Textbooks in Telecommunication Engineering

Series Editor
Tarek S. El-Bawab, PhD
Dean of Engineering and Applied Science
Nile University
Giza, Egypt

Telecommunications have evolved to embrace almost all aspects of our everyday life, including education, research, health care, business, banking, entertainment, space, remote sensing, meteorology, defense, homeland security, and social media, among others. With such progress in Telecom, it became evident that specialized telecommunication engineering education programs are necessary to accelerate the pace of advancement in this field. These programs will focus on network science and engineering; have curricula, labs, and textbooks of their own; and should prepare future engineers and researchers for several emerging challenges. The IEEE Communications Society's Telecommunication Engineering Education (TEE) movement, led by Tarek S. El-Bawab, resulted in recognition of this field by the Accreditation Board for Engineering and Technology (ABET), November 1, 2014. The Springer's Series Textbooks in Telecommunication Engineering capitalizes on this milestone, and aims at designing, developing, and promoting high-quality textbooks to fulfill the teaching and research needs of this discipline, and those of related university curricula. The goal is to do so at both the undergraduate and graduate levels, and globally. The new series will supplement today's literature with modern and innovative telecommunication engineering textbooks and will make inroads in areas of network science and engineering where textbooks have been largely missing. The series aims at producing high-quality volumes featuring interactive content; innovative presentation media; classroom materials for students and professors; and dedicated websites. Book proposals are solicited in all topics of telecommunication engineering including, but not limited to: network architecture and protocols; traffic engineering; telecommunication signaling and control; network availability, reliability, protection, and restoration; network management; network security; network design, measurements, and modeling; broadband access; MSO/cable networks; VoIP and IPTV; transmission media and systems; switching and routing (from legacy to next-generation paradigms); telecommunication software; wireless communication systems; wireless, cellular and personal networks; satellite and space communications and networks; optical communications and networks; free-space optical communications; cognitive communications and networks; green communications and networks; heterogeneous networks; dynamic networks; storage networks; ad hoc and sensor networks; social networks; software defined networks; interactive and multimedia communications and networks; network applications and services; e-health; e-business; big data; Internet of things; telecom economics and business; telecom regulation and standardization; and telecommunication labs of all kinds. Proposals of interest should suggest textbooks that can be used to design university courses, either in full or in part. They should focus on recent advances in the field while capturing legacy principles that are necessary for students to understand the bases of the discipline and appreciate its evolution trends. Books in this series will provide high-quality illustrations, examples, problems and case studies. For further information, please contact: Dr. Tarek S. El-Bawab, Series Editor, Dean of Engineering and Applied Sciences at Nile University (Egypt), telbawab@ieee.org; or Mary James, Senior Editor, Springer, mary.james@springer.com.

More information about this series at http://www.springer.com/series/13835

Tomasz P. Zieliński

Starting Digital Signal Processing in Telecommunication Engineering

A Laboratory-based Course

 Springer

Tomasz P. Zieliński
Department of Telecommunications
AGH University of Science and Technology
Kraków, Poland

ISSN 2524-4345   ISSN 2524-4353   (electronic)
Textbooks in Telecommunication Engineering
ISBN 978-3-030-49258-8   ISBN 978-3-030-49256-4   (eBook)
https://doi.org/10.1007/978-3-030-49256-4

This Springer imprint is published by the registered company Springer Nature Switzerland AG
The registered company address is: Gewerbestrasse 11, 6330 Cham, Switzerland

Textbooks in Telecommunication Engineering

Tomasz P. Zieliński

Starting Digital Signal Processing in Telecommunication Engineering

A Laboratory-based Course

Springer

To my dearest wife and parents, with many thanks for their love, priceless support in my long-distance, all-life, professional run, and great understanding.

And . . . to all readers who would like to live with DSP passion.

Foreword

When we were preparing sixteen years ago the first edition of professor Tomasz P. Zielinski's monograph, titled "**Digital Signal Processing. From Theory to Use,**" for publication in Poland in 2005, I did not think that this book, that treats very complex issues, would have such good readers' success. It had turned out that it was even more popular in our country than some world-renowned telecommunication titles we had translated. Why? Because the merits of severe mathematical problems were **explained in so accessible way** and it was directly shown how to use many **software solutions**, presented in the book, in a variety of fascinating, different applications. It was the first book of this type in Polish technical literature in the field of telecommunications. In this, I find the justification that this publication was appreciated by readers so much. The book had two editions and it was systematically reprinted (recently in 2014).

Apart from being appreciated by the reading market, in 2006 the book was awarded for the best academic book presented at the 13th National Fair of Academic Book ATENA 2006 in Poland and professor Tomasz P. Zieliński got the award of the Minister of Science and Higher Education in Poland as the book author.

Years go by. At present, 16 years later, professor Zieliński wrote a new book. In big part, it takes all the best from our previous, Polish monograph in its DSP part, but with significant modification and even further simplifications. A completely **new software-defined digital telecommunication section** has been added, containing about 350 pages. Knowing professor Zieliński writing style, I am fully convinced in good readers' reception of his new work.

Years go by and times are changing. The modern access to the information is different than it was years ago. The Internet is cutting our knowledge, and thoughts, into pieces. How to distinguish valuable ones? Where to find them? How to combine all pieces together? Questions, questions! May be it will be more effective to start our infinite queries from reading one good textbook?

I am proud that I can recommend this new book of such experienced Polish author.

Editor-in-Chief   Krzysztof Wiśniewski
Wydawnictwa Komunikacji i Łączności
Warszawa
21 April 2020

Preface

After almost 40 years' long friendship with Digital Signal Processing (DSP) and many, many years of teaching it, I have realized, finally, that most of my students, more than 90% of them, are not interested in DSP. They are obliged only to take DSP course and learn it in many different specializations, but they are more interested in some other topics. There is no sense to feel offended: the world surrounding is so fascinating and beautiful. What should result from this observation for me as a teacher? Of course, to fight for student attention and to try to make the course as attractive as it could be. At university practice, attractive means: short, comprehensive, fully understandable, and focused on applications, not on theoretical derivations.

I hope that this textbook fulfills the mentioned above, severe students conditions. I have tried to make it as compact as possible, but still presenting all the primary and important DSP concepts. The book is intended to cover in an attractive way introductory material for:

- the first student DSP laboratory, i.e. introduction to signals, primary spectral analysis, and digital filtering (Chaps. 1–9),
- the second student DSP laboratory, with important, special DSP topics and applications, including re-sampling, Hilbert transform and analytic signal, adaptive filters and modern spectral estimation methods as well as multimedia, speech–audio–video signal processing (Chaps. 10–16), and
- the first DSP-based digital telecommunication laboratory, with an introduction to software-defined radio technology, digital modulation basics, single-carrier and multi-carrier transmission fundamentals as well as digital receivers of FM/RDS radio, DAB radio, DSL modems and 4G/5G digital telephony (Chaps. 17–24).

Multimedia part (Chaps. 14–16) can be used as a bridge between general-purpose and strictly digital telecommunication DSP applications.

All the chapters end with exercises aimed at involving the student into her/his own funny *bungee-jump* experiments. Their difficulty is marked with "*, **, ***." Since "happy people do not count the time," an attempt is made to attract student attention and interest and reach teaching goals by project-oriented, case study-based self-education.

In order to make all the chapters *self-contained*, a short list of the most important references and further readings accompanies each of them. The short list of references offers students essential, DSP knowledge. The book intends to offer an easy explanation for the most common DSP problems and to involve students into the fun of DSP and not to lose them in the complex mathematical DSP at the beginning of its course in their education. When being interested in DSP, a contemporary student can find everything in the web. She/he even prefers to do it herself/himself—as their private investigation, not as an official obligation. This is the book well-known philosophy: *less is more*—less references, more joy and freedom, more self-studying, more unforeseen surprises and exciting adventures. In this aspect, all important papers and books are left for *future reading*—which is desirable and recommended but not obligatory.

In my opinion, a good explanation and understanding of a starting point is the most important in teaching and learning. This book aims at being a different type of introductory text on DSP with a more informal, friendly dialog.

My general intention was to make the book as short and simple as possible. Because none of us have time for technical epics—the world is changing too fast. Despite a very wide thematic scope, I had planned to write no more than 480 pages: 24 chapters with less than 20 pages each. Unfortunately, I did not succeed in realization of this goal. The final book manuscript had about 800 pages. But, in my opinion, the simplicity of book was not lost: all the chapters are really essential. So how has this happened? It is the presentation that is responsible for making the book immense. The book is full of diagrams, figures, exercises, and programs, which are the heart of DSP. These programs give the readers a chance to see all the math in action and help to understand definitively the underlying DSP concepts. Not all books allow this. All of them tell how to do it, but only a few show how to write a program solving a problem. Programs are used for the verification of a concept's correctness. Programs do not forgive mistakes. Everything should be considered in them: even one simple error can cause unpredictable consequences. Everything is explained by program in this book and this is its real strength, in my opinion.

Distinguishing between absolutely basic and complementary information is a crucial task in our everyday life. Between *what has to be done* and *what can be done*. Most students are obliged to take many courses, that are mainly focused on fundamental concepts. In this book all important issues are highlighted with a gray background and optional mathematical derivations are specially marked in order to be easily skipped during the first reading. Additionally, all chapters are proceeded by short introductions and ended with generalizing summaries—they help readers to select and remember fundamental problems discussed in each chapter. Last but not least, all new important terms appearing in the text that should be remembered are marked with different colors—we can say that the knowledge is well *labeled* and *annotated*. All of this, in my opinion, is a great help in fast acquiring and effective knowledge consolidation.

Finally, I would like to provide some *logistic* information which, in my opinion, should be very important for most readers.

1. **Book web page with programs**. All programs from this book, and some others, together with supported data, are available at the book web page: http://kt.agh.edu.pl/~tzielin/books/DSPforTelecomm/.

2. **Matlab/Octave version and toolboxes required**. Most of the programs do not use special Matlab functions and can be run in any Matlab/Octave version. However, the proposed solutions are very often compared with Matlab ones, and in such a situation the Signal Processing Toolbox is needed (or functions from the SPT should be commented). In image processing chapter, many Image Processing Toolbox are exploited. In analog filter design chapter, a few functions from the Control Toolbox are used but they are not necessarily required (`impulse()`, `step()`).

3. **Loading data to Matlab**. Due to changes in Matlab software, readers are asked to carefully exchange, when necessary, the following new Matlab functions `audioread()`, `audiowrite()`, `audioinfo()`, `audiorecorder()` to old ones: `wavread()`, `wavwrite()`, `wavrecord()`, etc.

4. **Auxiliary functions**. To allow readers to partially solve a problem of missing toolboxes, some very important auxiliary functions are given on the book web page in sub-folder/auxiliary. For example, very useful `spectrogram()` function for signal spectrogram calculation and visualization. Together, we could increase these sub-sets significantly.

We wish you an enjoyable and fruitful reading.

Kraków, Poland   Tomasz P. Zieliński
April 2020

Acknowledgements

In the first place, I would like to thank a lot all my colleagues with whom I had a pleasure to work during my whole, long, professional DSP life. We were discovering and enjoying the DSP world together. This book could not have been written without their inspiration and longstanding kind co-operation in different scientific projects. I would like to especially thank **Roman Rumian, Pawel Turcza, Krzysztof Duda, Tomasz Twardowski, Jarosław Bułat, Andrzej Skalski, Przemysław Korohoda, Dimitar Deliyski, Bogdan Matuszewski, Mirosław Socha, Jacek Stępień, Rafał Fraczek, Łukasz Zbydniewski, Marcin Wiśniewski, Artur Kos, and Grzegorz Cisek**.

In this aspect, I specially appreciate my last 10 years of active co-operation with **Krzysztof Duda**, my former Ph.D. student, at present Professor of the AGH University. Thank you, Krzysiek! It is a great pleasure for a teacher that his very talented pupil still wants to conquer DSP peaks together.

Considering telecommunication part of this book ...

I would like to thank warmly my co-student and colleague **Dr Roman Rumian**, the biggest hardware DSP guru at our AGH University. He had invited me to the MPEG audio, real-time, fixed-point DSP implementation project in the middle of 1990s of the twentieth century. For me, it was an unforgettable lesson of the DSP magic.

I would like to especially thank **Doctors: Paweł Turcza, Jarosław Bułat, and Tomasz Twardowski** for their very fruitful, stimulating, lasting 10-year co-operation in the field of Digital Subscriber Line modems (at the beginning of the twenty-first century).

I would like to express my special great gratitude to my friend **Dr Tomasz Twardowski** for introducing to me a field of software-defined radio in the year 2011. I could always count for his kind, patient, and comprehensive help during my first steps in programming digital receivers of analog FM/RDS radio as well as DAB digital radio. All the time, he is my high-prof telecommunication mentor. Thanks a lot, Dear Tom!

I would like to especially thank **Michael Häner-Höin** who has allowed me to use his Matlab programs of the DAB receiver as an example in this book. The code is a result of his M.Sc. Thesis entitled "SW-Realisierung eines DAB-Empfängers mit GNU Radio" defended in 2011 at School of Engineering (the Center for Signal Processing and Communications Engineering) of the ZHAW **Zurich University of Applied Science**, under the supervision of **Prof. Dr Marcel Rupf**. I would like to express my gratitude to the ZHAW for the same agreement.

I would like to thank **Dr Jarosław Bułat** for his longstanding, fantastic support for our all DSP laboratories and student projects and continuous generation of many fresh didactic ideas, especially, for his preparation and leading collective student projects on software real-time FM radio and DAB receivers.

I want to especially thank **Dr Marek Sikora** from my department for his priceless help, detailed explanation, and personal participation in the preparation of all Matlab programs dedicated for a digital single-carrier receiver, which are presented in this book. They represent at present a core of our laboratory on digital telecommunication fundamentals.

I would like also to highlight the great participation of my last Ph.D. student **Grzegorz Cisek** in high-tech part of the books. Thank you, Grzegorz, for inviting me for jointly discovering new wireless communication technologies, namely 4G-LTE and 5G-NR, and for writing demonstration programs for these two standards.

I would like also to thank **Joanna Marek**, my B.Sc. student, for interesting me in the analysis of amateur nano-satellite signals and supporting me with their recordings, done by her during Erasmus+ project realized at Institut Universitaire de Technologie 1 of **Universite Grenoble Alpes** in Grenoble, France.

Finally, last but not least, I want also to thank a lot ALL my colleagues from Telecommunication Department, AGH University, Kraków, Poland, for educating me for the last 15 years in different aspects of digital telecommunication systems. With a special gratitude to **Professors: Andrzej Dziech, Andrzej Jajszyk, Wiesław Ludwin, Andrzej Pach, and Zdzisław Papir and Doctors: Jarosław Bułat, Piotr Chołda, Michał Grega, Lucjan Janowski, Mikołaj Leszczuk, Marek Sikora, and Jacek Wszołek**.

Considering the book editing ...

First, I would like to appreciate my gratitude to **Krzysztof Wiśniewski**, editor-in-chief of Transport and Communication Publishers, Warsaw, Poland, for allowing me to use in this book figures for my Polish textbook "Cyfrowe przetwarzanie sygnałów. Od teorii do zastosowań" (Digital Signal Processing. From Theory to Applications), published till now many times from 2005, which were graphically prepared by myself.

I would like to express great thanks to my colleagues **Jarosław Bułat, Grzegorz Cisek, Piotr Chołda, Marek Sikora, and Jacek Wszołek** who have read selected book chapters and carefully reviewed them.

Finally, I would like also to sincerely thank a lot all my students who were my severe reviewers during final book editing. Thanks to their apt comments and suggestions large parts in this book were significantly improved. I specially appreciate the help of **Zuzanna Zajączkowska, Maciej Jankowski, Michał Markiewicz, Arkadiusz Pajor, Kamil Szczeszek, Maciej Hejmej, and Miłosz Sabadach**.

And finally, I would like to thank very, very much the **Reviewers** for their priceless help in enhancing the book quality!

The original version of this book was revised: The copyright year has been changed from 2020 to 2021. The correction to this book is available at https://doi.org/10.1007/978-3-030-49256-4_25

Contents

Contents

Chapter 1
Signals: Acquisition, Classification, Sampling

Let's try to climb mountain peaks! Do not send the SOS signal seeing the first obstacle!

1.1 Introduction

In digital signal processing (DSP) a signal is a vector (or matrix) of numbers taken from the real world. These numbers are acquired via sensors and Analog-to-Digital Converters (ADC). They are called the signal samples—samples of some known or unknown functions. During sampling a function value is discretized in time (in its argument) and quantized in amplitude (in its value). A sinusoid, e.g. power supply voltage, and damped sinusoids, e.g. generated in magnetic resonance, are the most frequently observed and the best known signals since they are solutions of second-order differential equations describing many existing physical phenomena. The other well-known signal are electrocardiograms and fingerprints.

In each signal the information is hidden about object which generated it. Most often signal analysis is focused on finding and extracting this information. In shape of the ECG signals there is the information about state of our heart. Fingerprint image allows to distinguish the person. Many signals are generated by people in some technical systems. Analysis of echoes of transmitted radar signal, penetrating our neighborhood, tells us about surrounding objects and their velocity. We have millions of sensors in the world and millions of phenomena to track and to analyze.

1.2 Digital Signal Processing Systems

In Fig. 1.1 a simple diagram of digital signal acquisition and processing systems is shown. Typically an observed real-world phenomena is continuous-time (analog), e.g. our heartbeat (ECG) or speech signals. The low-pass (LP) filter should pass only

© Springer Nature Switzerland AG 2021
T. P. Zieliński, *Starting Digital Signal Processing in Telecommunication Engineering*, Textbooks in Telecommunication Engineering, https://doi.org/10.1007/978-3-030-49256-4_1

signals' sinusoidal components having frequencies smaller than half of the sampling frequency (called sampling rate also, i.e. number of signal samples taken per one second). As a result each low-frequency sine, which is passed, after the sampling has more than two samples per period. Without usage of the LP filter, sinusoids with higher frequencies would be also present on the ADC input. They would be sampled to sparsely, i.e. their samples might be taken once for a few signal periods only. As a result, when no low-pass filter is used, high-frequency signals could look after sampling as low-frequency ones. It is creating for us a signal ambiguity problem: *is an observed signal really a low-frequency one or it only looks to be such?* From this reason usage of low-pass filter is obligatory.

Coming back to Fig. 1.1: the low-pass filter before the analog-to-digital (A/D) converter removes the unwanted high-frequency signal components, too *high* in regard to chosen sampling rate, while the low-pass filter after the digital-to-analog converter (D/A) smoothes jumps present in analog signal, resulting from converting real values into integer ones via rounding. The first filter is called an *anti-aliasing* one, while the second—a *reconstruction* one.

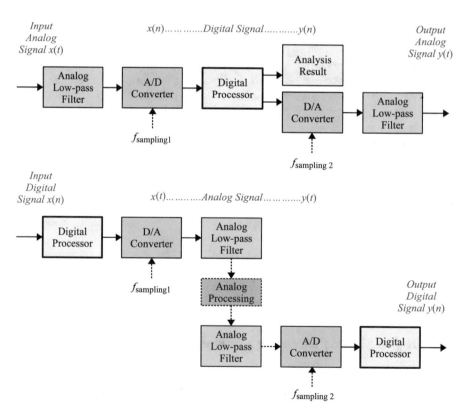

Fig. 1.1: Two digital paths of signal analysis and processing: (up) analog-digital-analog, (down) digital-analog-digital

The digital signal has a form of vector of numbers. These numbers are taken from sensors: sensor values are discretized in time and quantized in value by an A/D converter (ADC), like in computer sound cards. Next, they are processed by digital computer. Calculation results can be shown/stored or inversely converted to the analog world by the D/A converter (DAC). After the DAC the analog signals have step-wise shapes: they take only values from a predefined set of numbers, most values are missing. For example, the number "1" could represent the whole interval [0.5 ... 1.5), number "2"—interval [1.5...2.5) and so on. For this reason the low-pass filter, called a *reconstruction filter*, should smooth the synthesized signal and remove these "steps." Very often, in communication, the processing chain is inverted: at the input we have digital sinusoid. It takes different states, i.e. different values of amplitude, phase, or frequency and it is carrying bits in "state" numbers (e.g. state number 9, decimally, is equal to 1001, binary). In a transmitter the digital signal is converted to its analog form, then transmitted, passed through the communication channel (e.g. telephone/cable line, for DSL modems, or fiber), back converted into the digital form in the receiver, synchronized, and analyzed. This way the transmitted bits, 1001 in our example, are recovered in a receiver from the state number of acquired signal.

The simplest example of a DSP acquisition system is a personal computer equipped with microphone, AD converter, microprocessor, DA converter, and speaker. We can record our speech or environmental sound using it, do some digital processing (e.g. mixing and adding some special audio effects), and after that play our recordings. During recording the computer hardware is expected specification of the following parameters: sampling frequency/sampling rate (number of signal samples taken per one second), e.g.

8000, 11025, 16000, 22050, 24000, 32000, 44100, 48000, 96000, 192000, ...,

number of channels, one (mono) or two (stereo), and number of bits for one sample, for example 8 or 16. Changing pressure of the acoustical longitudinal wave is causing motion of the microphone membrane. This motion is transformed into voltage, also changing, which is processed by the low-pass filter removing signal components having frequency equal or higher than half of the sampling frequency. Then the AD converter is periodically taking samples of the signal (with earlier specified sampling frequency), rounding their values in amplitude to some predefined levels, and coding the level numbers as integer numbers with a sign. Then the stream of such numbers is processed by a computer, stored, transmitted or ... converted back to the analog form. In the last case the DA converter synthesizes analog voltages corresponding to integer numbers, interpreting them as numbers of voltage levels, and, finally, the analog signal is smoothed by the analog low-pass filter aiming at removal of output voltage switching ("jumps") resulting from the voltage quantization ("leveling") process.

There are plenty of DSP applications. For example, radar/sonar/USG echographic systems where technically generated electromagnetic or sound waves are emitted, reflected by different objects, and later analyzed in order to find object characterization. Or digital telecommunication systems in which single or multiple sines with different amplitudes, frequencies, and phase shifts are synthesized, transmitted, and analyzed in receivers, since bits are hidden (*carried*) in sine parameters. And so on, and so forth. After each A/D and before each D/A converter, typically the DSP *machinery* is used.

Rounding operation, or signal binarization is performed in the following way. Typically each AD converter has some predefined input voltage range, for example $[-5V, 5V]$. For example, having $N = 8$ output bits, the converter divides this range into $2^N = 256$ intervals, enumerating them from -128 ($= 10000000_2$ binary) to 127 ($= 01111111_2$ binary). In the C language it is a signed char variable, codded using U2 two complement notation: $1 = 00000001_2$, $-1 = 11111111_2 = -128 + 127$, the most significant bit is negative. In the discussed case the interval width in volts $\Delta V = 10/256V$. Let us assume that the input voltage is equal to $1.2345V$ or $-1.2345V$. It will be coded as (very, very roughly speaking)

$$\text{round}\left(\frac{1.2345V}{10V/256}\right) = \text{round}\,(31.6031999\ldots) = 32 = 00100000_2, \quad (1.1)$$

$$\text{round}\left(\frac{-1.2345V}{10V/256}\right) = \text{round}\,(\text{-}31.6031999\ldots) = -32 = 11100000_2. \quad (1.2)$$

One can say: binary representation is not important for me! I will do all calculation in Matlab/Octave or Python where very precise 32- or 64-bit floating-point number representation is used. Yes, it is true. All your calculation will be done really with high precision but they could be performed upon data which have lost already their precision during analog-to-digital conversion and have been rounded to nearest signed integers! The floating-point computing will help us only with precision of further computing.

For Inquisitive Readers It is important also to remark that the same bits after the ADC can be interpreted not as integers but also as fractional numbers lying in the range $[-1, 1)$. It can be beneficial during signal multiplications since multiplying integer numbers gives us a bigger number as a result (e.g. $2 \cdot 3 = 6$) while multiplying fractions gives us a smaller number (e.g. $0.2 \cdot 0.3 = 0.06$). This nice feature protects us against computational overflow. In fixed-point DSP processor the fractional number representation can be used.

From N_b-bit ADC, one can obtain only 2^{N_b} values being N_b-bit binary sequences of zeros and ones. For example, for 8-bit ADC we obtain $2^8 = 256$ different 8-bit patterns, changing from 00000000_2 to 11111111_2, and bits are numbered from 0 to 7 ($N_b - 1$) from right (the least significant) to left (the most significant). From these binary sequences one can obtain signed or unsigned integer values v_i or fractional values v_f, creating summations of 2's, taken to the positive or negative powers. For example in two's complement (U2) notation, the following values are obtained:

$$v_i = -b_{(N_b-1)} \cdot 2^{(N_b-1)} + \sum_{k=0}^{N_b-2} b_k \cdot 2^k, \quad v_f = \text{sign}(-b_{(N_b-1)}) + \sum_{k=0}^{N_b-2} b_k \cdot 2^{-((N_b-1)-k)}$$

(1.3)

For $N_b = 3$ bits and signed integers using U2 representation, the following bit interpretation is used:

Binary	100_2	101_2	110_2	111_2	000_2	001_2	010_2	011_2
Meaning	-4	$-4+1$	$-4+2$	$-4+2+1$ 0		1	2	2+1
Value v_i	-4	-3	-2	-1	0	1	2	3

while for signed fractional values the following interpretation is valid:

Binary	100_2	101_2	110_2	111_2	000_2	001_2	010_2	011_2
Meaning	-1	$-1+\frac{1}{4}$	$-1+\frac{1}{2}$	$-1+\frac{3}{4}$	0	$\frac{1}{4}$	$\frac{1}{2}$	$\frac{1}{2}+\frac{1}{4}$
Value v_f	-1	-0.75	-0.5	-0.25	0	0.25	0.5	0.75

If you are interested in a Matlab program doing the above calculations, look at lab01_ex_binary.m at book repository.

Exercise 1.1 (Recording and Playing Speech). Let us do the first experiment. We will record a few words, a fragment of our own speech, using computer sound card and Matlab environment. After that we will look at the acquired signal shape and check values of the signal samples. Run the below program 1.1. You should see a signal waveform more or less similar to the one presented in Fig. 1.2 in its upper part. Perform tasks listed below.

- Record some voiced phonemes (a, e, i, o, u) when vocal folds are working, opening and closing, and unvoiced ones (s, c, h) when vocal folds are open all the time. Check voiced plosives (d, g) and unvoiced ones (k, p). Test complicated words and longer sentences. Zoom different signal fragments. Look for periodic oscillatory fragments and noise-like speech intervals.
- Try to speak very loud (observe a clipping/saturation effect when signal is going beyond the allowed voltage limit) and very weak (observe clearly visible amplitude quantization levels).
- Record the same word using low sampling rate with small number of bits per sample and very high sampling rate with many bits per sample. Can you hear the difference? Hmm ... me not, unfortunately, ... "I am sixty four" (as in the Beatles' song). Do not worry: it was a joke.

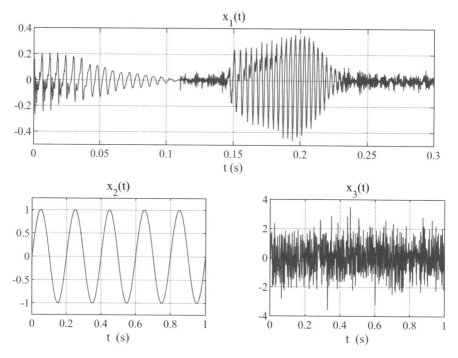

Fig. 1.2: Time waveforms of three signals: $x_1(t)$—a recorded speech signal, $x_2(t)$—a generated 5 Hz sinusoid $\sin(2\pi 5t)$ (in Matlab `t=0:0.001:1; sin(2*pi*5*t)`), $x_3(t)$—a generated Gaussian noise (using Matlab function `randn(1,1000)`), with mean value equal to 0 and standard deviation equal to 1, and assumed to have the same time support as the sinusoid. Notice that time at horizontal axis is scaled in seconds what is important for evaluation of signal duration and its frequency content

- Play recorded files with different frequencies allowed by the sound card. Why the speaking is faster or slower?
- Add different recordings sampled with the same or different frequencies. Listen to the result of your digital sound mixer.
- Try to find pitch frequency of your normal speech (frequency of vocal folds opening and closing) using visual observation of the signal plot. Measure the speech period in seconds and calculate its inverse. You should obtain a number approximately in the range [80–250] Hz. Check different voiced vowels (a, e, i, o, u). Do special recordings speaking vowel "a" as a singer: a bass, baritone, tenor, .. , soprano, alto, i.e. try to change pitch frequency from low to high value.

Listing 1.1: Recording and playing sounds

```
1   % File lab01_ex_audio.m
2   clear all; close all;
3
4   % Audio signal acquisition
5   fs = 8000;          % sampling frequency (samples per second):
6                       % 8000, 11025, 16000, 22050, 32000, 44100, 48000, 96000,
7   bits = 8;           % number of bits per sample: 8, 16, 24, 32
8   channels = 1;       % number of channels: 1 or 2 (mono/stereo)
9   recording = audiorecorder(fs, bits, channels); % create recording object
10  disp('Press any key and to record audio'); pause;
11  record(recording);     % start recording
12  pause(2);              % record for two seconds
13  stop(recording);       % stop recording
14  play(recording);       % listen
15  audio = getaudiodata( recording, 'single' ); % import data from the audio object
16
17  % Verification - listening, plotting
18  sound(audio,fs);            % play a recorded sound
19  x = audio; clear audio;     % copy audio, clear audio
20  Nx = length(x);             % get number of samples
21  n= 0:Nx-1;                  % indexes of samples
22  dt = 1/fs;                  % calculate sampling period
23  t = dt*n;                   % calculate time stamps
24  figure; plot(x,'bo-'); xlabel('sample number n'); title('x(n)'); grid;
25  figure; plot(t,x,'b-'); xlabel('t (s)'); title('x(t)'); grid;
26
27  % Write to disc and read from disc
28  audiowrite('speech.wav',x,fs,'BitsPerSample',bits); % write the recording
29  [y,fs] = audioread('speech.wav');                   % read it from file
30  sound(y,fs);                                        % play it again
```

1.3 Signal Classes

From functional analysis and signal theory point of view signals can be divided into
the following four groups (look at Fig. 1.3):

- **continuous-time and continuous-value (CT-CV)**, e.g. pure analog speech, audio, ECG signal, etc.,
- **discrete-time and continuous-value (DT-CV)**, CT-CV signal after discretization in time or space (sampling) but without value quantization, e.g. signal from
  single CCD camera capacitor, with its charge induced by light, before AD conversion, or set of signal samples generated in a computer with floating-point
  high precision from a mathematical function,
- **continuous-time and discrete-value (CT-DV)**, for example signal just after
  the DA converter, before smoothing by low-pass, reconstruction filter, being
  continuous but quantized in amplitude, e.g. digital music played from a compact
  disc or Internet,

- **discrete-time and discrete-value (DT-DV)**, for example signal after any AD converter, having specified number of bits (N-bits give 2^N states—quantization levels) and working with some sampling frequency, e.g. digital multimedia: speech, audio, image, video.

In Fig. 1.3 the CT-CV, DT-CV, CT-DV, and DT-DV versions of a pure sinusoidal signal are presented. We will be training generation of different signals in the next chapter. In this one we learn computer synthesis of the sine only, *the King of the Road* in the DSP world, and use it for practical demonstration of discussed concepts. The sine repeating f_0 times per second, for example 5 times, is given by the following math formula:

$$x(t) = \sin(2\pi f_0 t), \quad x(t) = \sin(2\pi 5t), \tag{1.4}$$

and generated by the following Matlab program (assumed: signal duration = 1 s, sampling rate = 1000 samples per second, i.e. sampling period equal to $1/1000 = 0.001$ s):

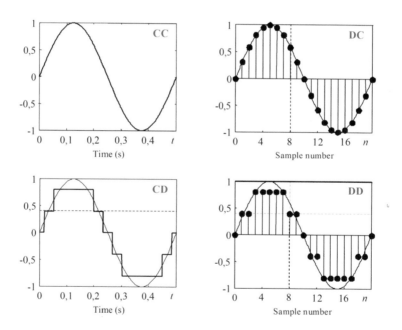

Fig. 1.3: Different representations of a sine signal repeating two times per second (period 0.5 s, frequency 2 Hz): **(CT-CV)** (left up) continuous-time–continuous-value $x(t)$, **(CT-DV)** (left down) continuous-time–discrete (quantized) value $x_q(t)$, **(DT-CV)** (right up) discrete-time–continuous-value $x(nT) = x(n) = x_n$ and **(DT-DV)** (right down) discrete-time–discrete (quantized) value $x_q(nT) = x_q(n)$. T denotes sampling period. In left plots—wide solid line represents a signal, in right plots—black dots represent a signal [15]

```
T=1; dt=0.001; t=0:dt:T; f0=5; x = sin(2*pi*f0*t); plot(t,x);
```

Let us check if the Eq. (1.4) defines the desired signal. Since for $t = 1$ s the sine angle is equal to f_0-th multiplicity of 2π—the sine period, i.e. sine repeats f_0 times per second, in our case 5 times.

Exercise 1.2 (Discretizing Signals). Try to generate exactly the same signal plots as presented in Fig. 1.3. Modify the Matlab code given in Listing 1.2. Generate and discretize a sum of two sinusoids with different frequencies and amplitudes not equal to 1.

Listing 1.2: Signal discretization

```
1   % File lab01_ex_discretizattion.m
2   clear all; close all;
3
4   T = 1;
5   fs1 = 1000;        fs2 = 50;              % sampling frequencies 1 & 2
6   dt1 = 1/fs1;       dt2 = 1/fs2;           % sampling periods 1 & 2
7   N1 = ceil(T/dt1); N2 = ceil(T/dt2);       % numbers of samples to generate
8   n1  = 0:N1-1;      n2  = 0:N2-1;           % vectors of sample indexes
9   t1 = dt1*n1;       t2 = dt2*n2;            % vectors of sampling time moments
10  x1 = sin(2*pi*1*t1);                      % sinusoid repeating 1 times per second
11  x2 = sin(2*pi*1*t2);                      % sampled with different frequencies
12
13  x_min=-1.5; x_max=1.5; x_minmax=x_max-x_min;  % ADC range in Volts
14  Nb=3; Nq=2^Nb;                            % number of bits, number of quantization levels
15  dx = x_minmax/Nq;                         % width of the quantization level
16  x2q = dx*round(x2/dx);                    % quantization of signal value
17  K = fs1/fs2; x2qt = [];
18  for k=1:N2, x2qt = [x2qt, x2q(k)*ones(1,K) ]; end
19
20  figure; plot(t1,x1,'b-',t2,x2,'bo',t2,x2q,'r*',t1,x2qt,'k-');
21  xlabel('t  (s)'); title('x(t)'); grid;
```

Additionally, having in mind signal generation, signals belong to the two main groups:

- **deterministic**—a function describing a signal waveform (curve, shape) exits, like sine, exponent, gaussoid, etc.,
- **random, stochastic**—a function describing a signal waveform is unpredictable.

In Fig. 1.4 detailed diagram of signal classification is presented.

The following signals belong to the deterministic class: periodic ones, quasi-periodic (having periodic components but as a mixture not periodic, e.g. when a ratio of periods of at least two signal components is an irrational number: $T_m/T_n \neq m/n$, where m, n are two integer numbers), modulated and impulsive with finite energy.

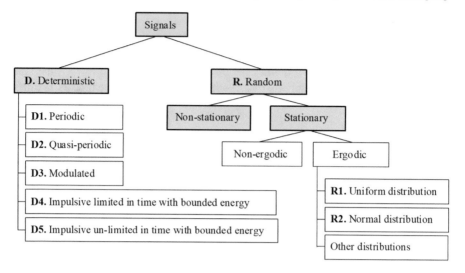

Fig. 1.4: Basic signal classification [15]

The random signals are divided into two groups: stationary and non-stationary. Let us assume that we have built a matrix putting into its rows different sequences of random samples taken from the same sensor. Figure 1.5 illustrates our explanation using analog signals: in rows we have 4 different signal realizations. If the signal is stationary, statistical signal parameters (like mean, variation, etc.), calculated for each instant across many signal realizations are the same, in our case—means and variations of the samples in each column are equal (in Fig. 1.5 parameters calculated vertically). For non-stationary signals they are different. Stationary signals are additionally ergodic when statistical parameters calculated for every instant over many signal realization are the same as for each signal realization in time. In our example, when statistical parameters of each column are the same as of each row (in Fig. 1.5 parameters calculated vertically and horizontally should be the same). In this situation, one signal realization, one matrix row, is sufficient for derivation of statistical description of the observed stochastic process.

Exercise 1.3 (Noise Stationarity/Ergodicity). Run the following Matlab code and verify stationarity of the generated Gaussian noise:

```
N=1000; X=rand(N,N); mcols=mean(X); mrows=mean(X.');
[ mcols.' mrows.']
```

(.') denotes matrix transposition: converting rows into columns. Any Matlab function performed upon a matrix is executed independently over each matrix column. Therefore the mean() function calculates mean value of each column of the matrix X. Exchange mean() function with var() function.

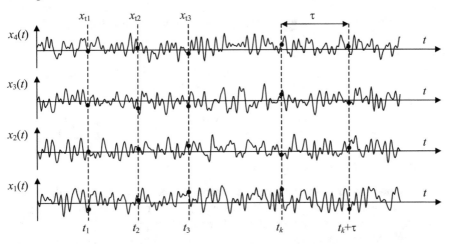

Fig. 1.5: Graphical illustration of random process stationarity and ergodicity. Four different realizations $x_k(t)$, $k = 1, 2, 3, 4$, of the same analog random process X are plotted horizontally. x_{t1}, x_{t2}, x_{t3} denote observed random variables in times t_1, t_2, t_3. The process is stationary if: (1) statistical parameters (e.g. means, variations,...) of random variables x_{t1}, x_{t2}, x_{t3} are the same and do not depend on t, (2) variable correlation $E[x_t x_{t+\tau}]) = R(\tau)$ depends only on time shift τ between variables but not on t. The process is ergodic when statistical parameters *across* process realizations (vertically) are the same as statistical parameters of each realization individually (horizontally) [15]

The shape in time domain of the random signals is unknown (unpredictable) but function describing probability of taken signal values (probability density function (PDF), marked as $P(x)$) can be known. The most popular random signal is a Gaussian noise (generated by Matlab function x=randn(1,N)). It is the most frequent signal disturbance with $P(x)$ described by the well-known Gaussian curve, having maximum for $x = 0$ (it is the most probable noise value) and decaying on both sides (depending on the noise variance). The uniform noise is an another noise example that has uniform distribution of values in some interval [a, b] (Matlab function x=rand(1,N) generates signal samples in the range $[0, 1]$). We will investigate both types of noises in the next chapter, do not worry about not seeing their figures at present.

Two of the most famous examples of a deterministic and a random signal, a sinusoid and Gaussian noise, are presented in Fig. 1.2, as signals $x_2(t)$ and $x_3(t)$, together with a fragment of real speech $x_1(t)$. In the real speech we can distinguish oscillatory/deterministic-like as well as noisy/random-like intervals. Consequently, the theory *works*.

Going further, considering dimensionality, signal can be treated as a set (collection) of some elements (consisting of a single or multiple values) captured in one or many dimensions. For example it can be

- one-dimensional (1D), e.g. mono speech or audio samples changing in time,
- two-dimensional (2D), e.g. one picture from camera, i.e. pixel values changing in *x-y* space coordinates,
- three-dimensional (3D), e.g. computer tomography (CT) data (so-called voxels) changing in *x-y-z* space coordinates or movie as a sequence of 2D *x-y* pixel-based pictures changing in time,
- four-dimensional (4D), e.g. 3D *x-y-z* field of temperature, pressure, or pollution changing in time, functional/dynamic 4D CT as a 3D CT *x-y-z* data changing in time,
- multi-dimensional.

Of course, no surprise, things can be becoming more and more complicated. In the above list I have tried to present only examples of one-value function with increasing number of arguments: time and space coordinates. But the function can be a multi-value one: for one set of arguments the function could take not only one but many values, which is the case in multi-channel sound systems (e.g. stereo, 5.1 or 7.1) or color images, e.g. RGB ones with Red, Green, and Blue components per one pixel. Multi-value signals can come from multi-value sensors, for example RGB color or NMR perpendicular x-y detectors, but can be also created artificially like telecommunication IQ signals (*In phase* and *Quadrature*) which are additionally interpreted as complex-value numbers for easier processing. Uff ... "What a wonderful world!"

In the next section we will investigate in detail the signal sampling problem.

1.4 Base-Band and Sub-Band Sampling

It is easy to predict and see that the signal should be sampled sufficiently dense in order not to lose its fast variations. How dense? The *base-band* signal theory says: more than two samples for one period of the fastest signal periodic component, i.e. its component having the highest frequency. In Fig. 1.6 we see example of *good* and *bad* sampling.

But the above rule represents only part of the truth. In the second, more general *sub-band* version it is said: if a signal of interest has only frequency components lying in some limited band, it is sufficient to isolate it from surrounding lower and higher frequency signals, and use sampling frequency two times bigger than the width of the signal frequency bandwidth. Information about the signal will not be lost. This phenomena will be shown in this section.

Till now we were treating signals as time functions. Therefore there is no problem to do the following assignments (Δt—sampling period, distance between samples in seconds ($= 1/f_s$), f_s—sampling frequency, number of samples per second ($= 1/\Delta t$)):

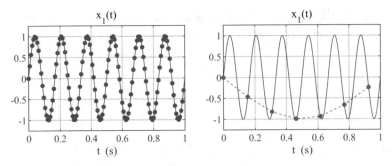

Fig. 1.6: Example of correct (left) and incorrect (right) sampling with *frequency aliasing*

$$t = n \cdot \Delta t = n \cdot \frac{1}{f_s} \qquad (1.5)$$

and to obtain discrete-time signals from all continuous-time signal functions presented till now in this chapter. For example, after these substitutions the signal (1.4) has a form:

$$x(t) = A \cdot \sin(2\pi f_0 t) \quad \rightarrow \quad x(n \cdot \Delta t) = A \cdot \sin\left(2\pi \frac{f_0}{f_s} n\right), \quad F_0 = \frac{f_0}{f_s}. \qquad (1.6)$$

Equation (1.6) tells us that continuous-time sinusoids having different frequencies f_k and different sampling frequencies f_{sk} but the same normalized frequency F_k:

$$\frac{f_1}{f_{s1}} = \frac{f_2}{f_{s2}} = \ldots = \frac{f_k}{f_{sk}} = F_k \qquad (1.7)$$

after time discretization have exactly the same sequence of samples, so they will be completely none-distinguishable. As a consequence signal processing algorithms designed for sinusoid with frequency f_1 and sampled with frequency f_{s1} can be used without no hesitation for processing of sinusoid with frequency f_2 and sampled with frequency f_{s2} (when Eq. (1.7) is fulfilled).

So now we can ask the question: is it all? Does the sampling represents only discretization of a function argument, a function of time in our example? The answer is: NO. In the real-world signal processing signal values are also discretized by AD converters: in this context we are saying that the signals values are *quantized*. Due to high-precision floating-point 32-bit or 64-bit number representation we generate in computers quasi-DT-CV signals and perform quasi-perfect calculations on them, concerning the signal values. But we can round them (as in Matlab Fixed-Point DSP Blockset) and test the influence of number representation upon signal processing results. In real-time DSP very often fixed-point processors are used with limited precision of data representation and arithmetic. Let us remember that different signal representations have been presented in Fig. 1.3.

Let us go back to signal sampling. Let us assume that a signal is a sum of sinusoids with different frequencies and f_{max} and f_{min} denote the highest and the lowest of them. As already stated the signal sampling frequency f_s should fulfill one of the following rules, so-called **sampling theorems (Nyquist theorems)**:

$$\text{Base-band version} \qquad f_s > 2f_{max}, \qquad\qquad (1.8)$$
$$\text{Sub-band version} \qquad f_s > \Delta f = f_{max} - f_{min}. \qquad (1.9)$$

Let us take the sine/cosine signal (1.6) having one frequency from the following set of frequencies: $f_0 + kf_s$ and $-f_0 + kf_s$ where k is any integer value: $k = 0, \pm 1, \pm 2, \pm 3, \dots$ and $0 \le f_0 < f_s/2$. Since the following trigonometric equivalences holds

$$\sin(k \cdot 2\pi) = 0, \quad \cos(k \cdot 2\pi) = 1, \qquad\qquad (1.10)$$
$$\sin(\alpha \pm \beta) = \sin(\alpha)\cos(\beta) \pm \sin(\beta)\cos(\alpha) \qquad (1.11)$$
$$\cos(\alpha \pm \beta) = \cos(\alpha)\cos(\beta) \mp \sin(\alpha)\sin(\beta) \qquad (1.12)$$

we can write the sine signal as

$$
\begin{aligned}
x(n) &= \sin\left(2\pi\frac{kf_s \pm f_0}{fs}n\right) = \sin\left(2\pi kn \pm 2\pi\frac{f_0}{fs}n\right) = \dots \\
&= \sin\left(2\pi\frac{\pm f_0}{fs}n\right) = \pm \sin\left(2\pi\frac{f_0}{fs}n\right)
\end{aligned}
\qquad (1.13)
$$

and similarly the cosine one:

$$
\begin{aligned}
x(n) &= \cos\left(2\pi\frac{kf_s \pm f_0}{fs}n\right) = \cos\left(2\pi kn \pm 2\pi\frac{f_0}{fs}n\right) = \dots \\
&= \cos\left(2\pi\frac{\pm f_0}{fs}n\right) = \cos\left(2\pi\frac{f_0}{fs}n\right).
\end{aligned}
\qquad (1.14)
$$

As we see all high-frequency signals ($kf_s + f_0$ and $kf_s - f_0, k = 1, 2, 3, \dots$), not fulfilling Eq. (1.8), after sampling have exactly the same samples as the signal with low frequency f_0. In order to avoid this ambiguity/aliasing (*So what do I see right now, a high or low frequency?*) we have to be sure that the signal being sampled fulfills the Nyquist theorem. Therefore the analog signal is low-pass filtered before the A/D converter as shown in Fig. 1.1, which removes all unwanted high-frequency signal components. The cut-off frequency of the filter should be smaller than half of the sampling frequency, i.e. the filter should pass only the components with frequencies

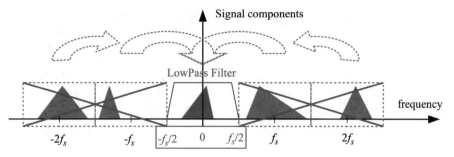

Fig. 1.7: Graphical illustration of a risk of signal components aliasing when sampling frequency f_s is too low in regard to all signal component frequencies. A low-pass filter has to be used that removes all signal components having frequencies higher than $\frac{f_s}{2}$ and lower than $\frac{-f_s}{2}$, which will be understandable after a few chapters. Otherwise they will be leaked to the band $\left[\frac{-f_s}{2}, \frac{f_s}{2}\right]$ and will be visible as low-frequency ones

lying in the band $[0, f_s/2)$ Hz. In Fig. 1.7 a risk of signal components misinterpretation is graphically illustrated. The problem of described above signal ambiguity is called a frequency aliasing phenomena in signal processing nomenclature.

The second possibility, used widely in telecommunication systems, is usage of the band-pass filter in the frequency range $[kf_s/2, (k+1)f_s/2)$, instead of the low-pass filter before the A/D converter, and application of the same low sampling frequency f_s as before. At present the sub-band sampling rule (1.9) is used. Of course, the signal observed in this case looks like a low-frequency one but knowing the exact frequency band of the filter we know its frequency shift in respect to 0 Hz (the DC constant value) and can correct the frequency measurement. Thanks to this effect, sub-band high-frequency telecommunication signals can be processed with lower sampling frequencies. That is a good news!

Remark: More on frequency Aliasing (Sampling Ambiguity) Effect The very good example of sampling ambiguity is a car wheel turning backward during film watching. In reality the wheel is turning forward but we are taking pictures too rarely and it looks on the movie that the wheel is turning back! One more example. Let us assume that we are taking a picture of a black and white chess-board with a low-resolution camera and we are frustrated obtaining completely white or completely black images and not *seeing* the chess-board pattern. The camera image sampling in space is too low: we have too less number of pixels per one chess-board square. The other example. We are walking on the pavement with many holes but somehow we are doing steps in such a way (*sampling the pavement*) that we are not falling into them, we do not *see* them. In case of signal we prefer to *see* everything. Not to observe any spurious/misleading effect like wheels "turning back" in a movie.

Exercise 1.4 (Good and Wrong Signal Sampling or Function Generation).
Try to do a figure similar to Fig. 1.6 showing the original high-frequency signal,
for example 9000 Hz sine, properly sampled by you at $f_s = 48$ kHz, and its
misleading low-frequency version obtained after wrong sampling with $f_s = 8$
kHz. Listen to both signals using the Matlab function sound(x,fs).

Exercise 1.5 (Wrong Re-sampling Music Files). Find in the Internet music file with high sampling rate, at least 48000 samples per second, and
with high-frequency content. Take every 6-th sample of it and play the resultant signal in Matlab/Octave environment at 8 kHz using the command
sound(x(1:6:end), 8000). Do you hear the difference? You should.

**Exercise 1.6 (Testing Sampling Theorem—On the Path to Sub-Band Sam-
pling).** For example let assume us that the sampling frequency $f_s = 1000$ Hz
and the analyzed (co)sinusoid has frequency $f_0 = 100$ Hz. The (co)sinusoids
with the following frequencies (written using bold characters) will look exactly
the same (with $\pm$ sinus sign exception which is impossible to catch):

k	$k \cdot f_s - f_0$	$k \cdot f_s$	$k \cdot f_s + f_0$
0	-100	0	**100**
1	**900**	1000	**1100**
2	**1900**	2000	**2100**
3	**2900**	3000	**3100**
4	**3900**	4000	**4100**
...	....	...	....

Run the program presented in Listing 1.3. 11 sines and 11 cosines with different
frequencies $kf_s \pm f_0$ are generated and displayed in one plot, separately sines
(up) and cosines (down). How many signals do we see and how many do we
expected to see? The signals perfectly overlay. Modify the program, changing
$f_0 = 100$ Hz to 200 Hz, then to 50 Hz. Observe how the figures have changed.

Listing 1.3: Sampling and danger of signal uncertainty

```
1   % File lab01_ex_sampling.m
2   % Sampling and signal aliasing (uncertainty)
3   clear all; close all;
4
5   fs = 2000;       % sampling frequency
6   dt = 1/fs;       % sampling period
7   Nx = 100;        % number of signal samples
8
9   n = 0 : Nx-1;    % vector of indexes of signal samples
10  t = dt*n;        % vector of sampling instants (in seconds)
11
12  fx = 50;         % frequency of power supply
13  figure;
14  for k = 0 : 10
15      k            % checking loop execution
16      x1 = 230*sqrt(2) * sin(2*pi*(+fx + k*fs)*t);   % freq = +fx + k*fs
17      x2 = 230*sqrt(2) * sin(2*pi*(-fx + k*fs)*t);   % freq = -fx + k*fs
18      x3 = 230*sqrt(2) * cos(2*pi*(+fx + k*fs)*t);   % freq = +fx + k*fs
19      x4 = 230*sqrt(2) * cos(2*pi*(-fx + k*fs)*t);   % freq = -fx + k*fs
20      subplot(211); plot(t,x1,'bo-',t,x2,'r*-'); hold on; title('Sines');
21      subplot(212); plot(t,x3,'bo-',t,x4,'r*-'); hold on; title('Cosines');
22  end
```

Sub-band Sampling Repeating: to avoid the signal ambiguity effect it is necessary to ensure that frequencies of all components of analog signal, being sampled, are lower than half of the sampling frequency, in our Exercise 1.6—in the range $[0, \ldots, 500)$ Hz. Or ... yes, yes, we can try to ride round this *drawback* and even make an advantage from it. Looking at the frequency values written in the Exercise 1.6 one can do very important conclusion about the so-called sub-band sampling: if one knows the bandwidth of the signal she/he could sample it with smaller frequency according to the rule (1.9) and adjusts the sampling frequency according to the bandwidth, not to the maximum frequency. Why is it possible? Because when one knows that her/his signal is in some known frequency range, e.g. [2000 2500) Hz, even observing low-frequency signal, e.g. 100 Hz, is not misleading because, knowing the frequency *shift* 2000 Hz, the true signal frequency can be calculated, e.g. (100+2000) Hz = 2100 Hz, and no ambiguity exists. This topic will be further discussed in next chapters. In such case sampling frequency should be two times bigger than the signal frequency band. This rule, the *killer* for most students, is known as **the general sub-band Nyquist sampling theorem**.

1.5 Analog Signal Reconstruction

Freedom. Everybody would like to have a chance to go back. Can we go back to the exactly the same analog signal from its samples, i.e. from discrete-time but continuous-value (DT-CV) signal?

The answer is YES if the sampling theorem frequency restriction (1.8) has been fulfilled! If we assume that our analog signal is a sum of sinusoidal components with different frequencies, and if the sampling frequency was more than two times bigger than the highest signal frequency—it is possible to *go back* to the analog world. Proof of this is not possible now (in some books it takes even several pages). But we can demonstrate experimentally that such reconstruction is in practice possible.

The analog signal $x(t)$ is reconstructed, using it non-quantized samples $x(nT)$, T—sampling period, from the following formula:

$$x(t) = \sum_{n=-\infty}^{\infty} x(nT) \frac{\sin\left(\frac{\pi}{T}(t-nT)\right)}{\frac{\pi}{T}(t-nT)}. \tag{1.15}$$

The output analog signal is a sum of many analog functions $\text{Sinc}(a) = \sin(a)/a, a = \frac{\pi}{T} \cdot (t-nT)$ (see Fig. 1.8), that are shifted in time by $n \cdot T$ and scaled in amplitude by $x(n \cdot T)$ (i.e. centered at each signal sample and multiplied by its value). $\text{Sinc}(a)$ function is a sinusoid divided by its angle changing from minus to plus infinity: $-\infty \leq a \leq \infty$. The function is equal to 1 for $a = 0$. It oscillates (as sine does) and it is periodically crossing zero for $a = k \cdot \pi$ (as a sine). It is decaying also with increase of the value of a. What is interesting, the $\text{Sinc}(a)$ centered at one signal sample has zero values at positions of all remaining samples. In consequence, in the sum it does not change signal values in these time points. Deeper practical explanation of the signal reconstruction from its samples offers the program 1.4.

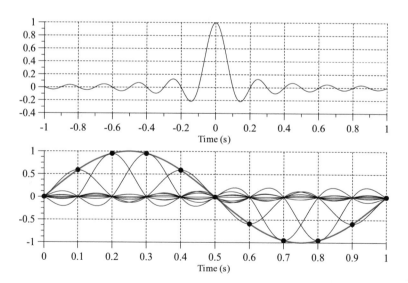

Fig. 1.8: Graphical illustration of analog signal reconstruction from its samples: (up) Sinc function, (down) the sine reconstruction process: addition of many shifted Sinc functions scaled in amplitude by sample values gives us a reconstructed signal plotted with the thick line. In the figure an analog sinusoid is reconstructed from its samples marked with "•" [15]

Exercise 1.7 (Illustration of Analog Signal Reconstruction). Analyze the Matlab code presented in program 1.4, illustrating the idea of analog (CT-CV) signal reconstruction from samples of discrete-time continuous-value (DT-CV) signal. We would like to go back from signal with coarse sampling to signal with finer sampling. Try to change values of program parameters in order to make the reconstruction error lower. What is a source of the error?

Listing 1.4: Signal reconstruction after sampling

```
1   % File lab01_ex_reconstruction.m
2
3   clear all; close all; hold off;
4
5   % Program parameters
6   f0 = 1;          % frequency of signal x(n) [Hz]
7   fs = 100;        % frequency of sampling [Hz]
8   N = 400;         % number of signal x(n) samples
9   D = 10;          % down-sampling ratio (D-times)
10
11  % Generation of sine x(n) with fx frequency
12  T = 1/fs;                              % sampling period for x(n)
13  t = 0 : T : (N-1)*T;                   % time instants for x(n)
14  x = sin(2*pi*f0*t);                    % generation of sine x(n)
15  figure; plot(t,x,'b-',t,x,'ro'); grid; title('Sine sampled with high frequency');
16
17  % Sine down-sampling: only every D-th sample is left
18  xD = x(1 : D : length(x));             % down-sampled sine x(n)
19  ND = length(xD); TD = D*T; tD = 0 : TD : (ND-1)*TD;
20  figure; plot(t,x,'b-',tD,xD,'ro'); grid; title('Sine sampled with low frequency');
21  figure; plot(tD,xD,'b-',tD,xD,'ro'); grid; title('Sine sampled with low frequency');
22  figure; stem(xD,'b'); title('Sine sampled with low frequency');
23
24  % Sinc(a) = sin(a)/a function
25  tt = - (N-1)*T : T : (N-1)*T;          % time support of Sinc function
26  fSinc = sin(pi/TD*tt)./(pi/TD*tt);     % Sinc function generation
27  fSinc(N) = 1;                          % its value for arg=0 (division 0/0)
28  tz = [ -tD(end:-1:1) tD(2:ND)];        % instants of zero-crossing plus 0
29  z = [zeros(1,ND-1) 1 zeros(1,ND-1)];   % Sinc values at time tz
30  figure; plot(tt,fSinc); %,'b',tz,z,'o'); grid; title('Sinc function');
31
32  % Reconstruction of the original signal from the decimated signal
33  figure;
34  y = zeros(1,N);
35  for k = 1 : ND
36      fSinc1 = fSinc( (N)-(k-1)*D : (2*N-1)-(k-1)*D );
37      y1 = xD(k) * fSinc1;
38      y = y + y1;
39      subplot(311); plot(t,fSinc1); grid; title('Next shifted Sinc function');
40      subplot(312); plot(t,y1);     grid; title('Next shifted and scaled Sinc function');
```

```
41    subplot(313); plot(t,y);        grid; title('Summed y(t) till now');
42    pause
43  end
44  figure; plot(t,y,'b',t,x,'r');     grid; title('Reconstructed signal');
45  figure; plot(t,x(1:N)-y(1:N),'b'); grid; title('Reconstruction error');
```

1.6 Summary

There is no doubt that good understanding of the *signal* concept is extremely important in contemporary *digital signal processing*. In this first introductory chapter we have tried to see signals diversity, their different origin, types, shapes, features, applications. We have learned a little bit about signal acquisition, including proper sampling. What is the most important? What should be remembered?

1. Real-world signals are continuous-time functions/waveforms which are sampled and after this, having already a form of vectors and matrices of numbers, they are analyzed and processed by digital computers.
2. Sampling rate has to be carefully chosen in order not to *lose* the original signal shape and, in consequence, information about its abrupt change or frequency of repeating.
3. In Matlab we generate signals discretizing time variable and putting the resultant vector of sampling instants into a signal function of interest. In this chapter we did it only for sine/cosine signals.
4. Discrete-time signals, as sequences of numbers, can have very different shapes and forms. They are classified as deterministic and random, one or more-dimensional, real and complex-value, periodic, quasi-periodic and non-periodic, impulsive or not, stationary or not, ...

1.7 Private Investigations: Free-Style Bungee Jumps

Exercise 1.8 (What a Wonderful World of ... Sounds!). Find in the Internet files of recorded sounds, read them into Matlab, observe their waveforms, and play them inside the Matlab environment. Many different sounds you can find on this web page: https://www.findsounds.com/. I am sure that you will be pleased.

Exercise 1.9 (* At Railway Station). Take your speech signal (a vector of speech samples) and make a few delayed copies of it. For example, delay the vector by 100, 250, 500, and 750 elements. Scale delayed copies by 1, 0.5, 0.25, and 0.1, respectively, add all of them, and listen to the result. Wow! Are you really at the railway station?

Exercise 1.10 (* BALANGA Records Studio). Read any real audio signal with music (a wav file). Zoom its different fragments. Observe changing frequency (oscillation) content. Play the signal using sound card. Observe that knowing sampling frequency (and in consequence sampling period, distance between the samples) is very important! Do specially some mistakes in specification of the sampling rate during signal reproduction (playing). Add a few signal recordings to themselves. Listen to the result. Cut and add different sound pieces. Try to obtain something your boyfriend, girlfriend, or grandfather, even the laboratory assistant, will enjoy.

Exercise 1.11 (* Is My Heart Broken?). Find in the Internet an ECG heart activity signal. E.g. take it from the page https://archive.physionet.org/cgi-bin/atm/ ATM. Choose, for example, MIT BIH Arythmia Database, Record: 100, Signals: all, Length: 1 min, Time format: seconds, Data format: standard, Toolbox Plot: waveform, Toolbox export signals as .mat. Download file xxx.mat and xxx.info (ASCII). Find signal periodicity looking at signal plot. How many heartbeats per second?

```
load ecg100.mat; whos;
fs=360; N=length(val(1,:)); dt=1/fs; t=dt*(0:N-1);
plot(t, val(1,:)); xlabel('t [s]'); title('ECG(t)'); grid; pause
```

References

1. L.F. Chaparro, *Signals and Systems Using Matlab* (Academic Press, Burlington MA, 2011)
2. M.H. Hayes, *Schaum's Outline of Theory and Problems of Digital Signal Processing* (McGraw-Hill, New York, 1999, 2011)
3. E.C. Ifeachor, B.W. Jervis, *Digital Signal Processing. A Practical Approach* (Addison-Wesley Educational Publishers, New Jersey, 2001)
4. V.K. Ingle, J.G. Proakis, *Digital Signal Processing Using Matlab* (PWS Publishing, Boston, 1997; CL Engineering, 2011)
5. R.G. Lyons, *Understanding Digital Signal Processing* (Addison-Wesley Longman Publishing, Boston, 1996, 2005, 2010)
6. J.H. McClellan, R.W. Schafer, M.A. Yoder, *DSP FIRST: A Multimedia Approach* (Prentice Hall, Upper Saddle River, 1998, 2015)
7. S.K. Mitra, *Digital Signal Processing. A Computer-Based Approach* (McGraw-Hill, New York, 1998)
8. A.V. Oppenheim, R.W. Schafer, *Discrete-Time Signal Processing* (Pearson Education, Upper Saddle River, 2013)
9. A.V. Oppenheim, A.S. Willsky, S.H. Nawab, *Signals & Systems* (Prentice Hall, Upper Saddle River, 1997, 2006)
10. A. Papoulis, *Signal Analysis* (Mc-Graw Hill, New York, 1977)

11. J.G. Proakis, D.G. Manolakis, *Digital Signal Processing. Principles, Algorithms, and Applications* (Macmillan, New York, 1992; Pearson, Upper Saddle River, 2006)
12. S.W. Smith, *The Scientist and Engineer's Guide to Digital Signal Processing* (California Technical Publishing, San Diego, 1997, 1999). Online: http://www.dspguide.com/
13. K. Steiglitz, *A Digital Signal Processing Primer: With Applications to Digital Audio and Computer Music* (Pearson, Upper Saddle River, 1996)
14. M. Vetterli, J. Kovacevic, V.K. Goyal, *Foundations of Signal Processing* (Cambridge University Press, Cambridge, 2014)
15. T.P. Zieliński, *Cyfrowe Przetwarzanie Sygnałów. Od Teorii do Zastosowań (Digital Signal Processing. From Theory to Applications)* (Wydawnictwa Komunikacji i Łączności (Transport and Communication Publishers), Warszawa, Poland, 2005, 2007, 2009, 2014)

Chapter 2
Signals: Generation, Modulation, Parameters

In high mountains all steps are very important, the second also.
Welcome in Chap. 2!

2.1 Introduction

In the first chapter we were dealing with proper signal acquisition, rules of good sampling, signal computer representation, and classification/interpretation. We were observing and enjoying the real-world signals diversity. Now we will do *the deeper dive*: we will generate the most important deterministic and random signals, including amplitude and frequency modulated ones. Doing this we will *feel* better the signal concept and will better understand the signal anatomy. After that we learn about quantities and functions describing signal features like its minimum, maximum and mean value, signal variance, energy, power, RMS value, signal-to-noise ratio (SNR), and signal(s) correlation function. The signal parameters (*descriptors*) are important because they summarize the signal *behavior* in a set of a few numbers. We can track change of these numbers in time. Having them in hand we can easily imagine the nature of signal variability.

The outline of this chapter is as follows. We will start with generation of deterministic signals, continue with random ones, than become familiar with a concept of signal instantaneous frequency (IF) and generate signals with given IF. Finally, we learn mathematical definition of the most important signal parameters and code for their calculation. Special attention will be given to auto- and cross-correlation functions and their applications.

2.2 Deterministic Signals

Signals can be described by many functions. But some of them are more important than the others. In Table 2.1 equations defining some of the most frequently occurring or specially used deterministic signals are given. Sinusoid, a signal of power

© Springer Nature Switzerland AG 2021
T. P. Zieliński, *Starting Digital Signal Processing in Telecommunication Engineering*, Textbooks in Telecommunication Engineering,
https://doi.org/10.1007/978-3-030-49256-4_2

Table 2.1: Basic functions defining deterministic signals with their Matlab code

Signal type	Math definition	Matlab code
Sine, cosine	$x_1(t) = A \cdot \sin(2\pi f_0 t + \phi)$	`x1=A*sin(2*pi*f0*t+fi);`
Cmplx harmonic	$x_2(t) = A \cdot e^{j(2\pi f_0 t + \phi)}$	`x2=A*exp(j*(2*pi*f0*` `t+fi));`
Exponential[a]	$x_3(t) = A \cdot e^{-\frac{1}{T}t}$	`x3=A*exp(-t/T);`
Damped sine[a]	$x_4(t) = A \cdot e^{-\frac{1}{T}t} \cdot \sin(2\pi f_0 t + \phi)$	`x4=x3.*sin(2*pi*` `f0*t+phi);`
Gaussoid[b]	$x_5(t) = A \cdot e^{-\frac{1}{2}\left(\frac{t-t_0}{T}\right)^2}$	`x5=A*exp(-(1/2)*` `((t-t0)/T).^2 );`
AM Modulated[c]	$x_6(t) = A(1 + k_A m_A(t)) \cdot \sin(2\pi f_c t)$	`x6=A*(1+kA*mA).*` `sin(2*pi*fc*t);`
FM Modulated[c]	$x_7(t) = A\sin\left(2\pi\left(f_c t + k_F \int_0^t m_F(t)dt\right)\right)$	`x7=A*sin(2*pi*` `(fc*t+kF*cumsum(mF)*dt));`

[a] Decaying exponent having value $\frac{1}{e}A$ for $t = T$
[b] Gaussoid having maximum A for $t = t_0$ and width approximately equal to $6T$ around it
[c] A, F—amplitude and frequency, k_A, k_F—amplitude and frequency modulation depths, $m_A(t), m_F(t)$—amplitude and frequency modulation functions, f_c—carrier frequency

supply, is generated in a natural way by different resonant oscillatory circuits or resonant mechanical objects, for example a pendulum. Damped sinusoids are observed in resonators with attenuation, e.g. in magnetic resonance. Signal modulated in amplitude, frequency, and phase are used in radio broadcasting and in many digital communication systems.

Simple signals can be summed to each other or multiplied by themselves giving as a result a new signal, more complicated, even a very sophisticated mixture of signals. The power supply voltage consists of fundamental frequency component (50 or 60 Hz) and its harmonics. Damped sine is an example of multiplication of sine and exponential signal: we say that exponent is modulating sine in amplitude. Gaussoids are used also for amplitude sine modulation: impulsive oscillatory signals transmitted in radar/sonar systems can be created this way. In turn, a received signal analyzed in radar/sonar systems is a sum of different copies of the transmitted signal, reflected from different targets. And so on, and so forth.

Exercise 2.1 (Generation of Deterministic Signals). We can record real-world signals like in Exercise 1.1 but it is also possible to generate them using functions defining them, i.e. from their mathematical *recipe*. Such skill is very important because in many applications, for example in radar echolocation or digital telecommunication, we are digitally generating signals that are next converted into analog form and transmitted. So, we can at present proudly say that, in this exercise, we will deal with programmable signal generators.

In most cases typical signal generation is as easy as going for lunch to the closest fast food bar: one should know the mathematical signal formula and put numbers into it, especially *time* vector with all moments of *sample taking* from the function. Listing 2.1 presents a Matlab code of such operation while Figs. 2.1 and 2.2—waveforms of signals generated with its help, of course, after slight modifications. As an exercise, please, modify the program and generate these signals yourself.

Listing 2.1: Generation of deterministic signals

```
1   % File: lab02_ex_deterministic.m
2   clear all; close all;
3
4   fs = 1000;        % sampling frequency (samples per second):
5   N = 1000;         % number of samples to generate
6   dt = 1/fs;        % time between samples
7   n = 0 : N-1;      % vector of sample indexes
8   t = dt * n;       % vector of sampling time moments
9   x1 = sin(2*pi*5*t);    % sinusoid repeating of 5 times per second
10  x2 = exp( -t/0.1 );              % exponent with T=0.1s
11  x3 = exp( -(1/2)*((t-0.5)/0.1).^2 );  % Gaussian centered at t=0.5s with T=0.1s
12  x4a = sin( 2*pi*(100*t+90*cumsum(x1)*dt) );  % signal modulated in frequency by x1
13  x4b = sin( 2*pi*(100*t-90/(2*pi*5)*cos(2*pi*5*t) ) );  % the same theoretically
14  x5 = 1/3*randn(1,N);         % NOBODY IS PERFECT! disturbing noise
15
16  figure;
17  subplot(6,1,1); plot(t,x1,'b.-');  grid; title('x(t)');
18  subplot(6,1,2); plot(t,x2,'b.-');  grid;
19  subplot(6,1,3); plot(t,x3,'b.-');  grid;
20  subplot(6,1,4); plot(t,x4a,'b.-'); grid;
21  subplot(6,1,5); plot(t,x4b,'b.-'); grid;
22  subplot(6,1,6); plot(t,x5,'b.-');  grid; xlabel('t [s]');
```

2.3 Random Signals

When a signal does not have a specific shape, very often it has a specific distribution of its values, i.e. some values are observed more frequently than the others and they are more probable than the other values. The probability density function (PDF) denoted as $p(x)$ characterizes this feature: it tells what is the probability that a signal will take a value belonging to a small interval around x. Therefore the PDF integral should be equal to 1: the signal for sure should have some value:

$$\int_{-\infty}^{\infty} p(x)dx = 1 \tag{2.1}$$

For: (1) Gaussian noise having mean value equal to $(\bar{x})$ and variance equal to (σ^2), and (2) noise with uniform distribution of values in the open interval $a < x < b$, the PDF is defined as, respectively:

$$p_G(x) = \frac{1}{\sqrt{2\pi\sigma^2}} \exp\left[-\frac{(x-\bar{x})^2}{2\sigma^2}\right], \qquad p_U(x) = \begin{cases} \frac{1}{b-a} & \text{for} \quad a < x < b \\ 0 & \text{for} \quad \text{other } x \text{ values} \end{cases} \tag{2.2}$$

In Fig. 2.3 both probability density functions defined in Eq. (2.2), the Gaussian and the uniform, are plotted for predefined values of their parameters. At this moment, it is important to stress that deterministic gaussoid, defined in Table 2.1, as a signal shape, and signal probability density functions with a Gaussian shape, defined in Eq. (2.2) and plotted in Fig. 2.3, concern and describe two completely different things.

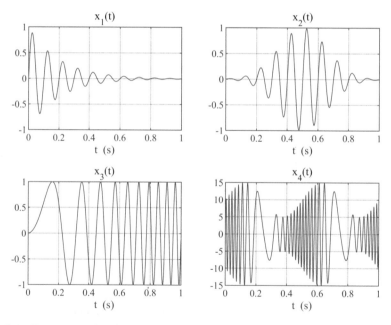

Fig. 2.1: Examples of mono-component deterministic signals: (left up) sinusoid damped (attenuated) by exponent ($T = 0.2$): $\exp(-5t)\sin(2\pi 10t)$, (right up) sinusoid with Gaussian *envelope* ($T = 0.1581$): $\exp(-20(t-0.5)^2)\sin(2\pi 10t)$, (left down) sinusoid with linearly increasing frequency (LFM): $\sin(2\pi \cdot 0.5 \cdot 20t^2)$, (right down) sinusoid being, both, modulated in amplitude and frequency: $10(1 + 0.5\sin(2\pi 2t))\sin(2\pi(30t + 25\sin(2\pi 2t)/(4\pi)))$

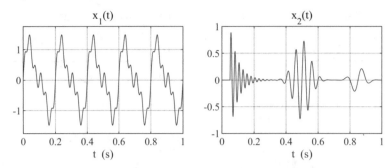

Fig. 2.2: Examples of multi-component signals: (left) multi-sine signal $x_1(t) = \sin(2\pi 5t) + 0.5\sin(2\pi 10t) + 0.25\sin(2\pi 30t)$, (right) summation $x_2(t)$ of three delayed oscillatory impulses with different frequencies and amplitude *envelopes*

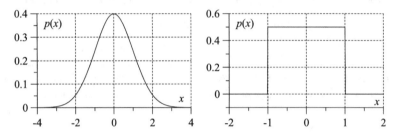

Fig. 2.3: Probability density functions for Gaussian noise with $\bar{x} = 0$, $\sigma^2 = 1$ (left), and uniform noise in the interval $[-1, 1]$ (for $a = -1, b = 1$) (right) [19]

In this book we will deal with discrete-time random signals $x(n)$, not $x(t)$. These signals are interpreted as a sequence of independent random variables, specified by their PDF functions. When these functions are identical, the signal generation process is called IID, Independent and Identically Distributed. In consequence, the process/signal is also stationary—see Fig. 1.5 and its description in Chap. 1, including definition of process stationarity. Finally, when one long discrete-time realization of a stationary process, as a sequence of realizations of IID random variables (signal samples), has a PDF identical to the PDF of all IID processes, then the process/signal is called ergodic—please, look once more the Fig. 1.5. In this case, one random signal realization only is sufficient for finding all statistical features of the signal. Such situation is assumed in the remaining part of this sub-chapter.

The central limit theorem specifies that sum of many independent random variables tends to the Gaussian (normal) PDF. For this reason such type of noise disturbance is the most frequently observed in real-world measurements and it is typically used during DSP algorithm testing. Estimation of signal mean value ($\bar{x}$), variance (σ_x^2), and standard deviation (σ_x), based on signal samples, is addressed in Table 2.2. Their definitions and Matlab implementations, which are given in it, can

Table 2.2: Basic signal features and their calculation in Matlab

Feature	Math definition	Matlab code
Mean	$\bar{x} = \frac{1}{N} \sum\limits_{n=1}^{N} x(n)$	`mean(x); sum(x)/length(x);`
Variance	$\sigma_x^2 = \frac{1}{N} \sum\limits_{n=1}^{N} (x(n) - \bar{x})^2$	`var(x); sum((x-mean(x)).^2)/N;`
Deviation	$std_x = \sqrt{\frac{1}{N-1} \sum\limits_{n=1}^{N} (x(n) - \bar{x})^2}$	`std(x); sqrt(sum((x-mean(x)).^2) /(N-1);`
Energy[a]	$E_x = T \cdot \sum\limits_{n=1}^{N} x^2(n)$	`T*sum(x.^2);`
Power[a]	$P_x = \frac{E_x}{N \cdot T} = \frac{1}{N} \sum\limits_{n=1}^{N} x^2(n)$	`sum(x.^2)/length(x);`
RMS value	$rms_x = \sqrt{P_x} = \sqrt{\frac{1}{N} \sum\limits_{n=1}^{N} x^2(n)}$	`sqrt(sum(x.^2)/length(x))`
SNR	$10 \cdot \log_{10} \frac{P_x}{P_n}$ (dB)	`10*log10(Px/Pn);`
Correlation[b]	$R_{xx}[k] = \frac{1}{C} \sum\limits_{n=1}^{N-k} x(n) y^*(n+k)$	`xcorr(x); xcorr(x,y);`

[a] T denotes sampling period, i.e. distance between signal samples (sampling frequency inverse $\frac{1}{f_s}$)
[b] C is a normalization constant equal to 1, N or $N - k$ deciding on estimator features

be used for finding real-world noise parameter values. Additionally, special tests should be performed for verification of the noise PDF type.

In Matlab pseudo-random sequences of numbers are generated by functions: `randn()` and `rand()`. The first of them returns the pseudo-Gaussian noise (*normal* noise) with $\bar{x} = 0$, $\sigma = 1$, while the second—the pseudo-uniform noise in the range $[0, 1]$. In turn, the function `px=hist(x,M)` is responsible in Matlab for the signal PDF estimation: it divides the signal value range $[\min(x), \max(x)]$ into desired number M of sub-intervals and calculated how many signal samples belong to each of them. When called by `hist(x,M)`, it only plots the estimated $p(x)$ shape. In Fig. 2.4 two noisy signals are presented, first having normal (Gaussian) PDF with mean equal to 0 and standard deviation equal to 1, and the second—having uniform PDF in the interval [0, 1]. In the figure histograms of both signals are shown also. In the left figure we see Gaussian-like bell around 0, spreading in horizontal axis from -3 to $+3$ (± 3 standard deviation equal to 1), and in the right—a rectangular *hat*, spreading in x axis from 0 to 1.

In Matlab language one can also very easily embed her/his signal in white Gaussian noise ensuring required signal-to-noise ratio (SNR), defined in Table 2.2. For this purpose, the function `awgn()`, Additive White Gaussian Noise (AWGN), is used: `x=awgn(x,SNR)`. Additive—because noise is added to our signal, white—because noise power is equally spread across (between) all frequencies, i.e. its frequency spectrum is *white*, Gaussian—because the function PDF has Gaussian shape. Here we can do short generalization of noise description: terms Gaussian/uniform

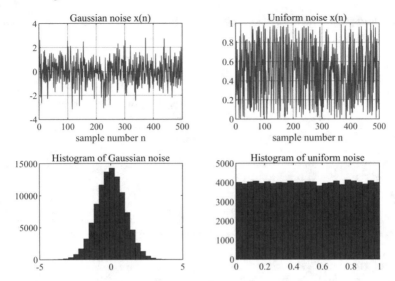

Fig. 2.4: Examples of pseudo-random signals and their histograms: (left) Gaussian noise (function randn()), (right) uniform noise (function rand())

describe the probability of noise value distribution, while terms white/pink/blue—stand for distribution of noise power between different frequencies (white—equal, pink—decreasing 6 decibels per octave, blue—increasing 6 decibels per octave). Increase/decrease 6 dB per frequency octave corresponds to 20 dB per frequency decade (decade = 10-times increase/decrease).

We encourage Reader to further reading one of many random signal theory books.

Remark: For ICT Funs In each computer language there are functions for pseudo-random number generation. Typically first congruent (additive + multiplicative) recursive number generator is used:

$$x_{n+1} = (a \cdot x_n + m) \bmod p \tag{2.3}$$

giving uniformly distributed values in the range $[1, p-1)$. In ANSI C the following parameters values are used: $p = 2^{32}, a = 1103515245, m = 12345$. After division by p numbers from the range $[0, 1)$ are obtained. Next these numbers are transformed to desired distribution of values—for example they are put into the inverse of cumulative distribution equation. In case of the normal distribution it is very easy to implement Box–Muller transformation of numbers $[0, 1)$. Look at Exercise 2.15.

Please, find additional information about the "Box–Muller transform" in . . . as usual
. . . the Internet.

Exercise 2.2 (Probability Study of Random Numbers). Make use of the
following Matlab code for obtaining plots very similar to these presented in
Fig. 2.4:

```
N=10000; M=25;
x1=randn(1,N); figure; plot(x1); figure; hist(x1,M);
x2=rand (1,N); figure; plot(x2); figure; hist(x2,M);
```

Please, observe consequences of changing values of N and M (histograms are
becoming smoother and closer to theoretical ones) with the increase of (N).
Run the program a few times and observe different signal shapes and different
values of signals parameters, for example minimum, maximum, and mean
values. Modify the code and generate: (1) pseudo-random Gaussian noise with
mean value equal to 10 and standard deviation equal to 3, (2) pseudo-random
uniform noise with values in the range $(-2, 2)$.

2.4 Sines and Instantaneous Frequency

As already mentioned in Sect. 1.3 sinusoid is the most popular signal. Analog sinusoid repeating f_0 times per second is given by the formula:

$$x(t) = \sin(2\pi f_0 t) = \sin(\omega_0 t), \quad \omega_0 = 2\pi f_0. \tag{2.4}$$

For example, sine $\sin(2\pi 10t)$ repeats 10 times per second since for $t = 1$ s the sine
argument (angle) is equal to the 10-th multiplicity of the 2π being the sine period.
The sinusoid (2.4) with amplitude A and phase ϕ is defined as

$$x(t) = A\sin(2\pi f_0 t + \phi) = A\sin(\varphi(t)), \tag{2.5}$$

where $\varphi(t)$, an argument of sine, is a function of time. When signal amplitude and
phase is changing in time $(A(t), \phi(t)$ instead of $A, \phi)$, the sinusoid is modulated
in amplitude and phase. Frequency modulation concept will be introduced later.
In many applications, for example in all damped resonance systems, sinusoid is
damped (attenuated) by exponential function:

$$x(t) = Ae^{-\lambda t}\sin(2\pi f_0 t + \phi). \tag{2.6}$$

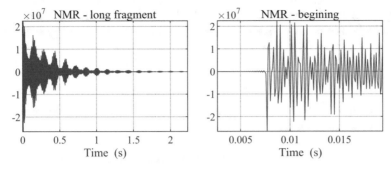

Fig. 2.5: One channel, real part, of recorded Nuclear Magnetic Resonance (NMR) signal being summation of many damped sines with different frequencies and different attenuation: a long fragment and its zoomed beginning

Sinusoids and damped sinusoids occur very often not alone but in linear superpositions (sums), like 50 Hz power supply voltage with its harmonics 100, 150, 200, 250, ... Hz:

$$x(t) = \sum_{k=1}^{K} A_k \sin(2\pi f_k t + \phi_k), \qquad (2.7)$$

or multi-component resonance signals:

$$x(t) = \sum_{k=1}^{K} A_k e^{-\lambda_k t} \sin(2\pi f_k t + \phi_k). \qquad (2.8)$$

In magnetic resonance imaging (MRI) and nuclear magnetic resonance (NMR) the damped components are even complex-value ones: real and imaginary parts are coming from two different perpendicular sensors:

$$x(t) = \sum_{k=1}^{K} A_k e^{-\lambda_k t} e^{j(2\pi f_k t + \phi_k)}. \qquad (2.9)$$

In Fig. 2.5 real part of a NMR signal (2.9), consisting of 15 damped complex oscillations, is presented.

In communication systems the signal (2.5) has varying (modulated) amplitude A and phase ϕ:

$$x(t) = A(t) \sin(2\pi f_0 t + \phi(t)) \qquad (2.10)$$

in order to encode the transmitted information in the changes of these parameters. The instantaneous frequency of the sinusoid is defined as

$$\omega_{inst}(t) = \frac{d\phi(t)}{dt}, \quad f_{inst}(t) = \frac{1}{2\pi} \frac{d\phi(t)}{dt}, \qquad (2.11)$$

i.e. it is a time-derivative of its angle, divided by 2π. Let us apply definition (2.11) to signal (2.4). The result is $f_{inst}(t) = f_0$—the correct value because signal frequency is constant and equal to f_0. Applying (2.11) to (2.10) gives

$$f_{inst}(t) = f_0 + \frac{1}{2\pi} \frac{d\phi(t)}{dt}. \tag{2.12}$$

Therefore in order to obtain from (2.10) a signal with the desired functional change $m_F(t)$ of instantaneous frequency around f_0:

$$f(t) = f_0 + k_F \cdot m_F(t) \tag{2.13}$$

one should set:

$$\phi(t) = 2\pi \cdot k_F \cdot \int_0^t m_F(t)dt, \tag{2.14}$$

since the derivative of definite integral is equal to the function being integrated:

$$f_{inst}(t) = f_0 + \frac{1}{2\pi} \frac{d}{dt}\left(2\pi \cdot k_F \cdot \int_0^t m_F(t)dt\right) = f_0 + k_F \cdot m_F(t). \tag{2.15}$$

Plots shown in the second row of Fig. 2.1 present two signals modulated in frequency. In the first of them, the instantaneous frequency starts from 0 Hz and it is linearly increasing 20 Hz per second: $f_{inst} = 0 + 20t$, while in the second one—it cosinusoidally oscillates around 30 Hz with depth 25 Hz, repeating up-down cycle two times per second: $f_{inst} = 30 + 25\cos(2\pi 2t)$. In the second case, the signal is additionally modulated in amplitude, i.e. its amplitude is changing in time according to the formula: $A(t) = 10 \cdot (1 + 0.5\sin(2\pi 2t))$.

Wow! *Is it an easy DSP starter?!* Yes, it is. It results from my teaching experience that understanding the concept of signal instantaneous frequency is the main key opening a door to understanding the concept of *signal* itself.

Exercise 2.3 (Generating Signals Modulated in Frequency). This is one of the most important exercise of this book since the frequency is the basic signal parameter. Use program from the Listing 2.1 and Matlab code from the Table 2.1, and generate different signals modulated in frequency, for example the following ones having:

- parameters as in two down plots of Fig. 2.1,
- frequency changing linearly from value f_1 to value f_2, for example from 0.01 Hz to 100 Hz in 1 s,
- frequency changing sinusoidally $\pm\Delta f$ hertz around some given f_0, for example $\pm$ 10 Hz around 10 Hz once per second,

- frequency change described by slowly varying function $f_{inst}(t)$ chosen yourself, for example $f_{inst}(t) = 10 \cdot e^{2t}$.

Assume, for example, sampling frequency equal to 1000 Hz. Check visually, whether the frequency requirements are fulfilled, observing the signal waveform/signature in each case and measuring period of oscillation (or counting changing number of samples per signal period).

Exercise 2.4 (** Generating FM Signals with Violation of the Sampling Theorem).** This is a wonderful exercise. Understanding it means that you are really very good in signal sampling and modulation, and you are ready for *climbing* next chapters of this book. Use the program 2.1 and the Table 2.1. Change sampling frequency to $f_s = 8000$ Hz and number of samples to be generated to $N = 5f_s$ (5 s). Generate a signal having frequency 0.01 Hz at $t = 0$ s and then increasing it linearly 4000 Hz per second. Plot the signal and hear it (plot(t,x); sound(x,fs);. Why the signal shape is repeating? Why do you hear the sound going periodically *up-and-down*? If it is too difficult for you, use a *Life-belt: a call to your friend* from the Wheel of Fortune—call the special Matlab function aimed at tracking of varying frequency content of a signal:

```
spectrogram(x,512,512-32,1024,fs,'yaxis');
```

You see 2D plot of frequency change in time. Observe the curve of frequency change. Explain origin of a *zig–zag* shape.

Now generate a signal with sinusoidal frequency modulation: carrier frequency $f_0 = 7500$ Hz, modulation depth $\Delta f = 500$ Hz, modulating frequency $f_m = 0.5$ Hz (once per two seconds). Is not the signal oscillating to slowly in the plot? Is not the played sound too low-frequency one? How do you explain this? Now change sampling frequency to $f_s = 44.1$ kHz and repeat last experiments. Why at present the sound is a high-frequency one?

2.5 Signal Parameters incl. Correlation Function

Having a signal we can calculate many numbers describing (characterizing) it features, like minimum, maximum, peak-to-peak (difference between maximum and minimum) and mean values, signal energy, power (energy in time), root-mean-square (RMS) value, variance, standard deviation (STD), and many others. Values of these parameters help us to *feel* the signal nature without observing its shape.

We are like a doctor who is looking at blood analysis results instead of observing a patient physiognomy. For more details see Table 2.2.

Signal-to-Noise Ratio (SNR), given in Table 2.2, is defined as a ratio of signal-to-noise power, expressed in decibels. It is used for characterization of signal strength in respect to embedded noise. SNR describes signal acquisition conditions and efficiency of performed, different DSP operations since we can compare the SNR before and after the signal processing. A ratio $R = \frac{y}{x}$ of two not-negative values y, x is expressed in decibels with the help of the following equation:

$$R(\mathrm{dB}) = 20 \cdot \log_{10}\left(\frac{y}{x}\right) = 10 \cdot \log_{10}\left(\frac{y^2}{x^2}\right). \tag{2.16}$$

In case of SNR, decimal logarithm of signal-to-noise power ratio is calculated and multiplied by 10.

Instead of simple one-value signal parameters, we can also associate with each signal a function having more values and describing it in some way. Signals can be self-similar (repetitive, periodic) or similar to each other. **Auto- and cross-correlation** functions are used for examination of this phenomena. Auto-correlation of one N-sample long $x(n)$ signal is defined in Table 2.2. Below definition of a cross-correlation of two different signals $x(n)$ and $y(n)$, both having N samples, is given:

$$R_{xy}(k) = \frac{1}{C}\sum_{n=1}^{N-k} x(n)y^*(n+k), \tag{2.17}$$

where a normalization constant C is equal 1, N or $N - k$ (for each k the exact number of samples taking part is accumulation is different and equal $N - k$). $()^*$ denotes complex conjugation if signal has complex values (otherwise it is omitted).

In the above definition the following operations are performed:

1. the first signal $x(n)$ is not moved,
2. the second signal shift value is initialized: $k = k_0$,
3. the second signal, $y^*(n)$ or $x^*(n)$ as a special case of $y^*(n)$, is shifted by k samples,
4. both signals are multiplied,
5. multiplication results are accumulated and the sum is stored as $R(k)$,
6. if necessary, value of k is changed and jump to step 3 is performed.

The distinctive maximum of the function $R(k)$ tells us that after the shift of k samples both signals $x(n)$ and $y^*(n)$ are similar to each other or one signal $x(n)$ is similar to complex conjugation of its own shift (i.e. it is periodic) and k is the period.

Correlation function is used, for example, in

- speech analysis where auto-correlation is used for finding pitch period of vocal cords opening and closing,
- radar systems where signal echoes (signal reflections coming back to transmitter) are cross-correlated with the sent signal penetrating the neighborhood, and moving object distance and velocity is found,

- telecommunication systems where receiver is cross-correlating the received signal with known pilot sequence, aiming at data synchronization and channel equalization.

Noise signal does not auto-correlate with itself for any $k \neq 0$ since there is no connection between its present value and past/future values. Noise auto-correlation function has a single peak for $k = 0$: the signal is only similar to itself with no shift.

Auto-correlation function of (co)sinusoid $x(n) = A \sin(2\pi(f_0/f_s)n)$ is equal to a cosine:

$$R_x(k) = \frac{A^2}{2} \cos\left(2\pi\frac{f_0}{f_s}k\right) \tag{2.18}$$

because (co)sinusoid is periodically self-similar to itself.

Exercise 2.5 (Calculating the Signal Features (Descriptors)). Calculate signal features defined in Table 2.2 for sinusoid (with integer and fractional number of periods), mixture of sinusoids, speech signal, and white Gaussian noise (with different lengths).

Exercise 2.6 (My AWGN: Embedding Signal in Gaussian Noise). In Octave there is no awgn() function for signal embedding in additive white Gaussian noise. Use SNR definition, presented in Table 2.2. Write a function adding to a given signal $x(n)$ the Gaussian noise, ensuring the SNR level requested by a user.

Exercise 2.7 (My XCorr and Its First DSP Mission). Write code of your own myxcorr() procedure, calculating auto- and cross-correlation function of two signals $x(n)$ and $y(n)$—use definition given in Table 2.2. Verify its correctness comparing the function output with output of the Matlab function xcorr(). Apply it to cosinusoid and verify validity of Eq. (2.18). Try to find your own speech period, auto-correlating the signal with itself.

OK. I have changed my mind: in the second chapter still more *father's help* is expected and needed, I will help you with calculation of the signal auto-correlation function. In Fig. 2.6 a fragment of voiced speech is shown, the waveform of the "a" vowel (240 samples for sampling rate 8000 samples per second), and plot of its auto-correlation function, calculated for the shift parameter k changing in the

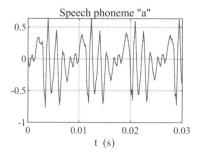
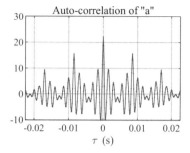

Fig. 2.6: Waveform of the vovel 'a' (left) and its auto-correlation (right)

range $-180, \ldots, 0, \ldots, 180$. One can observe that the signal is periodic with the period approximately equal to 0.01 s (80 samples). The auto-correlation function is symmetric (the same values are obtained for positive and negative values of k) since the shift left and right of the second signal give the same result when signals being correlated are the same. The auto-correlation function has the maximum value for $k = 0$—the signal is the most similar to itself in such case—no surprise. The highest side-maxima are observed for shift equal to $\approx \pm 0.01$ s which is true—it is the signal period. The Fig. 2.6 was generated with the help of below program which can be used for further experiments with correlation function testing in different applications.

Listing 2.2: Calculation of auto/cross-correlation function in Matlab

```
1  % File: Lab02_ex_correlation.m
2  clear all; close all;
3
4  [x,fs] = audioread('A8.wav',[1001,1240]);   % your waveform
5  x=x(:,1).';                        % first channel only
6
7  Nx = length(x); n=0:Nx-1;   % find number of samples, sample numbers
8  dt = 1/fs; t = dt*n;        % scale in seconds
9  K = 180;                    % how many shifts left and right
10
11 y = x;                      % more general, ready for cross-correlation
12 R(K+1) = sum( x .* conj(y) );                        % no shift, k=0
13 for k = 1 : K                                        % MAIN LOOP
14     R(K+1+k) = sum( x(1  : Nx-k) .* conj(y(1+k : Nx )) ); % shift left
15     R(K+1-k) = sum( x(1+k : Nx ) .* conj(y(1  : Nx-k)) ); % shift right
16 end                                                  %
17 k = -K : K;
18 figure
19 subplot(211); plot(t,x,'b.-');
20 xlabel('t (s)'); title('Signal fragment'); grid;
21 subplot(212); plot( k*dt, R, 'b.-');
22 xlabel('\tau (s)'); title('Signal auto-correlation'); grid;
```

Exercise 2.8 (Scaling Auto-Correlation Function). Add normalization of the auto-correlation function to the program 2.2, i.e. division by C value as in Eq. (2.17). Observe its influence upon the function shape.

2.6 Summary

In this chapter we did a longer visit in the signals' world, wonderful for me and a strange *Zoo*, or an old-fashion *museum*, for many of my students. We have generated different types of mono-component signals, deterministic and random ones, as well as their mixtures. We study more carefully amplitude and frequency signal modulation. We learn definition of sine instantaneous frequency and signal correlation. We become familiar with the most important signal parameters and their calculation. What should be remembered?

1. Deterministic signals have known shapes that are described by some mathematical function. Random signal have unknown shapes—they are unpredictable—but their probability distribution functions (PDFs), in Matlab histograms, are usually described by some specific functions, e.g. Gaussian curve.
2. The most important deterministic signal is a sinusoid and sinusoid having amplitude decreasing exponentially (so-called damped sinusoid). Such signals are generated by many real-world objects described by second-order differential equations.
3. The most important random signal is the Gaussian noise. The theory (central limit theorem) *says* that when there are many opposed factors, the Gaussian noise results.
4. Signals typically occur in mixtures: several components added together (e.g. radar echoes, many speaker talking simultaneously) or signal of interest and disturbances (like in telecommunication systems when each service is the disturber for all others).
5. A special very important class of signals represent modulated ones. Any signal can change (modulate) amplitude and/or frequency/phase of a sine. In telecommunication systems the sine is a carrier and the modulating signal—the information that is transmitted and should be correctly received.
6. It is extremely important to remember that instantaneous frequency of a single sine is equal to derivative of its angle divided by 2π. This knowledge is priceless in signal generators and transmitters.

7. Generation of deterministic signals is not difficult. We simply calculate values of functions that are defining the signals, e.g. sine, cosine, exponent, Gaussian, ... In Matlab the situation is even simpler: we send to a function a whole vector of argument values, typically time moments, and collect at once all function values—signal samples.

8. Having a signal we can calculate numbers describing it: its minimum, maximum and mean value, variance, deviation, energy, power, RMS value, signal-to-noise ratio (SNR), ... and many other. Each of them is a measure of one important signal feature. A set of them all is significantly smaller than a set of all signal samples but can give us an essential signal characterization.

9. The auto-correlation function specifies self-similarity of a signal. It is defined as an inner product (multiplication of corresponding elements and summation) of original signal samples and the shifted ones calculated for different shift values. Dominant values of the auto-correlation function for certain shift values tell us that some signal components repeat after these shifts, i.e. the shifts are their periods.

10. While introduced for the first time, the auto- and cross-correlation typically do not look for us as very important heroes of the DSP epic. But they really are! They play very important role in signal frequency analysis (power spectral density), signal detection (matched filter), and statistical signal processing (Wiener and Kalman filter). During any call each mobile phone calculates auto-correlation of the speech of any talker. How many auto-correlations are calculated all around the world now?

2.7 Private Investigations: Free-Style Bungee Jumps

Exercise 2.9 (* NMR as Sherlock-Holmes). Generate the complex-value signal (2.9) having now the form:

$$x(n) = \sum_{k=1}^{K} A_k e^{-d_k n} e^{j2\pi F_k n} + w(n), \quad F_k = \frac{f_k}{f_s}$$

which is exploited for testing algorithms for analysis of Nuclear Magnetic Resonance (NMR) signals [1]. Assume the following values of normalized frequencies, amplitudes, and damping factors:

F_k	$= -0.205$	-0.2	0.05	0.1	0.105	0.205
A_k	$= 1.0$	0.25	1.0	5.0	0.75	1
d_k	$= 0.005$	0.0	0.0	0.001	0.01	0.005

Assume $n = 0, 1, 2, \ldots, N - 1, N = 2048$. Add white Gaussian noise to the signal using Matlab functions: `x=x+std*randn(size(x))` or `x=awgn(x,SNR)` (additive white Gaussian noise with specified SNR). If, having a signal, we would find parameters of its all oscillatory components, we could deduce objects which have generated them. This way one can characterize ingredients of chemical compounds which can be safe or dangerous for human being. The method is called a spectroscopy.

Listing 2.3: Synthesis of NMR signal

```
 1 │ % File: Lab02_ex_nmr.m
 2 │ clear all; close all;
 3 │ Nx = 2^11;   % number of samples
 4 │ K  = 6;      % number of components
 5 │ F  = [ -0.205   -0.2    0.05    0.1     0.105    0.205  ];
 6 │ A  = [  1       0.25    1       5       0.75     1      ];
 7 │ d  = [  0.005   0       0       0.001   0.01     0.005  ];
 8 │ fi = zeros(1,K);
 9 │
10 │ % Signal generation
11 │ figure
12 │ x = zeros(1,Nx); n = 0:Nx-1;
13 │ for k=1:K
14 │     k
15 │     x1 = A(k) .* exp(-d(k)*n) .* exp(j*2*pi*F(k)*n);
16 │     x = x + x1;
17 │     subplot(211); plot(real(x1)); grid; title('x1 real'); axis tight;
18 │     subplot(212); plot(real(x)); grid; title('x real'); axis tight;
19 │     xlabel('sample number'); pause
20 │ end
21 │ x = awgn(x,40,'measured');  % x = x + 0.0187*randn(size(x));
22 │ figure;
23 │ subplot(211); plot(real(x)); grid; title('real(x)'); axis tight;
24 │ subplot(212); plot(imag(x)); grid; title('imag(x)'); axis tight;
25 │ xlabel('sample number'); pause
```

Exercise 2.10 (* KARAOKE Piano Bar). Visit one of Internet pages with virtual pianos (e.g. https://recursivearts.com/virtual-piano/, https://virtualpiano.net/) and play the instrument for a while. After this analyze the below Matlab code. Do you recognize the song? Now try to synthesize a musical track of your favorite song for the nearest Karaoke Bar. Musical frequency scale is defined as follows (see https://pages.mtu.edu/~suits/notefreqs.html):

- note A_k in k-th octave has frequency $f_k^A = 2^k \cdot 27.5$ Hz, where $k = 0, 1, 2, 3, \ldots 8$;
- 12 notes $\{A_k, B_k^{flat} = B_k^b, B_k, C_k, C_k^{sharp} = C_k^{\#}, D_k, D_k^{sharp} = D_k^{\#}, E_k, F_k, F_k^{sharp} = F_k^{\#}, G_k, A_k^{flat} = A_k^b\}$ of the k-th octave have the following frequencies, respectively, for $m = 0, 1, 2, 3, \ldots, 11 : f_{k,m} = f_k^A \cdot 2^{m/12}$.

Listing 2.4: Virtual piano program

```
1   % File: Lab02_ex_doremi.m
2   clear all; close all;
3
4   fs = 8000;  T = 0.5;
5   dt = 1/fs;  N = round(T/dt);  t = dt*(0:N-1);
6   damp = exp(-t/(T/2));
7
8   %          C     D     E     F     G     A     H
9   %          do    re    mi    fa    sol   la    si
10  freqs = [ 261.6, 293.7, 329.6, 349.6, 391.9, 440.0, 493.9 ];
11  kb = [ freqs; 2*freqs ]; % keybord;
12  temp = kb'; f=temp(:),
13
14  % Gama
15  mscale = [];
16  for k = 1 : length(f)
17      x = damp .* sin(2*pi*f(k)*t);
18      mscale = [ mscale x ];
19  end
20  soundsc(mscale,fs);
21  pause( T * (length(f)+1) );
22
23  % My song
24  myfreqs = [ kb(1,5) kb(1,5) kb(1,6) kb(1,5) kb(2,1) kb(1,7) ...
25              kb(1,5) kb(1,5) kb(1,6) kb(1,5) kb(2,2) kb(1,2) ...
26              kb(1,5) kb(1,5) kb(2,5) kb(2,3) kb(2,1) kb(1,7) kb(1,6) ...
27              kb(2,4) kb(2,4) kb(2,3) kb(2,1) kb(2,2) kb(2,1) ];
28  mysong = [];
29  for k = 1 : length(myfreqs)
30      x = damp .* sin(2*pi*myfreqs(k)*t);
31      mysong = [ mysong x ];
32  end
33  soundsc(mysong,fs);
```

Exercise 2.11 (* Fire! Fire!). Using elaborate frequency modulation patterns try to synthesize a very impressive alarm signal for your local fire brigade, your computer laboratory, or your boyfriend/girlfriend.

Exercise 2.12 (* My First Digital Modem). Generate a random sequence of *N* bits 0/1 using Matlab command: round(rand(1,N)). Modulate a sinusoid in amplitude or frequency according to the bit values 0/1, i.e. *high* or *low*. Write a program for bit recovery. Possibilities: AM: tracking local maximum/minimum, peak2peak value, FM: tracking local speed of zero-crossing.

Exercise 2.13 (* How My Vocal Cords Are Working? Part 2). Repeat the Exercise 1.1 but now find the pitch period for different voiced phonemes using the auto-correlation function. Calculate frequencies of vocal cords opening and closing.

Exercise 2.14 (* Is My Heart Still Broken? Part 2). Repeat the Exercise 1.11 but now find periodicity of the ECG signal using the auto-correlation function. Calculate the number of heartbeats per second.

Exercise 2.15 (* Generation of Random Numbers). Generate random numbers with uniform distribution in the interval $(0, 1)$ and then transform it into normal distribution using Box–Muller method. Test the program given below.

Listing 2.5: Random number generation by hand

```
 1  % File: Lab02_ex_BoxMuller.m
 2  clear all; close all;
 3
 4  % Uniform [0,1]
 5  r = rand_mult(10000,123);        % multiplicative generator
 6  %r = rand_multadd(10000,123);    % multiplicative + additive generator
 7  figure; plot(r,'bx');
 8  figure; hist(r,20);
 9
10  % Uniform [0,1] --> Normal(0,1)
11  N = 10000; r1 = rand(1,N);  r2 = rand(1,N);
12  n1 = sqrt(-2*log(r1)) .* cos(2*pi*r2);
13  n2 = sqrt(-2*log(r1)) .* sin(2*pi*r2);
14  figure;
15  subplot(211); hist(n1,20);
16  subplot(212); hist(n1,20);
17
18  %####################################
19  function s=rand_mult( N, seed )
20  a = 69069; p = 2^12; s = zeros(N,1);  % specially designed values
21  for i=1:N
22      s(i) = mod(seed*a,p); seed = s(i);
23  end
24  s = s/p;
25  end
26
27  %####################################
28  function s=rand_multadd( N, seed )
29  a = 69069; m = 5; p = 2^32; s = zeros(N,1); % specially designed values
30  for i=1:N
31      s(i) = mod(seed*a,p);
32      seed = s(i);
33  end
34  s = s/p;
35  end
```

References

1. E. Aboutanios, Y. Kopsinis, D. Rubtsov, Instantaneous frequency based spectral analysis of nuclear magnetic resonance spectroscopy data. Comput. Electr. Eng. **38**, 52–67 (2012)
2. J. Bendat, A. Piersol, *Random Data: Analysis and Measurement Procedures* (Wiley, New York, 1971; 4th 2010)

3. L.F. Chaparro, *Signals and Systems Using Matlab* (Academic Press, Burlington MA, 2011)
4. M.H. Hayes, *Statistical Digital Signal Processing and Modeling* (Wiley, New York, 1996)
5. M.H. Hayes, *Schaum's Outline of Theory and Problems of Digital Signal Processing* (McGraw-Hill, New York, 1999, 2011)
6. E.C. Ifeachor, B.W. Jervis, *Digital Signal Processing. A Practical Approach* (Addison-Wesley Educational Publishers, New Jersey, 2001)
7. V.K. Ingle, J.G. Proakis, *Digital Signal Processing Using Matlab* (PWS Publishing, Boston, 1997; CL Engineering, 2011)
8. S.M. Kay, *Fundamentals of Statistical Signal Processing: Estimation Theory* (PTR Prentice Hall, Englewood Cliffs, 1993)
9. R.G. Lyons, *Understanding Digital Signal Processing* (Addison-Wesley Longman Publishing, Boston, 1996, 2005, 2010)
10. J.H. McClellan, R.W. Schafer, M.A. Yoder, *DSP FIRST: A Multimedia Approach* (Prentice Hall, Englewood Cliffs, 1998, 2015)
11. S.K. Mitra, *Digital Signal Processing. A Computer-Based Approach* (McGraw-Hill, New York, 1998)
12. A.V. Oppenheim, R.W. Schafer, *Discrete-Time Signal Processing* (Pearson Education, Upper Saddle River, 2013)
13. A.V. Oppenheim, A.S. Willsky, S.H. Nawab, *Signals & Systems* (Prentice Hall, Upper Saddle River, 1997, 2006)
14. A. Papoulis, *Signal Analysis* (Mc-Graw Hill, New York, 1977)
15. J.G. Proakis, D.G. Manolakis, *Digital Signal Processing. Principles, Algorithms, and Applications* (Macmillan, New York, 1992; Pearson, Upper Saddle River, 2006)
16. S.W. Smith, *The Scientist and Engineer's Guide to Digital Signal Processing* (California Technical Publishing, San Diego, 1997, 1999). Online: http://www.dspguide.com/
17. K. Steiglitz, *A Digital Signal Processing Primer: With Applications to Digital Audio and Computer Music* (Pearson, Upper Saddle River, 1996)
18. M. Vetterli, J. Kovacevic, V.K. Goyal, *Foundations of Signal Processing* (Cambridge University Press, Cambridge, 2014)
19. T.P. Zieliński, *Cyfrowe Przetwarzanie Sygnałów. Od Teorii do Zastosowań (Digital Signal Processing. From Theory to Applications)* (Wydawnictwa Komunikacji i Łączności (Transport and Communication Publishers), Warszawa, Poland, 2005, 2007, 2009, 2014)

Chapter 3
Signal Orthogonal Transforms

Everything can be taken to pieces and assembled back. A signal also.

3.1 Introduction

This chapter is devoted to fundamental concept of signal decomposition into some simpler components (into some *smaller parts*). We are very close to mathematical theory of functional analysis, i.e. representation of one function as a result of weighted summation of some other *basic* functions. The basis functions should be orthogonal to each other: their inner products (*similarity* to each other) should be equal to 0. They are many sets of functions fulfilling this condition, in consequence there are many orthogonal signal transformations, for example discrete cosine transforms (DCTs), discrete sine transforms (DSTs), discrete Fourier (DFT), Hartley, Haar and Walsh–Hadamard transform. In this chapter we learn about general orthogonal signal analysis (looking for a signal *recipe/prescription*, i.e. calculation of signal similarity coefficients/weights in respect to some elementary signals) and orthogonal signal synthesis (summation of elementary signals scaled by calculated similarity weights). In computer implementation both operations are straightforward: first a vector of signal samples has to be multiplied by an *analysis* orthogonal matrix having different basis function in each row. Then the resultant vector of similarity coefficient is multiplied by a *synthesis* matrix being transposition and complex conjugation of the first. For real-value transformations only matrix transposition is done which shifts samples of basis functions from rows to corresponding columns. Wow! Yes! In discrete case both direct and inverse orthogonal signal transformation simplify to multiplication of a vector and by a rectangular matrix. When only a few similarity coefficients are significant for a given signal, we are telling that the transformation has compact *support*. It is the case when basis functions well fit to signal components. When we synthesize a signal from modified similarity coefficients, some filtering of signal content is done.

© Springer Nature Switzerland AG 2021
T. P. Zieliński, *Starting Digital Signal Processing in Telecommunication Engineering*, Textbooks in Telecommunication Engineering,
https://doi.org/10.1007/978-3-030-49256-4_3

3.2 Orthogonal Transformation by Intuition: From Points in 3D to Vector Spaces

And now slowly, from the beginning. Each color can be treated, roughly speaking, as a summation of three basics ones: red–green–blue (RGB). Each point in 3-dimensional space XYZ has three coordinates/components (x,y,z). Could any signal be represented as a linear superposition of some elementary (basic) components? The answer is YES and there is infinite number of such representations. The only question is how to choose *good* ones. For example, an RGB color can be coded using YIQ, YUV, or YCbCr components that could better fit to some specific applications. Similarly, using again the 3D space template, we could rotate our coordinate XYZ system in 3D and a point (x,y,z), not lying upon axes 0-X, 0-Y, 0-Z, could have simpler representation (xr, yr, 0) in the rotated system 0-Xr, 0-Yr, 0-Zr.

So, the question is what is used as RGB or XYZ in signal decomposition case? Since in real world, in circuits and systems engineering, different frequency components (sines) are treated (attenuated and delayed) in different manner, the most important is frequency content of any signal. Therefore, our elementary basis functions, playing a role of RGB colors in signal space, should be some frequency patterns. Since signals are treated by us as functions, this reasoning leads us to functional analysis—a branch of mathematics dealing with the problem of approximation of one function by linear superposition of some other functions, obligatory orthogonal. Orthogonality means that each two basic functions/signals are totally independent, in some way *perpendicular* to themselves as axes 0-X, 0-Y, 0-Z in 3D. The well-known Fourier series fulfills the orthogonality requirement and allows us to decompose any continuous differentiable periodic function into an infinite summation of sines and cosines having frequencies being an integer multiple of fundamental frequency, inverse of a signal period.

Digital signals are N-element vectors of samples: $[x_1, x_2, x_3, \ldots, x_N]$. Therefore vector spaces and linear algebra offer the best starting point for explanation and understanding the concept of orthogonal signal decomposition (by analogy to XYZ one).

Any signal, understood as a vector of samples, can be represented as a linear superposition (summation) of some orthogonal vectors, treated as elementary functions of the decomposition. Typically, the orthogonal vectors are obtained by sampling some oscillatory functions, sines and cosines, with different frequencies. The idea of orthogonal signal transformation is graphically explained in Fig. 3.1. Matrix **A**, having in rows samples of elementary functions, into summation of which our signal is decomposed, is multiplied by vector of signals samples $\bar{\mathbf{x}}, x(n), n = 1 \ldots N$.:

$$\bar{\mathbf{y}} = \mathbf{A} \cdot \bar{\mathbf{x}}. \tag{3.1}$$

Vector $\bar{\mathbf{y}}$ represents a decomposition result: a signal *recipe* called a signal spectrum. It consists of coefficients telling us how much our signal is similar to each

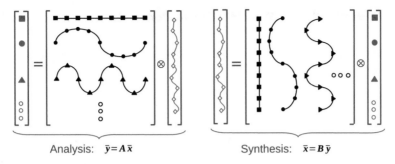

Fig. 3.1: Graphical explanation of orthogonal signal analysis/decomposition (left) and synthesis (right)

elementary function. When we take all elementary functions, scale them by these coefficients and add, the signal is perfectly reconstructed:

$$\bar{\mathbf{x}} = \mathbf{A}^{-1} \cdot \bar{\mathbf{y}} = (\mathbf{A}^*)^T \cdot \bar{\mathbf{y}}. \tag{3.2}$$

Equation (3.1) describes orthogonal signal transformation (i.e. the signal analysis), while Eq. (3.2) inverse orthogonal transformation (i.e. signal synthesis). Since inverse of an orthogonal matrix is obtained by complex conjugation of the matrix elements and the matrix transposition (rows becomes columns), during signal synthesis the same elementary vectors are used as during analysis: the signal is synthesized as a sum of weighted elementary functions. Only so much and so much.

When equality $\mathbf{A}^{-1} = (\mathbf{A}^*)^T$ holds, the matrix $\mathbf{A}$ is more than *orthogonal* only: it is orthogonal and the same time normalized, i.e. *orthonormal*. In such case $\mathbf{A}^{-1} \cdot \mathbf{A} = \mathbf{I}$, the identity matrix. In this chapter we will consider such case only. When matrix $\mathbf{A}$ is *only* orthogonal, i.e. its rows and column are mutually orthogonal, multiplication of the matrix and its transposed conjugation gives a diagonal matrix with not all elements equal to 1 on the main diagonal. In such case, some additional scaling of orthogonal forward (3.1) and backward (3.2) transformations is required. It will be introduced in next chapter on discrete orthogonal Fourier transform (DFT).

Many such signal decompositions (orthogonal matrices) exist. They are giving us information about the signal *content*. Knowing the signal decomposition result, we can remove (modify) some of its components and synthesize its *filtered* version. This is like with a soup: having its recipe we can modify it, for example remove cucumbers and increase number of tomatoes, and boil a different soup using the modified prescription.

3.3 Orthogonal Transformation Mathematical Basics

Let $\bar{\mathbf{x}}$ and $\bar{\mathbf{y}}$ denote N-element vertical vectors of signal samples, real- or complex-value ones:

$$\bar{\mathbf{x}} = \begin{bmatrix} x_1 \\ x_2 \\ \vdots \\ x_N \end{bmatrix}, \quad \bar{\mathbf{y}} = \begin{bmatrix} y_1 \\ y_2 \\ \vdots \\ y_N \end{bmatrix}, \tag{3.3}$$

They are orthogonal (*independent, perpendicular*) when their inner product $\langle \bar{\mathbf{x}}, \bar{\mathbf{y}} \rangle$, summation of multiplications of all corresponding elements is equal to zero:

$$\langle \bar{\mathbf{x}}, \bar{\mathbf{y}} \rangle = x_1 y_1^* + x_2 y_2^* + \ldots + x_N y_N^* = \begin{bmatrix} y_1^* & y_2^* & \cdots & y_N^* \end{bmatrix} \begin{bmatrix} x_1 \\ x_2 \\ \vdots \\ x_N \end{bmatrix} = \bar{\mathbf{y}}^H \bar{\mathbf{x}} = 0. \tag{3.4}$$

In Eq. (3.4) the mark "*" denotes complex conjugation, negation of imaginary part of complex-value data, and "$()^H$"—complex conjugation and additional vector transposition (changing orientation vertical $\leftrightarrow$ horizontal). Both vectors $\bar{\mathbf{x}}, \bar{\mathbf{y}}$ are orthonormal when additionally they have norm equal to 1 (perfect similarity to itself):

$$\|\bar{\mathbf{v}}\| = \langle \bar{\mathbf{v}}, \ \bar{\mathbf{v}} \rangle = 1, \tag{3.5}$$

Inner product can be treated as a measure of similarity: 1 = the same, 0 = completely different. Unitary vectors $\bar{\mathbf{e}}_x = [1,0,0]$, $\bar{\mathbf{e}}_y = [0,1,0]$, $\bar{\mathbf{e}}_z = [0,0,1]$, defining axes 0-X, 0-Y, 0-Z in 3D space are orthogonal because their mutual inner products are equal to 0 (1s are in different positions):

$$\langle \bar{\mathbf{e}}_x, \bar{\mathbf{e}}_y \rangle = 1 \cdot 0 + 0 \cdot 1 + 0 \cdot 0 = 0,$$
$$\langle \bar{\mathbf{e}}_x, \bar{\mathbf{e}}_z \rangle = 1 \cdot 0 + 0 \cdot 0 + 0 \cdot 1 = 0,$$
$$\langle \bar{\mathbf{e}}_y, \bar{\mathbf{e}}_z \rangle = 0 \cdot 0 + 1 \cdot 0 + 0 \cdot 1 = 0.$$

Exercise 3.1 (Checking Vector Orthogonality). In this short program we are checking orthogonality of two vector of numbers:

```
x = [ 1; -1; 2 ], y = [ 3; 1; -1 ], pause
o1 = sum(x .* conj(y)), o2 = (y') * x, pause
```

Change values of vector elements and run program a few times. Then make the vectors longer. At present vectors are vertical. Make them horizontal. Modify the program if it stops to work after the last change.

Let us choose N orthonormal, vertical, basis vectors $\bar{\mathbf{v}}_k, k = 1, 2, 3, \ldots, N$, N-element each, and build matrix $\mathbf{V}$, having conjugated vectors in rows (conjugation if complex-value):

$$
\mathbf{V} = \begin{bmatrix} \bar{\mathbf{v}}_1^H \\ \bar{\mathbf{v}}_2^H \\ \vdots \\ \bar{\mathbf{v}}_N^H \end{bmatrix} = \begin{bmatrix} v_{1,1}^* & v_{1,2}^* & \cdots & v_{1,N}^* \\ v_{2,1}^* & v_{2,2}^* & \cdots & v_{2,N}^* \\ \vdots & \vdots & \ddots & \vdots \\ v_{N,1}^* & v_{N,1}^* & \cdots & v_{N,N}^* \end{bmatrix},
\tag{3.6}
$$

where $v_{k,n}$ denotes n-th element of k-th vector. By definition it is an *orthogonal* matrix since it has *orthogonal* rows. When we multiply vector $\bar{\mathbf{x}}$ by this matrix we do its orthogonal transformation:

$$
\bar{\mathbf{X}} = \begin{bmatrix} X_1 \\ X_2 \\ \vdots \\ X_N \end{bmatrix} = \begin{bmatrix} <\bar{\mathbf{v}}_1^H, \bar{\mathbf{x}}> \\ <\bar{\mathbf{v}}_2^H, \bar{\mathbf{x}}> \\ \vdots \\ <\bar{\mathbf{v}}_N^H, \bar{\mathbf{x}}> \end{bmatrix} = \begin{bmatrix} v_{1,1}^* & v_{1,2}^* & \cdots & v_{1,N}^* \\ v_{2,1}^* & v_{2,2}^* & \cdots & v_{2,N}^* \\ \vdots & \vdots & \ddots & \vdots \\ v_{N,1}^* & v_{N,1}^* & \cdots & v_{N,N}^* \end{bmatrix} \begin{bmatrix} x_1 \\ x_2 \\ \vdots \\ x_N \end{bmatrix}
\tag{3.7}
$$

Vector $\bar{\mathbf{X}}$ consists of coefficients $<\bar{\mathbf{v}}_k^H, \bar{\mathbf{x}}>, k = 1, 2, \ldots, N$, specifying similarity of vector $\bar{\mathbf{x}}$ to the elementary vectors $\bar{\mathbf{v}}_k^H$. Equation (3.7) can be written as:

$$
\bar{\mathbf{X}} = \mathbf{V} \cdot \bar{\mathbf{x}}.
\tag{3.8}
$$

Multiplication of both sides of (3.8) by the inverse of matrix $\mathbf{V}$ results in

$$
\mathbf{V}^{-1} \bar{\mathbf{X}} = (\mathbf{V}^{-1} \mathbf{V}) \cdot \bar{\mathbf{x}} = \mathbf{I} \cdot \bar{\mathbf{x}} = \bar{\mathbf{x}}.
\tag{3.9}
$$

Therefore the vector $\bar{\mathbf{x}}$ can be always recovered from vector $\bar{\mathbf{X}}$ of similarity coefficients for any matrix $\mathbf{V}$ having an inverse:

$$
\bar{\mathbf{x}} = \mathbf{V}^{-1} \cdot \bar{\mathbf{X}}.
\tag{3.10}
$$

Exercise 3.2 (Forward and Backward: Solving Inverse Equation).
Let us do a very simple experiment: we will generate a vector and a rectangular matrix with random numbers. Then we will perform the vector transformation (3.8) and the inverse transformation (3.10), assuming that the matrix $\mathbf{V}$ has an inverse:

```
N=100; x=randn(N,1); V=randn(N,N);
X = V*x; xe = inv(V)*X;
error = max(abs( x - xe )),
```

What do we see? We came **back** perfectly! Concluding: there is no problem with doing perfect signal analysis and synthesis using any matrix and its inverse, if it exists. But interpretation of the similarity vector $\bar{\mathbf{X}}$ is difficult in this case: the signal is reconstructed as sum of columns of matrix $\mathbf{V}^{-1}$ scaled by coefficients $\bar{\mathbf{X}}$. If column of $\mathbf{V}^{-1}$ would be equal to rows of $\mathbf{V}$ the situation would be much more comfortable for us: the same functions would be used for the analysis and synthesis. We could easily interpreted elements of $\bar{\mathbf{X}}$ in such case. For the non-orthogonal matrix $\mathbf{V}$ it is impossible. Run the program a few times for different random data. Is the perfect signal reconstruction observed all the time?

But we have assumed that $\mathbf{V}$ is an **orthogonal matrix** and as such it not only has an inverse but this inverse is equal to transposition and conjugation of $\mathbf{V}$ (denoted by operator $(.)^H$):

$$\mathbf{V}^{-1} = \mathbf{V}^H = (\mathbf{V}^*)^T. \tag{3.11}$$

Only orthogonal matrix $\mathbf{V}$ has this nice feature! So we see that during signal synthesis from similarity coefficients $\bar{\mathbf{X}}$ we are using the same vectors as for the analysis, only transposed and conjugated:

$$\bar{\mathbf{x}} = (\mathbf{V}^*)^T \cdot \bar{\mathbf{X}}, \qquad \bar{\mathbf{x}} = \sum_{k=1}^{N} X_k \bar{\mathbf{v}}_k. \tag{3.12}$$

and the same but more *user-friendly*:

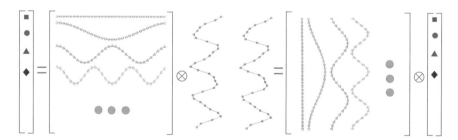

Fig. 3.2: Graphical illustration of signal analysis (left) and synthesis (right) by means of orthogonal transformations. During decomposition signal is multiplied by a matrix having reference oscillatory signals in its rows. Calculated similarity coefficients are used next for signal synthesis: vector with them is multiplied by a matrix having basis functions in its columns—the columns are scaled by transform coefficients and added

$$
\begin{bmatrix} x_1 \\ x_2 \\ \vdots \\ x_N \end{bmatrix} = \begin{bmatrix} v_{1,1} & v_{2,1} & \cdots & v_{N,1} \\ v_{1,2} & v_{2,2} & \cdots & v_{N,2} \\ \vdots & \vdots & \ddots & \vdots \\ v_{1,N} & v_{2,N} & \cdots & v_{N,N} \end{bmatrix} \begin{bmatrix} X_1 \\ X_2 \\ \vdots \\ X_N \end{bmatrix},
$$

$$
\begin{bmatrix} x_1 \\ x_2 \\ \vdots \\ x_N \end{bmatrix} = X_1 \begin{bmatrix} v_{1,1} \\ v_{1,2} \\ \vdots \\ v_{1,N} \end{bmatrix} + X_2 \begin{bmatrix} v_{2,1} \\ v_{2,2} \\ \vdots \\ v_{2,N} \end{bmatrix} + \ldots + X_N \begin{bmatrix} v_{N,1} \\ v_{N,2} \\ \vdots \\ v_{N,N} \end{bmatrix}
$$

Therefore these coefficients have a big sense: our signal is represented as a summation of basis orthogonal vectors taken with some weights. Basic concept of orthogonal signal transformation is once again repeated in Fig. 3.2 similar to Fig. 3.1. Like in the army: permanent repeating makes masters from us!

So doing orthogonal signal transformation we are looking for signal $\bar{\mathbf{x}}$ similarity to vectors $\bar{\mathbf{v}}_k^*$, $k = 1, 2, 3, \ldots, N$, and doing inverse transformation we are doing synthesis of $\bar{\mathbf{x}}$ using vectors $\bar{\mathbf{v}}_k$, $k = 1, 2, 3, \ldots, N$. Since the signal is represented as a sum of vectors $\bar{\mathbf{v}}_k$, we can say that during orthogonal transformation it is decomposed into these vectors (*spanned* by these vectors). And we can choose them (signal components we are looking for) in different ways. The best are vectors $\bar{\mathbf{v}}_k$ similar to real-life (real-world) components, e.g. (co)sine oscillations. In Fig. 3.2 graphical illustration of forward and backward orthogonal signal transformations is given.

In signal theory books matrix equations of direct and inverse orthogonal transformation of signals/vectors are typically written in the following form:

$$
\textbf{ANALYSIS}: \quad X_k = \langle \bar{\mathbf{x}}, \bar{\mathbf{v}}_k \rangle = \bar{\mathbf{v}}_k^H \cdot \bar{\mathbf{x}} = \sum_{n=1}^{N} x_n v_{k,n}^*, \quad k = 1, 2, 3, \ldots, N, \quad (3.13)
$$

$$
\textbf{SYNTHESIS}: \quad \bar{\mathbf{x}} = \sum_{k=1}^{N} X_k \bar{\mathbf{v}}_k. \quad (3.14)
$$

In the above discussion we have assumed more general case when vectors $\bar{\mathbf{v}}_k$ are complex-value. In such situation during analysis (3.7), (3.13) their conjugation is used. The reason of this is that in the next chapter we will discuss with more details the discrete Fourier transformations being a complex-value one. For real-value orthogonal basis vectors, the conjugation is not used (there is no imaginary part to be negated).

Soup Example Orthogonal transformations of signals and orthogonal signal decompositions are *unpleasant* concept for students. To make them more *tasty* I usually present to students *a soup example*. Let us assume that we have the soup components: potatoes, tomatoes, cucumbers, onions …, and that they are orthogonal. Then we go to a bar and we eat fantastic soup. Of course, we are interested in its recipe. So, what we are doing? We compare the soup as a whole with a reference potato (some basic pattern) and find the similarity coefficient for potatoes, i.e. how many they are in our soup, e.g. one and a half. Next, we repeat this operation for all soup ingredients, and find in this way (analysis) a complete soup prescription. Next, we go home and during a Sunday dinner we boil the soup (do it synthesis) for our guests using its know recipe. The same is with a signal: we are finding its ingredients and synthesizing it, even with slight modification of the recipe (e.g. with a little bit less or more amount of some signal frequency components).

Exercise 3.3 (Forward and Backward: Usage of Orthogonal Matrix).
At present we modify the program in Exercise 3.2. We use orthogonal transformation matrix **V** of the DCT-IV transform, which will be defined later, and in the backward signal synthesis we exchange its inverse with its joint transposition and conjugation, in Matlab denoted as ( ) ′ :

```
N=100; x=randn(N,1);
k=(0:N-1); n=(0:N-1);
V=sqrt(2/N)*cos( pi/N * (n'+1/2)*(k+1/2) );
X = V' * x; xe = V * X;
error = max(abs( x-xe )), ortho = sum( V(:,10).*conj(V(:,20)) ),
```

If the signal reconstruction error is on the level of computational accuracy, let us say 10^{-13}, we can conclude that time spent for reading equations in this chapter was not lost at least! We can find with ease *ingredients* of our signal and synthesize the signal back using them (i.e. *boil the soup*). In the last program line we have checked orthogonality of the 10-th and 20-th column of the matrix **V**. Please, check orthogonality of all pairs of matrix columns. Or alternatively, multiply the matrix **V** and its transposed conjugation V^H, in Matlab ortho = V′*V, and see whether the identity matrix is obtained.

Exercise 3.4 (Testing Simple 3D Orthogonal Transformation). Let us do a very simple experiment: we will rotate orthogonal unitary vectors of the 3D coordinate system and check orthogonality of the rotated ones. To do verification in one step we will build a matrix from input vectors (put them into matrix columns), multiply it by any 3D rotation matrix, and check orthogonality of the resultant matrix. For orthogonal matrix the matrix multiplication by its conjugated transposition should result in a diagonal matrix! In the case of orthonormality (additional normalization)—only "1s" on the main diagonal are allowed.

```
a = pi/4; c = cos(a); s = sin(a); % rotation angle a
Rx = [ 1 0 0; 0 c s; 0 -s c ]; % 3D 0-X rotation matrix
V = eye(3), Vx = Rx*V, % rotation, new 3D system vectors
ortho = Vx'*Vx, pause; % checking results: diagonal or not?
```

Modify the program: add rotation matrices around O-Y and 0-Z axes. Create more complex rotations multiplying rotation matrices by themselves:

Rxyz=Rx*Ry*Rz. Check that in each case you can go back doing inverse (backward) rotation, i.e. doing *inverse* orthogonal transformation with a matrix being conjugated transposition of the direct (forward) transformation matrix. At present let us do *the final cut*: let us transform any point $(u) = (x, y, z)$ from the first coordinate system to the second one and go back, i.e. do in chain the forward and backward orthogonal transformation of a point in 3D space, i.e. the vector **u**:

```
u = [ 1; 2; 3], pause % vector/signal
ux = (Vx') * u; [u ux], pause % forward transformation
ub = Vx * ux; [ux ub], pause % backward transformation
```

It is a magic! **Perfect reconstruction!** Now do it for any 3D vector (signal): u = randn(3,1). It works for any square orthonormal matrix of any size! But may be some matrices are *better*. Which ones and why? How to find them?

3.4 Important Orthogonal Transforms

Signal orthogonal transforms differ in selection of orthogonal functions that are used for signal decomposition. Many functions can be used for this purpose: sines, cosines, summation of sines and cosines, and different rectangular-shape sequences. In Fig. 3.3 function shapes for several orthogonal transformations are shown for $N = 8$. When basis functions are more similar to signal components, less number of decomposition coefficients have significant values. We are telling in such situation that a *sparse* signal representation is obtained and the decomposition functions offer *compact signal support*. In this case signal components are well represented by decomposition functions (since these functions are similar to signal components). And this fact apart from existence of a fast algorithm should be used for the transform selection. The most popular are discrete cosine transforms using sampled cosines functions as elementary signals.

Definitions of the most important **real-value discrete orthogonal transforms** are listed below where $v_{k,n}$ and $v_k(n)$ denote n-th sample of k-th basic function and $k, n = 0, 1, 2, \ldots, N-1$ for all transformations except DCT-I: $k, n = 0, 1, 2, \ldots, N$.

1) discrete cosine transforms DCT-I, DCT-II, DCT-III, DCT-IV:

$$\textbf{DCT-I:} \qquad v_{k,n} = v_k(n) = \sqrt{2/N} \cdot c(k)\, c(n) \cdot \cos\left[\frac{\pi kn}{N}\right], \qquad (3.15)$$

$$\textbf{DCT-II:} \qquad v_{k,n} = v_k(n) = \sqrt{2/N} \cdot c(k) \cdot \cos\left[\frac{\pi k(n+1/2)}{N}\right], \qquad (3.16)$$

$$\textbf{DCT-III:} \qquad v_{k,n} = v_k(n) = \sqrt{2/N} \cdot c(n) \cdot \cos\left[\frac{\pi (k+1/2)n}{N}\right], \qquad (3.17)$$

$$\textbf{DCT-IV:} \qquad v_{k,n} = v_k(n) = \sqrt{2/N} \cdot \cos\left[\frac{\pi (k+1/2)(n+1/2)}{N}\right], \qquad (3.18)$$

where

$$c(m) = \begin{cases} 1/\sqrt{2}, & m=0 \quad \text{or} \quad m=N \\ 1, & 0 < m < N \end{cases} \qquad (3.19)$$

2) discrete sine transform:

$$v_{k,n} = v_k(n) = \sqrt{\frac{2}{N+1}} \sin\left[\frac{\pi (k+1)(n+1)}{N+1}\right], \qquad (3.20)$$

3) discrete Hartley transform:

$$v_{k,n} = v_k(n) = \frac{1}{\sqrt{N}}\left(\cos\frac{2\pi}{N}kn + \sin\frac{2\pi}{N}kn\right) == \sqrt{\frac{2}{N}}\sin\left(\frac{2\pi}{N}kn + \frac{\pi}{4}\right), \quad (3.21)$$

4) Hadamard transform (only values 1 and -1 divided by $\sqrt{N}$):

$$v_{k,n} = v_k(n) = \frac{1}{\sqrt{N}}(-1)^{f(k,n)}, \qquad (3.22)$$

where

$$f(k,n) = \sum_{i=0}^{M-1} k_i n_i, \quad M = \log_2 N, \quad k_i,\, n_i = 0,\, 1,$$

$$k = k_0 + 2k_1 + \ldots + 2^{M-1}k_{M-1},$$

$$n = n_0 + 2n_1 + \ldots + 2^{M-1}n_{M-1},$$

5) Haar transform (only values $\pm 2^{p/2}$ divided by $\sqrt{N}$):

$$v_{0,n} = 1/\sqrt{N}, \quad v_{k,n} = \frac{1}{\sqrt{N}}\begin{cases} 2^{p/2}, & \frac{q-1}{2^p} \le \frac{n}{N} < \frac{q-1/2}{2^p} \\ -2^{p/2}, & \frac{q-1/2}{2^p} \le \frac{n}{N} < \frac{q}{2^p} \\ 0, & \text{other } n \end{cases}, \quad k = 1,\, 2,\ldots,N-1,$$

$$(3.23)$$

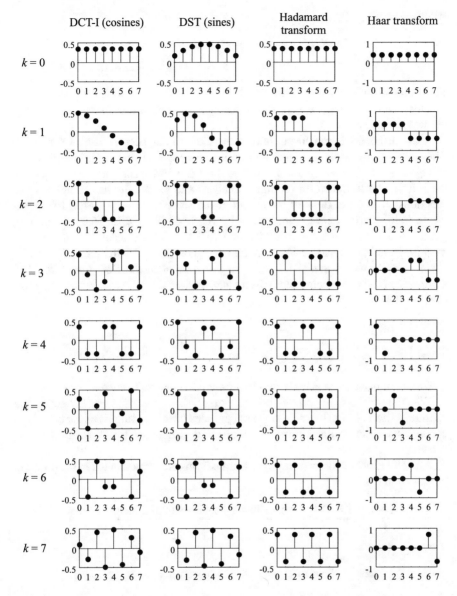

Fig. 3.3: Vectors $\bar{\mathbf{v}}_k, k = 0, 1, 2, \ldots, N - 1$, for some real-value orthogonal transformations: cosine, sine, Hadamard, and Haar (vertically from left to right). REMARK: for Hadamard transform vectors are presented in changed order: $\bar{\mathbf{v}}_0, \bar{\mathbf{v}}_4, \bar{\mathbf{v}}_6, \bar{\mathbf{v}}_2, \bar{\mathbf{v}}_3, \bar{\mathbf{v}}_7, \bar{\mathbf{v}}_5, \bar{\mathbf{v}}_1$, in order to obtain increasing frequency of oscillations [10]

where

$$N = 2^m, \quad k = 2^p + q - 1, \quad 0 \le p \le m - 1,$$
$$p = 0 \ \Rightarrow \ q = 0, 1, \quad p \ne 0 \ \Rightarrow \ 1 \le q \le 2^p.$$

For example for $N = 4$ one has: $k = 0, 1, 2, 3$; $p = 0, 0, 1, 1$; $q = 0, 1, 1, 2$.

Please, *do not worry, be happy* even seeing equations of the Haar transform. Not all days are rainy.

Most orthogonal transforms have real-value signal decomposition functions. They are mainly used in image and audio compression, like DCT in JPEG/M-PEG and AAC standards. But one of the most important and the most frequently used in the world transform for signal analysis and processing has complex-value functions! It is more *difficult* for us but at the same time very useful, for example in multi-carrier telecommunication (DAB, DVB-T, ADSL, Wi-Fi, LTE, 5G), psycho-acoustical sub-band audio coding (MP3 and AAC standards), and general-purpose signal analysis like in magnetic resonance (MRI, NMR) . This is the **discrete Fourier transform (DFT)**, *the biggest Animal in the DSP forest! the real King of the DSP road!* being time-discretized version of the Fourier series equations ($j = \sqrt{-1}$):

$$v_{k,n} = v_k(n) = \frac{1}{\sqrt{N}} \exp\left(j\frac{2\pi}{N}kn \right), \quad k, n = 0, 1, 2, \ldots, N-1, \qquad (3.24)$$

Due to its importance it is individually presented in the next chapter.

Why the DFT is so important? Because it is *shift invariant*: the original signal and its time-shifted version have the some magnitudes (absolute values) of the transform coefficients. This is not the case for real-value orthogonal transformation and is shown in next section.

All defined above discrete orthogonal transforms are implemented in the program presented in Listing 3.1.

Listing 3.1: Orthogonal sine/cosine-like signal transformations

```
1   % lab03_ex_transforms.m
2   clear all; close all;
3
4   N = 8; % for ortho checking N=8 (examples), for spectrum viewing e.g. N=100
5
6   % Orthogonal transform matrices (basis function in columns k, n - samples number)
7   k=0:N-1; n=0:N-1;        % k-basis function index, n-b.f. sample index
8                            % Comment the below two lines after understanding them
9   Indexes = n'*k,          % outer product of argument vectors
10  CosMatrix = cos(n'*k), pause   % function values for index matrix
11
12  k=(0:N-1); n=(0:N-1); nk = [1,N];
13  V1 = sqrt(2/(N-1)) * cos( pi*n'*k/(N-1));                      % DCT-I
14  V1(nk,:)=V1(nk,:)/sqrt(2); V1(:,nk)=V1(:,nk)/sqrt(2);         % DCT-I
15  V2 = sqrt(2/N) * cos( pi*(n+1/2)'*k/N ); V2(:,1)=V2(:,1)/sqrt(2); % DCT-II
16  V3 = sqrt(2/N) * cos( pi*n'*(k+1/2)/N ); V3(1,:)=V3(1,:)/sqrt(2); % DCT-III
```

```
17  V4 = sqrt(2/N) * cos( pi/N *(n+1/2)'*(k+1/2) );              % DCT-IV
18  V5 = sqrt(2/(N+1)) * sin(pi*(n+1)'*(k+1)/(N+1));             % DST
19  V6 = sqrt(1/N) * ( cos(2*pi/N*n'*k) + sin(2*pi/N*n'*k) );    % Hartley
20  V7 = sqrt(1/N) * exp(j*2*pi/N *n'*k);   % DFT - only info, real() below for DFT
21
22  V = V1;                      % our choice
23  ortho = V' * V,              % orthogonality test: 1s on main diag, 0s elsewhere?
24  figure;                      % open figure
25  for k = 1:N                  % comment these four lines
26      k, stem(real(V(:,k))), pause  % after seeing the transform
27  end                          % basis functions
28
29  % Choosing signal to be transformed forward and backward (vertical vector)
30  N1 = floor(N/2); N2=N-N1;
31  x1 = 5*V(:,2) + 10*V(:,7);        % #1: linear superposition of V columns
32  x2 = [ x1(N); x1(1:N-1) ];        % #2: first signal shifted by one sample
33  x3 = sin(2*pi/7*n');              % #3: pure sine
34  x4 = [ones(N1,1); -ones(N2,1)];   % #4: rectangular wave
35  x5 = randn(N,1);                  % #5: random Gaussian noise
36  % x6 = ...;                       % #6: now your turn
37  x = x1;                           % OUR CHOICE
38  figure; plot(n,real(x),'ro-'); title('x(n)');  % signal plot
39
40  % Transformations                 % (.)' - transposition and conjugation
41  X = V'*x;                         % direct transformation (analysis)
42  figure; stem(real(X)); title('X(k)');   % plot of transform coefficients
43
44  if(0)                             % possible "spectrum" modification
45   X(8) = 0;                        % remove unwanted components
46   figure; stem(real(X)); title('Xm(k)'); % plot of modified transform coefficients
47  end                               %
48
49  xs = V*X;                         % inverse transformation (synthesis)
50  error = max(abs(x-xs)),           % signal reconstruction error
51  figure; plot(n,real(x),'ro-',n,real(xs),'b*-'); title('xs(n)'); % synthesized signal
```

Exercise 3.5 (First ORTHO Tests). In the beginning set N=8. First, observe
how tricky generation of the transformation matrix is: first matrix of all pairs
of index values is calculated, then this matrix is put as an argument to a basis function and the whole orthogonal matrix is obtained at once. Then, check
matrix orthogonality (look at multiplication of the matrix and its transposition V'*V—the identity matrix should result). Next, set N=100 and observe
in the loop shape of the basis functions—columns of the matrix V. You should
see faster and faster oscillations. Comment the loop. Now observe shape of
the transform coefficients for different input signals (bigger values should be
concentrated around signal frequency components) and the shape of the reconstructed signal—without coefficients modification it should be exactly the same
as the input one. Now turn on the spectrum modification and try to remove some
signal components. It can be well done when the signal is summation of transformation matrix columns. Finally, implement $N = 8$ point Wash–Hadamard
transform interpreting properly Fig. 3.3.

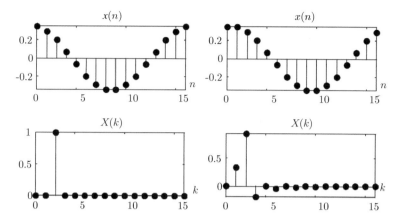

Fig. 3.4: Example of signal decomposition using discrete cosine transform (DCT-I) for $N = 16$. The analyzed signal is (left) perfectly equal to the third basis vector, (right) equal to the third basis vector but circularly shifted one sample right. As we see this very small shift has the dramatic influence upon the result: after shift the spectrum of similarity coefficients is smeared, more vectors have to be used for signal reconstruction. The DCT transform is not shift invariant. The DFT one, discussed in the next chapter, is! [10]

3.5 Transformation Experiments

In this section we will demonstrate the most important features of the orthogonal transform *machinery* using program 3.1.

Perfect Reconstruction   The orthogonal signal transformation is perfect reversible. Every signal, perfectly *smooth* like a sinusoid or perfectly *rough* like random noise, will go back to itself perfectly after the direct and inverse orthogonal transformation. It has been already presented in Exercises 3.3.

Perfect Signal Matching: Perfect Spectrum Compactness   Since after the orthogonal signal decomposition any signal is represented as summation of scaled orthogonal basis functions, it is no surprise that the signal spectrum is perfectly *sharp/compact* when the signal components are exactly equal to scaled basis functions only. In such case the orthogonal decomposition exhibits perfectly the signal content because basis functions perfectly fits to themselves. Such situation is presented in left part of Fig. 3.4. The signal has one component being exactly equal to the 3-rd basis function. Perfect fit! The audience applause!

Shift-Variance and Spectrum Smearing   But the situation is completely different when signal components are not equal to basis functions: the signal can consist of only one pure sinusoid being out of the basis set either of only one basis function but a little bit shifted in time and ... *the Happy End* disappears! The spectrum

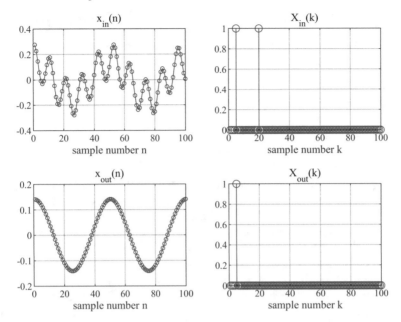

Fig. 3.5: Example of signal components separation in domain of transformation coefficients: (up left) input signal with two-components, low-frequency, and high-frequency ones, (up right) signal DCT-I spectrum, (down right) modified spectrum—the high-frequency component is removed, (down left) synthesized signal—only low-frequency component is synthesized

is smeared and analytic skills of orthogonal transforms are lost. In such case the orthogonal transformation still offers the perfect signal reconstruction but the signal spectrum (transform coefficients, *similarity* coefficients) is not compact. In right part of Fig. 3.4 the same signal is analyzed but it is circularly shifted one sample right. We observe that the spectrum compactness is completely lost. Seeing this the students are leaving the lecture in silence muttering under breath: *We lost our time again!*

Signal Components Separation in Domain of Transform Coefficients Having a signal with several components we can

- calculate orthogonal transform coefficients (i.e. the signal *spectrum*),
- identify which coefficients belong to components of interest and which to *disturbers*,
- set values of disturber coefficients to zero,
- synthesize signal from the modified signal *spectrum*.

This way we can separate some signal components from the rest, as a special case we can minimize noise embedded in the signal. Figure 3.5 illustrates this procedure.

Exercise 3.6 (Testing Orthogonal Transformations). Use the program 3.1 to test yourself all the listed above features of orthogonal transforms (in the beginning disable spectrum modification option):

- choose different matrices V=V1,V2,...,V7 and check their orthogonality (operation V' *V should result in identity matrix with 1s on the main diagonal and 0s elsewhere);
- observe that each testing signal x=x1,x2,...,x5, independently from its shape, is perfectly reconstructed;
- observe that for the first signal the spectrum is perfectly matched to signal components (no surprise! it is summation of scaled columns of the V matrix!) but for the second one, which is simply one-sample shifted version of the first signal, the spectrum *compactness* is completely lost (no surprise, these transforms are not *time-invariant*);
- observe that for the third signal, a pure sine, the spectrum has many nonzero coefficients: many basis functions are required for representation of signal having frequency not present in the basis functions set;
- now turn on the spectrum modification option and choose again the signal number one; observe that setting to zero the 8-th coefficient cause removing the basis function associated with it from the synthesized signal: filtration of signal components can be realized this way.
- now you are ready to *drive a car by yourself*: cut-and-paste the program code, computers allow a lot; for example increase value of N, analyze a speech signal read from disc, ...

Exercise 3.7 (More Fun: Audio Compression Using 1D DCT). Record one speech word, *reach* acoustically, with 11025 Hz sampling ratio, approximately 1–2 s. Perform the DCT upon it, DCT-IV manually or using the Matlab dct() function, then quantize the transform coefficients (see exercises in Chap. 1) and perform the inverse DCT. Compare the original and synthesized speech in one plot and by listening. Coefficients associated with lower frequencies could be quantized less.

Exercise 3.8 (Mega Fun: Image Compression Using the 2D DCT Transformation). 2D DCT image transformation is a cascade of 1D DCTs: first of each image row, then of each column of the matrix resulting from the first step. After the first operation image pixels are replaced by DCT coefficients. Read any image to Matlab using function [img,map]=imread('x.y'); and display it imshow(img,map). Convert it to gray scale if image is in color img=rgb2gray(img);, then transform it into double precision format: img=double(img)/255;. Set colormap(gray). Perform the 2D DCT coeffs=dct2(img) and display matrix of transform coefficients also as an image: imshow(abs(coeffs)). Modify the coefficient values, for example leaving ones in the left-up corner or taking only the biggest of them, and removing the rest. See result. Perform inverse DCT transformation img=idct2(coeffs); and compare synthesized image with the original one. Repeat this operation for different DCT coefficient selections. In case of any problems, look at the program lab03_ex_image.m.

2D DCT transformation of $M \times N$ image can be manually done in Matlab coeffs=U'*img*V where U and V are DCT matrices with dimensions $M \times M$ and $N \times N$, respectively, having basis functions in their column. Check it. How to implement the inverse transformation in matrix form?

Conclusion: Take Things as They Are, Do Not Expect Impossible When the analyzed signal is a linear superposition of some orthogonal vectors used by the transformation (i.e. $\bar{\mathbf{x}} = c_k \bar{\mathbf{v}}_k + c_l \bar{\mathbf{v}}_l + c_m \bar{\mathbf{v}}_m$), the similarity coefficients, i.e. the signal spectrum, give us perfect information about the signal content, i.e. coefficients c_k, c_l and c_m ("the spectrum is compact"). When signal components do not perfectly fit to the basis vectors used for the decomposition—the spectrum is *smeared*. Even slight signal shift in time leads to interpretation problems-mash. We can explain this fact using the 3D space analogy again: after some rotation of the coordinate system the unitary vector $\bar{\mathbf{e}}_x = [1, 0, 0]$ for sure will lose its perfect "energy" compactness and after rotation more coordinates are necessary to describe it, e.g. $\bar{\mathbf{e}}_x = [0.8, 0.6, 0]$. This is the same vector but represented in two different 3D coordinate systems (vector spaces)—one better and one worse. The DFT transform, which is discussed in the next chapter, is in this context more robust—it is signal time-shift invariant (in sense of spectrum magnitude).

3.6 Optimal Discrete Orthogonal Transforms

Important Mathematical Generalization This section has an *off-road* character and should be absolutely skipped by Readers without math interests. An interesting generalization of orthogonal signal decomposition is derived in it. Let us assume that the analyzed signal vector $\bar{\mathbf{x}}$ to be decomposed and orthonormal decomposition/basis vectors $\bar{\mathbf{v}}_k, k = 1, 2, 3, \ldots, N$, are vertical, as in Eq. (3.13). For random signals with mean value equal to 0, decomposition coefficients:

$$X_k = \bar{\mathbf{v}}_k^H \cdot \bar{\mathbf{x}} \tag{3.25}$$

should be completely independent (uncorrelated), i.e.

$$E\left[X_k \cdot X_l^H\right] = E\left[(\bar{\mathbf{v}}_k^H \bar{\mathbf{x}}) \cdot (\bar{\mathbf{v}}_l^H \bar{\mathbf{x}})^H\right] = \bar{\mathbf{v}}_k^H \cdot E\left[\bar{\mathbf{x}} \cdot \bar{\mathbf{x}}^H\right] \cdot \bar{\mathbf{v}}_l =$$
$$= \bar{\mathbf{v}}_k^H \left[\mathbf{R}_{xx}\right] \bar{\mathbf{v}}_l = \begin{cases} 0, & \text{for } k \neq l, \\ \lambda_k, & \text{for } k = l, \end{cases} \tag{3.26}$$

where $()^H$ denotes the Hermitian transpose, $E[.]$—the *statistically expected value*, λ_k—a constant, and $\mathbf{R}_{xx}$—the signal auto-correlation matrix with conjugate (Hermitian) symmetry:

$$\mathbf{R}_{xx} = E[\mathbf{x}\mathbf{x}^H] = \begin{bmatrix} E[x_1 x_1^*] & E[x_1 x_2^*] & \cdots & E[x_1 x_N^*] \\ E[x_2 x_1^*] & E[x_2 x_2^*] & \cdots & E[x_2 x_N^*] \\ \vdots & \vdots & \ddots & \vdots \\ E[x_N x_1^*] & E[x_N x_2^*] & \cdots & E[x_N x_N^*] \end{bmatrix}$$

$$= \begin{bmatrix} R_{xx}(0) & R_{xx}(-1) & \cdots & R_{xx}(-(N-1)) \\ R_{xx}(1) & R_{xx}(0) & \cdots & R_{xx}(-(N-2)) \\ \vdots & \vdots & \ddots & \vdots \\ R_{xx}(N-1) & R_{xx}(N-2) & \cdots & R_{xx}(0) \end{bmatrix} \tag{3.27}$$

consisting of auto-correlation values $R_{xx}(m)$. Signal samples x_k, elements of the vector $\bar{\mathbf{x}}$, are treated by us as independent and identically distributed discrete-time variables. Since the basis functions are orthonormal:

$$\bar{\mathbf{v}}_k^H \cdot \bar{\mathbf{v}}_l = \begin{cases} 0, & k \neq l, \\ 1, & k = l, \end{cases} \tag{3.28}$$

after left multiplication of Eq. (3.26) by $\bar{\mathbf{v}}_\mathbf{k}$, one obtains the following formula:

$$\mathbf{R}_{xx}\bar{\mathbf{v}}_k = \lambda_k \bar{\mathbf{v}}_k, \quad k = 1, 2, 3, \ldots, N. \tag{3.29}$$

informing us that vectors $\bar{\mathbf{v}}_k$ should be *eigenvectors* and values λ_k—*eigenvalues* of the matrix $\mathbf{R}_{xx}$, and can be found by its *eigenvalue decomposition* (EVD):

$$\mathbf{R}_{xx} = \sum_{k=1}^{N} \lambda_k \mathbf{v}_k \mathbf{v}_k^H, \quad \lambda_1 \geq \lambda_2 \geq \lambda_3 \geq \ldots \geq \lambda_N. \tag{3.30}$$

Eigenvectors are unitary (orthonormal) and they optimally span (adjust) the vector space for a concrete signal in directions of the signal energy concentration.

This result is very important. It tells us how to choose orthogonal decomposition vectors for a random signal, aiming at maximum concentration of signal energy in the smallest number of orthogonal transformation coefficients: as eigenvectors of the signal auto-correlation matrix $\mathbf{R}_{xx}$. Of course, if computational expense of such signal treatment is acceptable. Discrete orthogonal transform with *optimal* signal-driven basis vectors chosen this way is called the Karhunen–Loève transform. For auto-regressive first-order AR(1) signals, fulfilling the relation:

$$x(n) = ax(n-1) + noise(n)/\sqrt{1-a^2} \tag{3.31}$$

and having auto-correlation function of the form $R_{xx}(m) = a^m$, the DCT-II transform is a very close approximation of the described above optimal Karhunen–Loève signal decomposition for value of a close to 1. For this reason, DCT-II is widely used in image processing.

As a side conclusion, we can generalize this section saying that one should always try to choose such discrete orthogonal transformation which functions are the most similar (best fitted) to an analyzed signal. Such procedure guarantees the most *compact* signal decomposition, i.e. represented by left plot of Fig. 3.4, not the right one.

3.7 Summary

In this chapter we studied discrete orthogonal transformations of discrete-time signals. There are whole books dealing only with this one topic—so important it is for signal analysis and processing. Now we will summarize in points the most important things which should be remembered.

1. Two vectors are orthogonal when their inner product is equal to zero (sum of multiplied elements). A matrix is orthogonal when it is built from rows or columns which are all orthogonal to themselves. Inverse of an orthogonal matrix is equal to conjugation and transposition of its elements (the first row becomes the first column, the second row becomes the second column, and so on). Multiplication of the matrix and its inverse gives an identity matrix with ones on the main diagonal: $\mathbf{V}^H \cdot \mathbf{V} = \mathbf{I}$.

2. Orthogonal transformation of a vector $\bar{\mathbf{x}}$ of N signal samples is defined as its multiplication with a rectangular $N \times N$ orthogonal matrix $\mathbf{V}^H$:
$\bar{\mathbf{X}} = \mathbf{V}^H \cdot \bar{\mathbf{x}}$.
The matrix has in its rows so-called *basis signals/function* into which the analyzed signal is decomposed. The transformation result $\bar{\mathbf{X}}$ is a vector of coefficients telling us about quantity of each basis function presence in the analyzed signal. Thanks to this we are informed about a signal content: the signal becomes for us a summation of scaled *basis functions/signals*.

3. The transformation result can be used for signal synthesis/reconstruction. In this case the vector of similarity coefficients $\bar{\mathbf{X}}$ is multiplied by the matrix $\mathbf{V}$ giving in a result:
$\mathbf{V} \cdot \bar{\mathbf{X}} = \mathbf{V} \cdot \mathbf{V}^H \cdot \bar{\mathbf{x}} = \mathbf{I} \cdot \bar{\mathbf{x}} = \bar{\mathbf{x}}$.
This is a consequence of the fact that an orthogonal matrix is used and its inverse is equal to transposed conjugation.

4. Doing some modification of the transformation result before backward transformation we can change the signal content, i.e. separate signal components or reduce the noise.

5. There are many orthogonal discrete transformations: DCT, DST, DFT, ...
The transformation is good in our application when only a few similarity coefficients are significant, i.e. the vector of coefficient has *compact* form. It is achieved only in the situation when *basis functions* are perfectly matched to signal components. It almost never happens for real-word data. Hmm...Uff...

6. If decomposition coefficients have the same values after the signal shift, the transformation is *shift invariant*. Unfortunately, only the complex-value discrete Fourier transform has this nice feature (and only for magnitude of its coefficients).
7. The Karhunen–Loève transform is the optimal discrete orthogonal transform for random data—it maximizes compactness of the transform coefficients. The DCT-II is close to KL transform for AR(1) signals.

3.8 Private Investigations: Free-Style Bungee Jumps

Exercise 3.9 (Writing Program of a Universal Ortho-Screwdriver). Matlab functions are computer scripts having extension ".m" and starting with the reserved word `function`. After it, in square brackets there are listed variables computed inside the function while on the right, after the function name, in round brackets, there are listed variables sent to the function. The function task is to compute values of output variables using values of input variables. Matlab function recognizes each variable size. In program 3.2 an example of simple function call is presented.

Modify the function code from Listing 3.2 written above. Name it as `myortho.m`. The function should calculate different direct and inverse orthogonal transforms of any signal x of arbitrary length. You should check the perfect reconstruction feature for each transform. In case of DCT-IV and DFT compare your results with output of Matlab functions `dct()` and `fft()` (in the second one in Matlab no scaling of basis functions is done).

Listing 3.2: Example of writing and calling functions in Matlab

```
1    % File lab03_ex_function_call.m
2
3    % Main program - scripts myprog.m
4    x = [1; 2; 3];  trans = 1;  direction = -1;
5    [X, M] = myfun( x, trans, direction ),
6
7    % Function - script myfun.m
8    function [ Y, N ] = myfun( y, ortho, direct )
9    N = length(y);                % length of the input signal
10   if( ortho==1 ) A = eye(N);    % definition of the
11   else           A = rand(N,N); % transform matrix, basis functions in columns
12   end                           % end of if()
13   if(direct==1)  A = A' ; end   % for direct transform
14   Y = A*y;                      % signal transformation
15   end
```

Exercise 3.10 (* Periodicity/Frequency Hunter: Orthogonal Signal Analysis).
In Chaps. 1 and 2 we were reading and generating different signals, and trying to find (calculate) a period of signal repetition. But inverse of the signal period is equal

to repetition frequency. The orthogonal transformations can be used for frequency analysis. Looking at basis functions of DCT transform, presented in Fig. 3.3, we see that each next basis function (in the DCT matrix) has higher frequency of oscillations. Therefore if some transform coefficients are significantly bigger than the others, we can conclude that our signal is mostly built from functions associated with them, consequently it consists of frequencies of them. These way we can estimate frequency of oscillations present in our data. Try to use this approach to estimation of the following frequencies: generated artificially sinusoids (initial test), vocal cords opening and closing (for different isolated vowels—record your own speech) and heartbeat periodicity (file (ECG100.mat) from the first laboratory on signals). Use your own function OrthoScrewDriver() or the Matlab function dct(). Compare obtained results with visual finding of signal period from its time plot and with period values calculated using the auto-correlation function.

Exercise 3.11 (* Frequency Killer: Orthogonal Signal Tailoring/Filtering).
Knowing coefficients of orthogonal signal components one can manipulate them: set some of them to zero and increase values of some other ones. The same way as in the cinema: you are passing only when you have a ticket! This way after the orthogonal synthesis a different signal is obtained: it should be more useful for us since it was personally *tailored* by us. For example, noise or unwanted signal components can be removed or reduced this way. Generate a signal consisting of a few sinusoids and try to remove some of them in the domain of transform coefficients (setting their values to zero). Do the same with fragments of speech or audio as well as our ECG signals.

Exercise 3.12 (** Flip-Flop: Orthogonal Signal Transformations with Rectangular Shape Functions). Try to write a Matlab code implementing Hadamard (3.22) and Haar (3.23) orthogonal transforms for $N = 2^p$. Both of them have basis functions with rectangular shapes. Note that the basis functions of the first transform have a form of rectangular-shape oscillations having the same amplitude while in the second transform the amplitudes are different, additionally the functions are impulsive and try to detect not only frequencies of signal components but also their time localization. First, generate basis functions for $N = 8$ and compare them with these presented in Fig. 3.3. Then, in the second *high-mountains* task, try to write a program for calculation of the Wash–Hadamard transform for any $N = 2^p$ equal to a power of 2. Use the transform definition (3.22) or find in the Internet and apply recursive generation rule of the WH orthogonal matrix.

Exercise 3.13 (*** Mount Everest of Orthogonality: Optimal Orthogonal Transforms for Noisy Signals). What are optimal shapes of orthogonal basis functions for noisy signals? How to find their collection, which do *packing* of the signal energy to the smallest number of them (how to ensure *compactness* of random signal spectrum?) If you are really interested in this story, look for its continuation in many *optimum signal processing* books. May be the following Matlab functions R=xcorr(x); Rxx=toeplitz(R); [V,D] = eig(Rxx) can help you to find an answer?

References

1. V. Britanak, P.C. Yip, K.R. Rao, *Discrete Cosine and Sine Transforms: General Properties, Fast Algorithms and Integer Approximations* (Academic Press, Boston, 2006)
2. D.F. Elliott, K.R. Rao, *Fast Transforms. Algorithms, Analyses, Applications* (Academic Press, New York, 1982)
3. R.C. Gonzales, R.E. Woods, *Digital Image Processing* (Addison-Wesley Publishing Company, Reading, 1992; Pearson, Upper Saddle River, 2017)
4. N.J. Jayant, P. Noll, *Digital Coding of Waveforms* (Prentice-Hall, Englewood Cliffs, 1984)
5. J.S. Lim, *Two-Dimensional Signal and Image Processing* (Prentice Hall, Upper Saddle River NJ, 1990)
6. H.S. Malvar, *Signal Processing with Lapped Transforms* (Artech House, Norwood, 1992)
7. K. Rao, P. Yip, *Discrete Cosine Transform* (Academic Press, New York, 1990; Elsevier 2014)
8. G. Strang, The discrete cosine transform. SIAM Rev. **41**(1), 135–147 (1999)
9. R. Wang, *Introduction to Orthogonal Transforms with Applications in Data Processing and Analysis* (Cambridge University Press, Cambridge, 2012)
10. T.P. Zieliński, *Cyfrowe Przetwarzanie Sygnałów. Od Teorii do Zastosowań (Digital Signal Processing. From Theory to Applications)* (Wydawnictwa Komunikacji i Łączności (Transport and Communication Publishers), Warszawa, Poland, 2005, 2007, 2009, 2014)

Chapter 4
Discrete Fourier Transforms: DtFT and DFT

> *Always there are many runners but typically the winner is the same: DFT in the form of FFT!*

4.1 Introduction

This chapter is devoted to practical computer-based frequency analysis of discrete-time signals, i.e. vectors of signal samples, by means of Fourier transform-based methods. We are assuming now that the analyzed signal is a summation of different oscillatory components with different frequencies and we are interested in finding them. From the chapter about orthogonal transforms we remember that signal analysis (decomposition into simpler components) is performed by calculation of signal similarity to some reference oscillations. The similarity coefficients are calculated as inner products of the signal vector and some reference vectors (sum of products of corresponding elements). In analog signal theory the methodology is exactly the same but the inner product has a form of infinite integral of the product of signal and reference function, calculated for an infinite number of reference frequencies. One obtains this way a signal spectrum being a continuous function of frequency. When values of this function, i.e. signal similarity measures to some reference *oscillations*, are multiplied by these oscillations, and all oscillations are added together in infinite integral over frequency—the signal is synthesized (reconstructed) from its spectral description. The direct and inverse continuous Fourier transform (CFT) act the same way in the analog world as discrete orthogonal transforms in discrete-time world.

When the analog signal is periodic and repeats every T seconds, the signal integration in CFT can be limited to one signal period only because all information about the signal is in this time interval. Being periodic, the signal can have only components with frequencies being multiplicities of the signal repetition frequency $f_0 = 1/T$, i.e. $f_k = k \cdot f_0$. Thanks to this, the *analyzing* integration and final signal *synthesizing* integration from similarity coefficients are repeated not

© Springer Nature Switzerland AG 2021
T. P. Zieliński, *Starting Digital Signal Processing in Telecommunication Engineering*, Textbooks in Telecommunication Engineering, https://doi.org/10.1007/978-3-030-49256-4_4

for all frequencies. In such case the CFT is taking a form of Fourier series (FS), its special case.

When we are coming to discrete world, the CFT is changing to discrete-time Fourier transform (DtFT) and the FS are replaced with discrete Fourier transform (DFT).

In this chapter we learn only minimum amount of information concerning CFT, FS, DtFT, and DFT. We become familiar with their definitions and primary features. The main goal is to understand, from one side, the *broth-erhood* relation between DtFT and DFT, and, from the second side, differences in their practical usage. In frequency analysis performed on a digital computer, time-limited N-samples long signal has to be used in DtFT, similarly as in DFT. For this reason the only difference between both transforms relies on different frequency sets which are used by them. Let us assume that sampling frequency is equal to f_s and the analyzed signal has N samples. In DFT one can only calculate similarity coefficients for N frequencies equal to $f_k = k \cdot f_0, k = 0, 1, \ldots, N-1$, N multiplicities of $f_0 = f_s/N$, while in DtFT a user does not have any restrictions in her/his frequency choice. It is interesting that DSP users very often forget about DtFT which offers better spectrum inspection than DFT.

Finally, we will make a link to the previous chapter on orthogonal transforms. DFT is a special type of $N \times N$ orthogonal transform. In contrary to orthogonal transformations discussed before, it is using complex-value, not real-value, harmonic oscillations as orthogonal basis functions to which the signal is decomposed, cosine in the real part and sine in the imaginary part. Let us repeat the definition of normalized DFT basis functions for smoother continuation (k—function number and transformation matrix row number, n—sample number and transformation matrix column number, $k, n = 0, 1, 2, \ldots, N-1$):

$$v_{k,n} = v_k(n) = \frac{1}{\sqrt{N}} e^{j\frac{2\pi}{N}kn} = \frac{1}{\sqrt{N}} \left(\cos\left(\frac{2\pi}{N}kn\right) + j \cdot \sin\left(\frac{2\pi}{N}kn\right) \right). \quad (4.1)$$

Due to its complexity, the transformation result is robust to signal shift in time (delay), i.e. after this modification the absolute value of the signal transform coefficients does not change after the signal shift. In this chapter we will derive the DFT equation from the Fourier series analysis and show its relation to DtFT, its older *brother*.

4.2 Continuous Fourier Transform and Fourier Series

Let us start from the beginning, from an analog world description. The **continuous Fourier transforms** (CFT), direct and inverse, are defined as follows:

$$X(f) = \int_{-\infty}^{\infty} x(t) e^{-j2\pi ft} dt, \quad x(t) = \int_{-\infty}^{\infty} X(f) e^{j2\pi ft} df, \quad j = \sqrt{-1}. \quad (4.2)$$

Again, during analysis, the continuous-time signal $x(t)$ is compared with complex-conjugated continuous-time basis functions $e^{-j2\pi ft}$, now complex-value ones. It is done by performing integration (summation) of their product. The integration is convergent for limited energy signals only, for others—concept of generalized functions (distributions) should be applied. During synthesis all basis functions $e^{j2\pi ft}$ are scaled by corresponding, calculated spectral *coefficients* $X(f)$ and summed (integrated). Any signal is represented as infinite *summation/integration* of complex-value harmonic signals of the form:

$$e^{j2\pi ft} = \cos(2\pi ft) + j \cdot \sin(2\pi ft) \tag{4.3}$$

with different frequencies f. Pure cosine and sine signals with frequency f_0 have the following Fourier spectral decomposition (representation):

$$\cos(2\pi f_0 t) = \frac{e^{j2\pi f_0 t} + e^{-j2\pi f_0 t}}{2}, \qquad \sin(2\pi f_0 t) = \frac{e^{j2\pi f_0 t} - e^{-j2\pi f_0 t}}{2j}, \tag{4.4}$$

or:

$$\cos(2\pi f_0 t) = \frac{1}{2}e^{j2\pi f_0 t} + \frac{1}{2}e^{-j2\pi f_0 t}, \qquad \sin(2\pi f_0 t) = \frac{-j}{2}e^{j2\pi f_0 t} + \frac{j}{2}e^{-j2\pi f_0 t}. \tag{4.5}$$

They are summation of two harmonic signals (4.3), first with *positive* frequency f_0 and second with *negative* frequency $-f_0$. Fourier spectrum coefficients for cosine and sine are, respectively, equal to $[1/2, 1/2]$ and $[-0.5j, 0.5j]$, first for positive frequency, then for negative (*amount* of two basis signals, all remaining transform coefficients are equal to zero). We are doing here deliberately very big simplifications not mentioning the Dirac Delta functions but aiming at more intuitive, less formal presentation. Fourier spectra of pure cosine and sine with frequency f_0 are presented in Fig. 4.1.

For real-value signals, the CFT spectrum has conjugate (Hermitian) symmetry in respect to frequency $f = 0$ Hz, i.e. it is the same for positive and negative frequencies in its real part and negated in its imaginary part:

$$X(-f) = X^*(f). \tag{4.6}$$

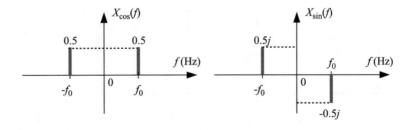

Fig. 4.1: Fourier spectrum of cosine (left) and sine (right)—see Eq. (4.5)

This feature is inherited from functions of $\cos()$ and $\sin()$:

$$X(f) = \int_{-\infty}^{\infty} x(t)e^{-j2\pi ft}\,dt = \underbrace{\int_{-\infty}^{\infty} x(t)\cos(2\pi ft)\,dt}_{X_{Re}(-f)=X_{Re}(f)} - j\underbrace{\int_{-\infty}^{\infty} x(t)\sin(2\pi ft)\,dt}_{X_{Im}(-f)=-X_{Im}(f)} \quad . \quad (4.7)$$

Spectra of pure cosine and sine signals, presented in Fig. 4.1, are the best examples of the CFT spectrum symmetry.

It is very informative to calculate the Fourier spectrum of a rectangular pulse equal to 1 in the interval $[-T, T]$ and zero elsewhere:

$$R_T(f) = \int_{-\infty}^{\infty} r_T(t)e^{-j2\pi ft}\,dt = \int_{-T}^{T} 1 \cdot e^{-j2\pi ft}\,dt = \frac{1}{-j2\pi f}e^{-j2\pi ft}\Big|_{-T}^{T} = \ldots$$

$$\frac{e^{-j2\pi fT} - e^{j2\pi fT}}{-j2\pi f} = \frac{-j2\sin(2\pi fT)}{-j2\pi f} = \frac{\sin(2\pi fT)}{\pi f} = 2T\,\mathrm{sinc}(2\pi fT). \quad (4.8)$$

Value for $f = 0$ we find calculating derivatives of nominator and denominator of the final formula in Eq. (4.8) in respect to f:

$$R_T(f)\big|_{f=0} = \frac{(2\pi T)\cos(2\pi fT)}{\pi}\bigg|_{f=0} = 2T. \quad (4.9)$$

Signal of rectangular pulse and its Fourier spectrum are presented in Fig. 4.2. The plots have been done using program 4.1.

Exercise 4.1 (Fourier Spectrum of the Rectangular Pulse). Run program 4.1 which is doing the Fourier spectrum visualization of a rectangular pulse. Observe oscillatory shape of this spectrum. Around $f = 0$ Hz the so-called spectral *main-lobe* of the oscillations is located. On both sides of it the so-called

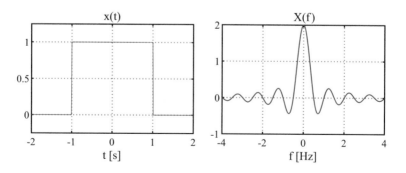

Fig. 4.2: Rectangular pulse and its Fourier spectrum

oscillatory spectral *side-lobes* are visible. Change the pulse duration. Observe that the shorter the pulse is, the wider is its spectrum. Note also that the spectrum has values equal to zero for frequencies being multiplicities of $1/(2T) : f = k \cdot (1/(2T))$.

Listing 4.1: Fourier spectrum of rectangular pulse

```
1  % lab04_ex_rectpulse.m
2  clear all; close all;
3
4  T = 1; t = -2*T : T/100 : 2*T;
5  x = zeros(1,length(t)); indx = find(abs(t)<=T); x(indx)=ones(1,length(indx));
6  figure; plot(t,x,'b-'); xlabel('t [s]'); title('x(t)'); grid;
7
8  f0 = 1/T; f= -4*f0 : f0/100 : 4*f0;
9  X = sin(2*pi*f*T) ./ (pi*f);
10 X( ceil(length(X)/2) ) = 2*T;
11 figure; plot(f,X,'b-'); xlabel('f [Hz]'); title('X(f)'); grid;
```

In DSP we deal with discrete-time signals taken from real-world objects. Therefore it is very important to know which are theoretical spectra of the most popular continuous-time signals. Why? Since the same spectra should be obtained during computer calculations performed upon discrete-time signal representations. In Table 4.1 some spectra examples (definitions) are given.

We should also know which are spectral consequences of different operations performed upon the signal. To find corresponding mathematical formulas one should put the modified signal into CFT integral (4.2) and calculate it. This is a routine exercise during analog circuits and signals (or signal theory) workouts. I recommend

Table 4.1: Continuous-time signals and their continuous-frequency CFT spectra

No	Signal name	Signal equation	Spectrum equation
1	Rectangular pulse	$r_T(t) = \begin{cases} 0 & \text{for } |t| > T \\ 1 & \text{for } |t| \leq T \end{cases}$	$X(\omega) = 2\frac{\sin \omega T}{\omega}$
2	Sign signal	$x(t) = \text{sign}(t)$	$X(\omega) = \frac{2}{j\omega}$
3	Gaussian function	$x(t) = e^{-at^2}$	$X(\omega) = \sqrt{\frac{\pi}{a}} e^{-\omega^2/(4a)}$
4	One-side exponential	$x(t) = \begin{cases} 0 & t < 0 \\ e^{-at} & t \geq 0 \end{cases}, \quad a > 0$	$X(\omega) = \frac{1}{a+j\omega}$
5	Damped sine	$x(t) = \begin{cases} 0 & t < 0 \\ Ae^{-at}\sin(\omega_0 t) & t \geq 0 \end{cases}$	$X(\omega) = \frac{A\omega_0}{(a+j\omega)^2+\omega_0^2}$
6	Damped cosine	$x(t) = \begin{cases} 0 & t < 0 \\ Ae^{-at}\cos(\omega_0 t) & t \geq 0 \end{cases}$	$X(\omega) = A\frac{a+j\omega}{(a+j\omega)^2+\omega_0^2}$
7	Cosine fragment	$x(t) = \cos(\omega_0 t) \cdot r_T(t)$	$X(\omega) = \frac{\sin((\omega-\omega_0)T)}{\omega-\omega_0} + \frac{\sin((\omega+\omega_0)T)}{\omega+\omega_0}$

Table 4.2: Basic CFT features: signal processing and its spectral consequence

No	Feature	Signal manipulation	Spectral consequence
1	Linearity	$ax(t)+by(t)$	$aX(f)+bY(f)$
2	Scaling	$x(at), \quad a>0$	$\frac{1}{a}X\left(\frac{f}{a}\right)$
3	Time reverse	$x(-t)$	$X(-f)$
4	Conjugation	$x^*(t)$	$X^*(-f)$
5	Time shift	$x(t-t_0)$	$e^{-j2\pi f t_0}X(f)$
6	Frequency Shift	$e^{\pm j2\pi f_0 t}x(t)$	$X(f\mp f_0)$
7	Multiplication	$x(t)\cdot y(t)$	$\int_{-\infty}^{\infty} X(v)Y(f-v)dv$
8	Complex modulation	$e^{\pm j2\pi f_0 t}x(t)$	$X(f\mp f_0)$
9	Cos() modulation	$x(t)\cos(2\pi f_0 t)$	$\frac{1}{2}\left[X(f-f_0)+X(f+f_0)\right]$
10	Sin() modulation	$x(t)\sin(2\pi f_0 t)$	$\frac{-j}{2}\left[X(f-f_0)-X(f+f_0)\right]$
11	Convolution	$\int_{-\infty}^{\infty} x(\tau)y(t-\tau)d\tau$	$X(f)\cdot Y(f)$
12	Correlation	$\int_{-\infty}^{\infty} x(t)y^*(t+\tau)dt$	$X(f)\cdot Y^*(f)$
13	Derivative	$\frac{d^n x(t)}{dt^n}$	$(j2\pi f)^n \cdot X(f)$
14	Energy—Parseval eq.	$\int_{-\infty}^{\infty} x(t)x^*(t)dt$	$\int_{-\infty}^{\infty} X(f)X^*(f)df$

to do it for one or two signal modifications. Examples could be found in many textbooks. The most important CFT features are listed in Table 4.2.

At present, as an example, we will derive a few important spectral relations which will be very often used later in this book ($\omega = 2\pi f$):

1. **signal time shift**—results only in signal spectrum phase change (after introducing new variable $\tau = t - t_0$, from where $t = \tau + t_0$):

$$\int_{-\infty}^{\infty} x(t-t_0)e^{-j\omega t}dt = \int_{-\infty}^{\infty} x(\tau)e^{-j\omega(\tau+t_0)}d\tau = e^{-j\omega t_0}\int_{-\infty}^{\infty} x(\tau)e^{-j\omega \tau}d\tau = e^{-j\omega t_0}X(\omega);$$

(4.10)

2. **complex modulation**—causes frequency shift of the signal spectrum to the modulation frequency:

$$\int_{-\infty}^{\infty} \left(e^{\pm j2\pi f_0 t}x(t)\right)e^{-j2\pi f t}dt = \int_{-\infty}^{\infty} x(t)e^{-j2\pi(f\mp f_0)t}dt = X(f\mp f_0); \quad (4.11)$$

3. **convolution of two signals**—results in multiplication of their spectra, which is extremely important in signal filtering (new variable $\lambda = t - \tau$, from where $t = \tau + \lambda$):

$$\int\limits_{-\infty}^{\infty}\left(\int\limits_{-\infty}^{\infty}x(\tau)y(t-\tau)d\tau\right)e^{-j\omega t}dt = \int\limits_{-\infty}^{\infty}\left(\int\limits_{-\infty}^{\infty}x(\tau)e^{-j\omega\tau}d\tau\right)y(\lambda)e^{-j\omega\lambda}\,(d\lambda+d\tau) = \dots$$

$$\left[\int\limits_{-\infty}^{\infty}x(\tau)e^{-j\omega\tau}d\tau\right]\cdot\left[\int\limits_{-\infty}^{\infty}y(\lambda)e^{-j\omega\lambda}d\lambda\right] = X(f)Y(f); \tag{4.12}$$

4. **multiplication of two signals**—results in convolution of their spectra, extremely important in spectral analysis (we show that inverse Fourier transform of convolution of two signal spectra is equal to multiplication of these signals, i.e. we will present an inverse proof; using new variable $u = f - v$, from where $f = v + u$):

$$\int\limits_{-\infty}^{\infty}\left(\int\limits_{-\infty}^{\infty}X(v)Y(f-v)dv\right)e^{j2\pi ft}df =$$

$$= \left[\int\limits_{-\infty}^{\infty}X(v)e^{j2\pi vt}dv\right]\cdot\left[\int\limits_{-\infty}^{\infty}Y(u)e^{j2\pi ut}du\right] = x(t)y(t); \tag{4.13}$$

5. **signal energy—Parseval's equation**—integration of squared signal in time domains is equivalent to the integration of its squared Fourier spectra in the frequency domain, important in signal power and spectral density analysis:

$$\int\limits_{-\infty}^{\infty}x(t)x^*(t)dt = \int\limits_{-\infty}^{\infty}\left(\int\limits_{-\infty}^{\infty}X(f)e^{j2\pi ft}df\right)x^*(t)dt =$$

$$= \int\limits_{-\infty}^{\infty}X(f)\left(\int\limits_{-\infty}^{\infty}x^*(t)e^{j2\pi ft}dt\right)df = \int\limits_{-\infty}^{\infty}X(f)X^*(f)df. \tag{4.14}$$

Fourier series use the same methodology as CFT but are dedicated to analysis and synthesis of periodic signals: only one signal period T is analyzed (multiplied with the reference and integrated, the result is divided by T) and only frequencies being multiplies of the signal repetition frequency $kf_0 = k\frac{1}{T}, k = -\infty, \dots, \infty$, are checked:

$$X_k = \frac{1}{T}\int\limits_0^T x(t)e^{-j2\pi(kf_0)t}dt, \quad x(t) = \sum_{k=-\infty}^{+\infty}X_k e^{j2\pi(kf_0)t}, \quad f_0 = \frac{1}{T}. \tag{4.15}$$

The Fourier series equations are written also in the so-called *trigonometric* version:

$$a_k = \frac{1}{T} \int_0^T x(t) \cos\left(2\pi(kf_0)t\right) dt, \quad b_k = \frac{1}{T} \int_0^T x(t) \sin\left(2\pi(kf_0)t\right) dt \qquad (4.16)$$

$$x(t) = a_0 + 2 \sum_{k=1}^{+\infty} \left[a_k \cos\left(2\pi(kf_0)t\right) + b_k \sin\left(2\pi(kf_0)t\right)\right] = \qquad (4.17)$$

$$= X_0 + 2 \sum_{k=1}^{+\infty} |X_k| \cos\left(2\pi(kf_0)t + \sphericalangle X_k\right)$$

$$X_k = a_k - jb_k, \quad |X_k| = \sqrt{a_k^2 + b_k^2}, \quad \sphericalangle X = \text{arctg}\left(\frac{-b_k}{a_k}\right).$$

4.3 Discrete-Time Fourier Transform: From CFT to DtFT

Presentation of the continuous Fourier transform, given above, is very important for us because in computer-based frequency analysis a discretized CFT version is very widely used. Let us rewrite the CFT into more computer-friendly form. Denoting sampling frequency as f_s, sampling period as $\Delta t = 1/f_s$, sampling time as $t = n \cdot \Delta t$, and exchanging infinite integral with infinite summation, Eq. (4.2) of the forward CFT takes the following form:

$$X(f) = \int_{-\infty}^{+\infty} x(t)e^{-j2\pi ft} dt \quad \Rightarrow \quad X(f) = \sum_{n=-\infty}^{+\infty} x(n \cdot \Delta t)e^{-j2\pi f(n \cdot \Delta t)}. \qquad (4.18)$$

Going further, we can write final equations for DtFT and its inverse as (defining $\Omega = 2\pi \frac{f}{f_s}$):

$$\text{Analysis:} \quad X\left(\frac{f}{f_s}\right) = \sum_{n=-\infty}^{+\infty} x(n)e^{-j2\pi \frac{f}{f_s}n} = \sum_{n=-\infty}^{+\infty} x(n)e^{-j\Omega n}, \qquad (4.19)$$

$$\text{Synthesis:} \quad x(n) = \frac{1}{f_s} \int_{-f_s/2}^{+f_s/2} X\left(\frac{f}{f_s}\right) \cdot e^{j2\pi \frac{f}{f_s}n} df. \qquad (4.20)$$

In (4.19) $X\left(\frac{f}{f_s}\right)$ can be calculated for any value of frequency f, being a continuous variable, but there is no need for this because the function $e^{-j2\pi \frac{f}{f_s}n}$ is periodic in respect to f and has period f_s:

$$e^{-j2\pi \frac{(f+k \cdot f_s)}{f_s}n} = e^{-j2\pi \frac{f}{f_s}n} \cdot e^{-j2\pi kn} = e^{-j2\pi \frac{f}{f_s}n}. \qquad (4.21)$$

Therefore, it is sufficient to calculate $X(\frac{f}{f_s})$ for $-f_s/2 \leq f < f_s/2$ or for $0 \leq f < f_s$. In the first case the inspection of the spectrum is more intuitive and easier for interpretation because pairs of positive and negative frequency components are visible in the spectrum. Going back to the sampling Exercise 1.4 presented in Chap. 1, for $f_s = 1000$ Hz and $f_x = 100$ Hz we see in the spectrum signal components -100 Hz and 100 Hz, not 100 Hz and 900 Hz (see equations (1.13) and (1.14)). In fact during discretization of CFT we are sampling not only analyzed signals but also the reference functions $\cos()$ and $\sin()$. When their frequency is too high, the sampling theorem is not fulfilled, and the high-frequency reference signals are under-sampled and look as low-frequency ones and, as such, they fit to the analyzed low-frequency signal again. From this reason the DtFT spectrum is periodic and there is no need for its whole computation.

When we have only N signal samples, after dividing (4.19) by N, one obtains the following equation:

$$X\left(\frac{f}{f_s}\right) = \frac{1}{N} \sum_{n=0}^{N-1} x(n)e^{-j2\pi\frac{f}{f_s}n}, \quad -f_s/2 \leq f < f_s/2, \qquad (4.22)$$

which offers properly scaled signal amplitude spectrum (for example, the cosine spectrum has two peaks equal to 1/2 for frequencies f_0 and $-f_0$). In DtFT (4.22) we can sample (discretize in frequency) the spectrum as dense as we want, significantly denser than in the DFT method, being discussed later in this chapter, where the frequency step $\Delta f = f_s/N$ is always used. From this reason (4.22) should be treated as a basic tool for spectral zooming and allows us to see details invisible in DFT. It is building a bridge between digital and analog signal theory.

It is very important also to note that, analogically to CFT and its Eqs. (4.6), (4.7), the DtFT spectrum $X() = X_{Re}() + X_{Im}()$ has conjugate symmetry also around the frequency $f = 0$ Hz—it is the same in its real part $X_{Re}()$ and negated in its imaginary part $X_{Im}()$:

$$X_{Re}\left(\frac{-f}{f_s}\right) = X_{Re}\left(\frac{f}{f_s}\right), \qquad X_{Im}\left(\frac{-f}{f_s}\right) = -X_{Im}\left(\frac{f}{f_s}\right). \qquad (4.23)$$

Fundamentals of frequency analysis of signals by means of DtFT, discretized in frequency, are summarized in Fig. 4.3. A pure cosine is analyzed in it. Signals are presented on the left side, while on the right their CFT and DtFT spectra. We see on the left side, one after the other: continuous-time cosine, a continuous-time rectangular *window*—a function, one of many possible, used for cutting a cosine fragment, result of their multiplication, i.e. the signal fragment to be analyzed, and, finally, its time-discretized version. The CFT spectrum of a cosine $\cos(2\pi f_0 t)$ is equal to $\frac{1}{2}$ for $-f_0$ and f_0 (see Eq. (4.5)). The CFT spectrum of the rectangular *cutting* function has an oscillatory shape described by $\frac{\sin(x)}{x}$ function (see Eq. (4.8)). The CFT spectrum of a cut cosine consists of two copies of the CFT spectrum of rectangular window, shifted to frequencies $-f_0$ and f_0 (due to modulation feature of the CFT

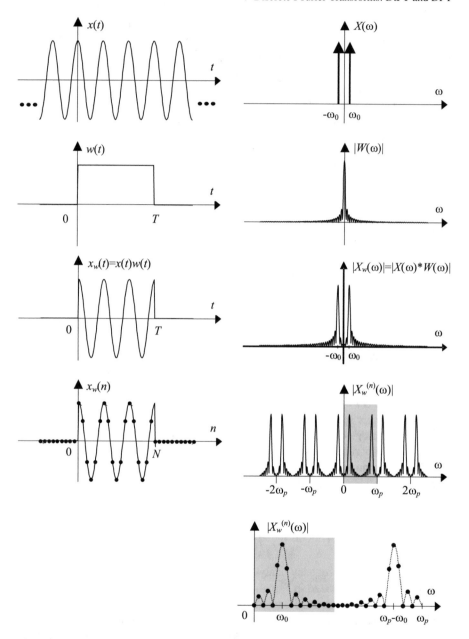

Fig. 4.3: Graphical illustration of fundamental principles of digital spectral analysis: (left) signals, (right) their spectra. In consecutive rows: (1) infinite cosine and its theoretical Fourier spectrum, (2) rectangular window and its theoretical Fourier spectrum, (3) multiplication of cosine and rectangular window and its spectrum (convolution of two above spectra marked with "*"), (4) sampled signal and its periodic spectrum, (5) sampled one period of the repeating spectrum [11]

transform—see feature 9 in Table 4.2). After signal discretization in time with sampling frequency f_s, the continuous-frequency DtFT spectrum is obtained in which the CFT spectrum is repeated with period f_s—due to Eq. (4.21). Since the DtFT spectrum is periodic, only its one period can be calculated. In Fig. 4.3 this is one DtFT spectrum period from $[0, f_s)$ Hz, the same as in the discrete Fourier transform (DFT) presented in the next section. In computer implementation some sampling of the frequency axis has to be chosen, which is presented also. When different *window* function is used for cutting a signal fragment, the observed signal spectrum has different shapes but its peaks are still located at signal frequency components. In Exercise 4.2 we will apply and test some exemplary window functions (rectangular, Hanning, and Chebyshev) during signal DtFT analysis.

Short Summary Since during DtFT-based frequency analysis, signal is multiplied with some *windowing* function, the DtFT spectrum of the signal fragment is a result of convolution of the signal spectrum and spectrum of the window—see Eq. (4.13). Signal discretization causes spectrum periodic repeating (sampling frequency is the period)—due to Eq. (4.21). Therefore, the DtFT spectrum of a discrete signal should be calculated only in the frequency range $\left[-\frac{f_s}{2}, \frac{f_s}{2}\right)$ for complex-value signals or in the range $[0, \frac{f_s}{2})$ for real-value signals—due to its conjugate symmetry (4.23).

In program 4.2 the Matlab code of the DtFT algorithm is presented. It will be used as a frequency detective for performing some initial experiments, recognition of DtFT features, and validation of the presented above mathematical material. We will start from the DtFT analysis of a pure cosine fragment cut by exemplary *window*: rectangular, Hanning, or Chebyshev. Do Exercise 4.2. Look at plots shown in Fig. 4.2, presenting DtFT spectra being solutions/results of the consecutive exercise tasks.

Listing 4.2: Matlab program for DtFT calculation

```
1   % lab04_ex_dft_dtft_analysis.m
2   clear all; close all;
3
4   N = 100;                          % number of samples: 100 --> 1000
5   fs = 1000; dt=1/fs; t=dt*(0:N-1);  % sampling ratio
6   df = 10;                          % sampling step in DtFT: 10 --> 1
7   fmax = 2.5*fs;                    % sampling range in DtFT: 2.5 --> 0.5
8   fx1 = 100;                        % frequency of signal component 1
9   fx2 = 250;   Ax2 =0.001;          % frequency and amplitude of signal component 2
10                                    % 250 --> 110, 0.001 --> 0.00001
11  % Signal
12  x1 = cos(2*pi*fx1*t);            % first component
13  x2 = Ax2*cos(2*pi*fx2*t);        % 250Hz --> 110Hz, 0.001 --> 0.00001
14  x = x1; % + x2;                  % x1, x1+x2, 20*log10(0.00001)=-100 dB
15  figure; stem(x); title('x(n)'); pause  % analyzed signal
16
```

```
17  % Windowing
18  w1 = boxcar(N)';                        % rectangular window
19  w2 = hanning(N)';                       % Hanning window
20  w3 = chebwin(N,140)';                   % Chebyshev window, 80, 100, 120, 140
21  w = w1;                                 % w1 --> w2, w3 (80, 100, 120, 140)
22  figure; stem(w); title('w(n)'); pause  % window
23  x = x .* w;                             % x = x, w, x.*w
24  figure; stem(x); title('xw(n)'); pause % windowed signal
25
26  % DFT - later in this chapter (red circles)
27  % k=0:N-1; n=0:N-1; F = exp(-j*2*pi*(k'*n)); X = (1/N)*F*x;
28  f0 = fs/N; f1 = f0*(0:N-1);   % DFT freq step = f0 = 1/(N*dt)
29  for k = 1:N
30      X1(k) = sum( x .* exp(-j*2*pi/N* (k-1) *(0:N-1) ) )/ N;
31      % X1(k) = sum( x .* exp(-j*2*pi/N* (f1(k)/fs) *(0:N-1) ) )/ N;
32  end
33  %X1 = N*X1/sum(w);                       % scaling for any window
34
35  % DtFT - already discussed (blue line)
36  f2 = -fmax : df : fmax;                  % df = 10 --> 1; first this freq. range
37  for k = 1 : length(f2)
38      X2(k) = sum( x .* exp(-j*2*pi* (f2(k)/fs) *( 0:N-1) ) ) / N;
39  end
40  %X2 = N*X2/sum(w);                       % scaling for any window
41
42  % Figures
43  figure; plot(f1,abs(X1),'ro',f2,abs(X2),'b-');
44  xlabel('f (Hz)'); grid; pause
45  figure; plot(f1,20*log10(abs(X1)),'ro',f2,20*log10(abs(X2)),'b-');
46  xlabel('f (Hz)'); grid; pause
```

Exercise 4.2 (DtFT of a Cosine with Rectangular Window). Use program 4.2 for computing DtFT spectrum of a simple cosine signal. Choose $f_{x1} = 100$ Hz as a signal frequency and $f_s = 1000$ Hz as sampling frequency. Generate $N = 100$ signal samples. Choose x=x1.

1. Analyze the signal using DtFT in the frequency range $[-f_{max}, \ldots, f_{max}]$, $f_{max} = 2.5 f_s$ with the frequency step $df = 10$ Hz, equal to DFT step $f_0 = \frac{f_s}{N}$. We are expecting two sharp peaks at frequencies $f = -100$ Hz and $f = 100$ Hz since, due to Eq. (4.5), after discretization our signal has the form:

$$\cos\left(2\pi \frac{f_{x1}}{f_s} n\right) = \frac{e^{j2\pi \frac{f_{x1}}{f_s} n} + e^{-j2\pi \frac{f_{x1}}{f_s} n}}{2}. \qquad (4.24)$$

Why we see only two peaks in DFT and much more peaks in DtFT, periodically repeating around multiplies of the sampling frequency f_s? Because DFT calculates the signal spectrum only in the range $[0, f_s)$ with the step f_0 and we see only two components of the cosine: $f_{x1} = 10f_0$ and $f_s - f_{x1} \doteq f_s - 10f_0$. In DtFT situation is different. Due to equations (4.21), the generated reference signals of *higher* frequencies $kf_s \pm f_{x1}$ have exactly the same samples as for the low frequencies $\pm f_{x1}$ and for them the *perfect fit* is valid also. Conclusion: the frequency range $[-0.5f_s, \ldots, 0.5f_s]$ is all we need. For real-value signals, having symmetrical spectra, even $[0, \ldots, 0.5f_s]$ is enough.

2. Now decrease the DtFT spectrum sampling 10 times, other words set $df = 1$ Hz. *Wow! What happens!* Take it easy: now you see repeating spectrum of rectangular pulse (4.8) shown in Fig. 4.2, at present its absolute value is calculated. *But why I did not see it before?!* Since the rectangular pulse spectrum is oscillatory, it is crossing through zero and we before, by chance, took only samples at those zeros and at spectral peaks. *But why the spectrum of rectangular pulse is present in the spectrum of the cosine? Where the rectangular pulse is hidden in math equations?* We analyze not the WHOLE cosine but its N-samples long FRAGMENT, cut by the rectangular pulse. Therefore two analog time-infinite signals are multiplied: cosine and rectangular pulse, and the resultant spectrum is equal to convolution of their individual spectra (see *multiplication* feature in Table 4.2 and Eq. (4.13)). For this reason we see spectrum of the rectangular pulse in the position of cosine spectral peaks. We can also apply in this case the *cosine modulation* feature from the same table: multiplying any signal $w(n)$ by cosine with frequency f_{x1} shifts the signal spectrum to cosine frequencies: f_{x1} and $-f_{x1}$ and scale them by $1/2$. In discrete-time case:

$$W\left(\frac{f}{f_s}\right) = \sum_{n=-\infty}^{+\infty} \left[w(n) \cos\left(2\pi \frac{f_{x1}}{f_s} t\right) \right] e^{-j2\pi \frac{f}{f_s} n} = \frac{1}{2} \sum_{n=-\infty}^{+\infty} w(n) e^{-j2\pi \frac{(f-f_{x1})}{f_s} n} + \ldots$$

$$\frac{1}{2} \sum_{n=-\infty}^{+\infty} w(n) e^{-j2\pi \frac{(f+f_{x1})}{f_s} n} = \frac{1}{2} W\left(\frac{f-f_{x1}}{f_s}\right) + \frac{1}{2} W\left(\frac{f+f_{x1}}{f_s}\right). \quad (4.25)$$

Because we sample the DtFT spectrum in wider frequency range then the DFT spectrum is sampled, we have many copies of the cosine spectrum, in consequence, we see many copies of the rectangular pulse spectrum.

3. *I do not want to watch the same film many times?* No problem. We are changing frequency range of interest to $[-0.5f_s, \ldots, 0.5f_s]$ remaining the small frequency step $df = 1$ Hz of DtFT spectrum sampling. Run the program. Are you satisfied? *Yes, but if I had the second very weak signal frequency component lying apart, I would not see it in the spectrum because it would be hidden by big spectral oscillations coming from the strong signal!*

4. Yes, you are right! Add a second weak cosine component to the signal: x=x1+x2, for example, with frequency $f_{x2} = 250$ Hz and very small amplitude $A_{x2} = 0.001$. Run the program. The second component is not visi-

ble! *I knew! I knew!* Yes, as usual, you knew how to complaint but I know
... how to solve the problem.

5. Multiply the two-component signal with samples of Hanning window function x=x.*w;, i.e. exchange the rectangular window with the Hanning
   window. In Matlab: w2=hanning(N); w=w2;. This function has lower
   level of spectral side-lobes than the rectangular window (look at Fig. 4.2)
   at the price of wider spectral main-lobe. Run the program. Now the second frequency component is visible. *But if the second component would be
   very, very weak indeed, having only* $A_2 = 0.00001$ (10^{-5}, *ten micro-volts)?*

6. No problem. We are choosing adjustable Chebyshev window function having side-lobes on the level of $A_{sl} = 10^{-7}$, $20\log_{10}(A_{sl}) = -140$ dB. In Matlab: w3=chewin(N,140); w=w3. Run the program. The second com-
   ponent is now seen. *Wow! But now the spectral peaks are very wide! If the
   second component had a frequency very close to the first one, for example,*
   $f_{x2} = 110$ *Hz, I would not see it!*

7. No problem. Let us make use of the scaling feature of the CFT given in
   Table 4.2, in consequence being the feature of the DtFT also. Making the
   signal longer (for $a < 1$) leads to its spectrum narrowing. For example,
   the window 10 times longer has the DtFT/DFT spectrum 10 times more
   narrow. Therefore, we will increase now the length of our signal vector 10
   times, setting $N = 1000$. Run the program. Yes, it works. *But now ... after
   the window usage, amplitudes of spectral peaks are not correct! For cosine*
   $1/2$ *is expected!*

8. Yes. I admire your curiosity! Now we have to change the spectrum normalization. Since windows are reducing amplitude of oscillations in signal
   fragment being analyzed, we should compensate this effect! We will exchange dividing the spectrum by N, which is correct for the rectangular
   window, by sum of window samples, which is correct for any window (for
   rectangular one we have N as before). In Matlab: uncomment the line: X =
   N*X/sum(w); You see! Now the spectrum scaling is corrected.

9. *But ...* Bang! Time is over! ... You are the game winner. The worst thing
   pupil can do is not asking questions!

4.4 Window Functions

4.4.1 Practical Summary

It turned out in the previous section how important are functions used for
cutting signal into fragments which are called windows! For signals being summations of pure tones we observe in their spectra only scaled and shifted copies

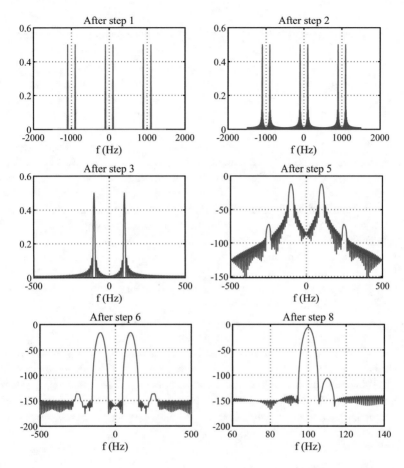

Fig. 4.4: DtFT spectra calculated for signals in consecutive steps (points) of Exercise 4.2: initially for cosine 100 Hz, $N = 100$ samples, $f_s = 1000$ Hz. After step 1: spectrum of rectangular window not visible, after step 2: window spectrum is visible after decreasing frequency step df from 10 Hz to 1 Hz, after step 3: reduction of frequency range due to spectrum periodicity, after step 5: addition of the second signal component with frequency 250 Hz and amplitude 0.001 and using Hanning window, after step 6: changing amplitude of the second component to 0.00001 and using Chebyshev window with side-lobes level of -140 dB, after step 8: changing second component frequency to 110 Hz and increasing number of samples to $N = 1000$, additionally improving spectrum scaling

of *windows* spectra (Fig. 4.4). Therefore one should choose *windows* very carefully: they should help us in spectral analysis, do not create troubles. The good window should have both: very narrow spectral main-lobe (similar to rectangular window) and a very big attenuation of the spectral side-lobes (in contrary to the rectangular window). The narrow spectral main-lobe allows to distinguish in the spectrum signal components having similar frequency values, while highly attenuated spectral side-lobes of the window spectrum makes possible to see in the spectrum, both, very strong and very weak components (with very large and very small amplitudes).

Window *tailoring* is a great DSP art! There are many window functions with precisely specified, fixed shapes: rectangular, triangular (Bartlett), Hamming, Hanning, Blackman, Blackman–Harris, and many others. There are also flexible windows with adjustable shapes: Chebyshev and Kaiser windows are the most popular among them. The latter allows to change the shape of the window and its spectrum in a controlled way and make a compromise between frequency spectrum resolution (width of the window main-lobe) and amplitude spectrum resolution (attenuation of the side-lobes). A special type of windows, flat-top ones, is designed to have a very flat main-lobe peak at the cost of increasing its width. Such windows allow very precise amplitude measurements of many signal components (for example, of power voltage supply harmonics) but they require their significant separation in frequency.

In this chapter the DtFT spectrum of the rectangular window is derived and it is shown how a big family of cosine-based windows is designed (Hamming, Hanning, Blackman, etc.), summarized in Table 4.3. Design equations of Chebyshev and Kaiser windows are presented also with an explanation of their usage. In Fig. 4.5 different window shapes (up) and their DtFT spectra (down) are compared. *Easy riders* can skip the mathematical part, which follows, and go directly to Exercise 4.3 using program from Listing 4.3.

4.4.2 Mathematical Description

Rectangular Window Let us start with the rectangular window.

$$w_R(n) = \begin{cases} 1, & n = 0,\ 1,\ 2,\ldots,\ N-1, \\ 0, & \text{other } n. \end{cases} \tag{4.26}$$

After introduction of a new variable—angular frequency:

$$\Omega = 2\pi f/f_s, \tag{4.27}$$

and putting Eq. (4.26) into the DtFT definition (4.19), we obtain

$$W_R(\Omega) = \sum_{n=-\infty}^{\infty} w_R(n)e^{-j\Omega n} = \sum_{n=0}^{N-1} w_R(n)e^{-j\Omega n} = \sum_{n=0}^{N-1} 1 \cdot e^{-j\Omega n}. \tag{4.28}$$

After multiplying both sides of Eq. (4.28) by $e^{-j\Omega}$ and rewriting the equation we have

$$e^{-j\Omega}W_R(\Omega) = \sum_{n=1}^{N} e^{-j\Omega n} = \sum_{n=0}^{N-1} e^{-j\Omega n} + e^{-j\Omega N} - e^{-j\Omega 0} = W_R(\Omega) + e^{-j\Omega N} - 1.$$

(4.29)

Now we can calculate the value of $W_R(\Omega)$:

$$W_R(\Omega) \cdot (1 - e^{-j\Omega}) = 1 - e^{-j\Omega N} \qquad (4.30)$$

$$W_R(\Omega) = \frac{1 - e^{-j\Omega N}}{1 - e^{-j\Omega}} = \frac{e^{-j\Omega N/2} \left(e^{j\Omega N/2} - e^{-j\Omega N/2}\right)}{e^{-j\Omega/2} \left(e^{j\Omega/2} - e^{-j\Omega/2}\right)} = \cdots$$

$$= e^{-j\Omega(N-1)/2} \frac{\sin(\Omega N/2)}{\sin(\Omega/2)}.$$

(4.31)

In a similar way it can be derived that DtFT spectrum of the odd-length rectangular window $w_{RS}(n), n = -M, \ldots, -1, 0, 1, \ldots, M, N = 2M + 1$, symmetrical around $n = 0$, is equal to:

$$W_{RS}(\Omega) = \frac{\sin\left(\Omega(2M+1)/2\right)}{\sin\left(\Omega/2\right)}.$$

(4.32)

Since $w_{RS}(n)$ is obtained by shifting $w_R(n)$ M samples left, e.g. $w_R(n+M)$, therefore $W_{RS}(\Omega)$ is equal to $W_R(\Omega)$ multiplied by $e^{j\Omega M}$. Of course, absolute values of both spectra are the same: $|W_R(\Omega)| = |W_{RS}(\Omega)|$. A main spectral lobe of the rectangular window (i.e. distance between first zero-crossings on both sides around $\Omega = 0$) is equal to $4\pi/N$, since from Eq. (4.31) we have (first zeros of the sin() function)

$$\begin{aligned} \Omega_1 N/2 &= \pi & \Rightarrow & \quad \Omega_1 = 2\pi/N \\ \Omega_2 N/2 &= -\pi & \Rightarrow & \quad \Omega_2 = -2\pi/N \\ & & & \Delta\Omega_R = \Omega_1 - \Omega_2 = 4\pi/N. \end{aligned}$$

In practice we analyze not infinite signals but their shorter or longer fragments. Some special functions are used for cutting long signal into fragments. They are called "window" functions because we are looking at signals "through" them. The window functions are extremely important in signal theory, especially in spectral analysis and filter design.

Cosine Windows In the beginning we ask fundamental question: what window functions are used and what are their spectra? A big family of so-called cosine-type windows is obtained by multiplication of N samples long rectangular window by sum of K cosines with different angular frequencies (Ω_k) and amplitudes (A_k) ($n = -\infty, \ldots, -1, 0, 1, \ldots, +\infty$):

$$w(n) = w_R(n) \sum_{k=0}^{K} A_k \cos\left(\Omega_k n\right) = w_R(n) \sum_{k=0}^{K} A_k \left(\frac{1}{2}e^{j\Omega_k} + \frac{1}{2}e^{-j\Omega_k}\right), \quad \Omega_k = \frac{2\pi k}{N-1}.$$

$$(4.33)$$

The DtFT (4.19) of (4.33) is equal to:

$$W(\Omega) = \sum_{k=0}^{K} \frac{A_k}{2} \left(\sum_{n=-\infty}^{+\infty} w_R(n) \cdot e^{-j(\Omega-\Omega_k)} + \sum_{n=-\infty}^{+\infty} w_R(n) \cdot e^{-j(\Omega+\Omega_k)} \right) =$$

$$\sum_{k=0}^{K} \frac{A_k}{2} \left(W_R(\Omega - \Omega_k) + W_R(\Omega + \Omega_k) \right) \quad (4.34)$$

due to transformation linearity: spectrum of sum of signals is equal to sum of their individual spectra. Since the window spectrum (4.34) is a sum of scaled (by A_k) and shifted ($\Omega_k = 2\pi k/(N-1)$) spectra of the rectangular window (4.31), (4.32), such weights (A_k) are chosen which minimizes side-lobe oscillations in final spectrum $W(\Omega)$. Definitions and spectral parameters of the most popular cosine windows are given in Table 4.3.

Table 4.3: Definitions and parameters of the most popular non-parametric digital window functions. Denotations: Δ_{ml}—width of the main-lobe of the spectrum around $\Omega = 0$ (rad/s), A_{sl}—relative attenuation of the highest side-lobe in relation to $W(0)$

Window name, Matlab function	Definition $w(n), n = 0, 1, 2, \ldots, N-1$	Δ_{ml}	A_{sl}
Rectangular, w=boxcar()	1	$\frac{4\pi}{N}$	13.3 dB
Triangular, bartlett()	$1 - \frac{2\lvert n-(N-1)/2\rvert}{N-1}$	$\frac{8\pi}{N}$	26.5 dB
Hanning (Hann), hanning()	$\frac{1}{2}\left(1 - \cos\left(\frac{2\pi n}{N-1}\right)\right)$	$\frac{8\pi}{N}$	31.5 dB
Hamming, hamming()	$0.54 - 0.46\cos\left(\frac{2\pi n}{N-1}\right)$	$\frac{8\pi}{N}$	42.7 dB
Blackman, blackman()	$0.42 - 0.50\cos\left(\frac{2\pi n}{N-1}\right) + 0.08\cos\left(\frac{4\pi n}{N-1}\right)$	$\frac{12\pi}{N}$	58.1 dB

Window spectral features are characterized by the shape of its DtFT spectrum. The best window should have a very narrow peak around frequency 0 Hz (the so-called *main-lobe*) and highly attenuated spectral side-peaks, lying elsewhere (the so-called *side-lobes*). The first feature is measured by Δ_{ml} (width of the main-lobe), the second by A_{sl} (attenuation of the highest spectral side-lobe). It is impossible to fulfill both criteria at the same time. The rectangular window has the sharpest main-lobe but, unfortunately, a very high level of spectral side-lobes. Different designers of other windows have tried to increase the side-lobes attenuation at the price of making the spectral main-lobe wider, but as least as possible. In Table 4.3 a few window functions are defined and values of their Δ_{ml} and A_{sl} are given. We see that the more cosine terms the window has, the larger its width Δ_{ml} is (multiplicity of $4\pi/N$—the size of the rectangular window is shifted K times) and the bigger

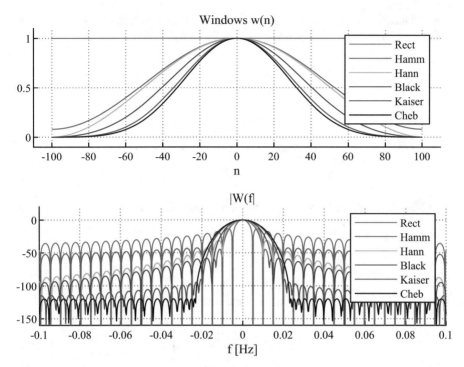

Fig. 4.5: Exemplary shapes (up) and their DtFT spectra (down) for the following windows (they are becoming more *peaky* in the upper plot): rectangular, Hamming, Hanning, Blackman, Kaiser with $\beta = 12$ and Chebyshev -120 dB. Windows are ordered from the lowest to the highest attenuation of spectral side-lobes (oscillatory ripples observed in window spectrum) obtained at the cost of widening the spectral main-lobe (spectral peak around 0 Hz)

attenuation A_{sl} is. Exemplary shapes of windows (up) and their DtFT spectra (down) are presented in Fig. 4.5. Since window functions are real-value ones their spectra are always symmetrical around $\Omega = 0$.

Special Adjustable Windows Apart from mentioned above fixed-shape non-parametric windows having fixed spectral features there are two very important parametric windows which can change their shape and spectral characteristics. The first one is the Dolph–Chebyshev window defined as follows ($N = 2M + 1$):

$$w_{DC}\left[m + (M+1)\right] = C\left[\frac{1}{\gamma} + 2\sum_{k=1}^{M} T_{N-1}\left(\beta\cos\frac{\pi k}{N}\right)\cos\frac{2\pi km}{N}\right], \quad -M \le m \le M,$$
(4.35)

where γ denotes the required relative height of maximum spectrum side-lode in relation to the height of the spectrum main-lobe (e.g. $\gamma = 0{,}01$ or $0{,}001$, which corresponds to relative attenuation of the side-lobe $A_{sl} = 20\log_{10}(\gamma) = 40$ dB or 60 dB) and parameter β, depending on γ, is given by

$$\beta = \cosh\left(\frac{1}{N-1}\cosh^{-1}\frac{1}{\gamma}\right) = \cosh\left(\frac{1}{N-1}\cosh^{-1}(10^{A_{sl}/20})\right). \quad (4.36)$$

$T_{N-1}(x)$ is Chebyshev polynomial of the $(N-1)$-th order:

$$T_N(x) = \begin{cases} \cos\left((N-1)\cos^{-1}x\right), & |x| \le 1 \\ \cosh\left((N-1)\cosh^{-1}x\right), & |x| > 1. \end{cases} \quad (4.37)$$

The Kaiser window, for N even or odd, is defined by formula:

$$w_K(n) = \frac{I_0\left(\beta\sqrt{1-\left(\frac{n-(N-1)/2}{(N-1)/2}\right)^2}\right)}{I_0(\beta)}, \quad 0 \le n \le N-1, \quad (4.38)$$

where $I_0(\beta)$ denotes Bessel function of the 0-th order:

$$I_0(x) = 1 + \sum_{k=1}^{\infty}\left[\frac{(x/2)^k}{k!}\right]^2. \quad (4.39)$$

In literature one can find equations connecting required values of Δ_{ml} and A_{sl} with values of Kaiser window parameters β and N:

$$\beta = \begin{cases} 0 & A_{sl} < 13.26 \text{ dB} \\ 0.76609(A_{sl}-13.26)^{0.4}+0,09834(A_{sl}-13.26), & 13.26 < A_{sl} < 60 \text{ dB} \\ 0.12438(A_{sl}+6.3) & 60 < A_{sl} < 120 \text{ dB} \end{cases} \quad (4.40)$$

$$N = \lceil K \rceil, \quad K = \frac{24\pi(A_{sl}+12)}{155\cdot\Delta_{ml}} + 1, \quad (4.41)$$

where $\lceil K \rceil$ denotes the smallest integer value equal to or greater than K. In Matlab we have functions `kaiser()` and `chebwin()`.

4.4.3 Application Example

Exercise 4.3 (DtFT of Different Windows). First, choose any window and calculate its DtFT spectra for different lengths, for example, $N = 50, 100, 200, 500,$

1000, 2000. Plot all spectra in decibels in one figure. What is your conclusion?
Next, calculate the DtFT spectra for Kaiser window for the value of β changing from 0 to 14 with step 2. Plot all spectra in decibels in one figure. What
is your conclusion? Finally, repeat Exercise 4.2 using in it the Chebyshev window. Change the signal length as well as frequencies and amplitudes of its two
components. Try to obtain a compromise between the Δ_{ml} and A_{sl}.

Listing 4.3: DtFT spectra of different windows

```matlab
1   % lab04_ex_windows.m
2   % DtFT of windows, window importance in frequency analysis
3
4   clear all; close all;
5
6   M = 100;        % one-side number of window samples
7   N = 2*M+1;      % all samples
8   n = -M : M;     % sample indexes
9   text={'Rect','Triang','Hamm','Hann','Black','Kaiser','Cheb'}
10
11  % Window functions, i.e. "windows" in columns of matrix W
12  w(:,1) = boxcar(N);         w(:,2) = bartlett(N);
13  w(:,3) = hamming(N);        w(:,4) = hanning(N);
14  w(:,5) = blackman(N);       w(:,6) = kaiser(N,10);
15  w(:,7) = chebwin(N,120);
16  [ N, Nw] = size(w);         % N window length, Nw number of windows
17
18      figure
19      plot(n,real(w)); xlabel('n'); title('Windows w(n)'); grid;
20      axis([-(M+10) M+10 -0.1 1.1]); legend(text); pause
21
22  % Normalization required for correct interpretation of amplitude spectrum
23  w = w ./ repmat(sum(w),N,1);    % normalization (in column) to sum(w)=1
24
25  % DtFT of windows
26  f = -1/10 : (1/N)/20 : 1/10;                % normalized freq ( f/fs )
27  for k = 1:length(f)
28      bk(1:N) = exp( -j*2*pi*f(k)*n );        % reference Fourier harmonics
29      W(k,1:Nw) = bk(1:N) * w(1:N,1:Nw);      % DtFT coeffs for all windows (in cols)
30  end
31      figure
32      subplot(211); plot(f,real(W)); title('Real(DtFT)'); grid;
33      subplot(212); plot(f,imag(W)); title('Imag(DtFT)'); xlabel('f [Hz]'); grid; pause
34      figure
35      subplot(211); plot(f,abs(W)); title('Abs(DtFT)'); grid;
36      subplot(212); plot(f,angle(W)); title('Angle(DtFT)'); xlabel('f [Hz]'); grid; pause
37      figure
38      subplot(111); plot(f,20*log10(abs(W))); xlabel('f [Hz]'); title('|W(f)|');
39      grid; axis([min(f) max(f) -160 20]);
40      legend(text); pause
41
```

```
42   % Windowed signal
43   x = 2*cos(2*pi*1/20*n);      % signal
44   x = w .* repmat(x',1,Nw);    % its multiplication with many different windows
45   for k = 1:length(f)
46       bk = exp( -j*2*pi*f(k)*n );
47       X(k,:) = bk * x;
48   end
49       figure
50       subplot(111); plot(f,20*log10(abs(X))); xlabel('f [Hz]'); title('|X(f|');
51       grid; axis( [min(f) max(f) -160 20]);
52       legend(text)
53
54   % Repeat the program for different values of N: 50, 100, 200, 500
```

Windows Application Summary The window spectrum should have narrow "main-lobe" to allow seeing in it two separate peaks for frequencies Ω_k and Ω_l, which can lie very close to each other. Otherwise, we might observe one broad peak instead of two narrow ones because of their fusion.

The window spectrum should also have high attenuation of side-lobes in the situation when amplitudes A_k and A_l of two cosine components differ a lot and spectral peak of the weaker component could be *lost/missed* in high spectral side-lobes (*in the grass*) of the stronger component.

4.5 Discrete Fourier Transform

DFT represents discretization result of the Fourier series (4.15) which is defined in analog world for periodic signals (fundamental frequency $f_0 = 1/T$, T—period, $T = N \cdot \Delta t$, $t = n \cdot \Delta t$):

$$
\begin{aligned}
X_k &= \frac{1}{T} \int_0^T x(t) e^{-j2\pi(k f_0)t} \, dt \approx \frac{1}{N \cdot \Delta t} \sum_{n=0}^{N-1} x(n \cdot \Delta t) e^{-j2\pi(k \frac{1}{N \cdot \Delta t})(n \cdot \Delta t)} \Delta t = \dots \\
&= \frac{1}{N} \sum_{n=0}^{N-1} x(n) e^{-j\frac{2\pi}{N} kn}, \quad f_k = k \cdot f_0 = k\frac{f_s}{N}, \quad k = 0, 1, 2 \dots, N-1. \quad (4.42)
\end{aligned}
$$

Equation (4.42) and its inverse can be written in matrix form as orthogonal transformation pair:

$$
\text{Analysis}: \qquad \bar{\mathbf{X}} = \frac{1}{N} \mathbf{F} \cdot \bar{\mathbf{x}}, \qquad (4.43)
$$

$$
\text{Synthesis}: \qquad \bar{\mathbf{x}} = \mathbf{F}^H \cdot \bar{\mathbf{X}} = (\mathbf{F}^*)^T \cdot \bar{\mathbf{X}}, \qquad (4.44)
$$

well-known for us from Chap. 2. The transformation matrix $\mathbf{F}$ is defined as, for example, also for $N = 4$:

$$
\mathbf{F}_N =
\begin{bmatrix}
1 & 1 & \cdots & 1 \\
1 & e^{-j\frac{2\pi}{N}1\cdot1} & \cdots & e^{-j\frac{2\pi}{N}1\cdot(N-1)} \\
\vdots & \vdots & \ddots & \vdots \\
1 & e^{-j\frac{2\pi}{N}(N-1)\cdot1} & \cdots & e^{-j\frac{2\pi}{N}(N-1)\cdot(N-1)}
\end{bmatrix}, \quad
\mathbf{F}_4 =
\begin{bmatrix}
1 & 1 & 1 & 1 \\
1 & e^{-j\frac{2\pi}{4}1\cdot1} & e^{-j\frac{2\pi}{4}1\cdot2} & e^{-j\frac{2\pi}{4}1\cdot3} \\
1 & e^{-j\frac{2\pi}{4}2\cdot1} & e^{-j\frac{2\pi}{4}2\cdot2} & e^{-j\frac{2\pi}{4}2\cdot3} \\
1 & e^{-j\frac{2\pi}{4}3\cdot1} & e^{-j\frac{2\pi}{4}3\cdot2} & e^{-j\frac{2\pi}{4}3\cdot3}
\end{bmatrix},
$$

$$(4.45)$$

so it has in its rows conjugated orthogonal harmonic vectors, with different scaling than in Eq. (4.1). In non-vector form and with changed normalization, equations (4.43) and (4.44) have the following form:

$$
\textbf{Analysis}: \quad X(k) = \frac{1}{N}\sum_{n=1}^{N} x(n)e^{-j\frac{2\pi}{N}kn}, \quad k = 0,\,1,\,2,\ldots,N-1, \quad (4.46)
$$

$$
\textbf{Synthesis}: \quad x(n) = \sum_{k=1}^{N} X(k)e^{j\frac{2\pi}{N}kn}, \quad n = 0,\,1,\,2,\ldots,N-1. \quad (4.47)
$$

In Eq. (4.46) the signal is compared (correlated) with conjugation of harmonic Fourier basis function, while in Eq. (4.47) it is represented as a sum of basic functions scaled by spectral ("similarity") coefficients $X(k)$, calculated in Eq. (4.46).

First, the main student problem, after calculation of the signal DFT spectrum using Eq. (4.46), is how to connect calculated spectral coefficients X_k with real-world frequencies when the frequency is missing in these equations! But we know how long the signals are (N samples) and which is the sampling frequency (f_s), so as a result we know also the time duration of signals: $T = N \cdot \Delta t = N/f_s$. In first row ($k = 0$) of matrix $\mathbf{F}$ we have only ones, in the second ($k = 1$)—one period of cos() in real part and one period of -sin() in imaginary part, in the third ($k = 2$)—two periods, later three, four, five, $\ldots$, $N-1$ periods. Therefore, since we know T, we should deduce that X_0 is a mean value of the signal, spectral coefficient X_1 corresponds to frequency $1 \cdot f_0 = 1/T$, X_2—to $2 \cdot f_0$, X_3—to $3 \cdot f_0$, and so on. This sounds reasonable since in Fourier series coefficients are also calculated for frequencies $k \cdot f_0$.

That is it! Now a reader should do some computer experiments and $\ldots$ find visually with surprise conjugate symmetry of the DFT spectrum: $X_k, k = 0, 1, 2, \ldots, N-1$. Yes, indeed, the spectrum of our speech has such symmetry! This is typically the second student surprise! We analyze, for example, a real-value signal having only one frequency component but in the spectrum we see two peaks: one in its first half and one in the second half. This phenomena results from the fact that for $k = 1, 2, 3 \ldots, N-1$ the following relations hold:

$$X_{N-k} = X_k^* \quad \Rightarrow \quad \text{real}(X_{N-k}) = \text{real}(X_k), \quad \text{imag}(X_{N-k}) = -\text{imag}(X_k) \quad (4.48)$$

Additionally the DFT spectrum always has zeros in imaginary part for $k = 0$ and $k = N/2$:

$$\text{imag}(X_0) = \text{imag}(X_{N/2}) = 0. \quad (4.49)$$

Why the DFT spectrum has conjugate symmetry? What is the sense of computing something twice? The first answer is that conjugated Fourier harmonics which are used for signal decomposition in Eq. (4.46) are the same for k and $N - k$, only conjugated. For the k-th harmonic we have

$$e^{-j\frac{2\pi}{N}kn}, \quad n = 0,1,2,\ldots,N-1, \quad (4.50)$$

while for the $(N - k)$-th:

$$e^{-j\frac{2\pi}{N}(N-k)n} = e^{-j2\pi n} \cdot e^{+j\frac{2\pi}{N}kn} = e^{+j\frac{2\pi}{N}kn}. \quad (4.51)$$

Therefore the calculated DFT coefficients have also the same values, only complex-conjugated:

$$X(k) = \frac{1}{N} \sum_{n=1}^{N} x(n) \cos\left(\frac{2\pi}{N}kn\right) - j\frac{1}{N} \sum_{n=1}^{N} x(n) \sin\left(\frac{2\pi}{N}kn\right) = \mathbf{a} - \mathbf{jb} \quad (4.52)$$

$$X(N-k) = \frac{1}{N} \sum_{n=1}^{N} x(n) \cos\left(\frac{2\pi}{N}kn\right) + j\frac{1}{N} \sum_{n=1}^{N} x(n) \sin\left(\frac{2\pi}{N}kn\right) = \mathbf{a} + \mathbf{jb}. \quad (4.53)$$

The second explanation of the spectrum (a)symmetry phenomena is that real-value cosine and sine functions can be expressed as a summation of two Fourier harmonics used for signal decomposition:

$$\cos\left(\frac{2\pi}{N}kn\right) = \frac{e^{j\frac{2\pi}{N}kn} + e^{-j\frac{2\pi}{N}(N-k)n}}{2}, \quad (4.54)$$

$$\sin\left(\frac{2\pi}{N}kn\right) = \frac{e^{j\frac{2\pi}{N}kn} - e^{-j\frac{2\pi}{N}(N-k)n}}{2j}, \quad (4.55)$$

therefore, since the DFT transform is linear, when analyzing real-value signals we have two symmetrical peaks in the DFT spectrum with complex-conjugated values.

Exercise 4.4 (DFT of a Cosine with Rectangular Window). Make use of the Matlab program 4.2 in which the DFT algorithm is also implemented. In figures generated by the program, the DFT spectra are marked with red circles and compared with the DtFT spectra, denoted by blue solid line. In the beginning, try to obtain the same plots as presented in Fig. 4.6. Set the following values of program parameters: sampling ratio $f_s = 1000$ Hz, number of signal samples $N = 50$, only component `x1`, in the beginning with frequency 100 Hz, than with 110 Hz. Check validity of the DFT spectrum (a)symmetry (Eq. (4.48)). Next, apply different windows to the signal. Again check the DFT spectrum (a)symmetry.

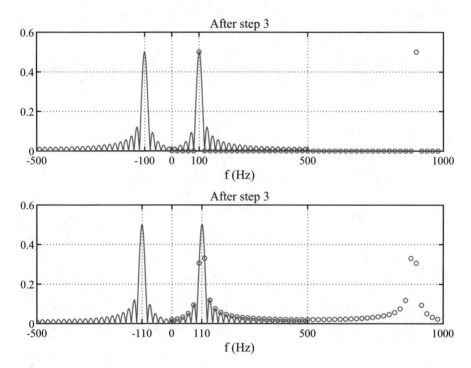

Fig. 4.6: DtFT and DFT spectra of two cosines, blue line and red dots, respectively: (up) signal with $f_0 = 100$ Hz: perfect DFT match to the cosine frequency, (down) signal with $f_0 = 110$ Hz: the worst DFT match to the cosine frequency. Values of parameters: sampling frequency $f_s = 1000$ Hz, $N = 50$ samples, DFT spectrum discretization step $\Delta f = f_0 = f_s/N = 20$ Hz. Observe symmetry of the DFT spectrum marked with red circles. Notice the different shapes of correct DtFT spectrum marked with blue solid line (two shifted spectra of the rectangular window)

4.6 Summary

Signal processing consists of two fundamental branches: frequency analysis
and signal filtering (noise reduction and rejections of some signal components).
This chapter has been focused on fundamentals of frequency analysis by means
of Fourier transform—it has a crucial significance in our DSP course. You
should understand everything in it. If you do not, I am very sorry, read this
chapter again and again ... before the examination. Yes, you can. Personally,
some papers I was reading repeatedly even 10 years. With final success. What
should be remembered?

1. Most frequency analysis methods are very similar: one compares a signal
   with reference frequency components (*oscillations*) multiplying it with the
   references and accumulating single products. This is valid in analog and
   digital world, done by Fourier integrals and summations.
2. DtFT is a discretized continuous Fourier transform and DFT is a result
   of the discretization of Fourier series. In computer implementation both
   transforms are very similar: they use the same signal samples and treat them
   in almost the same manner. The only difference is that in N-sample DFT the
   signal spectrum is computed for precisely specified set of N frequencies:
   $f_k = k \cdot f_0, k = 0, 1, 2, \ldots, (N-1)$, where $f_0 = 1/T$ is the inverse of the
   analyzed signal duration. In DtFT the choice of frequencies is completely
   free. For real-value signals we are typically choosing frequency values in
   the range $[0, f_s/2)$. The DtFT offers better visualization of the theoretical
   signal spectrum due to its possible more dense sampling.
3. The DtFT is more application flexible but the DFT is faster due to the existence of very fast DFT implementations (FFT algorithms are presented
   in the next chapter). DtFT calculation can be speed up by using fast implementations of the chirp-Z transform (usage of three FFTs, not discussed
   here)—never the less the method is slower than the DFT.
4. One should always remember that in practice frequency analysis is performed upon a signal fragment, not on the whole signal. What is the consequence of this fact? That a fragment *cutting* operation has a very big influence on the final result! Simply taking signal samples *from–to* means that
   we are multiplying a signal with an observation window function having a
   value equal to "1" *from–to* and "0" elsewhere.
5. If two signals are multiplied, the Fourier spectra are convoluted. Due to this,
   the signal windowing performed during signal fragment cutting causes that
   the theoretical spectrum of the time-infinite signal is convoluted with the
   window spectrum modifying it. Therefore, one should choose very carefully the window function during spectral analysis.
6. Increasing K-times the length of the analyzed signal, one improves K-times
   the DFT frequency resolution, no matter what the window function is used.

As a consequence, signal components having near frequencies are better distinguished.

7. Using windows with a low level of spectral side-lobes, one improves amplitude resolution of the spectrum by the cost of decreasing its frequency resolution. But the frequency resolution can be always improved by signal enlargement.

8. There are many window functions. The best should offer the most narrow spectral main-lobe (good frequency resolution) and the lowest level of spectral side-lobes (good amplitude resolution). In practice, Kaiser and Chebyshev windows are very often used due to their adjustable shapes and changeable spectral features.

4.7 Private Investigations: Free-Style Bungee Jumps

Exercise 4.5 (Am I An Enrico Caruso? Finding Frequency of Vocal Cords Oscillation). Use DtFT and DFT *detectives* for finding the frequency of your vocal cords opening and closing for different vowels.

Exercise 4.6 (In the Wonderful World of Sounds). From the Internet web page *FindSounds* (https://www.findsounds.com/) take 2–3 signals of different origin and use DtFT and DFT for finding their frequency content. Scale frequency axis in hertz. Overlay two spectra in one plot.

Exercise 4.7 (Did You Break My Heart? What Is the Frequency of My Heartbeat?). Take from the Internet an ECG heart activity signal, e.g. from the page https://archive.physionet.org/cgi-bin/atm/ATM. Calculate the frequency of the heartbeat using DtFT and DFT.

Exercise 4.8 (Steel Factory Secrets). Analyze signals of supply voltages and currents recorded for operating arc furnace. Take them from the file `load('UI.mat');` `whos`, given at the book web page. Do spectral DtFT analysis for interesting parts of the spectra. Estimate frequencies and amplitudes of fundamental frequency 50 Hz (close to 50) and its harmonics 100, 150, 200, 250, ... Hz.

Exercise 4.9 (Clear Water, Clear Power Supply). Please, analyze recorded power supply voltage signal `tu.dat` which is used for testing algorithms for monitoring electric power quality (https://grouper.ieee.org/groups/1159/). First, extract time and voltage values from matrix columns:
`load('tu.dat'); t=tu(:,1); u=tu(:,2); plot(t,u);).`
Then, estimate frequencies and amplitudes of fundamental frequency 50 Hz (close to 50) and its harmonics, if they are present.

Exercise 4.10 (Mysteries of NMR Laboratory). Do frequency analysis of pseudo-NMR signal synthesized in first additional exercise after Chap. 2.

References

1. L.F. Chaparro, *Signals and Systems Using Matlab* (Academic Press, Burlington MA, 2011)
2. F.J. Harris, On the use of windows for harmonic analysis with the discrete Fourier transform. Proc. IEEE **66**(1), 51–83 (1978)
3. V.K. Ingle, J.G. Proakis, *Digital Signal Processing Using Matlab* (PWS Publishing, Boston, 1997; CL Engineering, 2011)
4. R.G. Lyons, *Understanding Digital Signal Processing* (Addison-Wesley Longman Publishing, Boston, 1996, 2005, 2010)
5. A.V. Oppenheim, R.W. Schafer, *Discrete-Time Signal Processing* (Pearson Education, Upper Saddle River, 2013)
6. A.V. Oppenheim, A.S. Willsky, S.H. Nawab, *Signals & Systems* (Prentice Hall, Upper Saddle River, 1997, 2006)
7. A. Papoulis, *Signal Analysis* (McGraw Hill, New York, 1977)
8. J.G. Proakis, D.G. Manolakis, *Digital Signal Processing. Principles, Algorithms, and Applications* (Macmillan, New York, 1992; Pearson, Upper Saddle River, 2006)
9. S.W. Smith, *The Scientist and Engineer's Guide to Digital Signal Processing* (California Technical Publishing, San Diego, 1997, 1999). Online: http://www.dspguide.com/
10. M. Vetterli, J. Kovacevic, V.K. Goyal, *Foundations of Signal Processing* (Cambridge University Press, Cambridge, 2014)
11. T.P. Zieliński, *Cyfrowe Przetwarzanie Sygnałów. Od Teorii do Zastosowań (Digital Signal Processing. From Theory to Applications)* (Wydawnictwa Komunikacji i Łączności (Transport and Communication Publishers), Warszawa, Poland, 2005, 2007, 2009, 2014)

Chapter 5
Fast Fourier Transform

Salute! Your Majesty Fast Fourier Transform - the DSP King (Kong)!

5.1 Introduction

Discrete Fourier transform (DFT) is the orthogonal transform defined in previous chapter by Eqs. (4.46), (4.47) in which reference oscillatory signals are complex-value Fourier harmonics. The forward DFT (transforming signal samples $x(n)$ to spectral coefficients $X(k)$) and the inverse DFT (transforming back spectral coefficients to signal samples) are defined with very similar equations being different only in the sign of the exponent and $\frac{1}{N}$ scaling of the first sum:

$$X(k) = \frac{1}{N}\sum_{n=0}^{N-1} x(n)e^{-j(2\pi/N)kn}, \quad x(n) = \sum_{n=0}^{N-1} X(k)e^{+j(2\pi/N)kn}, \quad k,n = 0,1,\ldots,N-1.$$

(5.1)

Therefore fast algorithms, designed for the forward DFT, can be used also for the inverse DFT (IDFT) after change of the exponent sign and the overall scaling.

> **Warning!** It is important to note that scaling by $\frac{1}{N}$ is done inside Matlab function x=ifft(X) of the second IDFT equation, not in the first DFT one X=fft(x). We prefer putting scaling in the first of them since this is correct when DFT is derived from Fourier series—see Eq. (4.42). For such scaling, mathematically correct values of spectral coefficients are obtained. In this chapter the scale factor $\frac{1}{N}$ is completely neglected in order to unify algorithms for DFT and IDFT calculation.

© Springer Nature Switzerland AG 2021
T. P. Zieliński, *Starting Digital Signal Processing in Telecommunication Engineering*, Textbooks in Telecommunication Engineering, https://doi.org/10.1007/978-3-030-49256-4_5

Both algorithms, of DFT and IDFT, can be written using matrix notation typical for all orthogonal transforms (see previous chapter): the transform matrix with dimensions $N \times N$, having in rows sampled conjugated reference oscillatory basis functions, is multiplied by a vertical vector of signal samples. Therefore the DFT computation requires calculations of N^2 complex-value multiplications and $N \cdot (N-1)$ complex-value additions. As an example we will calculate DFT coefficients, without $\frac{1}{N}$ scaling, for signal $x(n) = [1,2,3,4]$ having $N = 4$ samples:

$$X(0) = \sum_{n=0}^{4-1} x(n)e^{-j(2\pi/4)0n} = x(0) + x(1) + x(2) + x(3) = \mathbf{10},$$

$$X(1) = \sum_{n=0}^{3} x(n)e^{-j(2\pi/4)1n} = x(0)e^0 + x(1)e^{-j(2\pi/4)} + x(2)e^{-j(2\pi/4)2}$$
$$+ x(3)e^{-j(2\pi/4)3} = 1 - j2 - 3 + j4 = \mathbf{-2+j2},$$

$$X(2) = \sum_{n=0}^{3} x(n)e^{-j(2\pi/4)2n} = x(0)e^0 + x(1)e^{-j(2\pi/4)2} + x(2)e^{-j(2\pi/4)2\cdot2}$$
$$+ x(3)e^{-j(2\pi/4)2\cdot3} = 1 - 2 + 3 - 4 = \mathbf{-2},$$

$$X(3) = \sum_{n=0}^{3} x(n)e^{-j(2\pi/4)3n} = x(0)e^0 + x(1)e^{-j(2\pi/4)3} + x(2)e^{-j(2\pi/4)3\cdot2}$$
$$+ x(3)e^{-j(2\pi/4)3\cdot3} = 1 + j2 - 3 - j4 = \mathbf{-2+j2}.$$

The resultant DFT spectrum is equal to $X(k) = [10, -2 + j2, -2, -2 - j2]$. The first value, equal to 10, is the sum of signal samples, the following ones are coefficients *measuring* the analyzed signal similarity to complex-value signals with reference frequencies (their real part specifies similarity to the cosine, while imaginary part to the sine). Let us write the same calculations in matrix form:

$$
\begin{bmatrix} X(0) \\ X(1) \\ X(2) \\ X(3) \end{bmatrix} =
\begin{bmatrix} 1 & 1 & 1 & 1 \\ 1 & -j & -1 & j \\ 1 & -1 & 1 & -1 \\ 1 & j & -1 & -j \end{bmatrix}
\begin{bmatrix} x(0) \\ x(1) \\ x(2) \\ x(3) \end{bmatrix} =
\begin{bmatrix} 1 & 1 & 1 & 1 \\ 1 & -j & -1 & j \\ 1 & -1 & 1 & -1 \\ 1 & j & -1 & -j \end{bmatrix}
\begin{bmatrix} 1 \\ 2 \\ 3 \\ 4 \end{bmatrix} =
\begin{bmatrix} 10 \\ -2+j2 \\ -2 \\ -2-j2 \end{bmatrix}.
$$

Fast DFT calculation algorithms make use of the observation that values in the transformation matrix are repeating (since sin() and cos() functions are periodic) and they are multiplied many times by the same signal samples. Such situations are identified and removed: multiplications are done only once and their results used a few times. For example, let us see how the fast algorithm works in the discussed case of $N = 4$ DFT:

$$
\begin{bmatrix} X(0) \\ X(1) \\ X(2) \\ X(3) \end{bmatrix} = \begin{bmatrix} 1 & 1 & 1 & 1 \\ 1 & -j & -1 & j \\ 1 & -1 & 1 & -1 \\ 1 & j & -1 & -j \end{bmatrix} \begin{bmatrix} x(0) \\ x(1) \\ x(2) \\ x(3) \end{bmatrix} = \tag{5.2}
$$

$$
= \begin{bmatrix} 1 & 1 \\ 1 & -1 \\ 1 & 1 \\ 1 & -1 \end{bmatrix} \begin{bmatrix} x(0) \\ x(2) \end{bmatrix} + \begin{bmatrix} 1 & 1 \\ -j & j \\ -1 & -1 \\ j & -j \end{bmatrix} \begin{bmatrix} x(1) \\ x(3) \end{bmatrix} = \tag{5.3}
$$

$$
= \begin{bmatrix} 1 & 1 \\ 1 & -1 \\ 1 & 1 \\ 1 & -1 \end{bmatrix} \begin{bmatrix} x(0) \\ x(2) \end{bmatrix} + \begin{bmatrix} 1 \\ -j \\ -1 \\ j \end{bmatrix} .* \begin{bmatrix} 1 & 1 \\ 1 & -1 \\ 1 & 1 \\ 1 & -1 \end{bmatrix} \begin{bmatrix} x(1) \\ x(3) \end{bmatrix} = \tag{5.4}
$$

$$
= \begin{bmatrix} \mathbf{1} & \mathbf{1} \\ \mathbf{1} & \mathbf{-1} \\ & \\ & \end{bmatrix} \begin{bmatrix} 1 \\ 3 \end{bmatrix} + \begin{bmatrix} 1 \\ -j \\ -1 \\ j \end{bmatrix} .* \begin{bmatrix} \mathbf{1} & \mathbf{1} \\ \mathbf{1} & \mathbf{-1} \\ & \\ & \end{bmatrix} \begin{bmatrix} 2 \\ 4 \end{bmatrix} = \tag{5.5}
$$

$$
= \begin{bmatrix} 4 \\ -2 \\ 4 \\ -2 \end{bmatrix} + \begin{bmatrix} 1 \\ -j \\ -1 \\ j \end{bmatrix} .* \begin{bmatrix} 6 \\ -2 \\ 6 \\ -2 \end{bmatrix} = \begin{bmatrix} 10 \\ \text{-2+2j} \\ \text{-2} \\ \text{-2+2j} \end{bmatrix} \tag{5.6}
$$

First, Eq. (5.3)—the signal samples are grouped into even- and odd-indexed ones and the 4×4 transformation matrix is replaced by two sub-matrices 4×2. Additionally, Eq. (5.4)—odd-indexed signal samples multiplication by the second sub-matrix is re-organized: first multiplication is done by the simplified matrix, then the obtained result is corrected: each calculated value is multiplied by different correction term. .* in Eq. (5.4) denotes, as in Matlab, multiplication of corresponding values of two vectors or matrices, first-by-first, second-by-second, ... and so on. Now we see that lower 2×2 matrices are the same as the upper ones (marked with bold font and blue color). Therefore it is sufficient to multiply signal samples by upper 2×2 matrices only (bold/blue ones), then copy the result and put it down. For this reason in Eq. (5.5) elements of lower matrices are missing and in Eq. (5.6) two calculated values (bold/blue) in the upper part are copied down and marked in red! Finally, the same result is obtained as in Eq. (5.1) but faster. In the original algorithm we do $4^2 = 16$ multiplication. In the fast version: $2 \cdot 2^2 = 8$ plus 4 multiplications with corrections, all together 12 operations.

This savings does not seem to be significant. However when the signal vector is long, for example, has $N = 1024$ samples, and the data partition into even- and odd-indexed signal samples is repeated many times, to the moment of obtaining vectors having two samples only, the savings are really impressive: about 200 TIMES FASTER (in general not $N \cdot N$ multiplications are performed but only $\frac{N}{2} \log_2(N)$). It is like to drive a car with speed 200 kilometers per hour instead of 1 kilometer per hour only. Or like pay only 1 euro/dollar instead of 200 euro/dollars. Who would like to move as a turtle? Or overpay. Do you not want?

Let us now look to the *heart* of the radix-2 FFT algorithm. After recursive multi-level partition of signal samples into even- and odd-indexed ones, two-element vectors are obtained and $\frac{N}{2}$ 2-point DFTs are performed upon them. Then: (1) 2-point DFT spectra are combined into 4-point ones, (2) 4-point DFT spectra into 8-point DFT spectra, (3) 8-point to 16-point, and so on, up to the reconstruction of the DFT spectrum of the whole signal.

So, ..., *the text of this chapter is yours.*

5.2 Radix-2 FFT Algorithm

Fast Fourier Transform (FFT) is a fast algorithm for computation of discrete Fourier transform (DFT) discussed in Chap. 4. In turn DFT represents itself an orthogonal transformation of the form:

$$\bar{\mathbf{X}}_{N\times 1} = \mathbf{W}_{N\times N} \cdot \bar{\mathbf{x}}_{N\times 1}, \tag{5.7}$$

where $\bar{\mathbf{x}}_{N\times 1}$ is a vertical vector of N signal samples, $\mathbf{W}_{N\times N}$ is an orthogonal $N \times N$ transformation matrix (with orthogonal rows), and $\bar{\mathbf{X}}_{N\times 1}$ is a vertical N-element vector of calculated DFT spectrum coefficients. In case of DFT, the matrix-based Eq. (5.1) has the following form:

$$\begin{bmatrix} X(0) \\ X(1) \\ \vdots \\ X(N-1) \end{bmatrix} = \begin{bmatrix} 1 & 1 & \cdots & 1 \\ 1 & e^{-j\frac{2\pi}{N}1\cdot 1} & \cdots & e^{-j\frac{2\pi}{N}1\cdot(N-1)} \\ \vdots & \vdots & \ddots & \vdots \\ 1 & e^{-j\frac{2\pi}{N}(N-1)\cdot 1} & \cdots & e^{-j\frac{2\pi}{N}(N-1)\cdot(N-1)} \end{bmatrix} \begin{bmatrix} x(0) \\ x(1) \\ \vdots \\ x(N-1) \end{bmatrix} \tag{5.8}$$

while the non-matrix one is defined as (we are at present neglecting division by N in comparison to Eq. (4.46):

$$X(k) = \sum_{n=0}^{N-1} x(n)e^{-j\frac{2\pi}{N}kn}, \quad k = 0, 1, 2, \ldots, N-1. \tag{5.9}$$

At the first look there is no possibility to do calculation faster: N-element vertical vector should be multiplied by $N \times N$-element matrix. This is the task! However the matrix elements (k—row/oscillation index, n—column/sample index):

$$W[k,n] = \exp\left(-j\frac{2\pi}{N}kn\right), \tag{5.10}$$

repeat (they have the same values due to periodicity of sin()/cos() functions) and some of them are multiplied by the same signals samples. These repeated multiplications should not be performed: the previous arithmetical results should be copied only. Savings offered by this strategy is very big. Instead of performing $N \cdot N$ multiplications (e.g. for $N = 1024$ we have $1024 \cdot 1024 = 1048576$), only 10240 multiplications have to be done (approximately) or even two times less if some additional trick is done. Therefore we are taking about decreasing the number of multiplications more than 100 times or even 200 times. This is like driving 200 km per hour, not 1 km per hour. Wow! So how is it done. This chapter is giving an answer to this question. To be precise, we are talking now about radix-2 decimation-in-time FFT algorithm. There are many other FFT algorithms, for example, radix-4, split-radix, prime-factor, ...

Let us introduce the following denotations:

$$W_N = e^{-j\frac{2\pi}{N}}, \quad (W_N)^{kn} = W_N^{kn} = e^{-j\frac{2\pi}{N}kn}. \tag{5.11}$$

Next, let us do in Eq. (5.9) summations of multiplications of W_N^{kn} with even ($n = 2r$) and odd ($n = 2r + 1$) samples of $x(n)$ separately:

$$X(k) = \sum_{r=0}^{N/2-1} x(2r)e^{-j\frac{2\pi}{N}k(2r)} + \sum_{r=0}^{N/2-1} x(2r+1)e^{-j\frac{2\pi}{N}k(2r+1)}, \quad k = 0, 1, 2, \ldots, N-1. \tag{5.12}$$

We can replace multiplication by 2 in exp(.) by dividing by $N/2$:

$$X(k) = \sum_{r=0}^{N/2-1} x(2r)e^{-j\frac{2\pi}{N/2}kr} + e^{-j\frac{2\pi}{N}k} \cdot \sum_{r=0}^{N/2-1} x(2r+1)e^{-j\frac{2\pi}{N/2}kr}, \quad k = 0, 1, 2, \ldots, N-1. \tag{5.13}$$

Now we see that we have obtained summation of $N/2$-point DFTs of even and odd signal samples (and the *odd* DFT result is multiplied by correction term $\exp(-j2\pi k/N)$). With the only difference that k is not changing in the range $0, 1, 2, \ldots, N/2 - 1$ as in $N/2$-point DFT but in the range $0, 1, 2, \ldots, N-1$ as in the N-point DFT. But we can observe that for values of k bigger then $N/2 - 1$ (i.e. $N/2 + k$, $k = 0, 1, 2, \ldots, N/2 - 1$), we obtain the same result as for values smaller then $N/2$ (i.e. k):

$$e^{-j\frac{2\pi}{N/2}(N/2+k)r} = e^{-j2\pi r}e^{-j\frac{2\pi}{N/2}kr} = e^{-j\frac{2\pi}{N/2}kr}, \quad k = 0, 1, 2, \ldots, \frac{N}{2} - 1. \tag{5.14}$$

What does result from this "discovery"? Is it a real "treasure" or not? Yes, it is! One can perform two DFTs upon two two-times shorter $N/2$-point vectors of samples (even and odd), copy each of calculated DFT vectors two times, making them this way two times longer (enlargement k from $N/2 - 1$ to $N - 1$), multiply the second vector associated with odd samples by the correction term $\exp(-j2\pi k/N)$, and do final summation of vectors (even and odd parts, respectively):

$$\bar{\mathbf{X}}^{(\mathrm{e})}_{N/2\times 1} = \mathbf{W}_{N/2\times N/2}\cdot\bar{\mathbf{x}}^{(\mathrm{e})}_{N/2\times 1}, \qquad \bar{\mathbf{X}}^{(\mathrm{o})}_{N/2\times 1} = \mathbf{W}_{N/2\times N/2}\cdot\bar{\mathbf{x}}^{(\mathrm{o})}_{N/2\times 1}, \qquad (5.15)$$

$$\bar{\mathbf{X}}_{N\times 1} = \begin{bmatrix}\bar{\mathbf{X}}^{(\mathrm{e})}_{N/2\times 1}\\ \bar{\mathbf{X}}^{(\mathrm{e})}_{N/2\times 1}\end{bmatrix} + \begin{bmatrix}e^{-j\frac{2\pi}{N}0}\\ e^{-j\frac{2\pi}{N}1}\\ \vdots\\ e^{-j\frac{2\pi}{N}(N-1)}\end{bmatrix}.*\begin{bmatrix}\bar{\mathbf{X}}^{(\mathrm{o})}_{N/2\times 1}\\ \bar{\mathbf{X}}^{(\mathrm{o})}_{N/2\times 1}\end{bmatrix}. \qquad (5.16)$$

.∗ in Eq. (5.16) denotes, as in Matlab, multiplication of corresponding values of two vectors or matrices. Let us calculate the obtained speed-up! As we see, now in Eq. (5.15) we multiply two times matrix with dimensions $N/2\times N/2$ by $N/2$-element vector, which gives $2N^2/4 = N^2/2$ multiplications, while the initial matrix equation (5.8) requires two times more, N^2 multiplications. Of course, we should not forget about N additional multiplications required for correction of the DFT of odd samples, however, in comparison with $N^2/2$ multiplications, the increase is insignificant.

Equations (5.15), (5.16) look serious and with dignity. But their graphical interpretation presented in Fig. 5.1 is very simple. It convincingly confirms our step-by-step calculations presented in Eqs. (5.2)–(5.6), leading to the fast FFT algorithm. Lower parts of *even* and *odd* matrices, in figure white, are exactly the same as their upper parts, therefore calculated upper results can be copied down and about 50 percents of multiplications are not performed. In Matlab program 5.1 we are performing operations presented in Fig. 5.1. Wow! How simple is the computational trick used in radix-2 DIT FFT!

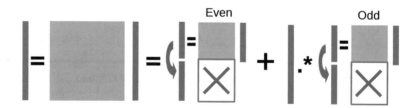

Fig. 5.1: Graphical illustration of reduction of number of multiplications in the DIT radix-2 FFT algorithm after one-level *even/odd* samples partitioning. Multiplication of even-/odd-indexed samples by "white" sub-matrices with red "×" is not performed since "white" sub-matrices are the same as upper "blue" ones—result of the upper matrix-vector multiplications is simply copied. ".∗" denotes element-by-element multiplications as in Matlab

Listing 5.1: First step in the radix-2 DIT FFT algorithm

```
1   % lab05_ex_partition1.m
2   % One-level FFT : first sample re-ordering
3   clear all; close all;
4
5   N = 100; x = rand(1,N);
6   Xm = fft(x);
7   Xe = fft(x(1:2:N));
8   Xo = fft(x(2:2:N));
9   X = [ Xe, Xe ] + exp(-j*2*pi/N*(0:1:N-1)) .* [Xo, Xo ];
10  error = max( abs( X - Xm ) ),
```

Exercise 5.1 (Matrix Interpretation of Speeding-Up Calculations in Radix-2 DIT FFT). In the introduction we have presented, as an example, matrix analysis of the 4-point DFT calculation. It was shown that after sample grouping into even- and odd-indexed ones, the square transformation matrix is replaced with two smaller matrices having the same elements in their lower parts as in upper parts. Set $N = 4, 8, 16$ in program 5.1 and verify this observation. Use the DFT transformation sub-matrices for calculation of spectra Xe and Xo. Check result correctness.

So, if approximately two times reduction of multiplication number is obtained only by simple separating of even and odd signal samples and performing two-times shorter DFTs on them, WHY not to continue and why not to divide signal samples once more: do partition of even samples into even and odd (even–even, even–odd) and odd samples into even and odd (odd–even, odd–odd). After that we will have four vectors four times shorter and four $N/4$-point DFTs will be perform with corrections. Such proceeding offers complexity of $4(N^2/16) = N^2/4$ plus $2N$ multiplications. So now not 2 but 4 times reduction of multiplication number is get in comparison to Eq. (5.7).

In radix-2 algorithm repetitive partitioning data to even/odd samples is repeated to the moment when $N/2$ vectors having only 2 samples are obtained. Then $N/2$ times 2-point DFT spectra are computed. Next $N/2$ 2-point spectra are combined with corrections and give $N/4$ 4-point spectra, then $N/4$ 4-point spectra are combined to $N/8$ 8-point spectra, 8-point to 16-point, ... etc., to the moment when the final N-point DFT spectrum is reconstructed. On each algorithm level one performs N complex multiplications (due to required correction of N samples) and in bottom level $N/2$ times 2-point DFTs are done:

$$\begin{bmatrix} X(0) \\ X(1) \end{bmatrix} = \begin{bmatrix} 1 & 1 \\ 1 & e^{-j\frac{2\pi}{2}1\cdot1} \end{bmatrix} \begin{bmatrix} x(0) \\ x(1) \end{bmatrix} = \begin{bmatrix} 1 & 1 \\ 1 & -1 \end{bmatrix} \begin{bmatrix} x(0) \\ x(1) \end{bmatrix}, \qquad (5.17)$$

representing only addition and subtraction of two samples with no multiplication. Summarizing the total number of multiplication, we have N corrections (multiplica-

tions) done on each spectra combining level. The number of such levels is equal to
the number of even/odd sample partitions minus 1, i.e. $\log_2(N) - 1$. Minus 1, since
the decomposition is stopped for vectors with 2 samples upon which 2-point DFTs
are executed. Therefore, when we neglect the simplicity of the 2-point DFT level
and assume the regular algorithm structure, $N \log_2(N)$ number of multiplication is
required. For $N = 1024$ the speed-up is more than 100 times while compared to N^2!

The following program 5.2 presents two-level fast radix-2 DFT computation and
it is an introduction to the divide even/odd philosophy of the FFT.

Listing 5.2: First two steps in the radix-2 DIT FFT algorithm

```
1    % lab05_ex_partition2.m
2
3    clear all; close all;
4
5    N=64;
6    x = randn(1,N);
7    X = fft(x);
8
9    % One-level decomposition
10   x1 = x(1:2:end);    % even samples
11   X1 = fft(x1);
12   x2 = x(2:2:end);    % odd samples
13   X2 = fft(x2);
14   Xr = [ X1 X1 ] + exp(-j*2*pi/N*(0:N-1)) .* [ X2 X2 ];
15   error_manual_1 = max( abs(X-Xr) ), pause
16
17   % Two-level decomposition
18   x11 = x1(1:2:end);  % even even samples
19   X11 = fft(x11);
20   x12 = x1(2:2:end);  % even odd samples
21   X12 = fft(x12);
22   X1 = [ X11 X11 ] + exp(-j*2*pi/(N/2)*(0:N/2-1)) .* [ X12 X12 ];
23
24   x21 = x2(1:2:end);  % odd even samples
25   X21 = fft(x21);
26   x22 = x2(2:2:end);  % odd odd samples
27   X22 = fft(x22);
28   X2 = [ X21 X21 ] + exp(-j*2*pi/(N/2)*(0:N/2-1)) .* [ X22 X22 ];
29
30   Xr = [ X1 X1 ] + exp(-j*2*pi/N*(0:N-1)) .* [ X2 X2 ];
31   error_manual_2 = max( abs(X-Xr) ), pause
32
33   % Calling our recursive multi-level FFT function
34   Xr = myRecFFT(x);
35   error_recursive = max( abs(X-Xr) ), pause
36
37   % Calling our non-recursive FFT function
38   Xr = myFFT(x);
39   error_nonrecursive = max( abs(X-Xr) ), pause
```

Finally, the program 5.3 presents a recursive implementation of the complete FFT algorithm with all even/odd decomposition levels. It should be used as a test for checking understanding of the radix-2 FFT philosophy based on even/odd partition of signal samples. It is important to notice how simple the program is (!) thanks to the fact that now not we but Matlab is responsible for memory management inside each function call and for exchange of results between the function successive calls.

Listing 5.3: Radix-2 DIT FFT algorithm implemented as recursive function

```
1  function X = myRecFFT(x)
2
3  % My recursive radix-2 FFT function
4  N = length(x);
5  if(N==2)
6      X(1) = x(1) + x(2);       % # 2-point DFT
7      X(2) = x(1) - x(2);       % # on the lowest level
8  else
9      X1 = myRecFFT(x(1:2:N));  % call itself on even samples
10     X2 = myRecFFT(x(2:2:N));  % call itself on odd samples
11     X = [ X1 X1 ] + exp(-j*2*pi/N*(0:N-1)).* [X2 X2];   % combine spectra
12 end
```

5.3 FFT Butterflies

So, is that all about the famous FFT? Absolutely NO. Two important issues still are not discussed. The first is an additional simple observation which can offer additional reduction of the FFT computational time by half. Does it sound impossible?

In the FFT algorithm one longer spectrum is reconstructed repeatedly from two two-times shorter spectra. Shorter *even* and *odd* spectra, in the first decomposition stage, were denoted in our programs as $X_1(k), X_2(k)$, $k = 0, 1, 2, \ldots, N/2 - 1$, while the spectrum reconstructed from them as $X(k)$, $k = 0, 1, 2, \ldots, N - 1$. Let us denote shorter spectra as $X_e(k)$ and $X_o(k)$. After this, Eq. (5.13) describing the final spectrum reconstruction can be rewritten as ($k = 0, 1, 2, \ldots, N/2 - 1$):

$$X(k) \quad\quad = X_e(k) + e^{-j\frac{2\pi}{N}k} \cdot X_o(k), \tag{5.18}$$

$$X(N/2+k) = X_e(k) + e^{-j\frac{2\pi}{N}(N/2+k)} \cdot X_o(k) = X_e(k) + e^{-j\pi}e^{-j\frac{2\pi}{N}k} \cdot X_o(k). \tag{5.19}$$

Since $e^{-j\pi} = -1$, the second correction (5.19) differs only from the first one (5.18) with sign:

$$X(N/2+k) = X_e(k) - e^{-j\frac{2\pi}{N}k} \cdot X_o(k). \tag{5.20}$$

Eureka! The second term calculated in Eq. (5.18) is added to the first term in Eq. (5.18) and subtracted in Eq. (5.19). Therefore only one multiplication can be done instead of two:

$$X_o^{(c)}(k) = e^{-j\frac{2\pi}{N}k} \cdot X_o(k), \tag{5.21}$$

and after its result should be added and subtracted from $X_e(k)$, $k = 0, 1, \ldots, N/2 - 1$:

$$X(k) \qquad = X_e(k) + X_o^{(c)}(k), \tag{5.22}$$

$$X(N/2 + k) = X_e(k) - X_o^{(c)}(k) \tag{5.23}$$

This way the overall number of multiplication is reduced by half to $(1/2)N \log_2(N)$ since $\frac{N}{2}$ not N corrections are done (one correction is used two times, one time negated). This twin operations is called the FFT *butterfly*: two numbers are coming into the module in which the second of them is corrected (Eq. (5.21)) and then added (Eq. (5.22)) and subtracted (Eq. (5.23)) from the first one, which gives two output numbers.

As an example let us write all butterfly equations of the one-level fast radix-2 DIT DFT algorithm:

$$\begin{cases} X(0) = X_e(0) + e^{-j\frac{2\pi}{N}0}X_o(0), \\ X(4) = X_e(4) + e^{-j\frac{2\pi}{N}4}X_o(4) = X_e(0) - e^{-j\frac{2\pi}{N}0}X_o(0), \end{cases}$$

$$\begin{cases} X(1) = X_e(1) + e^{-j\frac{2\pi}{N}1}X_o(1), \\ X(5) = X_e(5) + e^{-j\frac{2\pi}{N}5}X_o(5) = X_e(1) - e^{-j\frac{2\pi}{N}1}X_o(1), \end{cases}$$

$$\begin{cases} X(2) = X_e(2) + e^{-j\frac{2\pi}{N}2}X_o(2), \\ X(6) = X_e(6) + e^{-j\frac{2\pi}{N}6}X_o(6) = X_e(2) - e^{-j\frac{2\pi}{N}2}X_o(2), \end{cases}$$

$$\begin{cases} X(3) = X_e(3) + e^{-j\frac{2\pi}{N}3}X_o(3), \\ X(7) = X_e(7) + e^{-j\frac{2\pi}{N}7}X_o(7) = X_e(3) - e^{-j\frac{2\pi}{N}3}X_o(3), \end{cases}$$

The FFT butterfly computation is presented in Fig. 5.2. The name *butterfly* is used since in module figure/diagram we see characteristic butterfly contour/shape.

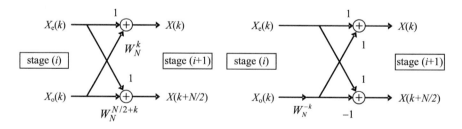

Fig. 5.2: Computational structure of the basic decimation-in-time (DIT) radix-2 FFT algorithm, the so-called *butterfly*. (left) full version with two multiplications, (right) simplified version with only one multiplication. Values close to arrows denote multiplication by them. W_N is defined by Eq. (5.11). Thanks to equality: $W_N^{(N/2+k)} = -W_N^k$ only one multiplication is done instead of two [11]

5.4 Fast Signal Samples Re-ordering

The second important issue in the FFT algorithm is changing samples position due to dividing them into two sets: even- and odd-indexed. Since moving data in computer memory requires also processor cycles, as multiplications, it may look that we won a lot, reducing the number of multiplications, but at the cost of significant increasing sample movement. If we could directly put the input signal sample to its target position (obtained after many even/odd sample partitions), we would benefit a lot. Therefore let us look more carefully at the partition process.

Let us assume that we are calculating 8-point FFT upon signal samples having values equal to sample numbers, and perform two-level even–odd sample sorting and four 2-point DFTs. Consecutive steps of samples re-ordering are presented in Fig. 5.3.

What do we observe? That the input sample position index, written binary, after many even/odd partitions, is equal to the original index but written in bit-reversed manner. For example, sample with binary position 011b (3) can go directly to binary position 110b (6) which is marked in figure with red dotted line. Therefore, only one move in computer memory is required for each signal sample which significantly simplifies the sample re-positioning. Look at Fig. 5.3. In the lowest level of the FFT algorithm the series of four 2-point DFTs takes place upon two neighboring data (short blue lines at the bottom of partition diagram are presented), next four 2-point DFT spectra are combined into two 4-point DFT spectra (two blue lines in the middle of the diagram), and, finally, 4-point spectra are combined into one 8-point DFT spectrum (one blue line in the upper part).

	Signal sample indexes							
Original (binary)	000	001	010	011	100	101	110	111
Original (decimal)	0	1	2	3	4	5	6	7
After 1-st even/odd (decimal)	0	2	4	6	1	3	5	7
After 2-nd even/odd (decimal)	0	4	2	6	1	5	3	7
After 2-nd even/odd (binary)	**000**	**100**	**010**	**110**	**001**	**101**	**011**	**111**

Fig. 5.3: Graphical illustration of two-level *even/odd* signal samples sorting before the 8-point radix-2 DIT FFT algorithm. It is assumed that signal samples have values equal to their indexes. Each sample goes finally to the new position having index with reversed bits, for example, the sample number 4 (100b) goes to the position 1 (001b) which is marked with red dotted line. Horizontal blue lines connect samples being in the same block of butterflies (the same sub-spectrum)

5.5 Example: 8-Point Radix-2 DIT FFT

In this section we will summarize our present knowledge about the FFT in *all-in-one* example. We will draw complete FFT block diagrams of the 8-point radix-2 decimation-in-time FFT algorithm and analyze programs implementing them. In Fig. 5.4 first two *even/odd* decomposition levels of the algorithm are presented. They lead to the final 8-point FFT computation shown in Fig. 5.5. The FFT program implementing it is written in two versions, as data block-oriented and single butterfly-based.

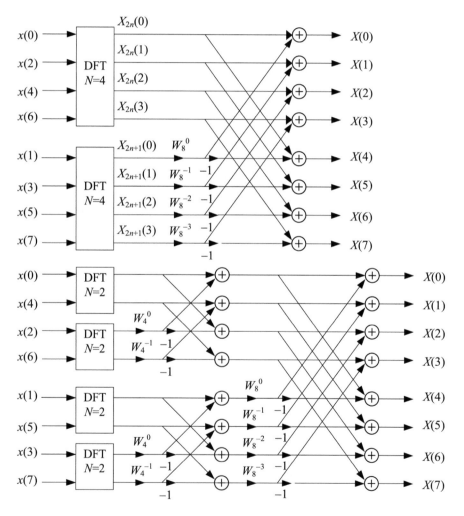

Fig. 5.4: Consecutive derivation of 8-point decimation-in-time radix-2 FFT algorithm leading to diagram presented in Fig. 5.5 [11]

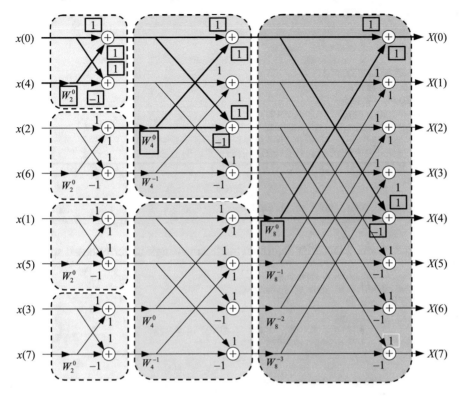

Fig. 5.5: Final block diagram for the decimation-in-time radix-2 FFT algorithm for $N = 8$ [11]

Listing 5.4: Radix-2 DIT FFT algorithm in Matlab—data blocks approach

```
1   % lab05_ex_blocks.m
2   % Radix-2 decimation in time (DIT) FFT algorithm
3     clear all; close all;
4
5     N=8;              % number of signal samples (power of 2 required)
6     x=0:N-1;          % exemplary analyzed signal
7     Nbit=log2(N);     % number of bits for sample indexes, e.g. N=8, Nbit=3
8
9   % Samples reordering (in bitreverse fashion)
10    n = 0:N-1;        % indexes
11    m = dec2bin(n);   % bits of indexes
12    m = m(:,Nbit:-1:1);  % reverse of bits
13    m = bin2dec(m);   % new indexes
14    y(m+1) = x(n+1);  % data reordering
15    y, pause          % check result
16
17  % All 2-point DFTs
18    y = [ 1 1; 1 -1] * [ y(1:2:N); y(2:2:N) ]; y=y(:)';
```

```
19
20   % Butterflies, i.e. spectra reconstruction
21      Nlev=Nbit;        % number of levels
22      Nfft=2;           % initial Nfft value after 2-point DFTs
23      for lev=2:Nlev % LEVELS
24          Nfft = 2*Nfft;
25          Nblocks = N/Nfft;
26          W = exp(-j*2*pi/Nfft*(0:Nfft-1));
27          for k = 1:Nblocks % butterflies
28              y1 = y( 1           + (k-1)*Nfft : Nfft/2 + (k-1)*Nfft );
29              y2 = y( Nfft/2+1 + (k-1)*Nfft : Nfft    + (k-1)*Nfft );
30              y(1 + (k-1)*Nfft : Nfft + (k-1)*Nfft ) = [ y1 y1 ] + W .* [ y2 y2 ];
31          end
32      end
33      ERROR = max( abs( fft(x) - y ) ), pause
```

Listing 5.5: Radix-2 DIT FFT algorithm in Matlab—single butterfly-based approach

```
1    % lab05_ex_butterfly_in_a_loop.m
2    % Radix-2 decimation in time (DIT) FFT algorithm
3    clear all; close all;
4
5       N=8;             % number of signal samples (power of 2 required)
6       x=0:N-1;         % exemplary analyzed signal
7       Nbit=log2(N);    % number of bits for sample indexes, e.g. N=8, Nbit=3
8
9    %Samples reordering (in bitreverse fashion)
10   for n=0:N-1
11       nc = n;                       % old sample position - copying
12       m = 0;                        % new sample position - initialization
13       for k=1:Nbit                  % check all bits
14           if(rem(n,2)==1)           % check low significant bit
15               m = m + 2^(Nbit-k);   % accumulate "m" using bitreversed weight
16               n = n - 1;            % decrement by 1
17           end
18           n=n/2;                    % shift all bits one position right
19       end
20       y(m+1) = x(nc+1);             % copy sample into new "bitreversed" position
21   end
22   y, pause                         % check result
23
24   % Butterflies
25   Nlev=Nbit; % number of levels
26   for lev=1:Nlev % LEVELS
27       bw=2^(lev-1);                 % butterflies width
28       nbb=2^(lev-1);                % number of butterflies per block
29       sbb=2^lev;                    % shift between blocks
30       nbl=N/2^lev;                  % number of blocks
31       W=exp(-j*2*pi/2^lev);         % correction coefficient
32       for bu=1:nbb % BUTTERFLIES
33           Wb=W^(bu-1);                                % correction for given butterfly
34           for bl=1:nbl % BLOCKS
35               up   = 1    + (bu-1) + (bl-1)*sbb;  % up sample index
```

```
36          down = 1+bw + (bu-1) + (bl-1)*sbb;   % down sample index
37          temp = Wb * y(down);                 % temporary value
38          y(down) = y(up) - temp;              % new down sample
39          y(up)   = y(up) + temp;              % new up sample
40        end
41      end
42    end
43    ERROR = max( abs( fft(x) - y ) ), pause
```

5.6 Efficient FFT Usage for Real-Value Signals

Fourier harmonic signals, being oscillatory basis functions in DFT, have complex-values. Our analyzed signal can be also a superposition (summation) of Fourier harmonics (e.g. like in NMR, in OFDM multi-carrier transmission). However, the most often, our signal has real-value samples and spectrum with conjugate symmetry in which half of the coefficients is repeated (with complex conjugation). The second half is calculated but is useless: the first half is for us sufficient, they contain the whole information about the signal.

We can exploit the DFT spectrum symmetry (valid for real-value data only) in two ways, perform one N-point DFT (FFT) but calculate:

1. two spectra of two real-value signals having N samples each,
2. one spectrum of a real-value signal having $2N$ samples.

Case 1: One FFT, Two Signals We have two real-value signals $x_1(n)$ and $x_2(n)$ having N samples. Their DFT spectra $X_1(k)$ and $X_2(k)$ are also N samples long. Let us create a complex signal having $x_1(n)$ in its real part and $x_2(n)$ in the imaginary part:

$$x(n) = x_1(n) + jx_2(n) \tag{5.24}$$

Due to transform linearity, the DFT of $x(n)$ is equal to:

$$X(k) = X_1(k) + jX_2(k) \tag{5.25}$$

Making use of the DFT spectrum symmetry (in respect to the $N/2$-th spectral coefficient) in real part and its asymmetry in the imaginary part, we can reconstruct the spectra $X_1(k)$ and $X_2(k)$ from $X(k)$ ($k = 1, 2, \ldots, N-1$):

$$X_1(k) = \frac{\text{Re}\,(X(k) + X(N-k))}{2} + j\frac{\text{Im}\,(X(k) - X(N-k))}{2}, \tag{5.26}$$

$$X_2(k) = \frac{\text{Im}\,(X(k) + X(N-k))}{2} - j\frac{\text{Re}\,(X(k) - X(N-k))}{2}. \tag{5.27}$$

The following *trick* is used: the symmetrical values are removed when *symmetrically* subtracted (values from the beginning and corresponding values from the end) and they are amplified by 2—when symmetrically added (values from the beginning and corresponding values from the end). The *inverse* rule is valid for asymmetrical data. Mean signal value is always in the real part of the $k = 0$ DFT coefficient: $X_1(0) = \text{Re}(X(0)), X_2(0) = \text{Im}(X(0))$.

Case 2: One FFT, Signal Two Times Longer In this case we divide a signal having $2N$ samples into even- and odd-indexed ones, exactly the same way as in the radix-2 DIT FFT algorithm discussed before, put them into real and imaginary parts of $x(n)$ as $x_1(n)$ and $x_2(n)$, and perform N-point FFT. After this we reconstruct spectra $X_1(k)$ and $X_2(k)$ using Eqs. (5.26), (5.27). Finally, the whole signal spectrum is reconstructed from its *even* and *odd* spectra using Eq. (5.13), the same way as in program 5.1.

Exercise 5.2 (Two-Times Faster FFT Spectra Calculation for Real-Value Signals). In program 5.6 we address computing two DFT spectra in one FFT call. Spectrum of the first signal has been already calculated with success. Finish the program, writing a code for computing spectrum of the second signal. After that add a new *functionality* to the program: assume that two input signals are even and odd samples of one two-times longer signal. Reconstruct spectrum of this signal.

Listing 5.6: Calculation of two DFT spectra using one FFT call

```
1   % lab05_ex_2in1
2   clear all; close all;
3   N=16;
4
5   x1 = randn(1,N);       % Signals
6   x2 = randn(1,N);
7   x3(1:2:2*N) = x1;      x3(2:2:2*N) = x2; % even and odd samples
8
9   X1 = fft(x1);          % Their DFT spectra
10  X2 = fft(x2);
11  X3 = fft(x3);
12
13  % Exploit this symmetry
14  x12 = x1 + j*x2;       % Artificial complex-value signal
15  X12 = fft(x12);        % Its DFT spectrum
16  X12r = real(X12);
17  X12i = imag(X12);
18
19  % Reconstruction of X1 from X12
20  X1r(2:N) = (X12r(2:N)+X12r(N:-1:2))/2;  % using symmetry  of Real(X1)
21  X1i(2:N) = (X12i(2:N)-X12i(N:-1:2))/2;  % using asymmetry of Imag(X1)
22  X1r(1) = X12r(1);
23  X1i(1) = 0;
24  X1rec = X1r + j*X1i;
25  error_X1 = max(abs( X1 - X1rec )), pause
```

```
26
27   % Reconstruction of X2 from X12
28   % ... to be done
29   % Reconstruction of X3 from X1 and X2
30   % ... to be done
```

5.7 FFT Algorithm with Decimation-in-Frequency

In the second big family of fast Fourier transform algorithms, not input signal samples are re-ordered as even-/odd-indexed ones but the calculated spectra coefficients.
Let us first calculate even spectral coefficients ($r = 0, 1, 2, \ldots, \frac{N}{2} - 1$):

$$X(2r) = \sum_{n=0}^{N-1} x(n) e^{-j\frac{2\pi}{N}(2r)n} = \tag{5.28}$$

$$= \sum_{n=0}^{N/2-1} x(n) e^{-j\frac{2\pi}{N}(2r)n} + \sum_{n=N/2}^{N-1} x(n) e^{-j\frac{2\pi}{N}(2r)n} = \tag{5.29}$$

$$= \sum_{n=0}^{N/2-1} x(n) e^{-j\frac{2\pi}{N}(2r)n} + \sum_{n=0}^{N/2-1} x\left(\frac{N}{2}+n\right) e^{-j\frac{2\pi}{N}(2r)\left(\frac{N}{2}+n\right)} = \tag{5.30}$$

$$= \sum_{n=0}^{N/2-1} x(n) e^{-j\frac{2\pi}{N/2}(r)n} + \sum_{n=0}^{N/2-1} x\left(\frac{N}{2}+n\right) e^{-j\frac{2\pi}{N/2}(r)n} e^{-j2\pi r} = \tag{5.31}$$

$$= \sum_{n=0}^{N/2-1} \left[x(n) + x\left(\frac{N}{2}+n\right)\right] e^{-j\frac{2\pi}{N/2}(r)n} \tag{5.32}$$

In Eq. (5.29) we divide summation in Eq. (5.28) into two parts (to sample $N/2 - 1$
and above). In Eq. (5.30) we change denotation of sample index in the second sum
and obtain the same summation limits as in the first sum. Then in Eq. (5.31) we
calculate the exponent argument in the second sum, exploiting equality $e^{-j2\pi r} = 1$.
Since it simplified to the exponent of the first sum, we combine two sums into one
in Eq. (5.32). As a result we obtain $N/2$-point DFT performed upon summation of
the first and the second half of the signal samples (having $N/2$ samples each).
   Now let us do calculation of the odd spectral coefficients:

$$X(2r+1) = \sum_{n=0}^{N-1} x(n) e^{-j\frac{2\pi}{N}(2r+1)n}, \quad r = 0, 1, 2, \ldots, \frac{N}{2} - 1 \tag{5.33}'$$

$$= \sum_{n=0}^{N/2-1} x(n) e^{-j\frac{2\pi}{N}(2r+1)n} + \sum_{n=N/2}^{N-1} x(n) e^{-j\frac{2\pi}{N}(2r+1)n} = \tag{5.34}$$

$$= \sum_{n=0}^{N/2-1} x(n)e^{-j\frac{2\pi}{N}(2r+1)n} + \sum_{n=0}^{N/2-1} x\left(\frac{N}{2}+n\right)e^{-j\frac{2\pi}{N}(2r+1)\left(\frac{N}{2}+n\right)} =$$

$$(5.35)$$

$$= \sum_{n=0}^{N/2-1}\left[x(n) - x\left(\frac{N}{2}+n\right)\right]e^{-j\frac{2\pi}{N}(2r+1)n} = \qquad (5.36)$$

$$= \sum_{n=0}^{N/2-1}\left[e^{-j\frac{2\pi}{N}n}\left(x(n) - x\left(\frac{N}{2}+n\right)\right)\right]e^{-j\frac{2\pi}{N/2}(r)n}. \qquad (5.37)$$

In Eq. (5.35) we have exploited the following substitution:

$$e^{-j\frac{2\pi}{N}(2r+1)\frac{N}{2}} = e^{-j2\pi r}e^{-j\pi} = -1. \qquad (5.38)$$

Now, the $X(2r+1)$ spectrum is obtained as a result of $N/2$-point DFT performed upon subtraction of the first and the second half of signal samples but initially multiplied by $\exp(-j2\pi/Nn)$, where $n = 0,1,2,\ldots,N/2-1$.

Concluding, in radix-2 decimation-in-frequency FFT algorithm we are performing two-times smaller $N/2$-point DFTs on summation and (corrected) subtraction of the first and the second half of the signal samples. Data re-ordering procedure is repeated recursively. In Fig. 5.6 first decomposition level of $N = 8$ DIF FFT algorithm is presented, while in Fig. 5.7—the whole algorithm.

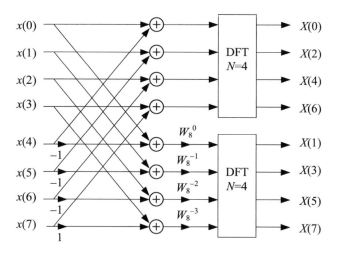

Fig. 5.6: First decomposition level of the $N = 8$-point decimation-in-frequency radix-2 FFT algorithm leading to diagram presented in Fig. 5.7 [11]

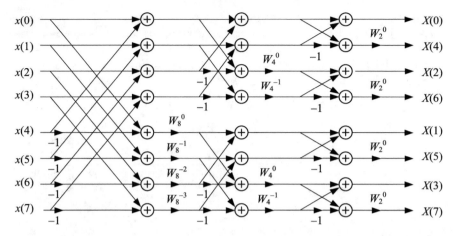

Fig. 5.7: Final block diagram for the decimation-in-frequency radix-2 FFT algorithm for $N = 8$ [11]

Exercise 5.3 (DIF FFT: With First Two Decomposition Levels). Write Matlab program implementing DIF FFT algorithm with first two decomposition levels only, for arbitrary signal having 2^p samples. Make use of diagrams presented in Figs. 5.6 and 5.7.

5.8 Summary

In this chapter we dealt with the fast Fourier transform algorithms. They are really very fast, not 5, 10, or even 100% (2-times) faster—they are about 200000% faster, YES! 200 TIMES! for signals having $N = 1024$ samples. Wow! How is it possible? During DFT calculation we multiply a vector of signal samples with matrix of samples of frequency basis/reference functions. How could any savings be done in such strictly defined mathematical operation?! Yes, it is possible to reduce the number of multiplications since the matrix elements are equal to sampled values of sine and cosine periodic functions, which are repeating and therefore they are repeatedly multiplied by the same signal samples. There is no sense to repeat some multiplications: it is more practical to copy the result calculated already. Exploiting this idea further leads to radix-K FFT algorithms in which divide-by-K concept is used: (1) signal is recursively decomposed into K fragments (e.g. even/odd samples for $K = 2$) until obtaining K-sample long sub-signals, (2) then many short K-point DFT spectra are computed, and (3) the spectra are combined in recursive way (e.g. 2-point to 4-point, 4-point to 8-point, etc.), up to the reconstruction of the whole signal DFT spectrum. So something that looks impossible, becomes possible. What we should remember about the FFT?

1. All decimation-in-time FFT algorithms exploit the *divide-and-conquer* methodology: (1) they recursively divide input samples into smaller groups, for example, by 2, by 2, by 2 . . . , or by 4, by 4 . . . , (2) when the further division is impossible they calculate DFT spectra of very short sample vectors (e.g. with 2 or 4 elements), and after that (3) they are recursively combining smaller spectra, e.g. initially 2- or 4-point, into longer ones, e.g. 4- or 8-point, and so on: 8-, 16-, 32-, 64-point, . . . , finally doing reconstruction of the whole signal DFT spectrum.

2. Samples partition can be fixed, e.g. into 2 or 4 groups at each sample decomposition level (radix-2, radix-4 algorithms) or can be changed from level to level as it is done in split-radix algorithms, for example, first into 2 groups, then into four groups, next into 3 groups, etc. Algorithms with bigger radix values offer better computational speed-up but restrict the signal to have a length being power of 4, 8, . . . Split-radix algorithms can be better adjusted to signal length.

3. In FFT algorithms are used the so-called *butterflies*—computation blocks with two inputs and outputs. Calculations are done *in-place*: input values are replaced by output values and no memory allocation problem exists. The second input number is first corrected in the block and then added and subtracted from the first one. The name *butterfly* is used since in module figure/diagram we see characteristic butterfly contour/shape.

4. In the simplest radix-2 N-point DIT FFT algorithm input samples are first re-ordered (result from recursive data partitioning) and then we have $\log_2(N)$ calculation levels. Each level consists of blocks of butterfly modules. In the beginning we have always $N/2$ blocks with one butterfly having width of one sample. In each next level: number of blocks is two times lower, number of butterflies in each block is two times bigger, and butterfly width is two times larger.

5. Implementation of radix-2 DIT FFT algorithm in any computer language is very simple: first we do sample re-ordering (in some DSP processors a special bit-reversed addressing mode should be turned on), then we have three nested loops (level, block, and butterfly number) inside which a single butterfly module is executed.

6. Apart from decimation-in-time (DIT) FFT algorithms, there are used also decimation-in-frequency (DIF) FFT algorithms. They offer the same computational speed-up but their philosophy is different: one recursively divides not signal samples but calculated spectral coefficients into groups, in the simplest case of radix-2 DIF—into even and odd numbered. In DIF FFTs the input signal samples have not to be initially re-ordered (*advan-*

tage) but, at the price of this, the FFT spectral coefficients are obtained in wrong sequence (*drawback*). But there are applications in which wrong order of calculated spectral coefficients is not important. Fast signal filtering and correlating in frequency domain are such examples explained in the next chapter.

7. As already stated in the motto of this chapter: the FFT is a King of a DSP highway! It is so fast that it is used not only for signal frequency analysis (in next chapter) but also it is exploited as a computational *hammer* in other DSP tasks, when-ever and where-ever it is possible. For example, in fast implementations of digital non-recursive signal filtering having a form of digital convolution of signal samples with filter weights or signal correlation. These two operations have similar representation: multiply-and-accumulate the result. Fast signal filtering or signal correlation in frequency domain are performed as follows: instead of time-consuming signal convolution/correlation in time domain three FFTs are performed, two direct done separately upon two signal vectors and one inverse performed upon the resultant multiplication of two signal FFT spectra.

5.9 Private Investigations: Free-Style Bungee Jumps

Exercise 5.4 (Need for Speed). In 99.9% Matlab is not your favorite computer language. You a master of Basic, C, C++, Fortran, Java, Julia, Pascal, Python,... For sure you have a *Need for Speed!* Therefore make yourself a little fun and implement any FFT program from this chapter in *your* language. At the end compare the calculated FFT spectrum with the Matlab result.

Exercise 5.5 (Dancing with the Stars: One Step Forward and One Step Backward). Performing in a cascade the direct (forward) and the inverse (backward) FFT algorithm, we should return perfectly to the same signal (with error on the level 10^{-14} of the signal amplitude). Write a program of inverse FFT modifying any Matlab code of direct FFT. FFT and inverse FFT programs should differ only in the exponent sign. Check your implementation using the Matlab function ifft(). Next, perform FFT and inverse FFT upon any signal and check the results: are both signals the same?

Exercise 5.6 (Radix-2 DIF FFT: On the Other Side of the Moon).** Write universal Matlab program for arbitrary $N = 2^P$ implementing radix-2 FFT algorithm with decimation-in-frequency (DIF) with all even/odd decomposition levels.

Exercise 5.7 (* DCT-II via FFT). Write Matlab program for calculation of orthogonal transform DCT-II presented in Chap. 2 ($c(k=0) = \sqrt{1/N}$, $c(k>0) = \sqrt{2/N}$):

$$X^{DCT}(k) = c(k) \cdot \sum_{n=0}^{N-1} x(n) \cos\left(\frac{\pi(2n+1) \cdot k}{2N}\right), \qquad 0 \leq k \leq N-1, \qquad (5.39)$$

with the FFT use. Equation connecting both transforms is as follows:

$$X^{DCT}(k) = \mathrm{Re}\left[c(k)e^{-j\frac{\pi k}{2N}} \cdot \mathrm{FFT}_N(\tilde{x}(n))\right], \tag{5.40}$$

where signal $\tilde{x}(n)$ is equal to:

$$\tilde{x}(n) = x(2n), \quad \tilde{x}(N-n-1) = x(2n+1), \quad n = 0,1,2,\ldots,N/2-1. \tag{5.41}$$

Check correctness of your implementation (outputs of both programs).

Further Reading

1. R.E. Blahut, *Fast Algorithms for Digital Signal Processing* (Addison-Wesley, Reading, 1985)
2. C.S. Burrus, T.W. Parks, *DFT/FFT and Convolution Algorithms. Theory and Implementation* (Wiley, New York, 1985)
3. P.M. Embree, *C Algorithms for Real-Time DSP* (Prentice Hall, Upper Saddle River, 1995)
4. H.K. Garg, *Digital Signal Processing Algorithms: Number Theory, Convolution, Fast Fourier Transforms, and Applications* (CRC Press, Boca Raton, 1998)
5. M.H. Hayes, *Schaum's Outline of Theory and Problems of Digital Signal Processing* (McGraw-Hill, New York, 1999, 2011)
6. V.K. Ingle, J.G. Proakis, *Digital Signal Processing Using Matlab* (PWS Publishing, Boston, 1997; CL Engineering, 2011)
7. R.G. Lyons, *Understanding Digital Signal Processing* (Addison-Wesley Longman Publishing, Boston, 1996, 2005, 2010)
8. A.V. Oppenheim, R.W. Schafer, *Discrete-Time Signal Processing* (Pearson Education, Upper Saddle River, 2013)
9. J.G. Proakis, D.G. Manolakis, *Digital Signal Processing. Principles, Algorithms, and Applications* (Macmillan, New York, 1992; Pearson, Upper Saddle River, 2006)
10. S.W. Smith, *The Scientist and Engineer's Guide to Digital Signal Processing* (California Technical Publishing, San Diego, 1997, 1999). Online: http://www.dspguide.com/
11. T.P. Zieliński, *Cyfrowe Przetwarzanie Sygnałów. Od Teorii do Zastosowań (Digital Signal Processing. From Theory to Applications)* (Wydawnictwa Komunikacji i Łączności (Transport and Communication Publishers), Warszawa, Poland, 2005, 2007, 2009, 2014)

Chapter 6
FFT Applications: Tips and Tricks

Having an F1 car or the X-Wing spacecraft it would be a big sin not to use it at full speed!

6.1 Introduction

This chapter aims to do comprehensive, short introduction to important FFT applications in field of digital signal processing. Since the FFT is really, really fast, about 100 times faster than regular DFT for 1000 signal samples, one should exploit its speed not only in spectral analysis, that is its primary usage, but also as a computational "hammer" for calculation of functions connected with signal spectrum. For example, for signal convolution and correlation. We start with demonstration of proper scaling of DFT amplitude spectrum and of its zooming possibilities, realized by appending additional zeros to an analyzed signal and performing the FFT. Zeros can be also appended to the signal spectrum—after IFFT we are obtaining interpolated signal. We will advise also to show only first half of the FFT spectrum for real-value signals since the second half is symmetric in its real part while asymmetric in the imaginary part. We will stress significance of the window function choice for the FFT spectrum amplitude and frequency resolution and show proper spectrum scaling when window is used.

In the first part of Chap. 4 on DtFT and DFT, we have discussed features of the discrete-time Fourier transform spectrum, result of discretization of the continuous Fourier transform. It was stressed that DtFT offers more flexible spectrum visualization (*from-to* any frequency with arbitrary frequency step) but at the cost of more time-consuming computing implementation. But a fast algorithm for DtFT calculation exists also. It makes use of three (I)FFTs and is known as chirp-Z transform. We learn about it.

When the signal is contaminated in noise or it is changing its deterministic content in random manner, like during bit transmission, we should calculate power spectra. How the power spectral density (PSD) function is defined? How

© Springer Nature Switzerland AG 2021
T. P. Zieliński, *Starting Digital Signal Processing in Telecommunication Engineering*, Textbooks in Telecommunication Engineering, https://doi.org/10.1007/978-3-030-49256-4_6

it can be calculated? We will answer these questions. It will turn out that the PSD calculation requires computation of correlation function which is widely used in statistical signal processing. We will calculate it fast by means of FFT.

Since the correlation function computation is almost the same as calculation of signal convolution, we will derive algorithms for fast signal convolution using the FFT. It will be used by us in next chapters for fast non-recursive FIR digital signal filtering. Wow!

Next, signal frequencies of signal components can vary in time, like in speech, audio, AM and FM modulated signals. Calculation of one FFT signal spectrum for many signal samples is not a good solution in this case. The spectrum will be smeared because, for sure, the changing signal components will not fit well to the constant-frequency Fourier harmonics. As a consequence, many spectral coefficients will be *turned on* and the spectrum will be smeared. Solution to this is assuming that signal components are not changing very fast, cutting signal into smaller fragments, calculating FFT spectra for them, and putting all spectra together into one matrix for precise inspection of the spectrum change. Such repetitive calculation of FFT on overlapping signal fragments is called short-time Fourier transform (performed on short consecutive signal parts). In Matlab the `spectrogram()` function implements the STFT.

Finally, we will end chapter with description of one interpolated DFT algorithm. What is the difference between DFT spectrum interpolation (zooming) by DtFT and interpolated IpDFT algorithms? In IpDFT one derives mathematical equations for DFT spectrum values. Then take a few of them around the spectral peak, solve set of some equations, and calculate frequency of one signal component. This is completely different approach than DtFT. Better frequency estimates are obtained for signals which mathematical model is known, in our case for summation of damped sinusoids. In this method not only sine frequency can be found but also its damping which is very important in different resonance spectroscopy methods.

So this is a chapter with a lot of *DSP cookies*.

6.2 FFT Usage Principles

Butterfly FFT algorithms, not very difficult, give the same result as slow direct multiplication of DFT matrix with a vector of signal samples but they are significantly faster. For this reason the FFT should replace DFT in all our programs. The classical TV advertisement says: "There is no sense to over-pay for the washing powder: buy cheaper one if you do not see the difference in action." But how to use FFT properly, efficiently, and flexible?

In this section, in the chapter beginning, we briefly summarize the general principles of proper FFT usage, on the base of [15]. In most cases we use DFT/FFT for obtaining signal *amplitude spectrum*, i.e. for measuring amplitudes of single frequency components of the signal. Having these amplitudes known, we can calculate later the signal *power spectrum*, its *power spectral density* and *amplitude spectral density*. In previous chapters, we have discussed only a problem how to calculate correct signal *amplitude spectrum* from the DFT/FFT result. At present we will deal with the other spectra also.

Repeating the DFT Definition Let us assume that an FFT procedure is implementing the DFT definition with correct amplitude scaling for the rectangular window (see Eq. (4.42)):

$$X(k) = \frac{1}{N} \sum_{n=0}^{N-1} x(n) e^{-j\frac{2\pi}{N}kn}, \quad k = 0,\ 1,\ 2,\ldots,N-1, \tag{6.1}$$

where $x(n), n = 0,1,2,\ldots,N-1$, is an analyzed signal, $X(k), k = 0,1,2,\ldots,N-1$— its amplitude FFT spectrum, and $e^{-j(2\pi/N)kn}$—the n-th sample of k-th Fourier basis function (reference oscillatory signals). When in Eq. (6.1) the following operations are performed: (1) $\frac{1}{N}$ scaling is removed, (2) positions of $X(k)$ and $x(n)$ are exchanged, (3) summation is performed over k for $n = 0,1\ldots N-1$, and (4) sign in the exponent is changed from "$-$" to "$+$," Eq. (6.1) becomes a definition of the inverse Fourier transform :

$$x(n) = \sum_{k=0}^{N-1} X(k) e^{j\frac{2\pi}{N}kn}, \quad n = 0,\ 1,\ 2,\ldots,N-1. \tag{6.2}$$

For this reason fast algorithms designed for direct FFT can also be used for inverse FFT after slight modifications. After performing Eqs. (6.1), (6.2) in a cascade, one obtains exactly the same signal (with negligible computational error). It is ensured by scaling the result by $\frac{1}{N}$, now performed in the first Eq. (6.1), however, typically done in the second equation in software implementations. Choice of the proper spectrum scaling, not only by number $\frac{1}{N}$, will be further discussed by us in this chapter.

Direct or Inverse FFT FFT is implemented in different languages and libraries. Very often names FFT and DFT are used in them interchangeably. Very often the same function can be used for direct (*forward*) and inverse (*backward*) FFT which differs only in sign of the exponent: -1 for direct and 1 for inverse. Typically, we are calling one FFT function in this way X=fft(x,direction) specifying whether direct or inverse FFT is to be computed. In Matlab there are two separate functions which are called as follows: X=fft(x) ; x=ifft(X) ; . Performing FFT and inverse FFT in cascade, one should return back the same signal. Try it in your favorite language or in Matlab.

Choosing FFT Length From one side, since frequency discretization (step) in direct DFT is equal to sampling frequency divided by number of analyzed samples N, i.e. $\Delta f = f_s/N$, we are interested in increasing the number of analyzed signal samples N in order to have better spectrum resolution, i.e. smaller Δf. From the other side, well remembering the most known radix-2 decimation algorithms, we feel that we are obligated to prefer FFT lengths N being powers of 2: 256, 512, 1024, 2048, ... What is the problem? If the signal sampling frequency is already given and the number of signal samples has to be power of 2, it is very difficult to obtain well-rounded frequency resolutions like 0.25, 0.5, 1, 2, 5, 10 Hz. Hmm ... But we should remember that there are many different data-partition FFT approaches, not only radix-2 one. Fast procedures from the most popular and frequently used FFTW library (http://www.fftw.org/) are optimized for different signal lengths, for example, ones being equal [15]:

$$N = 2^a \cdot 3^b \cdot 5^c \cdot 7^d \cdot 11^e \cdot 13^f, \tag{6.3}$$

where numbers a,b,c,d are arbitrary none-negative integers and the sum $e + f$ is equal 0 either 1. It is wise to know this.

Describing Frequency Axis When FFT spectral coefficients are already calculated, typically the first biggest problem rely on connecting frequency values with them. *What frequency is this and this and ... ?!.* Coefficients $X(1), X2), X(3), \ldots,$ $X(N)$ are associated with frequencies being multiplicities of the fundamental frequency $f_0 = f_s/N$, i.e. to: $0, f_0, 2f_0, 3f_0, \ldots, (N - 1)f_0$. For example, for $f_s = 1000$ Hz and $N = 100$ we have: $0, 10, 20, \ldots, 990$ Hz.

Meaning of Spectral Coefficients After euphoria of knowing what frequency components are present in my signal, the next question arises: *how much of each of them we have?* DFT/FFT is an orthogonal transformation. The signal is decomposed into summation of some elementary functions being references of certain frequencies. If properly scaled, the FFT coefficients are amplitudes of these reference signals. *Properly?* In most cases we should divide the FFT result by N, in Matlab also, obtaining: X=fft(x)/N; plot(fs/N*(0:N-1),abs(X)) .

Using Only First Half of the FFT Spectrum In case of real-value signal it is a good practice to display only first half of the calculated FFT spectrum since the second half have complex-conjugated values, which are confusing for non- experienced user. In Matlab: X=fft(x)/N; plot(fs/N*(0:N/2), abs(X(1:N/2+1)). In such case, it has a sense also to change the FFT scaling: divide the spectrum not by N but by $N/2$: in Matlab: X=2*fft(x)/N; . With this modification we will see only one peak for each sine/cosine being exactly equal to the signal amplitude, not half of it.

FFT Normalization for Rectangular Window Typically, FFT procedures return spectral coefficients $X(k)$ defined by Eq. (6.1), therefore, with no normalization. Alternatively, they divide $X(k)$ by $\sqrt{N}$, N either $N/2$. Performing a sequence of direct and inverse FFT we should go back to the same signal. If not, we should check the scaling implemented in our subroutines. In direct FFT there are three possibilities. In the first case, the inner product of signal and reference frequency function is not normalized: analyzing a cosine with amplitude equal to one we obtain two spectral peaks with height $N/2$. In this way work functions from FFTW package and math libraries of most languages, including Matlab using FFTW. In

the second case, mathematically the most correct, the inner product result is divided by $\sqrt{(N)}$ offering orthonormality of the DFT transform (inner product of each basis function with itself is equal to 1). Such normalization is used in Mathematica language. When analyzing a unitary cosine we obtain two spectral peaks with heights $\sqrt{N}/2$. In the third normalization strategy, the FFT coefficient can be divided by N and cosine has two spectral peaks equal to $1/2$ (for positive and negative frequency). This is mathematically correct and practically good result during signal analysis. Necessity of negative frequencies existence results from complexity of Fourier basis functions (Fourier harmonics). But during practical frequency analysis, for engineers, these cosine representation as summation of $1/2$ of the positive frequency harmonic and $1/2$ of the negative frequency can be treated as making things more difficult than they in reality are. Therefore in the fourth method, division by N can be replaced by division by $N/2$ and our cosine has in the spectrum two peaks with amplitudes 1—what is correct when we observe only one half of the FFT spectrum connected with positive frequencies. In Matlab:
`X=2*fft(x)/N; plot(fs/N*(0:N/2),abs(X(1:N/2+1)).`

FFT Normalization for Arbitrary Window In order to obtain amplitude signal spectrum, i.e. correct amplitudes of its frequency components, the FFT result should be scaled. In the simplest situation when no extra window function is used, only the rectangular one, we should simply divide the spectrum by N and multiply it by 2 in order to obtain correct amplitude spectrum in positive frequency range. But, as we remember from chapter on DtFT and DFT, one should multiply signal fragment with deliberately chosen window function $w(n)$ (and additionally use sufficiently long signal fragment, i.e. high value of N) in order to ensure required amplitude and frequency resolution of the spectrum. In such case the scaling of FFT amplitude spectrum has to be changed to general formula:
`Xampl=2*fft(x)/sum(w);.`

Power Spectrum and Power Spectral Density The last question to be answered before FFT usage is what kind of FFT spectrum we are interested in: amplitude, power, power spectral density, or amplitude spectral density [15]. In different applications and for different signals different FFT spectra are preferred. First, we should specify our expectations. Signal power and RMS value were defined in Table 2.2. One can analytically or computationally verify that sine with amplitude A has a power equal to $\frac{1}{2}A^2$ and the RMS value equal to the square root of the power $\frac{1}{\sqrt{2}}A$. Basic equalities and Matlab commands for primary spectral analysis are given in Table 6.1. We start from signal $x(n)$ windowing and computation of its FFT, then we take absolute value of the result and receive $|X(k)|$. Signal amplitude spectrum (AS) $X_{AS}(k)$ is obtained by multiplying calculated $|X(k)|$ by 2 (taking into account negative frequencies) and normalizing the result—division by summation of window $w(n)$ coefficients (division by N is correct only for rectangular window). The signal power spectrum $X_{PS}(k)$, as described above, is simply squared $X_{AS}(k)$ divided by 2.

Density spectra are calculated in similar way. The amplitude spectral density (ASD) $X_{ASD}(k)$ is only differently normalized than $X_{AS}(k)$, by square root of summation

Table 6.1: Mathematical definition and computation in Matlab of amplitude and power spectrum (AS, PS) as well as amplitude and power spectral density (ASD, PSD) for signal $x(n)$ multiplied by window function $w(n), n = 0, 1, 2, \ldots, N - 1$. Values of spectral coefficients computed in Matlab are valid for $k = 1, 2, 3, \ldots, N/2+1$ and they correspond to frequencies $kf_s/N, k = 0, 1, 2, \ldots, \frac{N}{2}$

Quantity	Math definition, $k = 0 \ldots \frac{N}{2}$	Matlab code
DFT Eq. (6.1), |DFT|	$X(k) = \sum\limits_{n=0}^{N-1} x(n)w(n)e^{-j\frac{2\pi}{N}kn}$	`X=fft(x.*w); Xa = abs(X);`
Window Sum	$S_1 = \sum\limits_{n=0}^{N-1} w(n)$	`S1=sum(w);`
Window Energy	$S_2 = \sum\limits_{n=0}^{N-1} w^2(n)$	`S2=sum(w.^2);`
Ampl Spectrum (AS)	$X_{AS}(k) = 2\frac{\|X(k)\|}{S_1}$ (V)	`Xas=2*Xa / S1;`
Power Spectrum (PS)	$X_{PS}(k) = \frac{X_{AS}^2(k)}{2}$ (V^2)	`Xps=Xas.^2 / 2;`
Ampl Spectral Density (ASD)	$X_{ASD}(k) = 2\frac{\|X(k)\|}{\sqrt{f_s S_2}}$ $\left(\frac{V}{\sqrt{Hz}}\right)$	`Xasd=2*Xa / sqrt(fs*S2);`
Power Spectral Density (PSD)	$X_{PSD}(k) = \frac{X_{ASD}^2(k)}{2}$ $\left(\frac{V^2}{Hz}\right)$	`Xpsd=Xasd.^2 / 2;;`

not of window coefficients but their squared values, the sum is additionally multiplied by the sampling frequency f_s. The power spectral density (PSD) $X_{PSD}(k)$ is a squared $X_{ASD}(k)$ divided by 2, as before.

A complete *all-in-one* program for computation of all signal spectra defined in Table 6.1 is presented in Listing 6.1. Signal components have amplitudes 1 and $\sqrt{2}$. Calculated spectra are shown in Fig. 6.1. The AS spectrum shows these values. The PS spectrum, as expected, gives values $\frac{1}{2}$ and 1. Correct are also plots of ASD and PSD.

Listing 6.1: Computation of different FFT spectra

```
1   % lab06_ex_fft_usage.m
2   clear all; close all;
3
4   % Signal
5   N=5000; fs=10000; xlsb=0.001; % N=5000=(2^3)*(5^4)=8*625 = fast alg.
6   dt=1/fs; t=dt*(0:N-1);
7   x = 1*sin(2*pi*1000*t) + sqrt(2)*sin(2*pi*3501.1234*t);
8
9   % Noise - alternative: SNR=80(?); x = awgn(x,SNR,'measured');
10  x = floor( x/xlsb + 0.5 )*xlsb; noise_level = xlsb/sqrt(6*fs),
11
12  % Windowing
13  w1 = rectwin(N)'; w2 = hann(N)'; w3 = kaiser(N,12)'; w4 = flattopwin(N)';
14  w  = w4;
```

```
15  x  = x .* w;
16
17  % Spectral estimation
18  X  = fft(x); Xa = abs(X);        % FFT without scaling, no division (by sqrt(N),N)
19  XN = X/N;                         % FFT with typical division by N (rectangular wind)
20  Xas = 2*Xa/sum(w);               % Ampl spectrum  (AS) - correct for any window
21  Xps = Xas.^2/2;                  % Power spectrum (PS) - correct for any window
22  Xasd = 2*Xa/sqrt(fs*sum(w.^2));  % Amplitude spectral density (ASD) - as above
23  Xpsd = Xasd.^2/2;                % Power spectral density (PSD) - as above
24
25  % PS = AS^2/2; ASD=AS/sqrt(ENBW); PSD = ASD^2/2 = PS/ENBW = AS^2/ENBW;
26  S1 = sum(w); S2 = sum(w.^2); ENBW = fs*S2/S1^2, NENBW = N*S2/S1^2,
27  err_asd = max(abs(Xasd-Xas/sqrt(ENBW))),
28  err_psd = max((abs(Xpsd)-Xas.^2/ENBW)), pause
29
30  % Figures
31  k = 1:N/2+1;                     % indexes of non-negative frequencies
32  f0 = fs/N, f = f0 * (k-1);       % none-negative frequencies (1st spectrum half)
33  figure;                          % change stem() to semilogy()
34  subplot(221); stem(f(k),Xas(k)); xlabel('f(Hz)'); ylabel('[V]'); title('AS');
35  subplot(222); stem(f(k),Xasd(k)); xlabel('f(Hz)'); ylabel('[V/\surdHz]'); title('ASD');
36  subplot(223); stem(f(k),Xps(k)); xlabel('f(Hz)'); ylabel('[V^2]'); title('PS');
37  subplot(224); stem(f(k),Xpsd(k)); xlabel('f(Hz)'); ylabel('[V^2/Hz]'); title('PSD');
38  pause
```

Exercise 6.1 (Standard FFT Usage for Different Signal Spectra Calculation). Carefully analyze code of the program 6.1. Read all comments. Run the program. You should see plots presented in Fig. 6.1. In the beginning make an extra copy of the last figure and replace in it the function stem() with the function semilogy() (only this). Run the program. Notice low level of noise. Next modify amplitudes and frequencies of two signal components. Calculate expected theoretical values (using Table 6.1) and check figure correctness. Increase value of xlsb—voltage corresponding to the low significant bit of the AD converter—and make this way level of noise higher. Observe spectra. Increase the xlsb once more, check result and return to the initial xlsb setting. Observe that till now you were using the Matlab flattopwin() function as a window. It has very wide spectral main-lobe and relatively low level of spectral side-lobes letting you, both, correct measurement of signal component amplitudes (reduction of spectral leakage effect), even in the case when they differ a lot (high dynamic range). Change the window to rectangular and Hann. Observe spectral peaks at present. Are their heights correct or not? Make one component being significantly weaker than the other, e.g. decrease its amplitude 1000 times. Observe spectra.

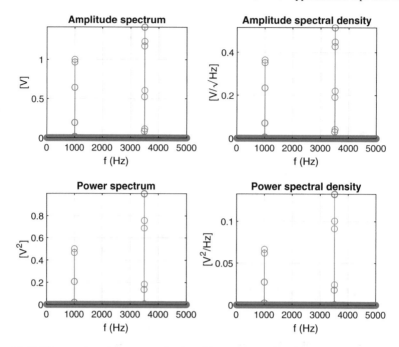

Fig. 6.1: Different signal spectra calculated in program 6.1, from (up-left) to (down-right): amplitude spectrum (AS), power spectrum (PS), amplitude spectral density (ASD), and power spectral density (PSD). Linear scale

6.3 Fast Frequency Resolution Improvement

We can very easily increase frequency resolution (*zoom*) of the DFT spectrum as a whole. How to do it? In discrete-time Fourier transform we analyze N signal samples assuming zeros before and after them. So the result will be the same when we perform FFT procedure on vector of N signal samples with some $(K-1)N$ zeros appended at the end. This way we increase the signal length artificially and the FFT function *sees* the $K \cdot N$ samples and build the DFT matrix with dimensions $KN \times KN$ instead of $N \times N$. So we have now KN frequency reference functions spanning the range $[0, f_s)$ which result in frequency resolution $f_s/(KN)$ instead of f_s/N, i.e. K times higher (with frequency step K times smaller). The spectral window, if necessary, should have the length N and do weighting of original signal samples only. The program 6.2 is demonstrating the described DFT zooming trick.Plots presented in

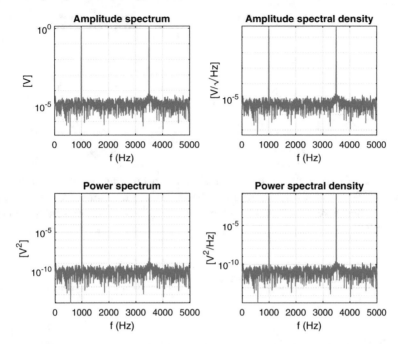

Fig. 6.2: Different signal spectra calculated in program 6.1, from (up-left) to (down-right): amplitude spectrum (AS), power spectrum (PS), amplitude spectral density (ASD), and power spectral density (PSD). Logarithmic scale

Fig. 6.2 have been generated with its use. In order to make things simpler the FFT is computed for a pure sine signals with two different frequencies: 20 Hz (left) and 22 Hz (right). Frequency discretization in original FFT is equal to $f_s/N = 5$Hz while with $K = 25$ times signal enlargement by zero appending—25 times less, i.e. 0.2Hz. Improvement in spectrum shape is tremendous: theoretical spectrum of rectangular window is clearly visible (blue small dots). Without zero appending the spectral result can look very different—compare FFT red dots in both figures.

It is important to remember that in the FFT Matlab function additional zeros are automatically appended at the signal end on request, i.e. when we specify FFT length M bigger then original signal length N. In such case, $M - N$ zeros are appended. In Matlab for rectangular window we should write
`X=fft(x,M)/N); plot(fs/M*(0:M-1),abs(X))`.

It is important to observe that all spectra calculated in the previous section can be *zoomed* since each of them represent a scaled version of the amplitude spectrum which can be zoomed, as shown in this section (Fig. 6.3).

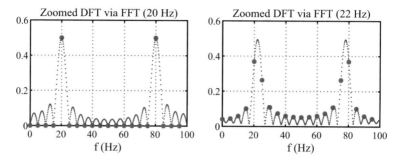

Fig. 6.3: DFT (red circles) and DtFT (blue dots) spectra of a signal having only one sinusoid: 20 Hz left and 22 Hz right. Higher frequency resolution DtFT spectrum is obtained after increasing signal length by appending zeros at the end and calculating longer FFT

Listing 6.2: FFT spectrum with denser frequency sampling

```
1    % lab06_ex_dft_zoom.m
2    clear all; close all;
3
4    N=20; K=25; fs=100;          % FFT length, resolution increase, sampling frequency
5    dt=1/fs; t=dt*(0:N-1);       % sampling period, sampling time moments
6    x = sin(2*pi*22*t);          % signal
7    w = rectwin(N)';             % window: rectwin(),hanning(),blackman(),flattopwin(),..
8    x = x.*w;                    % signal windowing
9    X = fft(x)/sum(w);           % FFT without appending zeros
10   Xz = fft( [x,zeros(1,(K-1)*N)] )/sum(w);  % with zeros; or X=fft(x,K*N)/sum(w)
11   f = fs/N*(0:N-1); fz = fs/(K*N)*(0:K*N-1);   % frequency axis
12   figure; plot(f,abs(X),'ro',fz,abs(Xz),'b.','MarkerFaceColor','r'); xlabel('f (Hz)');
13   title('Zoomed DFT via FFT'); grid; pause
```

Exercise 6.2 (Zoomed FFT by Appending Extra Zeros). Run the program 6.2. Choose different windows. Change N to 50 and 100. Add logarithmic scaling of spectral coefficients. Show only coefficients for non-negative frequencies. Multiply spectrum by 2 in this case. Add calculation of power spectral density—see Table 6.1.

6.4 FFT of Noisy Signals: Welch Power Spectral Density

In Sect. 6.2 general FFT application rules were presented. In part, computation of signal power spectrum and its power spectral density were described. But presented solutions were more *applied* than *statistically correct*. Only one FFT procedure was executed and all resulting amplitude and power spectra were calculated. Such approach is good for deterministic signals but not for noisy ones. In case of noise, individual spectra differ a lot, and several of them have to be computed and averaged in order to obtain reliable estimate of noise power or power spectral density. This section is devoted to this problem: how to apply FFT in case of noisy signals?

In *high math* theory, the signal power spectral density is defined as Fourier transform of the signal auto-correlation function. The cross-correlation function between two signals with finite energy is defined by formula:

$$R_{xy}(\tau) = \int_{-\infty}^{+\infty} x(t) y^*(t - \tau) \, dt \tag{6.4}$$

and their cross power spectral density is defined as Fourier transform of the first:

$$P_{xy}(f) = \int_{-\infty}^{\infty} R_{xy}(\tau) e^{-j2\pi f \tau} d\tau. \tag{6.5}$$

After setting $y(n) = x(n)$, the auto-correlation function of the signal $x(n)$ is obtained and its power spectral density. Considering two signals we are making discussion deliberately a little bit *deeper*. After time discretization the Eq. (6.5) takes the form:

$$P_{xy}(f/f_s) = \sum_{m=-\infty}^{\infty} R_{xy}(m) e^{-j2\pi(f/f_s)m}. \tag{6.6}$$

Let us assume that we have only one signal $x(n)$ with N samples. We can first calculate estimates of its auto-correlation function for $-N + 1 \leq m \leq N - 1$:

$$\hat{R}_{xx}^{(1)}(m) = \frac{1}{N} \sum_{n=0}^{N-1-|m|} x(n) x^*(n - m), \tag{6.7}$$

$$\hat{R}_{xx}^{(2)}(m) = \frac{1}{N - |m|} \sum_{n=0}^{N-1-|m|} x(n) x^*(n - m), \tag{6.8}$$

and then estimate of its PSD:

$$\hat{P}_{xx}^{(N)}(f/f_s) = \sum_{m=-(N-1)}^{N-1} \hat{R}_{xx}(m) e^{-j2\pi(f/f_s)m}. \tag{6.9}$$

Estimator $\hat{R}_{xx}^{(1)}(m)$ is *biased* (has an offset in respect to the true, correct value), while the estimator $\hat{R}_{xx}^{(2)}(m)$ is *un-biased*. However $\hat{R}_{xx}^{(2)}(m)$ has bigger variance than $\hat{R}_{xx}^{(1)}(m)$, i.e. its scatter around the *expected value* (mean) is bigger. In turn, the PSD estimator $\hat{P}_{xx}^{(N)}(f/f_s)$ is not *consistent* since its variance does not tend to zero with the increase of signal length N to infinity.

There are different methods coping with this drawback.

In Blackman–Tukey method the Fourier transform is performed upon the windowed auto-correlation function estimator, i.e. $\hat{R}_{xx}(m)$ multiplied by chosen window function $w(n)$ (Hamming, Hann, or some other):

$$\hat{P}_{xx}^{(N)}(f/f_s) = \sum_{m=-(N-1)}^{N-1} w(m)\hat{R}_{xx}(m)\, e^{-j2\pi(f/f_s)m}. \tag{6.10}$$

It is interesting to mark that this method, directly implementing the PSD definition, was in the past available in Matlab but at present it is not. Since it is computationally attractive and very educational we will discuss it later after introducing fast computation of convolution/correlation by FFT.

In Welch method, nowadays the most frequently used approach for PSD estimation, relation between the PSD and the so-called *periodogram* $1/(2N+1)|X_N(f/fs)|^2$, the squared DtFT, is exploited:

$$P_{xx}\left(\frac{f}{f_s}\right) = \lim_{N\to\infty} E\left[\frac{1}{2N+1} \left| \sum_{n=-N}^{N} x(n)e^{-j\frac{2\pi n f}{f_s}} \right|^2 \right] = \lim_{N\to\infty} E\left[\frac{1}{2N+1} X_N\left(\frac{f}{f_s}\right) X_N^*\left(\frac{f}{f_s}\right) \right]. \tag{6.11}$$

Please note that we were calculated periodograms, squared FFT spectra, in Sect. 6.2. In Welch approach many windowed periodograms (windowed, squared FFTs) are calculated and averaged. Input sequence of N signal samples $x(n)$ is divided into L fragments $x^{(l)}(n)$ with M samples, which overlay or not in dependence on the offset (step) D:

$$x^{(l)}(n) = x(n+lD), \quad 0 \le l \le L-1, \quad 0 \le n \le M-1. \tag{6.12}$$

Next each data fragment is multiplied with window function $w(n)$ (e.g. Hamming), the DtFT is computed, the result is squared and divided by sum of squared window coefficients:

$$\hat{P}_M^{(l)}(f) = \frac{1}{E_w} \left| \sum_{n=0}^{M-1} x^{(l)}(n)w(n)e^{-j2\pi(f/f_s)n} \right|^2, \quad E_w = \sum_{n=0}^{M-1} w^2(n) \tag{6.13}$$

The final PSD estimator is a mean value of calculated modified periodograms:

$$\hat{P}_{xx}^w(f) = \frac{1}{L} \sum_{l=0}^{L-1} \hat{P}_M^{(l)}(f). \tag{6.14}$$

When $D = M$ consecutive signal fragments do not overlay (Bartlett method), in turn for $D = M/2$ they overlay in fifty percents.

How to implement the Welch method? It is not difficult, especially its Bartlett version in Matlab. A code example is presented in Listing 6.3—being a continuation of the program 6.1. The signal amplitude and power spectral densities are calculated. Since in Matlab each operation (function) which is performed upon a matrix is executed over matrix columns, we can put consecutive (one-by-one) signal fragments into *signal* matrix X columns (line 7) using function reshape(), create a *window* matrix W having repeated window function $w(n)$ in each column using function repmat() (line 8), then multiply both matrices X.*W (line 9), then perform fft() on the matrices product (line 9) and, finally, square the result, multiply it by 2, normalize by doubled energy of window coefficients multiplied by sampling frequency f_s, and calculate mean value of all columns (spectra). Uff... Last operation is performed via matrix transposition (().'), calling function sum() and division by M. This way the signal amplitude spectral density is obtained. Next, it is squared and divided by 2 giving the signal power spectral density.

In Fig. 6.4 there are presented two calculated mean FFT spectra of signal amplitude and power. The Welch spectra averaging concept is applied. We can see with ease the difference between these mean spectra and spectra presented in Fig. 6.2— the noise floor is significantly smoother and on the expected level.

Listing 6.3: Calculation of noise-robust amplitude and power spectra using FFT

```
1   % lab06_ex_welch.m continuation of lab06_ex_fft_usage.m
2
3   M = 100; NM=N*M; n = 0 : NM-1; t=dt*(0:NM-1);
4   x = 1*sin(2*pi*1000*t) + sqrt(2)*sin(2*pi*3501.1234*t);  % two sines
5   %x = 1*sin(2*pi*(1000*t + 0.5*50*t.^2));  % LFM signal: increase of 50 Hz/s
6   x = floor( x/xlsb + 0.5 )*xlsb;      % noise addition
7   X = reshape(x,N,M);                  % matrix with signal fragments in M columns
8   W = repmat( w', 1, M);               % matrix with the same window in M columns
9   Xa = abs( fft( X.*W ) );             % FFT of each column of X.*W, then abs()
10  Xasd2 = 2*Xa/sqrt(fs*sum(w.^2));     % Amplitude Spectral Density (ASD)
11  Xpsd2 = Xasd2.^2/2;                  % Power Spectral Density (PSD)
12
13  Xasd2 = sum(Xasd2.')/M; Xpsd2 = sum(Xpsd2.')/M; % cols (spectra) --> rows, mean of
        rows
14
15  % Figures
16  % Mean amplitude spectrum and mean power spectrum (for many signal fragments)
17  figure
18  subplot(121); semilogy(f(k),Xasd(k),f(k),Xasd2(k),'g');
19  xlabel('f (Hz)'); ylabel('[V/\surdHz]]'); title('Amplitude SD');
20  subplot(122); semilogy(f(k),Xpsd(k),f(k),Xpsd2(k),'g');
21  xlabel('f (Hz)'); ylabel('[V^2/Hz]]'); title('Power SD');
22  pause
```

Exercise 6.3 (Welch PSD Calculation and Verification). Carefully analyze code of the program 6.3. At present signal amplitude and power spectral densities are calculated, i.e. ASD and PSD, respectively. Add calculation of average amplitude and power signal spectra, AS and PS. Become familiar with pwelch() or periodogram() Matlab functions. Use them, set their parameter values and try to obtain results that are similar to received in the program 6.3.

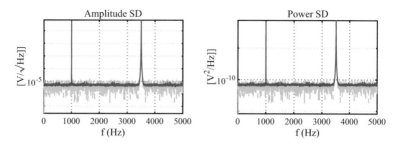

Fig. 6.4: Amplitude (left) and power (right) spectral densities after averaging of 100 FFT spectra (narrow blue solid curve). Noise level becomes significantly smoother after mean DFT spectrum calculation in comparison to the single DFT spectrum computing (light green dashed line)

6.5 FFT of Time-Varying Signals

6.5.1 Short-Time Fourier Transform

An analyzed signal can vary in time. For example, sine signal can have slowly changing amplitude, frequency, or phase, i.e. it can be modulated in amplitude, frequency, or phase, like telecommunication carriers. Our speech signal has permanently changing frequency content. In this case, only one spectrum calculated for the whole signal changing in time can be misleading! It offers mean values of signal components parameters inside the observation window—we cannot see in it trajectories of changes. In order to track frequency changes of the signal in time we should use the so-called short-time Fourier transform:

1. cut the signal into many, shorter overlapping fragments using any window,
2. calculate FFT spectrum of each fragment and collect all of them into one matrix,
3. display matrix values in 3D (mesh(), surf()) or as color/gray-scale image (imagesc() (gray-levels),
4. observe how the signal spectrum is changing in time—observe frequency and amplitude modulation curves of individual signal components.

The applied window should not be too long and too short since:

- for too long window more *average* than *instantaneous* spectrum is obtained, with smearing in time axis,
- for too short window the obtained instantaneous spectrum has too low-frequency resolution and visual smearing in frequency axis is observed.

In program 6.4 the above-described procedure is implemented in Matlab in simplified form: the signal is cut into non-overlapping fragments. In program 6.3, being a continuation of the programs 6.1, the signal was cut also into fragments, many spectra were computed and then they were averaged. We aimed at noise suppression then. At present all calculated spectra are important for us and we are not averaging them: we deliberately observe in them signal change in frequency domain. Figure 6.5 presents a STFT spectrum: a sequence of many amplitude spectra computed for shorter signal fragments and stored as a 2D time–frequency matrix. The STFT spectrum was computed for cosine signal changing its frequency linearly from 1000 Hz to 3500 Hz during 50 s of signal observation. In left figure the spectrum is plotted as a 3D mesh while in right one as color image. The signal frequency change is very well visible in both plots. It is not the case when only one spectrum of the whole signal is computed, which demonstrates Fig. 6.6. In the left plot one amplitude spectrum, calculated for the whole very long signal, is shown, while in the right one one mean amplitude spectrum of many consecutive signal fragments. In both cases the spectrum is wide and tells us that signal consists of many frequencies what is misleading. In fact the signal has only one sinusoidal component changing its frequency in time (Fig. 6.5).

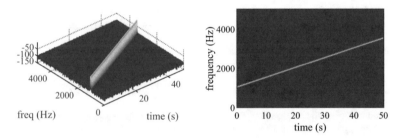

Fig. 6.5: Short-time Fourier transform spectrum $|X(t, f)|$ of cosine signal changing its frequency linearly from 1000 Hz to 3500 Hz during 50 s of signal observation, calculated using Hann window. (left) 3D mesh plot, (right) visualized as color image

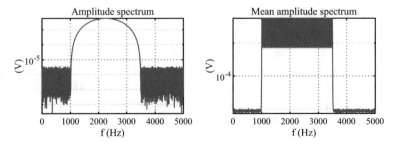

Fig. 6.6: Amplitude spectra calculated for cosine signal changing its frequency linearly from 1000 Hz to 3500 Hz during 50 s of signal observation, calculated using Hann window. (left) amplitude spectrum calculated for the whole long signal, (right) mean amplitude spectrum of many consecutive signal fragments

Listing 6.4: STFT calculation: sequence of short FFT amplitude spectra—continuation of Listing 6.4

```
1   % lab06_ex_stft.m continuation of lab06_ex_welch.m
2   % ...
3   X = reshape(x,N,M);              % put signal fragments into M columns of matrix X
4   W = repmat( w', 1, M);          % put the same window w into M columns of matrix W
5   Xa = abs( fft( X.*W ) );        % perform FFT of each column of X.*W, then abs()
6
7   figure
8   Xas = 2*Xa(1:N/2+1,1:M)/sum(w); % absolute values, positive frequencies, scaling
9   t = dt*(N/2 : N : N/2+(M-1)*N);
10  mesh(t,f,20*log10(Xas)); view(-40,70); axis tight;  % AS matrix as a 3D mesh
11  cb=colorbar('location','EastOutside'); set( get(cb,'Ylabel'),'String','V (dB)');
12  xlabel('time (s)'); ylabel('frequency (Hz)'); pause
13
14  figure
15  imagesc(t,f,20*log10(Xas));     % AS matrix as an image
16  cb=colorbar('location','EastOutside'); set( get(cb,'Ylabel'),'String','V (dB)');
17  xlabel('time (s)'); ylabel('frequency (Hz)'); pause
18
19  % For comparison - one mean spectrum of many short signal fragments
20  Xas = 2*Xa/sum(w);              % many amplitude spectra (AS)
21  Xasm = sum(Xas')/M;            % mean AS, cols --> rows, sum of rows
22  figure
23  subplot(111); semilogy(f(k),Xasm(k),'b'); xlabel('f (Hz)'); title('Mean AS');
24  pause
25
26  % For comparison - one spectrum of very long signal fragment
27  w = hann(NM)';                  % very long window
28  Xas1 = 2*abs(fft(x.*w))/sum(w); % one AS of the whole long signal
29  k = 1:1:NM/2+1; f=fs/NM*(0:NM-1); % frequencies of AS coefficients
30  figure
31  subplot(111); semilogy(f(k),Xas1(k),'b'); xlabel('f (Hz)'); title('One long AS');
```

Mathematically, for discrete-time signals, the short-time Fourier transform is defined as:

$$X(n,f) = \frac{2}{\sum_{n=0}^{N-1} w(n)} \sum_{m=-\infty}^{\infty} [x(m)w(m-n)] e^{-j2\pi \frac{f}{f_s}m}, \quad 0 \le f < f_s/2 \quad (6.15)$$

A window $w(m)$ has non-zeros values only for $m = 0, 1, 2, \ldots, N-1$. The value n denotes the window shift, i.e. after it the shifted window $w(m-n)$ has non-zero values for $m = n, n+1, n+2, \ldots, m+(N-1)$ and signal samples having these indexes are weighted by the window and transformed with FFT.

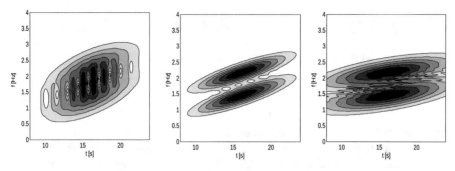

Fig. 6.7: Window length influence upon STFT time–frequency resolution: (left) window too short, (center) optimal window length, (right) window too long. Sum of two LFM signals with Gaussian time-envelopes is analyzed [27]

The FFT and window length N should be very carefully chosen. Too short or too long window causes unwanted STFT spectrum *smearing* in direction of the frequency axis (too short) or the time axis (too long), which is shown in Fig. 6.7.

Exercise 6.4 (STFT Calculation and Verification). Carefully analyze code of Listing 6.4 which is continuation of the program 6.3. At present a signal is cut into non-overlapped fragments and only signal amplitude spectrum is calculated. Become familiar with `spectrogram()` Matlab function. Use this function, set its parameter values, and try to get plots which are similar to ones obtained in the program 6.4. Next, try to add signal fragments overlapping—perform FFT not in the matrix form but in the loop, one-by-one over signal fragments. Change window length and its overlap. Observe changes in the STFT spectrum. Read any speech signal, then calculate and display its STFT spectrum.

6.5.2 Wigner–Ville Time–Frequency Signal Representation

The short-time Fourier transform is only one example of the so-called time–frequency distributions (TFD), intend to track the signal spectrum change. Most of them belong to the Cohen's TFD family [9]. A special role in it plays the Wigner–Ville TFD, which is a perfect tool for analysis of mono-component signals with linear frequency modulation. It was proposed by Wigner [25] for real-value signals, and extended to complex-value signals by Ville. Such signals are widely used in radar signal processing. The continuous-time Wigner–Ville TFD (WVD) is defined as a continuous-time Fourier transform, performed over variable τ, upon a WV signal kernel:

$$X(t, \omega) = \int_{-\infty}^{\infty} xx(t, \tau) e^{-j\omega\tau} d\tau, \quad xx(t, \tau) = x\left(t + \frac{\tau}{2}\right) \cdot x^*\left(t - \frac{\tau}{2}\right), \quad (6.16)$$

and repeated for many values of t. For a complex-value signal with linear frequency modulation, the WV kernel is equal to:

$$x(t) = e^{j(\omega_0 t + 0.5\alpha t^2)} \quad \Rightarrow \quad xx(t, \tau) = e^{j(\omega_0 + \alpha t)\tau}, \quad (6.17)$$

and the WVD magnitude has maximum at instantaneous angular signal frequency $(\omega_0 + \alpha t)$, regardless the t value. For this reason, the repeated WVD tracks perfectly the instantaneous signal frequency. In discrete-time implementation the WVD is computed as follows:

1. calculation of a complex-value, analytic signal corresponding to a real value, analyzed signal, using the Hilbert transform (discussed in the chapter on special filters; in Matlab: `xa=hilbert(x);`
2. cutting the signal into fragments N-samples long, in Matlab: `xf=xa(n:n+(N-1));`
3. calculation of the WV kernel of each signal fragment, in Matlab: `xx=xf.*conj( xf(end:-1:1) );`
4. performing FFT upon each WV kernel, in Matlab: `X=fft(xx)/N;`
5. collecting all FFT spectra into a matrix and displaying absolute values of matrix elements.

Mono-component signals different than the LFM ones, and multi-components signals, have the so-called *cross-terms* in the WVD spectrum. There are many methods for their suppression. If you are interested how it is done, take any book on time–frequency signal analysis.

Exercise 6.5 (Wigner–Ville Distribution of Signals Modulated in Frequency). Modify program `lab06_ex_wvd.m` from the book repository. Generate a signal with linear frequency modulation (LFM) and calculate its WVD. Check whether the frequency change is properly tracked. Propose correct scaling of the frequency axis—due to signal multiplication the observed frequency is two times higher. Repeat experiment for different speed of frequency change, starting from small values. Apply different windows to the WV kernel. Use different values of N. Compare obtained WVD spectra with STFT spectra. Now, calculate WVD for a signal with sinusoidal FM, i.e. SFM, and for a two-component signal: LFM plus SFM. Finally, return to the LMF signal and omit calculation of the analytic signal version. What has changed? Why?

6.6 Fast Convolution and Correlation Based on FFT

One of the most important laws in analog signal theory tells: "Considering time and frequency: convolution in one domain corresponds to multiplication in the other domain and vice versa." For example, convolution of two signals in time results in multiplication of their frequency spectra while multiplication of two signals causes convolution of their spectra. These relations hold also for discrete-time signals. Therefore, instead of convoluting two long signals in time domain, for example, input samples and weights of very long FIR digital filter, it is better to calculate their DFT spectra using fast FFT algorithms, multiply them and compute the inverse FFT. The result will be exactly the same but computational effort—significantly smaller. Such signal convoluting by their spectra multiplication is called *fast convolution* and it is widely used for fast implementation of FIR digital signal filtering (see Chap. 9). Since computationally correlation of two signals is very similar to convolution of two signals (with the only difference that the second signal is not time-reversed but complex conjugated), fast convolutions algorithms and programs can be with ease apply to efficient correlation computing. In this section we learn how all of this it is done, of course thanks to our sweet FFT.

6.6.1 Linear Convolution

Linear convolution of two signals $x(n)$ and $h(n)$, i.e. two sequences of samples, is defined as:

$$y(n) = \sum_{k=-\infty}^{\infty} x(k)h(n-k). \tag{6.18}$$

Graphical illustration of signal linear convolution is presented in left part of Fig. 6.8. Two signals are convoluted, $x(k) = [2,1,1,2]$, having $N = 4$ samples, and $h(k) = [1,2,3]$, consisting of $M = 3$ samples. Order of operations is as follows:

1. first, the second signal is reversed in time: $h(k) \to h(-k)$—third row from the top,
2. then, many times shifted right by 1 sample: $h(-k) \to h(1-k), h(1-k) \to h(2-k), h(2-k) \to h(3-k), \ldots$—the following rows are obtained,
3. after every shift, the second signal samples are multiplied with corresponding samples of the first signal $x(k)[2,1,1,2]$ (i.e. samples being in the same position) in the sample-by-sample manner: $x(k)h(n-k)$ for every value of k, e.g. for $n = 2$ and shifted signal $h[2-k]$ we have

$$x(2)h(0) = 1 \cdot 1 = 1, \quad x(1)h(1) = 1 \cdot 2 = 2, \quad x(0)h(2) = 2 \cdot 3 = 6, \tag{6.19}$$

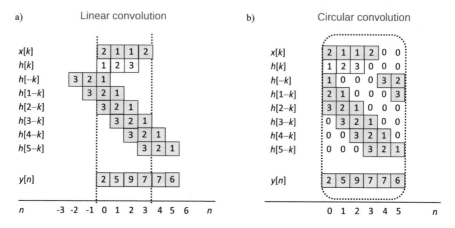

Fig. 6.8: Linear (left) and circular modulo-6 (right) convolution of two sequences of samples: $x(k) = [2,1,1,2], N = 4$, and $h(k) = [1,2,3], M = 3$. In both cases the convolution result is the same, has $K = N + M - 1 = 6$ samples, and is equal to $y(n) = [2,5,9,7,7,6]$. For clarity of presentation, we are neglecting signal synchronization problem

4. finally, all products are summed (accumulated) giving a convolution result $y(n)$ for each shift n of $h(k)$. This way one obtains $N + M - 1$ samples of $y(n)$. For example, for $n = 2$ we have (see Fig. 6.8):

$$x(2)h(0) + x(1)h(1) + x(0)h(2) = 1 + 2 + 6 = 9. \qquad (6.20)$$

Since linear convolution plays a very important role in digital signal filtering, its understanding is crucial in our course. For this reason, in Fig. 6.9 one additional, very simple graphical explanation is presented in which two vectors $x(k)$ and $h(k)$, consisting of three 1s, are convolved. The second is reversed, shifted forward, multiplied by the first, and *mult* results are accumulated. Only first three output samples $y(0)$, $y(1)$, and $y(3)$ are computed. They represent sum of one, two, and three 1s. *It is a last chance to catch a train to London. The train's going.*

6.6.2 Circular Convolution

Circular convolution modulo-K of two signals, $x(k)$ having N samples and $h(k)$ with M samples $(N > M)$, is defined as:

$$y(n) = \sum_{k=0}^{K-1} x(k)h((n-k) \bmod K), \qquad (6.21)$$

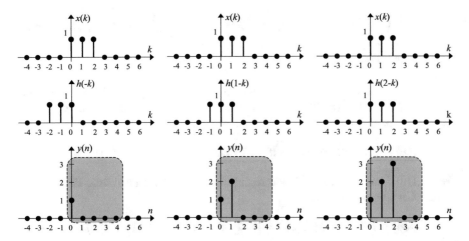

Fig. 6.9: Linear convolution of two vectors $\mathbf{x} = [1,1,1]$ and $\mathbf{h} = [1,1,1]$. Only three first output samples $y(0)$, $y(1)$, and $y(3)$ are computed [27]

where $h((n-k) \bmod K)$ denotes circular modulo-K shift n position right of the signal $h(-k)$. Let us assume that $K = N + M - 1$ and both signals $x(k)$ and $h(k)$ are initially enlarged with padding zeros, appended at their ends, to the length $N + M - 1$. Then they are convoluted in circular manner: the second signal is reversed in time (in circular modulo-K manner), shifted one sample right (also in modulo-K fashion), multiplied by the first signal, then all products are accumulated. This operation and its result are presented in Fig. 6.8 (right side). As we see, exactly the same values are obtained as for the linear convolution. If $N - M$ zeros are appended only to the second signal and the modulo-N convolution is performed, the first $M - 1$ samples of the signal $y(n)$ are wrong. But the remaining ones are correct.

Exercise 6.6 (Halo! My name Is Linear Convolution: Linear Not Circular!). Check manually calculation of linear and circular convolution of two signals, presented graphically in Fig. 6.8. Correct results are given in the bottom of the figure. Linear and circular convolutions are not well distinguished by students. This is like with my person: I am from Poland but recognized as coming from Holland.

6.6.3 Fast Linear Convolution

The DtFT of signal $y(n)$, result of convoluting two signals $x(n)$ and $h(n)$, is equal to (using denotation: $m = n - k$):

$$Y(\Omega) = \sum_{n=-\infty}^{\infty} \left(\sum_{k=-\infty}^{\infty} x(k)h(n-k) \right) e^{-j\Omega n} = \sum_{k=-\infty}^{\infty} x(k)e^{-j\Omega k} \sum_{m=-\infty}^{\infty} h(m)e^{-j\Omega m} =$$

$$= X(\Omega)H(\Omega), \tag{6.22}$$

i.e. to multiplication of DtFT spectra of convoluted signals $x(k)$ and $h(k)$. Therefore, instead of using Eq. (6.18), one can calculate vector of signal samples $\mathbf{y}(n)$ from Eq. (6.22), performing inverse DtFT upon product $X(\Omega)Y(\Omega)$:

$$\mathbf{y}(n) = \text{IDTFT}(\ \mathbf{Y}(\Omega)\ ) = \text{IDTFT}(\ \text{DFT}(\mathbf{x}(n)) \cdot * \text{DFT}(\mathbf{h}(n))\ ). \tag{6.23}$$

When DtFT and inverse DtFT are replaced by fast FFT and inverse FFT, the so-called *fast convolution* result:

$$\mathbf{y}(n) = \text{IFFT}(\ \mathbf{X}(k)\mathbf{H}(k)\ ) = \text{IFFT}(\ \text{FFT}(\mathbf{x}(n)) \cdot * \text{FFT}(\mathbf{h}(n))\ ) \tag{6.24}$$

where .* denotes, as in Matlab, multiplication of corresponding elements of two equal-length vectors, sample-by-sample. Fast convolution is beneficial for long signals: time-consuming signal convolving in time domain is replaced by three FFTs, two direct and one inverse. However the shorter signal should be appended with padding zeros at its end to the length of the longer signal. Or both signals have to be enlarged with zeros to the same length but bigger. What signal length is optimal?

Block of samples $\mathbf{y}(n)$ calculated in Eq. (6.24) using DFTs, represents result not of the linear convolution of signals $x(k)$ and $h(k)$ but of the circular one. When only $N - M$ zeros are appended to the second signal, the first $M - 1$ samples of the convolution result are wrong, because after initial time-reversion of the signal $h(k)$ its $M - 1$ samples *hits* to last samples of the signal $x(k)$ and are multiplied with them. In order to avoid this effect, both signal have to be appended with zeros minimum to the length $N + M - 1$, as explained above in Fig. 6.8. We will test in detail the described algorithm of fast convolution of two signals in Exercise 6.7. This way an FIR digital signal filtering, addressed in Chap. 9, can be implemented in a fast way. It is possible because such filtering has a form of convolution of signal samples $x(n)$ with specially designed filter weights $h(n)$. This operation, calculation of local weighted signal average, can cause removal of some *frequencies* from the processed signal.

6.6.4 Fast Overlap–Add and Overlap–Save Sectioned Convolution

In Eq. (6.24) FFTs are performed upon blocks of signal samples. There are two drawbacks of the data-block signal processing. From one side, it can cause a long time delay of out samples in respect to input samples. From the second side, in real-time signal filtering (convolving), when signal samples are permanently coming in, we would wait forever for end of data stream. Therefore cutting signal into smaller fragments and implementation of fast convolution in real-time over smaller vectors, *data pieces*, is necessary. In this section two fast sectioned convolution methods will be described, in which fast FFT-based convolution concept is applied, but for separate, consecutive signal fragments, not to the whole signal.

Fast Overlap–Add Method In this method zeros are appended to all, non-overlapping input signal sections, changing block-based circular convolution to the linear one. Due to this, the output signal sections are longer, they overlap and have to be added. The computational procedure is graphically illustrated in Fig. 6.10, implemented in program 6.5 and tested in Exercise 6.7. Let us assume that a shorter signal $h(k)$ has $M = 7$ samples. A longer signal $x(n)$, with $L = 46$ samples, is cut into non-overlapped fragments having $N = 10$ samples, its last 6 samples are neglected. In the beginning, $N - 1 = 9$ zeros are appended to the $h(k)$ end, the $h_z(n)$ is get, and $(N + M - 1 = 16)$-point FFT over $h_z(k)$ is performed. Then $M - 1 = 6$ zeros are appended to each, consecutive, N-samples long fragment of signal $x(n)$, and short $x_{1z}(n), \ldots, x_{5z}(n)$ sub-signals are obtained. Next $(L + M - 1) = 16$-point FFT is computed upon each short data block. Obtained results are multiplied with already computed FFT of $h_z(k)$ (in sample-by-sample manner, starting from the first samples) and inverse FFTs of the multiplication results are calculated for all sub-signals. $N + M - 1 = 16$ samples of the output sub-signals $y_{1z}(n), \ldots, y_{5z}(n)$ are obtained this way. Then they are combined (partially added): first $(M - 1) = 6$ of the next block are added to the last $(M - 1)$ samples of the previous block, already calculated, and the remaining next N samples of the next block are appended to the end of $y(n)$ calculated till now.

Fast Overlap–Save Method In this method zeros are not appended to the input signal sections and sectioned circular, not linear, convolutions are performed. For this reason, first $M - 1$ samples of each calculated output signal block are wrong! Because they should be taken from the end of a previous block, the input signal sections have to overlap. Graphical method illustration of the fast overlap–save convolution method is presented in Fig. 6.11. Let us assume that a filter $h(n)$ has M weights and it is artificially extended to the length N by appending $N - M$ zeros to its end. In turn, the filtered $x(n)$ consists of L samples and it is divided into fragments having N samples, but overlapping with $M - 1$ samples. Because we are not appending $M - 1$ zeros to signal fragments and applying the fast FFT-based convolution concept to signal sections, the first $M - 1$ samples of each partial convolution result are wrong, since circular—not linear—convolution is performed. The solution is to discard these samples and use the $M - 1$ last samples calculated for the previous signal fragment. But it is only possible when the processed input signal blocks overlap with $M - 1$ samples. Therefore, such overlapping is done.

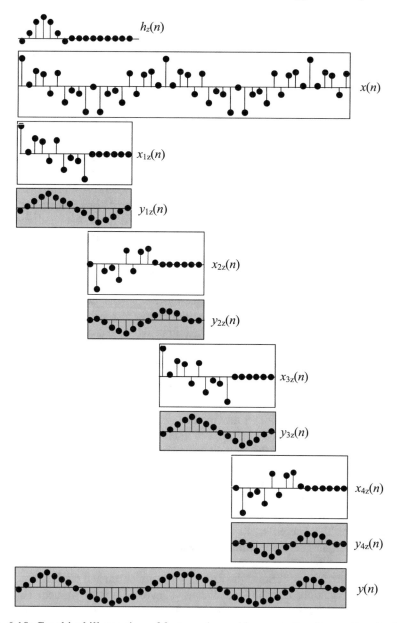

Fig. 6.10: Graphical illustration of fast overlap–add convolution (by sections) of two discrete-time sequences: $x(n)$, e.g. signal to be filtered, and $h(n)$, e.g. filter weights. Denotations: $M = 7$—length of $h(n)$, $N = 10$—length of $x(n)$ fragment (both signals without appended zeros), $h_z(n)$—filter weights appended with $N - 1 = 9$ zeros, $x_{1z}(n)$, $x_{2z}(n)$, $x_{3z}(n)$, $x_{4z}(n)$—consecutive $N = 10$-element fragments of input signal $x(n)$, appended with $M - 1 = 6$ zeros and overlapping with $M - 1 = 6$ samples, $y_{1z}(n)$, $y_{2z}(n)$, $y_{3z}(n)$, $y_{4z}(n)$—consecutive fragments of the output signal $y(n)$, all resulting from inverse FFT, which are combined (added when overlapping) [27]

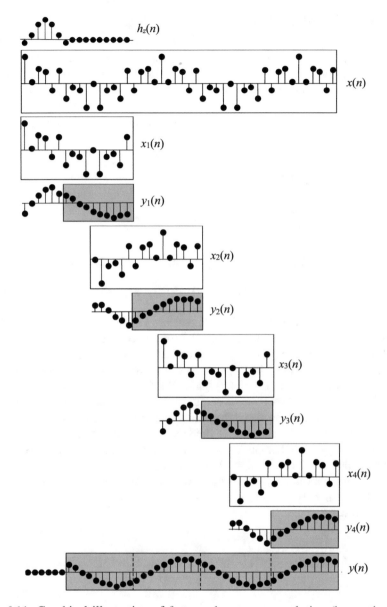

Fig. 6.11: Graphical illustration of fast overlap–save convolution (by sections) of two discrete-time sequences: $x(n)$, e.g. signal to be filtered, and $h(n)$, e.g. filter weights. Denotations: $M = 7$—length of $h(n)$, $N = 16$—length of $x(n)$ fragment (both without appended zeros), $h_z(n)$—filter weights appended with $N - M = 9$ zeros to the length N=16, $x_1(n)$, $x_2(n)$, $x_3(n)$, $x_4(n)$—consecutive $N = 16$-element fragments of input signal $x(n)$, overlapping with $M - 1 = 6$ samples, $y_1(n)$, $y_2(n)$, $y_3(n)$, $y_4(n)$—results of inverse FFTs ($y_k(n) = $ IFFT(FFT($x_k(n)$).* FFT($h_z(n)$))): their fragments, presented on gray background, are put one-by-one giving the output signal $y(n)$, the final filtering result [27]

6.6.5 Fast Signal Correlation

Correlation and convolution differ only in time-reversal and complex conjugation
of the second signal: in convolution (see (Eq. (6.18))) the time-reversal of the second signal is done—summation over k is performed—but its complex conjugation is
not applied, while in the correlation (see Eq. (6.25)) the time-reversal of the second
signal does not take place—summation over n is performed—but its conjugation is
applied (setting $n - k = m$):

$$r_{xy}(k) = \sum_{n=-\infty}^{+\infty} x(n) y^*(n-k), \tag{6.25}$$

$$R_{xy}(\Omega) = \sum_{k=-\infty}^{\infty} \left(\sum_{n=-\infty}^{+\infty} x(n) y^*(n-k) \right) e^{-j\Omega k} = \tag{6.26}$$

$$= \sum_{n=-\infty}^{\infty} x(n) e^{-j\Omega n} \sum_{m=-\infty}^{\infty} y^*(m) e^{j\Omega m} = X(\Omega) Y^*(\Omega), \tag{6.27}$$

$$\mathbf{r}_{xy}(k) = \text{IDtFT}(\mathbf{R}_{xy}(\Omega)) = \text{IDtFT}(\ \mathbf{X}(\Omega) \ .* \ \mathbf{Y}^*(\Omega) \ ), \tag{6.28}$$

where .* denotes element-by-element vector multiplication (as in Matlab).
In Eq. (6.25) we consider more general form of correlation of complex-value
signals—()* denotes complex conjugation of the second signal (such correlations
we perform, for examples, in telecommunication applications as complex-value
matched filters). We see in Eq. (6.27) that in the DtFT spectrum of the correlation
result, the second signal spectrum is conjugated while for signal convolution (6.22)
it is not. Conjugation of the signal frequency spectrum is equivalent to conjugation
of the signal itself and its reversion in time (see Table 4.2). Concluding: in order to
obtain correlation while computing convolution we should only do time-reversal
and conjugation of the second signal before the convolution. Therefore all fast algorithms designed for convolution can be used also for fast correlating complex-value
signals. User has only to do time-reversal and conjugation of the second signal
before the procedure call. That is it! In Matlab:

```
rxy = conv(x,conj(y(end:-1:1)));
```

6.6.6 Fast Convolution/Correlation Example and Program

All details discussed in this section concerning optimal FFT usage for fast signal
convolution and correlation are presented in program 6.5 written in Matlab. User
should read it very carefully. This section, albeit does not looking as being connected to the chapter subject, is very close connected to spectral analysis. Firstly,
fast convolution will be used in the next section to fast computing of DtFT offering

selective (*from-to*) frequency spectrum zooming. Secondly, the correlation function is widely used in spectral analysis of random (noisy) signals because power spectral density is defined as Fourier transform of it (see Eqs. (6.4), (6.5)). Later in this chapter we will calculate Blackman–Tukey PSD estimator exploiting the signal auto-correlation function.

Exercise 6.7 (Fast and Last ConvCorr Train to London!). Analyze the program 6.5. Choose short testing signal, setting `sig=1;`. Run the program, observe figures. The correct convolution result is $y(n) = [1, 2, 3, 3, 3, 2, 1]$. Note, that enlarging only the second signals with zeros does not work. Appending zeros to both signals to the length $(N + M - 1)$ offers good result. See, that even for so short signals fast correlation methods works also. Now set `sig=2`. For this option overlap–add convolution methods are switched on. Display signals `yy` (calculated new output fragment) and `y4` (the whole convolution result calculated so far) inside the loop. The overlap–save fast convolution method is not implemented in the program. Please, write its Matlab code. Check its correctness calculating an error in respect to the Matlab `conv()` function.

Listing 6.5: Fast convolution and correlation using FFT

```
1   % lab06_ex_fastconvcorr.m Fast signal convolution and correlation using FFT
2   clear all; close all
3
4   sig = 2;     % 1/2, signal: 1=short, 2=long
5   if(sig==1)  N=5;    M=3;   x = ones(1,N);  h = ones(1,M);  % signals to be
6   else        N=256;  M=32;  x = randn(1,N); h = randn(1,M); % convolved
7   end
8   n = 1:N+M-1; nn = 1:N; % sample indexes
9   figure;
10  subplot(211); stem(x); title('x(n)');
11  subplot(212); stem(h); title('h(n)'); pause
12
13  % Conv by Matlab function
14  y1 = conv(x,h);
15  figure; stem(y1); title('y1(n)'); pause
16
17  % Fast conv - WRONG!
18  hz = [ h zeros(1,N-M) ];        % append N-M zeros to the shorter signal only
19  y2 = ifft( fft(x) .* fft(hz) ); % fast conv, first M-1 samples are wrong
20  figure; plot(nn,y1(nn),'ro',nn,y2(nn),'bx'); title('y1(n) & y2(n)');
21  error2 = max(abs(y1(M:N)-y2(M:N))), pause
22
23  % Fast conv - GOOD!
24  hzz = [ h zeros(1,N-M) zeros(1,M-1) ];  % append zeros to the length N+M-1
25  xz = [ x zeros(1,M-1) ];                % append zeros to the length N+M-1
26  y3 = ifft( fft(xz) .* fft(hzz) );       % fast conv, all samples are good
27  figure; plot(n,y1,'ro',n,y3,'bx'); title('y1(n) & y3(n)');
```

```
28     error3 = max(abs(y1-y3)), pause
29
30  % Fast conv by pieces - OVERLAP-ADD method
31     if( sig > 1 )                        % only for long signal
32     L = M;                               % signal fragment length
33     K = N/L;                             % number of signal fragments
34     hzz = [ h zeros(1,L-M) zeros(1,M-1) ]; % append zeros to the shorter signal
35     Hzz = fft(hzz);                      % its FFT
36     y4 = zeros(1,M-1);                   % output signal initialization
37     for k = 1:K                          % LOOP
38         m = 1 + (k-1)*L : L + (k-1)*L;   % select samples indexes
39         xz = [ x(m) zeros(1,M-1) ];      % cut signal fragment
40         YY = fft(xz) .* Hzz;             % # fast convolution - spectra mult
41         yy = ifft( YY );                 % # inverse FFT
42         y4(end-(M-2):end) = y4(end-(M-2):end) + yy(1:M-1); % output overlap-add
43         y4 = [ y4, yy(M:L+M-1) ];                          % output append
44     end
45     figure; plot(n,y1,'ro',n,y4,'bx'); title('y1(n) & y4(n)');
46     error4 = max(abs(y1-y4)), pause
47     end
48
49  % Fast cross-correlation by fast convolution
50     R1 = xcorr( x, h );
51     R2 = conv( x, conj( h(end:-1:1) ) );
52     Kmax=max(M,N); Kmin=min(M,N); R2 = [ zeros(1,Kmax-Kmin) R2 ];
53     m = -(Kmax-1) : 1 : (Kmax-1);
54     figure; plot(m,R1,'ro',m,R2,'bx'); title('R1(n) & R2(n)');
55     error5 = max( abs( R1-R2 ) ), pause
56
57  % Fast computation of the first M coeffs of the auto-correlation function
58     R1 = xcorr( x, x );           % Matlab auto-correlation function
59     xz = [x zeros(1,M-1)];        % add M-1 zeros on the end of signal
60     X = fft(xz);                  % calculate (N+M-1)-point FFT
61     X = X.*conj(X);               % calculate |X(k)|^2
62     R2 = ifft(X);                 % calculate inverse (N+M-1)-point FFT
63     R2 = real(R2(1:M));           % choose correct values, scale
64     m=1:M; figure; plot(m,R1(N:N+M-1),'ro',m,R2(m),'bx'); grid;
65     title('Autocorr Rxx(m): Matlab (RED), WE (BLUE)'); xlabel('m'); pause
66     error6 = max( abs( R1(N:N+M-1) - R2(m) ) ), pause
```

Exercise 6.8 (Virtual Concert Hall). Read from disc one of acoustical impulse responses (room, bathroom, cathedral,...). Check sampling frequency. Record your speech, about 5–10 s, with the same sampling ratio. Perform fast convolution of both signals. Listen to the result. Wow!

6.7 Fast DtFT Calculation via Chirp-Z Transform

As demonstrated in Chap. 4, the discrete-time Fourier transform (DtFT) offers precise Fourier spectrum zooming for any frequency range $[f_1, f_2]$ with arbitrarily chosen step Δf but at the price of significant computational increase.

Let us assume that we would like to calculate DtFT coefficients $X(k)$ of signal $x(n)$, $n = 0, 1, 2, \ldots, N-1$, for M frequencies f_k, starting from f_0 and with step Δf:

$$X(k) = \sum_{n=0}^{N-1} x(n) e^{-j2\pi \frac{f_k}{f_s} n}, \quad f_k = f_0 + k \cdot \Delta f, \quad k = 0, \ldots, M-1. \quad (6.29)$$

Equation (6.29) requires NM complex multiplications. For big values of N and M it is beneficial to use the chirp-Z transform (CZT), exploiting FFT. After introduction of two new variables A and W:

$$A \equiv e^{-j2\pi \frac{f_0}{f_s}}, \quad W \equiv e^{-j2\pi \frac{\Delta f}{2 f_s}} \quad (6.30)$$

Eq. (6.29) takes the form of CZT:

$$X(k) = \sum_{n=0}^{N-1} x(n) A^n W^{2kn}, \quad k = 0, \ldots, M-1. \quad (6.31)$$

Since the following equality holds $2kn = n^2 + k^2 - (k-n)^2$, Eq. (6.31) can be written as:

$$X(k) = W^{k^2} \sum_{n=0}^{N-1} \underbrace{\left[x(n) A^n W^{n^2} \right]}_{y_1(n)} \underbrace{W^{-(k-n)^2}}_{y_2(n)}, \quad k = 0, 1, \ldots, M-1. \quad (6.32)$$

Since in Eq. (6.32) two sequences of samples:

$$y_1(n) = x(n) A^n W^{n^2}, \quad (6.33)$$

$$y_2(n) = W^{-n^2}, \quad (6.34)$$

are convolved, Eq. (6.32) can be calculated efficiently in frequency domain using three fast FFTs, two direct and one inverse:

$$\mathbf{X}(k) = W^{k^2} \cdot \text{IFFT} \left(\text{FFT} \left(\tilde{\mathbf{y}}_1(n) \right) . * \text{FFT} \left(\tilde{\mathbf{y}}_2(n) \right) \right). \quad (6.35)$$

Result of circular convolution (6.35) is the same as result of linear convolution (6.32) when in (6.35) not signals $y_1(n)$ and $y_2(n)$ are transformed but their modified versions $\tilde{y}_1(n)$ and $\tilde{y}_2(n)$. The first signal is calculated from Eq. (6.33) for $0 \le n \le N-1$ and $M-1$ zeros are appended at its end. It is obvious why it is done: standard signal *conditioning* before performing circular convolution which should result in linear convolution. The second signal is calculated from Eq. (6.34) for

$n = [0, 1, \ldots, (M-1), -(N-1), -(N-2), \ldots, -2, -1]$, so it is prepared for time-reversion after which its samples $n = 1, \ldots, M$ go to end and hit zeros added to the first signal. Remaining details are presented in program 6.6.

Listing 6.6: Fast DtFT calculation

```
1   % lab06_ex_dtft.m
2   clear all; close all
3
4   N = 256; M = 32;            % number of signal samples, number of frequency bins
5   fs = 128;                   % sampling ratio (Hz)
6   fd = 5; fu = 9;             % down and up frequency (from-to)
7   df = (fu-fd)/(M-1);         % step in frequency
8   f = fd : df : fu;           % frequencies of interest
9   x = rand(1, N);             % analyzed signal, random noise [0, 1]
10
11  NM1 = N+M-1;
12  A = exp( -j*2*pi * fd/fs );                       % for first frequency
13  W = exp( -j*2*pi * ((fu-fd)/(2*(M-1))/fs) );      % for frequency step
14  y1 = zeros(1,NM1); k=0:N-1; y1(k+1)=((A*W.^k).^k).*x(k+1); % init of y1
15  k = [ 0:M-1, -(N-1):1:-1]; y2 = W.^(-k.^2);               % init of y2
16  Y1 = fft(y1);              % #  fast circular convolution
17  Y2 = fft(y2);              % #  of signals y1 and y2
18  Y = Y1.*Y2;               % #
19  y = ifft(Y)/(N/2);        % #
20  k=0:M-1; XcztN(k+1) = y(k+1) .* (W.^(k.^2));      % phase correction
21
22  n = 0:N-1; Xref=x*exp(-j*2*pi*n(:)*f/fs)/(N/2); % reference - matrix def
23  error = max(abs( XcztN - Xref )),                % error
24
25  figure;
26  subplot(2,1,1); plot(f,real(XcztN),'r.-'); grid on; hold on;
27                  plot(f,real(Xref),'bo-');
28  subplot(2,1,2); plot(f,imag(XcztN),'r.-'); grid on; hold on;
29                  plot(f,imag(Xref),'bo-');  pause
```

Exercise 6.9 (Spectral Microscope). Apply fast DtFT algorithm to signal analyzed in program 6.2. Interpolate very dense its Fourier spectrum around the signal frequency peak. Add calculated fragment of the signal spectrum to figure plotted in Exercise 6.2. Mark spectrum samples with magenta circles.

6.8 Blackman–Tukey PSD Fast Calculation

Having fast algorithm for computation of signal auto-correlation, we can calculate efficiently the Blackman–Tukey PSD estimator defined by Eq. (6.10) and compare it with Welch PSD. This is done in program 6.7.

Listing 6.7: Fast calculation of Blackman–Tukey PSD estimator

```
1   % lab06_ex_psd.m
2   clear all; close all
3
4
5   % lab06_ex_psd.m  Blackman-Tukey PSD compared with Welch PSD
6
7     clear all; close all; subplot(111);
8
9   % Parameters
10    N = 256; fs = 1000;                    % number of samples, sampling ratio
11    M = N/4;                               % number of computed AutoCorr coeffs
12    df = fs/N, f=0:df:(N-1)*df;            % PSD spectrum frequencies
13    dt=1/fs; t=0:dt:(N-1)*dt;              % time stamps for signal
14    tR=-(N-1)*dt:dt:(N-1)*dt; k=1:N/2+1;   % time stamps for auto-correlation
15
16  % Generation of analyzed signal - sinusoid 100 Hz in noise
17    x = sin(2*pi*25.5*t) + 0.1*randn(1,N);     % signal + Gaussian noise
18    figure; plot(t,x); grid; axis tight; title('x(n)'); xlabel('time [s]'); pause
19
20  % Fast computation of the first M coeffs of the auto-correlation using FFT
21    xz = [x zeros(1,M-1)];                 % appending M-1 zeros
22    X = fft(xz); X = X.*conj(X); R = ifft(X);  % FFT, |X(k)|^2, IFFT
23    R = real(R(1:M))/N;                    % coeffs of interest, scaling
24
25  % PDS estimation - Blackman-Tukey method = Fourier transform of signal autocorr
26    w = hanning(2*M-1); w=w(M:2*M-1); w=w';    % choose window
27    Rw = R .* w;                           % multiply autocorr with window
28    s = [ Rw(1:M) zeros(1,N-2*M+1) Rw(M:-1:2)]; % input to FFT is symmetrical
29    P1  = real(fft(s))/fs;                 % the FFT result, real-value vector
30    figure; subplot(211); plot(f(k),P1(k),'b'); grid;
31    title('Blackman-Tukey estimation of PSD'); xlabel('f [Hz]'); ylabel('V^2 / Hz');
32
33  % PSD estimation - Welch method = averaging periodograms of signal fragments
34    Nfft = N; Nwind = 2*M; Noverlap = Nwind/2; % lengths: FFT, window, overlap
35    Nshift = Nwind-Noverlap;               % window shift
36    M = floor((N-Nwind)/Nshift)+1;         % number of signal fragments
37    w = hanning(Nwind)';                   % window choice
38    P2 = zeros(1,Nfft);                    % initialization
39    for m=1:M                              % number of the signal fragment
40       n = 1+(m-1)*Nshift : Nwind+(m-1)*Nshift; % which samples?
41       X = fft( x(n) .* w, Nfft);          % FFT of windowed fragment
42       P2 = P2 + abs(X).^2;                % accumulate
43    end                                    % end of loop
44    P2 = P2/(fs*M*sum(w.*w));              % PSD normalization
45    subplot(212); plot(f(k),P2(k)); grid; title('Welch estimation of PSD');
46    xlabel('f [Hz]'); ylabel('V^2 / Hz'); pause
```

Exercise 6.10 (Discovering Noisy Sounds). Analyze code of the program 6.7. Run it increasing level of noise up to the moment when you completely *loose* your signal. Observe both, signal and spectrum shape. Then, take from web page *FindSounds* some noisy sounds, read them to the program and analyze.

6.9 Fast Estimation of Damped Sinusoids by Interpolated DFT

If an analyzed signal is a mixture of cosines, like in Eq. (4.33), its DFT spectrum is specified by Eq. (4.34). Let us assume now more general and more universal signal model, than the above one, in which each cosine signal component has phase shifts ϕ_k and is attenuated by an exponent with damping factor d_k:

$$x(n) = \sum_{k=1}^{K} x_k(n), \quad x_k(n) = A_k \cos(\Omega_k n + \phi_k) e^{-d_k n}, \quad n = 0, 1, 2, \ldots, N-1. \quad (6.36)$$

After defining a complex-value signal $s_k(n)$ of the form:

$$s_k(n) = A_k e^{j\phi_k} e^{-d_k n} e^{+j\Omega_k n} = A_k e^{j\phi_k} \lambda_k^n, \quad \lambda_k = e^{-d_k + j\Omega_k}. \quad (6.37)$$

each signal component $x_k(n)$ in Eq. (6.36) can be expressed as summation of a corresponding signal $s_k(n)$ (6.37) and its complex conjugation, divided by two:

$$x_k(n) = \frac{1}{2} \left(s_k(n) + s_k^*(n) \right). \quad (6.38)$$

In consequence we can rewrite Eq. (6.36) into the following form:

$$x(n) = \sum_{k=1}^{K} \frac{1}{2} \left(s_k(n) + s_k^*(n) \right) = \sum_{k=1}^{K} \left(\frac{A_k}{2} e^{j\phi_k} \lambda_k^n + \frac{A_k}{2} e^{-j\phi_k} \lambda_k^{*n} \right). \quad (6.39)$$

In chapter on DFT we have derived spectrum of the rectangular window (equations from (4.28) to (4.31)). In similar way we can calculate now the DFT spectrum of complex signal $s_k(n)$ defined by Eq. (6.37), and obtain

$$S_k(\Omega) = A_k e^{j\phi_k} \frac{1 - \lambda_k^N}{1 - \lambda_k e^{-j\Omega}}. \quad (6.40)$$

I am leaving this derivation for an ambitious/interesting reader. As a consequence the DFT spectrum $X_k(\Omega)$ of each signal $x_k(n)$ (6.38) is equal to:

$$X_k(\Omega) = \frac{A}{2}\left(e^{j\phi_k}\frac{1-\lambda_k^N}{1-\lambda_k e^{-j\Omega}} + e^{-j\phi_k}\frac{1-\lambda_k^{*N}}{1-\lambda_k^* e^{-j\Omega}}\right). \tag{6.41}$$

At present we are in *the turning point* of the story. If maxima of the DFT spectrum of signal defined by Eq. (6.36) is well separated, it is possible to estimate Ω_k and d_k of each signal component from three DFT samples $X(m-1), X(m), X(m+1)$, corresponding to DFT angular frequencies $\Omega_{m-1}, \Omega_m, \Omega_{m+1}$, lying around the peak $|X(m)|$ in the DFT magnitude spectrum, i.e. close to Ω_k.

Let us define ratio R as:

$$R_k = \frac{X(m-1)-X(m)}{X(m)-X(m+1)}. \tag{6.42}$$

and put Eq. (6.41) into Eq. (6.42). When the second component in (6.41) is neglected (for the negative frequency), we obtain

$$R_k = \frac{1-\lambda_k e^{-j\Omega_{m+1}}}{1-\lambda_k e^{-j\Omega_{m-1}}}r, \quad r = \frac{-e^{-j\Omega_m}+e^{-j\Omega_{m-1}}}{-e^{-j\Omega_{m+1}}+e^{-j\Omega_m}}, \tag{6.43}$$

From Eq. (6.43) one can calculate λ_k as the only unknown value:

$$\lambda_k = e^{j\Omega_k}\frac{r-R}{re^{-j2\pi/N}-Re^{j2\pi/N}}, \tag{6.44}$$

and remembering its definition in (6.37)

$$\lambda_k = e^{-d_k+j\Omega_k} \tag{6.45}$$

we can calculate next values of Ω_k and d_k from λ_k:

$$d_k = -\,\mathrm{Re}\left[\ln(\lambda_k)\right], \quad \Omega_k = \mathrm{Im}\left[\ln(\lambda_k)\right]. \tag{6.46}$$

The above derivation was first done by Yoshida et al. [23].

Again, when the second component in Eq. (6.41) is neglected (for the negative frequency), signal amplitude A_k and phase ϕ_k are easy to be estimated:

$$A_k = |2X[m]/c|, \quad \phi_k = \mathrm{angle}\,(2X[m]/c), \quad \text{where} \quad c = \frac{1-\lambda_k^N}{1-\lambda_k e^{-j\Omega_m}}. \tag{6.47}$$

To Remember Algorithms similar to the described above, are called in literature interpolated DFTs. They have a very important feature that parameters of signals modeled by Eq. (6.36), i.e. being a sum of damped sinusoids, can be precisely recovered from them using a few FFT/DFT samples lying around spectral peaks of individual signal components. Before, in the beginning of this chapter, we have been interpolating shape of FFT spectrum by performing FFT upon signal with appended zeros. Now, a different type of *interpolation* is performed: we are exploiting some FFT samples for more precise calculation of the frequency value of one signal component using known signal model. The frequency lying in between the FFT spectrum bins of its peak. *In-between*. Interpolation!

There are many different IpDFT algorithms. The Bertocco–Yoshida one was described here. In other algorithms different numbers of DFT samples lying around spectral peaks are used. The DFT spectrum can be also calculated for windowed signals, e.g. using Hanning window. It is possible to estimate many signal components from their DFT/FFT peaks when signal component peaks are well separated in DFT spectrum. If not, in order to minimize influence of spectral leakage from one component to the other, an iterative leakage compensation can be applied [11, 26].

In program 6.8 Bertocco–Yoshida IpDFT algorithm without leakage correction is implemented. The analyzed damped cosine signal is presented on the left side of Fig. 6.12, while on the right side beginning of the signal DFT spectrum magnitude. Values of signal amplitude, damping, frequency, and phase are correctly estimated.

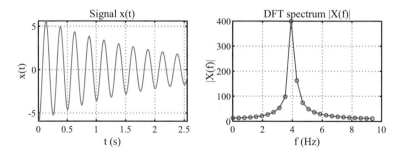

Fig. 6.12: Damped cosine signal analyzed by the IpDFT algorithm (left) and beginning of its DFT spectrum magnitude (right)

Listing 6.8: Matlab implementation of Bertocco–Yoshida interpolated DFT algorithm

```
1   % lab06_ex_ipdft.m  Interpolated DFT/FFT
2   clear all; close all;
3
4   % Test signal
5   N=256; fs=100;                 % number of samples, sampling ratio [Hz]
6   Ax=6; dx=0.5; fx=4; px=3;      % signal amplitude, damping, frequency, phase
7   dt=1/(fs); t=(0:N-1)*dt;       % sampling time
8   x = Ax*exp(-dx*t).*cos(2*pi*fx*t+px); figure; plot(t,x); pause % signal generation
9
10  % Interpolated DFT for maximum absolute DFT peak
11  Xw = fft(x); figure; plot(abs(Xw)), pause  % computation of DFT/FFT
12   [Xabs, ind] = max(abs(Xw(1:round(N/2))));  % find maximum and its index
13  km1 = ind-1; k=ind; kp1 = ind+1;           % three DFT samples around max
14  dw = 2*pi/N;                   % DFT frequency step
15  wkm1= (km1-1)*dw;              % angular frequency of DFT bin with index k-1
16  wk  = (k  -1)*dw;              % angular frequency of DFT bin with index k
17  wkp1= (kp1-1)*dw;             % angular frequency of DFT bin with index k+1
18  r = ( -exp(-j*wk)+exp(-j*wkm1) )/( -exp(-j*wkp1)+exp(-j*wk) ); % eq.(6.43)
19  R = ( Xw(km1)-Xw(k) )/( Xw(k)-Xw(kp1) );                     % eq.(6.42)
20  lambda = exp(j*wk)*(r-R)/( r*exp(-j*2*pi/N)-R*exp(j*2*pi/N) ); % eq.(6.44)
21  we = imag(log(lambda));       % estimated angular frequency
22  de = -real(log(lambda));      % estimated damping
23  fe = we*fs/(2*pi);            % angular frequency --> frequency
24  de = de*fs;                   % normalized damping (de/fs) --> damping (de)
25
26  if round(1e6*R)==-1e6         % COHERENT SAMPLING, dx=0
27     Ae = 2*abs(Xw(k))/N;       % estimated amplitude
28     pe = angle(Xw(k));         % estimated phase
29  else                          % NON-COHERENT SAMPLING
30     c = (1-lambda^N)/(1-lambda*exp(-j*wk));                   % eq.(6.47)
31     c = 2*Xw(k)/c;                                            % eq.(6.47)
32     Ae = abs(c);               % estimated amplitude
33     pe = angle(c);             % estimated phase
34  end
35
36  result = [ Ae, de, fe, pe ],            % display results
37  errors = [ Ae-Ax, de-dx, fe-fx, pe-px ], pause % show errors
```

Exercise 6.11 (Testing IpDFT Algorithm). Analyze code of the program 6.8. Run it. Observe figures. Are the estimated signal amplitude and frequency correct? They should be. Take a sound recording of single piano, guitar, or trumpet note from *FindSounds* web page. Read it into the program. Cut-off silence part from the recording beginning and end. Choose three FFT points lying around one spectrum peak. May be the first one? Or the highest one? What parameters has a damped sine you have selected? Do you trust in this result? How to verify its correctness?

Hmm... Hmm... This DSP course has been promised to be easy! Yes, it is. This was the most difficult derivation in this chapter. But it was extremely important for us, since in engineer's practice precise estimation of each damped sine parameters (A_k, f_k, ϕ_k, d_k) is very often priceless because such signals widely occurs in surrounding us world (due to physical damped resonance phenomena). We should also observe that IpDFT technique is different than DFT zooming realized by appending zeros to the signal end. In IpDFT we exploit the signal model, use it for a few DFT points (three in the described method), create a system of equations, and solve it. The solution obtained this way has very good accuracy and it is computationally fast due to the FFT usage. All four signal parameters are computed.

6.10 Summary

So our FFT cruiser, after a long voyage, after many intellectual storms, is finally sailing into a port. A lot places, views, and memories. What should stay *unforgettable*?

1. FFT has many faces. It can be used not only in spectral analysis but also as computational *hammer* in many other applications, for example, for fast calculation of signal convolution and correlation.
2. In the kingdom of spectral estimation, FFT can be applied to calculation of four different spectra types, appropriate for different usage, e.g. amplitude and power spectrum $((V)$ and (V^2), respectively), amplitude and power spectral density $((V/\sqrt{(Hz)})$ and $(V^2/Hz))$. Their definitions should be remembered.
3. Noisy random signals and signals with strong noise background require different FFT treatment. In their case, spectral density functions should be preferred and many FFT spectra of different signal fragments have to be calculated and averaged. Such signal processing minimizes variation of noise spectrum and ensures more reliable (consistent) estimation of noise level. PSD estimation using the Welch method, averaging windowed periodograms, is the most prominent example in this FFT application area.
4. Time-varying signals require also calculation of many FFT spectra of consecutive shorter signal fragments. However, not for the purpose of mean spectrum calculation. The spectra are stored one-by-one into *time–frequency* matrix and they are displayed all as a 3D mesh or color image which allows observation of the spectrum change in time.
5. The whole DFT spectrum of N-sample long signal can be zoomed K times after appending $K(N-1)$ zeros at the signal end and after performing longer, KN-point FFT.

6. We can zoom also not the whole DFT spectrum but its selected part only (from one frequency to the other frequency with a chosen step) using the chirp-Z transform exploiting three FFTs.
7. As already mentioned, the FFT is widely used for fast calculation of signal convolution and correlation. Both operations are extremely important in DSP: the first one describes always stable, non-recursive signal filtering, while the second one finds application in detection of one signal *hidden inside* the other signal (echo detectors, telecommunication receivers).
8. Interpolated FFT algorithms, exploiting damped sine signal model, should be used for fast and precise estimation of parameters of such signals.
9. Everybody should bless the FFT king!

6.11 Private Investigations: Free-Style Bungee Jumps

Exercise 6.12 (Looking for a Needle in a Bottle of Hay). Generate a sinus. Add a weak white Gaussian noise to it—use `randn()` function. Calculate and observe power FFT frequency spectrum. Then step-by-step increase the level of noise, stop when a sine peak is not visible in the spectrum. Then increase the signal length, calculate more power spectra and their average. Find how many spectra should be averaged in order to see again the signal peak.

Exercise 6.13 (Steel Factory Secrets: Revisited). Apply your present knowledge to spectral analysis of supply voltages and currents recorded for operating arc furnace, previously processed in Exercise 4.9. Read signals from files `load ('UI.mat'); whos`. Calculate the amplitude FFT spectrum of the whole signal, then its Welch PSD estimate (`psd()`) and time–frequency spectrogram (`spectrogram()`). Try to estimate frequencies and amplitudes of fundamental frequency 50 Hz (close to 50) and its harmonics 100, 150, 200, 250, ... Hz.

Exercise 6.14 (Piano Sound Spectrogram: Describing the Beauty). In one of previous exercises you generated a piano-like sound signal. At present use your own short-time Fourier transform program or the `spectrogram()` Matlab function and do *inverse engineering*: find sequence of frequencies hidden in your piano masterpiece. You can also take any recording from the Internet, for example, from *Find-Sounds* web page (https://www.findsounds.com/). Adjust properly: window length, window shift, FFT length (zero padding) in order to better see the frequency change.

Exercise 6.15 (Speech Spectrogram: Emotion Detector). Frequency of vocal cords opening and closing (pitch frequency) become higher during emotional speaking. Record a few times the same word, increasing level of excitement, astonishment, fear, ... Calculate and compare their spectrograms. Do you see no change? May be you are cold-blooded cad? Adjust properly: window length, window shift, FFT length (zero padding) in order to track better change of pitch frequency.

Exercise 6.16 (Is My Heart Still Broken? Part 3). Apply fast correlation algorithm, described in this section, for estimation of heartbeat periodicity in ECG recording analyzed in Chaps. 1 and 2 in Exercises 1.11 and 2.14.

References

1. F. Auger, P. Flandrin, Improving the readability of time-frequency and time-scale representations by the reassignment method. IEEE Trans. Signal Process. **43**(5), 1068–1089 (1995)
2. F. Auger, P. Flandrin, P. Goncalves, O. Lemoine, *Time-Frequency Toolbox*. Online: http://tftb.nongnu.org/
3. S. Bagchi, S.K. Mitra, *The Nonuniform Discrete Fourier Transform and Its Applications in Signal Processing* (Kluwer, Boston, 1999)
4. J. Bendat, A. Piersol, *Engineering Applications of Correlation and Spectral Analysis* (Wiley, New York, 1980, 1993)
5. R.E. Blahut, *Fast Algorithms for Digital Signal Processing* (Addison-Wesley, Reading, 1985)
6. B. Boashash (ed.), *Time-Frequency Signal Analysis and Processing. A Comprehensive Reference* (Elsevier, Oxford, 2003)
7. C.S. Burrus, T.W. Parks, *DFT/FFT and Convolution Algorithms. Theory and Implementation* (Wiley, New York, 1985)
8. W.H. Chen, C.H. Smith, S. Fralick, A fast computational algorithm for the discrete cosine transform. IEEE Trans. Commun. **25**, 1004–1009 (1977)
9. L. Cohen, *Time-Frequency Analysis* (Prentice Hall, Englewood Cliffs, 1995)
10. L. Cohen, P. Loughlin (eds.), *Recent Developments in Time-Frequency Analysis* (Kluwer, Boston, 1998)
11. K. Duda, T.P. Zieliński, Efficacy of the frequency and damping estimation of a real-value sinusoid. IEEE Mag. Instrum. Meas. 48–58 (2013)
12. D.F. Elliott, K.R. Rao, *Fast Transforms. Algorithms, Analyses, Applications* (Academic Press, New York, 1982)
13. P. Flandrin, *Time-Frequency/Time-Scale Analysis* (Academic Press, San Diego, 1999)
14. P. Flandrin, *Explorations in Time-Frequency Analysis* (Cambridge University Press, Cambridge, 2018)
15. G. Heinzel, A. Rudiger, R. Schilling, *Spectrum and spectral density estimation by the Discrete Fourier transform (DFT), including a comprehensive list of window functions and some new flat-top windows,* Technical Report, Max-Planck-Institut fur Gravitationsphysik (Albert-Einstein-Institut), Teilinstitut Hannover, 2002. Online: https://pure.mpg.de/rest/items/item_152164_1/component/file_152163/content
16. F. Hlawatsch, F. Auger (eds.), *Time-Frequency Analysis: Concepts and Methods* (Wiley, Chichester, 2008, 2013)

17. V.K. Ingle, J.G. Proakis, *Digital Signal Processing Using Matlab* (PWS Publishing, Boston, 1997; CL Engineering, 2011)
18. R.G. Lyons, *Understanding Digital Signal Processing* (Addison-Wesley Longman Publishing, Boston, 1996, 2005, 2010)
19. A.V. Oppenheim, R.W. Schafer, *Discrete-Time Signal Processing* (Pearson Education, Upper Saddle River, 2013)
20. J.G. Proakis, D.G. Manolakis, *Digital Signal Processing. Principles, Algorithms, and Applications* (Macmillan, New York, 1992; Pearson, Upper Saddle River, 2006)
21. D.S. Qian, D. Chen, *Joint Time-Frequency Analysis* (Prentice Hall, Upper Saddle River, 1996)
22. K. Shin, J.K. Hammond, *Fundamentals of Signal Processing for Sound and Vibration Engineers* (Wiley, Chichester, 2007)
23. S.W. Smith, *The Scientist and Engineer's Guide to Digital Signal Processing* (California Technical Publishing, San Diego, 1997, 1999). Online: http://www.dspguide.com/
24. P. Stoica, R. Moses, *Introduction to Spectral Analysis* (Prentice Hall, Upper Saddle River, 1997)
25. P. Wigner, Quantum-mechanical distribution functions revisited, in *Perspectives in Quantum Theory*, ed. by W. Yowgram, A. van der Merwe (Dover, New York, 1979)
26. R.C. Wu, C.T. Chiang, Analysis of the exponential signal by the interpolated DFT algorithm. IEEE Trans. Instrum. Meas. **59**(12), 3306–3317 (2010)
27. T.P. Zieliński, *Cyfrowe Przetwarzanie Sygnałów. Od Teorii do Zastosowań (Digital Signal Processing. From Theory to Applications)* (Wydawnictwa Komunikacji i Łączności (Transport and Communication Publishers), Warszawa, Poland, 2005, 2007, 2009, 2014)

Chapter 7
Analog Filters

Fresh salad versus old wine, or old Cadillac, or vinyl records.
There is no hard rule that newer things are always better!

7.1 Introduction

Talking about *old-fashioned* analog filters is always very difficult to me. Young audience expects the hottest news from first magazine pages. Unfortunately, analog filter theory is not the hottest topic. But this knowledge is still priceless, like old good wine, old good Cadillac, or old good vinyl records. Why?

Our world is analog. Our speech and our heart bit. When we are interfacing our DSP systems with a real world, we have to:

- firstly—on input, before the A/D converter—remove in analog way all frequency components lying *out-of-band* in our sampling scheme,
- secondly—on output, after the D/A converter—smooth in analog manner our continuous-time but discrete-value (step-like) signals.

Both operations require analog filters. Therefore, at least these two types of analog filters are absolutely necessary for us, DSP enthusiasts.

The second motivation for analog filter introducing in DSP book is that the theory of high-quality analog filter design is very well established and exist possibilities for very easy transformation of analog filter designs to digital ones, for example, using the bilinear transformation. Thanks to this analog filter design experience can be applied also in DSP core.

In this chapter only the simplest linear time-invariant analog systems/filters are presented. But such filters are used in more than 99% of all analog filter applications. Analog filters represent connections of passive elements (resistors, inductors, capacitors), like in *RLC* circuit, which can be additionally accompanied by electronic operational amplifiers. Each analog filter has the so-called *impulse response*, i.e. response to Dirac impulse occurring on its input. Theoretical Dirac impulse consists of all frequencies. In the filter impulse response

© Springer Nature Switzerland AG 2021
T. P. Zieliński, *Starting Digital Signal Processing in Telecommunication Engineering*, Textbooks in Telecommunication Engineering, https://doi.org/10.1007/978-3-030-49256-4_7

there are present only these frequencies that the filter can pass. The Fourier transform of the impulse response is called the filter frequency response—it tells us how much each input frequency component will be gained/attenuated by the filter and how much delayed.

Having the analog filter structure and its passive element values, we can write a mathematical equation describing the filter (consisting of derivatives and integrals of flowing currents) and then transform it into differential equation of higher order—calculating derivatives of its both sides. The final analog filter differential equation is connecting output filter voltage (and its derivatives) with the input voltage (and its derivatives). The Laplace transform allows us to change (transform) this differential equation into filter transfer function, while the Fourier transform into filter frequency response. Filter design task is to choose proper transfer function coefficients being also coefficients of the filter differential equation. Having values of these coefficients and knowing relation between them and passive RLC filter elements, one can solve set of equations and find values of the elements. Next, the only task is to go the nearest Electronic Store and to buy elements with the closest values to the designed ones.

Generally, in this chapter we make a short but consistent walk from analog LTI system theory to the design of concrete analog circuits. We become familiar with analog low-pass filters designed by Butterworth, Chebyshev, and Cauer (elliptic filter). We learn how to transform analog low-pass filter into another low-pass filter or high-pass, band-pass, or band-stop filter.

This will be *a simple story*. But I hope that different panorama views we will see during our *analog* trip will be *unforgettable* for some Readers.

7.2 Analog LTI Systems

Linear time-invariant (LTI) analog systems are specified by differential equations defining relation between input signal $x(t)$ and its derivatives (multiplied by coefficients b_m) and output signal $y(t)$ and its derivatives (taken with coefficients a_m):

$$x(t) \quad \Rightarrow \quad [\quad \text{Differential EQUATION} \quad] \quad \Rightarrow \quad y(t). \tag{7.1}$$

E.g.

$$b_0 x(t) + b_1 \frac{dx(t)}{dt} + \ldots + b_M \frac{dx^M(t)}{dt^M} = y(t) + a_1 \frac{dy(t)}{dt} + \ldots + a_N \frac{dy^N(t)}{dt^N}. \tag{7.2}$$

Coefficients $\{b_m, a_n\}$ depend on the system structure (its element connection) and values of passive elements used (R, L, C—resistance, inductance, capacitance). The Laplace transform (with complex variable s) is defined as:

$$L(x(t)) = X(s) = \int_{-\infty}^{+\infty} x(t)e^{-st}\,dt \tag{7.3}$$

and has two features very important for us:

$$L\left(\frac{dx^m(t)}{dt^m}\right) = s^m X(s), \qquad L\left(\int_{-\infty}^{t} x(t)dt\right) = \frac{1}{s}X(s). \tag{7.4}$$

The first of them is true for zero initial conditions.

Calculation of the Laplace transform of both sides of the differential equation of the analog system (7.2), having in mind features (7.4), results with the following system transfer function (TF):

$$\left[b_0 + b_1 s^1 + b_2 s^2 + \ldots + b_M s^M\right] X(s) = \left[1 + a_1 s^1 + a_2 s^2 + \ldots + a_N s^N\right] Y(s), \tag{7.5}$$

$$H(s) = \frac{Y(s)}{X(s)} = \frac{b_0 + b_1 s^1 + b_2 s^2 + \ldots + b_M s^M}{1 + a_1 s^1 + a_2 s^2 + \ldots + a_N s^N}. \tag{7.6}$$

Having differential equation describing the system, we can write with ease its transfer function, for example:

$$7x(t) + 6\frac{dx^2(t)}{dt^2} + 5\frac{dx^3(t)}{dt^3} = y(t) + 2\frac{dy^1(t)}{dt^1} + 3\frac{dy^2(t)}{dt^2} + 4\frac{dy^3(t)}{dt^3} \quad \Rightarrow$$

$$\Rightarrow \quad H(s) = \frac{7 + 6s^2 + 5s^3}{1 + 2s^1 + 3s^2 + 4s^3}$$

and, vice versa, from system transfer function deduce its differential equation:

$$H(s) = \frac{2 + 3s^1 + 4s^2}{1 + 5s^2 + 6s^4 + 7s^6} \quad \Rightarrow$$

$$\Rightarrow \quad 2x(t) + 3\frac{dx(t)}{dt} + 4\frac{dx^2(t)}{dt^2} = y(t) + 5\frac{dy^2(t)}{dt^2} + 6\frac{dy^4(t)}{dt^4} + 7\frac{dy^6(t)}{dt^6}.$$

Designing an analog filter relies on choosing values of $\{R,L,C\}$ elements and structure of their connection. After this we obtain a concrete differential equation with coefficients $\{b_m, a_n\}$. They precisely describe the filter and its filter transfer function. After setting:

$$s = j\omega = j2\pi f \tag{7.7}$$

the Laplace transform changes to continuous Fourier transform:

$$F(x(t)) = X(\omega) = \int_{-\infty}^{+\infty} x(t)e^{-j\omega t}\,dt \tag{7.8}$$

and the system TF to the system frequency response (FR):

$$H(\omega) = \frac{Y(\omega)}{X(\omega)} = \frac{b_0 + b_1(j\omega)^1 + b_2(j\omega)^2 + \ldots + b_M(j\omega)^M}{1 + a_1(j\omega)^1 + a_2(j\omega)^2 + \ldots + a_N(j\omega)^N}. \tag{7.9}$$

The complex number $H(\omega)$, $\omega = 2\pi f$:

$$H(\omega) = |H(\omega)| \cdot e^{j\angle H(\omega)} \tag{7.10}$$

is telling us what the system will do with a given frequency f. Its magnitude $|H(\omega)|$:

$$M(\omega) = |H(\omega)| = \sqrt{\mathrm{Re}\,(H(\omega))^2 + \mathrm{Im}\,(H(\omega))^2} \tag{7.11}$$

gives us information how the system will amplify/attenuate the signal with angular frequency $\omega = 2\pi f$, while its angle $\angle H(\omega)$:

$$\Phi(\omega) = \angle H(\omega) = \tan^{-1}\left(\frac{\mathrm{Im}(H(\omega))}{\mathrm{Re}(H(\omega))}\right) \tag{7.12}$$

gives information about the signal phase shift, i.e. its time delay on the system output.

Analog or digital frequency filter design relies on constructing a circuit or program which pass from its input to its output only selected band of frequencies. The filters, as frequency selectors, are classified into: low-pass (only low frequencies are passed), high-pass (high frequencies), band-pass (mid-range frequencies from-to), and band-stop (all frequencies except a specified band). In Fig. 7.1 all these filters types are shown as well as desirable (linear) and undesirable (non-linear) filter phase response. When filter phase response is a linear function of frequency in the pass-band, all frequencies which are passed by the filter are delayed by the same amount of time and signal shape in the pass-band is not changed (this issue is discussed below).

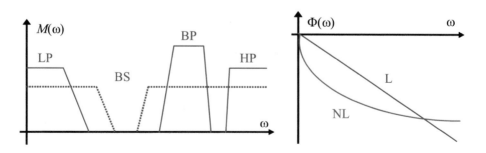

Fig. 7.1: (left)—frequency filter types in respect to frequencies which are passed: Low-Pass (LP), High-Pass (HP), Band-Pass (BP), and Band-Stop (BS), (right)—linear (L), desirable, and non-linear (NL), undesirable, filter phase response [7]

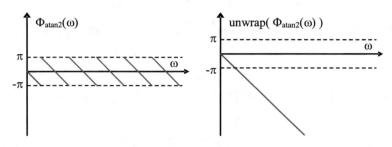

Fig. 7.2: Illustration of phase response calculation problem. Phase (angle of complex number) can be calculated only in the range $[-\pi, \pi)$, in Matlab using functions angle(), atan2()—then the phase unwrapping is necessary (unwrap()) [7]

Angle of a complex number can be computed only in the range $[-\pi, \pi)$. For this reason filter phase response, calculated using Matlab functions angle(), atan2(), is *wrapped* and the unwrap() function must be used—see Fig. 7.2.

Now let us look at analog systems from slightly different perspective. Let us deduce an equation relating output signal $y(t)$ with input signal $x(t)$ for the linear time-invariant (LTI) system. In system theory circuits are characterized by their impulse response, i.e. response to an ideal input impulse, e.g. Dirac delta function, which Fourier spectrum is equal to 1 for all frequencies. Therefore, when one does frequency analysis of the system response to perfect impulse, she/he knows what frequencies the system is passing and how they are deformed by the circuit. Let us now analyze step-by-step relation between an input signal (left, blue) and LTI system output (right, red):

$$\text{INPUT} \rightarrow \text{OUTPUT}$$
$$\delta(t) \rightarrow h(t) \qquad \text{(impulse response definition)}$$
$$\delta(t-\tau) \rightarrow h(t-\tau) \qquad \text{(delayed by } \tau \text{ due to time-invariance)}$$
$$x(\tau)\delta(t-\tau) \rightarrow x(\tau)h(t-\tau) \qquad \text{(delayed and scaled by constant } x(\tau))$$
$$\int_{-\infty}^{+\infty} x(\tau)\delta(t-\tau)d\tau \rightarrow \int_{-\infty}^{+\infty} x(\tau)h(t-\tau)d\tau \qquad \text{(sum of delayed and scaled signals)}$$
$$x(t) \rightarrow y(t) \qquad \text{(generalization of input and output)}$$

In summary:

$$y(t) = H[x(t)] = H\left[\int_{-\infty}^{\infty} x(\tau)\delta(t-\tau)d\tau\right] = \int_{-\infty}^{\infty} x(\tau)H[\delta(t-\tau)]\,d\tau = \int_{-\infty}^{\infty} x(\tau)h(t-\tau)d\tau.$$

$$(7.13)$$

Therefore, the LTI system output $y(t)$ is result of convolving input $x(t)$ with system impulse response $h(t)$:

$$y(t) = \int_{-\infty}^{+\infty} x(\tau)h(t-\tau)d\tau.$$

$$(7.14)$$

Performing continuous Fourier transform (7.8) on both sides of Eq. (7.14) and using setting $\xi = t - \tau$, we obtain the fundamental frequency consequence of signal convolution:

$$Y(\omega) = \int_{-\infty}^{\infty}\left(\int_{-\infty}^{\infty} x(\tau)h(t-\tau)d\tau\right)e^{-j\omega t}dt$$

$$= \int_{-\infty}^{\infty}\left(\int_{-\infty}^{\infty} x(\tau)e^{-j\omega\tau}d\tau\right)h(\xi)e^{-j\omega\xi}\,(d\xi + d\tau) = \dots$$

$$= \left(\int_{-\infty}^{\infty} x(\tau)e^{-j\omega\tau}d\tau\right)\left(\int_{-\infty}^{\infty} h(\xi)e^{-j\omega\xi}d\xi\right) = X(\omega)\cdot H(\omega).$$

$$(7.15)$$

In summary, Fourier transform $Y(\omega)$ of the LTI signal output is equal to multiplication of Fourier transform $X(\omega)$ of the input signal and system frequency response $H(\omega)$ (Fourier transform of the system impulse response):

$$Y(\omega) = X(\omega)\cdot H(\omega).$$

$$(7.16)$$

If $H(\omega_0) = 0$ for any angular frequency ω_0, this frequency is removed by the system: $Y(\omega_0) = 0$ despite value of $X(\omega_0)$. Similarly, if $H(\omega_0) = 1$ for any angular frequency ω_0, this frequency is passed unchanged by the system since for it $Y(\omega_0) = X(\omega_0)$, despite value of $X(\omega_0)$.

If harmonic signal:

$$x(t) = e^{j\omega_0 t}$$

$$(7.17)$$

is processed by the LTI system described by Eq. (7.14):

$$y(t) = \int_{-\infty}^{\infty} h(\tau)x(t-\tau)d\tau = \int_{-\infty}^{\infty} h(\tau)e^{j\omega_0(t-\tau)}d\tau = \left[\int_{-\infty}^{\infty} h(\tau)e^{-j\omega_0\tau}d\tau\right]e^{j\omega_0 t} = H(\omega_0)e^{j\omega_0 t}$$

$$(7.18)$$

the system output is equal to:

$$y(t) = H(\omega_0) \cdot e^{j\omega_0 t} = \left[M(\omega_0) \cdot e^{j\Phi(\omega_0)} \right] \cdot e^{j\omega_0 t} = M(\omega_0) \cdot e^{j(\omega_0 t + \Phi(\omega_0))}. \qquad (7.19)$$

It is the input signal scaled in amplitude by $M(\omega_0)$ and shifted in phase by $\Phi(\omega_0)$, i.e. shifted in time by $\Delta t = \frac{\Phi(\omega_0)}{\omega_0}$:

$$y(t) = M(\omega_0) \cdot e^{j\omega_0(t+\Delta t)}, \quad \Delta t = \frac{\Phi(\omega_0)}{\omega_0}. \qquad (7.20)$$

Assuming that $T = \frac{1}{f_0}$ is the sinusoid period, we can connect the phase shift $\Phi(\omega_0)$ of any sinusoid with its time delay Δt:

$$\Phi(\omega_0) = \Delta t \cdot \omega_0 = \Delta t \cdot 2\pi f_0 = \frac{\Delta t}{T} \cdot 2\pi. \qquad (7.21)$$

Coming back to our LTI system. The phase response is negative when the system is delaying an input. When it is additionally linear:

$$\Phi(\omega) = -\alpha \cdot \omega, \qquad (7.22)$$

Eq. (7.19) takes the following form:

$$y(t) = M(\omega_0) \cdot A e^{j\omega_0(t-\alpha)}, \qquad (7.23)$$

which is telling us that the input signal is delayed at output by α. This delay does not depend on signal frequency. So, if the system is linear, all his K components with different amplitudes and frequencies will be delayed by the same value α:

$$y(t) = \sum_{k=1}^{K} M(\omega_k) A_k e^{j\omega_k(t-\alpha)} \qquad (7.24)$$

and, for $M(\omega_k) = 1$, $k = 1, 2, \ldots, K$, the whole signal will be only delayed by α on the system output and its shape will not be changed. This signal processing feature is very important in many applications. For example in ECG analysis, because a medical doctor is specially interested in signal shape while investigating heart work anomalies, or in Hi-Fi acoustics where audiophiles do not want to lose space localization of sound sources (our ears localize sound source by triangulation and relative delay of the same sound in both ears).

Conclusion Linear phase response in the pass-band is a desirable feature of an LTI system. It guarantees that the passing signal is delayed only and does not have shape distortion.

Exercise 7.1 (LTI System: Frequency Response, Impulse Response). In program 7.1 a simple analog LTI system is analyzed with a= [1,2], b= [3,4,5] in Matlab:

$$H(s) = \frac{1s^1 + 2}{3s^2 + 4s^1 + 5}.$$

Different filter characteristics are computed and plotted: its frequency and phase response in frequency domain as well as its impulse and step response in time domain. In Fig. 7.3 only system frequency and phase responses are shown. They allow us to verify amplification/attenuation and phase shift of different signal frequency components on the system output. Slightly modify values of system transfer function $\{b,a\}$ coefficients and observe changing shapes of system responses (plot them in the same figure). Finally, concentrate on frequency system response and try to find such values of $\{b,a\}$ for which system is passing from input to output only high-frequency input components, e.g. higher than 1 Hz. Spend only 5 min on this task. You should accept an eventual failure. Nothing is perfect. This LTI system design method for sure is not.

Listing 7.1: Analysis of simple analog LTI system/filter

```
1  % lab07_ex_lti.m
2  clear all; close all;
3
4  b = [1, 2];         % coefficients {b} of the TF nominator   polynomial
5  a = [3, 4, 5];      % coefficients {a} of the TF denominator polynomial
6  f = 0 : 0.01 : 10;  % frequencies of interest
7  t = 0 : 0.01 : 10;  % time of interest
8  f0 = 1;             % radius (w0=2*pi*f0) for circle in s=j*w domain
9  [ H, h ] = AFigs(b,a,f,t,1);  % figures for analog filter
10
11 % ###############################
12 function [H,h] = AFigs(b,a,f,t,f0)
13
14 % Position of zeros and poles
15 z = roots(b)/(2*pi); p = roots(a)/(2*pi);  % scaling for frequency
16 phi = 0:pi/1000:2*pi; si = f0*sin(phi); co = f0*cos(phi);
17 figure; plot(real(z),imag(z),'ro',real(p),imag(p),'b*',co,si,'k-');
18 xlabel('real()'); xlabel('imag()'); title('TF Zeros (o) & Poles (*)'); grid; pause
19
20 % Frequency response
21 w = 2*pi*f; s = j*w;          % angular frequency, Laplace transform variable
22 H = polyval(b,s)./polyval(a,s);  % frequency response H(f) = H(s=j*2*pi*f)
23 % H = freqs(b,a,2*pi*f);       % Matlab function
24 % Figures
25 figure; plot(f,abs(H)); xlabel('f (Hz)'); title(' |H(f)|'); grid; pause
26 figure; plot(f,20*log10(abs(H))); xlabel('f (Hz)'); title(' |H(f)|  (dB)'); grid; pause
27 figure; semilogx(f,20*log10(abs(H))); xlabel('f (Hz)'); title(' |H(f)|  (dB)'); grid;
         pause
28 figure; plot(f,angle(H)); xlabel('f (Hz)'); title('\angle H(f)  (rd)'); grid; pause
```

```
29
30   % Impulse response - functions impulse() and step() from Control Toolbox
31   sys=tf(b,a);
32   figure
33   subplot(211); impulse(sys),       % plot impulse response h(t)
34   subplot(212); step(sys), pause    % plot step response u(t); test also:
35   h = impulse(sys,t);               % calculate impulse response h(t)
36   figure; plot(t,h); grid; xlabel('t (s)'); title('Impulse response h(t)'); pause
37   u = step(sys,t);                  % calculate step response u(t)
38   figure; plot(t,u); grid; xlabel('t (s)'); title('Step response u(t)'); pause
39
40   end
```

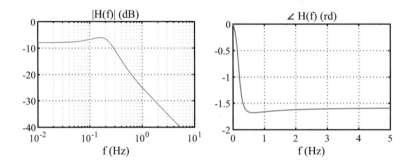

Fig. 7.3: Frequency response (left) and phase response (right) of an LTI system analyzed in Exercise 7.1

7.3 RLC Circuit Example

Let us analyze the RLC circuit—sequential connection of resistance R, inductance L, and capacitance C—presented in Fig. 7.4. The input voltage $u(t)$ causes flow of a current $i(t)$, which generates voltages $u_R(t)$, $u_L(t)$, and $u_C(t)$ upon elements R, L, C that are proportional: to the current for R, to the current derivative for L, and to the bounded integral from 0 to t for C. From Kirchhoff voltage law, the sum of this voltages is equal to the input voltage $u_{in}(t)$ while the output voltage $u_{out}(t)$ is equal to the capacitor voltage $u_C(t)$ only:

$$u_{in}(t) = u_R(t) + u_L(t) + u_C(t) = R \cdot i(t) + L\frac{di(t)}{dt} + \frac{1}{C}\int_0^t i(t)dt \qquad (7.25)$$

$$u_{out}(t) = u_C(t) = \frac{1}{C}\int_0^t i(t)dt. \qquad (7.26)$$

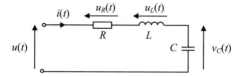

Fig. 7.4: Sequential connection of R, L, C elements, the RLC circuit [7]

After performing the Laplace transform (7.3) of both sides of Eqs. (7.25), (7.26) and after using Laplace transform features (7.4):

$$U_{in}(s) = R \cdot I(s) + sL \cdot I(s) + \frac{1}{Cs} \cdot I(s) = \left[R + sL + \frac{1}{Cs} \right] \cdot I(s) \qquad (7.27)$$

$$U_{out}(s) = U_C(s) = \left[\frac{1}{Cs} \right] \cdot I(s) \qquad (7.28)$$

one obtains the transfer function of *RLC* circuits:

$$H(s) = \frac{U_{out}(s)}{U_{in}(s)} = \frac{\frac{1}{Cs}}{R + Ls + \frac{1}{Cs}} = \frac{1}{1 + RC \cdot s + LC \cdot s^2} \qquad (7.29)$$

and its frequency response after using equality (7.7):

$$H(\omega) = \frac{U_{out}(\omega)}{U_{in}(\omega)} = \frac{1}{1 + RC \cdot (j\omega) + LC \cdot (j\omega)^2} = \frac{\frac{1}{LC}}{\frac{1}{LC} + \frac{R}{L}(j\omega) + (j\omega)^2}. \qquad (7.30)$$

Eq. (7.30) tells us what the circuit is doing with input signal component with angular frequency ω. We can vary value of ω in (7.30), calculate $H(\omega)$, and plot it.

After introduction of the following new variables: ω_0—circuit resonance angular frequency of un-damped oscillations and ξ—circuit damping, as well as A, d, ω_1—amplitude, damping and angular frequency of circuit impulse response, having a form of damped sinusoid:

$$\omega_0 = 1/\sqrt{LC}, \quad \xi = (R/L)/(2\omega_0), \quad A = \frac{\omega_0}{\sqrt{1 - \xi^2}}, \quad d = \xi \omega_0, \quad \omega_1 = \omega_0 \sqrt{1 - \xi^2}$$
$$(7.31)$$

we get from Eq. (7.31):

$$H(\omega) = \frac{\omega_0^2}{\omega_0^2 + 2\xi \omega_0 (j\omega) + (j\omega)^2} = \frac{A\omega_1}{(d + j\omega)^2 + \omega_1^2}. \qquad (7.32)$$

The following signal has the continuous Fourier transform (7.8) equal to (7.32):

$$h(t) = \begin{cases} Ae^{-d \cdot t} \sin \omega_1 t & \text{for } t \geq 0, \\ 0 & \text{for } t < 0, \end{cases} \qquad (7.33)$$

therefore (7.33) represents the impulse response of the *RLC* circuit. In Fig. 7.5 there are presented frequency and time characteristics for the RLC circuit from Fig. 7.4, having transfer function (7.32), un-damped resonance frequency $\omega_0 = 1$, and damping $\xi = 0.3$.

What the *RLC* filter will do with arbitrary one frequency component, for example, $f_0 = 10$ Hz one?

Input: $x(t) = \sin(2\pi \cdot f_0 \cdot t) = \sin(2\pi \cdot 10 \cdot t)$

Output: $y(t) = |H(f_0)| \cdot \sin(2\pi \cdot 10 \cdot t + \angle(H(f_0))) = A \cdot \sin(2\pi \cdot 10 \cdot t + \varphi)$

The output signal amplitude is $A = |H(f_0)|$ and the new phase is $\varphi = \angle(H(f_0))$.

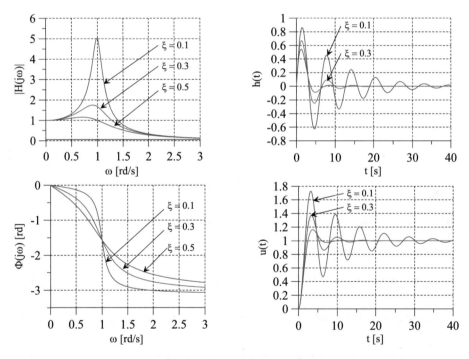

Fig. 7.5: Frequency and time characteristics of *RLC* circuit from Fig. 7.5, with transfer function (7.32), having un-damped resonance frequency $\omega_0 = 1$ and damping $\xi = 0.1, 0.3, 0.5$. From left to right, up to down: magnitude response (amplification/attenuation), response for Dirac impulse function, phase response (delay), response for unitary step excitation [7]

Exercise 7.2 (RLC Circuit). In program 7.2 there are computed and visualized different characteristics of an exemplary RLC circuit. Values of its elements are chosen arbitrarily. Run the program and observe figures. Calculate the angular frequency ω_0 and ω_1 as well as circuit damping ξ and signal impulse response damping d. Change *RLC* values in order to obtain lower and higher circuit damping ξ. Plot figures similar to Fig. 7.5. Finally, try to change $\{R, L, C\}$ values and to obtain a low-pass filter with cut-off angular frequency $\omega_{3dB} = 1$ rad/s.

Listing 7.2: Analysis of analog RLC circuit

```
1   % lab07_ex_rlc.m
2   clear all; close all;
3
4   R = 10;                      % resistance in ohms
5   L = 2*10^(-3);               % inductance in henrys
6   C = 5*10^(-6);               % capacitance in farads
7
8   w0  = 1/sqrt(L*C);      f0 = w0/(2*pi),      % undumped resonance frequency
9   ksi = (R/L)/(2*w0),                          % should be smaller than 1
10  w1  = w0*sqrt(1-ksi^2); f1 = w1/(2*pi), pause % damped resonance frequency
11
12  b = [ 1 ];                   % coeffs of nominator polynomial
13  a = [ L*C, R*C, 1 ];         % coeffs of denominator poly (from the highest order)
14  %z = roots(b), p = roots(a), gain = b(1)/a(1),    % coeffs --> roots, gain
15  %[z,p,gain] = tf2zp(b,a),    % the same in one Matlab function
16
17  f=0:1:10000; t=0:0.000001:2.5e-3; f0=0;
18  [ H, h ] = AFigs(b,a,f,t,f0); % figures for analog filter
```

7.4 Analog Filter Design by Zeros and Poles Method

In analog filter design we would like to obtain filter frequency response characterized by

- good *flatness* in the filter pass-band, i.e. gain close 1,
- good *sharpness* of the filter transition from pass-band to stop-band,
- very high attenuation in the stop-band, i.e. gain ≈ 0.

However, it is difficult to propose "good/compact" equations for choosing such values of polynomial coefficients $\{b_m, a_n\}$, which ensure required frequency features of the filter. It is much easier to design system with a required $H(f)$, designing roots $\{z_m, p_n\}$ of the transfer function (TF) polynomials:

$$H(s) = \frac{Y(s)}{X(s)} = \frac{b_M \cdot (s-z_1)(s-z_2) \cdot \ldots \cdot (s-z_M)}{a_N \cdot (s-p_1)(s-p_2) \cdot \ldots \cdot (s-p_N)}. \tag{7.34}$$

Why? Because system transfer function $H(s)$, given by Eq. (7.34), becomes the system frequency response (system frequency characteristics) after setting (7.7) $s = j\omega = j2\pi f$:

$$H(\omega) = \frac{Y(\omega)}{X(\omega)} = \frac{b_M \cdot (j\omega - z_1)(j\omega - z_2) \cdot \ldots \cdot (j\omega - z_M)}{a_N \cdot (j\omega - p_1)(j\omega - p_2) \cdot \ldots \cdot (j\omega - p_N)}. \tag{7.35}$$

The analog filter design by TF zeros-poles placement is illustrated in Fig. 7.6. For given value of ω we have in Eq. (7.35) many complex-value vectors:

$$j\omega - z_m = B_m e^{j\theta_m}, \quad j\omega - p_n = A_n e^{j\phi_n}, \tag{7.36}$$

and we can express magnitude and phase of the filter frequency response Eq. (7.35) as follows:

$$H(j\omega) = M(\omega)e^{j\Phi(\omega)} = \frac{b_M \prod\limits_{m=1}^{M} B_m e^{j\theta_m}}{a_N \prod\limits_{n=1}^{N} A_n e^{j\phi_n}} = \frac{B(j\omega)}{A(j\omega)}. \tag{7.37}$$

From Eq. (7.37) the frequency response magnitude and phase are equal to:

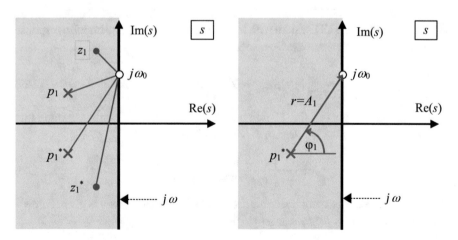

Fig. 7.6: Illustration of analog filter frequency response $H(\omega)$ calculation in "s" plane: (left) magnitude response, (right) phase response. The filter transfer function $H(s)$ has zeros marked with blue dots (•) and poles marked with red crosses (×) [7]

$$M(\omega) = \frac{b_M \prod\limits_{m=1}^{M} B_m}{a_N \prod\limits_{n=1}^{N} A_n}, \qquad \Phi(\omega) = \sum_{m=1}^{M} \theta_m - \sum_{n=1}^{N} \phi_n. \qquad (7.38)$$

Therefore, the filter amplification/attenuation for a given frequency is proportional to the ratio of two products of complex vector magnitudes: product of distances of $j\omega$ to all TF zeros z_m divided by product of distances of $j\omega$ to all TF poles p_n. In turn, filter phase shift for a given frequency is difference of two sums of complex vector angles: sum of angles of TF zeros minus sum of angles of TF poles.

When frequency is changing, the $j\omega = j2\pi f$ is *moving* along the imaginary axis in the complex-value plane. Therefore, it is easy to put *on its way* zero $z_k = j2\pi f_k$ or pole $p_l = j2\pi f_l$ and turn to zero any term in the nominator and denominator of (7.35): $(j\omega - z_k) = 0$ or $(j\omega - p_l) = 0$. This causes complete removal of signal with frequency f_k on the system output and infinite amplification by the system (division by zero) of signal with frequency f_l. Since we want to avoid infinite amplification, we put the TF pole close to imaginary axis at left half-plane: $p_l = -\delta_k + j2\pi f_l$, where δ_k is a small number, and we divide in (7.35) by small number: as a result the TF value is high (amplification!) but not infinite. The analog system stability requires placement of the TF poles in the left half-plane of complex variable s ($-\delta_k$, negative real part!) because only in such situation the system impulse response decays to zero after some time. From the system stability point of view, position of TF zeros is arbitrary.

For Questioning The LTI system with transfer function (7.34) has impulse response of the form:

$$h(t) = \sum_{k=1}^{N} h_k(t), \qquad h_k(t) = \begin{cases} 0 & \text{dla } t < 0, \\ c_k e^{p_k t} & \text{dla } t \geq 0, \end{cases} \qquad (7.39)$$

where p_k denotes roots of the TF denominator polynomial. When we assume that they are complex-value, i.e. $p_k = \sigma_k + j\omega_k$, we obtain

$$h_k(t) = \begin{cases} 0 & \text{dla } t < 0, \\ c_k e^{(\sigma_k + j\omega_k)t} & \text{dla } t \geq 0. \end{cases} \qquad (7.40)$$

Because roots occur in complex-conjugated pairs (explanation why is given below), we have (for $c_k = u_k + jv_k, \beta_k = \text{atan}\left(\frac{v_k}{u_k}\right)$):

$$h_k(t) + h_k^{(*)}(t) = c_k e^{(\sigma_k + j\omega_k)t} + c_k^* e^{(\sigma_k - j\omega_k)t} = 2\sqrt{u_k^2 + v_k^2} \cdot e^{\sigma_k t} \cdot \cos(\omega_k t + \beta_k). \qquad (7.41)$$

Each pair of impulse response components $h_k(t)$ and $h_k^{(*)}(t)$ is decaying only when $\sigma_k < 0$. Therefore TF roots p_k can lie only in left half-plane of Laplace variable s.

Single Frequency Attenuate/Amplify Analog Filter Design If we want to suppress signal with frequency f_1 on the system output, it is sufficient to set in Eqs. (7.34), (7.35):

$$z_1 = j2\pi f_1, \quad z_2 = z_1^* \tag{7.42}$$

since for $j\omega = j2\pi f_1$ the term $(j\omega - z_1) = 0$. If we want to amplify signal with frequency f_2 on the system output, it is sufficient to set:

$$p_1 = -\delta + j2\pi f_2, \quad p_2 = p_1^* \tag{7.43}$$

since for $j\omega = j2\pi f_2$ the term $(j\omega - p_1) = \delta$ and we will be dividing in Eq. (7.35) by a small number δ causing that the TF value for frequency f_2 will be high, the higher the smaller δ is. The very important question is why in Eqs. (7.42), (7.43) $z_2 = z_1^*$ and $p_2 = p_1^*$, so why we need a pair of conjugated zeros and a pair of conjugated poles to remove (f_1) and amplify (f_2), single frequency components? Because practical system realization requires polynomial with real-values coefficients $\{b_m, a_n\}$ and this is obtained when zeros and poles occur in conjugated pairs:

$$(s - (c + jd))(s - (c - jd)) = s^2 - s(c - jd) - s(c + jd) + (c + jd)(c - jd) = \ldots \tag{7.44}$$
$$= 1 \cdot s^2 - 2c \cdot s + (c^2 + d^2). \tag{7.45}$$

We see that roots $c \pm jd$ of the polynomial (7.45) of complex variable s are complex, but polynomial coefficients $[1, -2c, c^2 + d^2]$ are real.

Simple Band-Pass Filter Design Example Having in mind all above recommendations, let us try to propose by TF zeros & poles placement of an analog filter, amplifying signal components having angular frequency close to $\omega_0 = 10$ rd/s and attenuating the remaining ones. To achieve this goal, we can place three TF poles close to ω_0 in the left half-plane of complex variable s and one TF zero left and right to them:

$$p_{1,2} = -0.5 \pm j9.5; \quad p_{3,4} = -1 \pm j10; \quad p_{5,6} = -0.5 \pm j10.5$$
$$z_{1,2} = \pm j5, \quad z_{3,4} = \pm j15$$

and obtain the following transfer function:

$$H(s) = \frac{(s - j5)(s + j5)}{(s + 0.5 - j9.5)(s + 0.5 + j9.5)(s + 1 - j10)(s + 1 + j10)} \cdots$$
$$\cdots \frac{(s - j15)(s + j15)}{(s + 0.5 - j10.5)(s + 0.5 + j10.5)}.$$

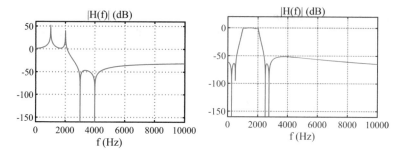

Fig. 7.7: Frequency responses of two filters designed in Exercise 7.3: (left) the amplifier (1000 and 2000 Hz) and attenuator (3000 and 4000 Hz) of selected frequencies, and (right) band-pass filter in the frequency range from 1000 Hz to 2000 Hz [7]

Exercise 7.3 (Design of Analog Filters by TF Zeros & Poles Placement). In program 7.3 three different analog filters are designed using the ZP method. The first of them is simple analog amplifier and attenuator of selected frequencies. The second and the third have ambitions to be a low-pass and band-pass analog filters. Obtained frequency responses of the first and second filter are shown in Fig. 7.7. Run the program. Do recommended exercises. Finally, try to design a good low-pass filter for cut-off angular frequency $\omega_{3dB} = 1$ rad/s.

Listing 7.3: Design of analog filters by appropriate placement of transfer function zeros and poles in complex s-plane

```
1   % lab07_ex_zp_design.m
2   clear all; close all;
3
4   task = 1;    % 1=Remove/Amplify, 2=LP, 3=BP
5   if(task==1) % Simple ZP design: remove 3000, 4000 Hz, amplify 1000, 2000 Hz
6       z = j*2*pi*[ 3000 4000 ];                z = [ z conj(z) ];
7       p = [ -10 -10 ] + j*2*pi*[ 1000 2000 ]; p = [ p conj(p) ];
8       b = poly(z);   a = poly(p);
9       b = 100*b / abs(polyval( b,j*2*pi*2000) / polyval( a,j*2*pi*2000) );
10  end
11  % Design filter rejecting 1500 Hz and amplifying 3000 Hz.
12
13  %###############################
14  if(task==2) % ZP design of LowPass filter [ 0 - 1000 Hz ]
15      z = j*2*pi*[ 2000 3000  ];                z = [ z conj(z) ];
16      p = [ -2000 -2000 -400 ] + j*2*pi*[ 400 700 1000 ];   p = [ p conj(p) ];
17      b = poly(z);   a = poly(p);
18      b = b / abs(polyval( b,0) / polyval( a,0) );
```

```
19    end
20    % Using ZP method design a good analog HighPass filter for [ 2000 ... ] Hz.
21
22    %##################################
23    if(task==3) % ZP design of Band-Pass [ 1000 Hz - 2000 Hz ]
24        z = j*2*pi*[ 250 500  2500 2750  ];
25        z = [ z conj(z) ];
26        p = [ -450 -2000 -3500 -2000 -400 ] + j*2*pi*[ 1000 1250 1500 1750 2000 ];
27        p = [ p conj(p) ];
28        b = poly(z);  a = poly(p);
29        b = b / abs(polyval( b,j*2*pi*1500) / polyval( a,j*2*pi*1500) );
30    end
31
32    % Figures
33    f = 0 : 1 : 10000;          % frequencies of interest
34    t = 0 : 1e-5 : 50e-3;       % time of interest
35    f0 = 1000;                  % radius (w0=2*pi*f0) for circle in s=j*w domain
36    [ H, h ] = AFigs(b,a,f,t,1); % figures for analog filter
```

Conclusions Proper placement of roots "z_k" of the nominator polynomial (ze-ros of transfer function) is used for frequency attenuation. Their position is not limited. Proper placement of roots "p_k" of the denominator polynomial (poles of transfer function) is used for frequency amplification. Position of TF poles is allowed only in left half-plane because then the system impulse response de-cays to zero. Manual "zeros & poles placement" design method of the $H(s)$ is simple but very time-consuming.

Uff! We have just finished the introduction.

7.5 Butterworth, Chebyshev, and Elliptic Analog Filters

Standard analog filters are designed to pass only signal components with frequencies (see Fig. 7.8): lower than f_0 (low-pass, LP), higher than f_0 (high-pass, HP), only in the range from frequency f_1 to frequency f_2 (band-pass, BP), or out of this frequency

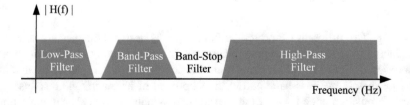

Fig. 7.8: Standard analog filter types: what frequencies will be passed?

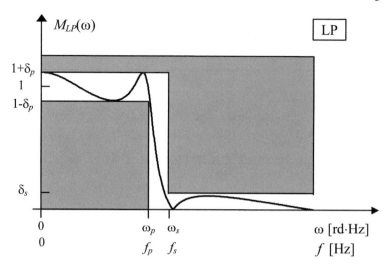

Fig. 7.9: Technical specification of requirements in analog filter design: δ_p, δ_s— allowed level of ripples in the pass-band and in the stop-band, ω_p, ω_s—angular frequencies of the pass-band end stop-band beginning (giving filter transition width) [7]

range (band-stop, BS, in some frequency range). Typically a so-called *normalized low-pass prototype filter* is designed first (always for $\omega_0 = 1$), then it is transformed to any other filter type using frequency transformations. Using one more transformation, e.g. the bilinear one, an analog filter can be transformed into a digital filter.

Typical specifications used in analog filter design are presented in Fig. 7.9. The magnitude frequency response of the filter should fits into the *tunnel* of requirements, concerning allowed oscillations in the pass-band (δ_p) and stop-band (δ_s) as well as width of the transition band ($\omega_s - \omega_p$).

Exist special mathematical rules for appropriate placement of TF zeros & poles in low-pass ($\omega_0 = 1$) prototype filters. **We can choose the following prototype filters:**

- Butterworth—only poles (on circle), no oscillations in $|H(f)|$, not sharp,
- Chebyshev type 1—only poles (on ellipse), oscillations in pass-band, sharper,
- Chebyshev type 2—poles & zeros, oscillations in stop-band, sharper,
- Cauer–Elliptic—poles & zeros, oscillations in pass-band and stop-band, very sharp.

In Fig. 7.10 different LP prototype filter ($\omega_0 = 1$) design strategies are shown in consideration of TF zeros and poles placement. In turn, LP prototype filter frequency responses, designed by different methods, are shown in Fig. 7.11. More oscillations the filter has in the pass-band and in the stop-band of its magnitude response, the sharper transition edge it has. Butterworth (no oscillations, non-sharp) and elliptic (oscillations everywhere, very sharp) filters are mutual reverses.

The Analog Filter Design Procedure Consists of the Following Steps

1. filter specification: first, we choose a Low-Pass filter prototype (Butterworth, Chebyshev 1/2 or Cauer—elliptic), its parameters (e.g. number N of TF poles and oscillation levels), and target frequency characteristics (Low-Pass, High-Pass, Band-Pass, Band-Stop),
2. design of prototype filter zeros & poles: then, we design $\{z_m, p_n\}$ of a Low-Pass prototype filter having $\omega_0 = 1$,
3. frequency transformation of prototype zeros & poles: next, $\{z_m, p_n\}$ of the LP prototype filter are transformed into zeros and poles of the target filter: LP (with different ω_0), HP, BP or BP,
4. target TF coefficients calculation: then coefficients $\{b_m, a_n\}$ of target TF polynomials are calculated using $\{z_m, p_n\}$,
5. RLC values calculation: finally, knowing $\{b_m, a_n\}$ we calculate values of RLC from set of equations.

Only low-pass Butterworth, Chebyshev, and Cauer/elliptic filters are designed in the method described above. They are called **prototype filters** and have features summarized in Table 7.1. High-pass, band-pass, and band-stop filters are obtained via frequency transformation of the prototypes. Designed

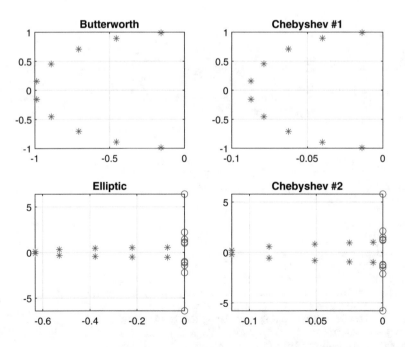

Fig. 7.10: Comparison of zeros and poles placement for different low-pass prototype filters ($\omega_0 = 1$), from left-to-right, from top-to-bottom: Butterworth (only poles on circle), Chebyshev type 1 (only poles on ellipse), elliptic (Cauer) (zeros and poles), and Chebyshev type 2 (zeros and poles)

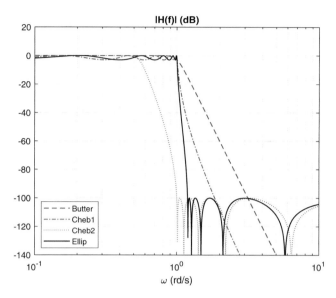

Fig. 7.11: Comparison of frequency responses of different low-pass prototypes filters ($\omega_0 = 1$) having $N = 10$ poles (their zeros and poles) placement is shown in Fig. 7.10

filters inherit prototype features—quantity of oscillations and sharpness—what is shown in Fig. 7.12 presenting results of band-pass filter design. Increasing number of prototype TF poles results in sharper transition edges of the target filter. The most useful Matlab functions for analog filter design are given in Listing 7.4.

Listing 7.4: Matlab functions useful for analog filter design

```
1  % lab07_ex_functions.m - program to be finished by Reader
2  clear all; close all;
3
4  % Polynomial roots to polynomial coefficients
5  b = poly(z);  a = poly(p);
6  % Polynomial coefficients to polynomial roots
7  z = roots(b); p = roots(a);
8  % Required order of transfer function (N = number of poles)
9  [N,Wn] = buttord(.,'s'), cheby1ord(.,'s'), cheby2ord(.,'s'), ellipord(.,'s'),
10 % Analog prototype (Rp - ripples in passband, Rs - ripples in stopband)
11 [b,a] = buttap(N), cheby1ap(N,Rp), cheby2ap(N,Rs), ellipap(N,Rp,Rs),
12 % Frequency transformation (required W0, Wn=[W1,W2], Rp, Rs)
13 [bt,at] = lp2lp(b,a,.), lp2hp(b,a,.), lp2bp(b,a,.), lp2bs(b,a,.),
14 % All-in-one analog filter design (required N, W0, Wn=[W1,W2], Rp, Rs)
15 [b,a] = butter(N,.,'s'), cheby1(N,.,'s'), cheby2(N,.,'s'), ellip(N,.,'s');
```

```
16   % Frequency response - magnitude, phase as a function of frequency
17   f=?; H = freqs(b,a,f)
18   f=?; H = polyval(b,j*2*pi*f)./polyval(a,j*2*pi*f);
19   figure; plot(f,20*log10(abs(H))); figure; plot(f,unwrap(angle(H)));
20   % Impulse response, step response - functions from Control Toolbox
21   figure; impulse(b,a);
22   figure; step(b,a);
```

Exercise 7.4 (My First Professional Analog Filter). Use functions from the Listing 7.4 as well as previous programs of this chapter and design transfer function of a band-stop or band-pass filter in the range $[10, ..., 20]$ kHz with allowed 3 dB ripples in the pass-band and -100 dB ripples in the stop-band (Fig. 7.12).

Table 7.1: Analog filters comparison

Name	Types	Oscillations pass-band	Oscillations stop-band	Edges	Non-linear phase
Butterworth	LP, HP, BP, BS	No	No	Flat	Very small
Chebyshev 1	LP, HP, BP, BS	Yes	No	Sharp	Small
Chebyshev 2	LP, HP, BP, BS	No	Yes	Sharp	Small
Cauer (elliptic)	LP, HP, BP, BS	Yes	Yes	Very sharp	Very big
Bessel	LP	No	No	Very flat	No

7.6 Frequency Transformation

In analog filter design, first, normalized ($\omega_0 = 1$) low-pass prototype filter is designed (of Butterworth, Chebyshev, Cauer, ...) and, then, it is transformed into target filter of any frequency characteristics (low-pass, high-pass, band-pass, and band-stop). In transfer function $H(s)$ of the prototype low-pass normalized ($\omega_0 = 1$) filter, variable s is substituted with a special function of s' ($s = g_{xx}(s')$) and polynomials of s are changed to polynomial of s':

$$H_{LP}^{(\omega_0=1)}(s) \underset{s = g_{xx}(s')}{\longrightarrow} H_{xx}(s'), \quad s = j\omega, \quad s' = jv, \quad xx = LP, HP, BP, BS. \quad (7.46)$$

In consequence $H(s)$ is transformed to $H(s')$ of a different filter, LP, HP, BP, or BS. The frequency filter transformation idea is explained in Fig. 7.13 while the mapping functions $s = g_{xx}(s')$ are given in Table 7.2.

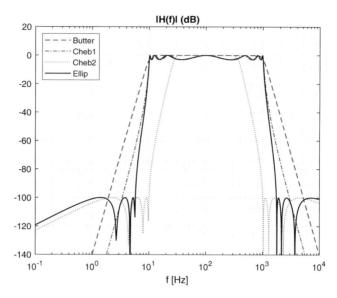

Fig. 7.12: Comparison of different analog filter designs on the example of a band-pass filter $[10, 1000]$ Hz. $N = 7$. The elliptic filter offers the fastest transition from pass-band to the stop-band

In practical realization, each single zero and pole of the prototype TF are mapped into possible multiple zeros and poles of the target filter TF. For example, let us do mapping of one TF zero in case of $LP \to BS$ transformation:

$$(s - z_m) \quad \to \quad \left(\frac{\Delta \omega \cdot (s')}{(s')^2 + \omega_0^2} - z_m \right) \quad \to \quad (-z_m) \frac{(s')^2 - \frac{\Delta \omega}{z_m}(s') + \omega_0^2}{(s')^2 + \omega_0^2}. \quad (7.47)$$

Doing transformation of the whole TF, all M zeros and N poles have to be mapped. In the discussed case of $LP \to BS$ mapping we have

$$H_{BS}(s') = \frac{b_M}{a_N} \cdot \frac{\displaystyle\prod_{m=1}^{M} (-z_m)}{\displaystyle\prod_{n=1}^{N} (-p_n)} \cdot \frac{\left[(s')^2 + \omega_0^2 \right]^{N-M} \displaystyle\prod_{m=1}^{M} \left[(s')^2 - (z_m^{-1} \Delta \omega)s' + \omega_0^2 \right]}{\displaystyle\prod_{n=1}^{N} \left[(s')^2 - (p_n^{-1} \Delta \omega)s' + \omega_0^2 \right]}. \quad (7.48)$$

Ufff! Take a breath! Do not worry! The program implementing this, shown in Listing 7.5, is not so difficult. Code of other transformations will be presented in the last listing of this chapter.

Having functions specifying relations between $s = j\omega$ and $s' = j\nu$ for different frequency mapping, we can find relations between ω and ν. Plotting them gives us

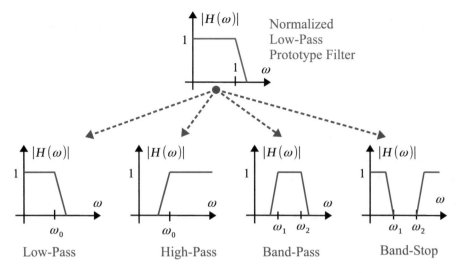

Fig. 7.13: Explanation of frequency transformation of analog filters: normalized low-pass prototype filter (B, C1, C2 or E) with cut-off angular frequency 1 is transformed to low-pass, high-pass, band-pass, or band-stop filters with arbitrary angular frequencies ω_1 and ω_2

Table 7.2: Frequency transformations of analog low-pass filter prototypes: $H_{LP}^{\omega_0=1}(s) \rightarrow H_{LP,HP,BP,BS}(s')$

No	Type	Function $s = g_{xx}(s')$	Transformation result of $(s - z_m)$
1	$LP \rightarrow LP$	$s = s'/\omega_0$ [a]	$\dfrac{s' - z_m \omega_0}{\omega_0}$
2	$LP \rightarrow HP$	$s = \omega_0/s'$ [a]	$(-z_m)\dfrac{(s') - \frac{\omega_0}{z_m}}{(s')}$
3	$LP \rightarrow BP$	$s = \dfrac{s'^2 + \omega_0^2}{\Delta\omega \cdot s'}$ [b]	$\dfrac{(s')^2 - z_m \Delta\omega \cdot (s') + \omega_0^2}{\Delta\omega \cdot (s')}$
4	$LP \rightarrow BP$	$s = \dfrac{\Delta\omega \cdot s'}{s'^2 + \omega_0^2}$ [b]	$(-z_m)\dfrac{(s')^2 - \frac{\Delta\omega}{z_m}(s') + \omega_0^2}{(s')^2 + \omega_0^2}$

[a]—$\omega_0 = \omega_{pass}$

[b]—$\omega_0 = \sqrt{\omega_{pass1}\,\omega_{pass2}}$, $\Delta\omega = \omega_{pass2} - \omega_{pass1}$

a very good graphical illustration how frequency response $H(\omega)$ of the prototype low-pass filter is transformed into frequency response $H(v)$ of filter with different type. Look at Fig. 7.14.

Listing 7.5: Matlab code of two frequency transformation functions

```
1   %--------------------------------------------
2   function [zz,pp,gain] = lp2hpTZ(z,p,gain,w0)
3   % LowPass to HighPass TZ
4
5   zz = []; pp = [];
6   for  k=1:length(z) % transformation of zeros
7       zz = [ zz w0/z(k) ];
8       gain = gain*(-z(k));
9   end
10  for  k=1:length(p) % transformation of poles
11      pp = [ pp w0/p(k) ];
12      gain = gain/(-p(k));
13  end
14  for k=1:(length(p)-length(z))
15      zz = [ zz 0 ];
16  end
17
18  %--------------------------------------------
19  function [zz,pp,gain] = lp2bsTZ(z,p,gain,w0,dw)
20  % LowPass to BandStop TZ
21
22  zz = []; pp = [];
23  for  k=1:length(z) % transformation of zeros
24      zz = [ zz roots([ 1 -dw/z(k) w0^2 ])' ];
25      gain = gain*(-z(k));
26  end
27  for  k=1:length(p) % transformation of poles
28      pp = [ pp roots([ 1 -dw/p(k) w0^2 ])' ];
29      gain = gain/(-p(k));
30  end
31  for k=1:(length(p)-length(z))
32      zz = [ zz roots([ 1 0 w0^2 ])' ];
33  end
34  %--------------------------------------------
```

Exercise 7.5 (Frequency Transformations in Analog Filter Design). Analytically check correctness of equations written in last column of Table 7.2. Next, check whether the Matlab code of $LP \to HP$ and $LP \to BS$ frequency transformations, presented in Listing 7.5, gives the same results as Matlab functions lp2hp() and lp2bs(). Finally, write code for $LP \to LP$ and $LP \to BP$ transformations.

Exercise 7.6 (Frequency Transformation of RLC Circuit Transfer Function). Choose such $\{R,L,C\}$ values in RLC circuit that it became a low-pass filter with cut-off angular frequency $\omega_{3dB} = 1$ rad/s (or design such low-pass $\omega_{3dB} = 1$ rad/s filter using the TF zeros-poles placement method). Then transform the filter TF into high-pass,

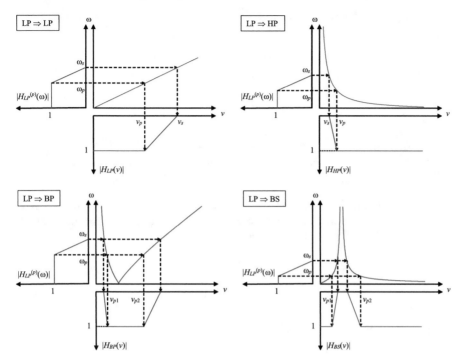

Fig. 7.14: Graphical explanation of frequency transformations of analog filter prototypes, in consecutive rows: $LP \rightarrow LP$, $LP \rightarrow HP$, $LP \rightarrow BP$, $LP \rightarrow BS$ [7]

band-pass, and band-stop analog filters. Cut-off frequencies are arbitrary, e.g. 100, 200, 500, 1000, 2000, 5000, 10000 Hz.

7.7 Butterworth Filter Design Example

In this chapter a very short example of low-pass and high-pass Butterworth filter design is presented. The low-pass filter has only N poles on the circle with radius ω_{3dB} and constant in the nominator (product of negated poles):

$$H_{B,N}^{(LP)}(s) = \prod_{k=1}^{N} (-p_k) / \prod_{k=1}^{N} (s - p_k). \tag{7.49}$$

The high-pass Butterworth filter has the same N poles on the circle and additionally N-multiple zero in 0, i.e. $(s-0)^N$:

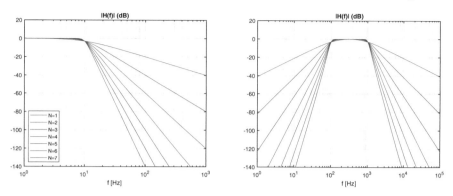

Fig. 7.15: Graphical illustration of the fact that Butterworth filter having $N = 1, 2, 3, ..., 7$ poles has frequency response decaying in frequency $-20N$ decibels per decade (decade = 10 times increase/decrease of frequency)

$$H_{B,N}^{(HP)}(s) = (s-0)^N / \prod_{k=1}^{N} (s - p_k). \tag{7.50}$$

In both filters poles are designed as:

$$p_k = \omega_{3\mathrm{dB}} e^{j\phi_k} = \omega_{3\mathrm{dB}} \exp\left[j\left(\frac{\pi}{2} + \frac{\pi}{2N} + (k-1)\frac{\pi}{N} \right) \right], \quad k = 1, 2, 3, ..., N. \tag{7.51}$$

One pole in the low-pass Butterworth filter causes decrease of magnitude frequency response -20 dB per decade, i.e. for frequency 10 times higher than the cut-off one. Therefore, for example, a filter having $N = 5$ poles offers falling of frequency response -100 dB per decade. Figure 7.15 illustrates this Butterworth filter feature.

Butterworth designed a low-pass filter which TF has only poles (no zeros!) placed on a circle. When two specific Butterworth filters with poles located on two circles (bigger and smaller) are designed and their poles combined, the Chebyshev low-pass filter type 1 is obtained. Again, the filter TF has no zeros but now its poles lie on an ellipse, not on a circle. Details can be found in Listing 7.7. In turn, transfer function of Chebyshev type 2 low-pass filter is modification of Chebyshev type 1 but it has also zeros apart from poles. Similarly placed poles and zeros have Cauer (elliptic) filter. However, in contrary to Chebyshev 2 filter, it allows oscillation in the pass-band, offering thanks to this sharper transition edge than the Chebyshev type 2 filter.

Step-by-step design of the Butterworth low-pass and high-pass filter is presented in Listing 7.6.

Listing 7.6: Analog low-pass and high-pass Butterworth filter design

```
1   % lab07_ex_butterworth.m
2   clear all; close all;
3
4   N=6;                              % number of poles
5   f0 = 100;                         % 3 dB cut-off frequency
6   dfi = pi/N;                       % angle of ''piece of cake''
7   fi = pi/2 + dfi/2 + (0 : N-1)*dfi; % all angles of poles
8   p = w0*exp(j*fi);                 % all poles
9   if(1)                             % LOW-PASS
10      z = [];                       % no zeros
11      gain = real( prod(-p) );      % gain
12  else                             % HIGH-PASS
13      z = zeros(1,N);               % zeros
14      gain = 1;                     % gain
15  end
16  a = poly(p);          % poles to coeffs of polynomial of denominator A(z)
17  b = poly(z);          % poles to coeffs of polynomial of nominator B(z)
18  f = 0:1:1000;         % frequencies of interest
19  w = 2*pi*f;           % angular frequencies
20  s = j*w;              % Laplace transform ''s'' variable
21  H = gain*polyval(b,s) ./ polyval(a,s);            % ratio of two polynomials
22  ang=0:pi/100:2*pi; co=w0*cos(ang); si=w0*sin(ang); % circle for poles
23  figure; plot(real(z),imag(z),'bo',real(p),imag(p),'r*',co,si,'k-'); pause
24  figure; plot(f,20*log10(abs(H))); grid; xlabel('f [Hz]'); title('|H(f)| (dB)'); pause
25  figure; semilogx(f,20*log10(abs(H))); grid; xlabel('f [Hz]'); title('|H(f)| (dB)');
            pause
```

Exercise 7.7 (Do It Yourself: Designing a Band-Pass and Band-Stop Butterworth Filters). Use programs from Listings 7.5 and 7.6, modify them and design yourself a band-pass and band-stop Butterworth filters, for example, for frequency range [10, ..., 20] kHz. Compare your filters with Matlab ones.

7.8 All Together Now: Unified Program for Analog Filter Design

Since appetite is increasing during eating, at the end of this chapter, for readers that are *still hungry*, a unified analog filter design program is given for self-studying. It allows design of Butterworth, Chebyshev, and elliptic filters of four types: LP, HP, BP, and BS. All operations are done by hand using the simplest Matlab functions.

Exercise 7.8 (Alice in Wonderland! Discovering Chebyshev Filter Prototypes). Analyze code of the program 7.7. In frequency transformations some lines are marked as comments, because the transfer function has the same values of zeros and poles and they should be removed—we do not trust Matlab and do it manually. After the loop we are adding zeros or poles which have not been rejected. Run the program for different input options. Observe figures. At the end, concentrate on design of Chebyshev filter prototypes. Replace the program code with dedicated Matlab functions cheby1ap(), cheby2ap(). Compare results.

Listing 7.7: Analog low-pass and high-pass Butterworth filter design

```
1   % lab07_ex_all_in_one.m
2   clear all; close all;
3
4   proto = 3;    % 0=Butt, 1=Cheb1, 2=Cheb2, 3=Ellip  (prototypes)
5   ftype = 4;    % 1=LP,   2=HP,    3=BP,    4=BS      (filter types)
6   N = 10;       % number of poles in the LowPass analog prototype filter
7
8   if(ftype<3)                           % LP, HP
9       f0 = 1000; w0 = 2*pi*f0;          % frequency for LP and HP
10  else                                  % BP, BS
11      f1 = 100;  f2 = 1000;             % frequencies for BP and BS
12      f0 = sqrt(f1*f2); w0 = 2*pi*f0;   %
13      dw = 2*pi*(f2-f1);                %
14  end
15
16  % ###############
17  % ANALOG PROTOTYPE
18  % ###############
19
20  if(proto==0) % Butterworth (NO oscillations), only poles, on circle, left half s-plane
21  % [ z,p,gain ] = buttap(N);
22      w00=1;
23      dfi=pi/N;
24      fi=pi/2+dfi/2+dfi*(0:N-1);
25
26      p=w00*exp(j*fi); % poles on a circle
27      z=[];
28      gain=prod(-p);
29  end
30
31  if(proto==1) % Chebyshev 1 (oscillations in Pass), only poles, on ellipse, left half s-plane
32  % [ z,p,gain ] = cheb1ap(N,Apass); % Apass=Rp ripples in Pass in dB
33      w00=1;
34      dfi=pi/N;
35      fi=pi/2+dfi/2+dfi*(0:N-1);
36
37      Apass = 3;
38      epsi = sqrt(10^(Apass/10)-1);
39      D = asinh(1/epsi)/N;
40      R1 = sinh(D);
41      R2 = cosh(D);
```

```
42      p1 = R1*exp(j*fi);
43      p2 = R2*exp(j*fi);
44      p = real(p1)+j*imag(p2);  % poles on an ellipse
45
46      z = [];
47      if(rem(N,2)==0) % N even
48          gain = 10^(-Apass/20)*prod(-p);
49      else            % N odd
50          gain = prod(-p);
51      end
52   end
53
54   if(proto==2) % Chebyshev 2 (oscillations in Pass and Stop)
55    % [ z,p,gain ] = cheb2ap(N,Astop); % Astop=Rs ripples in Stop
56      w00=1;
57      dfi=pi/N;
58      fi=pi/2+dfi/2+dfi*(0:N-1);
59
60      %Apass = 3;
61      %epsi = sqrt(10^(Apass/10)-1);
62      %D = asinh((1/epsi)/N);
63
64      Astop=120;
65      gamma = 1/sqrt(10^(Astop/10)-1);
66      D = asinh(1/gamma)/N;
67
68      R1 = sinh(D);
69      R2 = cosh(D);
70      p1 = R1*exp(j*fi);
71      p2 = R2*exp(j*fi);
72      p = real(p1)+j*imag(p2);
73      z = j*sin(fi);
74
75      gain = prod(z./p);
76      p = 1./p;
77      z = 1./z;
78   end
79
80   if(proto==3) % Elliptic from Matlab (oscillations in Pass and Stop)
81      [ z,p,gain ] = ellipap(N,3,120);  % Apass, Astop (Rp,Rs)
82   end
83
84   b = poly(z);
85   a = poly(p);
86
87   % CHECKING FREQUENCY RESPONSE OF THE ANALOG PROTOTYPE
88
89   alf = 0:pi/100:2*pi; c=cos(alf); s=sin(alf);
90   figure;
91   plot(real(z),imag(z),'ro',real(p),imag(p),'b*',c,s,'k-'); grid; pause
92
93   w  = 0 : 0.01 : 100;
94   s = j*w;
95   H = gain*polyval(b,s)./polyval(a,s);
```

```
 96    figure; plot(w,abs(H)); grid; xlabel('w (rd*Hz)'); title('|H(w)|'); pause
 97    figure; semilogx(w,20*log10(abs(H))); grid; xlabel('w (rd*Hz)');
 98    title('|H(w)| (dB)'); pause
 99    figure; semilogx(w,unwrap(angle(H))); grid; xlabel('w (rd*Hz)');
100    title('\angle H(w) (rad)'); pause
101
102    % ##########################################################
103    % FREQUENCY TRANSFORMATION LP (norm) -> { LP, HP, BP, BS }
104    % ##########################################################
105
106    Nz=length(z);
107    Np=length(p);
108    zt=[];
109    pt=[];
110    gaint=gain;
111
112    if(ftype==1) % LP-->LP
113        for k=1:Np              % from poles
114            pt = [ pt p(k)*w0 ];
115            gaint = gaint*w0;
116        end
117        for k=1:Nz              % from zeros
118            zt = [ zt z(k)*w0 ];
119            gaint = gaint/w0;
120        end
121    end
122    if(ftype==2) % LP->HP
123        for k=1:Np
124            pt = [ pt w0/p(k) ];
125            % zt = [ zt 0 ];            % zeros at s=0
126            gaint = gaint/(-p(k));
127        end
128        for k=1:Nz
129            zt = [ zt w0/z(k) ];
130            % pt = [ pt 0 ];            % poles at s=0
131            gaint = gaint*(-z(k));
132        end
133        zt = [ zt zeros(1,Np-Nz)];     % left zeros at s=0
134    end
135    if(ftype==3) % LP->BP
136        for k=1:Np
137            pt = [ pt roots([1,-p(k)*dw,w0^2])' ];
138            % zt = [ zt 0];
139            gaint = gaint*dw;
140        end
141        for k=1:Nz
142            zt = [ zt roots([1,-z(k)*dw,w0^2])' ];
143            % pt = [ pt 0];
144            gaint = gaint/dw;
145        end
146        zt = [ zt zeros(1,Np-Nz)];
147    end
148    if(ftype==4) % LP->BS
149        for k=1:Np
```

```
150        % zt = [ zt roots([1,0,w0^2])'] ;
151          pt = [ pt roots([1,-dw/p(k),w0^2 ])' ] ;
152          gaint = gaint/(-p(k));
153      end
154      for k=1:Nz
155        % pt = [ pt roots([1,0,w0^2])'] ;
156          zt = [ zt roots([1,-dw/z(k),w0^2 ])' ] ;
157          gaint = gaint*(-z(k));
158      end
159      for k=1:Np-Nz
160          zt = [ zt roots([1,0,w0^2])'] ;
161      end
162    end
163
164    z = zt;
165    p = pt;
166    gain = gaint;
167
168    b = poly(z);
169    a = poly(p);
170
171    % #####################################
172    % FINAL CHECKING OF DESIGNED ANALOG FILTER
173    % #####################################
174
175    if(ftype<3) fR=f0; else fR=f2; end
176    alf = 0:pi/100:2*pi; c=fR*cos(alf); s=fR*sin(alf);
177    figure;
178    z = z /(2*pi); p = p / (2*pi);
179    plot(real(z),imag(z),'ro',real(p),imag(p),'b*',c,s,'k-'); grid; pause
180
181    if(ftype<3) fmax=10*f0; else fmax=10*f2; end
182    f  = 0 : fmax/10000 : fmax;
183    s = j*2*pi*f;
184    H = gain*polyval(b,s)./polyval(a,s);
185    figure;
186    plot(f,abs(H)); grid; xlabel('f (Hz)'); title('|H(f)|'); pause
187    figure;
188    semilogx(f,20*log10(abs(H))); grid; xlabel('f (Hz)'); title('|H(w) (dB)'); pause
189    figure;
190    semilogx(f,unwrap(angle(H))); grid; xlabel('f (Hz)'); title('\angle H(w) (rad)'); pause
```

7.9 Example of Hardware Design of Analog Filters

Designed transfer function of analog filter should be mapped into appropriate hardware, allowing its implementation. In Fig. 7.16 four circuits with operational amplifiers are shown capable of implementing low-pass (left) and high-pass (right) analog-Butterworth filters of the first order (top) and second order (down). They have the following transfer function, respectively:

$$H_{LP}^{(1)}(s) = \frac{\frac{1}{RC}}{s + \frac{1}{RC}}, \qquad H_{HP}^{(1)}(s) = \frac{s}{s + \frac{1}{RC}}, \tag{7.52}$$

$$H_{LP}^{(2)}(s) = \frac{K/R_1 R_2 C_1 C_2}{s^2 + \left[\frac{1}{R_1 C_1} + \frac{1}{R_2 C_1} + \frac{1-K}{R_2 C_2}\right] s + \frac{1}{R_1 R_2 C_1 C_2}}, \tag{7.53}$$

$$H_{HP}^{(2)}(s) = \frac{Ks^2}{s^2 + \left[\frac{1}{R_2 C_2} + \frac{1}{R_2 C_1} + \frac{1-K}{R_1 C_1}\right] s + \frac{1}{R_1 R_2 C_1 C_2}}, \tag{7.54}$$

where $k = \frac{R_A + R_B}{R_A} = 1 + \frac{R_B}{R_A}$. Having a concrete transfer function equation, for example (7.52), (7.53), or (7.54), and values of its coefficients $\{b_k\}, \{a_k\}$, designed in Matlab, one should solve set of algebraic equations, like $b_1 = \text{function}(R_1, R_2, ..., C_1, C_2, ...)$, and find values of resistors and capacitors. For example, for $R_1 = R_2 = R$ and $C_1 = C_2 = C$ both second-order transfer functions in Eqs. (7.53), (7.54) have the same polynomial in TF denominator. Using the following qualities:

$$a_2 = 1, \qquad a_1 = \frac{3 - K}{RC}, \qquad a_0 = \frac{1}{R^2 C^2} \tag{7.55}$$

values $R, K = 1 + R_B/R_A$ can be calculated when C is arbitrarily chosen:

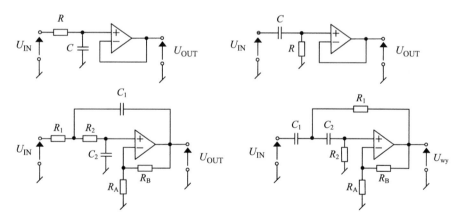

Fig. 7.16: Examples of low-pass (left) and high-pass (right) filter circuits, using operational amplifiers and passive RC elements. Their transfer functions are defined by Eqs. (7.52), (7.53), or (7.54). Up—first-order sections, down—so-called second-order bi-quadratic sections (they implement transfer functions being a ratio of two second-order polynomials of s). Higher order filters are obtained by cascading filters of the first and second order. Output impedance of each section is very low and allows their cascading. In first-order sections operational amplifier works as simple voltage repeater [7]

$$R = \frac{1}{C\sqrt{a_0}}, \qquad K = 3 - \frac{a_1}{\sqrt{a_0}}. \qquad (7.56)$$

When a transfer function is of higher order, i.e. has polynomials of higher order, it should be represented as a result of multiplication of simpler TFs with polynomials of the second order (so-called bi-quadratic sections using one operational amplifier only) and eventually one TF of the first order:

$$H(s) = \frac{b_{1,0}(s - z_0)}{a_{1,0}(s - p_0)} \cdot \frac{b_{2,1}(s - z_1)(s - z_1^*)}{a_{2,1}(s - p_1)(s - p_1^*)} \cdot \ldots \cdot \frac{b_{2,M}(s - z_M)(s - z_M^*)}{a_{2,M}(s - p_M)(s - p_M^*)}. \qquad (7.57)$$

In such case, in hardware design, bi-quadratic TF sections should be cascaded and accompanied, eventually, by one first order TF. In Fig. 7.17 there is shown a design example confirming the above statement. It is an analog low-pass Butterworth filter fulfilling the requirements: $A_{pass} = 2$ dB; $f_{pass} = 8000$ Hz, $A_{stop} = 40$ dB; $f_{stop} = 22050$ Hz, built from two bi-quadratic sections and one first-order section. It has the following transfer function:

$$H(s) = \frac{3\,042\,184\,930}{(s^2 + 34088.3s + 3\,042\,184\,930)} \cdot \frac{3\,042\,184\,930}{(s^2 + 89244.3s + 3\,042\,184\,930)} \cdot$$
$$\frac{55156}{(s + 55156)}, \qquad (7.58)$$

and frequency response presented in Fig. 7.18. Note phase response is approximately linear function of frequency in the filter pass-band therefore the signal shape will not be deteriorated a lot. Required values of output resistors R_x and R_y are calculated from equations specifying an overall, requested circuit gain G and an overall, desired, output circuit resistance:

$$G = K \cdot \frac{R_y}{R_x + R_y}, \qquad R_{out} = \frac{R_x \cdot R_y}{R_x + R_y}. \qquad (7.59)$$

Exercise 7.9 (Analog Filter Hardware). Become familiar with the Matlab program `lab07_ex_hardware.m`, available in the book archive, designing filter presented in Fig. 7.17. Analyze its code, run it, observe plotted figure and displayed values. Knowing transfer function (7.58) of the circuit from Fig. 7.17 and having Eqs. (7.52), (7.53), (7.54), verify whether the circuit RC values are calculated properly, i.e. whether they ensure obtaining correct values of the TF polynomial coefficients (7.58). Calculate frequency response of the circuit from Fig. 7.17 and frequency response of the analog system having TF (7.58). Display them in one figure, compare results with Fig. 7.18.

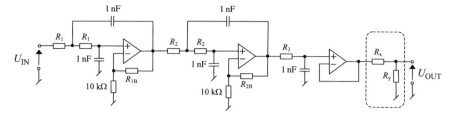

Fig. 7.17: Exemplary realization of 5-th order low-pass filter having transfer function described by Eq. (7.58). Values of remaining elements are as follows: $R_1 = R_2 = R_3 = 18.13\,\text{k}\Omega$, $R_{1B} = 13.82\,\text{k}\Omega$, $R_{2B} = 3.8197\,\text{k}\Omega$, $K_1 = 2.3820$, $K_2 = 1.3820$, $K_3 = 1$, $K = K_1 K_2 K_3 = 3.2918$. For requested overall gain $G = 1$ and output resistance $R_{out} = 10\,\text{k}\Omega$, we have $R_x = \frac{K}{G} R_{out} = 32.92\,\text{k}\Omega$ and $R_y = \frac{K}{K-G} R_{out} = 14.36\,\text{k}\Omega$ [7]

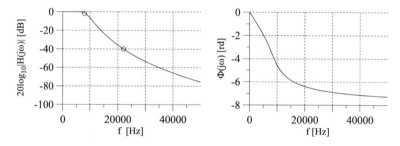

Fig. 7.18: Frequency response of the hardware analog filter from Fig. 7.17, having transfer function (7.58): (left) magnitude, (right) phase [7]

7.10 Summary

In this chapter we learn a little about analog filters. I heard already many nervous questions: *Why analog, why old-fashioned technology of the past? Why?* Because modern DSP has to live in symbiosis with surrounding world which is mechanically and electrically analog. Digital signals are transmitted nowadays as analog ones, which are changing their states. DSP systems require analog filters at their inputs (anti-aliasing filter) and outputs (reconstruction filter). Apart from this, designed analog filters can be easily changed into digital ones.

So what should be remembered?

1. Analog filters theory is like an old wine or old good Cadillac. It represents well-known old engineering achievement and nothing better has been proposed till now.
2. Analog filters are necessary elements in contemporary digital signal processing systems at their front-ends, inputs, and outputs. They are used for

removal of *out-of-band* disturbing signal from the point of view of Nyquist sampling theorem. At present they use state-of-the art electronic technology and are designed for work with GHz signals.

3. Analog filters are also very important for us because their transfer functions and frequency responses can be easily changed into transform functions and frequency responses of corresponding digital filters, using, for example, the bilinear transformation. Thanks to this very high-quality digital filters can be designed without application of elaborate time-consuming optimization schemes.

4. Filters are divided into four primary types (from the point of view what frequencies are passed by the filter): low-pass, high-pass, band-pass, and band-stop.

5. Prototype low-pass filters are aristocrats in the analog filter theory. They are very smart mathematically designed and offer different shape of filter frequency response in respect to its oscillations and sharpness. The well-known are: Butterworth, Chebyshev type 1 and 2, as well as Cauer (elliptic) prototype. Frequency response of Butterworth filter has no oscillations, Chebyshev type 1—oscillations only in the pass-band, Chebyshev type 2—oscillations only in the stop-band, Cauer (elliptic)—oscillations in the pass-band and stop-band. The more oscillations the filter frequency response has in its pass-band and stop-band, the sharper is transition band of the filter, i.e. switching from signal passing to rejecting.

6. Design of any analog filter starts from design of a low-pass normalized *prototype filter*. Normalized because for $\omega_{3dB} = 1$ rad/s. Then this filter is transformed into target low-pass, high-pass, band-pass, or band-stop filter. Transformation changes the prototype transfer function from $H(s)$ to $H(s')$. It is realized by substituting (s) by some function of new variable (s').

7.11 Private Investigations: Free-Style Bungee Jumps

Exercise 7.10 (Calculating RLC Circuit Frequency Response from Its Impulse Response). Use program 7.2 of the RLC circuit. Acquire impulse response samples from the function step(). Having it try to calculate the circuit frequency response as a Fourier transform of its impulse response. Compare obtained result with the circuit frequency response calculated from its transfer function as a ratio of two polynomials of variable $s = j\omega$.

Exercise 7.11 (Calculating Frequency and Damping of an RLC Circuit Impulse Response). Use program 7.2 of the RLC circuit. Acquire impulse response samples of the circuit from the function step(). Then use the interpolated DFT algorithm (IpDFT) and program, described in the previous chapter, for finding angular frequency ω_1 and damping of the impulse response signal. Compare obtained results with correct values.

Exercise 7.12 (Analog Filter Tutorial). Become familiar with the program `lab07_demo_all.m`. Analyze the Matlab code. Run program for six different options (`task = 1,2,3,4,5,6,7`). Observe figures.

Exercise 7.13 (RLC Circuit). Cut RLC fragment from the program `lab07_demo_all.m` (`task=1`). Change values of RLC parameters (ξ should be smaller than 1). Plot a few frequency responses in one figure—use `semilogx(f,20 × log10(abs(H)))`.

Exercise 7.14 (Zeros & Poles—*love and hate* Analog Filter). Set `task=2` in the program `lab07_demo_all.m`. Design an analog filter rejecting 1500 Hz and amplifying 3000 Hz.

Exercise 7.15 (Zeros & Poles: Analog High-Pass Filter with Flat Pass-Band). Set `task = 3, 4` in the program `lab07_demo_all.m`. Using *zeros & poles* method try to design good analog high-pass filter with cut-off frequency 2000 Hz (passing frequencies above 2000 Hz) having flat frequency response.

Exercise 7.16 (Butterworth Band-Pass Filter). Set `task=5` in the program `lab07_demo_all.m`. Design analog band-pass Butterworth filter for $N = 4,5,6,7,8$ for frequency range $[100,...,1000]$ Hz. Observe pole placement. Calculate $|H(f)|$ for all cases. Plot them in one figure using `semilogx( f, 20*log10(abs(H)))`. What is an influence of higher N value for the $|H(f)|$ shape?

Exercise 7.17 (Zeros & Poles Placements in Different Analog Filters). Set `task=6` in the program `lab07_demo_all.m`. Compare transfer function zero & poles placement for the same low-pass filter, e.g. having frequency range $[0...1000]$ Hz, but designed using different methodology (Butterworth, Chebyshev, Cauer).

Exercise 7.18 (Analog Filter Sharpness). Set `task=6` in the program `lab07_demo_all.m`. Calculate $[b,a]$ for the same band-stop filter $[f1...f2]$ using Matlab functions:

Listing 7.8: Matlab functions to be used

```
1   [b,a] = cheby1(N, Rp,    2*pi*[f1,f2], 'stop', 's');  % Cheb1 BS
2   [b,a] = cheby2(N, Rs,    2*pi*[f1,f2], 'stop', 's');  % Cheb2 BS
3   [b,a] = ellip (N, Rp, Rs,2*pi*[f1,f2], 'stop', 's');  % Ellip BS
```

and plot their $|H(f)|$ in one figure using `semilogx(f,20*log10(abs(H)))`. Which filter has the sharpest edge in frequency domain?

Exercise 7.19 (Analog Filter Impulse Response). Set `task=1,5,6` in the program `lab07_demo_all.m`. Acquire impulse response of analog filter using: `[h,t]=impulse(b,a,Tmax)`. Calculate its DtFT and compare with filter $|H(f)|$ in one figure.

Exercise 7.20 (** Mount Everest of Elliptic Filter Design). If you are still *hungry* try to design your own code for the elliptic filter prototype. Find yourself information about it. Add your code to program 7.7.

Exercise 7.21 (*** Hardware Design Verification). Download and install analog circuits simulator LTSpice (http://www.linear.com/designtools/software/)—freeware license. Implement in it analog filter from Fig. 7.17. Do simulations and find circuit frequency response. Compare it with response presented in Fig. 7.18.

Use AC source 10 kHz with amplitude 10 V as input signal (Edit/Components/Voltage). Use universal operational amplifiers: go to Edit/Components/, then choose UniversalOamp2 from folder /Opamps and supply it with DC ±15 V. Do AC analysis: Simulation/Edit Simulation Command/AC Analysis.

References

1. A. Ambardar, *Analog and Digital Signal Processing* (PWS Publishing, Boston, 1995)
2. H. Baher, *Analog and Digital Signal Processing* (Wiley, Chichester, 2001)
3. K.G. Beauchamp, *Signal Processing Using Analog and Digital Techniques* (Allen and Unwin, London, 1993)
4. L.F. Chaparro, *Signals and Systems Using Matlab* (Academic Press, Burlington MA, 2011)
5. J.G. Proakis, D.G. Manolakis, *Digital Signal Processing. Principles, Algorithms, and Applications* (Macmillan, New York, 1992)
6. L.D. Thede, *Analog and Digital Filter Design Using C* (Prentice Hall, Upper Saddle River, 1996)
7. T.P. Zieliński, *Cyfrowe Przetwarzanie Sygnałów. Od Teorii do Zastosowań (Digital Signal Processing. From Theory to Applications)* (Wydawnictwa Komunikacji i Łączności (Transport and Communication Publishers), Warszawa, Poland, 2005, 2007, 2009, 2014)

Chapter 8
IIR Digital Filters

Mothers and Fathers, Sisters and Brothers. It is a short story about Johnny Digital, a successor of Bartholomew Analog.

8.1 Introduction

Digital spectral analysis and digital filtering are two basic parts of each digital signal processing course. A time has come to open a new coffer with Aladdin's digital treasures, this time with digital filtering masterpieces.

Digital filter is a number processing module that obtains some signal samples $x(n)$ on its input and calculates output samples $y(n)$, having two sets of coefficients $\{b_k\}$ and $\{a_k\}$, that should be designed, see Fig. 8.1.

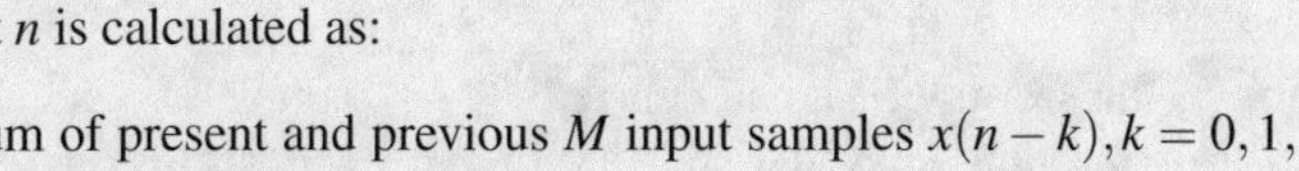

$$x(n) \longrightarrow \boxed{\{b_k\}, \{a_k\}} \longrightarrow y(n) = \sum_{k=0}^{M} b_k x(n-k) - \sum_{k=1}^{N} a_k y(n-k)$$

Fig. 8.1 The simplest block diagram of a digital filter

Digital filtering can be interpreted as running weighted "summation" of last input and output samples of a digital filter. The output filter sample $y(n)$ at time moment n is calculated as:

- a sum of present and previous M input samples $x(n-k), k = 0, 1, 2, ..., M$, taken with weights b_k,
- minus a sum of previous N output samples $y(n-k), k = 1, 2, 3, ..., N$, taken with weights a_k,

in summary:

$$y(n) = \sum_{k=0}^{M} b_k x(n-k) - \sum_{k=1}^{N} a_k y(n-k), \tag{8.1}$$

© Springer Nature Switzerland AG 2021
T. P. Zieliński, *Starting Digital Signal Processing in Telecommunication Engineering*, Textbooks in Telecommunication Engineering, https://doi.org/10.1007/978-3-030-49256-4_8

for example:

$$y(n) = \; b_0 x(n) + b_1 x(n-1) \qquad - \; a_1 y(n-1) + a_2 y(n-2),$$
$$y(n) = \; 0.1 x(n) + 10 x(n-1) \qquad - \; 1 y(n-1) + \; 2 y(n-2),$$

The following introductory specifications are worth to be remembered.

- Design of weights $\{b_k\}$ and $\{a_k\}$ is called a filter design.
- Calculation of $y(n)$ having $x(n)$ is called digital filtering.
- When all coefficients a_k are equal to zero (no recursion exists—the present output sample does not depend on previous outputs), the filter has finite impulse response (FIR) to Kronecker input impulse. *Finite* means equal to zero after some time. For this reason such filter is called an FIR filter.
- When some coefficients a_k are different from zero, there is a feedback in output samples calculation and filter has infinite impulse response (IIR). *Infinite* means does not decaying to zero on filter output. Such filter is called an IIR filter.

In digital filter design the Z-transform (ZT) plays a very significant, prestigious role. Due to it, transfer functions and frequency responses of digital filters can be easily calculated and set of filter coefficients can be found without problems. The ZT plays in digital world the same role as the Laplace transform in analog world. The Z-transform is defined as follows:

$$X(z) = \sum_{n=-\infty}^{\infty} x(n) z^{-n}, \tag{8.2}$$

where z is a complex variable similar to s in the Laplace transform. The Z-transform has two features which are very important for us:

$$Z(x(n-n_0)) = X(z) \cdot z^{-n_0}, \quad Z(c_1 \cdot x_1(n) + c_2 \cdot x_2(n)) = c_1 \cdot X_1(z) + c_2 \cdot X_2(z). \tag{8.3}$$

Firstly, the Z-transform of signal delayed by n_0 samples is equal to the Z-transform of the non-delayed signal multiplied by z^{-n_0}. Secondly, the Z-transform of the sum of signals is equal to the sum of these signals Z-transforms. Therefore, when the Z-transform is performed upon the digital filter Eq. (8.1), the digital filter transfer function (TF) $H(z)$ is obtained:

$$H(z) = \frac{Y(z)}{X(z)} = \frac{b_0 + b_1 z^{-1} + b_2 z^{-2} + \ldots + b_M z^{-M}}{1 + a_1 z^{-1} + a_2 z^{-2} + \ldots + a_N z^{-N}}. \tag{8.4}$$

The coefficients b_k and a_k are the same as in the filtering equation (8.1). What is extremely important? After using the setting (f_s—sampling frequency):

$$z = e^{j2\pi \frac{f}{f_s}} \tag{8.5}$$

the Z-transform (8.2) takes a form of the discrete-time Fourier transform:

$$X\left(\frac{f}{f_s}\right) = \sum_{n=-\infty}^{+\infty} x(n) e^{-j2\pi \frac{f}{f_s} n}, \tag{8.6}$$

and the digital filter transfer function $H(z)$ (8.4) becomes a function of frequency. It is changing into digital filter frequency response $H(f)$ which is telling us what the filter will do with signal component having frequency f. To find this, we have to put only values of frequency of interest into $H(f)$ and chosen/designed values of filter weights $\{b_k\}$ and $\{a_k\}$.

For different values of coefficients $\{b_k\}$ and $\{a_k\}$ digital filters have different:

- frequency characteristics, e.g. frequency magnitude and phase responses,
- time characteristics, e.g. impulse and step responses.

In this lecture and laboratory we will become familiar with:

- program implementation of digital filtering algorithm described by Eq. (8.1),
- digital filter design method in which roots of the filter transfer function polynomials (8.4) are specially chosen (zeros and poles placement method),
- designing digital filters by means of bilinear transformation of analog filters,
- designing digital Butterworth, Chebyshev, and Cauer (elliptic) filters this way,
- zeros and poles placements in the mentioned above three types of filters, as well as shapes of frequency responses of these filters, i.e. the existence of oscillations in them and sharpness of their transition edges.

8.2 Discrete-Time LTI Systems

Derivation of LTI Systems Input–Output Equation Digital filters designed in this chapter are discrete-time linear time-invariant systems since they fulfill the following two features:

- system response to linear superposition of different inputs is equal to superposition of individual system responses to each input separately (c_1, c_2—arbitrary constants):

$$
\begin{matrix} x_1(n) & \to & y_1(n) \\ x_2(n) & \to & y_2(n) \end{matrix} \quad \Rightarrow \quad c_1 x_1(n) + c_2 x_2(n) \to c_1 y_1(n) + c_2 y_1(n), \quad (8.7)
$$

- system response to the delayed input is the same as before but delayed:

$$
x(n) \to y(n) \quad \Rightarrow \quad x(n - n_0) \to y(n - n_0). \tag{8.8}
$$

Let us derive input–output relation of a discrete-time LTI system. The Kronecker delta impulse is defined as:

$$
\delta(n) = \begin{cases} 1, & n = 0, \\ 0, & n \neq 0. \end{cases} \tag{8.9}
$$

Making use of the LTI system features (8.7), (8.8) the following relations are valid between LTI system input and output :

$$
\begin{array}{lrcl}
(1) & \delta(n) & \to & h(n) \\
(2) & \delta(n - k) & \to & h(n - k) \\
(3) & x(k)\delta(n - k) & \to & x(k)h(n - k) \\
(4) & x(n) = \sum_{k=-\infty}^{\infty} x(k)\delta(n - k) & \to & y(n) = \sum_{k=-\infty}^{\infty} x(k)h(n - k).
\end{array}
$$

Relation (1) is a definition of an impulse response of a discrete-time system. Relation (2) is a consequence of system time invariance. Relations (3) and (4) are consequences of assumed system linearity: system response to summation of many delayed and scaled Kronecker delta impulses is equal to summation of scaled individual impulse responses.

As a result the following LTI input–output relation is achieved, having the written immediately below frequency consequence (spectra multiplication):

$$
y(n) = \sum_{k=-\infty}^{\infty} x(k)h(n - k) = x(n) \circledast h(n), \tag{8.10}
$$

$$
Y(\Omega) = X(\Omega)H(\Omega). \tag{8.11}
$$

It is a convolution of two signals $x(n)$ and $h(n)$ (denoted by $\circledast$). The second signal is reversed in time, shifted n samples and multiplied by the first signal $x(k)$ in sample-by-sample manner. Then all multiplication results are accumulated and obtained one value is treated as an output filter sample $y(n)$ for time stamp n. Single

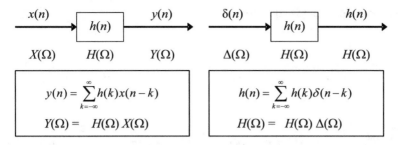

$$y(n) = \sum_{k=-\infty}^{\infty} h(k)x(n-k)$$

$$Y(\Omega) = H(\Omega) X(\Omega)$$

$$h(n) = \sum_{k=-\infty}^{\infty} h(k)\delta(n-k)$$

$$H(\Omega) = H(\Omega) \Delta(\Omega)$$

Fig. 8.2: Block diagrams and equations for discrete-time linear time-invariant (LTI) systems: (left) arbitrary input $x(n)$, (right) Kronecker delta input $\delta(n)$ having DtFT Fourier spectrum $\Delta(\Omega)$ (equal to 1 for all frequencies) [10]

time-reversion and shifting are relative operations; therefore, they can be performed upon the first signal and the result will be the same:

$$y(n) = \sum_{k=-\infty}^{\infty} h(k)x(n-k) = h(n) \circledast x(n). \tag{8.12}$$

Concluding, the output signal in LTI systems is equal to the convolution of its input with the system impulse response. LTI system impulse response is fully specifying the filter output for known input. The DtFT spectrum of Kronecker delta impulse consists of all frequencies ($\Delta(\Omega) = 1$ for all Ω). In the LTI system impulse response $h(n)$ there are only frequencies which the system can pass. The DtFT spectrum of $h(n)$ characterizes the LTI system behavior for each input frequency. Our present knowledge about the discrete-time LTI systems is summarized in Fig. 8.2. From Chaps. 4 and 6 we know that if two signals are convoluted, their DtFT spectra are multiplied. Therefore equality $|H(\Omega)| = 0$, valid for some angular frequency Ω, causes that this frequency component will not be passed by the LTI system from its input to output.

Frequency Response of LTI Systems Let us calculate the discrete-time Fourier transform of the LTI system impulse response $h(n)$:

$$H(\Omega) = \sum_{n=-\infty}^{\infty} h(n)e^{-j\Omega n}, \qquad \Omega = 2\pi\frac{f}{f_s}. \tag{8.13}$$

As a result the system frequency response $H(\Omega)$ is obtained, being a complex-value function of angular frequency Ω. It has magnitude $M(\Omega)$) and angle $\Phi(\Omega)$:

$$H(\Omega) = |H(\Omega)|e^{j\sphericalangle H(\Omega)} = M(\Omega)e^{j\Phi(\Omega)}. \tag{8.14}$$

When we assume the following input signal:

$$x(n) = e^{j\Omega n}, \tag{8.15}$$

the LTI system output is equal to:

$$y(n) = \sum_{k=-\infty}^{\infty} h(k)x(n-k) = \sum_{k=-\infty}^{\infty} h(k)e^{j\Omega(n-k)} = e^{j\Omega n}\sum_{k=-\infty}^{\infty} h(k)e^{-j\Omega k} =$$
$$= e^{j\Omega n}H(\Omega) = e^{j\Omega n}M(\Omega)e^{j\Phi(\Omega)} = M(\Omega)e^{j[\Omega n+\Phi(\Omega)]}, \qquad (8.16)$$

i.e. the input signal is multiplied in amplitude by $M(\Omega)$ and shifted in phase by $\Phi(\Omega)$. Therefore, knowing the system impulse response we have complete knowledge about the LTI system treatment of different frequency components.

Further Specifications of LTI Systems The LTI system is *causal* when its impulse response is equal to zero for $t < 0$: first the system excitation appears at $t = 0$, then the system responds. Taking this into account, the causal input–output LTI system Eq. (8.10) is equal to:

$$y(n) = \sum_{k=0}^{\infty} h(k)x(n-k) = \sum_{k=0}^{\infty} h_k x(n-k). \qquad (8.17)$$

The LTI system is *stable* when

$$\sum_{n=-\infty}^{\infty} |h(n)| < \infty. \qquad (8.18)$$

Resultant, joint impulse response of *parallel* and *serial* connection of two LTI systems (see Fig. 8.3) is equal, respectively:

Parallel LTI: $h(n) = h_1(n) + h_2(n), \qquad (8.19)$

Serial LTI: $h(n) = \sum_{k=-\infty}^{\infty} h_1(k)h_2(n-k) = h_1(n) \circledast h_2(n). \qquad (8.20)$

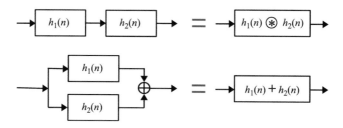

Fig. 8.3: Serial (up) and parallel (down) connection of two LTI systems

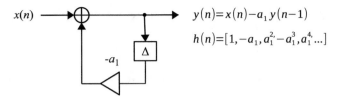

Fig. 8.4: Block diagram of the simplest digital recursive filter

From LTI Systems to Digital Filters Now we look at LTI systems from slightly different perspective. Let us consider a system having very long impulse response described by equation:

$$h(n) = (-a_1)^n, \quad \text{for} \quad n \geq 0. \tag{8.21}$$

Such impulse response has the system presented in Fig. 8.4 and described by the equation:

$$y(n) = x(n) - a_1 y(n-1). \tag{8.22}$$

The last equation can be extended to more general one:

$$y(n) = x(n) - \sum_{k=1}^{\infty} a_k y(n-k) \tag{8.23}$$

in which many previous system outputs are weighted in the process of calculation of the present output.

It can be proofed that the following equation:

$$y(n) = \sum_{k=0}^{\infty} b_k x(n-k) - \sum_{k=1}^{\infty} a_k y(n-k) \tag{8.24}$$

is the most general description of LTI systems as systems performing weighting averaging of input and output samples. Now h_k is replaced by b_k in order to have more consistent denotations.

The digital filters averaging is limited to orders M and N:

$$y(n) = \sum_{k=0}^{M} b_k x(n-k) - \sum_{k=1}^{N} a_k y(n-k). \tag{8.25}$$

In Fig. 8.5 block diagrams of typical digital filters are shown. Three main architectures can be distinguished:

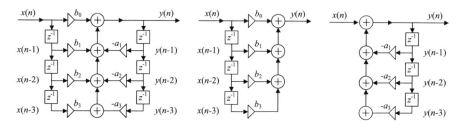

Fig. 8.5: Block diagrams of the digital filter structures: (left) full IIR (AR+MA), (center) FIR (MA), (right) reduced IIR (AR) [10]

- input and output averaging, and two buffers for samples, one on filter input and one on output (auto-regressive (AR)–moving average (MA) case),
- only input averaging and input buffer (MA case),
- only output averaging and output buffer (AR case).

8.3 Digital Signal Filtering

In this section we will write a first program for digital filtration and use it.

An algorithm for on-line digital signal filtering is summarized in Table 8.1. In DSP processors shift buffers are replaced with circular buffers and modulo-M and modulo-N sample addressing is used. Additionally, calculations are organized in an in-place manner: the newest sample, coming in or out, is put into an appropriate buffer into the position of the oldest sample. If you are interested in this topic, look at manuals of DSP processor vendors.

Matlab code implementing the described above filtering procedure is given in Listing 8.1. Results obtained with its use are presented in Fig. 8.6: input and output signals and their FFT spectra.

Listing 8.1: Matlab code of digital signal filtration

```
1   % lab08_ex_filtering.m
2   clear all; close all;
3
4   % Input signal x(n) - two sinusoids: 50 and 500 Hz
5   fs = 2000;                    % sampling ratio
6   Nx = 4000;                    % number of samples
7   dt = 1/fs; t = dt*(0:Nx-1);   % sampling moments
8   %x = zeros(1,Nx); x(1) = 1;   % Kronecker delta impulse
9   x = sin(2*pi*50*t+pi/3) + sin(2*pi*500*t+pi/7);  % sum of 2 sines
10
11  % Digital filter coefficients [ b, a] = ?
12  b = [ 1, 0, 2, 0, 1 ];                        % [ b0, b1, b2, b3, b4]
13  a = [ 1.0000, -3.2199, 3.9203, -2.1387, 0.4412 ]; % [a0=1, a1, a2, a3, a4]
14
```

```
15  │ % Different design - our future work ...
16  │ % #########################################
17  │ % b = ?, a= ?, a_compare=[...], b_compare=[...],
18  │ % #########################################
19  │
20  │ % Digital filtration: x(n) --> [ b, a ] --> y(n)
21  │ M = length(b);                    % number of {b} coefficients
22  │ N = length(a); a = a(2:N); N=N-1;  % number of {a} coeffs, remove a0=1, weight of y(n)
23  │ bx = zeros(1,M);                   % buffer for input samples x(t)
24  │ by = zeros(1,N);                   % buffer for output samples y(t)
25  │ for n = 1 : Nx                     % MAIN LOOP
26  │     bx = [ x(n) bx(1:M-1) ];       % put new x(n) into bx buffer
27  │     y(n) = sum( bx .* b ) - sum( by .* a );  % do filtration, find y(n)
28  │     by = [ y(n) by(1:N-1) ];       % put y(n) into by buffer
29  │ end
30  │
31  │ % Comparison of input and output
32  │ figure;
33  │ subplot(211); plot(t,x); grid;      % input  signal x(n)
34  │ subplot(212); plot(t,y); grid; pause  % output signal y(n)
35  │ figure; % signal spectra of the second halves of samples (transients are removed!)
36  │ k=Nx/2+1:Nx;  f0 = fs/(Nx/2);  f=f0*(0:Nx/2-1);
37  │ subplot(211); plot(f,20*log10(abs(2*fft(x(k)))/(Nx/2))); grid;
38  │ subplot(212); plot(f,20*log10(abs(2*fft(y(k)))/(Nx/2))); grid; pause
```

Table 8.1: Digital filtering algorithm

No	Operation name	Operation code	Matlab code
1	Correct coeffs a	$k = 1...N-1 : a_k = a_{k+1}, N = N-1$	`a=a(2:N); N=N-1;`
2	Initialize buffers	$k = 1,...,M : bx[k] = 0$	`bx=zeros(1,M);`
		$k = 1,...,N : by[k] = 0$	`by=zeros(1,N);`
3	REPEAT, $n = 1, 2, ...$		
3a	Shift input buffer bx	$k = M,...,2 : bx[k] = bx[k-1]$	`bx=[ 0 x(1:M-1)];`
3b	Put new $x(n)$ into bx	$bx[1] = x(n)$	`bx(1)=x(n);`
3c	Do filtration (8.25)	$y(n) = \sum\limits_{k=1}^{M} b_k \cdot bx[k] - \sum\limits_{k=1}^{N} a_k \cdot by[k]$	`y(n)=sum(bx.*b) -`
			`-sum(by.*a);`
3d	Shift output buffer by	$k = N,...,2 : by[k] = by[k-1]$	`by=[ 0 by(1:N-1)];`
3e	Put new $y(n)$ into by	$by[1] = y(n)$	`by(1)=y(n);`

ad. 1) coefficient a_0, if present, is removed from $\{a_k\}$ and N is decreased by 1
ad. 2) two buffers $bx[.]$ and $by[.]$ for input $x(.)$ and output $y(.)$ samples are initialized with 0s
ad. 3) in a loop, the following operations are performed
ad. 3a) samples in the input buffer $bx[.]$ are shifted one position right
ad. 3b) new sample $x(n)$ is put into the first, most left position in $bx[.]$
ad. 3c) samples in both buffers are multiplied by corresponding filter weights and sum/subtracted as in Eq. (8.25), new output filter value $y(n)$ is calculated
ad. 3d) samples in the output buffer $by[.]$ are shifted one position right
ad. 3e) value $y(n)$ is put into the first, most left position in $by[.]$

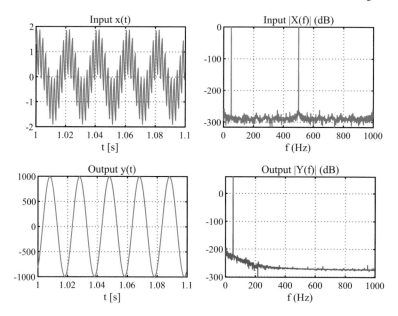

Fig. 8.6: Illustration of digital signal filtering done in Exercise 8.1: (left) input and output signals of a digital filter, (right) FFT spectra of both signals

Exercise 8.1 (My First Digital Filtration). Run program 8.1. Compare input and output signals and their spectra. Observe that the first signal component having 50 Hz is amplified 1000 times, while the second with 500 Hz is completely rejected (!). Modify the program code in order to obtain exactly the same plots as in Fig. 8.6. Modify frequencies of input signal components and try to design new weights of digital filter which should amplify the high-frequency component and reject the low-frequency one. Do not spend more than 15 min on this task.

8.4 Z-Transform and Its Features

The last exercise should show us that digital filters can be extremely effective but their design is not very easy, at least at present level of our knowledge. How digital filters are professionally designed? This will be the subject of our interest in the remaining part of this chapter and in the next few chapters.

First, we need a *transform* which will be eliminating delays in the digital filter equation (8.25) in a similar way the Laplace transform removes signal derivatives in differential equations of analog circuits. Such transform exists and is called the Z-transform:

$$X(z) = \sum_{n=-\infty}^{\infty} x(n)z^{-n}, \qquad (8.26)$$

where z is a complex-value variable, similar to variable s in the Laplace transform. Equation (8.26) defines Laurent polynomial of variable z: signal Z-transform is summation of signal samples, multiplied by z taken to the power of negated samples indexes.

Exercise 8.2 (From Signal Samples to Signal Z-Transform). In the beginning we will look at the Z-transform of a very simple signal having non-zero the following samples only:

$$x(-2) = -20, \quad x(-1) = -10, \quad x(0) = 0.1, \quad x(1) = 1, \quad x(2) = 2.$$

Its Z-transform exists only for $z \neq \infty$ and $z \neq 0$ and is equal to:

$$X(z) = -20z^2 - 10z^1 + 0.1 + 1z^{-1} + 2z^{-2}.$$

Exercise 8.3 (From Signal Z-Transform to Signal Samples). Conversely, knowing the signal Z-transform we have knowledge about signal samples values, for example:

$$X(z) = 5 + 4z^{-1} + 3z^{-2} + 2z^{-3} + 1z^{-4},$$
$$X(z) = x(0) + x(1)z^{-1} + x(2)z^{-2} + x(3)z^{-3} + x(4)z^{-4},$$
$$x(0) = 5, \quad x(1) = 4, \quad x(2) = 3, \quad x(3) = 2, \quad x(4) = 1.$$

From the point of view of our application, the most important are the following **Z-transform features**:

- signal shift:

$$\sum_{n=-\infty}^{\infty} x(n-n_0)z^{-n} = \sum_{m=-\infty}^{\infty} x(m)z^{-(m+n_0)} = z^{-n_0} \sum_{m=-\infty}^{\infty} x(m)z^{-m} = z^{-n_0}X(z),$$
$$(8.27)$$

- signal superposition (c_1, c_2—arbitrary constants):

$$\sum_{n=-\infty}^{\infty} [c_1 x_1(n) + c_2 x_2(n)] z^{-n} = c_1 \sum_{n=-\infty}^{\infty} x_1(n) z^{-n} + c_2 \sum_{n=-\infty}^{\infty} x_2(n) z^{-n} =$$
$$c_1 X_1(z) + c_2 X_2(z), \quad (8.28)$$

- signal convolution:

$$\sum_{n=-\infty}^{\infty} \left(\sum_{k=-\infty}^{\infty} x(k) y(n-k) \right) z^{-n} = \sum_{m=-\infty}^{\infty} \left(\sum_{k=-\infty}^{\infty} x(k) y(m) \right) z^{-(m+k)} =$$
$$\left(\sum_{k=-\infty}^{\infty} x(k) z^{-k} \right) \left(\sum_{m=-\infty}^{\infty} y(m) z^{-m} \right) = X(z) Y(z). \quad (8.29)$$

In many textbooks one can find more Z-transforms features and Z-transforms of many signals. For us the following pair {signal, its Z-transform} is important:

$$x(n) = a^n u(n) \quad \Leftrightarrow \quad X(z) = \frac{1}{1 - az^{-1}}, \quad |z| > |a|, \quad (8.30)$$

where $u(n)$ denotes the unitary step function:

$$u(n) = \begin{cases} 1, & n \geq 0 \\ 0, & n < 0 \end{cases}. \quad (8.31)$$

The $X(z)$ is calculated as a limit value of the Laurent polynomial:

$$\sum_{k=0}^{\infty} \left(\frac{a}{z} \right)^n. \quad (8.32)$$

Correspondence (8.30) $x(n) \Leftrightarrow X(z)$ can be used for finding impulse response $h(n)$ of a digital filter when its transfer function $H(z)$ has been already designed: $H(z) \Rightarrow h(n)$. In order to do this, however, it should be possible to represent/rewrite $H(z)$ as a summation of low-order transfer functions (8.30). An example presenting this is shown in Sect. 8.6.

8.5 Digital Filter Transfer Function and Frequency Response

Let us repeat the input–output equation of the discrete-time LTI system (8.10):

$$y(n) = \sum_{k=-\infty}^{\infty} x(k)h(n-k). \tag{8.33}$$

Performing the Z-transform on both sides of Eq. (8.33), due to Eq. (8.29) and system causality, we obtain

$$Y(z) = H(z)X(z) \quad \Rightarrow \quad H(z) = \frac{Y(z)}{X(z)} = \sum_{n=-\infty}^{\infty} h(n)z^{-n} = \sum_{n=0}^{\infty} h_n z^{-n}. \tag{8.34}$$

Doing the same upon LTI input–output system description (8.25):

$$y(n) = \sum_{k=0}^{M} b_k x(n-k) - \sum_{k=1}^{N} a_k y(n-k), \tag{8.35}$$

we get

$$Y(z) = \left[\sum_{k=0}^{M} b_k z^{-k} \right] X(z) - \left[\sum_{k=1}^{N} a_k z^{-k} \right] Y(z). \tag{8.36}$$

From Eq. (8.36) the digital LTI system transfer function is derived:

$$H(z) = \frac{Y(z)}{X(z)} = \frac{b_0 + b_1 z^{-1} + \ldots + b_M z^{-M}}{1 + a_1 z^{-1} + \ldots + a_N z^{-N}} = \frac{b_0}{1} \frac{(1 - z_1 z^{-1})\ldots(1 - z_M z^{-1})}{(1 - p_1 z^{-1})\ldots(1 - p_N z^{-1})}. \tag{8.37}$$

Finally, after multiplication of the transfer function (8.37) by $\frac{z^M}{z^M}$ and $\frac{z^N}{z^N}$ and doing some math rearrangements, polynomials of positive powers of variable z are obtained:

$$H(z) = \frac{z^N}{z^M} \frac{b_0 z^M + b_1 z^{M-1} + \ldots + b_M}{a_0 z^N + a_1 z^{N-1} + \ldots + a_N} = z^{N-M} \frac{b_0}{a_0} \frac{(z - z_1)\ldots(z - z_M)}{(z - p_1)\ldots(1 - p_N)}. \tag{8.38}$$

Example The digital filter having the following transfer function:

$$H(z) = \frac{3 + 2z^{-1} + z^{-2}}{1 + 2z^{-1} + 3z^{-3}} \tag{8.39}$$

is described by the following input–output equation:

$$y(n) = [3x(n) + 2x(n-1) + x(n-2)] - [2y(n-1) + 3y(n-3)]. \tag{8.40}$$

Concluding, having a digital filter input–output equation we can write its transfer function, and vice versa, knowing filter transfer function we can deduce the filter time equation.

The last two questions to answer are as follows:

- how to find quickly what a given digital filter with weights $\{b_k\}, \{a_k\}$ will do with any arbitrary frequency component,
- how to choose filter coefficients (weights) in order to design a filter with the desired treatment of different frequencies.

Since after setting:

$$z = e^{j\Omega} = e^{j2\pi \frac{f}{f_s}}, \tag{8.41}$$

the Z-transform is changing to the discrete-time Fourier transform and obtains its frequency interpretation (meaning), the same setting is used to change the filter transfer function $H(z)$ (8.37) into the filter frequency response $H(\Omega) = H(f/f_s)$:

$$H(\Omega) = \left(e^{j\Omega} \right)^{N-M} \cdot \frac{b_0}{a_0} \cdot \frac{(e^{j\Omega} - z_1)(e^{j\Omega} - z_2)...(e^{j\Omega} - z_M)}{(e^{j\Omega} - p_1)(e^{j\Omega} - p_2)...(e^{j\Omega} - p_N)}. \tag{8.42}$$

Let us introduce the following denotations:

$$e^{j\Omega} - z_m = B_m e^{j\theta_m}, \qquad e^{j\Omega} - p_n = A_n e^{j\varphi_n}, \tag{8.43}$$

where

$$B_m = \left| e^{j\Omega} - z_m \right|, \quad A_n = \left| e^{j\Omega} - p_n \right|, \quad \theta_m = \sphericalangle \left(e^{j\Omega} - z_m \right), \quad \varphi_n = \sphericalangle \left(e^{j\Omega} - p_n \right). \tag{8.44}$$

Now we can rewrite (8.42) into very compact form:

$$H(\Omega) = M(\Omega) e^{j\Phi(\Omega)} = \left(e^{j\Omega} \right)^{N-M} \frac{b_0}{a_0} \frac{\prod\limits_{m=1}^{M} B_m e^{j\theta_m}}{\prod\limits_{n=1}^{N} A_n e^{j\varphi_n}}, \tag{8.45}$$

where

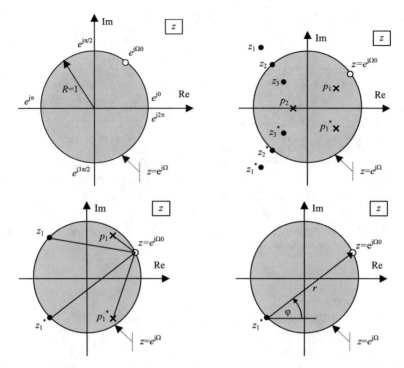

Fig. 8.7: Graphical illustration of digital filter design by appropriate placement of its transfer function zeros and poles, in rows from top to bottom: (1) unitary circle of $z = \exp\left(j2\pi\frac{f}{f_s}\right)$, (2) stability requirements: zeros can lie everywhere, while poles only inside the unitary circle, (3) calculation of filter magnitude response, see Eqs. (8.44), (8.45), (4) calculation of filter phase response, see Eqs. (8.44), (8.45) [10]

$$M(\Omega) = \frac{b_0}{a_0} \frac{\prod_{m=1}^{M} B_m}{\prod_{n=1}^{N} A_n}, \qquad \Phi(\Omega) = \Omega(N-M) + \sum_{m=1}^{M} \theta_m - \sum_{n=1}^{N} \phi_n \qquad (8.46)$$

are magnitude (amplitude) response and phase response of a digital filter. If $B_m = 0$ for some angular frequency Ω, this frequency component will be removed from the filter output. When A_n is very small for some angular frequency Ω, this frequency component will be amplified at the filter output. Appropriate placement of zeros z_m and poles p_n of the filter TF has influence upon B_m and A_n and decides about the filter frequency behavior. This method, called the zeros-and-poles one, is the simplest method for designing digital filters aimed at rejection and amplification of selected frequency signal components. It is graphically explained in Fig. 8.7.

To Remember The main rules for designing filter magnitude response $M(\Omega)$ are as follows:

- in order to remove f_1 frequency component, set TF zero $z_1 = \exp\left(j2\pi\frac{f_1}{f_s}\right)$, i.e. place it on the unitary circle at angle $2\pi\frac{f_1}{f_s}$; remember to add z_1 conjugation to set of TF zeros;
- in order to amplify f_2 frequency component, set TF pole $p_2 = 0.99 \cdot \exp\left(j2\pi\frac{f_2}{f_s}\right)$, i.e. place it inside the unitary circle but very close to it at angle $2\pi\frac{f_2}{f_s}$; remember to add p_2 conjugation to set of TF poles; value 0.99 can be changed: amplification is bigger when the pole is closer to the unitary circle;
- in order to ensure digital filter stability, its zeros can lie everywhere, while poles only inside the unitary circle.

Exercise 8.4 (Digital Filter Design by Its TF Zeros and Poles Placement. Calculation of Digital Filter Frequency Response). Become familiar with program 8.2, calculating and displaying digital filter frequency response as well as showing placement of its TF zeros and poles. Filter from Exercise 8.1 is analyzed. Run program 8.2. Copy fragments of its code to program 8.1. Your task is to add frequency response magnitude curve $M(\Omega)$ to the figure of the output signal spectrum and to check whether input signal components are correctly amplified and attenuated.

Listing 8.2: Matlab program for calculation of digital filter frequency response

```
1   % lab08_ex_freq_response.m
2   clear all; close all;
3
4   % Digital filter coefficients from program lab08_ex_filtering.m
5   bb = [ 1, 0, 2, 0, 1 ];
6   aa = [ 1.0000, -3.2199, 3.9203, -2.1387, 0.4412 ];
7
8   % [bb,aa] values were found by TF zeros & poles placement:
9   fs = 2000;    % sampling ratio
10  f1 = 50;      % frequency to be amplified
11  f2 = 500;     % frequency to be removed
12  p = 0.815*exp(j*2*pi*[f1 f1]/fs); p = [ p conj(p)];
13  a = poly(p);
14  z = exp(j*2*pi*[f2 f2]/fs); z = [ z conj(z)];
15  b = poly(z);
16  [a' aa'], pause, [b' bb'], pause  % comparison
17
```

```
18  % Position of zeros & poles in respect to unitary circle
19  fi = 0  : pi/1000 : 2*pi; si=sin(fi); co=cos(fi);
20  figure; plot(co,si,'k-',real(z),imag(z),'bo',real(p),imag(p),'r*');
21  title('ZP'); pause
22
23  % IIR filter frequency response, magnitude and phase
24  f = 0 : 0.5 : fs/2;            % frequencies of interest
25  zz = exp(j*2*pi*f/fs);         % eq. (8.41), zz - Z-transform variable
26  H = zz.^(length(a)-length(b)) .* polyval(b,zz) ./ polyval(a,zz);    % eq. (8.38)
27  % H = polyval(b(end:-1:1),conj(zz)) ./ polyval(a(end:-1:1),conj(zz));  % eq. (8.37)
28  Hm = freqz(b,a,f,fs);          % Matlab function for (8.40)
29  error = max(abs(H-Hm)), pause  % error in comparison to Matlab
30
31  figure; plot( f, 20*log10(abs(H))); xlabel('f (Hz)'); title('|H(f)| (dB)'); grid; pause
32  figure; plot( f, unwrap(angle(H))); xlabel('f (Hz)'); title('angle |H(f)| (rad)'); grid
        ; pause
```

8.6 Example: Digital Filter Design by TF Zeros and Poles Placement

Simple examples showing everything *in action* are priceless! In this section we will design a very simple digital filter amplifying frequency $f_1 = 10$ Hz and rejecting frequency $f_2 = 50$ Hz for sampling frequency $f_s = 1000$ Hz. Below step-by-step all calculations are given. Plots describing the designed digital filter are presented in Figs. 8.8, 8.9, and 8.10.

Choosing a TF Pole

$$p_1 = 0.98e^{j\Omega_1} = 0.98e^{j2\pi(f_1/f_s)} = 0.98e^{j2\pi(10/1000)} = 0.98e^{j\pi/50}.$$

Choosing a TF Zero

$$z_2 = e^{j\Omega_2} = e^{j2\pi(f_2/f_s)} = e^{j2\pi(50/1000)} = e^{j\pi/10}.$$

Calculation of Filter Transfer Function (TF)

$$H(z) = \frac{(1 - z_2 z^{-1})(1 - z_2^* z^{-1})}{(1 - p_1 z^{-1})(1 - p_1^* z^{-1})} = \frac{(1 - e^{j\pi/10} z^{-1})(1 - e^{-j\pi/10} z^{-1})}{(1 - 0.98 e^{j\pi/50} z^{-1})(1 - 0.98 e^{-j\pi/50} z^{-1})},$$

$$H(z) = \frac{1 - (e^{j\pi/10} + e^{-j\pi/10}) z^{-1} + z^{-2}}{1 - 0.98(e^{j\pi/50} + e^{-j\pi/50}) z^{-1} + z^{-2}} = \frac{1 - 2\cos(\pi/10) z^{-1} + z^{-2}}{1 - 0.98 \cdot 2\cos(\pi/50) z^{-1} + 0.9604 z^{-2}},$$

$$H(z) = \frac{1 - 1.9021 z^{-1} + z^{-2}}{1 - 1.9561 z^{-1} + 0.9604 z^{-2}}.$$

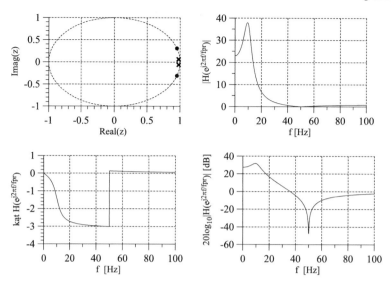

Fig. 8.8: Characterization of designed digital filter passing frequency 10 Hz and removing frequency 50 Hz. In rows: (1) zeros ($\circ$) and poles ($\times$) placement, (2) magnitude response in linear scale, (3) phase response, (4) magnitude response in decibels [10]

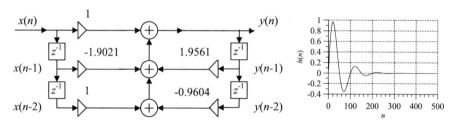

Fig. 8.9: Block diagram of the designed filter (left) and its impulse response, calculated by Matlab function `impulse(b,a)` [10]

Calculation of Filter Frequency Response

$$z = e^{j\Omega} = e^{j2\pi(f/f_s)},$$

$$H(\Omega) = \frac{1 - 1.9021e^{-j\Omega} + e^{-2j\Omega}}{1 - 1.9561e^{-j\Omega} + 0.9604e^{-2j\Omega}}.$$

Filter Behavior for Frequencies f_1 and f_2

$$H\left(2\pi\frac{f_1}{f_s}\right) = H\left(\frac{\pi}{50}\right) = Ge^{j\varphi} = 37.23 \cdot e^{-j1.40}, \quad H\left(2\pi\frac{f_2}{f_s}\right) = H\left(\frac{\pi}{10}\right) = 0.$$

Filter Input–Output Equation

$$\left[1 - 1.9561z^{-1} + 0.9604z^{-2}\right] Y(z) = \left[1 - 1.9021z^{-1} + z^{-2}\right] X(z),$$

$$y(n) - 1.9561y(n-1) + 0.9604y(n-2) = x(n) - 1.9021x(n-1) + x(n-2),$$

$$y(n) = x(n) - 1.9021x(n-1) + x(n-2) + 1.9561y(n-1) - 0.9604y(n-2).$$

Filter Impulse Response the filter transfer function is rewritten into summation of two simpler ratios of z-polynomials—it turns out that they are Z-transforms of known signals, see Eq. (8.30):

$$H(z) = \frac{0.7669 \cdot e^{-j1.5977}}{1 - 0.98e^{j\pi/50}z^{-1}} + \frac{0.7669 \cdot e^{j1.5977}}{1 - 0.98e^{-j\pi/50}z^{-1}},$$

$$h(n) = (0.7669 \cdot e^{-j1.5977})(0.98e^{j\pi/50})^n u(n) + (0.7669 \cdot e^{j1.5977})(0.98e^{-j\pi/50})^n u(n),$$

$$h(n) = h_0(n) + h_0^*(n) = 2\,\mathrm{Re}\,(h_0(n)) = 2\,\mathrm{Re}\left(0.7669e^{-j1.5977} \cdot (0.98)^n e^{j\pi n/50} \cdot u(n)\right),$$

$$h(n) = 2 \cdot 0.7669 \cdot (0.98)^n \cos(\pi n/50 - 1.5977),$$

$$h(0) = 1.$$

Filter Input $x(n)$ and Output $y(n)$

$$x(n) = \sin\left(2\pi\frac{f_1}{f_s}n\right) + \sin\left(2\pi\frac{f_2}{f_s}n\right) = \sin\left(\frac{\pi}{50}n\right) + \sin\left(\frac{\pi}{10}n\right),$$

$$y(n) = G\sin\left(\frac{\pi}{10}n + \phi\right) = 37.23 \cdot \sin\left(\frac{\pi}{10}n - 1.4\right).$$

Exercise 8.5 (Digital Filter Design Using the Zeros-and-Poles Method). Use programs 8.1 and 8.2, adjust values of their parameters to example presented in this section. Now coefficients b and a of the transfer function polynomials are calculated from the following polynomial roots:

```
p = 0.98*exp(j*2*pi*[f1]/fs); p = [p conj(p)];
z = (j*2*pi*[f2]/fs); z = [z conj(z)];
b = poly(z); a = poly(p);
```

Compare your input and output signals with the signals shown in Fig. 8.10. Find impulse response of the filter, using the Kronecker delta impulse as the filter excitation (input). Compare your result with the signal presented in Fig. 8.9.

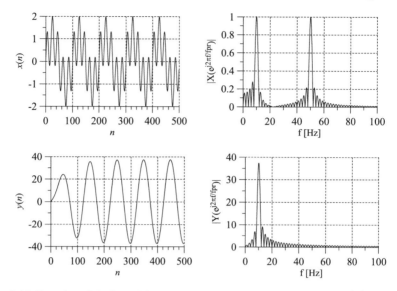

Fig. 8.10: Results of designed filter testing. In rows: (1) input signal $x(n)$, (2) magnitude of $x(n)$ FFT spectrum, (3) output signal $y(n)$, (4) magnitude of $y(n)$ FFT spectrum [10]

8.7 Digital Filter Design Using Bilinear Transformation

In Chap. 7, during *unforgettable* presentation of analog filters in the book on DSP, I argued that analog filter theory is not only used for the design of circuits for analog signal filtering on DSP front-ends, but also it is very widely used for the design of a very good analog transfer functions which can be very easily converted into discrete-time transfer functions of digital filters. Yes, it is true. Even during my last DSP lecture, I asked the Matlab environment about routines for IIR digital filter design and the `yulewalk()` function was the only fully digital optimization procedure which was offered. The other recommended functions, `butter()`, `cheby1()`, `cheby2()` and `ellip()`, are based on designing analog filters $H(s)$ and transforming them into digital ones $H(z)$.

So, how this *magic* transformation is done? By exchanging "s" variable in analog filter transfer function $H(s)$ with variable "z" using the so-called bilinear setting:

$$s = \frac{2}{T}\frac{z-1}{z+1} = 2f_s\frac{z-1}{z+1}, \tag{8.47}$$

where T is the sampling period (inverse of the sampling ratio f_s) in discrete-time system being designed. Thanks to this the $H(s)$, ratio of two polynomials of variable s, is changed into digital filter transfer function $H(z)$, ratio of two polynomials of variable z:

$$s = \frac{2}{T}\frac{z-1}{z+1} \qquad z = \frac{1+sT/2}{1-sT/2}$$
$$H_a(s) \qquad \rightarrow \qquad H_d(z) \qquad \rightarrow \qquad H_a(s) \tag{8.48}$$

In the last equation possibility of coming in the reverse direction is shown also. Having $H(z)$, we can take coefficients of its polynomials and put them into digital filter input–output equation, as we did in Sect. 8.5.

Knowing relation (8.47) between variables s and z, and remembering definitions:

$$s = j\omega = j2\pi f, \qquad z = e^{j\Omega} = e^{j2\pi \frac{f}{f_s}}, \tag{8.49}$$

we can find their non-linear relation, resulting from bilinear transformation:

$$j\omega = \frac{2}{T}\frac{e^{j\Omega}-1}{e^{j\Omega}+1} = \frac{2}{T}\frac{e^{j\Omega/2}\left(e^{j\Omega/2}-e^{-j\Omega/2}\right)/2}{e^{j\Omega/2}\left(e^{j\Omega/2}+e^{-j\Omega/2}\right)/2} = \frac{2}{T}\frac{j\sin\left(\Omega/2\right)}{\cos\left(\Omega/2\right)}, \tag{8.50}$$

in short:

$$\omega = \frac{2}{T}\tan\left(\Omega/2\right), \qquad \Omega = 2\,\mathrm{atan}\left(\omega T/2\right). \tag{8.51}$$

Graphical illustration of bilinear transformation is presented in Fig. 8.11. Zeros and poles of an analog transfer function $H(s)$, designed is space of the complex variable s of the Laplace transform, are mapped into zeros and poles of a digital transfer

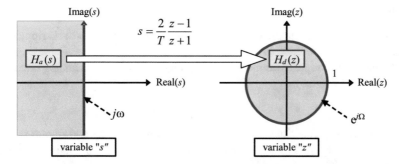

Fig. 8.11: Graphical illustration of the bilinear transformation of analog filter with transfer function $H_a(s)$ into digital filter with transfer function $H_d(z)$. Line $s = j\omega$ is transformed into unitary circle $z = e^{j\Omega}$. Left half-plane of variable s is transformed into interior of unitary circle, i.e. stable analog filters (poles in left half-plane) are transformed into stable digital filters (poles inside the unitary circle) [10]

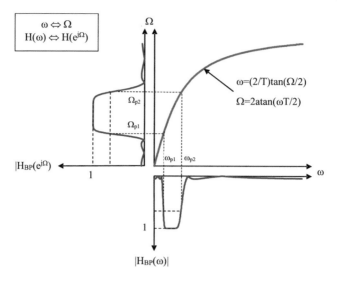

Fig. 8.12: Graphical illustration showing how analog filter frequency response (left) is mapped into digital filter frequency response (bottom) through the non-linear atan() function curve of bilinear transformation [10]

function $H(z)$, lying in space of the complex variable z of the Z-transform. Left half-plane of s is transformed into the interior of unitary circle in the plane of z and digital filter stability is guaranteed. In turn, in Fig. 8.12, it is shown how designed analog filter frequency response (bottom) is mapped into digital filter frequency response (left) by non-linear atan() function.

In the bilinear transformation method digital IIR filters are designed using the following steps:

1. required cut-off frequencies and allowed pass-band and stop-band ripples (oscillations) are specified first,
2. digital cut-off angular frequencies Ω_{3dB} are transformed into analog ones ω_{3dB} using the formula:

$$\omega_{3dB} = \frac{2}{T} \tan\left(\Omega_{3dB}/2\right), \qquad (8.52)$$

3. analog transfer functions $H_a(s)$, fulfilling analog frequency and ripples requirements, are designed (it is important to stress once more that analog design frequencies differ from digital ones),
4. the analog filter is transformed into the corresponding digital filter:

$$H_d\left(z\right) = H_a\left(s\right)\Big|_{s=(2/T)\frac{z-1}{z+1}}, \qquad (8.53)$$

5. during the analog-to-digital system mapping each root of analog transfer function (TF) polynomials is generating zeros and poles of the digital TF and mod-

ifying its gain according to the following formula, at present, as an example, given only for exemplary zero of the analog TF:

$$s - z_k = 2f_s \frac{z-1}{z+1} - z_k = (2f_s - z_k) \frac{z - \frac{2f_s+z_k}{2f_s-z_k}}{z+1}. \tag{8.54}$$

In Listing 8.3 the Matlab code is shown for bilinear mapping of analog TF zeros, poles, and gain into digital TF zeros, poles, and gain. In program 8.4 the whole procedure of digital IIR filter design is implemented.

Exercise 8.6 (Digital Filter Design Using the Bilinear Transformation). Try to design any digital IIR filter using skeleton of program 8.4. Feel free to use fragments of code from programs 8.1 and 8.2. Add plot of zeros and poles. Calculate and display yourself digital filter frequency response. Check it correctness. Do filtration of some signals. Display input and output signals and their FFT spectra.

Listing 8.3: Matlab function for bilinear transformation of analog filter into digital one

```
1  function [zz,pp,ggain] = bilinearTZ(z,p,gain,fs)
2  % Bilinear transformation: H(s) (analog filter) --> H(z) digital filter
3  % zeros, poles, gain (z,p,gain) --> zeros, poles, gain (zz,pp,ggain)
4
5  pp = []; zz = []; ggain = gain;
6  for k=1:length(z)    % transforming zeros
7      zz = [ zz (2*fs+z(k))/(2*fs-z(k)) ];
8      ggain = ggain*(2*fs-z(k));
9  end
10 for k=1:length(p)    % transforming poles
11     pp = [ pp (2*fs+p(k))/(2*fs-p(k)) ];
12     ggain = ggain/(2*fs-p(k));
13 end
14 if (length(p)>length(z)) zz = [ zz -1*ones(1,length(p)-length(z)) ]; end
15 if (length(p)<length(z)) pp = [ pp -1*ones(1,length(z)-length(p)) ]; end
```

Listing 8.4: Matlab code for IIR digital filter design using bilinear transformation of analog filters

```
1  % lab08_ex_iir_design.m
2
3  % Digital requirements
4  fs = 2000;        % sampling ratio
5  f1 = 400;         % minimum frequency of band-pass filter
6  f2 = 600;         % maximum frequency of band-pass filter
7  N  = 8;           % number of poles
8
```

```
 9    % Analog requirements --> digital requirements
10    f1 = 2*fs*tan(pi*f1/fs) / (2*pi); % f1: analog --> digital
11    f2 = 2*fs*tan(pi*f2/fs) / (2*pi); % f2: analog --> digital
12    w0 = 2*pi*sqrt(f1*f2);            % pass-band center
13    dw = 2*pi*(f2-f1);                % pass-band width
14
15    % Analog filter design
16    [z,p,gain] = cheb2ap(N,100);            % low-pass Chebyshev 2 prototype filter
17    [z,p,gain] = lp2bpTZ(z,p,gain,w0,dw);   % frequency transformation: LP to BP
18    b = real(gain*poly(z)); a = real(poly(p)); % analog zeros&poles --> [b,a] coeffs
19    f = 0 : fs/2000 : fs;                   % frequencies of interest
20    H = freqs(b,a,2*pi*f);                  % frequency response of analog filter
21    figure; plot(f,20*log10(abs(H))); xlabel('f (Hz)'); title('Analog |H(f)| (dB)'); pause
22
23    % Conversion of analog filter to digital
24    [z,p,gain] = bilinearTZ(z,p,gain,fs); % bilinear transformation
25    b = real(gain*poly(z)); a = real(poly(p));       % analog zeros&poles --> [b,a] coeffs
26    figure; fvtool(b,a), pause            % displaying digital filter
27
28    % Add plot of zeros & poles
29    % Calculate and display yourself filter frequency response
30    % Do filtration of some signals
```

Example of Digital IIR Filter Design   Required digital cut-off frequency:

$$\Omega_{3\text{dB}} = 2\pi \frac{f_{3\text{dB}}}{f_s} = 2\pi \frac{0.25}{1} = \frac{\pi}{2}. \tag{8.55}$$

Resulting analog cut-off frequency:

$$\omega_{3\text{dB}} = \frac{2}{T}\tan\left(\Omega_{3\text{dB}}/2\right) = 2\tan\left(\pi/4\right) = 2. \tag{8.56}$$

Analog filter:

$$H_a(s) = \frac{-p_1 p_2 p_3}{(s-p_1)(s-p_2)(s-p_3)} = \frac{\omega_{3\text{dB}}^3}{(s+\omega_{3\text{dB}})\left(s^2+s\omega_{3\text{dB}}+\omega_{3\text{dB}}^2\right)}, \tag{8.57}$$

where

$$p_1 = p_3^* = \omega_{3\text{dB}}\left[\left(-1+j\sqrt{3}\right)/2\right], \; p_2 = -\omega_{3\text{dB}}. \tag{8.58}$$

After taking into account that $\omega_{3\text{dB}} = 2$:

$$H_a(s) = \frac{8}{(s+2)(s^2+2s+4)}. \tag{8.59}$$

Digital filter:

$$H_d\left(z\right) = H_a\left(\frac{2}{T}\frac{z-1}{z+1}\right) = \frac{(z+1)^3}{2z\left(3z^2+1\right)} = \frac{\left(1+z^{-1}\right)^3}{2\left(3+z^{-2}\right)}. \tag{8.60}$$

Its frequency response:

$$H_d\left(\Omega\right) = \frac{\left(1+e^{-j\Omega}\right)^3}{2\left(3+e^{-2j\Omega}\right)} \tag{8.61}$$

8.8 Digital IIR Butterworth, Chebyshev, and Elliptic Filters

Matlab language has many ready-to-use functions for IIR digital filter design from analog prototypes. They are summarized in Listing 8.5.

Listing 8.5: Matlab functions useful for IIR digital filter design

```
1  % lab08_ex_functions.m - program to be finished and tested by Reader
2  clear all; close all;
3
4  % Iterative digital filter design
5  fdatool; filterDesigner;
6  % Required order of transfer function (N - number of poles), Wn=2*pi*fcut/fs
7  [N,Wn] = buttord(.), cheby1ord(.), cheby2ord(.), ellipord(.),
8  % All-in-one analog filter design (required N, W0, Wn=[W1,W2], Rp, Rs); R ripples
9  [b,a] = butter(N,Wn,.), cheby1(N,Rp,Wn,.), cheby2(N,Rs,Wn,.), ellip(N,Rp,Rs,Wn,.);
10 % Frequency response - magnitude, phase as a function of frequency
11 f=0:fs/2000:fs/2; H = freqz(b,a,f,fs)
12 f=0:fs/2000:fs/2; H = polyval(b,exp(j*2*pi*f/fs))./polyval(a,exp(j*2*pi*f/fs));
13 figure; plot(f,20*log10(abs(H))); figure; plot(f,unwrap(angle(H)));
```

Exercise 8.7 (Using Matlab Functions for IIR Filter Design). Select Matlab functions from Listing 8.5 and design an IIR digital filter of your dream. Display its magnitude and phase response. Apply it to any test signal, maybe to speech or music. For example, design a band-pass filter in the frequency range $[400, 600]$ Hz for sampling frequency $f_s = 2000$ Hz, having attenuation in the stop-band bigger than 60 dB. Going further, you can design Butterworth, Chebyshev type 1, Chebyshev type 2, and Cauer (elliptic) digital band-pass filters of the same order and display their magnitude responses in one figure. You should obtain figure similar to Fig. 7.12.

From our present perspective, having the method transforming analog filters into digital ones, the all-in-one Matlab program 7.7 for analog filter design should be extremely precious for us. It offers design "by hand" of analog Butterworth, Chebyshev type 1&2 as well as Cauer (elliptic) filters, low/high/band-pass and band-stop, which can be very easily transformed into fantastic IIR digital filters. One should only remember about transforming digital frequency requirements into analog ones by Eq. (8.52) before an analog filter design. Matlab code for changing analog TF into digital TF is given in Listing 8.4. What can I say more? *Run Forest, run!*

Exemplary placements of zeros and poles as well magnitude and phase responses of many digital filters, designed by bilinear transformation method, are shown below in Figs. 8.13, 8.14, 8.15, 8.16, 8.17. Our design requirements are marked in them with circles: we would like to obtain attenuation not bigger than -3 dB in the pass-band and not smaller than -60 dB in the stop-band, for different frequencies: 200, 300, 400, 600, and 700 hertz. We can observe that filters which allow oscillations in their magnitude response are *sharper*, i.e. have more narrow pass-to-stop frequency transition zones (among the filters with the same number of TF poles). We should prefer them when we aim at the minimization of filter order. See Fig. 7.12 for additional illustration/explanation of this phenomena.

The figures are many, maybe too many, but personally I admire the beauty of nice, symmetric zeros and poles patterns (looking like *stars in the sky*) as well mathematically rigorous shapes of frequency responses, theoretically derived. *Viva IIR digital filters!*

Exercise 8.8 (Beautiful IIR Filters). Use program 8.4. Try to design yourself some of IIR filters whose characteristics are shown in Figs. 8.13, 8.14, 8.15, 8.16, 8.17. Use also Matlab functions for designing digital elliptic filters.

8.9 IIR Filter Structures: Bi-quadratic Sections

IIR digital filters are described by time equations (in which scaled delayed input and output samples are added/subtracted) and by corresponding transfer functions. Both, time equations and transfer functions, can be written in forms differing in number of required arithmetic operations and memory data/coefficients locations. Because IIR digital filters are recursive (they have feedback loop), they can become unstable when calculation errors are accumulated. Errors occur when filters coefficients are represented with limited bit precision and calculations are done with fixed point arithmetic. Therefore suitable computational IIR filter structures should be used which are robust to limited precision arithmetic. For this reason the IIR digital filters are typically implemented as a cascade of digital second-order bi-quadratic

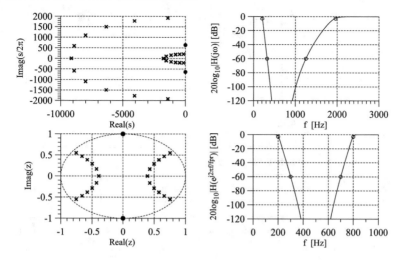

Fig. 8.13: Results of bilinear transformation of band-stop analog Butterworth filter (up) to corresponding digital filter (down): (left) zeros and poles placement, (right) magnitude frequency response [10]

sections/filters (transfer functions of the second order). It is significantly easier to control filter stability of smaller recursive systems.

Let us look at the transfer function decomposition problem. The following second-order TF:

$$H^{(1)}(z) = \frac{1 + 2z^{-1} + z^{-2}}{1 - 0.9z^{-1} + 0.2z^{-2}} \tag{8.62}$$

can be rewritten as:

$$H^{(2)}(z) = 5 + \frac{-4 + 6.5z^{-1}}{1 - 0.9z^{-1} + 0.2z^{-2}}, \tag{8.63}$$

$$H^{(3)}(z) = 5 - \frac{49}{1 - 0.4z^{-1}} - \frac{45}{1 - 0.5z^{-1}}, \tag{8.64}$$

$$H^{(4)}(z) = \left(\frac{1 + z^{-1}}{1 - 0.4z^{-1}} \right) \left(\frac{1 + z^{-1}}{1 - 0.5z^{-1}} \right), \tag{8.65}$$

$$H^{(5)}(z) = \left(1 + z^{-1}\right) \left(1 + z^{-1}\right) \left(\frac{1}{1 - 0.4z^{-1}} \right) \left(\frac{1}{1 - 0.5z^{-1}} \right). \tag{8.66}$$

Block diagrams of all filters are presented in Fig. 8.18. As we can see, even in this simple case, possibilities are many.

As already mentioned, an IIR digital filter having high-order transfer function is typically represented as a cascade connection (multiplication) of simpler second-order TF systems (SOS):

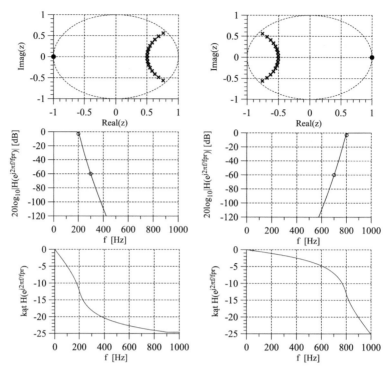

Fig. 8.14: Zeros and poles placement (up), magnitude frequency response (center), and phase frequency response (down) of digital Butterworth low-pass filter (left) and high-pass filter (right) [10]

$$H(z) = \frac{b_{k,0}}{a_{k,0}} \cdot \frac{(1 - z_k z^{-1})(1 - z_k^* z^{-1})}{(1 - p_k z^{-1})(1 - p_k^* z^{-1})}. \qquad (8.67)$$

In Matlab there are special functions doing this. Each second-order system can be implemented in one of the four forms: direct type I and II and their transposed version (see Fig. 8.19). Direct form type II is obtained by changing order of (MA) and (AR) parts of full type IIR filter (see Fig. 8.5). Transposed filter versions are derived by changing direction flow of samples, i.e. from output to input, and reversing each operation in only-MA and only-AR filters. Then the obtained two simple filters are cascaded, first transposed MA, then transposed AR, and vice versa. In transposed filter versions delay-one-sample blocks are shifted into different positions.

Exercise 8.9 (IIR Digital Filter Structures). Try to write Matlab functions of second-order IIR filter sections. Check them *in action*, i.e. doing some signal filtering.

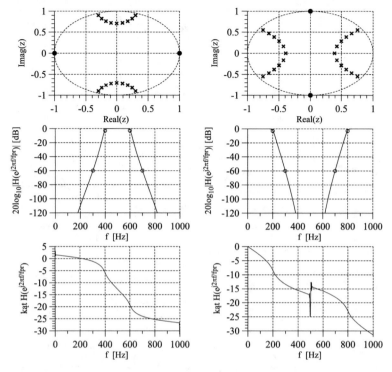

Fig. 8.15: Zeros and poles placement (up), magnitude frequency response (center), and phase frequency response (down) of digital Butterworth band-pass filter (left) and band-stop filter (right) [10]

8.10 Summary

This is a very important chapter. The fundamental one in this book. Why? Because the DSP core consists of two main topics: frequency analysis and digital filtering. In this chapter we learn the digital filtering basics. What should be remembered?

1. Digital filtering can be interpreted as simple local weighted averaging of last filter input samples (as well as already computed last output samples).
2. Filter design methods aim at selection of appropriate filter coefficients (weights). Till now we learn about filter transfer function zeros and poles placements and bilinear transformation of analog filters.

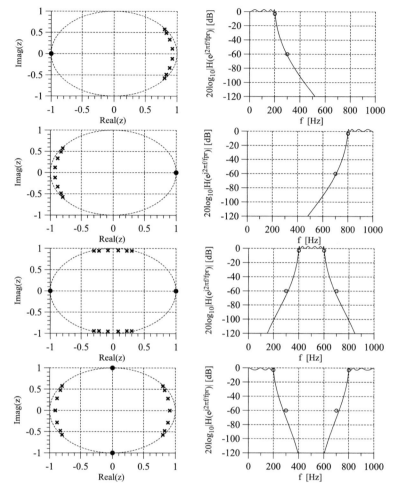

Fig. 8.16: Zeros and poles placement (left) and magnitude frequency response (right) of digital Chebyshev type 1 filter, from top to bottom: low-pass, high-pass, band-pass, and band-stop [10]

3. Z-transform of an input–output digital filtering equation (weighted summation of delayed input and output filter samples) gives a filter transfer function $H(z)$ which helps us in designing the filter weights. It has the same coefficients as the filter equation.

4. After setting $z = e^{j2\pi(f/f_s)}$, the filter transfer function $H(z)$ changes into filter frequency response $H(f/f_s)$ and tells us what the filter will do with any arbitrary frequency component. After appropriate adjusting the filter coefficients in filter frequency response, we can copy them to the input–output filter equation and use.

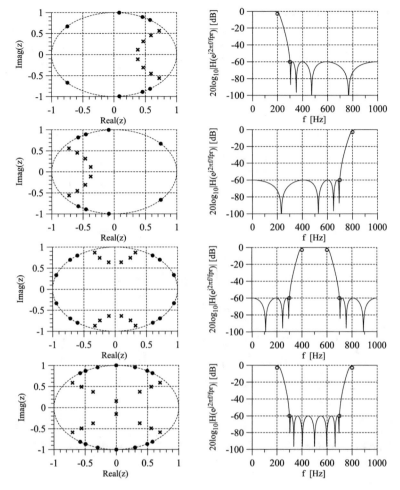

Fig. 8.17: Zeros and poles placement (left) and magnitude frequency response (right) of digital Chebyshev type 2 filter, from top to bottom: low-pass, high-pass, band-pass, and band-stop [10]

5. Digital filters are divided into two main groups: (1) recursive ones having infinite impulse response (IIR), performing weighting of input and output filter samples, (2) non-recursive ones having finite impulse response (FIR), performing weighting of input filter samples only. FIR filters are sub-class of the IIR filters.

6. Digital IIR filters are stable when its TF poles lie inside the unitary circle.

7. Digital IIR filters require small number of weights to be sharp and very frequency selective, they are fast. However, they should be very carefully

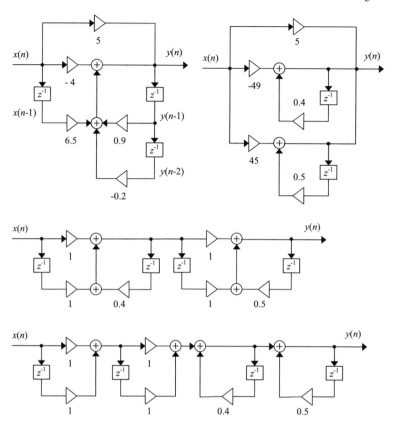

Fig. 8.18: Alternative implementation structures of the IIR digital SOS filter (8.62), described by equations (8.63)–(8.66)—presented in consecutive rows [10]

designed due to possible instability and do not have a perfectly linear-phase response—can change slightly the signal shape.

8. IIR filters are mainly obtained by means of bilinear transformation of analog filters: Butterworth, Chebyshev, and Cauer (elliptic).

9. Non-recursive digital FIR filters (finite impulse response ones), calculating a local mean average of filter input samples only, will be described in the next chapter.

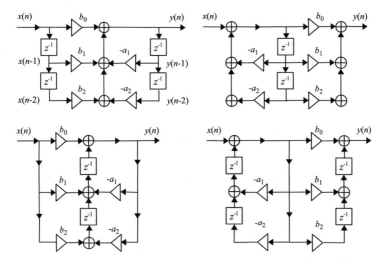

Fig. 8.19: Possible implementations of a second-order section (SOS) (8.67) of IIR digital filters. In rows: direct form type I and II, transposed form type I and II [10]

8.11 Private Investigations: Free-Style Bungee Jumps

Exercise 8.10 (Optimization of Digital Signal Filtering Loop). Look at digital signal filtering loop presented in Table 8.1 and program 8.1. Try to rewrite it: remove time-consuming operation of buffer shifting. Consider using circular buffers instead of shift buffers.

Exercise 8.11 (Separation of Real-World Signal Components). Record some speech signals or use signals from Internet, for example, take them from *FindSounds* web page. Calculate their spectra using Matlab or our `spectrogram()` function designed in Chap. 6 (see program 6.4). Try to separate signal frequency components via digital filtering, for example, pass only: (1) frequencies lower than 2000 Hz, (2) higher than 2000 Hz. Plot signals before and after the filter and listen to them. You can try to separate components of different technical sounds (like a police car alarm) or animal sounds (for example, frog and canary singing together).

Exercise 8.12 (Disturbance Reduction in Real-World Signals). Record a speech signal. Generate and add to it a disturbing sinusoid with a constant frequency. Design and apply a digital filter trying to remove the disturbance. Plot spectrogram of original speech signal, after the addition of disturber and after filter application. Listen to all signals. You can take a more complicated disturber from the *FindSounds* web page.

References

1. A. Antoniou, *Digital Filters. Analysis, Design, Applications* (McGraw-Hill, New York, 1993)
2. H. Baher, *Analog and Digital Signal Processing* (Wiley, Chichester, 2001)
3. L.B. Jackson, *Digital Filters and Signal Processing with Matlab Exercises* (Kluwer Academic Publishers, Boston, 1996; Springer, New York, 2013)
4. R.G. Lyons, *Understanding Digital Signal Processing* (Addison-Wesley Longman Publishing, Boston, 1996, 2005, 2010)
5. A.V. Oppenheim, R.W. Schafer, *Discrete-Time Signal Processing* (Prentice Hall, Englewood Cliffs, 1989)
6. T.W. Parks, C.S. Burrus, *Digital Filter Design* (Wiley, New York, 1987)
7. B. Porat, *A Course in Digital Signal Processing* (Wiley, New York, 1997)
8. J.G. Proakis, D.G. Manolakis, *Digital Signal Processing. Principles, Algorithms, and Applications* (Macmillan, New York, 1992)
9. L.D. Thede, *Analog and Digital Filter Design Using C* (Prentice Hall, Upper Saddle River, 1996)
10. T.P. Zieliński, *Cyfrowe Przetwarzanie Sygnałów. Od Teorii do Zastosowań (Digital Signal Processing. From Theory to Applications)* (Wydawnictwa Komunikacji i Łączności (Transport and Communication Publishers), Warszawa, Poland, 2005, 2007, 2009, 2014)

Chapter 9
FIR Digital Filters

Mirror, mirror on the wall, who's the prettiest of them all? The FIR digital filters are, My Cinderella.

9.1 Introduction

There are people that are talking a lot and therefore thinking that they are *pulling the strings and pressing the buttons*. But very often, the other persons, men of few words are taking decisions and solving things in silence. The same is with FIR digital filters: there are a lot of more attractive DSP solutions but in need the FIR filters are used.

FIR filters are special case of IIR filters. The FIR digital filtering represents a simple weighted averaging of last N samples of input signal $x(n)$ using filter coefficients $\{h_k\}$:

$$y(n) = \sum_{k=0}^{N-1} h(k)x(n-k) = \sum_{k=0}^{N-1} h_k x(n-k), \qquad (9.1)$$

$$y(n) = h_0 \cdot x(n-0) + h_1 \cdot x(n-1) + h_2 \cdot x(n-2) + \dots + h_{N-1} \cdot x(n-(N-1)), \qquad (9.2)$$

for example:

$$y(n) = 5 \cdot x(n) + 10 \cdot x(n-1) + 20 \cdot x(n-2) + 30 \cdot x(n-3). \qquad (9.3)$$

In Fig. 9.1 block diagram of the very short FIR filter is shown having only four weights $\{h_0, h_1, h_2, h_3\}$.

© Springer Nature Switzerland AG 2021
T. P. Zieliński, *Starting Digital Signal Processing in Telecommunication Engineering*, Textbooks in Telecommunication Engineering,
https://doi.org/10.1007/978-3-030-49256-4_9

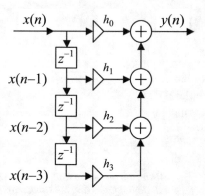

Fig. 9.1: Block diagram of an FIR digital filter calculating weighted average of last four input samples using four weights $h_k, k = 0, 1, 2, 3$

Design of weights $\{h_k\}$ is called an FIR filter design. Calculation of output signal samples $y(n)$, having input signal samples $x(n)$ and filter coefficients $\{h_k\}$, is called an FIR digital filtering.

Making connection to the previous chapter on IIR digital filtering, we can briefly characterize the FIR filters as digital filters having the following transfer function $H(z)$ and resulting from it frequency response $H(\Omega)$, after the setting $z = e^{j\Omega}$, $\Omega = 2\pi \frac{f}{f_s}$:

$$H(z) = \frac{Y(z)}{X(z)} = \sum_{k=0}^{N-1} h_k z^{-k} = h_0 \prod_{k=1}^{N-1} (1 - z_k z^{-1}) = h_0 z^{-(N-1)} \prod_{k=1}^{N-1} (z - z_k)$$

(9.4)

$$H(\Omega) = |H(\Omega)| e^{j\angle H(\Omega)} = M(\Omega) e^{j\Phi(\Omega)} = \frac{Y(\Omega)}{X(\Omega)} = \sum_{k=0}^{N-1} h_k e^{-j\Omega k}, \quad \Omega = 2\pi \frac{f}{f_s}.$$

(9.5)

The filter length N typically should be large in order to obtain sharp filter transition from its pass-band to stop-band. In different applications values of N vary, approximately, from 10 to 1000. Uhh ... 1000! Yes. In such case a design method could not be a *try and check* one: some mathematical foundation/derivation is required.

There are many methods for FIR digital filter design. One of the most simple is a window method. The filter weights are found in it in analytic way: inverse discrete-time Fourier transform of the desired filter frequency response is calculated. The theoretically derived filter weights are next multiplied by chosen window function and a compromise is done between the filter frequency response: flatness in the pass-band, attenuation in the stop-band, and sharpness in the transition band. Despite its simplicity, the window method offers a very

efficient low-pass, high-pass, band-pass, and band-stop FIR filters that are sufficient in most of DSP applications. In Matlab the function `fir1()` implements
them.

In the second very popular inverse DFT method the filter weights are calculated not analytically but computationally, performing the inverse DFT of the
specially shaped samples of the required filter frequency response. Then again,
a window function is applied upon calculated filter weights. In comparison to
the window method, a non-standard filter frequency characteristic can be designed, for example, having values different from 1 for selected frequencies in
the pass-band, e.g. 0.9 or 1.2 (i.e. only correcting the signal processing path
characteristics). This way signal frequency equalizers can be designed (even
the digital music correctors) or multi-pass-band filters, not only typical one-
band-pass/stop (1/0) LP, HP, BP, and BS FIR filters. Such filters are designed
by the `fir2()` Matlab functions.

Techniques relying on finding optimal filter weights by means of numerical
minimization of some cost functions represent the next qualitative level of FIR
filter design methods. The weighted least-squares (WLS) method as well as
Chebyshev min-max equiripple approximation method of the filter frequency
response shape are the well-known examples of such methods. In Matlab one
should use functions `firls()` and `firpm()`, respectively.

What are advantages of FIR non-recursive digital filers in comparison with
IIR recursive digital filters, discussed in the previous chapter? For limited number of finite-value weights they are always stable. Additionally, it is very easy to
design filters with linear-phase response in which all signal components, which
are passed by the filter, are delayed by the same amount of time. Thanks to
this, shape of the passed signal is not changed what is very important in many
applications. Linear-phase filter characteristics are guaranteed by symmetry or
asymmetry of filter weights according to their center.

As a main drawback of FIR filters, we have to mention, is a fact that long
filters with hundreds of weights are required to ensure very sharp filter transition
from its frequency pass-band to stop-band. However, large filter size does not
create nowadays a problem for filter practical implementation since specialized
very fast processor exists. Nevertheless, application of long FIR filters is always
more power consuming than of significantly shorter IIR digital filters.

Simple weighted running averaging of signal samples allows not only frequency low-high-band-stop filtration but also designing digital phase shifters
and digital differentiation filters. However, these types of filters, due to their
significance, especially in telecommunications, will be presented in separate
chapter.

In my opinion it will be an interesting story.

9.2 FIR Filter Description

The FIR digital filter is described by the following input–output relation ($x(n) \rightarrow y(n)$):

$$y(n) = \sum_{k=0}^{N-1} h(k)x(n-k) \quad \longleftrightarrow \quad Y(\Omega) = H(\Omega)X(\Omega), \quad \Omega = 2\pi\frac{f}{f_s}, \quad (9.6)$$

where $h(k)$ denotes the filter impulse response, also marked as h_k or b_k, as in the previous chapter, and f_s is a sampling frequency. The summation is limited only to $k = 0, 1, 2, ..., N-1$, since it is assumed that the filter is causal ($h(k) = 0$ for $k < 0$) and has only N non-zero weights $h(k)$ for non-negative values of k. FIR signal filtering has a form of input samples convolution with filter weights (left part of Eq. (9.6)). When two signals are convoluted, their DtFT spectra are multiplied (right part of Eq. (9.6)).

The discrete-time Fourier transform (DtFT) of the time-limited signal $h(k)$ and the inverse DtFT are defined as ($\Omega = 2\pi\frac{f}{f_s}$):

$$H(\Omega) = \sum_{k=0}^{N-1} h(k)e^{-j\Omega k}, \quad h(k) = \frac{1}{2\pi}\int_{-\pi}^{+\pi} H(\Omega)e^{j\Omega k}d\Omega = \frac{1}{f_s}\int_{-f_s/2}^{+f_s/2} H(f)e^{j2\pi\frac{f}{f_s}k}df.$$

$$(9.7)$$

$H(\Omega)$ is a complex-value number for each Ω:

$$H(\Omega) = \mathrm{Re}(H(\Omega)) + j\,\mathrm{Im}(H(\Omega)) = |H(\Omega)| \cdot e^{j\angle H(\Omega)}, \quad (9.8)$$

where

$$|H(\Omega)| = \sqrt{\mathrm{Re}^2(H(\Omega)) + \mathrm{Im}^2(H(\Omega))}, \quad (9.9)$$

$$\angle(H(\Omega)) = \mathrm{atan}\left(\frac{\mathrm{Im}(H(\Omega))}{\mathrm{Re}(H(\Omega))}\right). \quad (9.10)$$

$|H(\Omega)|$ tells us about filter amplification of radial frequency Ω while $\angle(H(\Omega))$ specifies the frequency phase shift (in consequence time delay—see beginning of the chapter on IIR digital filters). When the filter has a linear-phase response $\angle(H(\Omega)) = -\alpha\Omega$, its answer to the complex-value harmonic excitation $x(n) = e^{j\Omega n}$ is equal to (see Eq. (8.16)):

$$y(n) = |H(\Omega)|e^{j\angle(H(\Omega))} \cdot x(n) = |H(\Omega)|e^{-j\alpha\Omega} \cdot e^{j\Omega n} = |H(\Omega)|e^{j\Omega(n-\alpha)}, \quad (9.11)$$

i.e. each signal component, regardless its frequency value, is delayed by α samples at filter output. As a result, shape of the signal passing through the filter is not changed, the signal is only delayed α samples.

Let us write Eq. (9.7) in alternative form (for some integer constant M):

$$H(\Omega) = e^{-j\Omega M} \cdot \left[\sum_{k=0}^{N-1} h(k)e^{j\Omega(M-k)} \right] = e^{-j\Omega M} \cdot A(\Omega). \qquad (9.12)$$

It results from Eq. (9.12) that the filter phase response (phase shift) is linear function of Ω when the complex-value function $A(\Omega)$ takes real values only either imaginary values only. This is the case when filter weights are symmetrical or asymmetrical. Because phase response linearity is the biggest advantage of FIR digital filters in comparison with IIR filters, it will be proved below.

Proof. For ambitious Readers. Let us assume that M denotes the index of the central filter weight: $M = (N-1)/2$. When the number of the filter weights is odd $N = 2L+1$, we obtain $M = L$. In such case $A(\Omega)$ in Eq. (9.12) is equal to:

$$A(\Omega) = \sum_{k=0}^{L-1} (h(k) + h(N-1-k)) \cos(\Omega(M-k)) +$$

$$+ j \sum_{k=0}^{L-1} (h(k) - h(N-1-k)) \sin(\Omega(M-k)) + h(L). \qquad (9.13)$$

Therefore when filter impulse response is symmetrical ($h(k) = h(N-1-k), k = 0,1,2,...,N-1$), we have

$$A(\Omega) = \sum_{k=0}^{L-1} 2h(k) \cos(\Omega(M-k)) + h(L), \qquad (9.14)$$

while for the asymmetrical filter impulse response ($h(k) = -h(N-1-k), k = 0,1,2,...,N-1$) and for $h(L) = 0$, one obtains

$$A(\Omega) = j \sum_{k=0}^{L-1} 2h(k) \sin(\Omega(M-k)). \qquad (9.15)$$

In similar way derivation is done when filter has even number of weights $N = 2L$ (what is left for Reader as a homework).

$\square$

In Table 9.1 it is shown what filter length (even/odd) or what filter weights symmetry is required for design of linear-phase FIR digital filters of different types. Transfer function $H(z)$ of different FIR linear-phase filter types has obligatory different positions of its zeros (without proof!)—information about this is also included

Table 9.1: Specification of filter length (even $N = 2L$ or odd $N = 2L + 1$) and filter weights symmetry (symmetry $h(k) = h(N - k)$ or asymmetry $h(k) = -h(N - k)$ for $k = 0, 1, 2, ..., N - 1$)) which are required for design of linear-phase FIR digital filters of different types: LP—low-pass, HP—high-pass, BP—band-pass, BS—band-stop, H—Hilbert $-\pi/2$ (rad) phase shifter, D—differentiation filter

Type	Symmetry	Length	$A(\Omega)$	$H(z)$ zeros	$H(0) = 0$	$H\left(\frac{f_s}{2}\right) = 0$	Filter type
I	Symmetrical	Odd	Real	–	–	–	LP, HP, BP, BS
II	Symmetrical	Even	Real	$z = -1$	–	Yes	LP, BP
III	Asymmetrical	Odd	Imag	$z = \pm 1$	Yes	Yes	BP, H, D
IV	Asymmetrical	Even	Imag	$z = +1$	Yes	–	HP, BP, H, D

in the table. In Table 9.1 Hilbert (H) and differentiation (D) filters are considered also. They are discussed in next chapters.

Let us remember that equation $z = e^{j2\pi f/f_s}$ draws unitary circle for changing frequency f. The following partial cases can be distinguished:

- when polynomial of the variable z of the transfer function $H(z)$ has zero for $z = -1 = e^{j\pi}$, i.e. for $f = f_s/2$, it means that $H(f_s/2) = 0$, which allows only obtaining LP or BP filter,
- when $H(z)$ has zero for $z = 1$, i.e. for $f = 0$Hz, it means that $H(0) = 0$, which allows only designing HP and BP filters,
- finally, zeros $z = \pm 1$ allow only building BP filters.

In order to make *things* shorter and simpler, if not otherwise stated, in this chapter we will be designing only FIR filters $h(k)$ having odd number of weights $N = 2L + 1 = 2M + 1$ being symmetrical or asymmetrical around its central point.

Exercise 9.1 (Filter Impulse Response: Symmetric or Asymmetric, Even or Odd Length). Become familiar of program 9.1. Observe displayed plots: exemplary ones are shown in Fig. 9.2. Run it. Set h as h1, h2, h3, h4 for two versions of h0. Check what filters have always roots of their transfer function polynomials at points $z = \pm 1$. Observe magnitude response (responsible for filter gain) and phase response (responsible for time delay) of all filters. Note $\pm \pi$ jumps in phase response for frequencies of roots, lying on the unitary circle. Observe linear-phase characteristics of all filters in frequency-passing regions. Set h=h5, i.e. to low-pass filter impulse response generated by Matlab. A magic occurs! At present frequency response of the filter is very good: magnitude response (filter gain) is equal to 0 dB for low frequencies and more than -60 dB for high frequencies. The phase linearly decreases with frequency in filter pass-bands.

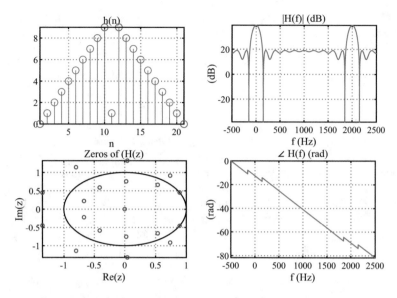

Fig. 9.2: Characteristics of a test FIR filter, designed in the program 9.1, in rows from left to right: symmetrical weights of odd-length filter, magnitude response, zeros placement in $H(z)$, filter phase response. $f_s =2000$ Hz

Listing 9.1: Testing symmetry and number of FIR filter weights

```
1   % lab09_ex_symmetry.m
2   clear all; close all;
3
4   % Impulse response - filter weights
5   L = 10;          % half of the filter length
6   h0 = (0:L-1); % h0 = randn(1,L);    % half of the filter weights
7   h1 = [ h0(1:L) 1  h0(L:-1:1) ];   % symmetry,  odd  length
8   h2 = [ h0(1:L)    h0(L:-1:1) ];   % symmetry,  even length, z=-1
9   h3 = [ -h0(1:L) 0 h0(L:-1:1) ];   % asymmetry, odd  length, z=+/-1
10  h4 = [ -h0(1:L)   h0(L:-1:1) ];   % asymmetry, even length, z=+1
11  h5 = fir1( 200, 0.5 );            % symmetry,  odd  length, Matlab example
12  h = h1;
13  figure; stem(h); title('h(n)'); grid; pause
14
15  % Roots (zeros) of the filter transfer function H(z)
16  z = roots(h); z = z( find( abs(z) < 10 ) );
17  alpha = 0 : pi/1000 : 2*pi; co=cos(alpha); si=sin(alpha);
18  figure; plot(co,si,'k-',real(z),imag(z),'bo'); title('TF Roots');
19  grid; pause
20
21  % Filter frequency response H(f)
22  fs = 2000; f = -0.25*fs : fs/10000 : 1.25*fs;
23  zz = exp(-j*2*pi*f/fs); % z^(-1)
24  H = polyval(h(end:-1:1),zz); % H = freqz(h,1,f,fs);
25  figure; plot(f,20*log10(abs(H))); xlabel('f (Hz)'); title(' |H(f)|');     grid; pause
26  figure; plot(f,unwrap(angle(H))); xlabel('f (Hz)'); title('phase H(f)'); grid; pause
```

Exercise 9.2 (Testing FIR Digital Filters Designed by Matlab Functions).
Become familiar with program 9.2. First, program parameters are set and signal to be filtered is generated. Then, weights of HP, LP, BP, and BS filters are designed by Matlab function `fir1()`. Next, roots of the filter transfer function $H(z)$ and filter frequency response are plotted. Finally, the input signal is filtered and filtration result is analyzed. Exemplary plots displayed by the program are shown in Fig. 9.3. Run the program for `task=1,2,3,4`. Observe placement of $H(z)$ roots in relation to unitary circle. Note linear-phase response in filter pass-band. Notice that $\pm\pi$ jumps occur in phase response for frequencies of $H(z)$ roots lying on the unitary circle. Change the filter length L: observe sharper filter transition from frequency *passing* to *stopping* for bigger values of L. Look carefully at comparison of input and output signals in one figure. Remember that the first valid output sample $y(n)$ has index N, due to computational transition effect resulting from filling the input buffer with samples $x(n)$. The sample $y(N)$ is delayed by $M = (N-1)/2$ in relation to the input signal $x(n)$. For $N = 2L+1$ and $M = (N-1)/2 = L$, the $y(N)$ corresponds to the input sample $x(N-L)$ while for $N = 2L$ and $M = (N-1)/2 = L-1/2$ to input signal value lying between samples $x(N-L)$ and $x(N-(L-1))$.

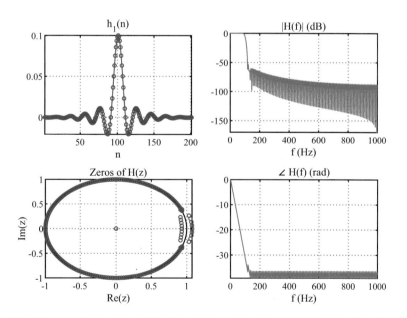

Fig. 9.3: Characteristics of an FIR filter designed in the program 9.2 using the window method, in rows from left to right: symmetrical weights of odd-length filter, filter magnitude response, zeros placement in $H(z)$, filter phase response

Listing 9.2: Testing FIR filters designed by Matlab functions

```
1  % lab09_ex_filtering.m
2  clear all; close all;
3
4  % Parameters
5  task=1;                 % 1=LP, 2=HP, 3=BP, 4=BS - filter type
6  fs=2000;                % sampling ratio
7  f0=100; f1=200; f2=300; % cut-off frequencies
8  L=100;                  % half of the filter length
9  N=2*L+1                 % filter length: 2*L or 2*L+1
10 M= (N-1)/2;             % central point of weights
11
12 % Signal
13 Nx=2000; n=0:Nx-1; dt=1/fs; t=dt*n;
14 x1 = cos(2*pi*10*t); x2 = cos(2*pi*250*t);
15 x  = x1 + x2;
16 if(task==1 | task==4) xref = x1; else xref = x2; end
17
18 % Matlab functions - design of filter impulse responses
19 h1=fir1(N-1,f0/(fs/2),'low');         % low-pass
20 h2=fir1(N-1,f0/(fs/2),'high');        % high-pass
21 h3=fir1(N-1, [f1,f2]/(fs/2),'bandpass'); % band-pass
22 h4=fir1(N-1, [f1,f2]/(fs/2),'stop');  % band-stop
23 if    (task==1) h=h1;
24 elseif(task==2) h=h2;
25 elseif(task==3) h=h3;
26 else            h=h4;
27 end
28 figure; stem(h); title('h(n)'); grid; pause
29
30 % Roots of filter transfer function
31 z = roots(h); z = z( find( abs(z) < 10 ) );
32 alpha = 0 : pi/1000 : 2*pi; co=cos(alpha); si=sin(alpha);
33 figure; plot(co,si,'k-',real(z),imag(z),'bo'); title('TF Zeros'); grid; pause
34
35 % Filter frequency response
36 f = 0 : fs/2000 : fs/2;
37 z = exp(-j*2*pi*f/fs);    % z^(-1) of the Z transform
38 H = polyval(fliplr(h),z); % H = freqz(h,1,f,fs);
39 figure
40 subplot(211); plot(f,20*log10(abs(H))); xlabel('f (Hz)'); title(' |H(f)|'); grid;
41 subplot(212); plot(f,unwrap(angle(H))); xlabel('f (Hz)'); title('Phase H(f)'); grid;
       pause
42
43 % FIR digital filtering
44 %y = conv(x,h); y = filter(h,1,x);
45 N = length(h);
46 bx = zeros(1,N);
47 for n = 1 : Nx
48     bx = [ x(n) bx(1:N-1) ];
49     y(n) = sum( bx .* h );
50 end
51
```

```
52  % Figures
53  n=0:Nx-1;
54  figure; % observe that first N-1 samples of y(n) are erroneous (transient)
55  subplot(211); plot(n,x,'.-'); xlabel('n'); title('x(n)'); grid;
56  subplot(212); plot(n,y,'.-'); xlabel('n'); title('y(n)'); grid; pause
57  figure; nx = L+1: Nx-L; ny = N:Nx;
58  subplot(111); plot(t(nx),xref(nx),'b.-',t(nx),y(ny),'r.-'); xlabel('n');
59  title('x(n), y(n)'); grid; pause
60  figure; % signal spectra of the second halves of samples (transients out!)
61  k=Nx/2+1:Nx;  f0 = fs/(Nx/2);  f=f0*(0:Nx/2-1);
62  subplot(211); plot(f,20*log10(abs(2*fft(x(k))))/(Nx/2)); grid;
63  subplot(212); plot(f,20*log10(abs(2*fft(y(k))))/(Nx/2)); grid; pause
```

9.3 Window Method

In the window method, first, theoretical filter impulse response $h(k)$ is derived analytically: the inverse discrete-time Fourier transform (DtFT) (9.7) is calculated upon
the required filter frequency response equal to 1 in pass-band and 0 in stop-band:

$$h(k) = \frac{1}{2\pi} \int_{-\pi}^{\pi} H_{XX}(\Omega) e^{j\Omega k} d\Omega, \quad -M \le k \le M, \quad (9.16)$$

where XX denotes: LP, HP, BP, or BS, and $H_{XX}(\Omega)$ is user-defined required filter
frequency response symmetrical or asymmetrical around $\Omega = 0$ (rad/s). Next, calculated values $h(k)$ are multiplied by carefully chosen window function $w(k)$, deciding
about filter frequency response oscillations in the pass-band and stop-band as well
as about the sharpness of filter transition band ($-M \le k \le M$):

$$h_w(k) = h(k)w(k) \quad \leftrightarrow \quad H_w(\Omega) = \frac{1}{2\pi} \int_{-\pi}^{\pi} H(\Theta) W(\Omega - \Theta) d\Theta. \quad (9.17)$$

Finally, the filter weights, cut by the window, are shifted M samples right ($0 \le k \le 2M$):

$$h_w^{(M)}(k) = h_w(k - M) \quad \leftrightarrow \quad H_w^{(M)}(\Omega) = e^{-j\Omega M} H_w(\Omega). \quad (9.18)$$

This operation makes the filter causal but delays the filter output by M samples.
Since for $N = 2M + 1$, the $H_w(\Omega)$ is a real-value function, we can conclude from
Eq. (9.18) that the filter phase response linearly decreases with frequency. This
means that shape of the passed signal is not changed on the output.

The filter design using the window method is a big pleasure: a little effort and
a very good effect. Only impulse response equation of the low-pass filter has to
be derived analytically, the other filters are represented as linear super-positions of

low-pass and all-pass filters working in parallel. Their outputs are subtracted from each other.

The impulse response of the low-pass filter with cut-off frequency f_0 (Hz) and cut-off angular frequency $\Omega_0 = 2\pi f_0/f_s$ (rad · Hz/Hz) is analytically calculated as follows:

$$
h_{LP}^{\Omega_0}(k) = \frac{1}{2\pi} \int_{-\pi}^{\pi} H_{LP}(e^{j\Omega}) e^{j\Omega k} d\Omega = \frac{1}{2\pi} \int_{-\Omega_0}^{\Omega_0} 1 \cdot e^{j\Omega k} d\Omega = \frac{1}{2\pi} \frac{1}{jk} e^{j\Omega k} \Big|_{-\Omega_0}^{\Omega_0} =
$$

$$
= \frac{1}{j2\pi k} \left[e^{j\Omega_0 k} - e^{-j\Omega_0 k} \right] = \frac{2j\sin(\Omega_0 k)}{j2\pi k} = \frac{\sin(\Omega_0 k)}{\pi k} = 2\frac{f_0}{f_s} \frac{\sin(\Omega_0 k)}{\Omega_0 k}. \quad (9.19)
$$

Equation (9.19) is only valid for $k \neq 0$ since for $k = 0$ division 0 by 0 takes place. Value $h_{LP}^{\Omega_0}(0)$ is found by calculating derivatives over k of nominator and denominator in Eq. (9.19) and computing value of the result for $k = 0$:

$$
h_{LP}^{\Omega_0}(0) = 2\frac{f_0}{f_s} \frac{d(\sin(\Omega_0 k))/dk}{d(\Omega_0 k)/dk} \Big|_{k \to 0} = 2\frac{f_0}{f_s} \cdot \frac{\Omega_0 \cos(\Omega_0 k)}{\Omega_0} \Big|_{k \to 0} = 2\frac{f_0}{f_s}. \quad (9.20)
$$

Having formulas (9.19), (9.20) for an impulse response $h_{LP}^{\Omega_0}(k)$ of the low-pass filter (LP), we can find with ease impulse responses of HP, BP, and BS filters. All equations are given in Table 9.2. One all-pass filter and different low-pass filters are working in parallel and their outputs are subtracted. The all-pass filter impulse response $h_{ALL}(k)$ is equal to Kronecker delta impulse $\delta(k)$: 1 only for $k = 0$, otherwise 0. Obtaining HP, BP, and BS filters from low-pass ones and the all-pass filter is presented in Fig. 9.4. For example, a high-pass filter is obtained by subtracting a low-pass filter from the all-pass one (first column of plots in the figure), while the band-pass filter is a result of subtracting two low-pass filters: the filter with wider pass-band minus the filter with narrower pass-band (second column of plots in the figure).

Exercise 9.3 (Theoretical Filter Impulse Responses). Analyze code of the program 9.3, implementing in Matlab impulse response equations from Table 9.2. Run the program, change filter parameters, observe different filter weights (shapes)–see Fig. 9.5. Add this code to the program 9.2. Observe placements of $H(z)$ roots as well as filter magnitude response (Fig. 9.6) and phase response. Do signal filtering. Compare input and output signals.

Table 9.2: Theoretical impulse responses $h(k), k = -M, ..., 0, ..., M$ of non-causal FIR digital filters derived analytically—by performing inverse DtFT upon required frequency response of the filter. $\delta(k)$—Kronecker impulse equal to 1 for $k = 0$ and 0 otherwise. Cut-off frequencies f_0, f_1, f_2 are selected by a user

Type	Denotation	Relation to LP/BP	$h(k), k \neq 0$	$h(0)$
LP	$h_{LP}^{\Omega_0}(k)$	–	$2\frac{f_0}{f_s}\frac{\sin(\Omega_0 k)}{\Omega_0 k}$	$2\frac{f_0}{f_s}$
HP	$h_{HP}^{\Omega_0}(k)$	$\delta(k) - h_{LP}^{\Omega_0}(k)$	$-2\frac{f_0}{f_s}\frac{\sin(\Omega_0 k)}{\Omega_0 k}$	$1 - 2\frac{f_0}{f_s}$
BP	$h_{BP}^{[\Omega_1,\Omega_2]}(k)$	$h_{LP}^{\Omega_2}(k) - h_{LP}^{\Omega_1}(k)$	$2\frac{f_2}{f_s}\frac{\sin(\Omega_2 k)}{\Omega_2 k} - 2\frac{f_1}{f_s}\frac{\sin(\Omega_1 k)}{\Omega_1 k}$	$2\frac{f_2}{f_s} - 2\frac{f_1}{f_s}$
BS	$h_{BS}^{[\Omega_1,\Omega_2]}(k)$	$\delta(k) - h_{BP}^{[\Omega_1,\Omega_2]}(k)$	$-\left(2\frac{f_2}{f_s}\frac{\sin(\Omega_2 k)}{\Omega_2 k} - 2\frac{f_1}{f_s}\frac{\sin(\Omega_1 k)}{\Omega_1 k}\right)$	$1 - \left(2\frac{f_2}{f_s} - 2\frac{f_1}{f_s}\right)$

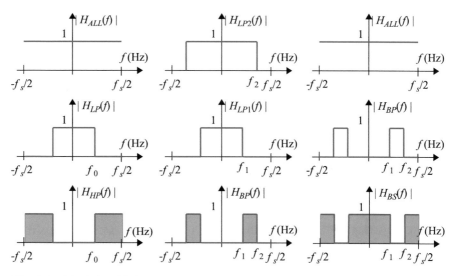

Fig. 9.4: Obtaining high-pass, band-pass, and band-stop filters (from left to right in columns) by subtracting outputs of working in parallel low-pass and all-pass filters. Filters presented in the middle row are subtracted from corresponding filters presented in the first row: as a result filters shown in the last row are get. For first column: all-pass filter minus low-pass filter gives a high-pass filter

Listing 9.3: Generation of theoretical impulse responses of LP/HP/BP/BS filters

```
1   % lab09_ex_impulse_response.m
2
3   % Filter parameters
4   fs = 1000;        % sampling frequency
5   f0 = 100;         % cut-off frequency for low-pass and high-pass filters
6   f1 = 200;         % low frequency for band-pass and band-stop filters
7   f2 = 300;         % high frequency for band-pass and band-stop filters
8   M = 100;          % half of the filter weights
9   N = 2*M + 1;      % number of weights {h(n)} (b(n) in IIR filters) - always odd
```

```
10  n = -M :1: M;    % non-causal weights indexes
11
12  % Impulse responses - weights
13  hALL = zeros(1,N); hALL(M+1)=1;                        % AllPass
14  hLP = sin(2*pi*f0/fs*n)./(pi*n); hLP(M+1) = 2*f0/fs;   % LowPass f0
15  hHP = hALL - hLP;                                      % HighPass
16  hLP1 = sin(2*pi*f1/fs*n)./(pi*n); hLP1(M+1) = 2*f1/fs; % LowPass f1
17  hLP2 = sin(2*pi*f2/fs*n)./(pi*n); hLP2(M+1) = 2*f2/fs; % LowPass f2
18  hBP = hLP2 - hLP1;                                     % BandPass [f1,f2]
19  hBS = hALL - hBP;                                      % BandStop [f1,f2]
20  hH = (1-cos(pi*n))./(pi*n); hH(M+1)=0;  % Hilbert filter -90deg phase shifter
21  hD = cos(pi*n)./n; hD(M+1)=0;           % differentiation filter
22  h = hLP;         % choose filter: hLP, hHP, hBP, hBS, hH, hD
23  figure; stem( n, h ); title('h(n)'); grid; pause
```

The only parameters which can be changed in the window method are:

- impulse response equation with specified cut-off frequencies $f_{0,1,2}, \Omega_{0,1,2}$ (Table 9.2),
- filter length $N = 2M + 1$ deciding about filter sharpness,
- window type deciding about filter flatness in the pass-band and attenuation in the stop-band as well as about width of the transition band.

Window choice has a crucial impact on filter quality because:

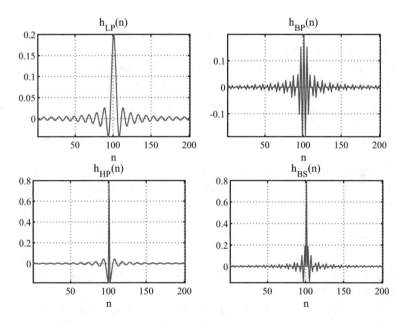

Fig. 9.5: Theoretical un-windowed impulse responses of FIR filters, designed in the program 9.3, in columns: LP, HP, BP, and BS filter

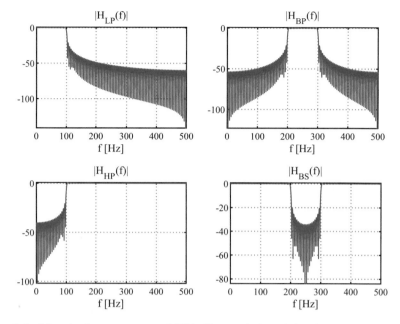

Fig. 9.6: Magnitude responses of FIR filters, designed in the program 9.3, in columns: LP, HP, BP, and BS filter. They correspond to impulse responses presented in Fig. 9.5

- the more narrow is its spectral main-lobe, the faster/sharper is filter transition from pass-band to stop-band,
- the bigger is attenuation of its spectral side-lobes, the better is filter flatness in the pass-band, and the higher is filter attenuation in the stop-band.

There are many window functions. Each of them could be used. But typically they have fixed shapes and fixed spectral features. For this reason the Chebyshev and Kaiser window usage is recommended because their shapes can be changed and their spectral features adjusted. For the Chebyshev window, first, one can choose directly the spectral side-lobes attenuation, and, then, change the window length in order to obtain a desired filter sharpness: fast transition from signal passing to stopping.

The following **filter design procedure** is widely used for the Kaiser window.

1. Choose filter type (LP, HP, BP, or BS) and cut-off frequencies of its $H(\Omega)$.
2. Compute samples of theoretical filter impulse response $h(n)$ using one equation from Table 9.2.
3. Decide about the filter quality, see Fig. 7.9 in chapter on analog filters:

   a. specify allowed level of oscillations δ_{pass} in the filter pass-band around gain equal to 1 (e.g. 0.01 around 1) and gain δ_{stop} in the filter stop-band (e.g. 0.0001), i.e. the filter attenuation,

b. specify Δf (in Hz) of the widest transition band (from pass-band to stop-band) of the filter edge.

4. **Design Kaiser window:**

   a. Calculate A using equation:

   $$\delta = \min\left(\delta_{pass}, \delta_{stop}\right), \quad A = -20\log_{10}\left(\delta\right). \tag{9.21}$$

   b. Calculate required β:

   $$\beta = \begin{cases} 0, & \text{for} & A < 21 \text{ dB}, \\ 0.5842(A-21)^{0.4} + 0.07886\,(A-21), & \text{for} & 21 \text{ dB} \le A \le 50 \text{ dB}, \\ 0.1102\,(A-8.7), & \text{for} & A > 50 \text{ dB}. \end{cases} \tag{9.22}$$

   c. Calculate required length of the filter (f_s—sampling frequency):

   $$N = 2M + 1 \ge \frac{A - 7.95}{14.36}\frac{f_s}{\Delta f}. \tag{9.23}$$

   d. Calculate the Kaiser window for given β and M, where $I_0(x)$ denotes the modified Bessel function of the zeroth order:

   $$w_K(k) = \begin{cases} \dfrac{I_0\left[\beta\sqrt{1-(n/M)^2}\right]}{I_0(\beta)}, & -M \le k \le M, \\ 0, & \text{elsewhere.} \end{cases} \tag{9.24}$$

5. **Window the impulse response:** multiply $h(k)$ of the filter and Kaiser window $w_K(k)$:

   $$h_w(k) = h(k) \cdot w_K(k), \quad -M \le k \le M. \tag{9.25}$$

6. **Shift impulse response** right by M samples:

   $$h_w^{(M)}(k) = h_w(k-M), \quad 0 \le k \le N-1. \tag{9.26}$$

7. **Use weights $\{h_k\}$** in filtering equation for any n:

   $$y(n) = \sum_{k=0}^{2M} h_w^{(M)}(k)x(n-k). \tag{9.27}$$

Exercise 9.4 (FIR Filter Design Using Window Method with Kaiser Window). Check whether Matlab code in Listing 9.4 correctly calculates Kaiser window parameters from Eqs. (9.21)–(9.24). Compare generated Kaiser window with output of the Matlab function `kaiser()`. Use calculated window for windowing the FIR filter impulse responses computed in the program 9.3. Verify the FIR digital filters designed this way using the program 9.2: check whether obtained filter frequency responses are correct (Fig. 9.7), estimate from them expected attenuation of different signal frequency components at filter output and observe their practical attenuation at the filter output as well as in FFT spectra of output signals.

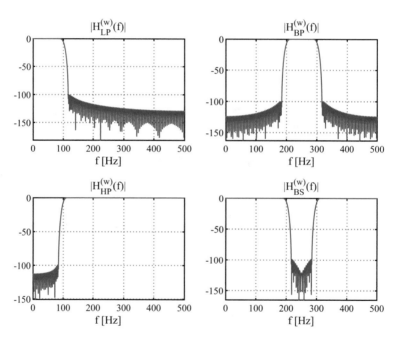

Fig. 9.7: Magnitude responses of FIR filters designed by window method, corresponding to frequency responses of un-windowed filters presented in Fig. 9.6. Theoretical impulse responses calculated in the program 9.3 have been multiplied by Kaiser windows designed in the program 9.4. In Columns: LP, HP, BP, and BS filter

Listing 9.4: Choosing Kaiser window for FIR filter design

```
1   % lab09_ex_kaiser.m
2   clear all; close all;
3
4   % Choose filter parameters
5   fs = 2000;        % sampling frequency (Hz)
6   fa1 = 150;        % low frequency  1   (Hz),  LP, BP, BS
7   fa2 = 200;        % low frequency  2   (Hz),  LP, BP, BS
8   fb1 = 300;        % high frequency 1   (Hz),  HP, BP, BS
9   fb2 = 350;        % high frequency 2   (Hz),  HP, BP, BS
10  if(1)
11    dpass = 0.001;  % allowed oscillations in passband, e.g. 0.1, 0.01, 0.001
12    dstop = 0.0001; % allowed oscillations in stopband, 0.001, 0.001, 0.0001
13  else % Alternative: from Apass and Astop in (dB) to dpass and dstop (linear)
14    Apass = 3;  Astop = 80;
15    dpass = ((10^(Ap/20))-1)/((10^(Ap/20))+1);
16    dstop = 10^(-As/20);
17  end
18
19  type = 'bp';      % lp=LowPass, hp=HighPass, bp=BandPass, bs=BandStop
20
21  % Calculate Kaiser window parameters
22  if (type=='lp') % Filter LP
23      df=fa2-fa1; fc=((fa1+fa2)/2)/fs; wc=2*pi*fc;
24  end
25  if (type=='hp') % Filter HP
26      df=fb2-fb1; fc=((fb1+fb2)/2)/fs; wc=2*pi*fc;
27  end
28  if (type=='bp' | type=='bs') % Filter BP or BS
29      df1=fa2-fa1; df2=fb2-fb1; df=min(df1,df2);
30      f1=(fa1+(df/2))/fs; f2=(fb2-(df/2))/fs;
31      w1=2*pi*f1; w2=2*pi*f2;
32  end
33  % beta = ?
34  d=min(dpass,dstop); A=-20*log10(d);
35  if(A>=50)             beta=0.1102*(A-8.7); end
36  if(A>21 & A<50)       beta=(0.5842*(A-21)^0.4)+0.07886*(A-21); end
37  if(A<=21)             beta=0; end
38  % D = ?
39  if(A>21)  D=(A-7.95)/14.36; end
40  if(A<=21) D=0.922; end
41  % N = ?
42  N=(D*fs/df)+1; N=ceil(N); if(rem(N,2)==0) N=N+1; end
43  N, pause
44  M = (N-1)/2; m = 1 : M; n = 1 : N;
45
46  % Generate Kaiser window
47  % e.g. w=hanning(N)'; w=hamming(N)', w=blackman(N)'; w=chebwin(N,Astop)'
48  % w=Kaiser(N,beta)';
49  w = besseli( 0, beta * sqrt(1-((n-1)-M).^2/M^2) ) / besseli(0,beta);
50  figure; plot(n,w,'bo-'); title('Window function w(n)'); grid; pause
```

9.4 Inverse DFT Method

In the inverse DFT method of FIR filter design, the filter impulse response samples $h(n)$ (index k is replaced by n in this section) is not derived analytically but calculated by a computer program. The inverse DFT (FFT) algorithm is executed:

$$h(n) = \sum_{k=0}^{N-1} H(k)e^{j\left(\frac{2\pi}{N}k\right)n}, \quad n = 0, 1, 2, ..., N-1 \tag{9.28}$$

directly upon user-selected samples of the required filter frequency response:

$$H(k) = H(\Omega_k) = H\left(\frac{2\pi}{N}k\right) = H\left(2\pi\frac{k \cdot f_0}{f_s}\right), \quad f_0 = \frac{f_s}{N}. \tag{9.29}$$

In order to obtain real-value, not complex-value, filter weights, the DFT spectrum (a)symmetry should be fulfilled: $H(N-k) = H^*(k)$. The calculated filter coefficients (9.28) are multiplied with window samples, as before, in order to improve filter features: linearity in the passband and attenuation in the stop-band.

Why is the windowing necessary? To answer this question let us perform the DtFT over calculated filter weights (9.28):

$$H(\Omega) = \frac{1}{N}\sum_{n=0}^{N-1} h(n)e^{-j\Omega n} = \frac{1}{N}\sum_{n=0}^{N-1}\left[\sum_{k=0}^{N-1} X(k)e^{j\left(\frac{2\pi}{N}k\right)n}\right]e^{-j\Omega n}. \tag{9.30}$$

After changing order of summations in Eq. (9.30), we have

$$H(\Omega) = \frac{1}{N}\sum_{k=0}^{N-1} X(k)\left[\sum_{n=0}^{N-1} e^{j\left(\frac{2\pi}{N}k\right)n}e^{-j\Omega n}\right] = \frac{1}{N}\sum_{k=0}^{N-1} X(k)\left[\frac{1 - e^{-j\Omega N}}{1 - e^{j(2\pi/N)k}e^{-j\Omega}}\right], \tag{9.31}$$

where the following equality and derivation were exploited:

$$\sum_{n=0}^{N-1} r^n = \frac{1 - r^N}{1 - r} \tag{9.32}$$

$$\sum_{n=0}^{N-1} e^{j\left(\frac{2\pi}{N}k\right)n}e^{-j\Omega n} = \sum_{n=0}^{N-1} e^{j\left[\frac{2\pi}{N}k-\Omega\right]n} = \frac{1 - e^{j\left[\frac{2\pi}{N}k-\Omega\right]N}}{1 - e^{j\left[\frac{2\pi}{N}k-\Omega\right]}} = \frac{1 - e^{-j\Omega N}}{1 - e^{j\left[\frac{2\pi}{N}k-\Omega\right]}} \tag{9.33}$$

Equation (9.31) represents summation of shifted oscillatory functions:

$$\left[\frac{1 - e^{-j\Omega N}}{1 - e^{j(2\pi/N)k}e^{-j\Omega}}\right] = e^{-j\Omega(N-1)/2}e^{-j\pi k/N}\frac{\sin\left(\Omega N/2\right)}{\sin\left(\Omega/2 - \pi k/N\right)}. \tag{9.34}$$

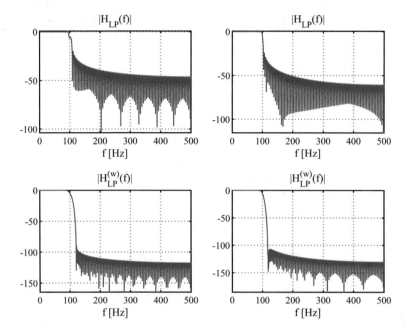

Fig. 9.8: Magnitude responses of two FIR filters designed by inverse DFT method using the program 9.5: (left) frequency oversampling $K = 1$, (right) $K = 4$, (top) without Kaiser window, (down) with Kaiser ($\beta = 10$)

Therefore the frequency response $H(\Omega)$ of the designed filter (9.31) is oscillatory between points $H(k)$. In order to improve the designed filter and to reduce the oscillations, calculated $h(n)$ is multiplied with a chosen spectral window.

Exercise 9.5 (FIR Filter Design Using Inverse DFT). Matlab code implementing the FIR filter design method using inverse DFT is given in Listing 9.5– see Fig. 9.8. Calculated weights of the FIR filter are not windowed in it. Do the windowing and plot the magnitude and phase filter response. Compare them in one figure with characteristics of the filter designed using the window method and the same window. Test different oversampling orders in frequency domain, set K=1,2,4. Compare your filters with filters generated by the Matlab function fir2(). Calculate and plot magnitude of the function (9.34) of variable Ω.

Listing 9.5: Matlab code of inverse DFT method of FIR filter design

```
1   % lab09_ex_inv_dft.m
2   clear all; close all;
3
4   % Parameters
5   fs = 2000;                   % sampling frequency (Hz)
6   f0=100; f1=200; f2=300;      % cut-off frequencies
7   M = 100;                     % half of the filter weights
8   K = 4;                       % oversampling order in frequency domain
9   N = 2*M + 1;                 % number of weights {h(n)} (b(n) in IIR filters) - always odd
10  n = -M :1: M;                % non-causal weights indexes
11  NK = N*K; if(rem(NK,2)==1) NK=NK+1; end   % after oversampling in frequency
12  df = fs/NK; f = df*(0 : NK/2);            % frequencies
13  H0 = zeros(1,NK/2+1);        % initialization
14  H1 = ones(1,NK/2+1);         % initialization
15  % Low-Pass
16  ind = find( f<=f0 );
17  HLP = H0; HLP(ind) = ones(1,length(ind)); HLP(ind(end))=0.5;
18  % High-Pass
19  ind = find( f>=f0 );
20  HHP = H0; HHP(ind) = ones(1,length(ind)); HLP(ind(1))=0.5;
21  % Band-Pass and Band-Stop
22  ind1 = find( f< f1 );
23  ind2 = find( f<=f2 );    %
24  ind  = find( ind2 > ind1(end) );
25  HBP = H0; HBP(ind) = ones(1,length(ind));  HBP(ind(1))=0.5; HBP(ind(end))=0.5;
26  HBS = H1; HBS(ind) = zeros(1,length(ind)); HBS(ind(1))=0.5; HBS(ind(end))=0.5;
27
28  % Our choice: LP, HP, BP, BS
29  H = HBS;                                  % copy
30  H(NK:-1:NK/2+2) = H(2:NK/2);              % by symmetry
31  h = ifft( H );                            % inverse DFT
32  h = [ h(M+1:-1:2) h(1:M+1) ];             % selection of 2M+1 weights
33  figure; stem(n,h); title('h(n)'); grid; pause   % figure
```

9.5 Weighted Least-Squares Method

In this section we will describe FIR filter design method in which filter weights $h(n)$ result from least-squares fitting of their DtFT $H(\Omega_k)$ to the desired filter frequency response $H_d(\Omega_k)$, specified by a user. We limit our discussion to the design case described by Eq. (9.14), i.e. symmetrical $N = 2M + 1$ weights: $h(N-k) = h(k), k = 0, 1, 2, ..., 2M$.

In the beginning, let us assume that Ω_k of our requirements result from DFT frequency sampling grid:

$$\Omega_k = \frac{2\pi}{L}k, \quad k = 0, 1, 2, ..., L-1. \tag{9.35}$$

We would like to find such weights $h(n)$, fulfilling Eq. (9.14), for which the following LS error is minimal:

$$E_1 = \sum_{k=0}^{L-1} |H(\Omega_k) - H_d(\Omega_k)|^2 = \sum_{k=0}^{L-1} |M(\Omega_k) - M_d(\Omega_k)|^2. \tag{9.36}$$

From Parseval's equality telling about energy preservation by the Fourier transform (see Table 4.2)), minimization of the cost function E_1 (9.36) corresponds to minimization of a cost function E_2:

$$E_2 = \sum_{n=-(L-1)/2}^{(L-1)/2} |h(n) - h_d(n)|^2. \tag{9.37}$$

The error (9.36) is minimal when $H(\Omega_k) = H_d(\Omega_k)$ for all Ω_k, therefore the optimal filter weights are equal to the inverse DFT of our requirements, similarly as in the previous section.

The situation is different when number of points L (requirements), specified in the frequency domain, is bigger than number of filter weights $N = 2M + 1$ to be found. In this situation the error sum (9.37) can be divided into two parts:

$$E_2 = \sum_{n=-M}^{M} |h(n) - h_d(n)|^2 + 2 \sum_{n=M+1}^{(L-1)/2} |h_d(n)|^2 \tag{9.38}$$

and the cost function E_2 reaches minimum when the first term in (9.38) is the smallest, i.e. as designed filter weights are taken only samples $h_d(n), n = -M, ..., 0, ..., M$, resulting from inverse DFT of the $H_d(\Omega_k)$.

Finally, let us consider the most general case when radial frequencies $\Omega_k, k = 0, 1, 2, ... L - 1$, are sampled non-uniformly and we are specifying L values $A(\Omega_k)$ in Eq. (9.14). Our goal is to find the $M + 1$ weights of the FIR filter (half of them) by means of solving the following over-determined set of equations ($L > M + 1$) in least-squares sense:

$$\begin{bmatrix} 2\cos(\Omega_0 M) & 2\cos(\Omega_0(M-1)) & \cdots & 2\cos(\Omega_0) & 1 \\ 2\cos(\Omega_1 M) & 2\cos(\Omega_1(M-1)) & \cdots & 2\cos(\Omega_1) & 1 \\ \vdots & \vdots & \ddots & 1 & 1 \\ 2\cos(\Omega_{L-1} M) & 2\cos(\Omega_{L-1}(M-1)) & \cdots & 2\cos(\Omega_{L-1}) & 1 \end{bmatrix} \cdot \begin{bmatrix} h(0) \\ h(1) \\ \vdots \\ h(M) \end{bmatrix} = \begin{bmatrix} A(\Omega_0) \\ A(\Omega_1) \\ \vdots \\ A(\Omega_{L-1}) \end{bmatrix} \tag{9.39}$$

The matrix in Eq. (9.39) has dimensions $L \times (M+1)$. Therefore for $L > M+1$ we have more equations than unknowns and we are finding the solution in the least-squares sense.

After introducing a vector $\mathbf{e}$ of the LS approximation error, the Eq. (9.39) can be rewritten as follows:

$$\mathbf{Fh} \cong \mathbf{a}_d, \quad \mathbf{Fh} = \mathbf{a}_d + \mathbf{e}. \tag{9.40}$$

After *left* multiplication of both sides of Eq. (9.40) by the matrix $\mathbf{F}$ transposition, we have

$$\mathbf{F}^T \mathbf{F} \mathbf{h} = \mathbf{F}^T \mathbf{a}_p + \mathbf{F}^T \mathbf{e}. \tag{9.41}$$

For the optimal LS solution the error vector is orthogonal to rows of the matrix $\mathbf{F}^T$. Therefore the last term in Eq. (9.41) is equal to zero, which gives

$$\mathbf{F}^T \mathbf{F} \mathbf{h} = \mathbf{F}^T \mathbf{a}_d, \tag{9.42}$$

and after multiplication of both sides of Eq. (9.42) by the matrix $(\mathbf{F}^T \mathbf{F})^{-1}$, the designed filter weight is obtained:

$$\mathbf{h} = \left(\mathbf{F}^T \mathbf{F}\right)^{-1} \mathbf{F}^T \mathbf{a}_d = \mathrm{pinv}(\mathbf{F}) \mathbf{a}_d. \tag{9.43}$$

The function pinv() calculates the matrix $\mathbf{F}$ pseudo-inverse.

The weighted LS solution is obtained when the following weighted error function is used, with coefficients w_k specifying the difference significance:

$$E = \sum_{k=0}^{L-1} w_k |A(\Omega_k) - A_d(\Omega_k)|^2. \tag{9.44}$$

When these weights are put one-by-one on the main diagonal of the square matrix $\mathbf{W}$, the weighted LS solution is defined by following equations:

$$\mathbf{F}^T \mathbf{W} \mathbf{F} \mathbf{h} = \mathbf{F}^T \mathbf{W} \mathbf{a}_p \tag{9.45}$$

$$\mathbf{h} = \left(\mathbf{F}^T \mathbf{W} \mathbf{F}\right)^{-1} \mathbf{F}^T \mathbf{W} \mathbf{a}_p. \tag{9.46}$$

Exercise 9.6 (FIR Filter Design Using Weighted Least-Squares (WLS) Approach). Matlab code implementing the FIR filter design method using the WLS technique is presented in Listing 9.6 but only for an exemplary low-pass filter. Modify the program and allow calculation of HP, BP, and BS filter weights also. Add to the program a figure of $H(z)$ roots placements and plots of filter magnitude and phase responses. Compare your filters with filters calculated by the Matlab function firls(). See Fig. 9.9.

Listing 9.6: Matlab code of weighted least-squares (WLS) method of FIR filter design

```
1   % lab09_ex_wls.m
2     clear all; close all;
3
4   % Parameters
5   fs = 2000;               % sampling frequency (Hz)
6   f0 = 100;                % cut-off frequency
7   M = 100;                 % half of the filter weights, N=2M+1
8   K = 4;                   % oversampling order in frequency domain
9   % P points of frequency response Ad() for ang. frequencies 2*pi*p/P, p=0,1,...,P-1
10  P = K*2*M;               % number of points (even; P >= N=2M+1)
```

```
11   L1 = floor(f0/fs*P),        % number of first FR samples equal gain=1
12   Ad = [ ones(1,L1) 0.5 zeros(1,P-(2*L1-1)-2) 0.5 ones(1,L1-1)];
13   Ad = Ad';
14   % Chose weighting coeffs w(p), p=0,1,2,...,P-1, for Pass/Trans/Stop regions
15   wp = 1;                      % weight for PassBand
16   wt = 1;                      % weight for TransientBand
17   ws = 10000;                  % weight for StopBand
18   w = [ wp*ones(1,L1) wt ws*ones(1,P-(2*L1-1)-2) wt wp*ones(1,L1-1) ];
19   W = zeros(P,P);              %
20   for p=1:P                    % matrix with weights on the main diagonal
21       W(p,p)=w(p);             %
22   end                          %
23   % Find matrix F being solution of the matrix equation  W*F*h = W*(Ad + err)
24   F = [];
25   n = 0 : M-1;
26   for p = 0 : P-1
27       F = [ F; 2*cos(2*pi*(M-n)*p/P) 1 ];
28   end
29   % Find h[n] minimizing error of W*F*h = W*Ad
30   % h = pinv(W*F)*(W*Ad);      % method #1
31   h = (W*F)\(W*Ad);            % method #2
32   h = [ h; h(M:-1:1) ]';
33   figure; stem(-M:M,h); title('h(n)'); grid; pause   % figure
```

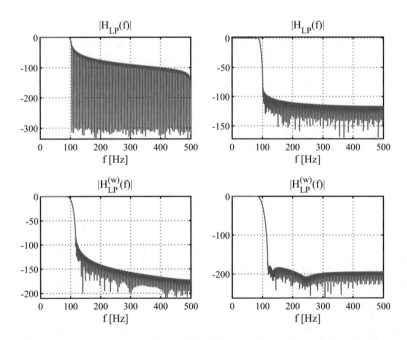

Fig. 9.9: Magnitude responses of two FIR filters designed by weighted least-squares method using the program 9.6: (left) frequency oversampling $K = 1$, (right) $K = 4$, (top) without Kaiser window, (down) with Kaiser ($\beta = 10$)

9.6 Min-Max Equiripple Chebyshev Approximation Method

In this section we learn one of the most powerful FIR filter design methods and the most frequently used in professional applications. It is doing Chebyshev min-max approximation of the filter frequency response, i.e. filter weights optimization aiming at minimization of the maximum filter frequency response error. The Parks-McClellan implementation of the Remez algorithm is presented here. In this algorithm the designed filter frequency response $A_d(\Omega)$ (Eq. (9.14)) is approximated by a sum of cosines taken with coefficients c_n. The following cost function $E(\Omega)$ is minimized:

$$E(\Omega) = W(\Omega) \left[\sum_{k=0}^{M} c_k \cos(\Omega k) - A_d(\Omega) \right], \tag{9.47}$$

where $W(\Omega)$ denotes non-negative weighting function, stressing significance of user expectations for different frequency regions. The Remez algorithm is based on the theorem specifying that a set of frequencies $\Omega_m, m = 1, 2, ..., M+2$, always exists for which the error function (9.47) gives only values $\pm\varepsilon$:

$$W(\Omega_m) \left[\sum_{k=0}^{M} c_k \cos(\Omega_m k) - A_d(\Omega_m) \right] = (-1)^m \varepsilon, \quad m = 1, 2, ..., M+2 \tag{9.48}$$

which are extreme. Let us assume that frequencies of these extremes are known. In such situation, we obtain from Eq. (9.48) a set of $M+2$ equations with $M+2$ unknowns:

$$
\begin{bmatrix}
1 & \cos(\Omega_1) & \cdots & \cos(M\Omega_1) & 1/W(\Omega_1) \\
1 & \cos(\Omega_2) & \cdots & \cos(M\Omega_2) & -1/W(\Omega_2) \\
\vdots & \vdots & \ddots & \vdots & \vdots \\
1 & \cos(\Omega_{M+1}) & \cdots & \cos(M\Omega_{M+1}) & (-1)^{M+1}/W(\Omega_{M+1}) \\
1 & \cos(\Omega_{M+2}) & \cdots & \cos(M\Omega_{M+2}) & (-1)^{M+2}/W(\Omega_{M+2})
\end{bmatrix}
\cdot
\begin{bmatrix}
c_0 \\ c_1 \\ \vdots \\ c_M \\ \varepsilon
\end{bmatrix}
=
\begin{bmatrix}
A_d(\Omega_1) \\ A_d(\Omega_2) \\ \vdots \\ A_d(\Omega_{M+1}) \\ A_d(\Omega_{M+2})
\end{bmatrix}
$$
$$\tag{9.49}$$

Solving it, we obtain $M+1$ approximation coefficients c_k and amplitude of oscillation ε. For consecutive angular frequencies Ω_m of the error extreme, the approximation error is equal to: $-\varepsilon, +\varepsilon, -\varepsilon, +\varepsilon, ...$ The angular frequencies Ω_m are not known. The Remez algorithm aims at their iterative approximate finding and, then, at the end, solving the Eq. (9.49). The algorithm consists of the following steps:

1. initialization of $M+2$ values of Ω_m,
2. solving the Eq. (9.49): finding coefficients $c_k, k = 0, 1, 2, ..., M$ and ε,
3. checking whether the amplitude of error function $E(\Omega)$ for Ω from the interval $[0, \pi]$ is bigger than calculated ε; if not, then STOP,

4. otherwise finding $M+2$ angular frequencies for which the error function $E(\Omega)$ has extreme values, setting them as Ω_m and returning to step 2.

Finding new extreme of the error function can be done by Lagrange interpolation technique.

For filter length $N = 2M + 1$ and symmetrical weights $h(N - k) = h(k)$, we have from Eq. (9.14):

$$A(\Omega) = \sum_{k=0}^{M-1} 2h(k)\cos(\Omega(M-k)) + h(M) = \sum_{k=0}^{M} h_k \cos(\Omega(M-k)) = \sum_{k=0}^{M} h_{M-k}\cos(\Omega k).$$

$$(9.50)$$

where

$$h_k = \begin{cases} 2h(k), & k = 0, 1, ..., M-1, \\ h(k), & k = M. \end{cases} \tag{9.51}$$

When we set $c_k = h_{M-k}$, $k = 0, 1, 2, ..., M$, then the Eqs. (9.47)–(9.49) directly describe the problem of filter frequency response $A_d(\Omega)$ approximation by the filter fulfilling condition (9.50). Therefore, after calculation of coefficients c_k by Remez algorithm, knowing that $c_k = h_{M-k}$ and remembering Eq. (9.51), we finally obtain

$$\mathbf{h} = \left\{ \frac{c_M}{2}, ..., \frac{c_2}{2}, \frac{c_1}{2}, c_0, \frac{c_1}{2}, \frac{c_2}{2}, ..., \frac{c_M}{2} \right\}. \tag{9.52}$$

Exercise 9.7 (Designing Min-Max Equiripple FIR Filters). The last exercise in this chapter is very easy. Since success of the Parks-McClellan implementation of the Remez algorithm strongly depends on the quality of frequency response interpolation, there is no possibility of taking a shortcut. Therefore it is better, in my opinion, to test existing good algorithm implementations, then spend hours on writing our own code offering, with big difficulties, worse results than other methods. Please, use Matlab function firpm(), generate filter weights for the same design tasks as before in Exercises 9.4, 9.5, and 9.6, and compare frequency responses of min-max filters (Fig. 9.10) with solutions offered by methods already discussed. Extend this Matlab code:

```
M=100; N=2*M+1; % filter length
fs=2000; f0=100; fpass1 = f0-10; fstop1 = f0+10; % frequencies
wpass = 1; wstop = 1; % weights for passband and stopband
h = firpm(N-1, [0 fpass1 fstop1 fs/2]/(fs/2), [1 1 0 0],...
[wpass wstop]); % function call
```

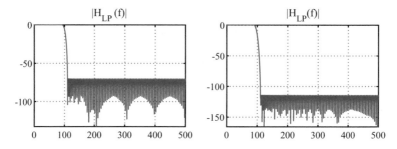

Fig. 9.10: Magnitude responses of two FIR filters designed by the Matlab function `firpm()`, implementing the Parks-McClellan algorithm. Generated by code from example 9.7: (left) `wpass=1, wstop=1`, (right) `wpass=1; wstop=10000`

9.7 Efficient FIR Filter Implementations

When talking about efficient FIR filter implementations, two things should be mentioned.

1. In order to ensure the desirable linear-phase response, the FIR filter weights should have specific symmetry, i.e. be "the same" in absolute value sense: $h(N-k) = h(k)$ or $h(N-k) = -h(k)$, where $k = 0...N$ and $N+1$ denotes number of filter coefficients. This means that number of performed multiplications can be reduced approximately by half by, first, doing addition (or subtraction) of the corresponding signal samples associated with "the same" filter weights, and, then, doing multiplication of the addition/subtraction results by half of the number of weights only. For example, LP, HP, BP, and BS filters with $N = 2*M$ we have

$$y(n) = h(M)x(n) + \sum_{1}^{M} h(M+k)[x(n+k) \pm x(x-k)]. \qquad (9.53)$$

2. Since FIR digital filtering has a form of linear convolution of two discrete-time signals, fast convolution algorithms, described in Sect. 6.6, can be applied in this case. Of course, one should remember to choose appropriate modification of the FFT-based circular convolution changing it into the linear one. A Reader, interested in this subject, is asked to read again more carefully the Sect. 6.6 and run programs presented in it.

When implementing FIR digital filters by means of fast convolution methods, a special care has been taken during implementation of on-line real-time FIR filtering using overlap methods, in particular when the filter impulse response is long and should be cut into sections too, similarly to the incoming signal samples. Now we will concentrate only on this subject and investigate it in Exercise 9.8.

Exercise 9.8 (Fast FFT-Based Sectioned Convolution in Matlab). In Sect. 6.6 fast convolution algorithms were described. Among others, the sectioned overlap-add and overlap-save methods. At present, we will concentrate only on the program of the overlap–save algorithm which was skipped before. Its Matlab code is presented in Listing 9.7. Analyze it and run. Compare its output with output of the Matlab function conv(). Specially analyze the second part of the program in which partition of filter weights is done also. Thanks to this operation the processing delay is significantly decreased which is beneficial when filter impulse response is very long. Find in the Internet samples of impulse response (IR) of any large acoustical object like cave or cathedral. Record your speech and convolve it with the downloaded IR using different methods.

Listing 9.7: Fast FFT-based sectioned convolution in Matlab

```
1   % lab09_ex_fastconv.m - fast signal convolution (FIR filtering) using FFT
2   clear all; close all
3
4   N=2048; M=128; x = randn(1,N); h = randn(1,M);
5   y1=conv(x,h);
6
7   % continuation of lab06_ex_fastconv.m
8
9   % Method #5: Fast convolution by pieces - OVERLAP-SAVE method
10  N1 = 2*M;                                   % length of one signal fragment
11  Lb = floor( (N-(M-1))/(N1-(M-1)) );         % number of signal fragments (blocks)
12  hz = [ h zeros(1,N1-M) ];                    % appending zeros to h(n)
13  Hz = fft(hz);                                % FFT of h(n)
14  for k = 0 : Lb-1                             % LOOP: fast conv of signal fragments
15      n1st = 1+k*(N1-(M-2)-1);                 % index of next signal fragment beginning
16      bx = x( n1st : n1st + N1-1 );            % cutting new samples of x(n) with overlap
17      Bx = fft( bx );                          % fft of the x(n) fragment
18      by = real( ifft( Bx .* Hz ) );          % fast convolution in frequency domain
19      y5(n1st+M-1+0 : n1st+M-1+(N1-M)) = by(M:N1); % remove first M-1 samples, store
20  end
21  Ny5 = length(y5); n5 = M:Ny5;
22  figure; plot(n5,y1(n5),'ro',n5,y5(n5),'bx'); title('y1(n) & y5(n)');
23  error5 = max(abs( y1(n5)-y5(n5) ))
24
25  % Method #6: Fast convolution by pieces - OVERLAP-SAVE method plus h(n) partition
26  Nx = N; Nh = M;                             % signal length, filter length
27  Nb = 32;                                     % partition length of h(n)
28  Lb = ceil(Nx/Nb);                            % number of signal fragments
29  Lh = ceil(Nh/Nb);                            % number of filter fragments
30  for k=1:Lh                                   % FFTs of filter weight fragments
31      H(k,1:2*Nb) = fft( [h((k-1)*Nb+1:k*Nb) zeros(1,Nb)] ); % with zeros
32  end
33  bx = zeros(1,2*Nb); X = zeros(Lh,2*Nb); y6 = []; 
34  for k = 0 : Lb-1                             % FOR ALL SIGNAL FRAGMENTS
```

```
35        bx = [ bx(Nb+1:2*Nb) x(k*Nb+1:k*Nb+Nb) ]; % take next signal fragment
36        X = [ fft(bx); X(1:Lh-1,:) ];              % calculate its FFT and store
37        XH = X.*H;                                  % # fast
38        by = real( ifft(sum(XH)) );                % # convolution
39        y6 = [ y6 by(Nb+1:2*Nb) ];                  % remove beginning, store result
40     end
41     Ny = min( length(y1), length(y6) ); n6 = Nh+1:Ny;
42     y1 = y1(n6); y6 = y6(n6);
43     figure; plot(n6,y1,'ro',n6,y6,'bx'); title('y1(n) & y6(n)');
44     error6 = max(abs(y1-y6)), pause
```

9.8 Summary

FIR digital filters are one of the most frequently applied. They are ready-to-use off-the-shelf solutions that are always present close to cash register at each DSP mini-mark just near the corner. They are simple and effective like rapid-drying glue. What is the most important about the FIR filters?

1. FIR digital filters are very simple: they calculate running weighted average of input signal samples. They do not have a loop back and they are always stable (finite input causes finite output). It is straightforward to guarantee their linear-phase response: the filter weights should be symmetrical or asymmetrical with respect to their center. The FIR filter is frequency selective when many signal samples, typically hundreds of them, are multiplied by corresponding filter coefficients and this is their main drawback. However, for many low-frequency kHz applications, like acoustical ones, this requirement is not computationally demanding even for low-cost DSP controllers and processors. And for high-speed tasks very fast GHz processors are available.

2. There are many methods for FIR filter design but even very simple ones, like the window method, allow to design very good filters with negligible oscillations in the pass-band and very strong signal attenuation in the stop-band. It is achieved by using appropriate windows for smooth tapering the shape of theoretically derived weights, as in window method, or by using special optimization procedures, like minimization of maximum error of fitting the desired filter frequency response (Chebyshev approximation of filter frequency response using Remez/Parks-McClellan algorithm). The sharpness of filter transition edge is always very easily obtained by increasing the number of filter weights.

3. FIR digital filters are mainly used for modification of signal frequency content (*pass* or *stop*). But they can be also used as *special task* filters, for

example, as special phase shifters (Hilbert filter) and differentiation filters. Due to significance of this topic, it will be discussed in a separate chapter.

4. Inverse DFT method allows for design of FIR digital filters for amplitude and phase correction (equalization) of non-ideal signal processing path, for example, in measurement, acoustical and telecommunication systems.

5. FIR filtering has a form of discrete-time convolution of signal samples and filter weights. For this reason, fast FFT-based convolution algorithms can be used for its fast implementation. Their application is especially beneficial for very long FIR filters. In case of very long signals and on-line filtering, fast sectioned convolution methods should be used: the overlap–add or the overlap–save. In order to decrease the delay introduced by them, a long filter impulse response can be also divided into sections.

9.9 Private Investigations: Free-Style Bungee Jumps

Exercise 9.9 (Mastering the Perfectness: Do the Deeper Dive). Apply any type of FIR digital filter in any practical signal processing example discussed so far in these books, i.e. to speech, music, ECG, power supply voltage, ... everything you want.

Exercise 9.10 (Flight to Unknown? What Is on the Other Side of the Moon?). Find in Internet any recorded signal, do frequency analysis of it using DFT/FFT and try to filter out some of signal components. In case of speech, music, or different sounds hearing effects of filtration might be interesting. Dear Reader, the floor is yours!

References

1. A. Antoniou, *Digital Filters. Analysis, Design, Applications* (McGraw-Hill, New York, 1993)
2. H. Baher, *Analog and Digital Signal Processing* (Wiley, Chichester, 2001)
3. L.B. Jackson, *Digital Filters and Signal Processing with Matlab Exercises* (Kluwer Academic Publishers, Boston, 1996; Springer, New York, 2013)
4. R.G. Lyons, *Understanding Digital Signal Processing* (Addison-Wesley Longman Publishing, Boston, 1996, 2005, 2010)
5. A.V. Oppenheim, R.W. Schafer, *Discrete-Time Signal Processing* (Prentice Hall, Englewood Cliffs, 1989)
6. T.W. Parks, C.S. Burrus, *Digital Filter Design* (Wiley, New York, 1987)
7. B. Porat, *A Course in Digital Signal Processing* (Wiley, New York, 1997)
8. J.G. Proakis, D.G. Manolakis, *Digital Signal Processing. Principles, Algorithms, and Applications* (Macmillan, New York, 1992)

9. L.D. Thede, *Analog and Digital Filter Design Using C* (Prentice Hall, Upper Saddle River, 1996)

10. T.P. Zieliński, *Cyfrowe Przetwarzanie Sygnalów. Od Teorii do Zastosowań (Digital Signal Processing. From Theory to Applications)* (Wydawnictwa Komunikacji i Łączności (Transport and Communication Publishers), Warszawa, Poland, 2005, 2007, 2009, 2014)

Chapter 10
FIR Filters in Signal Interpolation, Re-sampling, and Multi-Rate Processing

Who would not like to have a perfect micro-macro-scope for seeing a single electron and the whole solar system? This is an introduction to our dream of multi-resolution signal processing.

10.1 Introduction

In this chapter we discuss techniques of changing sampling ratio of signals that have been already sampled by analog-to-digital converters, i.e. they have a form of vectors of numbers. We are interested in having *more* samples per second, i.e. in signal *up-sampling*—increasing the sampling ratio, and in having *less* samples per second, i.e. in signal *down-sampling*—decreasing the sampling ratio. Typically, integer-value *K*-times up-sampling and integer-value *L*-times down-sampling are done. Let describe separately in the beginning these two DSP blocks.

In a digital signal *K-times up-sampler* of integer order, we are calculating new signal samples lying between samples that we already have. The old samples remain unchanged. The interpolation is done this way that, first, $K-1$ zeros are put between each two old signal samples, and, then, the signal undergoes low-pass smoothing operation, causing that the inserted zeros are adjusted to the signal *envelope*. The low-pass filter cut-off frequency is equal to one-half of the original signal sampling frequency $\left(\frac{f_s}{2} \right)$, no surprise, the same time being equal to $\frac{1}{K}$ of the new sampling frequency, being K time bigger: $\left(\frac{K \cdot f_s}{2} \right)$.

In a digital signal *L-times down-sampler* of integer order, we are leaving every *L*-th signal sample, removing the remaining ones. Very simple, but risky. The Nyquist sampling theorem specifies that the highest signal frequency component should have more than two samples per period. We cannot violate (breach) this rule during signal down-sampling, otherwise high-frequency signal components will look as low-frequency ones. In order to avoid this effect, the signal frequency content should be reduced to the new sampling ratio be-

© Springer Nature Switzerland AG 2021
T. P. Zieliński, *Starting Digital Signal Processing in Telecommunication Engineering*, Textbooks in Telecommunication Engineering,
https://doi.org/10.1007/978-3-030-49256-4_10

ing L-times lower. This task performs our *super-hero*, this time having a form of low-pass filter with 3 dB cut-off frequency $\frac{f_s}{2L} - f_{back}$. The slight frequency value decrease by f_{back} is needed since, in order to avoid aliasing, the filter stop-band should start exactly from the frequency $\frac{f_s}{2L}$ and it is reached after a filter some transition band. The down-sampling module is called a decimator in DSP world.

When a cascade of K-times signal up-sampler and L-times signal down-sampler is required, it looks initially that two signal processing scenarios are possible but in fact only one of them is allowed. Let us explain it on the following example in which digital audio from DAB or DVB broadcast with 48 ksps (kilo-samples per second) should be changed into digital signal for analog FM broadcast transmission, having 32 ksps. In this case $K = 2$ and $L = 3$. In the first, correct method, signal is initially 2-times up-sampled from 48 to $= 2 \cdot 48 = 96$ ksps and then 3-times down-sampled from 96 to $96/3 = 32$ ksps. In the second, wrong approach, the signal is first 3-times down-sampled from 48 to $48/3 = 16$ ksps and, then, 2-times up-sampled from 16 to $2 \cdot 16 = 32$ ksps. One should remember that in this cascade, an interpolator should always proceed a decimator! Why? Because in decimator, in signal down-sampler, the signal frequency band is reduced in order not to violate Nyquist sampling theorem, as stated above. And after the low-pass filter in decimator some signals frequency components are lost for ever. The second important remark is that in cascade of two low-pass filters, one in interpolator and one in decimator, only the filter with more narrow bandwidth can be left, the second one is redundant and can be removed.

We are very close to the introduction end. Is any further explanation still required? Yes, it is. An issue of required interpolation/decimation low-pass filter length. There is no problem when interpolation and decimation ratios are small, like 2, 3, 4, 5. The filter bandwidth is relatively high in such case and the filter frequency response flatness in the pass-band and sharpness in the transition band can be ensured with ease by short filters. However, we have a problem when interpolation and decimation ratios are very high, because in this case long filters are required. Let us explain the problem on example of changing sampling rate of CD disc $f_s = 44,100$ Hz to sampling rate of DAB radio $f_s = 48,000$ Hz. The common multiple of these frequencies is equal to 7,056,000, i.e. we should up-sample 44.1 kHz signal $K = 160$ times to 7.056 MHz and then down-sample the signal $L = 147$ times to 48 kHz. The up-sampling/down-sampling ratios of high and low-pass filters of interpolator and decimator are high and both filters have to be long. So what to do? Firstly, we can resign from perfect frequency change and accept frequency very close to 48 kHz, when such decision allows usage of smaller up/down re-sampling ratios. Secondly, instead of using one long filter we could use cascade of a few shorter filters and obtain the same result. For example, we can exchange interpolation filters with up-sampling $K = 160$ by cascade of shorter filters with up-sampling $K_1 = 4, K_2 = 4, K_3 = 5$, and $K_4 = 2$, requiring less weights and less computations.

The last hero of this chapter is *polyphase filtering* structure, very well-known by everybody in the *DSP City*. In digital up-samplers, very often, carelessly, we are multiplying filter weights by zero samples put between original signal samples. What is a sense of multiplying any number by zero? None. In turn, in digital down-samplers, very often, carelessly, we are low-pass filtering a signal and leaving every L-th sample only, removing the rest of calculated samples. What is a sense to calculate something and do not use it? None. In polyphase filtering we do only calculations which are needed! The filter weights are decomposed into modulo K or L sequences, they are separately convoluted with signal samples and individual results are combined together. This will be really a big fun to see it in work! And we will have this pleasure in this chapter.

In Matlab we have the following functions for signal re-sampling: `iterp()`, `decimate()`, `resample()`, and `upfirdown()`. It is a good idea to do first some home experiments with them.

10.2 Signal 1:K Up-Sampling—Simple Interpolation

Signal K-times up-sampling aims at having K times samples more, i.e. K instead of one. During up-sampling the sampling ratio is changed from f_{s1} to $f_{fs2} = Kf_{s1}$. How it is done? The computational procedure is presented in Fig. 10.1. First, $K-1$ zeros (filled red dots in the figure) are inserted between every two original signal samples and the such modified signal is processed by a low-pass filter with cut-off frequency equal to the original signal frequency bandwidth, i.e. half of the original sampling frequency $f_0 = f_{s1}/2$. From the chapter on FIR filters and description of the window method, we know that weights of such filter are equal to:

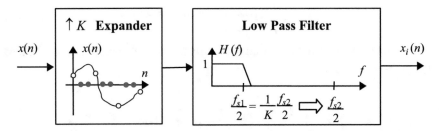

Fig. 10.1: Block diagram of signal interpolator increasing K-times signal sampling ratio. First, $K-1$ zeros are appended between every two signal samples by an expander, then the signal is smoothed with a low-pass filter, passing only the original signal bandwidth (having cut-off frequency equal to half of the original sampling frequency $f_{s1}/2$ and to $\frac{1}{K}$ of $f_{s2}/2 = \frac{Kf_{s1}}{2}$)

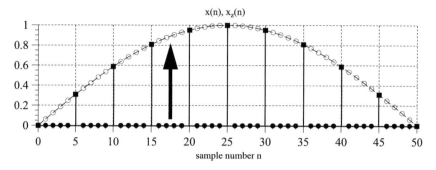

Fig. 10.2: Graphical illustration of signal smoothing operation, performed by the low-pass filter present in digital interpolator: inserted zeros are shifted up-or-down to the signal *envelope* [21]

$$h(n) = K \frac{\sin\left(2\pi(\frac{f_0}{f_{s2}})n\right)}{\pi n}, \quad f_0 = \frac{f_{s1}}{2} = \frac{\frac{f_{s2}}{2}}{K}, \quad n \in \mathbb{Z}. \tag{10.1}$$

Of course, the number of filter coefficients has to be limited. Therefore, in order to ensure flatness of the filter frequency response in the pass-band and good filter attenuation in the stop-band, the filter coefficients should be multiplied by a window function $w(n)$:

$$h_w(n) = h(n) \cdot w(n), \quad n = -M, ..., 0, ..., M, \tag{10.2}$$

for example, a Kaiser one, having frequency spectrum with low side-lobes level. The low-pass filter is needed for smoothing the signal, which becomes impulsive after zero insertion, and for restoring its original shape. Thanks to filtration the inserted zeros are shifted up-or-down to the signal *envelope* which is presented in Fig. 10.2.

At present we try to justify the choice of the LP filter. Let us denote the signal, with zeros added by an expander, as:

$$x_z(n) = \begin{cases} x(\frac{n}{K}), & n = 0, \pm K, \pm 2K, ... \\ 0, & \text{otherwise} \end{cases} \tag{10.3}$$

Now we calculate the DtFT spectrum of this signal, sampled with frequency K times higher, i.e. $f_{s2} = K f_{s1}$:

$$X\left(\frac{f}{f_{s2}}\right) = \sum_{n=-\infty}^{\infty} x_z(n) e^{-j2\pi \frac{f}{Kf_{s1}} n} = \sum_{m=-\infty}^{\infty} x_z(m \cdot K) \cdot e^{-j2\pi \frac{f}{Kf_{s1}}(m \cdot K)} =$$

$$\sum_{m=-\infty}^{\infty} x(m) \cdot e^{-j2\pi \frac{f}{f_{s1}}(m)} = X\left(\frac{f}{f_{s1}}\right). \tag{10.4}$$

Taking only every mK-th sample in summation (10.4) we avoid multiplication of Fourier harmonics by zeros. We see that the spectrum is periodic with period f_{s1}. In

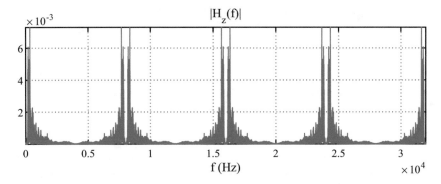

Fig. 10.3: Spectrum of 8 kHz speech after $K = 4$ expander, i.e. after zero insertion only. New sampling ratio is 4-time higher and the speech spectrum repeats $K = 4$ times in the range 0–32 kHz

Fig. 10.3 a spectrum of 8 kHz speech signal after $K = 4$ time expander is shown. The new frequency is equal to 32 kHz and in frequency range 0–32 kHz the signal spectrum repeats $K = 4$ times. Therefore its *mirror copies/images* have to be removed by a low-pass filter with cut-off frequency $\frac{f_{s1}}{2}$, i.e. $\frac{1}{K}$ of the new Nyquist frequency $\frac{f_{s2}}{2} = \frac{K f_{s1}}{2}$. Why the filter should have to reach the stop-band at the frequency $\frac{f_{s1}}{2}$? Because the original signal was sampled properly and there is no signal component with frequency higher than $\frac{f_{s1}}{2}$.

A program implementing digital up-sampler (interpolator), in simple but not fast way, is presented in Listing 10.1.

Exercise 10.1 (Testing Digital 1:K Up-Sampler). Use program 10.1. Generate different sums of sines and cosines and observe original and interpolated signals. Knowing FIR filter delay, equal to half of its length, try to synchronize input and output signals—display them in one figure. Use different values of: interpolation order K, filter length M, and Kaiser window parameter β (function firl() allows it). Compare results with the Matlab function interp(). Do one or two experiments with interpolation of real-world signals like speech or music. Read them to Matlab using functions audioread() or wavread(). For example, up-sample 6 times your speech (or singing) recorded at 8 kHz to 48 kHz, and then add it to music sampled at 48 kHz. This is an example of digital sound mixing: a Karaoke Bar example from the book beginning. Add also samples of speech (or singing) and music without speech interpolation.

Listing 10.1: Demonstration of simple signal up-sampling

```
1  % lab10_ex_interp.m - signal interpolation
2  clear all; close all;
3
4  K=5; M=50; N=2*M+1; Nx=200; ni = 1:200;     % parameters
5  x = sin(2*pi*(0:Nx-1)/20); figure; plot(x); pause  % signal and its plot
6  xz = zeros(1,K*Nx);                          % # zero
7  xz(1:K:end) = x;                             % # insertion
8  h = K*fir1(N, 1/K);                          % filter design
9  xi = filter(h,1,xz); % xi = conv(xz,h);      % filtering
10 figure; plot(xi); pause                      % display result
```

10.3 Signal L:1 Down-Sampling—Simple Decimation

Signal down-sampling operation is very easy to explain from practical point of view: if you want to have L-times samples less of the signal that has been already sampled and to take only every L-th sample you have, i.e. to decrease L-times its sampling ratio from f_{s1} to $f_{s2} = \frac{f_{s1}}{L}$, you should first reduce the signal bandwidth L-times according to the sampling theorem. Therefore, you should first use a low-pass filter leaving only $\frac{1}{L}$-th of the original signal frequency band equal to $\frac{f_{s1}}{2}$. After that, you can take only every L-th sample and remove the rest. Block diagram of the digital down-sampler is shown in Fig. 10.4.

The crucial difference between the low-pass filter used in interpolation and decimation is that the decimation low-pass filter should reach its stop-band at the frequency $\frac{f_{s2}}{2}$ while the interpolation one—reach 3 dB decrease of its pass-band at frequency $\frac{f_{s1}}{2}$. Therefore the decimation filter weights $g(n)$ should be designed for the slightly lower cut-off frequency $\frac{f_{s1}/2}{L} - f_{back}$:

$$g(n) = \frac{\sin\left(2\pi(\frac{f_0}{f_{s1}})n\right)}{\pi n}, \quad f_0 = \frac{\frac{f_{s1}}{2}}{L} - f_{back}, \quad n \in \mathbb{Z}. \tag{10.5}$$

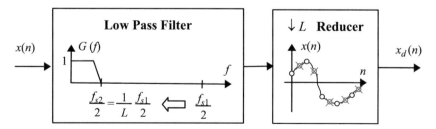

Fig. 10.4: Block diagram of signal decimator, decreasing L-times signal sampling ratio from f_{s1} to $f_{s2} = \frac{f_{s1}}{L}$. First, the signal bandwidth is reduced L-times (i.e. adjusted to the new sampling ratio according to the Nyquist theorem), then, only every L-th signal sample is left

Of course, weights from Eq. (10.5) have to be multiplied by a window function, the same way as in case of interpolation filter (Eq. (10.2)): $g_w(n) = g(n) \cdot w(n)$.

Program in Listing 10.2 shows the simplest version of digital down-sampler.

Exercise 10.2 (Testing Digital L:1 Down-Sampler). Use program 10.2. Repeat all tasks from Exercise 10.1 exchanging $1 : K$ interpolation with $L : 1$ decimation. At present, in Karaoke Bar, down-sample the music from 48 kHz to 8 kHz and add to your singing recorded at 8 kHz.

Listing 10.2: Demonstration of simple signal down-sampling

```
1  % lab10_ex_decim.m - signal decimation
2  clear all; close all;
3
4  L=3; M=50; N=2*M+1; Nx=L*500; nd = 1:200;        % parameters
5  x = sin(2*pi*(0:Nx-1)/100); figure; plot(x); pause   % signal and its plot
6  h = fir1(N, 1/L - 0.1*(1/L));                    % filter design
7  xd=filter(h,1,x); % xd = conv(x,h);              % filtering
8  xd = xd(1:L:end);                                % decimation
9  figure; plot(xd); pause                          % display result
```

10.4 Signal $K : L$ Re-sampling

Till now we know *only* how to increase or decrease signal sampling ratio by integer number of times. Why only *only*? We are also *only* doing single steps but can overcome the distance of kilometers! The same is with interpolation and decimation: you will *only* up-sample $K = 160$ times a song from your CD player and *only* down-sample the result $L = 147$ times, and will change the sampling frequency of the song $\frac{160}{147}$-times: from 44.1 kHz to 48 kHz! Now the recording is ready to be broadcast in DAB digital radio or added to the sound track of your favorite movie. In Fig. 10.5 there is presented a serial connection of K-order digital interpolator and L-order digital decimator, allowing implementation of many different signal re-sampling tasks.

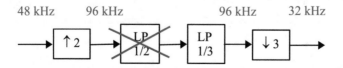

Fig. 10.5: Block diagram of signal re-sampler, changing signal sampling ratio K/L-times. In our example $K = 2$ and $L = 3$. It is a serial connection of K-times up-sampler (interpolator) and L-times down-sampler (decimator). From two filters working in a cascade, only the one with lower pass-band can be used [21]

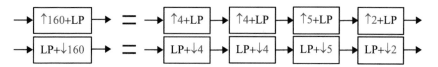

Fig. 10.6: Exemplary implementation of a high-order up-sampler (up) and down-sampler (down) as a serial connection of several low-order devices: $160 = 4 \cdot 4 \cdot 5 \cdot 2$. LP denotes a low-pass filter following the expander in the up-sampler and preceding the reducer in the down-sampler [21]

What should be remembered? In a cascade connection an interpolator should always proceed a decimator. When decimator is working first, its low-pass anti-aliasing filter is always removing some high-frequency components of the signal. They are lost but could *survive* in the second scenario when the signal is interpolated first. In such case, the interpolator LP filter, which is working first, is removing only high-frequency copies of the signal spectrum resulting from insertion of zero-value samples. The second fundamental rule is that only one low-pass filter is sufficient, the one with the more narrow pass-band. The second filter can be removed as shown in Fig. 10.5.

There is still one important issue to be addressed: the low-pass filter length. When interpolation or decimation order is very high, and the LP filter pass-band, $\frac{1}{K}$ or $\frac{1}{L}$, is very narrow, the FIR filter length has to be large in order to fulfilling frequency response requirements (very sharp transition at low-frequency). Therefore it is computationally more efficient to do interpolation/decimation in simple steps and use a cascade of lower order interpolators/decimators with short filter. This approach is illustrated in Fig. 10.6.

Exercise 10.3 (Testing Digital K:L Re-sampler). Use programs 10.1 and 10.2. Cascade the interpolator and decimator. Up-sample to 48 kHz the first signal, originally sampled at 32 kHz, and down-sample to 32 kHz the second signal, originally sampled at 48 kHz. Add signals at 32 kHz and 48 kHz. Try to do the same with signals sampled at 44.1 kHz (160/147 up-sampler) and 48 kHz (147/160 down-sampler). Is the re-sampling very slow? Why? Use the Matlab function `resample()`. What is going on?!

10.5 Matlab Functions for Signal Re-sampling

In this short section we learn and test Matlab functions dedicated for signal interpolation, decimation, and re-sampling. We apply them to real-world audio signals of speech and music.

Exercise 10.4 (Matlab Functions for Signal Re-sampling). Use program 10.3 for processing a reach sound music. Compare in one plot signal waveforms, original and modified. Do you hear the difference between them? Calculate noise level (SNR) introduced by signal filtering.

Listing 10.3: Re-sampling in Matlab

```matlab
% lab10_ex_reasampling_demo1.m
clear all; close all; subplot(111);

K = 3;    % UP-sampling K times
L = 2;    % DOWN-sampling L times

disp('Speech reading from HD ...');
%[x,fs,Nb ] = wavread('speech.wav'); x=x';
[x,fs ] = audio read('speech.wav'); x=x';

Nx=length(x);
dt=1/fs; t=dt*(0:Nx-1);
figure; plot(t,x); xlabel('t [s]'); title('x(t)'); grid; pause
soundsc(x,fs); pause

fsUP = K*fs;     % new K-times higher sampling frequency
fsDOWN = fsUP/L; % new L-times lower sampling frequency

% Interpolation via appending zeros in FFT spectrum: M times UP
disp('Speech interpolation via FFT ...'); pause
K = 3;                          % UP-sampling order
N = 2^14;                       % number of signal samples
Nz = (K-1)*N+1;                 % number of zeros to be added
y = x(1:N);                     % cut signal fragment
Y = fft(y);                     % perform its FFT
Y = [ Y(1:N/2) zeros(1,Nz) Y(N/2+2:N)]; % append zeros in the middle, ensure symmetry
yUP = K*real(ifft(Y));          % perform IFFT
figure; plot(dt*(0:N-1),y,'r.-',dt/K*(0:K*N-1),yUP,'b.-');
grid; xlabel('t [s]'); title('x(t) and xUP(t) - for FFT WITH ZEROS'); pause
soundsc(yUP,fsUP); pause

% Interpolation (K times UP) via Matlab function
disp('Speech interpolation via Matlab interp() ...');
xUP = interp(x,K);
figure; plot(t,x,'r.-',dt/K*(0:K*Nx-1),xUP,'b.-');
```

```
36   grid; xlabel('t [s]'); title('x(t) and xUP(t) - for INTERP()');
37   soundsc(xUP,fsUP); pause
38
39   % Decimation (L times DOWN) via Matlab function
40   disp('Speech decimation via Matlab decimate() ...');
41   xDOWN1 = decimate(xUP,L);
42   figure; plot(t,x,'r.-',dt/K*(0:K*Nx-1),xUP,'b.-',L*dt/K*(0:K*Nx/L-1)+dt/K,xDOWN1,'g.-')
       ;
43   grid; xlabel('t [s]'); title('x(t),  xUP(t),  xDOWN1(t)');
44   soundsc(xDOWN1,fsDOWN); pause
45
46   % Re-sampling (K times UP and L times DOWN) via Matlab function
47   disp('Speech resampling via Matlab resample() ...');
48   xDOWN2 = resample(x,K,L);
49   figure; plot(L*dt/K*(0:K*Nx/L-1)+dt/K,xDOWN1,'g.-',L*dt/K*(0:K*Nx/L-1),xDOWN2,'r.-');
50   grid; xlabel('t [s]'); title('xDOWN1(t),  xDOWN2(t)');
51   soundsc(xDOWN2,fsDOWN); pause
```

10.6 Fast Polyphase Re-sampling

Every mountain has its own Mount Everest. In signal re-sampling the polyphase
(PP) signal decomposition and the polyphase implementation of interpolator and
decimator filters are widely used. They are the highest peaks of the *Resampling*
mountains. Why? Because formalism of their mathematical description is rather
high and typically not tolerated by "people from flatland." Therefore our explanation
will be in half—mathematical and in half—experimental.

10.6.1 Polyphase Signal Decomposition

In polyphase K-order decomposition, signal samples $x(n)$ are divided into K sequences $x_k(n)$, starting from different sample number $k = 0, 1, 2, ..., K - 1$, and containing every K-th signal sample:

$$x_k(n) = x(k + m \cdot K), \quad k = 0, 1, 2, ..., K - 1, \quad -\infty < m < \infty. \qquad (10.6)$$

For example, for $K = 4$ decomposition we have (starting from $m = 0$):

$$\begin{aligned}
x_0(n) &= x(0), x(4), x(8), x(12), x(16), x(20), ... \qquad (10.7)\\
x_1(n) &= x(1), x(5), x(9), x(13), x(17), x(21), ...\\
x_2(n) &= x(2), x(6), x(10), x(14), x(18), x(22), ...\\
x_3(n) &= x(3), x(7), x(11), x(15), x(19), x(23), ...
\end{aligned}$$

Let us denote as $x_k^{(z)}(n)$ the polyphase sequences with appropriate number of zeros inserted between their elements. In our example we obtain

$$x_0^{(z)}(n) = x(0),\ 0\ ,0\ ,0\ ,x(4),\ 0\ ,0\ ,0\ ,x(8),\ 0\ ,0\ ,0\ ,x(12),...\qquad (10.8)$$
$$x_1^{(z)}(n) = x(1),\ 0\ ,0\ ,0\ ,x(5),\ 0\ ,0\ ,0\ ,x(9),\ 0\ ,0\ ,0\ ,x(13),...$$
$$x_2^{(z)}(n) = x(2),\ 0\ ,0\ ,0\ ,x(6),\ 0\ ,0\ ,0\ ,x(10),\ 0\ ,0\ ,0\ ,x(14),...$$
$$x_3^{(z)}(n) = x(3),\ 0\ ,0\ ,0\ ,x(7),\ 0\ ,0\ ,0\ ,x(11),\ 0\ ,0\ 0\ ,x(15),...$$

Additionally, let $x_k^{(zz)}(n)$ denote the polyphase sequence $x_k^{(z)}(n)$ with appropriate number of leading zeros, appended in the beginning, in our example:

$$x_0^{(zz)}(n) = x(0),\ 0\ ,0\ ,0\ ,x(4),\ 0\ ,0\ ,0\ ,x(8),\ 0\ ,0\ ,0\ ,x(16),...\qquad (10.9)$$
$$x_1^{(zz)}(n) = \ 0\ ,x(1),\ 0\ ,0\ ,0\ ,x(5),\ 0\ ,0\ ,0\ ,x(9),\ 0\ ,0\ ,0\ ,x(13),...$$
$$x_2^{(zz)}(n) = \ 0\ ,\ 0\ ,x(2),\ 0\ ,0\ ,0\ ,x(6),\ 0\ ,0\ ,0\ ,x(10),\ 0\ ,0\ ,0\ ,x(14),...$$
$$x_3^{(zz)}(n) = \ 0\ ,\ 0\ ,\ 0\ ,x(3),\ 0\ ,0\ ,0\ ,x(7),\ 0\ ,0\ ,0\ ,x(11),\ 0\ ,0\ 0\ ,x(15),...$$

Any signal $x(n)$, and filter weights $h(n)$ also, is a summation of its polyphase sequences:

$$x(n) = \sum_{k=0}^{K-1} \sum_{m=-\infty}^{\infty} x(k+m\cdot K) = \sum_{k=0}^{K-1} x_k^{(zz)}(n). \qquad (10.10)$$

The DtFT spectrum of the PP-decomposed signal is equal to:

$$X\left(\frac{f}{f_s}\right) = \sum_{n=-\infty}^{\infty} x(n)e^{-j2\pi\frac{f}{f_s}n} = \sum_{k=0}^{K-1}\left(\sum_{m=-\infty}^{\infty} x(k+m\cdot K)e^{-j2\pi\frac{f}{f_s}(k+m\cdot K)}\right) =$$
$$= \sum_{k=0}^{K-1} e^{-j2\pi\frac{f}{f_s}(k)}\left(\sum_{m=-\infty}^{\infty} x(k+m\cdot K)e^{-j2\pi\frac{f}{f_s}(m\cdot K)}\right) = \sum_{k=0}^{K-1} e^{-j2\pi\frac{f}{f_s}(k)}X_k\left(\frac{f}{f_s}K\right),$$

$$(10.11)$$

i.e. it is a sum of DtFT spectra $X_k(Kf/f_s)$ shifted in phase. That is why the decomposition is called a *polyphase* one. Spectra $X_k(Kf/f_s)$ consist of the original signal spectrum and its $K-1$ copies, similarly as in Fig. 10.3. However, after their addition all copies are canceled, which is shown in Exercise 10.5.

From now on, we are going down from mathematical mountain peaks to the math "flatland." Why the PP decomposition, with so *strange* equations, is so important for us? Because thanks to it, in an easy way, we will avoid multiplication of filter weights by zeros inserted to signal during up-sampling as well as calculation of signal values which are removed at decimator output.

Exercise 10.5 (DFT Spectra of Polyphase Signal Components). Use program 10.4. Analyze its code. Observe the DFT spectrum of each PP signal component and result of the spectra accumulation. Note that the final summation result is correct since the *inner* copies of the signal spectrum lying at frequencies $\frac{kf_s}{K}$, $k = 1...K - 1$ disappeared in result of spectra accumulation. Repeat experiments for different test signals, also for speech and music.

Listing 10.4: Testing DFT spectra of polyphase signal components

```
1   % lab10_ex_pp_spectrum.m - DFT spectra of poly-phase (PP) signal components
2   clear all; close all; figure;
3
4   K=5; N=K*256;                    % parameters: PP order, signal length
5   x=cos(2*pi/32*(0:N-1));          % analyzed signal
6
7   % Signal spectrum calculation
8   zer=zeros(1,N); Xsum=zer;        % initialization
9   z=exp(j*2*pi/N*(0:N-1));         % correction
10  for k=1:K
11      k
12      if(1) % PP components have inserted zeros and start from the x(k) sample
13          xpp = zer; xpp(k:K:N)=x(k:K:N);    % PP signal component
14          Xpp = fft(xpp);                    % its DFT/FFT
15          Xsum = Xsum + Xpp;                 % spectrum accumulation
16      else % PP components have inserted zeros and start from the x(1) sample
17          xpp = zer; xpp(1:K:N)=x(k:K:N);    % PP signal component
18          Xpp = fft(xpp);                    % its DFT/FFT
19          Xsum = Xsum + z.^(-k).*Xpp;        % spectrum accumulation
20      end
21      subplot(211); stem(abs(Xpp)); hold on;
22      subplot(212); stem(abs(Xsum)); pause
23  end
```

10.6.2 Fast Polyphase Interpolation

In this subsection we learn fast polyphase up-sampler implementation. An attempt of its easy explanation is done in Fig. 10.7. FIR digital filtering relies on multiplication of signal samples with filter weights and on summation of multiplication results. This is done by the filter in left up corner of the Fig. 10.7. The filter weights $h(n)$ can be represented as a summation of K polyphase sequences $h_k^{zz}(n)$, as in

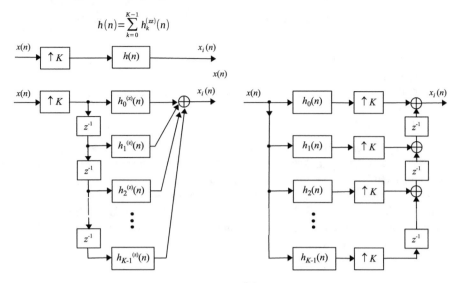

Fig. 10.7: Block diagram of signal polyphase interpolator increasing K-times signal sampling ratio: (left) the LP filter weights are decomposed into polyphase (PP) components, (right) sample expander (embedding $K-1$ zeros between every two signal samples) can be shifted right after the PP filtering [21]

Eq. (10.10). Recalling, Eq. (10.10) is a consequence of signal decomposition presented in Eq. (10.9). Therefore, we can implement the filtering as: (1) a parallel filtration of input signal by several PP sub-filters $h_k^{zz}(n)$, and (2) summation of individual filtration results. No unitary delays are required in such computational structure. When, during filter $h(n)$ partitioning, we use not PP components $h_k^{zz}(n)$ but $h_k^{z}(n)$, an extra unitary delays (blocks $[z^{-1}]$) have to added to the filter implementation, what is presented in left part of Fig. 10.7. No surprise.

In digital interpolator, samples of the input signal $x(n)$ come through expanders where $K-1$ zeros are inserted between each two original samples. Next, the zero-expended input signal is filtered in parallel by K PP sub-filters $h_k^{z}(n)$ (preceded with delays) and sub-filter outputs are added. Since zeros were inserted into the input data stream and polyphase sub-filters, it is obvious that most of the multiplications (by 0) are useless. How to avoid them? It can be strictly shown mathematically and informally graphically that expanders, inserting zeros between input signal samples, can be shifted to the right side, after the polyphase sub-filters. This is shown on right side of Fig. 10.7. Why is it possible? Because after each PP sub-filter, only each K-th sample is significant, the rest of them is equal to zero. Additionally, the significant samples after the PP sub-filters lie in different time instants.

At present we will interpret computationally the diagram presented on the right side of the Fig. 10.7. Each polyphase sub-filter $h_k(n)$ (without inserted zeros!) multiplies its weights with consecutive signal samples. Since these components are delayed by one sample in relation to each other, the following operations are performed:

1. the $K-1$ zeros are inserted between each two sub-filter outputs (note K-time expanders),
2. obtained signals are appropriately delayed with respect to each other (note unitary delay blocks $[z^{-1}]$),
3. all output sub-signals are summed.

Everything is done without multiplication by zeros! In alternative implementation, one can only circularly take outputs of the PP sub-filters without any zeros insertion. The above explanation is typical for most of textbooks.

Let us do something different. Let us assume that $K = 4$ and filter weights are equal to $h(k) = k$, $k = 0,1,2,...,23$, i.e. $h = [0,1,2,3,...,23]$. In the beginning, at time instants $n = 20,21,22,23$, we have the following organization of calculations, confirming their polyphase structure:

$$0x(5) + 4x(4) + \ 8x(3) + 12x(2) + 16x(1) + 20x(0) = y(20),$$
$$1x(5) + 5x(4) + \ 9x(3) + 13x(2) + 17x(1) + 21x(0) = y(21),$$
$$2x(5) + 6x(4) + 10x(3) + 14x(2) + 18x(1) + 22x(0) = y(22),$$
$$3x(5) + 7x(4) + 11x(3) + 15x(2) + 19x(1) + 23x(0) = y(23).$$

All sub-filters have calculated their valid outputs. They all make use of the same input samples $[x(5),x(4),x(3),x(2),x(1),x(0)]$, and they are *looking back*. They are using their own polyphase filter weights. $y(20)$ is the first valid output sample. Output of polyphase sub-filters is circularly switched.

And one more explanation. Let us look at the interpolation filtering in the simplest possible way: as a convolution of two signals, one with inserted zeros. Our problem is graphically illustrated in Fig. 10.8. Original signal samples are marked as dark blue squares, inserted zeros—as white squares. Filter weights $[8,7,6,5,4,3,2,1,0]$ are shifted right. Its polyphase components are marked with different background. We see that in each filter position only one set of polyphase filter weights (PP sub-filters) is used periodically/circularly: $h_0(n)[0,3,6], h_1(n)[1,4,7], h_2(n)[2,5,8]$. In consequence, it is possible to convolve the original signal samples (without inserted zeros!) independently with three sets of polyphase weights. And then build the output signal taking circularly samples from polyphase filter outputs, i.e. first sample from sub-filter $h_0(n)$, first from $h_1(n)$, and first from $h_2(n)$, then second samples from them, then third ones, and so on.

It is very interesting that after *tones of explanations*, the polyphase interpolation program is one of the shortest in this book. It is presented in Listing 10.5.

Exercise 10.6 (Testing Polyphase Interpolator). Use programs 10.1 and 10.5. Compare their outputs using some synthetic signals and some speech or

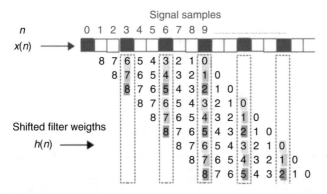

Fig. 10.8: Convolution-based interpretation of filtering problem occurring during signal interpolation—multiplications of filter weights with zeros inserted to the signal (white square boxes) should be avoided. Only multiplications with filter weights marked with color background (inside boxes marked with dashed lines) are performed. Weights of different PP filters are distinguished with different background colors

audio recordings. Using Matlab functions `tic()`,`toc()` check computational speed of both approaches. Overlay in one figure the original and the interpolated signal. Compare our design with Matlab function application: `xi=interp(x,K)` and `xi=resample(x,K,1)`.

Listing 10.5: Fast polyphase up-sampler implementation

```
1  % lab10_ex_iterp.m - continuation
2
3  % INTERPOLATION: Fast polyphase (PP) implementation - mults by zeros are removed
4  % K convolutions of PP filter weights with the original signal
5  xipp = zeros(1,length(x));                  % output initialization
6  for k=1:K                                    % LOOP START: PP filtering
7      xipp(k:K:K*Nx) = filter( h(k:K:end), 1, x );  % k-th PP component filtering
8  end                                          % LOOP END
9  MaxERR_INTERPpp = max(abs(xi-xipp)),         % error
10 figure; plot(ni,xi(ni),'ro-',ni,xipp(ni),'bx-'); grid; xlabel('n');
11 title('Interpolated signals - slow and fast (PP)'); pause
```

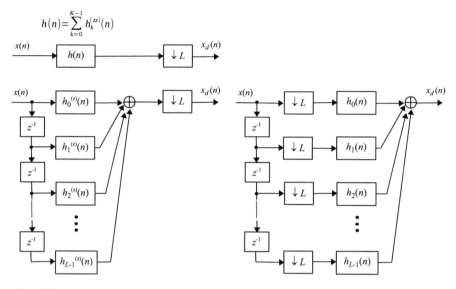

Fig. 10.9: Block diagram of signal polyphase decimator decreasing L-times signal sampling ratio: (left) the LP filter weights are decomposed into polyphase (PP) components $h_l^z(n)$, (right) sample reducer (only every L-th sample is left, $L-1$ samples lying in between are removed) can be shifted left before the PP filtering [21]

10.6.3 Fast Polyphase Decimation

In this subsection we learn fast polyphase down-sampler implementation. We had already our *Battle of Thermopylae*—the description of fast polyphase up-sampler! For this chapter one battle is enough. Now we will discuss the computational problem in more synthetic way. In case of signal down-sampling we would like to avoid calculation of low-pass filter outputs which are rejected (decimated) on decimator output.

In Fig. 10.9 the existing polyphase solution to the problem is explained. In the upper-left part a standard, simple down-sampler is shown: first classical one-sample-in one-sample-out low-pass filter is used which is followed by reducer leaving only every L-th sample and removing all the others. Since the filter weights $h(n)$ can be expressed as summation of its L polyphase components $h_l^{zz}(n), l = 0, 1, ..., L-1$ (see Eq. (10.10)), the filter can be implemented with ease as a summation of L PP parallel h_l^{zz} sections. When we use the PP filter components $h_l^z(n)$, instead of $h_l^{zz}(n)$, i.e. without preceding zeros, we have to add unitary delay elements $[z^{-1}]$ at the input to parallel branches of the PP filter implementation. As a result we obtain computation structure presented in left down diagram of the Fig. 10.9. Its drawback is that filter computes samples that are removed by the reducer at the decimator (down-sampler) output. However, thanks to mathematical discussion neglected here and the parallel PP filter implementation, we can shift the L-time reducer to the

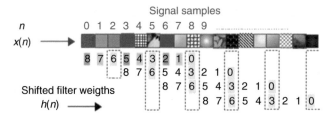

Fig. 10.10: Convolution interpretation of filtering problem during signal decimation—calculation of filtered values neglected on decimator output are avoided due to filter weights shifts by $L-1$ samples, not by only one. $L=3$ in our example. Weights of different PP filters are distinguished with different background colors

left before the PP filters. Thanks to this all sub-filters are working on decimated input samples and only output signal samples needed on the decimator outputs are calculated: all outputs of PP filters are added.

In turn, in Fig. 10.10 very simple convolutional interpretation of the problem is presented. Since after low-pass filtering, aimed at signal bandwidth reduction, only each L-th sample is left, there is no sense to calculate samples that are removed. Therefore the filter weights should be shifted not by one sample but at once by L samples, which is straightforward. Thanks to this calculation of unused samples is avoided. However, due to filter weight shifts bigger than 1, simple usage of fast FFT-based convolution algorithms becomes impossible in such case. The problem can be solved when we decompose signal samples and filter weights into PP sections, perform PP fast sub-filtering using FFT-based methods, and add outputs of PP filters. Hurrah! In our example PP filters $h_0(n)[0,3,6], h_1(n)[1,4,7], h_2(n)[2,5,8]$ are used and their weights are distinguished with different background colors.

Program of fast polyphase filtering is given in Listing 10.6. In order to obtain the same result as for decimator from the program 10.2, $L-1$ zeros are appended to the input signal beginning, and vector x is replaced with xz. In consequence, the k-th polyphase filter section h(k:L:end), k=1...K, is convoluted with the polyphase signal components starting from the sample L-k+1, i.e. the following filtration is done:

```
filter( h(k:L:end), 1, xz(L-k+1:L:end) ).
```

Exercise 10.7 (Testing Polyphase Decimator). Use programs 10.2 and 10.6. Compare their outputs using some synthetic signals and some speech or audio recordings. Using Matlab functions tic(), toc() check computational speed of both approaches. Overlay in one figure the original and the decimated signal. Compare our design with Matlab function application: xi=decimate(x,L) and xi=resample(x,1,L).

Listing 10.6: Fast polyphase down-sampler implementation

```
1   % lab10_ex_decim.m - continuation
2
3   % DECIMATION: Fast polyphase (PP) implementation - samples to be removed are not
        computed
4   % L convolutions of PP components of the original signal and the filter weights
5   xz = [ zeros(1,L-1) x(1:end-(L-1)) ]; % initial insertion of L-1 preceding zeros
6   for k=1:L                              % LOOP START: separate PP filtering
7       xdppm(k,:) = filter( h(k:L:end), 1, xz(L-k+1:L:end) ); % k-th PP section
8   end                                    % LOOP END
9   xdpp = sum(xdppm);                     % summation of PP components
10  MaxERR_DECIMpp = max(abs(xd-xdpp)),    % error
11  figure; plot(nd,xd(nd),'ro-',nd,xdpp(nd),'bx-'); xlabel('n');
12  title('Decimated signals - slow and fast (PP)'); pause
```

10.7 Filters with Fractional Delay

In digital telecommunication systems digital-to-analog converters are used in trans-mitters, analog signals taking some defined, strictly numbered states are passed through different channels, wired or wireless, and finally are converted back to digital form by analog-to-digital converters. The AD and DA converters, in TX in RX, respectively, are not synchronized. They can work with slightly different *clocks* (quartz oscillators) and they can have sampling time moments delayed with respect to each other. In such situation signal interpolation should be performed in a receiver: signal values in proper time moments should be calculated. During digital symbol synchronization procedures, we are interested also in taking samples in symbol centers since thanks to this we increase symbol strength and decrease the error rate.

In receivers different filters are used, typically band-pass (for service isola-tion/extraction) and low-pass (for removing high-frequency components after signal shifting to the base-band around 0 Hz). During filtering operation one can do also fractional signal delay and interpolation, i.e. signal delay by a fraction of sample number, for example, 0.25. In such case signal values are calculated not for indexes: $\{n = 0, 1, 2, 3, ...\}$ but for indexes $\{n = 0.25, 1.25, 2.25, 3.25, ...\}$. Is this difficult to do? No, it is not. During designing a digital filter we should put only filter indexes shifted appropriately.

It is important to notice that any FIR filter having odd number of samples $n = 0, 1, 2, ..., 2M$ has a central weight for index M and it delays output signal by M samples. In contrary, any FIR filter with even number of weights $n = 0, 1, 2, ..., 2M - 1$ does not have a central weight. Its axis of symmetry is between points $M - 1$ and M, in *virtual* sample with index $M - 0.5$, and such filter delays the signal by $M - 0.5$ samples, i.e. it is performing signal filtering and 0.5 interpolation the same time. Its output samples lie 0.5 between original samples. This nice phenomenon is further exploited in filters having a fractional delay, i.e. interpolating ones. For example in telecommunications during applications of raised cosine pulse-shaping filters described in Chaps. 20 and 21.

We will not discuss this issue long. Let us give one simple example. We can use theoretical impulse response $h(k)$ of any FIR filter designed by the window method from (Table 9.2), add to its index k a fractional shift parameter $-0.5 \le \Delta < 0.5$ and obtain this way a filter doing its job but the same time performing the interpolation task. The Δ-shift should be done also upon the window used. For example, an interpolating low-pass filter with Kaiser window is defined as $-M \le n \le M$:

$$h(n) = 2\frac{f_0}{f_s}\frac{\sin(\Omega_0(n-\Delta))}{\Omega_0(n-\Delta)} \cdot \frac{I_0\left[\beta\sqrt{1-((n-\Delta)/M)^2}\right]}{I_0(\beta)}. \tag{10.12}$$

In the described above low/band-pass filters with fractional delay, the filter impulse response is calculated once for a chosen delay value and then used for joint signal filtering and re-sampling. Filtered signal values are calculated in different time stamps, sampling frequency is not modified. Of course, delay value can be changed during filtering but the filter impulse response should be recalculated in such situation.

Exercise 10.8 (Testing Filters with Fractional Delay). Analyze the program 10.7. Run it. Observe in one plot output signals for different values of D delay. Check magnitude and phase responses of filters with different fractional delays—uncomment lines with plots and do required program modifications. Add to the program possibility of testing discrete raised cosine low-pass filter, defined by Eq. (20.17), and its square root modification (20.20), both defined in Chap. 20 on digital single carrier transmission.

Listing 10.7: Signal interpolation during filtering

```
1   % lab10_ex_fractional_delay
2   % Filtering and interpolation the same time
3   clear all; close all;
4
5   N = 1000; M=50;              % signal length, half of the fiter length
6   fs = 200; f0 = 50; fx = f0/10;   % frequencies: sampling, filter cutoff, signal
7
8   dt = 1/fs; t = dt*(0:N-1);      % time
9   x = sin(2*pi*fx*t);          % input signal
10  n = 0 : 2*M; Nc = M;         % filter indexes, central point
11
12  figure;
13  for D = -0.5 : 0.1 : 0.4       % we are checking different fractional delays
14      hLP = sin(2*pi*f0/fs*(n-(Nc+D)))./(pi*(n-(Nc+D)));   % low-pass filter
15      if( D==0 ) hLP(Nc+1) = 2*f0/fs; end
```

```
16   % figure; stem(hLP); title('h(n)'); pause
17     beta=10; w = besseli( 0, beta * sqrt(1-(n-(Nc+D)).^2/M^2) )/besseli(0,beta);
18   % figure; plot(n,w,'bo-'); title('Window function w(n)'); grid; pause
19     hD = hLP .* w;                              % h windowing
20   % f = 0:0.25:fs/2; z = exp(-j*2*pi*f/fs); H = polyval(fliplr(hD),z);  % freq
          response
21   % figure; plot(f,20*log10(abs(H))); grid; xlabel('f (Hz)'); title('|(H(f))|'); pause
22   % figure; plot(f,unwrap(angle(H))); grid; xlabel('f (Hz)'); title('ang(H(f))'); pause
23     y = conv(x,hD);                             % Matlab filtering function
24     y = y( 2*M+1 : end-2*M );                   % cutting transients
25     ty = t(M+1:N-M)-D*dt;                        % sampling times of output signal
26   % plot(t(M+1:N),x(M+1:N),'k-',ty,y,'o'); xlabel('t [s]'); title(' y(t)');
27   % axis([(M+1)*dt,1,-1.25,1.25]); grid; hold on; pause
28   end
```

10.8 Farrow Interpolator

In the previous section we dealt with taking signal samples in different time moments at output of FIR digital filters. In this section we will learn very popular Farrow structure in which signal resampling is performed separately, without filtering.

Describing this method, we start with polynomial interpolation of a function and Lagrange interpolation method. Let us assume that our discrete-time signal at each moment has a value described by a polynomial:

$$x(t) = a_0 + a_1 t + a_2 t^2 + \ldots + a_{N-1} t^{N-1}, \tag{10.13}$$

and that we have N signal samples:

$$x(t_n) = x_n, \quad n = 1, 2, 3, \ldots, N. \tag{10.14}$$

We can combine Eq. (10.13) written for all N signal samples (10.14) into one matrix equation:

$$\begin{bmatrix} 1 & t_1 & \cdots & t_1^{N-1} \\ 1 & t_2 & \cdots & t_2^{N-1} \\ \vdots & \vdots & \cdots & \vdots \\ 1 & t_N & \cdots & t_N^{N-1} \end{bmatrix} \begin{bmatrix} a_0 \\ a_1 \\ \vdots \\ a_{N-1} \end{bmatrix} = \begin{bmatrix} x_1 \\ x_2 \\ \vdots \\ x_N \end{bmatrix}, \quad \mathbf{Ta = x}, \tag{10.15}$$

and solve it with respect to vector of polynomial coefficients $\{a_n\}, n = 0, 1, 2, \ldots, N-1$: $\mathbf{a = T^{-1}x}$, in Matlab a=inv(T)*x. Then we can use coefficients $\{a_n\}$ for calculation of a signal value for any time t, using Eq. (10.13).

Polynomial (10.13) can be written as a summation of continuous-time Lagrange polynomials $L_n(t)$ of order $N - 1$, multiplied by corresponding signal samples $x_n = x(t_n)$, $n = 0, 1, 2, \ldots, N$:

$$x(t) = \sum_{n=1}^{N} x_n L_n(t), \qquad L_n(t) = \frac{\prod_{k=1,\dots,N,k \neq n}(t - t_k)}{\prod_{k=1,\dots,N,k \neq n}(t_n - t_k)}. \tag{10.16}$$

Example: Quadratic Lagrange Interpolation Using Farrow Filters Let us assume that $t_1 = -1, t_2 = 0, t_3 = 1$, i.e. we want to interpolate signal around $t = 0: -1 < t < 1$. Lagrange polynomials are in this case as follows:

$$L_1(t) = \frac{(t-0)(t-1)}{(-1-0)(-1-1)} = \frac{1}{2}t^2 - \frac{1}{2}t, \tag{10.17}$$

$$L_2(t) = \frac{(t+1)(t-1)}{(0-(-1))(0-1)} = -t^2 + 1, \tag{10.18}$$

$$L_3(t) = \frac{(t+1)(t-0)}{(1-(-1))(1-0)} = \frac{1}{2}t^2 + \frac{1}{2}t, \tag{10.19}$$

and signal interpolation is performed using the following equation:

$$x(t) = x(-1)\left[\frac{1}{2}t^2 - \frac{1}{2}t\right] + x(0)\left[-t^2 + 1\right] + x(1)\left[\frac{1}{2}t^2 + \frac{1}{2}t\right] =$$

$$= \left[\frac{1}{2}x(-1) - x(0) + \frac{1}{2}x(1)\right] \cdot t^2 + \left[-\frac{1}{2}x(-1) + \frac{1}{2}x(1)\right] \cdot t + [x(0)].$$

$$\tag{10.20}$$

The interpolation is correct in the neighborhood of a signal sample $x(0)$. Generalizing Eq. (10.20), we can write interpolation formula of any signal sample $x(n)$ in the neighborhood $-1 < t < 1$:

$$x(n+t) = \left[\frac{1}{2}x(n-1) - x(n) + \frac{1}{2}x(n+1)\right]t^2 + \left[-\frac{1}{2}x(n-1) + \frac{1}{2}x(n+1)\right]t + [x(n)],$$

$$\tag{10.21}$$

$$h_2(n) = \left[\frac{1}{2}, -1, \frac{1}{2}\right], \quad h_1(n) = \left[-\frac{1}{2}, 0, \frac{1}{2}\right], \quad h_0(n) = [0, 1, 0].$$

$$\tag{10.22}$$

We see that one can use three digital filters defined by Eq. (10.22) and exploit their outputs for computing signal values around the n-th signal sample.

Coefficients for digital filters, which are used for linear, quadratic, and cubic signal interpolation using Lagrange polynomials, are given in Table 10.1. They can be numerically calculated using program presented in Listing 10.8. Their usage is demonstrated in program 10.9.

Table 10.1: Coefficients of Farrow filters used in linear, quadratic, and cubic signal interpolation by means of Lagrange polynomials

Interp. order	Sample indexes	Digital filter coefficients
1	$[0,1]$	$h_1(n) = [-1,1], \quad h_0(n) = [1,0],$
2	$[-1,0,1]$	$h_2(n) = [\frac{1}{2},-1,\frac{1}{2}], \quad h_1(n) = [-\frac{1}{2}],0,\frac{1}{2}], \quad h_0(n) = [0,1,0],$
3	$[-1,0,1,2]$	$h_3(n) = \frac{1}{6}[-1,3,-3,1], \quad h_2(n) = \frac{1}{6}[3,-6,3,0],$
		$h_1(n) = \frac{1}{6}[-2,-3,6,-1], \quad h_0(n) = \frac{1}{6}[0,6,0,0].$

Exercise 10.9 (Signal Interpolation Using Farrow Filters). Use program 10.8 for calculation of Lagrange polynomials for signal interpolation of order $N = 5$. Find coefficients of associated Farrow filters. Add them to the program 10.9 and compare with other methods.

Listing 10.8: Calculation of Lagrange polynomials and Farrow filter coefficients

```
1   % lab10_ex_lagrange.m
2   % Interpolation using Lagrange polynomial
3   clear all; close all;;
4
5   % ORIGINAL
6   %x = [ 0 1 ];  y=[0 0.5 ]; Sc=1;              % for linear interpolation
7    x=[ -1 0 1 ]; y=[ 0 0.5 0.75 ]; Sc=1;        % for quadratic interpolation
8   %x=[-1 0 1 2]; y=[0 0.5 0.75 0.5 ]; Sc=6;     % for cubic interpolation
9   xi = -1 : 0.1 : 3;                            % function arguments of interest
10  [p,L,yi1] = lagrangeTZ(x,y,xi);              % Lagrange interpolation function
11  yi2 = polyval(p,xi);                         % classical polynomial interpolation
12  figure; plot(xi,yi1,'b-',xi,yi2,'k-',x,y,'ro'); title('y(k) and y(n)'); % result
13  Lagrange = L,  pause          % coefficients of Lagrange polynomials in rows
14  Farrow = L', pause            % coefficients of Farrow filters in rows
15
16  %###################################
17  function [p,L,xi] = lagrangeTZ(t,x,ti)
18  % Lagrange interpolation
19  % Input: points (t(n),x(n)), required points for ti(k)
20  % Output: p - main polynomial, L - matrix with Lagrange polynomials,
21  %         xi - interpolated values.
22
23  % Calculate polynomial coefficients
24  N=length(t); L = zeros(N,N); p = zeros(1,N);
25  for n = 1:N
26      if(n==1)    prodt(1) = prod(t(1)-t(2:N));   proots = t(2:N);
27      elseif(n==N) prodt(N) = prod(t(N)-t(1:N-1)); proots = t(1:N-1);
28      else        prodt(n) = prod(t(n)-[t(1:n-1) t(n+1:N)]); proots=[t(1:n-1) t(n+1:N)];
29      end
```

```
30    Ln = poly(proots)/prodt(n);    % Lagrange polynomial coefficients for n
31    L(n,1:N) = Ln;                  % storing
32    p = p + x(n) * Ln;             % calculation of polynomial coefficients
33  end
34  % Calculate interpolated function values using Lagrange polynomials
35  Ni=length(ti);
36  for k=1:Ni
37    xi(k)=0;
38    for n=1:N
39        if(n==1)    dti = ti(k)-t(2:N);
40        elseif(n==N) dti = ti(k)-t(1:N-1);
41        else        dti = ti(k)-[t(1:n-1) t(n+1:N)];
42        end
43        xi(k) = xi(k) + x(n)*prod(dti)/prodt(n);
44    end
45  end
```

Listing 10.9: Using Farrow filters for Lagrange signal interpolation

```
1   % lab10_ex_farrow.m
2   clear all; close all;
3
4   dx = pi/5;
5   xRef = 0 : pi/1000 : 10*2*pi; yRef = sin(xRef);   % reference signal
6   xDec = 0 : dx : 10*2*pi;      yDec = sin(xDec);   % decimated signal
7
8   % Linear interpolation
9   D = 0.25; % delay
10  h1 = [ -1  1 ]; h1 = h1(end:-1:1);   y1 = filter(h1,1,yDec);
11  h0 = [  1  0 ]; h0 = h0(end:-1:1);   y0 = filter(h0,1,yDec);
12  yL = D*y1 + y0;
13  xL = xDec - dx + D*dx;
14  % Quadratic interpolation
15  D = 0.5;
16  h2 = [ 1/2   -1   1/2 ]; h2 = h2(end:-1:1);  y2 = filter(h2,1,yDec);
17  h1 = [ -1/2   0   1/2 ]; h1 = h1(end:-1:1);  y1 = filter(h1,1,yDec);
18  h0 = [   0    1    0  ]; h0 = h0(end:-1:1);  y0 = filter(h0,1,yDec);
19  yQ = D*D*y2 + D*y1 + y0;
20  xQ = xDec - dx + D*dx;
21  % Cubic interpolation
22  D = 0.75;
23  h3 = 1/6 * [-1 +3 -3  1 ]; h3 = h3(end:-1:1);  y3 = filter(h3,1,yDec);
24  h2 = 1/6 * [ 3 -6  3  0 ]; h2 = h2(end:-1:1);  y2 = filter(h2,1,yDec);
25  h1 = 1/6 * [-2 -3 +6 -1 ]; h1 = h1(end:-1:1);  y1 = filter(h1,1,yDec);
26  h0 = 1/6 * [ 0  6  0  0 ]; h0 = h0(end:-1:1);  y0 = filter(h0,1,yDec);
27  yC = D*D*D*y3 + D*D*y2 + D*y1 + y0;
28  xC = xDec - 2*dx + D * dx;
29
30  figure; plot(xRef,yRef,'k-',xDec,yDec,'bo-',xL,yL,'gs-',xQ,yQ,'r*-',xC,yC,'mx-');
31  grid; pause
```

10.9 Asynchronous Sampling Rate Conversion

In real-time applications very often it is required to change signal re-sampling *on the run*, i.e. asynchronously. Typically it is done as follows: first, a signal is very densely interpolated, for example, 50 times. Then local three or four point Lagrange polynomial interpolation is adapted in real-time.

> **Exercise 10.10 (** ** Hybrid Interpolation).** Write a program for changing sampling ratio from 44,100 samples per second to 48,000 samples per second (160 times up and 147 times down). First interpolate signal 10, 20 and 40 times using classical DSP up-sampling procedure (zero insertion and filtering—remember that fast polyphase version of this operation exists) and, then, locally interpolate obtained signal using the Farrow filter technique.

10.10 Multi-Resolution Signal Decomposition and Wavelet Filters

Filters and re-sampling can be used for decomposition of an *all-band* signal, potentially having all frequency components, into summation of many *sub-band* signals, having only components, belonging to separate frequency bands. For extraction (isolation) of these sub-band components, many digital filters are used which work in parallel or in cascade. They are called, as a *team*, a filter bank.

10.10.1 Multi-band Filter Banks

Let us assume that we have L separate band-pass *analysis* filters, each responsible for different frequency range. Each filter reduces L times the signal bandwidth and leaves only each L-th signal sample. In consequence, we have L sub-band signals, having $\frac{N}{L}$ samples, all together $L \cdot \frac{N}{L} = N$ samples: the same as before but decomposed into frequency sub-bands. In each sub-band time-resolution of the signal is lower. Each sub-band signal can be further decomposed into sub-sub-band. Repeating this decomposition scheme, a multi-resolution signal representation is obtained. When we: (1) insert zeros into the position of removed samples (using sample expanders), (2) filter the zero-inserted sub-band signals by appropriate *synthesis* filters, associated with the *analysis* ones, and (3) sum all obtained sub-signals, the original signal is reconstructed. The signal was first decomposed into many sub-band *parts*, then composed/built back. Since the L-channel analysis-synthesis filter banks are discussed in chapter on audio processing/compression (precisely in Sect. 15.4), we will not continue this story now.

10.10.2 Two-Band Wavelet Filter Banks and Wavelet Transform

However, dyadic filter banks are worth to tell a few words more, because they lead us to wavelet filters and wavelet transform—a very impressive mathematical construction. In this section we will follow description presented in [11]. Let us assume that we have a special pair of low-pass $h_0(k)$ and high-pass $h_1(k)$ filters, both *half-band* ones, and perform a three-level, sub-band signal decomposition, according to scheme presented in Fig. 10.11. Ω_3 denotes the original signal bandwidth. After the first filtering and down-sampling, we have a high-frequency sub-signal with spectrum Π_2 (upper) and low-frequency sub-signal with spectrum Ω_2 (lower). Next, the low-band sub-signal is filtered once more by the same filter pair, and sub-sub-signals with spectra Π_1 and Ω_1 are obtained. The low-frequency signal Ω_1 is filtered next and signals Π_0 and Ω_0 result. In consequence, we have got four sub-signals, having spectra presented on upper part of the Fig. 10.12. It is important to observe that: $\Omega_3 = \Pi_2 + \Omega_2$, $\Omega_2 = \Pi_1 + \Omega_1$, $\Omega_1 = \Pi_0 + \Omega_0$. On the lower part of the Fig. 10.12, we see the signal *sub-space* interpretation of the performed decomposition: a better signal representation with more details is obtained, e.g. Ω_1, when high-frequency signal sub-space (with details), e.g. Π_0, is added to low-frequency signal sub-space (with smooth approximation), e.g. Ω_0. When we take into account number of samples, having by signals in different sub-bands, the signal is decomposed into the time–frequency (TF) grid presented in Fig. 10.13: lower frequency signal sub-bands are more narrow and have less samples, while higher frequency sub-bands are wider and have more samples. With each TF rectangular box in Fig. 10.13 only one sub-band signal sample is associated. When: (1) it is multiplied by the corresponding basis function of the wavelet decomposition, (2) this operation is repeated for all sub-band samples, and (3) all scaled *wavelets* are summed, the original signal is reconstructed. In Fig. 10.14 exemplary wavelet functions are shown associated with the TF wavelet grid, presented in Fig. 10.13. This is the fundamental difference between Fourier and wavelet signal analysis (representation): in the first method a signal is decomposed into sum of permanently oscillating sines and cosines, while in the second one into sum of oscillatory *wavelets* having limited time-support, i.e. placed in different time instants.

As already mentioned above, the decomposed signal can be reconstructed from sub-band sub-samples, i.e. wavelet coefficients. When we: (1) put zeros into positions of removed samples, (2) filter all sub-band signals back by the same filters, but with weights time-reversed, and (3) add resultant samples, the original signal is perfectly reconstructed (neglecting border samples). Dyadic synthesis filter bank is presented in Fig. 10.15. Filters $h_0(-k)$ and $h_1(-k)$ from Fig. 10.11 are the same as $h_0(k)$ and $h_1(k)$, their weights have only reversed order.

The two-tap Haar wavelet filters, called also the Daubechies D2, are the simplest ones $(k = 0, 1)$:

$$h_0(k) = \left\{ \frac{1}{\sqrt{2}}, \frac{1}{\sqrt{2}} \right\}, \quad h_1(k) = \left\{ \frac{1}{\sqrt{2}}, -\frac{1}{\sqrt{2}} \right\}. \tag{10.23}$$

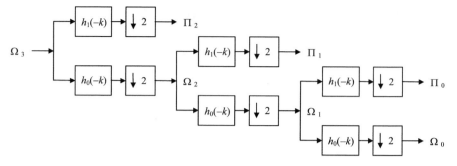

Fig. 10.11: Dyadic analysis filter bank. Input signal is filtered in a cascade of half-band low-pass ($h_0(-k)$) and high-pass ($h_1(-k)$) filters ("$-k$" denotes reversing order of filter weights) [21]

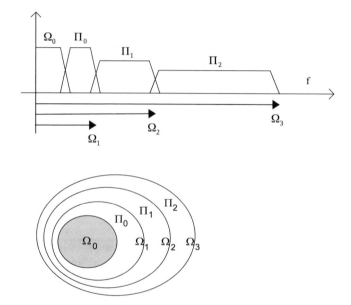

Fig. 10.12: Resultant frequency response of filter bank from Fig. 10.11 (up) and corresponding *signal space* interpretation (down) [21]

In turn, four-tap the Daubechies D4 filters are defined as follows ($k = 0,1,2,4$):

$$h_0(k) = \left\{ \frac{1+\sqrt{3}}{4\sqrt{2}}, \frac{3+\sqrt{3}}{4\sqrt{2}}, \frac{3-\sqrt{3}}{4\sqrt{2}}, \frac{1-\sqrt{3}}{4\sqrt{2}} \right\}, \tag{10.24}$$

$$h_1(k) = \left\{ \frac{1-\sqrt{3}}{4\sqrt{2}}, -\frac{3-\sqrt{3}}{4\sqrt{2}}, \frac{3+\sqrt{3}}{4\sqrt{2}}, -\frac{1+\sqrt{3}}{4\sqrt{2}} \right\}. \tag{10.25}$$

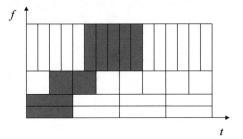

Fig. 10.13: Time–frequency grid of signal decomposition using dyadic wavelet filter bank: less samples for low-frequency narrow-band sub-band signals and more samples for high-frequency wide-band sub-band signals [21]

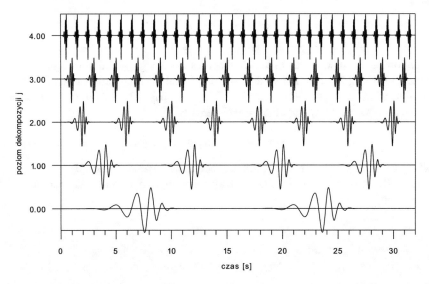

Fig. 10.14: Wavelet decomposition basis functions, corresponding to time–frequency grid of wavelet coefficients, presented in Fig. 10.13. The analyzed signal is summation of all *wavelets* scaled by sub-band signal samples, i.e. the wavelet coefficients. To improve figure readability only every 16-th wavelet is shown [21]

In general, K weights of the low-pass wavelet filter $h_0(k)$ are solution of the following set of $K/2 + 1$ equations with K unknowns (only for K even):

$$\sum_{k=0}^{K-1} h_0(k) = \sqrt{(2)}, \tag{10.26}$$

$$\sum_{k=0}^{K-1} h_0(k)h_0(k-2l) = \delta(l) \qquad l = 0, 1, 2, ..., K/2 - 1. \tag{10.27}$$

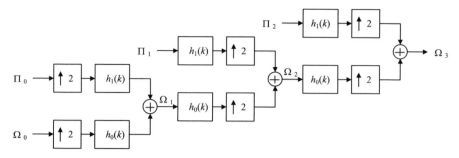

Fig. 10.15: Dyadic synthesis filter bank. Sub-band input signals are filtered in a cascade of half-band low-pass ($h_0(k)$) and high-pass ($h_1(k)$) filters, and summed [21]

For $K = 2$, the Haar filters are the only choice. For $K = 4$ we have $K/2 + 1 = 3$ equations and 4 unknowns. In the Daubechies D4 design, an additional equation is added, ensuring the best smoothness of the signal decomposition basis functions, i.e. *approximation* and *detail* wavelets. The high-pass filter coefficients $h_1(k)$ are calculated from the low-pass filter coefficients $h_0(k)$:

$$h_1(k) = \pm(-1)^k h_0(K - 1 - k), \tag{10.28}$$

where $k = 0, 1, ..., K - 1$ and K, even number, denotes order of the designed wavelet system. Readers interested in details are kindly requested to follow the mathematical derivation, presented below.

Exercise 10.11 (First Lesson on Signal Decomposition Using Wavelet Filter Bank). Try to write personally a Matlab program for one-level wavelet analysis and synthesis of arbitrary signal via signal filtering described above:

1. LP/HP analysis filters and $\downarrow 2$ ($2\times$) reducers,
2. followed by $\uparrow 2$ ($2\times$) expanders and LP/HP synthesis filters.

Make use of given weights of D2/D4 wavelet filters. Compare input and output signal samples: is the wavelet-based signal decomposition fully reversible? Find more wavelet filter coefficients in the Internet and use them also.

If you have a Wavelet Toolbox, become familiar with Matlab functions dwt() and idwt(), performing one-level signal analysis and synthesis using dyadic wavelet filter bank, with user-selected pair of LP and HP filters. Exploit both functions for decomposition and reconstruction of an arbitrary signal. Compare restored signal with output of your program.

10.10.3 Derivation of Wavelet Filter Equations

This section is *only for the brave!* We will derive in it the wavelet filter equations (10.26), (10.27) from mathematical wavelet signal decomposition assumptions. We will follow math presented in [11]. In my opinion it is one of the most fascinating and educating theoretical *climbing* in all DSP theory.

Wavelet Signal Decomposition as a Dyadic Filter Bank Let us assume that our continuous-time signal $x(t)$ can be approximated on the 0-th decomposition level by a sum of time-shifted orthonormal functions $\phi(t)$:

$$x(t) \approx \sum_n c_{0,n} \phi\left(2^0 t - n\right) = \sum_n c_{0,n} \phi\left(t - n\right), \qquad c_{0,n} = \int_{-\infty}^{+\infty} x(t) \phi^*\left(t - n\right) dt.$$

(10.29)

On the 1-st signal decomposition level, the approximating functions are two times *faster/shorter*, due to applied $2\times$ time scaling, and the signal approximation is two-times more detailed:

$$x(t) \approx \sum_n c_{1,n} \sqrt{2}\phi\left(2^1 t - n\right) = \sum_n c_{1,n} \sqrt{2}\phi\left(2t - n\right), \quad c_{1,n} = \int_{-\infty}^{+\infty} x(t) \sqrt{2}\phi^*\left(2t - n\right) dt.$$

(10.30)

Using denotations already used by us during description of cascade dyadic signal filtering, we can say that functions $\phi(2^0 t)$ and $\sqrt{(2)}\phi(2^1 t)$ span, respectively, the signal space Ω_0 and Ω_1, and Π_0 is an orthogonal complement of Ω_0 to Ω_1: $\Pi_0 = \Omega_1 - \Omega_0$, what is presented in Fig. 10.12. Let us denote functions spanning the space Π_0 as $\psi(2^0 t)$. They are orthogonal to $\phi(2^0 t)$. The first are high-frequency ones (signal (details)), while the second—low-frequency ones (signal smooth (approximation)).

After performing signal decomposition further, the signal on the k-th decomposition level ($k > m_0$, m_0—initial signal representation) can be expressed as:

$$x(t) = \sum_n c_{m_0,n} \varphi_{m_0,n}(t) + \sum_{m=m_0}^{k-1} \sum_n d_{m,n} \psi_{m,n}(t), \qquad (10.31)$$

where

$$c_{m,n} = \int_{-\infty}^{+\infty} x(t) 2^{m/2} \phi^*\left(2^m t - n\right) dt, \qquad d_{m,n} = \int_{-\infty}^{+\infty} x(t) 2^{m/2} \psi^*\left(2^m t - n\right) dt.$$

(10.32)

Coefficients $c_{m,n}$ describe the signal *smooth approximation* while $d_{m,n}$—the signal *details*. In wavelet signal representation concept, a signal is the summation of many shifted in time and frequency local *approximation* $\phi()$ and *detail* $\psi()$ functions, with finite time and frequency support, what is presented in Figs. 10.13 and 10.14.

In multi-resolution dyadic signal decomposition, approximation wavelet of the lower level has to be linear superposition of shifted *approximation* wavelets of the higher level:

$$\varphi(t) = \sum_k h_0(k) \sqrt{2} \varphi(2t - k). \tag{10.33}$$

The similar equation has to hold for the *detail* wavelet:

$$\psi(t) = \sum_k h_1(k) \sqrt{2} \varphi(2t - k). \tag{10.34}$$

Filter $h_0(k)$ in Eq. (10.33) is a low-frequency one, while filter $h_1(k)$ in Eq. (10.34) is a high-frequency one. Using settings $t = 2^m t$ and $l = 2n + k$ in Eq. (10.33), one obtains

$$\varphi(2^m t - n) = \sum_k h_0(k) \sqrt{2}\, \varphi(2(2^m t - n) - k) = \sum_l h_0(l - 2n) \sqrt{2} \varphi\left(2^{m+1} t - l\right). \tag{10.35}$$

Taking into account definition (10.32) of coefficients $(c_{m,n} = c_m(n))$, we get

$$c_m(n) = \int_{-\infty}^{+\infty} x(t) 2^{m/2} \left[\sum_l h_0(l - 2n) \sqrt{2} \phi^* \left(2^{m+1} t - l\right) \right] dt =$$

$$= \sum_l h_0(l - 2n) \left[\int_{-\infty}^{+\infty} x(t) 2^{(m+1)/2} \phi^* \left(2^{m+1} t - l\right) dt \right] = \sum_l h_0(l - 2n) c_{m+1}(l). \tag{10.36}$$

In similar way the following relation is derived for coefficients $d_{m,n} = d_m(n)$:

$$d_m(n) = \sum_l h_1(l - 2n)\, c_{m+1}(l). \tag{10.37}$$

The last two equations (10.36), (10.37) confirm that the wavelet transform of a signal, i.e. its decomposition (10.31) into superposition of shifted *approximation* and *detail* wavelets, can be realized in practice in dyadic filter bank with $h_0(n)$ and $h_1(n)$

filters (Fig. 10.11): we treat original signal samples as its approximation coefficients $c_m(n)$ at the high-frequency approximation level and then filter them by LP and HP filters and go to lower levels.

Finding Wavelet Filter Design Equations When one additionally assumes that:

$$\int_{-\infty}^{+\infty} \varphi(t)dt = 1,$$ (10.38)

and put Eq. (10.33) into Eq. (10.38) with setting $\tau = 2t - n$, she/he gets

$$\sum_n h_0(n) = \sqrt{2}.$$ (10.39)

It means that the filter has to pass the constant signal value. If additionally:

$$\sum_n (-1)^n h_0(n) = 0,$$ (10.40)

the filter frequency response has 0 for $\Omega = \pi$. Last two equations are fulfilled when:

$$\sum_n h_0(2n) = \sum_n h_0(2n+1) = \frac{\sqrt{2}}{2},$$ (10.41)

what tells us that the filter should have even number of taps. When we additionally request orthogonality of *approximation* and shifted *approximation* wavelets:

$$\int_{-\infty}^{+\infty} \varphi(t)\varphi(t-n)dt = \delta(n),$$ (10.42)

and use Eq. (10.33), after setting $\tau = 2t - m$ we obtain from (10.42) :

$$\sum_m h_0(m)h_0(m-2n) = \delta(n).$$ (10.43)

Finally, when we request zero value of the *detail* wavelet integral (and high-pass nature of the $h_1(k)$ filter):

$$\int_{-\infty}^{+\infty} \psi(t)dt = 0$$ (10.44)

after *some* derivations, which are ... skipped (*Hurrah!*), we get

$$h_1(n) = \pm(-1)^n h_0(N-1-n). \qquad (10.45)$$

Conclusion Thanks to the above mathematical explanation of the multi-resolution wavelet-based signal decomposition, we should be convinced that dyadic filter banks with appropriate filters let us calculate in practice wavelet transform coefficients of a signal. Additionally, we have derived Eqs. (10.39), (10.43), (10.45), which are obligatory for design of arbitrary wavelet system. So now, please, dear Reader, design a system of your dream.

Exercise 10.12 (Second Lesson on Signal Decomposition Using Wavelet Filter Bank).

1. Become familiar with the program 10.10, performing by-hand multi-level wavelet signal analysis and synthesis. Try to understand its every line. Find in the Internet coefficients of more wavelet filters and add them to the program, for example, higher order Daubechies and Coifman filters. Run the program: observe the reconstructed signal shape and note value of signal reconstruction error. Add new test signals to the program, may be speech or music? Listen to signals at different wavelet decomposition levels.

2. Note, how the *approximation* wavelet/function $\phi()$ from Eq. (10.31) is synthesized: an unitary impulse is passed through a cascade of 10 interpolators, each consisting of the "↑ 2" expander and the low-pass filter $h_{0s}(k)$. Our signal is expressed as a summation of its shifted copies. Modify the program and synthesize the *detail* wavelet/function $\psi()$ from Eq. (10.31). Compare calculated shapes of wavelet functions $\phi()$ and $\psi(n)$ for different wavelet filters. Find in literature, whether these shapes are correct. Since they are *wavelet-like*, the described signal decomposition/approximation is called the wavelet transform.

Listing 10.10: Discrete wavelet transform in Matlab

```
1   % lab11_ex_wavelets.m
2
3   % Parameters
4     niter = 3;          % decomposition level = number of iterations
5     nx = 2^niter*32;    % length of test signal
6   % LP synthesis filter coefficients h0s, e.g. Db4
7     h0s = [ (1+sqrt(3))/(4*sqrt(2)) (3+sqrt(3))/(4*sqrt(2)) ...
```

```
8            (3-sqrt(3))/(4*sqrt(2)) (1-sqrt(3))/(4*sqrt(2)) ];
9    % Calculate remaining filters
10     N = length(h0s); n = 0:N-1;
11     h1s = (-1).^n .* h0s(N:-1:1);        % HP synthesis filter
12     h0a = h0s(N:-1:1); h1a=h1s(N:-1:1);  % LP and HP analysis filters
13   % Synthesis of "approximation" wavelets - given
14   % Synthesis of "detail" wavelets - your homework!
15     c = 1;
16     for m = 1:10                         % synthesis levels
17         c0=[];
18         for k = 1:length(c)
19             c0(2*k-1)=c(k); c0(2*k)=0;   % 2-time expander
20         end
21         c = conv(c0,h0s);                % LP wavelet filter
22     end
23     figure; plot( c ); title('Approximation function/wavelet'); pause
24   % Test signal
25   % x=sin(2*pi*(1:nx)/32);
26     x=rand(1,nx);
27   % Wavelet analysis, discrete wavelet transform
28     cc = x;
29     for m = 1:niter
30         c0 = conv(cc,h0a);               % LP filtration
31         d0 = conv(cc,h1a);               % HP filtration
32         k=N:2:length(d0)-(N-1); kp=1:length(k); ord(m)=length(kp); dd(m,kp) = d0( k );
33         k=N:2:length(c0)-(N-1); cc=c0( k );
34     end
35   % Wavelet synthesis, inverse discrete wavelet transform
36     c=cc;
37     for m = niter:-1:1
38         c0=[]; d0=[];
39         for k = 1:length(c)
40             c0(2*k-1)=c(k); c0(2*k)=0;       % 2-times expander
41         end
42         c = conv(c0,h0s); nc=length(c);      % LP wavelet filter
43         for k = 1:ord(m)
44             d0(2*k-1) = dd(m,k); d0(2*k) = 0;  % 2-times expander
45         end
46         d = conv(d0,h1s); nd=length(d);      % HP wavelet filter
47         c = c(1:nd);
48         c = c + d;
49     end
50   % Final figures
51     n = 2*(N-1)*niter : length(c)-2*(N-1)*niter+1;
52     figure; plot(x); title('Input'); pause
53     figure; plot(n,c(n)-x(n)); title('Output-Input'); pause
```

10.11 Summary

1. Signal interpolation (up-sampling) and decimation (down-sampling) have extraordinary significance in our multi-task multi-resolution style of life and problem solving. We would like to observe data in different scales, switch between them in real-time, adapt scales. We are surrounded by multi-resolution multimedia: speech, audio, images, video, ... We are using multi-rate (multi-speed) data transmission.

2. Interpolation and decimation do not look difficult at a first glance. Simple zero-insertion and low-pass data smoothing is done in the first of them. Simple signal bandwidth reduction by low-pass filtering and removal of some samples is performed in the second case. But when we start thinking about computational optimization of these operations, things start to become very difficult. Severe mathematical notation of polyphase signals and systems are used. Problems are growing further when sub-band signals decomposition and processing is addressed.

3. What to repeat once more? When used in a cascade, the interpolator is always first before the decimator. Why? Because decimator is always filtering out high-frequency signal components which are lost *forever*.

4. Application of fast polyphase versions of digital up-samplers and down-samplers should be preferred. Why? Because there is no sense to multiply filter weights by zeros or filter data and not to exploit results of calculations.

5. When signal re-sampling is of high-order, applied filters have to be long. Therefore it is profitable to use a cascade of shorter filters. Computational expense of such up/down-sampling is typically significantly lower.

6. One should remember that signal interpolation can be realized cost-less during digital FIR filtration by slight shifting in time weights of the filter impulse response. Additionally, in asynchronous sampling rate conversion (ASRC), very important in synchronization parts of communication systems, the Lagrange interpolation polynomials are exploited in the form of Farrow filter—the state-of-the-art solution in signal re-sampling.

7. Finally, we become familiar in this chapter with main concept of signal analysis and synthesis by multi-channel (multi-band) filter banks, i.e. the sub-band signal decomposition and its down-sampling in separate sub-bands, as well as sub-band signal up-sampling and the original, all-band signal reconstruction. As a concrete example, we have studied with more details 2-band (dyadic) analysis and synthesis wavelet filter banks: we have derived design equations of wavelet filters and we have found their connection with the wavelet signal decomposition (transform).

8. However, in order to become familiar with modern implementation of re-sampling in modern telecommunication systems, further reading is recommended focused on usage of cascaded integrator-comb (CIC) **recursive filters** [3, 9] for this purpose.

10.12 Private Investigations: Free-Style Bungee Jumps

Exercise 10.13 (UP-DOWN with Only One Low-Pass Filter). During two-level combined UP and DOWN signal re-sampling we have been using a cascade of two low-pass filters. But only one is enough, this having more narrow pass-band. Verify this observation experimentally on one real and one synthetic signal. Compare in one figure two re-sampling results which are obtained with the usage of two filters and only one filter.

Exercise 10.14 (Mixing Digital Signals Sampled with Different Frequencies). Record 3–5 sec of your speech with sampling frequency equal to 11.025 kHz. Record 3–5 sec of music with sampling frequency equal to 44.1 kHz. Up-sample speech to 32 kHz. Use Matlab function rat() to find required Up/Down ratio. Down-sample music to 32 kHz. Use Matlab function rat() to find required Up/-Down ratio. Add up-sampled speech to down-sampled music. Adjust gains to hear both signals of them clearly.

References

1. A.N. Akansu, R.A. Haddad, *Multiresolution Signal Decomposition: Transforms, Subbands and Wavelets* (Academic Press, San Diego, 1992, 2000)
2. C.S. Burrus, R.A. Gopinath, H. Guo, *Introduction to Wavelets and Wavelet Transforms. A Primer* (Prentice Hall, Upper Saddle River, 1998)
3. C.K. Chui, *An Introduction to Wavelets* (Academic Press, Boston, 1992)
4. R.E. Crochiere, L.R. Rabiner, *Multirate Digital Signal Processing* (Prentice Hall, Englewood Cliffs, 1983)
5. I. Daubechies, *Ten Lectures on Wavelets* (SIAM, Philadelphia, 1992)
6. I. Daubechies, W. Sweldens, Factoring wavelet transforms into lifting steps. J. Fourier Anal. Appl. **4**(3), 245–267 (1998)
7. N.J. Fliege, *Multirate Digital Signal Processing. Multirate Systems, Filter Banks, Wavelets* (Wiley, Chichester, 1994)
8. F.J. Harris, Multirate filters for interpolating and desampling, in *Handbook of Digital Signal Processing*, ed. by F.D. Elliott (Academic Press, San Diego, 1987), pp. 173–287
9. F.J. Harris, *Multirate Signal Processing for Communication Systems* (Prentice Hall, Upper Saddle River, NJ, 2004)

10. R.G. Lyons, *Understanding Digital Signal Processing* (Addison-Wesley Longman Publishing, Boston, 1996, 2005, 2010)
11. S.G. Mallat, A theory for multiresolution signal decomposition: The wavelet representation. IEEE Trans. Pattern Anal. Mach. Intell. **II**(7), 674–693 (1989)
12. S. Mallat, *A Wavelet Tour of Signal Processing* (Academic Press, San Diego, 1998)
13. D.G. Manolakis, V.K. Ingle, *Applied Digital Signal Processing* (Cambridge University Press, Cambridge, 2011)
14. A. Mertins, *Signal Analysis: Wavelets, Filter Banks, Time-Frequency Transforms and Applications* (Wiley, Chichester, 1999)
15. L.D. Milic, *Multirate Filtering for Digital Signal Processing: MATLAB Applications* (Information Science Reference/IGI Publishing, Hershey PA, 2008)
16. G. Strang, T. Nguyen, *Wavelets and Filter Banks* (Wellesley-Cambridge Press, Wellesley, 1996)
17. W. Sweldens, P. Schröder, Building your own wavelets at home, in *Wavelets in Computer Graphics*. ACM SIGGRAPH Course Notes, pp. 15–87, 1996
18. A. Teolis, *Computational Signal Processing with Wavelets* (Birkhauser, Boston, 1998)
19. P.P. Vaidyanathan, *Multirate Systems and Filter Banks* (Prentice Hall, Englewood Cliffs, 1993)
20. G.W. Wornell, *Signal Processing with Fractals: A Wavelet-Based Approach* (Prentice Hall, Upper Saddle River, 1995)
21. T.P. Zieliński, *Cyfrowe Przetwarzanie Sygnałów. Od Teorii do Zastosowań (Digital Signal Processing. From Theory to Applications)* (Wydawnictwa Komunikacji i Łączności (Transport and Communication Publishers), Warszawa, Poland, 2005, 2007, 2009, 2014)

Chapter 11
FIR Filters for Phase Shifting and Differentiation

To be or not be? To be universal or very narrowly specialized?
A short story about special task filters.

11.1 Introduction

In FIR digital filtering any output sample is calculated as running weighted average of some last input samples. This operation is not only very simple but also very effective in signal components attenuation when filter weights are designed properly. A filter can have a few or even hundreds or thousands of weights. Most often their values should ensure special filter *behavior* in frequency domain, i.e. from its input to output the filter should pass only signal components belonging to specified frequency bands. In previous two chapters we were designing FIR low-pass, high-pass, band-pass, and band-stop filters only. Why? Because they are the most frequently used, for example for signal de-noising (as in bio-medical applications where information signals are very weak) or for separation of different signal frequency components (like in telecommunication or radar signal receivers).

Different frequencies can be amplified/attenuated and delayed by a digital filter in different way. We can exploit this *nice* feature and design filters passing only signals lying in predefined frequency bands (LP, HP, BP, BS). And we widely do it. But we can also design digital filters having some specific frequency responses and thanks to this performing some specific tasks which are very useful in practice, for example, shifting signal components in phase or differentiating a signal. This chapter is exactly devoted to these two types of FIR filters. Weights of both of them are designed using the window method: first filter frequency response is specified, then it is transformed to time-domain using the discrete-time Fourier transform (DtFT), next the calculated filter weights are windowed and, finally, they are convoluted with a signal being processed.

© Springer Nature Switzerland AG 2021   293
T. P. Zieliński, *Starting Digital Signal Processing in Telecommunication Engineering*, Textbooks in Telecommunication Engineering,
https://doi.org/10.1007/978-3-030-49256-4_11

The Hilbert filter has a frequency response $H(\Omega) = -j = e^{-j\pi/2}$ for $\Omega > 0$ and $H(\Omega) = j = e^{j\pi/2}$ for $\Omega < 0$. It is an all-pass filter (except the DC component), shifting all frequency components in phase -90 degrees. The filter allows very easy amplitude and phase/frequency signal demodulation and is widely used. How it works? Thanks to the Hilbert filter (transform) a complex-value signal is created, called an analytic signal, having in its real part the original signal and in the imaginary part the Hilbert filter output. Absolute value and angle of each analytic signal sample are equal, respectively, to instantaneous signal amplitude and instantaneous signal phase. In turn, signal frequency is derived as numerical derivative of the calculated phase.

The differentiation filter has a frequency response equal to $H(\Omega) = j\Omega$, i.e. it is, both, a plus 90 degree phase shifter ($j = e^{j\pi/2}$) and an amplifier with gain proportional to signal frequency. Signal differentiation is needed, for example, in discrete-time automatic control systems and in telecommunication receivers.

11.2 FIR Hilbert Filter

11.2.1 Hilbert Filter Basics

Frequency response of the digital Hilbert filter is defined as (see Fig. 11.1):

$$H_H(\Omega) = \begin{cases} -j, & \Omega > 0, \\ 0, & \Omega = 0, \\ j, & \Omega < 0, \end{cases} \tag{11.1}$$

where $\Omega = 2\pi \frac{f}{f_s}$, f_s—sampling frequency. Since the following identities holds

$$-j = e^{-j\pi/2}, \qquad j = e^{j\pi/2} \tag{11.2}$$

the complex-value input harmonic signals $x(n)$ with positive and negative angular frequencies Ω_0 and $-\Omega_0$ are shifted by this filter by $-\pi/2$ radians:

$$e^{j\Omega_0 n} \quad \rightarrow \quad e^{j\Omega_0 n} e^{-j\frac{\pi}{2}} = e^{j\left(\Omega_0 n - \frac{\pi}{2}\right)}, \tag{11.3}$$

$$e^{j(-\Omega_0)n} \quad \rightarrow \quad e^{j(-\Omega_0)n} e^{+j\frac{\pi}{2}} = e^{-j\left(\Omega_0 n - \frac{\pi}{2}\right)}. \tag{11.4}$$

In consequence a cosine with angular frequency Ω_0 is also shifted in phase by the same angle:

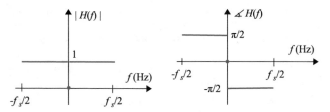

Fig. 11.1: Hilbert filter frequency response: (left) magnitude, (right) phase. f_s denotes sampling frequency

$$\cos(\Omega_0 n) = \frac{e^{j\Omega_0 n} + e^{j(-\Omega_0)n}}{2} \quad \rightarrow$$

$$\frac{e^{j\left(\Omega_0 n - \frac{\pi}{2}\right)} + e^{-j\left(\Omega_0 n - \frac{\pi}{2}\right)}}{2} = \cos\left(\Omega_0 n - \frac{\pi}{2}\right) = \sin(\Omega_0 n). \quad (11.5)$$

Therefore we can conclude that Hilbert filter is minus 90 degree phase shifter, transforming cosine signal into a sine one, independently of the cosine frequency. The Hilbert filter magnitude and phase frequency responses are presented in Fig. 11.1.

Now let us calculate the inverse DtFT of the filter frequency response given by Eq. (11.1):

$$h_H(n) = \frac{1}{2\pi} \int_{-\pi}^{\pi} H_H(\Omega) e^{j\Omega n} d\Omega = \frac{1}{2\pi} \int_{-\pi}^{0} j e^{j\Omega n} d\Omega + \frac{1}{2\pi} \int_{0}^{\pi} (-j) e^{j\Omega n} d\Omega =$$

$$= \frac{j}{2\pi} \left[\frac{1}{jn} e^{j\Omega n} \Big|_{-\pi}^{0} - \frac{1}{jn} e^{j\Omega n} \Big|_{0}^{\pi} \right] = \frac{1}{2\pi n} \left[(e^{j0} - e^{-j\pi n}) - (e^{j\pi n} - e^{j0}) \right] =$$

$$= \frac{1}{2\pi n} [2 - 2\cos \pi n] = \frac{1}{\pi n} [1 - \cos \pi n] = \frac{\sin^2(\pi n/2)}{\pi n/2}. \quad (11.6)$$

The FIR Hilbert filter, as each FIR digital filter, is described by the following input–output relation:

$$x(n) \quad \rightarrow \quad [\text{ filter weights } h_H(n)] \quad \rightarrow \quad y(n) = \sum_{k=-\infty}^{\infty} h_H(k) x(n-k) \quad (11.7)$$

i.e. output signal samples $y(n)$ are result of convolution of input signal samples $x(n)$ with filter weights $h_H(n)$, calculated in Eq. (11.6):

$$h_H(n) = \begin{cases} \frac{1-\cos(\pi n)}{\pi n} = \frac{\sin^2(\pi n/2)}{\pi n/2}, & n \neq 0, \\ 0, & n = 0. \end{cases} \quad (11.8)$$

Conclusion The Hilbert filter is not a typical *frequency-passing* and *frequency-stopping* filter. It follows from Eq. (11.1) that it is the minus 90 degree phase shifter. Theoretically, it should pass without amplitude change all signal components with different frequencies (since $|H(f)| = 1$ for $f \neq 0$ Hz), shifting them only in phase by $-\pi/2$ radians. Only the signal mean value, the DC one, is removed by the Hilbert filter (because $|H(f)| = 0$ for $f = 0$ Hz).

Exercise 11.1 (Hilbert Filter Impulse and Frequency Response). Use Matlab program 11.1. Choose Hilbert filter. Observe its impulse response and frequency response. Is the calculated frequency response similar to the theoretical one which is presented in Fig. 11.1. Increase value of M. Is the filter magnitude response less oscillatory now? Explain origin of the oscillations?

Listing 11.1: Generation of Hilbert and differentiation filter impulse response and calculation of the filter frequency response

```
1  % lab11_ex_impulse_freq_response.m
2    clear all; close all;
3
4  % Generate and plot the Hilbert filter impulse response
5    M=20; n=-M:M; hH=(1-cos(pi*n))./(pi*n); hH(M+1)=0;
6    figure; stem(n,hH); title('hH(n)'); grid; pause
7  % Generate and plot the differentiation filter impulse response
8    M=20; n=-M:M; hD=cos(pi*n)./n; hD(M+1)=0;
9    figure; stem(n,hD); title('hD(n)'); grid; pause
10 % Our choice: hH or hD
11   h = hH;
12 % Windowing
13 % w = blackman(2*M+1)'; figure; stem(w); pause, h = h .* w; figure; stem(h); pause
14 % Calculate filter frequency response::
15   Om = 0 : pi/1000 : pi; z=exp(-j*Om); H=polyval(h(end:-1:1),z);
16 % Plot its magnitude and phase response:
17   figure; plot(Om,abs(H)); title('abs( H(Om) )'); grid; pause
18   figure; plot(Om,unwrap(angle(H))); title('angle( H(Om) )'); grid; pause
19 % Correct the phase shift resulting from impulse response shift by M samples right
20   figure; plot(Om,unwrap( angle( H.*exp(j*Om*M) ) )); title('angle( H(Om) )');
21   grid; pause
```

11.2.2 Analytic Signal and AM-FM Demodulation Principle

Combining input and output of the Hilbert filter, the so-called analytic signal is obtained (see Fig. 11.2):

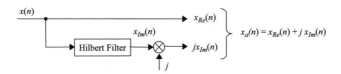

Fig. 11.2: Computation of analytic signal using Hilbert filter

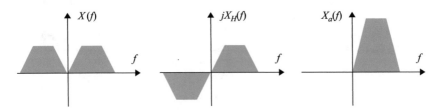

Fig. 11.3: Graphical illustration of the fact that analytic signal has spectrum equal to zero for negative frequencies. Spectra denotation: $X(f)$—for the original signal, $X_H(f)$—for the Hilbert filter output, and $X_a(f) = X(f) + jX_H(f)$—for the analytic signal

$$x_a(n) = x(n) + j \cdot H[x(n)] = x(n) + j \cdot x_H(n). \tag{11.9}$$

Remembering that the Hilbert transform of $\cos(\Omega_0 n)$ is $\sin(\Omega_0 n)$:

$$H[\cos(\Omega_0 n)] = \cos(\Omega_0 n - \pi/2) = \sin(\Omega_0 n), \tag{11.10}$$

and due to Euler equation $e^{j\alpha} = \cos(\alpha) + j\sin(\alpha)$, the analytic signal (11.9) associated with the cosine is equal to the complex-value Fourier harmonic signal,

$$x_a(n) = \cos(\Omega_0 n) + j \cdot \sin(\Omega_0 n) = e^{j\Omega_0 n}. \tag{11.11}$$

DtFT spectrum of the analytic signal is equal:

$$X_a(\Omega) = X(\Omega) + j \cdot X_H(\Omega) = X(\Omega) + j \cdot X(\Omega)H_H(\Omega) =$$
$$= X(\Omega) \cdot (1 + j \cdot H_H(\Omega)). \tag{11.12}$$

For positive angular frequencies $\Omega > 0$ we obtain

$$X_a(\Omega) = X(\Omega) \cdot (1 + j \cdot (-j)) = X(\Omega) \cdot (1 + 1) = 2X(\Omega), \tag{11.13}$$

while for negative ones $\Omega < 0$ we have

$$X_a(\Omega) = X(\Omega) \cdot (1 + j \cdot (j)) = X(\Omega) \cdot (1 - 1) = 0. \tag{11.14}$$

This is a very important result which should be remembered! Analytic, complex-value version of a real-value signal does not have negative frequency components, i.e. "mirror," complex-conjugated part is removed from the spectrum of a real-value signal. This phenomena is shown in Fig. 11.3. We can very easily check it using Matlab `fft()` function.

Exercise 11.2 (Analytic Signal Spectrum). Use Matlab program 11.2. A cosine with sinusoidal frequency modulation is generated in it: carrier 1000 Hz, modulation index 200 Hz, modulating frequency 10 Hz. Note the difference between the spectrum of the original signal and its analytic version: in the second negative frequency components are missing. Observe that the complex-value signal xm, which imaginary part is negated in comparison to the analytic signal, has a spectrum containing only negative frequency components. Test this feature on different signals, also with speech signals sampled with frequency 8000 Hz.

Listing 11.2: Generation of analytic signal and checking its frequency spectrum

```
1    % lab11_ex_analytic.m
2    clear all; close all;
3
4    fs = 8000;                          % sampling frequency
5    Nx = 8000;                          % number of samples
6    dt = 1/fs; t = dt*(0:Nx-1);         % sampling period, sampling moments
7    fc = 1000; kf = 200; fm = 10;       % carrier, modulation index, modulation frequency
8    x = cos(2*pi*(fc*t + kf/(2*pi*fm)*sin(2*pi*fm*t)));  % SFM signal
9    xa = hilbert( x );                  % analytic signal, positive frequencies
10   xp = x + j*imag(xa);                % analytic signal, the same as above
11   xm = x - j*imag(xa);                % negative frequencies
12   % Spectra
13   X  = fftshift( abs(fft(x))  )/Nx;   % for original signal
14   Xa = fftshift( abs(fft(xa)) )/Nx;   % for analytic signal with positive freqs only
15   Xm = fftshift( abs(fft(xm)) )/Nx;   % for analytic signal with negative freqs only
16   % Figures
17   f0 = fs/Nx; f=f0*(-Nx/2:Nx/2-1);
18   figure
19   subplot(311); plot(f,X ); title('|X(f)|');  xlabel('(Hz)'); grid;
20   subplot(312); plot(f,Xa); title('|Xa(f)|'); xlabel('(Hz)'); grid;
21   subplot(313); plot(f,Xm); title('|Xm(f)|'); xlabel('(Hz)'); grid;
22   pause
```

The Hilbert filter (Hilbert transform) is mainly used for signal demodulation. Let us assume that we have, for example, a signal as follows:

$$s(n) = A \cdot (1 + a(n)) \cdot \cos\left(2\pi \frac{f_c}{f_s} n + \varphi(n) \right), \tag{11.15}$$

being modulated in amplitude by signal $a(n)$ ($|a(n)| < 1$) and in phase by signal $\varphi(n)$ (both modulating signals are low-frequency ones in comparison to the signal carrier frequency f_c). Let us calculate analytic signal for (11.15):

$$s_a(n) = A \cdot (1 + a(n)) e^{j \cdot \left(2\pi \frac{f_c}{f_s} n + \varphi(n) \right)}. \tag{11.16}$$

We see that absolute value $|s_a(n)|$ and angle $\angle s_a(n)$ of $s_a(n)$ are equal:

$$|s_a(n)| = A \cdot (1 + a(n)), \tag{11.17}$$

$$\angle s_a(n) = 2\pi \frac{f_c}{f_s} n + \varphi(n), \tag{11.18}$$

so they allow finding $a(n)$ and $\varphi(n)$ and, next, instantaneous signal frequency $f_{inst}(n)$ as the derivative of the found angle:

$$f_{inst}(n) = \frac{f_s}{2\pi} \frac{\Delta(\angle s_a(n))}{\Delta n} = f_c + \frac{f_s}{2\pi} \frac{\Delta \varphi(n)}{\Delta n}. \tag{11.19}$$

For frequency modulation with signal $x_f(n)$ and modulation index k_f, the phase $\varphi(n)$ is equal:

$$\varphi(n) = \frac{2\pi}{f_s} k_f \sum_{k=0}^{n} x_f(k). \tag{11.20}$$

Therefore Eq. (11.19) takes the following form (derivative of the sum in last equation is equal to the last element of this sum):

$$f_{inst}(n) = f_c + k_f \cdot x_f(n), \tag{11.21}$$

allowing for finding $x_f(n)$ when f_c and k_f are known. EUREKA! Signal demodulation using Hilbert filter does not look very difficult.

Exercise 11.3 (Hilbert Filter Signal Demodulation). Use Matlab program 11.3. Signal modulated jointly in amplitude and frequency is generated in it. In the demodulation, first, analytic complex-value signal version is calculated using the Matlab function `hilbert()`. AM demodulation makes use of instantaneous value of analytic signal magnitude, while FM demodulation exploits unwrapped analytic signal angle and its derivative. Analyze the program code, admire its simplicity, note very small demodulation errors. Run program for different values of AM and FM modulation parameters. Does the program work correctly for all of them? Optional: change sampling frequency to 44.1 kHz and try to do AM and FM modulation of the carrier 10 kHz using speech signal sampled at 44.1 kHz. Listen to demodulated signals.

Listing 11.3: Demonstration of signal demodulation using Hilbert filter

```
1    % lab11_ex_hilbert_demod_primer.m
2      clear all; close all;
3
4      fs = 8000;                          % sampling frequency
5      Nx = 8000;                          % number of samples
6      dt = 1/fs; t = dt*(0:Nx-1);         % sampling period, sampling moments
7      fc = 1000; Ac = 10;                 % carrier frequency and amplitude
8      kA = 0.5; fA = 5;                   % AM modulation: index, frequency
9      kF = 200; fF = 10;                  % FM modulation: index, frequency
10     xA = cos(2*pi*fA*t);               % signal for AM
11     xF = cos(2*pi*fF*t);               % signal for FM
12   % Modulation
13     s =  Ac*(1 + kA*xA) .* cos( 2*pi*( fc*t + kF*cumsum(xF)*dt ) ); % SFM signal
14   % Demodulation
15     sa = hilbert( s );                  % analytic signal, positive frequencies
16     xA_est = ( abs(sa)/Ac - 1 ) / kA;
17     xF_est = ( diff( unwrap( angle( sa ) ) ) /(2*pi) - fc*dt ) / (kF*dt);
18   % Figures
19     figure
20     subplot(211); plot(t,xA,'r',t,xA_est,'b'); title('A(t)');
21     subplot(212); plot(t(2:end),xF(2:end),'r',t(2:end),xF_est,'b'); title('f(t)'); pause
22     k = round(0.25*Nx) : round(0.75*Nx);
23     error_a = max( abs( xA(k) - xA_est(k) )),   % amplitude demodulation error
24     error_f = max( abs( xF(k+1) - xF_est(k) )), % frequency demodulation error
```

11.2.3 Hilbert Filter Implementations

Hilbert phase shifter can be implemented in time domain as a digital filter, IIR or FIR, or in frequency domain using signal spectrum modification. At present we will discuss the FIR filter implementation and the spectrum modification methods.

Time-Domain FIR Filter This implementation directly follows the methodology of filter weights design used in the window method:

1. the Hilbert filter weights $h_H(n)$ are calculated analytically in Eq. (11.6),
2. they are multiplied by selected window function, deciding about the flatness of the filter magnitude response in the pass-band and width of the filter transition band, and weights $h_H^{(w)}(n)$ are obtained,
3. the filter length is adjusted to ensure very wide pass-band and very sharp and short filter transition bands, close to frequencies 0 Hz and $f_s/2$,
4. the filter weights are shifted M samples right giving $h_H^{(w)(M)}(n)$,
5. the designed filter weights are convoluted with input signal,
6. samples of the input and output signals are synchronized.

Figure 11.4 illustrates the design issues for filter having $N = 2M + 1 = 21$ and 41 weights. Results for un-windowed theoretical Hilbert filter impulse response are presented in the left column, while in the right column—results for the theoretical impulse response multiplied with Blackman window. In consecutive rows we see: filter weights shifted right by M samples, filter magnitude response and filter phase response (corrected after the right shift, i.e. for $H_H^{(w)}(n)$). We see that usage of window with bigger spectral side-lobes attenuation (the Blackman one) improves linearity of the filter magnitude response in the pass-band (reduces oscillations), at the cost of making the filter pass-band more narrow. Enlarging the filter length helps to reduce this effect and increases the filter working frequency band. The filter phase shift is equal to minus 90 degree, as expected.

Exercise 11.4 (Hilbert Filter in Time Domain). Modify program from Listing 11.1: uncomment lines 12 and 13, i.e. add to it the following Matlab commands:

```
w = blackman(2*M+1)'; stem(w); pause, h = h .* w;
stem(h); pause
```

Try to obtain the same results as presented in Fig. 11.4. Then, increase filter length (change value of M): observe that for longer filters the frequency response becomes sharper but level of oscillations does not change. Change window to kaiser() with $\beta = 10$ and chebwin() with $R = 120$. Note flatness of the filter magnitude response. The phase response is not changing. Do some experiments with values of β and R. Analyze Listing 11.4 with Matlab code of Hilbert filter implementation and analytic signal calculation in time domain. Call the function with cosines with different frequencies and compare in one figure real and imaginary parts of the function output. Cosine should be transformed into a sine. Draw a plot plot(real(xa),imag(xa)). You should see a circle when cosine frequency lies in the filter pass-band. Pay attention to synchronization of real and imaginary parts of the analytic signal, performed inside the function. It is necessary due to delay of the Hilbert filter output caused by the input buffer as well as to delay resulting from the filter weights shift. This issue is addressed in next paragraph.

Listing 11.4: Hilbert filter in time domain

```
1   function xa = hilbertTZ1(x,M,beta)
2     Nx = length(x);
3   % Generation of the Hilbert filter impulse response
4     n=-M:M; hH=(1-cos(pi*n))./(pi*n); hH(M+1)=0;
5   % Windowing using Kaiser window
6     w = kaiser(2*M+1,beta)'; hH = hH .* w;
7   % Filtration
8     xi = conv(x,hH);
9   % Analytic signal with synchronization and removing transient states
10    xa = x(M+1:Nx-M) + j*xi(2*M+1:Nx);
11  end
```

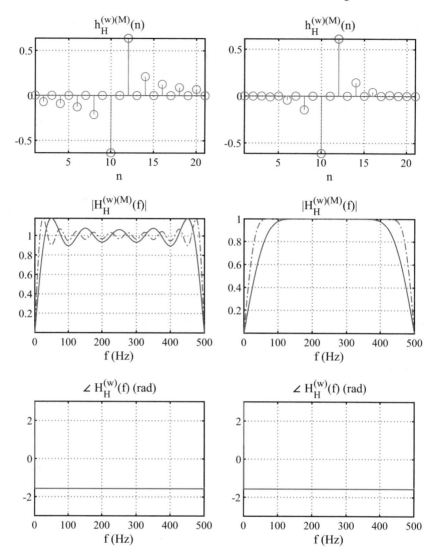

Fig. 11.4: FIR Hilbert filter characteristics, (left)—for rectangular window, (right)—for Blackman window. Up-down: filter impulse response $h_H^{(w)(M)}(n)$ (theoretical, multiplied with the window function), filter magnitude response $|H_H^{(w)(M)}(f)|$, filter phase response $\angle(H_H^{(w)}(f))$—for the non-shifted filter weights or shifted but with filter output synchronized with its input. Filter length $N = 2M + 1 = 21$—solid blue line, $N = 2M + 1 = 41$—dashed red line. Sampling frequency is equal to $f_s = 1000$ Hz

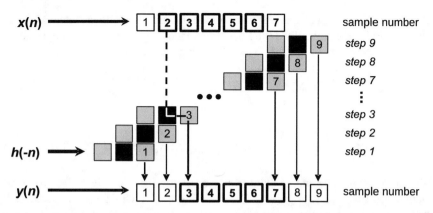

Fig. 11.5: Graphical illustration of *synchronization* of input and output samples of an odd-length FIR filter. Denotations: $x(n)$—input signal ($N_x = 7$), $h(-n)$—filter weights ($N = 2M + 1 = 3$), $y(n)$—output signal ($N_y = N_x + N - 1 = 9$). Corresponding input–output samples: $\{x(2), x(3), ..., x(6)\}$ and $\{y(3), y(4), ..., y(7)\}$, or $\{x(M+1), ..., x(N_x - M)\}$ and $\{y(N), ..., y(N_x)\}$

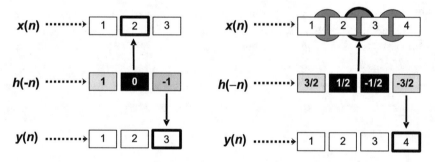

Fig. 11.6: Graphical illustration of *synchronization* of input and output samples of an FIR filter for: (left) odd number of filter weights ($N = 3$), (right) even number of filter weights ($N = 4$)

Synchronization of Filter Output with Input FIR filtering with $N = 2M + 1$ weights, implemented as convolution, introduce delay of M samples. Therefore input and output samples of the FIR Hilbert filter should be *synchronized/paired* very carefully. The first valid output sample has index N, since this is the first sample for which the inner filter buffer is completely filled with input data. But due to the performed M samples right shift of the theoretical Hilbert filter impulse response (initially non-causal), the output is delayed M samples in relation to the input. Therefore the $y(N)$ output sample corresponds to the $x(N - M)$ input sample, and so on ... The synchronization of input and output samples of any odd-length FIR filter is presented in two figures, Figs. 11.5 and 11.6.

Frequency-Domain Hilbert Filtering One can implement Hilbert filter in frequency-domain modifying signal DFT/FFT spectrum according to Eq. (11.1), i.e. multiplying its fragments by $j, -j$ either 0 and calculating the inverse DFT/FFT:

$$x(n) \xrightarrow{N-FFT} X(\Omega_k) \xrightarrow{\cdot H(\Omega_k)} H(\Omega_k) \cdot X(\Omega_k) \xrightarrow{N-FFT^{-1}} x_H(n), \qquad (11.22)$$

where

$$n = 0,1,2,...,N-1, \qquad k = 0,1,2,...,N-1, \qquad \Omega = \frac{2\pi}{N}k. \qquad (11.23)$$

We can also compute the analytic signal $x_a(n)$ for $x(n)$ directly in frequency domain using Eqs. (11.12)–(11.14): 1) calculating the DFT/FFT signal spectrum, 2) multiplying its appropriate fragments by weights $W(\Omega_k)$, equal to 2 either 0, and 3) performing the inverse DFT/FFT:

$$x(n) \xrightarrow{N-FFT} X(\Omega_k) \xrightarrow{\cdot W(\Omega_k)} W(\Omega_k)X(\Omega_k) \xrightarrow{N-FFT^{-1}} x_a(n). \qquad (11.24)$$

Exercise 11.5 (Hilbert Filter in Frequency Domain). Computation of analytic signal in frequency domain, described by Eqs. (11.22) and (11.24), is implemented in two functions presented in Listing 11.5. Using the following Matlab code:

```
x=rand(1,100); error = max( abs( hilbert(x) - hilbertTZ(x)))
```

check whether they give the same results as the Matlab function hilbert().

Listing 11.5: Matlab implementation of Hilbert filter in frequency domain

```
1   function xa = hilbertTZ2(x)
2   N = length(x);                      % input signal length
3   X = fft(x);                         % signal FFT, then its modification:
4   X(1)=0; X(N/2+1)=0;                 % # 0 for 0 Hz and fs/2
5   X(2:N/2)    = -j*X(2:N/2);          % # (-j) for positive frequencies
6   X(N/2+2:N)  = j*X(N/2+2:N);         % # (+j) for negative frequencies
7   xi = ifft(X);                       % inverse FFT of the modified spectrum
8   xa = x+j*xi;                        % creation of analytic signal
9   end
10
11  function xa = hilbertTZ3(x)
12  N = length(x);                      % input signal length
13  X = fft(x);                         % signal FFT, then its modification:
14  X(1)=X(1); X(N/2+1)=X(N/2+1);       % # 1 for 0 Hz and fs/2 (unchanged)
15  X(2:N/2)    = 2*X(2:N/2);           % # 2 for positive frequencies
16  X(N/2+2:N) = zeros(1,N/2-1);        % # 0 for negative frequencies
17  xa = ifft(X);                       % inverse FFT of the modified spectrum
18  end
```

11.2.4 Hilbert Filter Applications: AM/FM Demodulation

In the beginning we do short recapitulation. Equations (11.17), (11.18), (11.19) have fundamental importance for amplitude and phase signal demodulation. We will analyze signal having the following form:

$$x(n) = A(n)\cos(\varphi(n)). \tag{11.25}$$

Doing signal demodulation we, first, calculate its analytic version of $x(n)$:

$$x_a(n) = A(n)\cdot e^{j\,\varphi(n)}, \tag{11.26}$$

and, then, recover from it signal amplitude and angle:

$$A(n) = |x_a(n)|, \tag{11.27}$$

$$\varphi(n) = \angle x_a(n). \tag{11.28}$$

Knowing $\varphi(n)$ we can estimate instantaneous signal frequency, i.e. do it frequency demodulation:

$$f_{inst}(n) = \frac{f_s}{2\pi}\frac{\Delta\varphi(n)}{\Delta n}. \tag{11.29}$$

The Hilbert filter can be used for the following specific AM and FM signal demodulations.

AM-DSB-LC Signal Demodulation   In case of double-side-band (DSB) amplitude modulation (AM) with a large carrier (LC), a cosine with carrier angular frequency Ω_c is modulated in amplitude by a real-value slowly varying signal $(1+x(n))$:

$$y(n) = (1+x(n))\cos(\Omega_c n). \tag{11.30}$$

After calculation of analytic signal for $y(n)$, one obtains

$$y_a(n) = (1+x(n))\,e^{j\Omega_c n}. \tag{11.31}$$

Computing absolute value of $y_a(n)$ allows to reconstruct $(1+x(n))$ and find $x(n)$:

$$|y_a(n)| = 1+x(n) \quad \rightarrow \quad x(n) = |y_a(n)| - 1. \tag{11.32}$$

AM-DSB-SC Signal Demodulation   In case of DSB-AM with a suppressed carrier (SC), a cosine with carrier angular frequency Ω_c is modulated in amplitude by a real-value slowly varying signal $x(n)$. Therefore the modulated signal and its analytic versions are equal:

$$y(n) = x(n)\cos(\Omega_c n), \tag{11.33}$$

$$y_a(n) = x(n)e^{j\Omega_c n}. \tag{11.34}$$

In this case the demodulation is done in two steps. First, the signal carrier $e^{j\Omega_c n}$ is reconstructed from $y_a(n)$ (solving this *carrier recovery task* is presented in chapter on AM modulation). Next, $y_a(n)$ is multiplied by the estimated carrier but with negative angle:

$$y_a(n) \cdot e^{-j\Omega_c n} = \left[x(n)e^{j\Omega_c n}\right] \cdot e^{-j\Omega_c n} = x(n) \tag{11.35}$$

and signal $x(n)$ is restored.

AM-SSB-U or AM-USB Signal Demodulation In case of single-side-band (SSB) amplitude modulation using upper (U) side-band (denoted also as AM upper side-band AM-USB), the analytic version (11.9) of the modulating signal $x_a(n) = x(n) + jx_H(n)$ is directly generated in a transmitter. Such signal does not have negative frequency components, according to Eqs. (11.12)–(11.14), therefore it uses only one, right, positive side-band. Next, the signal is multiplied by a complex-value harmonic carrier $e^{j\Omega_c n}$, i.e. it is shifted-up in frequency to the angular frequency Ω_c, and, finally, its real part is left only:

$$y(n) = \mathrm{Re}\left\{(x(n) + jx_H(n)) \cdot e^{j\Omega_c n}\right\} = x(n)\cos(\Omega_c n) - x_H(n)\sin(\Omega_c n). \tag{11.36}$$

The right part of the last equation can be used directly, not the left part, in order to reduce the number of performed calculations: there is no sense to calculate imaginary part of the multiplication in the left equation part and after that neglect it. In a receiver of the AM-SSB-U (AM-USB) modulated signal, first, analytic signal $y_H(n)$ of the received signal $y(n)$ is calculated, i.e. an imaginary part of the signal is restored. Then, the signal carrier $e^{j\Omega_c n}$ is reconstructed and, after its angle negation, it is multiplied by $y_H(n)$ (the signal is shifted down in frequency to 0 Hz), and, finally, the real part of the result is taken only:

$$\mathrm{Re}\left\{(y(n) + jy_H(n)) \cdot e^{-j\Omega_c n}\right\} = \mathrm{Re}\left\{(x(n) + jx_H(n)) \cdot e^{j\Omega_c n} \cdot e^{-j\Omega_c n}\right\} = x(n). \tag{11.37}$$

AM-SSB-L or AM-LSB Signal Demodulation Demodulation of signal with single side-band (SSB) amplitude modulation using lower (L) side-band (denoted also as AM lower side-band AM-LSB), is exactly the same as for the upper side band with the only one small difference: the Hilbert filter output is put into imaginary part of the created complex-value signal with the negative sign, not the positive one: $x_b(n) = x(n) - jx_H(n)$. Due to this, the signal spectrum has only negative frequency components (look at Exercise 11.2 and program 11.2).

FM Signal Demodulation Let us assume that we have the following signal $s(n)$ (f_c—carrier frequency, f_s—sampling frequency):

$$s(n) = \cos\left(2\pi\left(\frac{f_c}{f_s} \cdot n + \frac{k_f}{f_s}\sum_{k=0}^{n} x(k)\right)\right),\tag{11.38}$$

modulated in frequency by signal $x(n)$ around the carrier frequency f_c and with modulation index k_f. Its demodulation consists of the following steps. First, the analytic signal of $s(n)$ is calculated:

$$s_a(n) = e^{j\cdot 2\pi\left(\frac{f_c}{f_s}\cdot n + \frac{k_f}{f_s}\sum_{k=0}^{n} x(k)\right)} = e^{j\cdot 2\pi\left(\frac{f_c}{f_s}\cdot n\right)} \cdot e^{j\cdot 2\pi\left(\frac{k_f}{f_s}\sum_{k=0}^{n} x(k)\right)}.\tag{11.39}$$

Next, the carrier $e^{j2\pi\frac{f_c}{f_s}n}$ is recovered (topic discussed in next chapters), negated in angle and multiplied by the $s_a(n)$:

$$s_a^{BB}(n) = s_a(n) \cdot e^{-j\cdot 2\pi\left(\frac{f_c}{f_s}\cdot n\right)} = e^{j\cdot 2\pi\frac{k_f}{f_s}\sum_{k=0}^{n} x(k)}.\tag{11.40}$$

This way signal down-conversion in frequency to the so-called *base band* is done. Then, angle of $s_a^{BB}(n)$ is calculated:

$$\varphi(n) = \angle s_a^{BB}(n) = 2\pi\frac{k_f}{f_s}\sum_{k=0}^{n} x(k),\tag{11.41}$$

and, finally, the modulating signal $x(n)$ is found as a scaled phase derivative (Eq. (11.19)):

$$x(n) = \frac{1}{2\pi}\frac{f_s}{k_f}\frac{\Delta\varphi(n)}{\Delta n}.\tag{11.42}$$

Since angle calculation of any complex-value number is limited to the interval $-\pi \leq \alpha < \pi$ and the angle in Eq. (11.41) can go beyond the limits, it preferable to exchange equations (11.40), (11.41) with the following ones ($a(n)$—possible slowly changing signal amplitude):

$$y(n) = s_a^{BB}(n)\left(s_a^{BB}(n-1)\right)^* = a(n)a(n-1)e^{j(\varphi(n)-\varphi(n-1))},\tag{11.43}$$

$$\angle y(n) = \Delta\varphi(n) = \varphi(n) - \varphi(n-1).\tag{11.44}$$

This method is better since using the function `angle()` in Matlab we calculate directly the phase difference, which is changing now in smaller range, and uncertainty concerning $\pm 2\pi$ jumps in the function `angle()` (or `atan2()`) is avoided.

The Hilbert–Huang Signal Decomposition In publication [3] a new, very *fresh* application of the Hilbert transform (filter) was proposed for tracking of time-varying spectral content of multi-components signals, i.e. for time–frequency anal-

ysis. First, signal is decomposed into many, individual, AM-FM modulated components, using empirical mode decomposition method. Then, analytic, complex-value
version of each k-th signal component is found using the Hilbert filter. Finally,
each k-th analytic signal is AM-FM demodulated: $AM_k(t), FM(t)$. Trajectories of
$AM_k(t)$ and $FM_k(t)$ are called a Hilbert–Huang (HH) spectrum. In Matlab the function hht(.) can be used. In fact, in Exercise 11.3 we have performed the HH
analysis of one AM-FM modulated signal component.

**Exercise 11.6 (Testing Prototypes of AM Demodulators Using the Hilbert
Filter).** The described above AM demodulation schemes are implemented in
program 11.6. Run the program for three modulation options and observe the
signal spectra. Add to the program the fourth possible AM modulation type,
i.e. the single side-band left (SSB-L, LSB), making use of code of the program 11.2. Record your own speech with sampling frequency 44.1 kHz and
repeat all tests. Observe carefully spectra of the modulated signals.

Exercise 11.7 (Testing Prototype of FM Demodulator Using Hilbert Filter).
Modify program 11.3. Set AM modulation index to 0: $k_A = 0$. Change sampling
frequency to $f_s = 44.1$ kHz and modulate in frequency a cosine carrier $f_c =$
10 kHz using the same speech signal as in Exercise 11.6. Observe spectra for
different values of modulation index.

Listing 11.6: AM demodulators using Hilbert filter

```
1   % lab11_ex_hilbert_am.m
2   clear all; close all;
3
4   fc = 10000;   % carrier frequency
5   typeAM = 3;   % 1=DSB-C, 2=DSB-SC, 3=SSB-U=USB
6   kA = 0.5;     % modulation index for DSB-C
7
8   % Read or generate modulating signal
9   [x,fs] = audioread( 'speech44100.wav', [1,1*44100] );   % samples from-to
10
11  Nx = length(x); x = x=x.';
12  dt=1/fs; t=dt*(0:Nx-1);
13  df=1/(Nx*dt); f=df*(-Nx/2:1:Nx/2-1);
14  figure; plot(t,x); grid; xlabel('t (s)'); title('x(t)');
15  soundsc(x,fs); pause
16
17  % Carrier amplitude calculation
18  if( typeAM == 1)   a = (1+kA*x);    end
19  if( typeAM == 2)   a = x;           end
```

```
20   if( typeAM == 3)   a = hilbert(x); end
21   % Carrier generation with AM
22   if( typeAM ==1 )  y = a .* cos(2*pi*fc*t); end
23   if( typeAM ==2 )  y = a .* cos(2*pi*fc*t); end
24   if( typeAM ==3 )  y = a .* exp(j*2*pi*fc*t); y = real(y); end
25
26   % AM Demodulation of a cosine
27   ya = hilbert(y); % calculation of analytic signal
28   if( typeAM == 1) xest =  (abs(ya)-1)/kA; end
29   if( typeAM == 2) xest = real( ya.*exp(-j*2*pi*fc*t) ); end
30   if( typeAM == 3) xest = real( ya.*exp(-j*2*pi*fc*t) ); end
31
32   % Results
33   figure; plot(t,x,'r',t,xest,'b'); grid; xlabel('t (s)'); title('x(t) and xest(t)');
34   soundsc(xest,fs)
35   n=1000:Nx-1000; max_err = max( abs( x(n) - xest(n) ) ), pause % error
36
37   figure; plot(f,fftshift(20*log10(abs(fft(x))/Nx))); grid; title(' |X(f)|'); pause
38   figure; plot(f,fftshift(20*log10(abs(fft(a))/Nx))); grid; title(' |A(f)|'); pause
39   figure; plot(f,fftshift(20*log10(abs(fft(y))/Nx))); grid; title(' |Y(f)|'); pause
40   figure; plot(f,fftshift(20*log10(abs(fft(ya))/Nx))); grid; title(' |Ya(f)|'); pause
41   figure; plot(f,fftshift(20*log10(abs(fft(xest))/Nx))); grid; title(' |Xest(f)|'); pause
```

11.3 FIR Differentiation Filters

Theoretical differentiation filter impulse response is calculated as inverse DtFT of the required frequency response $H(\Omega) = j\Omega$ (because Fourier transform of a signal derivative is equal to the Fourier transform of the signal itself, multiplied by $j\Omega$, in analog world $F\left(\frac{dx(t)}{dt}\right) = j\omega \cdot X(\omega)$:

$$h_D(n) = \frac{1}{2\pi} \int_{-\pi}^{\pi} (j\Omega)(e^{j\Omega n})d\Omega = \frac{2}{2\pi} \int_{0}^{\pi} (j\Omega)(j\sin(\Omega n))d\Omega =$$

$$-\frac{1}{\pi} \int_{0}^{\pi} \Omega \sin(\Omega n)d\Omega. \quad (11.45)$$

Frequency response $H(\Omega)$ is asymmetrical around $\Omega = 0$ and for its frequency approximation in Eq. (11.45) only asymmetrical sinus functions are required. Additionally, since the integrated function has the same values for negative and positive angular frequencies, it is possible to integrate only in the range $[0, \pi]$ and multiply the obtained result by 2. In Eq. (11.45) integration by parts can be applied:

$$\int_a^b u \cdot dv = u \cdot v \big|_a^b - \int_a^b v \cdot du, \tag{11.46}$$

$$u = \Omega, \quad dv = \sin(\Omega n)d\Omega \quad \rightarrow \quad v = \int dv = \int \sin(\Omega n)d\Omega = -\frac{1}{n}\cos(\Omega n). \tag{11.47}$$

from where we have

$$h_D(n) = -\frac{1}{\pi}\left[\Omega\left(-\frac{1}{n}\cos(\Omega n)\right)\Big|_0^\pi - \int_0^\pi \left(-\frac{1}{n}\cos(\Omega n)\right)d\Omega \right] =$$

$$= -\frac{1}{\pi}\left[\pi\left(-\frac{1}{n}\cos(\pi n)\right) + \frac{1}{n}\left(\frac{1}{n}\sin(\Omega n)\right)\Big|_0^\pi \right] = \frac{\cos(\pi n)}{n}. \tag{11.48}$$

One should apply window $w(n), n = -M, \ldots, 0, \ldots, M$ to the calculated $h_D(n)$ and then shift the result M samples right, exactly the same way we did in the case of Hilbert filter time-domain implementation. In there are presented impulse and frequency responses of two differential filters with length $N = 2M + 1 = 21$, are presented in Fig. 11.7: on the left side for the rectangular window and on the right side for the Blackman window. For Blackman window the magnitude response is less oscillatory but it has more narrow frequency range of the correct filter work (linear characteristic is required). Increasing the filter length from $N = 21$ to $N = 41$ makes the filter better in this aspect.

Exercise 11.8 (Testing Differentiation Filter). Use the program 11.1. Set h=hD;. Use different values of M and different windows. Observe changes in differentiation filter frequency response. Next generate a sine signal and do it differentiation using function presented in Listing 11.7. Signal xD is a differentiation result while signal xS is the input signal synchronized with it. Check a few points of the filter frequency response.

Listing 11.7: Differentiation filter in time domain

```
1   function [xS,xD] = diffTZ(x,M,beta)
2     Nx = length(x);
3   % Generate the differentiation filter impulse response
4     n=-M:M; hD=cos(pi*n)./n; hD(M+1)=0;
5   % Window it using the Kaiser window
6     w = kaiser(2*M+1,beta)'; hD = hD .* w;
7   % Filter the signal
8     xD = conv(x,hD);
9   % Synchronize and remove transient states
10    xS =   x(M+1:Nx-M)
11    xD = xD(2*M+1:Nx);
12  end
```

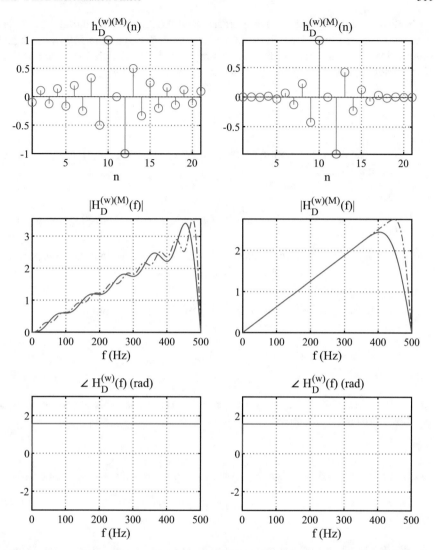

Fig. 11.7: FIR differentiation filter characteristics, (left) for rectangular window, (right) for Blackman window. Up-down: filter impulse response $h_D^{(w)(M)}(n)$ (theoretical, multiplied with the window function, shifted M samples right), filter magnitude frequency response $|H_D^{(w)(M)}(\Omega)|$, filter phase frequency response $\angle(H_D^{(w)}(\Omega))$—with window but without shift of the weights (or with shift correction). Filter length $N = 2M + 1 = 21$—solid line, $N = 2M + 1 = 41$—dashed line

Now we will analyze interesting practical application of the digital FIR differentiation filter. In digital receivers of FM-modulated signals the following signal demodulation method can be used. Let us repeat the FM modulation equation. Cosine carrier modulated in frequency by signal $x(t)$ is given by

$$c(t) = \cos(2\pi f_c t + \phi(t)), \quad \phi(t) = 2\pi k_f \int_0^t x(t)dt, \tag{11.49}$$

where k_f the denotes modulation index. To recover $x(t)$ having $c(t)$, it is possible to differentiate the signal $c(t)$:

$$\frac{dc(t)}{dt} = -\left(2\pi f_c + 2\pi k_f x(t)\right) \cdot \sin(2\pi f_c t + \phi(t)) = -e(t) \cdot \sin(2\pi f_c t + \phi(t)) \tag{11.50}$$

and detect the slow-changing envelope $e(t)$ of fast-changing sine. For this purpose we can

1. calculate the signal derivative (Eq. (11.50)) and square it, obtaining

$$e^2(t) \cdot \sin^2(a) = e^2(t) \cdot 0.5(1 - \cos(2a)) = 0.5e^2(t) - 0.5e^2(t) \cdot \cos(2a); \tag{11.51}$$

2. perform low-pass filtration of the result (11.51) and left this way only the first low-frequency component:
$$0.5e^2(t); \tag{11.52}$$

3. multiply by 2 and calculate the square root of the result (11.52), obtaining:
$$e(t) = 2\pi(f_c + k_f x(t)); \tag{11.53}$$

4. divide (11.53) by 2π, subtract f_c, divide by k_f and ... get the modulated signal $x(t)$.

In the above algorithm two filters are needed: a band-pass differential filter around $f_c \pm \Delta/2f$ and a low-pass filter with the cut-off frequency $\Delta f/2$. Δf should be calculated using the Carson's rule:

$$\Delta = 2f_{max}\left(\frac{k_f}{f_{max}} + 1\right). \tag{11.54}$$

Exercise 11.9 (Testing Prototype of FM Demodulator Using Differentiation Filter). Analyze program 11.8. Speech signal, sampled at 8 kHz, is up-sampled K-times and then it modulates in frequency the 8 kHz cosine carrier. Then, the carrier is demodulated using differentiation filter method and down-sampled. Compare the program code with equations presented above. Set different values of modulation index k_f. Observe width of the modulated signal spectrum. Test whether the Carson's rule is fulfilled. Remember to adjust cut-off frequency of the low-pass filter. Design differentiation filter using Matlab functions `firpm()` and `firls()`.

Listing 11.8: FM demodulator using differentiation filter

```
1    % lab11_ex_differentiation_fm1.m
2    clear all; close all;
3
4    Nx = 8000;          % number of input 8 kHz speech samples to be processed
5    M = 200; N=2*M+1;   % N=2M+1: length of used FIR filters (to be designed)
6    K  = 5;             % speech up-sampling ratio
7    fc = 8000;          % carrier frequency (to be changed by speech signal)
8    fs = K*fc;                       % sampling frequency after up-sampling
9    Nfigs = 2^14; nn = 1:Nfigs;      % for figures
10
11   % GENERATION OF FM-MODULATED CARRIER
12
13   % Read speech 8kHz signal to be FM-broadcasted
14     [x, fx ] = audioread('speech.wav', [1,Nx] ); x=x.';
15   soundsc(y,fx); figure; plot(x); tittle('x(n)'); pause
16   % Up-sampling signal to frequency fs
17   x = interp(x,K); Nx = length(x);
18   % Final frequency modulation of the carrier by x(n)
19     fmax = 4000;            % maximum signal frequency of speech (Hz)
20     DF = 5000;             % 2*DF = required bandwidth of FM modulated signal (Hz)
21     kf = (DF/fmax-1)*fmax;  % modulation index from Carson's rule
22     x = cos( 2*pi*( fc/fs*(0:Nx-1) + kf*cumsum(x)/fs ) ); x=x.'; clear n;
23   % FM DEMODULATION USING DIFFERENTIATION FILTER
24     y = x; clear x;
25   % Designing impulse response of the FIR differentiator using window method
26     n=-M:M; hD=cos(pi*n)./n; hD(M+1)=0; w = kaiser(N,10)'; hD = hD .* w;
27   % Differentiation
28     y = filter(hD, 1, y);  y = y(N:end);
29   % Power of 2
30     y = y.^2;
31   % Designing impulse response of the low-pass FIR filter
32     n=-M:M; hLP=sin(2*pi*4000/fs*n)./(pi*n); hLP(M+1)=2*(4000)/fs; hLP = hLP .* w;
33   % Low-pass filtration
34     y = filter(hLP, 1, y ); y = y(N:end);
35   % Decimation and square root
36     y = real( sqrt( 2 * y(1:K:end) ) );
37     y = (y - 2*pi*fc/fs)/(2*pi*kf/fs);
38   % Let the music play
39     soundsc(y,fx); figure; plot(y); tittle('x(n)'); pause
```

11.4 Summary

Simple weighted running averaging of signal samples allows not only frequency low-high-band-stop filtration but also designing digital phase shifters and digital differentiation filters. They were presented in this chapter. What should be remembered?

1. Discussed filters have weights designed using the *window method* methodology: their impulse responses are calculated analytically and after that they are appropriately windowed. The longer the filter is, the wider is frequency band of filter operation, i.e. signal phase shifting or signal differentiating. The window choice decides about the level of oscillations present in the filter frequency response.
2. The Hilbert filter is passing all frequencies apart from a DC signal component and is delaying them by minus $\pi/2$ radians. It is used for creation of the so-called analytic complex-value signal, having in its real part the original signal and in its imaginary part the Hilbert filter output. When the original signal is modulated/changed (*slowly*) in amplitude and in frequency, it can be very easily demodulated using its analytic version. For amplitude demodulation we should simply track absolute values of the analytic signal complex numbers, while for frequency demodulation track their phase, calculate its derivative, and scale the result.
3. Differentiation is very important mathematical operation. In discrete-time automatic control of dynamic systems, know-how of its robust to noise, real-time calculation is priceless. In courses on numerical analysis different methods of data differentiation are discussed, typically making use of difference equations and data approximation polynomials (e.g. in direct, Lagrange or Newton form) or local data extensions (e.g. Taylor one). Typically as the easiest differentiation solution such weights are used: $[-1, 1]$, $[-1, 0, 1]/2$ or $[1, -8, 0, 8, -1]/12$. But in DSP special differentiation filter are designed having significantly more weights.

11.5 Private Investigations: Free-Style Bungee Jumps

Exercise 11.10 (FM Demodulation with Differentiation Filter Revisited). Modify program 11.8. Increase up-sampling order and modulate separately two or three different cosine carriers with two or three independent speech signals. Before the FM demodulation use should filter out the FM radio services which you are not interested in. Therefore a band-pass filter should be applied, leaving only an FM radio station you would like to listen to.

Exercise 11.11 (FM Demodulation with Hilbert Filter Revisited). Modify the program 11.8 the same way as described in Exercise 11.10. After the separation of only one FM broadcast by a band-pass filter, do the FM signal demodulation using the Hilbert filter method—like in the program 11.3.

Exercise 11.12 (* Traditional and Amateur AM Radio Receivers). Find in Internet any signal of traditional AM-DSB radio broadcast or amateur AM-SSB radio transmission. Try to demodulate it using program 11.6.

References

1. S.C. Chapra, *Applied Numerical Methods with MATLAB for Engineers and Scientists* (McGraw-Hill, New York, 2012)
2. S.L. Hahn, *Hilbert Transforms in Signal Processing* (Artech House, London, 1996)
3. N.E. Huang, Z. Shen, S.R. Long, M.C. Wu, H.H. Shih, Q. Zheng, N.-C. Yen, C.C. Tung, H.H. Liu, The empirical mode decomposition and the Hilbert spectrum for nonlinear and non-stationary time series analysis. Proc. R. Soc. Lond. A **454**(1), 903–995 (1998)
4. R.G. Lyons, *Understanding Digital Signal Processing* (Addison-Wesley Longman Publishing, Boston, 1996, 2005, 2010)
5. A.V. Oppenheim, R.W. Schafer, *Discrete-Time Signal Processing* (Pearson Education, Upper Saddle River, 2013)
6. T.W. Parks, C.S. Burrus, *Digital Filter Design* (Wiley, New York, 1987)
7. J.G. Proakis, D.G. Manolakis, *Digital Signal Processing. Principles, Algorithms, and Applications* (Macmillan, New York, 1992; Pearson, Upper Saddle River, 2006)
8. R. Qu, A new approach to numerical differentiation and integration. Math. Comput. Modell. **24**(10), 55–68 (1996)
9. D.E.T. Romero, G.J Dolecek, Digital FIR Hilbert transformers: Fundamentals and efficient design methods, chapter 19 in *MATLAB - A Fundamental Tool for Scientific Computing and Engineering Applications*, ed. by V. Katsikis (IntechOpen, 2012), online available
10. T.P. Zieliński, *Cyfrowe Przetwarzanie Sygnałów. Od Teorii do Zastosowań (Digital Signal Processing. From Theory to Applications)* (Wydawnictwa Komunikacji i Łączności (Transport and Communication Publishers), Warszawa, Poland, 2005, 2007, 2009, 2014)

Chapter 12
FIR Adaptive Filters

Dinosaurs failed to survive because they could not adapt to changing world. In contrary to our adaptive filter which knows how to do it.

12.1 Introduction

There is no universal screwdriver but it is very annoying to have a drawer full of different types of them and still could not find the right one. Some compromise is needed. An adaptive filter represents such a compromise: it is only one filter but capable to adjust its weights according to changing *circumstances*, changing signal features.

The classical non-adaptive FIR digital filtration can be interpreted as a local weighted averaging some last input samples $x(n)$:

$$y(n) = \sum_{k=0}^{M} h_k x(n-k) = \mathbf{h}^T \mathbf{x}(n), \quad y(n) = h_0 x(n) + h_1 x(n-1) + ... + h_M x(n-M),$$

(12.1)

where $\mathbf{h}$ and $\mathbf{x}(n)$ represent, respectively, vertical vector of filter weights and vertical vector of last $M+1$ input signal samples:

$$\mathbf{h}^T = [h_0, \ h_1, \ ..., \ h_M], \quad \mathbf{x}^T(n) = [x(n), \ x(n-1), \ ..., x_M(n-M)]. \quad (12.2)$$

In standard FIR filters, weights $h_k, k = 0...M$, are fixed during usage and specially designed before their application (for example: LP, HP, BP, BS filters, Hilbert phase shifter, digital differentiator).

© Springer Nature Switzerland AG 2021
T. P. Zieliński, *Starting Digital Signal Processing in Telecommunication Engineering*, Textbooks in Telecommunication Engineering,
https://doi.org/10.1007/978-3-030-49256-4_12

ADAPTIVE FILTER

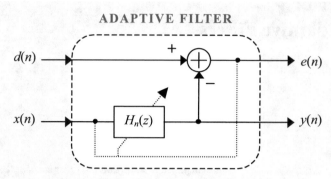

Fig. 12.1: Block diagram of the adaptive filter having two inputs: $d(n), x(n)$, and two outputs: $e(n), y(n)$ [24]

Block diagram of an adaptive filter is shown in Fig. 12.1. Adaptive filter is not a traditional filter having only one input and one output. It has two inputs $x(n)$ and $d(n)$:

– $x(n)$—a signal to be filtered,
– $d(n)$—a reference (desired) signal, not processed by the filter,

and two outputs $y(n)$ and $e(n)$:

– $y(n)$—result of signal $x(n)$ filtration,
– $e(n)$—a difference between signals $d(n)$ and $y(n)$: $e(n) = d(n) - y(n)$.

Only the input signal $x(n)$ is filtered by the filter $H_n(z)$. Goal of filter weights adaptation is to make the filtering result $y(n) = H[x(n)]$ equal to the second filter input $d(n)$, called the desired signal. Or at least make it similar as much as possible. Difference between the expected signal $d(n)$ and $y(n)$, being filtered $x(n)$, represents the second filter output, so-called *error* signal $e(n) = d(n) - y(n)$. During its work, adaptive filter is changing the filter weights **h** from sample to sample and minimizes a chosen cost function, having present and past error values as its arguments. In the simplest case of an LMS filter, the cost function has a form of squared momentum error $e^2(n)$.

FIR adaptive filter is described by Eq. (12.1) but its weights **h**, in contrary to Eq. (12.1), are time-dependent: they can be different for different time index n. Therefore we can add the time index n to the vector **h**:

$$\mathbf{h}^T(n) = [h_0(n), \ h_1(n), \ ..., \ h_M(n)], \qquad (12.3)$$

and still use the classical FIR filter equation (12.1) for FIR adaptive filter description. Vector of filter weights $h_k(n)$ is adapted/changed during the filter work:

$$\mathbf{h}(n+1) = \mathbf{h}(n) + \Delta\mathbf{h}(n), \qquad \text{new}(n+1) = \text{old}(n) + \text{change}(n), \qquad (12.4)$$

attempting to ensure minimization of some chosen cost function $J(n)$, associated with the error signal $e(n)$. Choice of the function $J(n)$ and its computer implementation decides about the filter type: (normalized) least mean square (NLMS), weighted least squares (WLS), and recursive weighted least squares (RWLS)(RLS).

Adaptive filters are widely used in very important applications, for example in: adaptive interference canceling (AIC), adaptive echo canceling (AEC), adaptive noise canceling (ANC), adaptive signal/line enhancement (ASE)(ALE), and adaptive telecommunication channel identification and equalization.

When one reads about adaptive filters for the first time, immediately the following question appears: what is the profit from making equal (or similar) one signal $(x(n))$ to the other one $(d(n))$? But very, very wide field of adaptive filter applications (presented wider in the next section) should confirm us that this is a fruitful idea. The merit of the adaptive filtering relies on the concept of adaptive correlation canceling (ACC) between two signals. In the most often application scenario we have

- a signal of interest with additive disturbance, i.e. $d(n) = s(n) + z(n)$, for example, airplane pilot speech plus engine purr acquired by microphone located close to pilot lips,
- and a deformed copy of the disturbance $x(n) = \tilde{z}(n)$, acquired by a different sensor, for example, only motor purr acquired by a second microphone put in the airplane cockpit but far away from the pilot.

Any digital filter is processing signals amplifying/attenuating and delaying its frequency components. The role of the adaptive filter is to do the same with a disturbance *copy*, i.e. filtering $x(n) = \tilde{z}(n)$ and making the signal $y(n)$ similar as much as possible to the disturbance $z(n)$ present in the signal $d(n)$: $y(n) \rightarrow \hat{z}(n)$. Then, after subtracting $y(n)$ from $d(n)$ we are reducing the disturbance present in the signal $d(n)$:

$$e(n) = d(n) - y(n) = (s(n) + z(n)) - H[\tilde{z}(n)] = s(n) + (z(n) - \hat{z}(n)). \quad (12.5)$$

This is an adaptive filter *magic*. There are plenty of applications fitting to this scenario. Adaptive interference (correlation) canceling can be applied in many important real-world application scenarios. Examples of some other signals $d(n)$ are as follows:

- signal from the telecommunication receiver antenna containing an echo from the transmitter antenna of the same device,
- signal from the microphone in the hands-free telephony system having an echo of the signal from the loudspeaker,

- ECG of a still unborn child with ECG of the mother, making the prenatal diagnosis more difficult,
- speech of a diver with a sound of its heartbeat.

Are you already interested in adaptive filters? Very well!

12.2 Adaptive Filter Application Scenarios

Adaptive filters have a lot of applications, to mention only a few of them: adaptive noise canceling headphones, adaptive echo canceling car speakerphone systems and conference hand-free systems, adaptive echo canceling in telecommunication receivers (e.g. signal of transmitting antenna is captured by a receiving antenna), disturbance of mother heartbeat present in an ECG signal of the baby during making prenatal medical examinations, signal of a diver heartbeat present in its speech, burr of pneumatic hammer adding itself to a speech of a worker. Adaptive processing is also used for adaptive impulse response identification and correction, for example in adaptive communication channel estimation and equalization, as well as for adaptive noise canceling (ANC) and speech enhancement in digital telephony (ALE—Adaptive Line Enhancement).

Applications are many. How they are classified? For example this way.

Adaptive Interference Canceling (AIC) This is the most frequently used application of adaptive filter (Fig. 12.2). The input signal $d(n)$ consists of information part $s(n)$ and disturbing interference $i(n)$ (*burr, purr, wir, wirr*), while the input signal $x(n)$ represents an available interference reference pattern $i_{ref}(n)$:

$$d(n) = s(n) + i(n), \tag{12.6}$$
$$x(n) = i_{ref}(n). \tag{12.7}$$

We would like to remove from $d(n)$ the oscillatory interference $i(n)$ which has added to our signal $s(n)$. It is assumed in the method that a reference pattern $i_{ref}(n)$ of the interference is available. It is similar to $i(n)$, because it is deliberately acquired (recorded) the same time as $i(n)$ was, but separately by another sensor (microphone). The role of the filter is to adjust/fit $i_{ref}(n)$ to $i(n)$, i.e. delay in time and scale in amplitude, and then to subtract it from signal $d(n)$:

$$e(n) = d(n) - y(n) = (s(n) + i(n)) - H[i_{ref}] = s(n) + (i(n) - H[i_{ref}(n)]). \tag{12.8}$$

The better the adaptation procedure of the filter $H[.]$ is, the smaller is the last term $i(n) - H[i_{ref}(n)]$ in Eq. (12.8) and the better is the interference canceling. Adaptive noise canceling (ANC) is a special case of AIC.

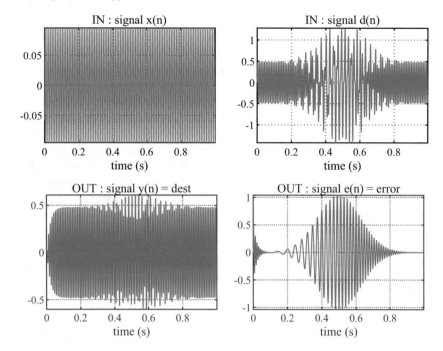

Fig. 12.2: Example of adaptive interference canceling (AIC), from left to right up-down: $x(n)$—reference pattern of an interference—a sine, $d(n)$—impulsive oscillatory signal disturbed by interference copy (delayed sine, scaled in amplitude), $y(n)$—filtered reference pattern of the interference, correlated with interference disturbing our impulsive signal, $e(n) = d(n) - y(n)$ error signal, i.e. our impulse with partially subtracted disturbance. In the beginning, filter adaptation is visible

Adaptive Echo Canceling (AEC) This is a special case of AIC. At present interference $i(n)$ is an echo $i_{ref}^{(echo)}(n)$ of some other signal $i_{ref}()$ which is available for us:

$$d(n) = s(n) + i_{ref}^{(echo)}(n), \tag{12.9}$$

$$x(n) = i_{ref}(n). \tag{12.10}$$

The best example is hands-free car-phone system where microphone is capturing not only a voice of a driver but also sound from loudspeakers with voice of his/her interlocutor. The loudspeaker signal is known.

Adaptive Signal/Line Enhancement (ASE)(ALE) In this example the adaptive filter is working as a linear predictor: input signals $d(n)$ and $x(n)$ are copies of the same noisy signal $s(n)$ but $x(n)$ is delayed in relation to $d(n)$ by one or a few samples:

$$d(n) = s(n+P),\qquad\qquad\qquad (12.11)$$

$$x(n) = s(n).\qquad\qquad\qquad (12.12)$$

Trying to minimize the error between the reference signal $d(n)$ and the filtered $x(n)$, the filter has to predict the next sample $s(n+P)$ on the base of previous known samples $s(n), s(n-1), s(n-2), ..., s(n-M)]$. Only deterministic, sine-like signals can be predicted. Each sample of a sinusoid can be calculated as a linear superposition of its two previous samples taken with some weights with fixed values. In these weights are *hidden* values of sinusoid parameters (amplitude, frequency, and phase). If signal contains K sines, the predicting filter should have $2K$ weights. Summarizing, the adaptive filter working in ASE/ALE scenario adjusts its weights to the linear prediction coefficients and the filter frequency response has peaks around sine frequencies which were *found* in a signal by the filter. As a result the signal components are amplified: the signal-to-noise ratio becomes higher (see Fig. 12.3). Having estimation of the LP coefficients, associated with the signal being processed, we can compute and track its frequency spectrum. Why? Because after setting $\Omega = 2\pi\frac{f}{f_s}$, a frequency response of the recursive IIR digital filter having coefficients $\mathbf{b} = [1]$ and $\mathbf{a} = [1, \mathbf{h}]$:

$$H_n(z)\big|_{z=\exp(j\Omega)} = H_n(\Omega) = \frac{1}{1 + h_1(n)e^{-j\Omega} + h_2(n)e^{-j2\Omega} + ... + h_{2K}(n)e^{-j2K\Omega}},$$
$$(12.13)$$

can be used for estimation of the signal spectrum shape. This spectrum can be estimated from sample to sample.

Adaptive Channel/Object Identification and Equalization (ACI/ACE) In this application the signal $d(n)$ is equal to output signal from some object, e.g. telecommunication channel, and signal $x(n)$ is equal to the signal exciting the object/channel, e.g. pilot sequence $p(n)$ in telecommunication systems:

$$d(n) = G[p(n)],\qquad\qquad\qquad (12.14)$$

$$x(n) = p(n).\qquad\qquad\qquad (12.15)$$

Adaptive filter $H[x(n)]$ tries to minimize the difference between the filtered $x(n)$ and $d(n)$:

$$e(n) = d(n) - y(n) = G[\,p(n)\,] - H[\,p(n)\,].\qquad\qquad\qquad (12.16)$$

This is obtained when $H[.]$ is the same as $G[.]$. Therefore the filter is adapting its weights to the object/channel weights: $H[.] \to G[.]$ and object/channel identification is performed. Knowing the transfer function $G[.]$ of the channel we can built the channel equalizer simply reversing the channel transfer function.

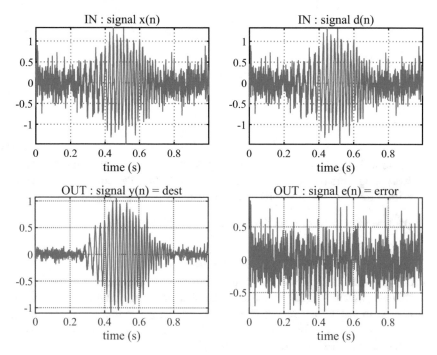

Fig. 12.3: Example of adaptive signal/line enhancement (ASE)(ALE) using filter configuration as linear predictor. Impulsive oscillatory signal, the same as in Fig. 12.2, contaminated with Gaussian noise, is de-noised by the filter. From left to right up-down: $x(n)$—noisy signal delayed one sample, $d(n)$—original noisy signal, $y(n)$—filtered $x(n)$ being prediction of $d(n)$, $e(n) = d(n) - y(n)$ prediction error signal

12.3 Adaptive Filter Types

In adaptive filters, one input signal $x(n)$ is filtered and subtracted from the other input signal $d(n)$, called the desired one:

$$e(n) = d(n) - H[x(n)]. \tag{12.17}$$

The difference signal is called an error signal. Filter is changing its weights in different way depending on the cost function which it minimizes. The following cost functions are used, incorporating in different way the momentum error values $e(n)$:

- the squared momentum error only:

$$J(n) = e(n)^2, \tag{12.18}$$

   leading in *long time horizon* to decreasing least mean squared error (LMS);
- cumulative least squared (LS) error from the filter start:

$$J(n) = \sum_{k=0}^{n} e^2(k), \tag{12.19}$$

- cumulative weighed least squared (WLS) error, forgetting past error values:

$$J(n) = \sum_{k=0}^{n} \lambda^{n-k} e^2(k) = \lambda^0 e^2(n) + \lambda^1 e^2(n-1) + \lambda^2 e^2(n-2) + \dots + \lambda^n e^2(0), \tag{12.20}$$

where λ denotes a forgetting factor $0 \ll \lambda \le 1$. For $\lambda = 1$ the filter remembers everything from the start and reduces to the LS filter. The smaller λ, the shorter is filter memory and faster its reactions, but lower noise robustness.

Adaptive filter names origin from criteria they minimize:

- LMS—the most popular, not demanding computationally but having slow convergence when adaptation parameter μ value is set in risk-averse way;
- NLMS—normalized LMS is faster, due to normalization of filter weights change by energy of last signal $x(n)$ samples; not demanding computationally;
- LS—with faster convergence than NLMS but being significantly more complex computationally; it is used only in stationary case for adjusting filter weight values from initial ones to optimal ones,
- WLS—LS with forgetting factor $0 \ll \lambda < 1$, which allows the filter to forget past errors, i.e. to have *shorter memory about the history*, and thanks to this to track an optimal local solution which can be slowly changing; the filter is demanding computationally; for $\lambda = 1$ we have standard LS filter;
- RWLS or simply RLS—recursive, fast, computationally *lighter* version of WLS adaptive filter—recommended for usage.

12.4 LMS Adaptive Filters

Derivation of Adaptation Rule In case of the LMS filters, the minimization of the squared instantaneous error value $e(n)$ is performed (Eq. (12.18)) which in longer *time horizon* leads to the minimization of the mean squared error value (therefore expectation operator $E[.]$ is removed):

$$J(n) = e(n)^2 = \left[d(n) - \sum_{k=0}^{M} h_k(n) x(n-k) \right]^2. \tag{12.21}$$

Since we are looking for a minimum of a non-negative convex function $J(n)$, it is sufficient to calculate its partial derivatives in respect to all coefficients $h_k(n)$, $k = 0\dots M$, i.e. gradients—directions of a function growth:

$$\frac{dJ(n)}{dh_k(n)} = 2 \cdot e(n) \cdot \frac{d}{dh_k(n)} \left[d(n) - h_0(n) \cdot x(n-0) - \ldots - h_M(n) \cdot x(n-M) \right],$$

$$(12.22)$$

$$\frac{dJ(n)}{dh_k(n)} = -2 * e(n) \cdot x(n-k),$$

$$(12.23)$$

and then to go in the opposite direction to the calculated gradient (minus sign) during adaptation of filter weights:

$$h_k(n+1) = h_k(n) - \frac{dJ(n)}{dh_k(n)},$$

$$(12.24)$$

$$h_k(n+1) = h_k(n) + 2 \cdot \mu \cdot e(n) \cdot x(n-k),$$

$$(12.25)$$

where μ denotes a speed adaptation coefficient/factor. Equation (12.25) can be written in a matrix form, which is giving better insight into computational structure of the filter weights update (at present neglecting scaling by 2 or incorporating it in μ coefficient):

$$\begin{bmatrix} h_0(n+1) \\ h_1(n+1) \\ \vdots \\ h_M(n+1) \end{bmatrix} = \begin{bmatrix} h_0(n) \\ h_1(n) \\ \vdots \\ h_M(n) \end{bmatrix} + \mu \cdot e(n) \begin{bmatrix} x(n) \\ x(n-1) \\ \vdots \\ x(n-M) \end{bmatrix}$$

$$(12.26)$$

or simply:

$$\mathbf{h}(n+1) = \mathbf{h}(n) + \mu \cdot e(n) \cdot \mathbf{x}(n),$$

$$(12.27)$$

In the classical LMS adaptive filter μ has fixed value.

For complex-value signals the LMS filter equations are defined as (denotations: $()^H$—conjugation and transposition of vector of filter weights, $()^*$—only conjugation):

$$e(n) = d(n) - \mathbf{h}^H(n)\mathbf{x}(n)$$

$$(12.28)$$

$$(1) \quad \mathbf{h}(n+1) = \mathbf{h}(n) + \mu_{complex} e^*(n)\mathbf{x}(n)$$

$$(12.29)$$

$$(2) \quad \mathbf{h}(n+1) = \mathbf{h}(n) + \mu_{complex} e(n)\mathbf{x}^*(n).$$

$$(12.30)$$

Optimal Solution (Wiener Filter) For stationary signals, one can directly calculate optimal weights $\mathbf{h}_{opt}$ to which the filter should converge. It can be done by (at present, in order to simplify notation, (n) in $\mathbf{h}(n)$ is neglected):

1. rewriting the cost function into the matrix form:

$$
\begin{aligned}
J(\mathbf{h}) &= E\left[\left(d(n) - \mathbf{h}^T \mathbf{x}(n)\right)^2\right] = E\left[d^2(n)\right] - 2E\left[d(n)\mathbf{h}^T \mathbf{x}(n)\right] + E\left[\left(\mathbf{h}^T \mathbf{x}(n)\right)^2\right] = \\
&= E\left[d^2(n)\right] - 2\mathbf{h}^T E\left[d(n)\mathbf{x}(n)\right] + E\left[\mathbf{h}^T \mathbf{x}(n)\mathbf{x}^T(n)\mathbf{h}\right] = \\
&= E\left[d^2(n)\right] - 2\mathbf{h}^T \mathbf{r}_{dx}^{(n)} + \mathbf{h}^T E\left[\mathbf{x}(n)\mathbf{x}^T(n)\right]\mathbf{h} = \\
&= E\left[d^2(n)\right] - 2\mathbf{h}^T \mathbf{r}_{dx}^{(n)} + \mathbf{h}^T \mathbf{R}_{xx}^{(n)} \mathbf{h}, \quad (12.31)
\end{aligned}
$$

where

- $\mathbf{h}$—vector of filter weights,
- $\mathbf{r}_{dx}$—vector of cross-correlation between signals $d(n)$ and $x(n)$,
- $\mathbf{R}_{xx}$—matrix of auto-correlation of signal $x(n)$

2. calculation of $J(\mathbf{h})$ derivative in respect to vector $\mathbf{h}$ and setting it to zero:

$$
\frac{dJ(\mathbf{h}))}{d\mathbf{h}} = -2 \cdot \mathbf{r}_{dx} + 2 \cdot \mathbf{R}_{xx} \cdot \mathbf{h} = 0. \quad (12.32)
$$

3. solving the last equation is respect to $\mathbf{h}$ (so-called optimal Wiener filter):

$$
\mathbf{h}_{opt} = \mathbf{R}_{xx}^{-1} \cdot \mathbf{r}_{dx}. \quad (12.33)
$$

Stability Condition The filter adaptation is convergent when the following condition is fulfilled:

$$
|1 - \mu \lambda_k| < 1, \qquad k = 0, 1, 2, ..., M, \quad (12.34)
$$

where λ_k denotes the k-th eigenvalue of the auto-correlation matrix $\mathbf{R}_{xx}$. Value of speed adaptation coefficient μ has to fulfill the equations:

$$
0 < \mu < \frac{2}{\lambda_{max}}, \qquad 0 < \mu < \frac{2}{\text{trace}(\mathbf{R}_{xx})}, \quad (12.35)
$$

where λ_{max} is the maximum eigenvalue of the auto-correlation matrix $\mathbf{R}_{xx}$ of signal $x(n)$. Larger values of μ offer faster adaptation convergence but they are more risky from the point of view of filter stability, and make the filter more sensitive to noise (the filter is fast and attempts to track a noise, offering noisy outputs).

During derivation of filter stability conditions the following terms are obtained:

$$
v_k(n+1) = (1 - \mu \lambda_k)^{n+1} v_k(0), \quad k = 0, 1, 2, ..., M. \quad (12.36)
$$

The slowest convergent term (the closest to 1) is associated with the smallest eigenvalue λ_{min}. After setting $\mu = 1/\lambda_{max}$ one obtains $(1 - \lambda_{min}/\lambda_{max})^{(n+1)}$. The smallest the difference between the minimum and maximum eigenvalue is,

the closer to 1 is the ratio $\lambda_{min}/\lambda_{max}$ and the closer to 0 is $(1 - \lambda_{min}/\lambda_{max})$—
in such situation filter convergence is very fast. If $\lambda_{min} = 0$, the filter is not
convergent at all. This tells that the convergence is faster for $x(n)$ being a noisy
(random) signal, because for such signal the ratio $\lambda_{min}/\lambda_{max}$ is close to 1 (a
signal has a lot of equally important components, not only a few strong ones).
Having this in mind, we should de-correlate (whiten) signal $x(n)$ and make the
matrix $(R)_{xx}$ orthogonal.

Other LMS Filters The classical equation (12.27) of the LMS filter can be modified in different ways. For example, this is a more universal filter description:

$$\mathbf{h}(n+1) = \mathbf{h}(n) + \mu(n) \cdot \mathbf{W}(n) \cdot e(n) \cdot \mathbf{x}(n). \qquad (12.37)$$

As especially marked, the speed factor μ can be also a function of time index n and
adapted, as in the normalized LMS algorithm. The $\mathbf{W}(n)$ denotes a matrix which
was introduced in order to increase the adaptation speed, as in the LMS-Newton
algorithm. Specific proposed changes of the standard LMS filter are briefly listed
below.

Normalized LMS Filter The μ is changed during filter work, i.e. divided by energy of last $M + 1$ samples of the signal $x(n)$:

$$\mu(n) = \frac{\mu}{\gamma + \|\mathbf{x}(n)\|^2} = \frac{\mu}{\gamma + \mathbf{x}^H(n)\mathbf{x}(n)} = \frac{\mu}{\gamma + \sum_{k=0}^{M} x^*(n-k)x(n-k)}, \qquad (12.38)$$

which makes the adaptation faster and convergence more stable (the second stability
condition in Eq. (12.35) is always fulfilled).

Sign-LMS Filter In adaptation equations only sign of the error signal $e(n)$ is used.

Leaky-LMS Filter When some eigenvalues of the auto-correlation matrix $\mathbf{R}_{xx}$ are
equal to zero, stability condition (12.34) is not fulfilled and the filter can be unstable.
Remedy for this is changing the adaptation equation to the following one:

$$\mathbf{h}(n+1) = (1 - \mu\gamma)\mathbf{h}(n) + \mu e(n)\mathbf{x}(n), \quad 0 < \gamma \ll 1, \qquad (12.39)$$

where γ denotes very small *leakage* coefficient (e.g. equal to 0.05).

Partial-Update-LMS Filter Not all filter weights are updated in time n or update
does not take place at all for some value of n.

Set-Membership-LMS Filter Filter weights update is done only when error $e(n)$
exceeds some threshold.

Frequency-Domain-LMS Filter Adaptation of filter weighs is done using not signal samples but their discrete Fourier transform coefficients. Signal samples are de-correlated by the DFT and adaptation is faster. While FFT is used, this approach is also computationally efficient and allows to adapt long filters. Therefore it is capable of modeling long impulse responses of real-world object, what is especially important in room acoustics.

Sub-band-LMS Filter Signals $d(n)$ and $x(n)$ are decomposed into frequency sub-bands and separate LMS filters are used in each sub-band.

Newton-LMS Filter Equation (12.37) is used with matrix $\mathbf{W}(n)$, being estimation of the matrix $\mathbf{R}_{xx}^{-1}$. We can say that the optimal Wiener solution (12.33) is approximated this way. Newton-LMS filters are very close conceptually to RLS adaptive filters, which are presented in the next section. The RLS filters are significantly better known and widely used than the Newton-LMS filter.

Exercise 12.1 (How Can I Work in This Whir!). The Matlab function `adaptTZ()`, presented in Listing 12.1, implements standard and normalized LMS adaptive filter (as well as RLS filter but do not interest in it now). Write a main program calling this function. Try to implement any application scenario of adaptive filtering. If you have no idea, record 3–4 s of your own speech, sampled at 8 kHz, and generate a sinusoid $s(n)$ with frequency 1 kHz, sampled also at 8 kHz. Add sinusoid to the speech and use this signal as $d(n)$. Multiply sinusoid by 0.25 and delay it by 4 samples and use this signal as $x(n)$: `x=[0 0 0 0 0.25*s(1:end-4)]`. Apply adaptive filter as correlation/interference canceler to reduce the sinusoidal disturbance. Add figure to the function `adaptTZ()`, displaying filter frequency response in the adaptation loop (`f=0:fs/2000:fs/2; plot(abs(freqz(h,1,f,fs)))`). Exchange sinusoid with SFM modulated signal: $f_c = 1000\,\text{Hz}$, $\Delta f = 500\,\text{Hz}$, $f_m = 0.5\,\text{Hz}$ (look at Chap. 2 on signal generation). Exchange SFM signal with recording of burr of any working machine, e.g. hairdryer, vacuum cleaner, etc. You can change filter length, μ value, use LMS or NMLS algorithm.
If you have problems with good value selection of filter length M and adaptation coefficients, LMS: `mi`, NLMS: `mi,gamma`, please, look to Listing 12.2 and find inspiration in it (even copy some fragments of it). You can also first do the Exercises 12.5 or 12.5 before realization of the present task.

> **Exercise 12.2 (Does Anybody Understand What I Am Telling?).** Record
> your own speech and add noise to it using the Matlab function `randn()`. Then
> use the function `adaptTZ()` from Listing 12.1 and apply adaptive signal en-
> hancement strategy, i.e. use the filter as linear predictor: `d=[s(2:end)];`
> `x=s(1:end-1);`. Listen to signal before and after filtration. Calculate signal-
> to-noise ratio before and after the filter. SNR is defined in Chap. 2 in Table 2.2.
> Measure SNR after the convergence of the adaptation filter (for this purpose ob-
> serve adaptation of filter weights). Try to find values of NLMS filter parameters
> (filter length M and adaptation speed `mi`) offering the highest SNR value.
>
> If you have problems with good value selection of filter length M and adaptation
> coefficients, LMS: `mi`, NLMS: `mi`, `gamma`, please, look to Listing 12.2 and
> find inspiration in it (even copy some fragments of it). You can also first do
> Exercises 12.8 or 12.9 before realization of the present task.

Listing 12.1: Matlab function implementing NLMS and RLS adaptive filters

```
1   function [y, e, h] = adaptTZ(d,x,M,mi,gamma,lambda,delta,ialg)
2   % M-filter length, LMS: mi, NLMS: mi, gamma, RLS: lambda, delta
3   % Initialization
4   h  = zeros(M,1);          % filter weights
5   bx = zeros(M,1);          % input buffer for x(n) samples
6   Rinv = delta*eye(M,M);    % inverse of auto-correlation matrix Rxx^{-1}
7   for n = 1 : length(x)            % adaptive filtering loop
8       bx = [ x(n); bx(1:M-1) ];    % take new sample of x(n) into the buffer
9       y(n) = h' * bx;              % filter x(n): y(n) = sum( h .* bx )
10      e(n) = d(n) - y(n);          % calculate error e(n)
11      if(ialg==1)      % LMS
12          h = h + mi * e(n) * bx;                      % update filter weights
13      elseif(ialg==2)  % NLMS
14          energy = bx' * bx;                           % energy of signal x(n)
15          h = h + mi/(gamma+energy) * e(n) * bx;  % update filter weights
16      else             % RLS
17          Rinv = (Rinv - Rinv*bx*bx'*Rinv/(lambda+bx'*Rinv*bx))/lambda; % update
18          h = h + Rinv * bx * e(n);                    % update
19      end
20  end
```

12.5 RLS Adaptive Filters

Recursive least squares (RLS) adaptive filters are used for minimizing the weighted
least squares (WLS) cost function (12.20) in very fast, recursive way. Forgetting
factor values, which are used, are in the range $0 \ll \lambda \leq 1$, typically slightly smaller
than 1, e.g. $\lambda = 0.99$. When $\lambda = 1$, the WLS filters give the best, robust to noise
LS estimate of $\mathbf{h}(n)$, based on the observation time interval from 0 to n. The smaller

λ is, the WLS filters are more forgetting the past and more *local*. They can fastly adapt to signal changes at the cost of being more sensitive to noise (output signals have bigger variance in stationary states).

Since for $\lambda = 1$ the cost function (12.20) simplifies to (12.19), we will derive adaptation equations for (12.20) as more universal.

Let us calculate derivative of (12.20) over single filter weight $h_k(n)$:

$$\frac{\partial J(n)}{\partial h_k(n)} = 2\sum_{i=0}^{n} \lambda^{n-i} e[i] \frac{\partial e(i)}{\partial h_k(n)} = -2\sum_{i=0}^{n} \lambda^{n-i} e(i)x(i-k) = 0. \tag{12.40}$$

After writing in (12.40) equation of the adaptive filter error signal $e(n)$, one gets

$$\sum_{i=0}^{n} \lambda^{n-i} \left\{ d(i) - \sum_{j=0}^{M} h_k(j)x(i-j) \right\} x(i-k) = 0. \tag{12.41}$$

After changing summation order and simple mathematical transformations, we have

$$\sum_{j=0}^{M} h_k(j) \left\{ \sum_{i=0}^{n} \lambda^{n-i} x(i-j)x(i-k) \right\} = \sum_{i=0}^{n} \lambda^{n-i} d(i)x(i-k). \tag{12.42}$$

Since the last equation is valid for $k = 0, 1, 2, ..., M$, it can be written in the following matrix form:

$$\mathbf{R}_{xx}(n)\mathbf{h}(n) = \mathbf{r}_{dx}(n) \tag{12.43}$$

where

$$\mathbf{R}_{xx}(n) = \sum_{i=0}^{n} \lambda^{n-i} \mathbf{x}(i)\mathbf{x}^T(i), \qquad \mathbf{r}_{dx}(n) = \sum_{i=0}^{n} \lambda^{n-i} d(i)\mathbf{x}(i). \tag{12.44}$$

We can summarize that in the weighted LS filter one should first calculate estimates of $\mathbf{R}_{xx}(n)$ and $\mathbf{r}_{dx}(n)$ using Eqs. (12.44), and then filter weights in moment n as a solution of the Eq. (12.43):

$$\mathbf{h}(n) = \mathbf{R}_{xx}^{-1}(n)\mathbf{r}_{dx}(n). \tag{12.45}$$

Equation (12.45) has the same form as the optimal Wiener solution (12.33). Of course, for next time stamp $(n+1)$ filter weights are equal to:

$$\mathbf{h}(n+1) = \mathbf{R}_{xx}^{-1}(n+1)\mathbf{r}_{dx}(n+1). \tag{12.46}$$

The WLS adaptive filter is very demanding computationally, due to immense matrix arithmetic growing in time when time index n becomes bigger and bigger. The only solution is to organize calculations in recursive way, i.e. not to calculate auto-correlation matrix and cross-correlation vector in Eqs. (12.44), starting from $n = 0$, but to update their values which are already known for the time stamp (n):

$$\mathbf{R}_{xx}(n+1) = \lambda \mathbf{R}_{xx}(n) + \mathbf{x}(n+1)\mathbf{x}^T(n+1), \tag{12.47}$$

$$\mathbf{r}_{dx}(n+1) = \lambda \mathbf{r}_{dx}(n) + d(n+1)\mathbf{x}(n+1). \tag{12.48}$$

During filter initialization for $n = 0$, the matrix $\mathbf{R}_{xx}(0)$ is set as a diagonal matrix with power of signal $x(n)$ upon the main diagonal, and vectors $\mathbf{h}(0)$ and $\mathbf{r}_{rd}(0)$ have elements equal to zero. Next, Eqs. (12.47) and (12.48) are used.

In matrix algebra the following matrix theorem holds for a square matrix $\mathbf{A}$ and vectors $\mathbf{u}$ and $\mathbf{v}$ with the same dimensions:

$$\left(\mathbf{A} + \mathbf{u}\mathbf{v}^T\right)^{-1} = \mathbf{A}^{-1} - \frac{\mathbf{A}^{-1}\mathbf{u}\mathbf{v}^T\mathbf{A}^{-1}}{1 + \mathbf{v}^T\mathbf{A}^{-1}\mathbf{u}}. \tag{12.49}$$

After setting $\mathbf{A} = \lambda \mathbf{R}_{xx}(n)$ and $\mathbf{u} = \mathbf{v} = \mathbf{x}(n+1)$, and using Eqs. (12.47) and (12.49), we obtain

$$\begin{aligned}
\mathbf{R}_{xx}^{-1}(n+1) &= \left[\lambda \mathbf{R}_{xx}(n) + \mathbf{x}(n+1)\mathbf{x}^T(n+1)\right]^{-1} = \\
&= \lambda^{-1}\mathbf{R}_{xx}^{-1}(n) - \frac{\lambda^{-2}\mathbf{R}_{xx}^{-1}(n)\mathbf{x}(n+1)\mathbf{x}^T(n+1)\mathbf{R}_{xx}^{-1}(n)}{1 + \lambda^{-1}\mathbf{x}^T(n+1)\mathbf{R}_{xx}^{-1}(n)\mathbf{x}(n+1)} = \\
&= \lambda^{-1}\mathbf{R}_{xx}^{-1}(n) - \frac{\lambda^{-1}\mathbf{R}_{xx}^{-1}(n)\mathbf{x}(n+1)\mathbf{x}^T(n+1)\mathbf{R}_{xx}^{-1}(n)}{\lambda + \mathbf{x}^T(n+1)\mathbf{R}_{xx}^{-1}(n)\mathbf{x}(n+1)}. \tag{12.50}
\end{aligned}$$

Thanks to Eq. (12.50) complicated matrix inversion in (12.46) is avoided. The matrix $\mathbf{R}_{xx}^{-1}(-1)$ is set with a small value on its main diagonal. Matrices $\mathbf{R}_{xx}^{-1}(0)$, $\mathbf{R}_{xx}^{-1}(1)$, $\mathbf{R}_{xx}^{-1}(2)$ are calculated recursively.

And *the final cut*. Now, we will show how to calculate new filter weight values at time $(n+1)$ as modification of previous values from time (n). Using Eq. (12.46) in Eq. (12.46), we get

$$\mathbf{h}(n+1) = \mathbf{R}_{xx}^{-1}(n+1) \cdot (\lambda \mathbf{r}_{dx}(n) + d(n+1)\mathbf{x}(n+1)). \tag{12.51}$$

At present we want to remove $\mathbf{r}_{dx}(n)$ from the last equation. We calculate $\mathbf{R}_{xx}(n)$ from Eq. (12.47) and put the result to Eq. (12.43), obtaining:

$$\mathbf{r}_{dx}(n) = \frac{1}{\lambda}\left[\mathbf{R}_{xx}(n+1) - \mathbf{x}(n+1)\mathbf{x}^T(n+1)\right]\mathbf{h}(n). \tag{12.52}$$

Finally, we use Eq. (12.52) in (12.51):

$$\begin{aligned}
\mathbf{h}(n+1) &= \mathbf{h}(n) + \mathbf{R}_{xx}^{-1}(n+1)\mathbf{x}(n+1)\left[d(n+1) - \mathbf{x}^T(n+1)\mathbf{h}(n))\right] = \\
&= \mathbf{h}(n) + \mathbf{R}_{xx}^{-1}(n+1)\mathbf{x}(n+1)[d(n+1) - y(n)] = \\
&= \mathbf{h}(n) + \mathbf{R}_{xx}^{-1}(n+1)\mathbf{x}(n+1)e(n). \tag{12.53}
\end{aligned}$$

As we see, the final recursive equation (12.53) of the WLS adaptive filter is very similar to the Eqs. (12.26) and (12.27) of the LMS adaptive filter, with matrix $\mathbf{W}(n)$

added for increasing convergence speed of the filter. In the RLS filter the matrix $\mathbf{W}(n)$ is replaced with $\mathbf{R}_{xx}^{-1}(n)$ and $\mu = 1$. What has a big sense, because the increasing and decreasing the adaptation speed is NOW input signal dependent.

The following two equations summarize the RLS adaptive filter weights update:

$$\mathbf{R}_{xx}^{-1}(n+1) = \lambda^{-1}\mathbf{R}_{xx}^{-1}(n) - \frac{\lambda^{-1}\mathbf{R}_{xx}^{-1}(n)\mathbf{x}(n+1)\mathbf{x}^T(n+1)\mathbf{R}_{xx}^{-1}(n)}{\lambda + \mathbf{x}^T(n+1)\mathbf{R}_{xx}^{-1}(n)\mathbf{x}(n+1)}, \quad (12.54)$$

$$\mathbf{h}(n+1) = \mathbf{h}(n) + \mathbf{R}_{xx}^{-1}(n+1)\mathbf{x}(n+1)e(n). \quad (12.55)$$

I know that this section was extremely boring to most Readers. But the RLS adaptive filters are too important in real-world applications and should not be described shorter only by simplifying *rule of a thumb*.

Exercise 12.3 (RLS vs. SledgeHammer! Fight of the Evening). Do the RLS adaptive filters really track faster and more accurate signal changes than the NLMS filter? Repeat experiments done in Exercises 12.1 and 12.2, but making use of the RLS algorithm instead of the (N)LMS one. The RLS filter is implemented also in the function adaptTZ() from Listing 12.1. Test different values of forgetting factor $\lambda \leq 1$ and initialization constant δ. How to choose them? Try to find inspiration in the all-in-one program 12.2. Filter with value of λ close to 1 (e.g. 0.999) is changing its weights very slowly. If this problem is too difficult for you, then compare RLS and (N)LMS filters adaptation speed solving easier problems 12.5 and 12.8, or slightly more difficult problems 12.6 and 12.9.

12.6 RLS Filters as Dynamic System Observers

A story which is just beginning is very important also. We can approximate real-world dynamic systems as piece-wise linear time-invariant systems. Typically, we know system input $x(n)$ and system output $d(n)$ samples, and assume that output samples are result of convolution of input samples with samples of the system impulse response $h(n)$. Therefore we assume that the following matrix equation is approximately valid:

$$\begin{bmatrix} x_M & x_{M-1} & \cdots & x_0 \\ x_{M+1} & x_M & \cdots & x_1 \\ \vdots & \vdots & \ddots & \vdots \\ x_{M+P} & x_{M+P-1} & \cdots & x_P \end{bmatrix} \begin{bmatrix} h_0 \\ h_1 \\ \vdots \\ h_M \end{bmatrix} = \begin{bmatrix} d_M \\ d_{M+1} \\ \vdots \\ d_{M+P} \end{bmatrix}, \tag{12.56}$$

in matrix form:

$$\mathbf{Xh} = \mathbf{d}. \tag{12.57}$$

The notation $x(n) = x_n$ was used. We have more equations than variables: $P > M$ which helps us decreasing error of $\mathbf{h}$ estimation. We solve the above equation in respect to $\mathbf{h}$ in least squares (LS) sense. First, we multiply both sides of (12.57) by the $\mathbf{X}^T$:

$$\left(\mathbf{X}^T\mathbf{X}\right)\mathbf{h} = \mathbf{X}^T\mathbf{d}. \tag{12.58}$$

The matrix $\mathbf{X}^T\mathbf{X}$ is a square one and has an inverse. Now, we multiply both sides of (12.58) by this inverse matrix:

$$\left(\mathbf{X}^T\mathbf{X}\right)^{-1}\mathbf{X}^T\mathbf{Xh} = \left(\mathbf{X}^T\mathbf{X}\right)^{-1}\mathbf{X}^T\mathbf{d},$$
$$\mathbf{I}\cdot\mathbf{h} = \left(\mathbf{X}^T\mathbf{X}\right)^{-1}\mathbf{X}^T\mathbf{d},$$
$$\mathbf{h} = \left(\mathbf{X}^T\mathbf{X}\right)^{-1}\mathbf{X}^T\mathbf{d}, \tag{12.59}$$

where $\mathbf{I}$ is the square identity matrix with ones on its main diagonal.

The LS solution (12.59) has one very big drawback: when P value grows, multiplications of big matrices have to be done. Therefore initial solving the Eq. (12.57) for $P = M$ is recommended (number of equations and unknowns is the same). And then only upgrading this solution should be done, after obtaining a new pair of values $\{x(n+1), d(n+1)\}$ (first $\{x(M+1), d(M+1)\}$, then $\{x(M+2), d(M+2)\}$, and so on) and having an additional equation. In fact, this is a computational strategy of the RWLS (RLS) adaptive filter algorithm (12.54), (12.55) with $\lambda = 1$. Therefore this filter can be used for recursive calculation of $\mathbf{h}$ in the Wiener object identification scenario (12.33), which is described also by Eqs. (12.56), (12.57).

Solution (12.59) minimizes the LS error:

$$J = (\mathbf{d} - \mathbf{Xh})^T(\mathbf{d} - \mathbf{Xh}) = \sum_{k=0}^{P} e^2(k). \tag{12.60}$$

If we are interested in minimization of the weighted LS error:

$$J = (\mathbf{d} - \mathbf{Xh})^T\mathbf{W}(\mathbf{d} - \mathbf{Xh}) = \sum_{k=0}^{n} \lambda^{n-k}e^2(k), \qquad \mathbf{W} = \begin{bmatrix} \lambda^n & 0 & \cdots & 0 \\ 0 & \lambda^{n-1} & \cdots & 0 \\ \vdots & \vdots & \ddots & 0 \\ 0 & 0 & 0 & \lambda^0 \end{bmatrix} \tag{12.61}$$

and in forgetting the oldest system equations, Eqs. (12.58) and (12.59) are replaced with, respectively:

$$(\mathbf{X}^T\mathbf{W}\mathbf{X})\mathbf{h} = (\mathbf{X}^T\mathbf{W})\mathbf{d} \quad \leftrightarrow \quad \mathbf{R}_{xx}\mathbf{h} = \mathbf{r}_{dx}, \tag{12.62}$$

$$\mathbf{h} = \left(\mathbf{X}^T\mathbf{W}\mathbf{X}\right)^{-1}\mathbf{X}^T\mathbf{W}\mathbf{d}. \tag{12.63}$$

Such solution is found and track by RWLS adaptive filter described by Eqs. (12.54), (12.55) with $\lambda < 1$.

Further elaboration of the dynamic system observer/tracker problem would lead us to Kalman filter theory. For ambitious reader a demonstration program lab12_ex_kalman.m is *deeply hidden* in the book archives, presenting our *hero* in action.

Exercise 12.4 (My Name Is Holmes: Sherlock Holmes). Use RLS adaptive filter algorithm from the function adaptTZ()—Listing 12.1. Generate $N_x =1000$ samples of random noise sequence using the Matlab function: x=randn(1,N). Use it as signal $x(n)$ in adaptive filter. Write your own loop for FIR signal filtering or use the Matlab function d=filter(h,1,x). Filter the generated noise, i.e. the signal $x(n)$. For the first $N_x/2$ samples use filter weights $h = [1,-0.5]$, for the second half—reversed weights $h = [-1,0.5]$. Use the filter output as the signal $d(n)$ in adaptive filter. Modify function adaptTZ()—it should return also some chosen filter weights, allowing observation of their change during adaptation, or plot selected filter weights in the adaptation loop inside the function. Call the modified function adaptTZ(). Set M=2, 5, 10. Try different values of λ. Did the filter recognize properly the filter $h = [\pm 1, \mp 0.5]$ used for filtering the signal $x(n)$? Solving this problem you can copy some code from the Listing 12.2.

12.7 Exercises

When one understands philosophy of adaptive filter work, it is time to find an application well fitting to adaptive filter usage. There are many such applications.

Adaptive filtering programs are not very long but they are extremely condensed: a few line doing a lot. Typical are students questions: *why this? and this?* It is very difficult to answer them without *a piece of math* because adaptive filters implement in practice some complex mathematical optimization strategies which are severely derived. But a typical final solution has a form of a few *golden lines* of code.

Therefore in this section, we will concentrate not on writing our own programs (because the most probable is that they will not be better than existing ones), but on solving some interesting problems with the use of well-known and very good adaptive filter algorithms. In the program 12.2 classical (N)LMS and RLS adaptive filters are implemented and used for adaptive interference (correlation) canceling as well as for adaptive signal enhancement by means of adaptive linear prediction. It is possible to use synthetic, generated signals (pure sine, LFM and SFM signals with optional envelopes: rectangular, linearly increasing, exponentially decaying, impulsive Gaussian) as well as real signals read from disc, at present only some speech recordings are offered. A Reader is encouraged to do 2-3 from 5 exercises proposed below.

Exercise 12.5 (Interference Canceling for Generated Signals). Analyze code of the program 12.2. Run it for the following settings:
```
isig=1; itask=1, env = 0,3, ialg = 1,2,3.
```
Observe results. Try to interpret them. Adaptive filter is working in this case with synthetic signals as adaptive interference canceler:
```
d(n) = s(n) + osc2(n);
x(n) = osc1(n);
```
We would like to remove from d(n) the oscillatory interference osc2(n), which is adding to our signal s(n). The signal osc1(n) is a reference interference pattern, similar to osc2(n), acquired (recorded) separately by another sensor (microphone). The role of the filter is to adjust osc1(n) to osc2(n), delay it in time and scale it in amplitude, and then to subtract it from d(n):
```
e(n) = d(n) - y(n) = (s(n) + osc2(n)) - filtered(osc1(n));
```
What is the maximum permitted values of mi? It is calculated using the Wiener filter theory. Try different values of mi in the program, smaller and bigger. Is the filter always stable? Are the final filter weights h(n) similar to the optimal Wiener filter weights hopt(n)?

Exercise 12.6 (Interference Canceling for Real-World Signals). In this exercise we will apply the adaptive interference canceler to our own speech signal. We will make the following settings in the program 12.2:
```
isig=2; itask=1, ialg=1,2,3.
```
Record separately your own speech s(n) and disturbing sound of *burr, purr, whir, whirr* osc1(n) of any machine/engine. Add a deformed (filtered) burr to your speech. Try to reduce interference using adaptive filter. Modify the program 12.2. Code for preparation of input signals for the adaptive filter could have such form:
```
h = [ 0, 0.75, 0, 0.25 ];
x = vibro;
d = speech + filter(h,1,vibro);
```
Does the adaptive filter work properly? Change filtering algorithms and speed adaptation parameters mi and lambd.

Exercise 12.7 (Echo Canceling for Real-World Signals). In this exercise we will test adaptive echo canceling (AEC), being a special form of adaptive interference canceling (AIC). Working scenario is the same as above in Exercises 12.5 and 12.6:

```
d(n) = speakerA(n) + echoB(n);
x(n) = speakerB(n);
```

which is typical for teleconference systems and hand-free phone systems, for example, in a car. `speakerA=sA` denotes our speech, `speakerB=sB` is a speech of a person we are taking to, our interlocutor, sent to a loudspeaker, `echoB` is an echo of `speakerB` signal captured by a microphone.

Record two clear speech fragments `sA(n)` and `sB(n)` lasting approx. 5–10 sec or take them from the Internet. Create two signals:

```
d(n) = delayed_1sec( sA(n) ) + filtered( sB(n) );
x(n) = sB(n);
```

Delaying `sA(n)` is recommended to give a filter time for finding unknown impulse response of a room. As filter weights use, for example, the following values `h = [ 0, 0, 1, 0, 0, 0.5, 0, 0, 0.25 ]`. Has the adaptive filter converged to `h`?

Try to find in the Internet impulse response of any room or use the supported impulse_resp_11kHz.wav as `h`. Is the filter working properly for it?

Exercise 12.8 (Signal Enhancement for Generated Signals). Adaptive signal enhancement (ASE) is used for years for enhancement of telecommunication speech signals. It is known also as adaptive line enhancement (ALE). Its goal is signal de-noising: the adaptive filter adjusts its frequency response to sinusoids present in the signal, decreasing this way level of noise and increasing the signal-to-noise ratio. Run the program 12.2. Make the following settings:

```
isig=1; itask=2; env=1; ialg=1,2,3;.
```

Observe filtering results. Try to understand them and interpret. In this case we have

```
d(n) = s(n);
x(n) = s(n-1);
```

and would like to predict `s(n)` having `s(n-1)`. Since only deterministic components of `s(n)` can be predicted, the filter adjusts peaks of its frequency response to these components and improve this way the signal SNR (signal-to-noise ratio).

Un-comment two figures in the adaptation loop:

```
subplot(211); stem(h); xlabel('n'); title('h(n)');
subplot(212); plot(f,abs(freqz(h,1,f,fs)));
title('|H(f)|');
```

Observe how the filter h(n) weights are changing and how the filter frequency response moves its peaks towards the frequencies of signal components. How much was the signal SNR improved by the ASE/ALE algorithm?

Exercise 12.9 (Signal Enhancement for Real Signals). Run the program 12.2. Make the following settings:
    isig=2; itask=2; ialg=1,2,3;.
Observe filtering results. Try to understand them and interpret. Apply the adaptive signal enhancement algorithm to recorded your own noisy speech signals. You can also add noise in Matlab to clear noise-free recordings.

Listing 12.2: Program for testing adaptive (N)LMS and RLS filters

```
1  % lab12_ex_adapt.m
2  clear all; close all;
3
4  % Task (AIC, ANC) and algorithm selection (LMS, NLMS, RLS)
5  isig  = 1;  % 1=synthetic or 2=real signal
6  itask = 1;  % 1=adaptive interference canceling (e.g. cross-talk)
7            % 2=adaptive signal de-noising using linear prediction method
8  ialg  = 1;  % adaptation algorithm: 1=LMS, 2=NLMS (normalized LMS), 3=RLS
9
10 % LMS filter (Least Mean Squares)
11 M = 50;                % number of adaptive filter coefficients (weights)
12 mi = 0.1;              % adaptation speed coefficient ( 0<mi<1)
13 % NLMS filter (normalized LMS), faster convergence
14 gamma = 0.001;         % not divide by zero constant, e.g. = 0.001
15 % RLS filter (recursive LS) - more-complex, faster convergence
16 lambd = 0.999;         % RLS - forgetting factor for Least-Squares cost function
17 Rinv = 0.5*eye(M,M);   % RLS - inverse of Rxx matrix
18
19 % Generation of an LFM (linear frequency modulation) test signal
20 if(isig == 1) % SYNTHETIC SIGNALS ========================================
21 env  = 1;            % choose information signal envelope:
22                      % 0=rectangular, 1=alfa*t, 2=exp(-alfa*t), 3=Gauss,
23 fs = 1000;           % sampling frequency
24 Nx = 1*fs;           % number of samples
25 A = 1;               % signal amplitude
26 f0 = 0;              % LFM: initial signal frequency
27 df = 100;            % LFM: frequency increase [Hz/s], SFM: modulation index [Hz]
28 fc = 1;              % SFM: carrier frequency
29 fm = 1;              % SFM: modulating frequency
30 f=0:fs/500:fs/2;                  % frequencies for plots
31 dt=1/fs; t=0:dt:(Nx-1)*dt;        % time points for plots
32 s = A*cos( 2*pi*(f0*t + 0.5*df*t.^2) );              % LFM SIGNAL
33 %s = A*cos( 2*pi*(fc*t + df/(2*pi*fm)*cos(2*pi*fm*t) );  % SFM SIGNAL
```

```
34                                                    % ENVELOPE choice:
35   if (env==0) w = boxcar(Nx)' ; end              % 0 = rectangular
36   if (env==1) alfa=5; w=alfa*t; end              % 1 = alfa*t
37   if (env==2) alfa=5; w=exp(-alfa*t) ; end       % 2 = exp(-alfa*t)
38   if (env==3) alfa=10; w=exp(-alfa*pi*(t-0.5).^2); end % 3 = Gauss
39
40   s = s .* w;                                     % SIGNAL WITH ENVELOPE
41   if (itask==1)                                   % TEST 1 - interference canceling
42     P = 0;                                        % no prediction
43     x = 0.1*sin(2*pi*200*t-pi/5);  % interference delayed and attenuated
44     d = s + 0.5*sin(2*pi*200*t);   % signal + interference
45   end
46   if (itask==2)                                   % TEST 2 - de-noising by linear prediction
47     P = 1;                                        % prediction order set to 1,2,3,...)
48     x = s + 0.25*randn(1,Nx);      % signal + noise
49     d = [ x(1+P:length(x)) 0 ];    % signal x(n) speed-up by P samples (earlier)
50   end
51   else % REAL SIGNALS ========================================
52     [s, fs] = audioread('speech8000.wav'); s=s' ;
53     [sA,fs] = audioread('speakerA.wav');   sA=sA' ;
54     [sB,fs] = audioread('speakerB.wav');   sB=sB' ;
55     P = 1; % delay in samples
56     if(itask==1)
57       s = sA;                                     % reference for comparison
58       x = sB; Nx = length(x);                     % original echo signal
59       d = sA + 0.25*[ zeros(1,P) sB(1:end-P) ]; % added echo copy:
60     end                                           % weaker (0.25), delayed (P)
61     if(itask==2)
62       x = s; Nx = length(x);                      % original noisy speech
63       d = [ x(1+P:length(x)) zeros(1,P) ]; % signal x speed-up by P samples
64     end
65     f=0:fs/500:fs/2;                              % frequencies for plots
66     dt = 1/fs; t = dt*(0:Nx-1);                   % time for plots
67   end %========================================
68
69   % Figures - input signals
70   figure;
71   subplot(211); plot(t,x); grid; title('IN : signal x(n)');
72   subplot(212); plot(t,d); grid; title('IN : signal d(n)'); xlabel('time (s)'); pause
73
74   % Calculation of optimal Wiener filter and limits for mi
75   for k = 0 : M
76     rxx(k+1) = sum( x(1+k:Nx) .* x(1:Nx-k) )/(Nx-M);  % auto-correlation of x(n)
77     rdx(k+1) = sum( d(1+k:Nx) .* x(1:Nx-k) )/(Nx-M); % cross-correlation of d(n) and x(n
       )
78   end
79   Rxx = toeplitz(rxx,rxx);          % symmetrical autocorrelation matrix of x(n)
80   h_opt = Rxx\rdx';                 % weights of the optimal Wiener filter
81   lambda = eig( Rxx );              % eigenvalue decomposition of Rxx
82   lambda = sort(lambda,'descend'); % sorting eigenvalues
83   disp('Suggested values of mi')
84   mi1_risc = 1/lambda(1),          % limit #1 - inverse of max eigen-value
85   mi2_risc = 1/sum(lambda), pause  % limit #2 - inverse of sum of all eigen-values
86   figure;
```

```
 87   subplot(211); stem( h_opt ); grid; title('Optimal Wiener filter h(n)');
 88   subplot(212); stem( lambda ); grid; title('Eigenvalues of matrix Rxx');
 89   % mi = mi2_risc/20;
 90
 91   % Adaptive filtration
 92   bx = zeros(M,1);                % initialization of buffer for input signal x(n)
 93   h  = zeros(M,1);                % initialization of filter weights (coefficients)
 94   y = [];
 95   e = []; figure;
 96   for i = 1 : length(x)           % Main loop
 97      bx = [ x(i); bx(1:M-1) ];    % put new sample of x(n) into the buffer
 98      dest = h' * bx;              % filtration of x(n) = prediction of d(n)
 99      err = d(i) - dest;           % prediction error
100      if (ialg==1) % LMS #########
101         h = h + ( 2*mi * err * bx );      % LMS  - weights adaptation
102      end
103      if (ialg==2) % NLMS ########
104         eng = bx' * bx;                              % signal energy in bx
105         h = h + ( (2*mi)/(gamma+eng) * err * bx ); % weights adaptation
106      end
107      if (ialg==3) % RLS #########
108         Rinv = (Rinv - Rinv*bx*bx'*Rinv/(lambd+bx'*Rinv*bx))/lambd; % new Rinv
109         h = h + (err * Rinv * bx );                         % new weights
110      end
111      if(0) % Observation of filter weights and filter frequency response
112         subplot(211); stem(h); xlabel('n'); title('h(n)'); grid;
113         subplot(212); plot(f,abs(freqz(h,1,f,fs))); xlabel('f (Hz)');
114         title('|H(f)|'); grid; % pause(0.25)
115      end
116      y = [y dest];
117      e = [e err];
118   end
119
120   % Figures - output signals
121   figure;
122   subplot(211); plot(t,y); grid; title('OUT : signal y(n) = dest');
123   subplot(212); plot(t,e); grid; title('OUT : signal e(n) = err');
124   xlabel('time [s]'); pause(0.25)
125   if (itask==1)
126      figure; subplot(111); plot(t,s,'r',t,e,'b'); grid; xlabel('time [s]');
127      title('Original (RED), filtration result (BLUE)'); pause(0.25)
128   end
129   if (itask==2)
130      n=Nx/2+1:Nx;
131      SNR_in_dB  = 10*log10( sum( s(n).^2 ) / sum( (d(n)-s(n)).^2 ) ),
132      SNR_out_dB = 10*log10( sum( s(n).^2 ) / sum( (s(n)-y(n)).^2 ) ),
133      n=1:Nx-P;
134      figure; subplot(111); plot(t(n),s(n),'k',t(n),d(n),'r',t(n),y(n),'b');
135      grid; xlabel('time (s)');
136      title('Reference (BLACK), Noisy (RED), Filtered (BLUE)'); pause(0.25)
137   end
138   figure; subplot(111); plot(1:M+1,h_opt,'ro-',1:M,h,'bx-'); grid;
139   title('h(n): Wiener (RED), our (BLUE)');   xlabel('n'); pause(0.25)
```

12.8 Summary

What we should remember about FIR adaptive filters?

1. The adaptive FIR filter is a moving average filter but with weights being changed during filter work, not fixed. It has two inputs and two outputs, in contrary to the standard filter with only one input and one output. The filtering goal is to adaptively process the first input signal $x(n)$ and to make it equal (or very similar) to the second input signal $d(n)$, desired one.

2. Adaptive filter works as correlation canceler: it tries to adjust (correlate) its first input signal $x(n)$ to its deformed *copy* present in the second one $d(n)$ and subtract it, leaving only in the second signal its part which is uncorrelated with the first input. In typical application, we have a signal with additive disturbance and disturbance reference, acquired by some other sensor. For example, speech of an airplane pilot with engine *burr1* and engine *burr2* recorded alone. The filter is adjusting *burr2* to *burr1* and subtracting it from disturbed pilot speech, reducing the engine interference.

3. Adaptive filters performing correlation canceling can be exploited as linear predictors. It is sufficient to delay signal $x(n)$ by one sample in relation to signal $d(n)$ and the filter, trying correlate both signal, has to predict the sample $d(n) = x(n)$ having $x(n-1)$. Since noise could not be predicted, the filter adjusted its frequency response peaks to sine-like components of $x(n)$ and amplify them, improving the signal-to-noise ratio (SNR). This way adaptive signal enhancement is done, also in telecommunication lines—adaptive line enhancement (ALE). Additionally, having linear prediction (LP) filter, one can calculate signal $x(n)$ spectrum since it is equal to frequency response of the LP filter. Not only calculate but also track the spectrum shape, because LP coefficients are updated for every new input sample of $x(n), d(n)$ signals. Bravo! Possibility of spectral analysis of time-varying signals is obtained.

4. Adaptive filters use different strategies for their weight correction, because they minimize different measures of the error between the filtered signal $x(n)$ and the desired signal $d(n)$: momentum, cumulative or weighted cumulative error. For this reason, different types of adaptive filters are used: least mean squares LMS (momentum), least squares LS (cumulative), and weighted least squares WLS (weighted cumulative). In practice two types of filters are the most important: LMS, eventually normalized, and RLS, being fast version of the WLS adaptive filter.

5. In stationary conditions adaptive filter weights tend to optimal Wiener solution: $\mathbf{h}_{opt} = \mathbf{R}_{xx}^{-1} \cdot \mathbf{r}_{xd}$, where $\mathbf{R}_{xx}^{-1}$ denotes inverse of the auto-correlation matrix of the signal $x(n)$ and $\mathbf{r}_{xd}$ is vector of cross-correlation values between signals $d(n)$ and $x(n)$.

6. Adaptive filters are changing their weights and try to reach the cost function minimum in the steepest descent way. The weight update is in the direction opposite to the local cost function gradient. But when the weights modification is too big, the filter cannot reach the minimum and *is jumping above it*. There are special constraints put on input signal $x(n)$ and values of filter parameters (e.g. μ in LMS filter) which ensure the filter convergence.

7. Adaptive filters are very widely used, to mention only: noise canceling headphones, acoustic noise control systems, echo canceling in: hand-free car-phone systems and combined telecommunication transmitters-receivers, telecommunication channels identification and equalization.

12.9 Private Investigations: Free-Style Bungee Jumps

Exercise 12.10 (Frequency Tracking Using Adaptive Linear Predictor). Use program 12.2. Set parameters: `isig=2`; `itask=2`; `ialg=1,2,3`, i.e. use the adaptive filter as a linear predictor. Set filter length `M=10`. Filter your own recorded speech signal. Linear prediction coefficients $\mathbf{a} = [1, \mathbf{h}]$ store into a special matrix. Then, outside the filtering loop calculates consecutive momentum signal spectra using the calculated LP vectors $\mathbf{a}$: `X=freqz(1,a,f,fs)`. Display matrix of calculated spectra (their magnitudes!) and compare the plot with figure generated by the Matlab function `spectrogram()`. Try to explain origin of their difference (in spectrogram we have detail FFT-based spectrum showing harmonics of vocal folds oscillations and vocal tract resonances, while in our method we are tracking only change of spectrum envelope, i.e. vocal tract resonances—see chapter on speech compression). You can also add the following line into the adaptation loop:
`plot(f,abs(freqz(1,a,f,fs))); title('|X(f)|');`
and display the momentum signal spectrum inside the loop. Try to apply the method to AM-FM multi-components signals.

Exercise 12.11 (Removing Power Supply (Mains) Interference from ECG Signal). Use program 12.2. Take from the Internet an ECG heart activity signal, e.g. from the page https://archive.physionet.org/cgi-bin/atm/ATM. Add to it a 50 Hz or 60 Hz sine, simulating a power supply (mains) interference. Try to remove the disturbance using adaptive filter working as correlation canceler.

Exercise 12.12 (Removing Muscle Interference from ECG Signal). Use program 12.2. Take from the Internet an ECG heart activity signal, e.g. from the page https://archive.physionet.org/cgi-bin/atm/ATM. Add to it a low-pass filtered Gaussian noise (`randn()`) simulating a disturbing muscles signal. Find in the Internet information about the highest frequency of the muscles interference. Exploit adaptive filter as a correlation canceler and a linear predictor, and try to suppress the interference. What adaptive filter structure is better in this application?

Exercise 12.13 (Removing Disturbances from Animal Sounds). Repeat exercise (12.11) or (12.12) for any animal sound found in the Internet, for example, for birds song (canary.wav in Windows), blue whale sound (bluewhale.au in Matlab), dog or wolf howl, etc.

References

1. M. Bellanger, *Adaptive Digital Filters* (Marcel Dekker, New York, Basel, 2001)
2. C. Breining, et al., Acoustic echo control - An application of very high order adaptive filters. IEEE Signal Process. Mag. **16**(4), 42–69 (1999)
3. C.F.N. Cowan, P.M. Grant, *Adaptive Filters* (Prentice Hall, Englewood Cliffs, 1985)
4. P.S.R. Diniz, *Adaptive Filtering: Algorithms and Practical Implementation* (Springer, New York, 2002, 2010)
5. B. Farhang-Boroujeny, *Adaptive Filters: Theory and Applications* (Wiley, New York, 2000)
6. G.-O. Glentis, K. Berberidis, S. Theodoridis, Efficient least squares adaptive algorithms for FIR transversal filtering. IEEE Signal Process. Mag. **16**(4), 12–41 (1999)
7. E. Hansler, G. Schmidt, *Acoustic Echo and Noise Control - A Practical Approach* (Wiley, Chichester, 2004)
8. S. Haykin, *Adaptive Filter Theory* (Prentice Hall, Upper Saddle River, 1996, 2001)
9. S. Haykin, *Modern Filters* (Macmillan, New York, 1990)
10. G.H. Hostetter, Recursive estimation, in *Handbook of Digital Signal Processing*, ed. by F.D. Elliott (Academic Press, San Diego, 1987), pp. 899–940
11. N. Kalouptsidis, S. Theodoridis, *Adaptive System Identification and Signal Processing Algorithms* (Prentice Hall, New York, 1993)
12. S.M. Kay, *Fundamentals of Statistical Signal Processing: Estimation Theory* (PTR Prentice Hall, Englewood Cliffs, 1993)
13. S.M. Kuo, D.R. Morgan, *Active Noise Control Systems: Algorithms and DSP Implementations* (Wiley, New York, 1996)
14. K.-A. Lee, W.-S. Gan, S.M. Kuo, *Subband Adaptive Filtering: Theory and Implementation* (Wiley, Hoboken NJ, 2009)
15. D.G. Manolakis, V.K. Ingle, S.M. Kogon, *Statistical and Adaptive Signal Processing: Spectral Estimation, Signal Modeling, Adaptive Filtering and Array Processing* (McGraw-Hill, Boston, 2000; Artech House, Boston, London, 2005)
16. S.J. Orfanidis, *Optimum Signal Processing. An Introduction* (Macmillan, New York, 1988)
17. A.D. Poularikas, Z.M. Ramadan, *Adaptive Filtering Primer with Matlab* (CRC, Boca Raton, 2006)
18. A.H. Sayed, *Adaptive Filters* (Wiley-IEEE Press, New York, 2011)
19. A.H. Sayed, *Fundamentals of Adaptive Filtering* (Wiley, New York, 2003)

20. J. Shynk, Frequency-domain and multirate adaptive filtering. IEEE Signal Process. Mag. **10**(1), 14–37 (1992)
21. S. Theodoridis, Adaptive filtering algorithms, in *Proc. IEEE Instrumentation and Measurement Technology Conference*, Budapest, 2001, pp. 1497–1501
22. J.R. Treichler, C.R. Johnson, M.G. Larimore, *Theory and Design of Adaptive Filters* (Wiley, New York, 1987)
23. B. Widrow, S. Stearns, *Adaptive Signal Processing* (Prentice Hall, Englewood Cliffs, 1985)
24. T.P. Zieliński, *Cyfrowe Przetwarzanie Sygnałów. Od Teorii do Zastosowań (Digital Signal Processing. From Theory to Applications)* (Wydawnictwa Komunikacji i Łączności (Transport and Communication Publishers), Warszawa, Poland, 2005, 2007, 2009, 2014)

Chapter 13
Modern Frequency and Damping Estimation Methods

To be or not be as a biological object described by the second-order ordinary differential equation? Would you like to emotionally oscillate like a damped sine?

13.1 Introduction

The need for frequency and damping measurement appears in many fields: electrical (e.g., transients estimation in electronic circuits, analysis of magnetic resonance signals, power quality measurement), mechanical, electromechanical, optical, biological, human, economical and social, geophysical, astrophysical, and chemical. The literature on this subject is extensive, and a significant amount of time is required to get a complete and consistent view in this field based on numerous different sources. There are many specialized computational environments with implemented standard estimation methods. Matlab is a prime example of such a sophisticated program. However, using toolbox functions for frequency and damping estimation is not always straightforward.

This chapter aims to provide a "Reader's digest" of joint frequency and damping estimation methods. It is constructed as a practical overview and tutorial and should be useful in selecting a suitable method for the case investigated by the Reader. For this reason, Matlab implementations of all described methods are given. The chapter is based on the method survey presented in [25]. For further reading about spectral analysis of signals, for example, the following interesting books can be recommended [5, 7, 10–13, 19].

The chapter is structured as follows. In Sect. 13.2, the measurement model of a signal of interest is formulated. In Sect. 13.3, definitions of eigenvalue and singular value matrix decompositions are given. In Sects. 13.4 to 13.6, the application of the Hilbert transform, linear prediction (auto-regressive) parametric modeling, and the interpolated discrete Fourier transform (DFT) for frequency and damping estimation is presented, respectively.

© Springer Nature Switzerland AG 2021
T. P. Zieliński, *Starting Digital Signal Processing in Telecommunication Engineering*, Textbooks in Telecommunication Engineering,
https://doi.org/10.1007/978-3-030-49256-4_13

13.2 Signal Model

Since in this chapter several matrix equations with large matrices will be presented, in order to simplify the notation, all signals will be denoted in it using two alternative ways, i.e. as x_n, y_n and $x(n), y(n)$. The assumed signal model, used in this chapter, is as follows:

$$x_n^{(k)} = A_k e^{-D_k n} \cos\left(\Omega_k n + \varphi_k\right), \quad n = 0,\ 1,\ 2, \ldots,\ N-1, \tag{13.1}$$

where (k) denotes signal component number, $A > 0$ signal amplitude, $0 < \Omega < \pi$ signal angular frequency in radians ($\Omega_k = 2\pi \frac{f_k}{f_s}$ and $\Omega = 2\pi$ rad corresponds to the sampling frequency f_s in hertz), $\pi < \varphi \le \pi$ phase angle in radians, n index of the sample, N the number of samples, and $D \ge 0$ damping of the digital signal. Signal of the form (13.1) represents an impulse response of object described by second-order differential equation. Since many physical phenomena are described by it (for example, RLC circuits, magnetic resonance), this chapter concerns many different application fields. The signal Eq. (13.1) can be rewritten into the following form:

$$x_n^{(k)} = \frac{A_k}{2} e^{j\varphi_k} \lambda_k^{\,n} + \frac{A_k}{2} e^{-j\varphi_k} \lambda_k^{*n}, \tag{13.2}$$

where

$$\lambda_k = e^{-D_k + j\Omega_k}, \tag{13.3}$$

and $(.)^*$ denotes complex conjugation. In general, in real-world measurements summation of signals (13.1) appears

$$x_n = \sum_{k=1}^{K} x_n^{(k)} = \sum_{k=1}^{K} \left(\frac{A_k}{2} e^{j\varphi_k} \lambda_k^{\,n} + \frac{A_k}{2} e^{-j\varphi_k} \lambda_k^{*n} \right) = \sum_{k=1}^{K} \left(c_k \lambda_k^{\,n} + c_k^* \lambda_k^{*n} \right), \tag{13.4}$$

where c_k is defined as

$$c_k = \frac{A_k}{2} e^{j\varphi_k}. \tag{13.5}$$

In some applications, such as in nuclear magnetic resonance (NMR), signals are recorded in space by two perpendicular sensors and they are combined together into one complex-value signal:

$$x_n = \sum_{k=1}^{K} A_k e^{j\varphi_k} \lambda_k^{\,n}, \quad \lambda_k = e^{-D_k + j\Omega_k}. \tag{13.6}$$

We assume also that the analyzed signal can be embedded in the possible measurement disturbance ε_n (e.g., noise, drift, interference), i.e.

$$y_n = x_n + \varepsilon_n, \quad n = 0,\ 1,\ 2, \ldots,\ N-1. \tag{13.7}$$

In summary, in this chapter we will analyze signals modeled by Eq. (13.4), but we will begin our investigation with the one-component signal (13.1). We will estimate values of Ω_k, D_k, A_k, and φ_k based on the measured signal y_n. We will concentrate only on the damped signal, as undamped signal is a special case with $D_k = 0$. When Ω_k and D_k are known, finding values of A_k and φ_k is straightforward (which is shown below); therefore, in most methods we only find values of Ω_k and D_k. In turn, in parametric modeling methods Ω_k and D_k can be calculated from coefficients of the linear self-prediction model; therefore, further discussion is terminated once those coefficient are calculated. Efficiency of different frequency and damping estimators can be compared with known Cramèr–Rao lower bounds that were calculated for this measurement problem in [22].

At present, we show how to calculate values $\{A_k, \phi_k\}, k = 1 \ldots K$, having signal samples $x_n, n = 0, 1, \ldots, N - 1$, and knowing values of $D_k, \Omega_k, k = 1 \ldots K$. Writing Eq. (13.4) in matrix form for each signal sample, we obtain

$$
\begin{bmatrix}
(\lambda_1)^0 & (\lambda_1^*)^0 & \cdots & \cdots & (\lambda_K)^0 & (\lambda_K^*)^0 \\
(\lambda_1)^1 & (\lambda_1^*)^1 & \cdots & \cdots & (\lambda_K)^1 & (\lambda_K^*)^1 \\
(\lambda_1)^2 & (\lambda_1^*)^2 & \cdots & \cdots & (\lambda_K)^2 & (\lambda_K^*)^2 \\
\vdots & \vdots & \ddots & \ddots & \vdots & \vdots \\
(\lambda_1)^{N-2} & (\lambda_1^*)^{(N-2)} & \cdots & \cdots & (\lambda_K)^{N-2} & (\lambda_K^*)^{(N-2)} \\
(\lambda_1)^{N-1} & (\lambda_1^*)^{(N-1)} & \cdots & \cdots & (\lambda_K)^{N-1} & (\lambda_K^*)^{(N-1)}
\end{bmatrix}
\begin{bmatrix}
c_1 \\
c_1^* \\
c_2 \\
\vdots \\
c_K \\
c_K^*
\end{bmatrix}
=
\begin{bmatrix}
x_0 \\
x_1 \\
x_2 \\
\vdots \\
x_{N-2} \\
x_{N-1}
\end{bmatrix}.
\qquad (13.8)
$$

Now it is sufficient to solve the above equation in least-squares sense in regard to **c**:

$$
\mathbf{Lc} = \mathbf{x} \quad \rightarrow \quad \widehat{\mathbf{c}} = (\mathbf{L}^T \mathbf{L})^{-1} \mathbf{L}^T \mathbf{x}
\qquad (13.9)
$$

and calculate amplitudes and phases of signal components from the following equation:

$$
\widehat{A}_k = 2|\widehat{c}_k|, \quad \widehat{\varphi}_k = \angle \widehat{c}_k,
\qquad (13.10)
$$

resulting from Eq. (13.5).

Exercise 13.1 (Sum of Damped Sine Signal Model). I can imagine faces of most Readers. In order to show that it is not a *scientific thriller*, let us do the following exercise. A program generating summation of 3 damped sinusoids and finding its parameters from signal samples is presented in Listing 13.1. Amplitudes and phases of the components are reconstructed assuming that dampings and frequencies are known. But they are also computed using a method which will be presented later. Please, analyze the program code, run it, and change signal parameters. Among others, this is a problem of guitar tuning or magnetic resonance imaging. Interesting, is not it?

Listing 13.1: Signal model of sum of damped sinusoids

```
1   % lab13_ex_model.m
2   clear all; close all;
3
4   SNR = 120;                               % signal-to-noise ratio
5
6   % One damped sine component - testing equations
7   N = 100; n = 0:N-1; n=n';                % number of signal samples
8   A = 10; D = 0.1; Om = 0.5; phi = 0.2;    % values of signal parameters
9   x = A*exp(-D*n) .* cos(Om*n+phi);        % first signal equation (A,D,Om,phi)
10  figure; plot(x); title('Single-damped'); pause % signal shape
11
12  c = A/2 * exp(j*phi);                    % c value
13  lamb = exp(-D+j*Om);                     % lambda value
14  xm = c * lamb.^n + conj(c)*conj(lamb).^n; % second signal equation (c,lambda)
15  error = max( abs( x-xm ) ),              % error between signals x and xm
16  figure; plot(x-xm); pause                % error
17
18  x = awgn( x, SNR);                       % optional noise addition
19  L = [ lamb.^n conj(lamb).^n ];           % matrix generation
20  c = pinv( L )*x;                         % solving matrix equation
21  A = 2*abs(c),                            % found A
22  phi = angle(c), pause                    % found phi
23
24  % Many components - at present three
25  A =    [ 3,     2,    1   ];             % their: amplitudes
26  D =    [ 0.1,   0.2,  0.3 ];             % dampings
27  Om =   [ 0.25,  0.5,  0.75 ];            %  frequencies
28  phi = [ 0.1,   0.2,  0.3 ];             % phase shifts
29  K = length(A);                          % number of components
30  x = zeros(N,1);
31  L = zeros(N,2*K);
32  for k = 1:K                              % signal generation
33      x = x + A(k)*exp(-D(k)*n) .* cos(Om(k)*n+phi(k));
34  end
35  figure; plot(x); title('Multi-damped'); pause % signal shape
36  x = awgn( x, SNR);                       % optional noise addition
37  [Om, D] = fMatPen(x,K), pause            % optional Om and D calculation
38  for k = 1:K                              % matrix L generation
39      L(:,2*k-1) = exp(-D(k)+j*Om(k)).^n;
40      L(:,2*k)   = conj( exp(-D(k)+j*Om(k)) ).^n;
41  end
42  disp('Results:');
43  c = pinv( L )*x;                         % solving matrix equation
44  A = 2*abs(c), pi = angle(c), D, Om, pause % results
```

Exercise 13.2 (Real-World   Damped   Oscillations).   In   programs
`lab13_ex_real_spectroscopy.m` and `lab13_ex_real_sdr.m`,
two damped oscillatory signals are simulated, originated from different
real-world measurement scenarios. The first signal comes from low-frequency
mechanical spectroscopy applied in material science for testing just manu-
factured new materials, while the second signal from software-defined radio,
where new carrier frequency oscillates after switching. Analyze code of both
programs and run them. In Fig. 13.1, both signals are shown.

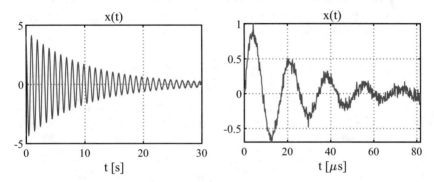

Fig. 13.1: Damped oscillatory signals originated from low-frequency mechanical
spectroscopy of materials (left) and software-defined radio—carrier frequency offset
after carrier switching (right)

13.3 Eigenvalue and Singular Value Matrix Decomposition

In this chapter we will use two fundamental matrix decompositions, the eigenvalue
and singular value ones. Therefore we start with the presentation of their definitions.

Eigenvalue Matrix Decomposition Let us assume that $\mathbf{A}$ is a square $M \times M$ matrix
[12]. An M-element vertical vector $\mathbf{u}$ is its eigenvector and γ associated eigenvalue
if they both fulfill the following equation:

$$\mathbf{Au} = \gamma \mathbf{u}. \tag{13.11}$$

The matrix multiplication is transforming the vector $\mathbf{u}$ into the same vector, only
scaled by γ. The matrix $\mathbf{A}_{M \times M}$ has maximally M different eigenvectors $\mathbf{u}_k$ and connected with them eigenvalues γ_k. The vectors defines only directions and have ar-
bitrary lengths, and therefore they are normalized to 1. Why? Because if vector $\mathbf{u}_k$

fulfills (13.11), it is done also by all its multiplies. Eigenvalues are called a matrix **A** spectrum. If Eq. (13.11) is written for all eigenvectors $\mathbf{u}_k, k = 1, 2, \ldots, M$, and if all equations are combined together, one obtains the following matrix formula (we are assuming: $\gamma_1 \geq \gamma_2 \geq \gamma_3 \geq \ldots \geq \gamma_M$):

$$\mathbf{AU} = \mathbf{UD}, \quad \mathbf{U} = \begin{bmatrix} \mathbf{u}_1 \ \mathbf{u}_2 \ \cdots \ \mathbf{u}_M \end{bmatrix}, \quad \mathbf{D} = \mathrm{diag}\,(\gamma_1, \gamma_2, \ldots, \gamma_M), \tag{13.12}$$

where a square matrix **U** has vectors $\mathbf{u}_k$ in its columns. Using Eq. (13.12), we can now represent matrix **A** as:

$$\mathbf{A} = \mathbf{UDU}^{-1}. \tag{13.13}$$

If matrix **A** is a Hermitian matrix (with complex-conjugated symmetry: $a_{i,j} = a_{j,i}^*$), its eigenvalues are real and eigenvectors are orthonormal (orthogonal and normalized to 1). In this case the matrix inverse $\mathbf{A}^{-1}$ is equal to $\mathbf{A}^H = (\mathbf{A}^T)^*$, i.e. to its transposition and conjugation. Therefore, from (13.13) we have

$$\mathbf{A} = \mathbf{UDU}^H = \sum_{k=1}^{M} \gamma_k \mathbf{u}_k \mathbf{u}_k^H. \tag{13.14}$$

In Eq. (13.14) the matrix **A** is a summation on first rank matrices $\mathbf{u}_k \mathbf{u}_k^H$, taken with weights γ_k. If we limit the summation only to the first K, the biggest (principal, dominant) eigenvalues:

$$\widetilde{\mathbf{A}} = \sum_{k=1}^{K<M} \gamma_k \mathbf{u}_k \mathbf{u}_k^H, \tag{13.15}$$

we obtain the best least-squares approximation of **A** using a lower-rank matrix (the Frobenius norm of their difference is the smallest).

Having EVD decomposition of the matrix **A**, it is very easy to calculate (find) its inverse and approximation of this inverse:

$$\mathbf{A}^{-1} = \mathbf{UD}^{-1}\mathbf{U}^H = \sum_{k=1}^{M} \frac{1}{\gamma_k} \mathbf{u}_k \mathbf{u}_k^H, \quad \widetilde{\mathbf{A}^{-1}} = \sum_{k=1}^{K<M} \frac{1}{\gamma_k} \mathbf{u}_k \mathbf{u}_k^H. \tag{13.16}$$

In Matlab we perform EVD matrix decomposition calling function [U,D]=eig(A).

In signal processing applications, the EVD decomposition is typically applied to symmetrical auto-correlation signal matrices, for example, used in LPC speech codecs (14.9), RLS adaptive filters (12.45) and spectrum estimation methods using signal auto-correlation function (for example, Blackman-Tukey power spectral density estimate (6.10)). In order to reduce noise influence on these matrices, they are approximated using only dominant (principal) components in Eq. (13.15), associated with the biggest eigenvalues γ_k (with the smallest for matrix inversion). If signal is a summation of K pure complex-value harmonics, only first K sub-matrices should be used, while for $2K$ real-value sines, $2K$ first sub-matrices.

Matrix Pencil of Two Matrices If we have two square matrices $\mathbf{A}$ and $\mathbf{B}$ with dimensions $M \times M$, their generalized eigenvalue γ and generalized eigenvector $\mathbf{u}$ have to fulfill the following equation [12]:

$$\mathbf{Au} = \gamma\mathbf{Bu} \quad \rightarrow \quad (\mathbf{A} - \gamma\mathbf{B})\mathbf{u} = \mathbf{0} \quad \rightarrow \quad \mathbf{A} - \gamma\mathbf{B} = \mathbf{0} \quad \rightarrow \quad \mathbf{B}^{pinv}\mathbf{A} - \gamma\mathbf{I} = \mathbf{0},$$
(13.17)

where matrix $\mathbf{B}^{inv}$ is a pseudo-inverse Moore–Penrose matrix to $\mathbf{B}$

$$\mathbf{B}^{pinv} = \left(\mathbf{B}^H\mathbf{B}\right)^{-1}\mathbf{B}^H.$$
(13.18)

Set of matrices $\mathbf{A} - \gamma\mathbf{B}$ is called the *Matrix Pencil*. In Matlab generalized eigenvectors $\mathbf{u}_k$ and eigenvalues γ_k for two matrices $\mathbf{A}$ and $\mathbf{B}$ are calculated as [U,D]=eig(A,B). We will use this method in damped sinusoid estimation problem.

Singular Value Matrix Decomposition [12] Let us assume that $\mathbf{A}$ is a rectangular matrix with dimensions $M \times N$ and real-value or complex-value elements. Its singular value decomposition (SVD) is defined as:

$$\mathbf{A} = \sum_{k=1}^{K=\min\{M,N\}} \gamma_k \mathbf{u}_k \mathbf{v}_k^H = \mathbf{USV}^H,$$
(13.19)

$$\widetilde{\mathbf{A}} = \sum_{k=1}^{L<\min\{M,N\}} \gamma_k \mathbf{u}_k \mathbf{v}_k^H.$$
(13.20)

If we limit summation in (13.19) only to the first $L < \min(M,N)$ components, the best least-squares approximation (13.20) of $\mathbf{A}$ is obtained by a lower-rank matrix. In Matlab we call the function: [U,S,V]=svd(A,0). As a result, we obtain

1. Unitary matrix $\mathbf{U}$ ($\mathbf{U}^H\mathbf{U} = \mathbf{I}$) with dimensions $M \times M$, having in columns left orthonormal (orthogonal and normalized) singular vectors $\mathbf{u}_k, k = 1,2,\ldots,M$;
2. Unitary matrix $\mathbf{V}$ ($\mathbf{V}^H\mathbf{V} = \mathbf{I}$) with dimensions $N \times N$, having in columns right orthonormal singular vectors $\mathbf{v}_k, k = 1,2,\ldots,N$;
3. Matrix $\mathbf{S}$ with dimensions $M \times N$ that can have non-zero and non-negative elements only on its main diagonal ($\sigma_1 \geq \sigma_2 \geq \sigma_3 \geq \ldots \geq \sigma_K \geq 0, K = \min(M,N)$).

The pseudo-inverse $\mathbf{A}^{pinv}$ of matrix $\mathbf{A}$ and its best LS approximation $\widetilde{\mathbf{A}^{pinv}}$ can be calculated using the SVD decomposition:

$$\mathbf{A}^{pinv} = \left(\mathbf{A}^H\mathbf{A}\right)^{-1}\mathbf{A}^H = \sum_{k=1}^{K=\min\{M,N\}} \frac{1}{\sigma_k}\mathbf{v}_k\mathbf{u}_k^H,$$
(13.21)

$$\widetilde{\mathbf{A}}^{pinv} = \sum_{k=1}^{L<\min\{M,N\}} \frac{1}{\sigma_k}\mathbf{v}_k\mathbf{u}_k^H.$$
(13.22)

We will use these nice features in this chapter.

Exercise 13.3 (SVD of an Image as a Breather). An interesting exercise is to perform the SVD decomposition of any image and do its reconstruction from the singular values and vectors, which is presented in the program 13.3. It gives us a fruitful insight into the merit of EVD- and SVD-based approximation of matrices using Eqs. (13.15), (13.20). Decompose your own image. Observe image improvement after addition of successive order-one matrices $\sigma_k \mathbf{v}_k \mathbf{u}_k^T$.

Listing 13.2: SVD of an image matrix

```
1   % lab13_ex_svd_image.m - SVD-based image decomposition
2   clear all; close all;
3
4   [X,map] = imread('lena512.bmp');             % read image
5   X = double(X);                               % change number represetation
6   image(X); title('Our image');                % display image
7   colormap(map); axis image off; pause         %
8
9   [U,S,V] = svd(X);                            % perform image SVD
10  image( U*S*V' ); title('SVD');               % show reconstructed result
11  colormap(map); axis image off; pause         % color map
12  mv = [1, 2, 3, 4, 5, 10, 15, 20, 25, 50];    % number of added rank-one images
13  for i = 1 : length(mv)                       % add approximation matrices in a loop
14      mv(i)                                    % how many rank-one images to add
15      mask = zeros(size(S));                   % masking matrix
16      mask( 1:mv(i), 1:mv(i) ) = 1;            % with 1s only on main diag beginning
17      image( U * (S.*mask) * V' );             % masking, reconstruction, image display
18      axis image off;
19      pause
20  end
```

13.4 Discrete Hilbert Transform Method

Applying the Hilbert transform (HT) to estimating frequency and damping has been reported in many publications; however, the HT was mainly used for measuring damping, for example, in [1]. The HT shifts the signal in phase by $-\frac{\pi}{2}$ rad, e.g. $\cos(\Omega n)$ becomes $\sin(\Omega n)$. An analytic signal (AS) is defined as a complex signal with the input samples in its real part and their HT in the imaginary part, e.g. for input $\cos(\Omega n)$, we have $e^{j\Omega n} = \cos(\Omega n) + j\sin(\Omega n)$. For discrete time signal of the form (13.1), its AS version is (we are neglecting the index (k) since the Hilbert method can be used only for one-component signals)

$$x_n^{(a)} = Ae^{-Dn}e^{j\Omega n + \varphi} \tag{13.23}$$

and damping and frequency are given by (approximation of derivative):

$$D_n = -\operatorname{diff}\left(\ln\left(\left|x_n^{(a)}\right|\right)\right), \quad D = \operatorname{mean}(D_n), \tag{13.24}$$

$$\Omega_n = \operatorname{diff}(\angle x_n^{(a)}), \quad \Omega = \operatorname{mean}(\Omega_n), \tag{13.25}$$

where "diff" denotes the difference between two consecutive values as implemented in the Matlab function `diff()`, and $|x|$ and $\angle(x)$ denote magnitude and angle of a complex number, respectively. In computer implementation, analytic signal can be calculated by convolution or modification of DFT coefficients. When Ω and D are known, we find values of A and φ using Eqs. (13.8), (13.9), (13.10).

Matlab implementation of the AS (or Hilbert) estimation method of frequency and damping is given in the function `fHilbert()` presented in Listing 13.3. The calculated instantaneous frequency and damping have strong oscillations at the beginning and the end of the observation interval. Therefore 1/4 of samples is discarded at the beginning and the end of the computed data. In the last program line, variable temp is used for damping estimation via line fitting, which is an option for computing the mean value. The method is dedicated for a single damped sine. In case of summation of several damped sine components, they should be first separated by band-pass filters and then analyzed individually using the below function.

Listing 13.3: Simple damped sinusoid parameters estimation using Hilbert filter (transform)

```
1  function [Om, D]=fHilbert(x,W);
2  % x = A*cos(Om*n+p).*exp(-n*D)
3  % W window 1-Hanning, 2-Blackman
4
5  N = length(x);
6  if(W==1) win = hanning(N)'; end        % window choice
7  if(W==2) win = blackman(N)'; end       %
8  Xh = hilbert(win.*x)./win;             % Hilbert filter/transform
9  Om = diff( unwrap( angle(Xh) ) );      % angle differentiation
10 D = -diff( log(abs(Xh)) );             % logarithm of abs(), differentiation
11 ind = round( N/2-N/4 : N/2+N/4 );      % removing begin/end transients
12 Om = mean(Om(ind));                    %  frequency
13 D = mean(D(ind));                      % damping
14 % temp = polyfit(ind,log(abs(Xh(ind))),1); D=-temp(1);
```

Exercise 13.4 (Application of Hilbert Method in Mechanical Spectroscopy). Apply the Hilbert method for estimation of parameters of the damped sine obtained in mechanical spectroscopy—use program `lab13_ex_real_spetroscopy.m`. Repeat experiments for different signal parameters.

13.5 Parametric Modeling Methods: Solving Linear Equations

13.5.1 Initial Problem Setup

Let us assume in the beginning that we have only one damped sine component, neglect the index (k), and write Eq. (13.1) for 3 consecutive signal samples:

$$\begin{cases} x(n) = A\cos(\Omega n + \varphi)e^{-Dn} \\ x(n-1) = A\cos(\Omega(n-1) + \varphi)e^{-D(n-1)} \\ x(n-2) = A\cos(\Omega(n-2) + \varphi)e^{-D(n-2)} \end{cases} \tag{13.26}$$

After multiplication of the second equation by $2\cos(\Omega)e^{-D}$, the third equation by e^{-2D}, and after subtraction of the modified third equation from the modified second equation, one obtains (after simple trigonometric transformations)

$$x(n) = -a_1 x(n-1) - a_2 x(n-2), \tag{13.27}$$

where

$$a_1 = -2\cos(\Omega)e^{-D}, \quad a_2 = e^{-2D}. \tag{13.28}$$

It results from Eq. (13.27) that actual signal sample x_n can be calculated as linear prediction of two previous samples x_{n-1} and x_{n-2}. In order to obtain the correct value of amplitude, the following initial condition should be used:

$$x(-1) = 0, \quad x(-2) = -Ae^D \sin(\Omega). \tag{13.29}$$

In general, the signal x_n (13.1) is an impulse response h_n of the following linear digital filter ($u(n)$—input, $v(n)$—output):

$$v(n) = b_1 u(n-1) - a_1 v(n-1) - a_2 v(n-2), \tag{13.30}$$

$$b_1 = Ae^{-D}\sin(\Omega), \quad a_1 = -2e^{-D}\cos(\Omega), \quad a_2 = e^{-2D}, \tag{13.31}$$

having a transfer function $H(z)$ with complex poles (zeros of denominator) exactly equal to λ and λ^* of the signal (13.2):

$$H(z) = \frac{b_1 z^{-1}}{1 + a_1 z^{-1} + a_2 z^{-2}} = \frac{b_1 z^{-1}}{(1 - \lambda z^{-1})(1 - \lambda^* z^{-1})}, \quad \lambda = e^{-D+j\Omega}. \tag{13.32}$$

In order to estimate frequency Ω and damping D, one should find such values of coefficients a_1 and a_2 for given N noisy measurements $\{y_0, y_1, \ldots, y_{N-1}\}$, where $y_n = x_n + \varepsilon_n$, that the linear prediction model $y_n = -a_1 y_{n-1} - a_2 y_{n-2}$ fits best to the measurement data y_n.

Many solutions for this optimization task are briefly summarized below. In general, the typical calculation path for algorithms covered in this section is as follows:

1. Calculate coefficients a_1, a_2 of the signal self-prediction model,
2. Find complex roots $\lambda = e^{-D}e^{j\Omega}$ and λ^* of the polynomial $1 + a_1 z^{-1} + a_2 z^{-2}$, describing the denominator of the transfer function of the second-order digital system (13.32); in Matlab function `lambda=roots([1, a1, a2]);`
3. Calculate Ω and D of the discrete-time signal:

$$\Omega = \angle(\lambda), \quad D = -\ln(|\lambda|), \quad (13.33)$$
$$\Omega = |\mathrm{Im}(\ln(\lambda))|, \quad D = -\mathrm{Re}(\ln(\lambda)), \quad (13.34)$$

4. Estimate signal amplitude and phase (using Eqs. (13.8), (13.9), (13.10)).

Having the transfer function $H(z)$, one can calculate the frequency response setting $z = e^{j\Omega}, 0 \le \Omega \le 2\pi$, and find its greatest magnitude corresponding to angular frequency Ω of the analyzed signal. However, estimation of damping is not possible using just the magnitude of frequency response.

13.5.2 Generalization for Multi-component Signals

When the analyzed signal consists of K damped sine components, as in Eq. (13.3), the same methodology as described above can be used (without a proof!). The next signal sample is predicted by previous $2K$ samples:

$$x(n) = -a_1 x(n-1) - a_2 x(n-2) - a_3 x(n-3) - \ldots - a_{2K} x(n-2K), \quad (13.35)$$

and the signal is generated by the following filter ($u(n)$—input, $v(n)$—output):

$$v(n) = b_0 u(n) + \ldots + b_{2K-1} u(n-(2K-1)) - a_1 v(n-1) - \ldots - a_{2K} v(n-2K), \quad (13.36)$$

having transfer function:

$$H(z) = \frac{b_0 + b_1 z^{-1} + \ldots + b_{2K-1} z^{-(2K-1)}}{1 + a_1 z^{-1} + a_2 z^{-2} + a_3 z^{-3} \ldots + a_{2K} z^{-2K}}. \quad (13.37)$$

When we find coefficients $\{a_1, a_2, a_3, \ldots, a_{2K}\}$ of the signal linear self-prediction model (13.35), next we can calculate from them values of parameters $\lambda_k = D_k + j\Omega_k$, describing components of our signal, as roots of the following polynomial:

$$1 + a_1 z^{-1} + a_2 z^{-2} + a_3 z^{-3} + \ldots + a_{2K} z^{-2K} =$$
$$(1 - \lambda_1 z^{-1})(1 - \lambda_1^* z^{-1}) \ldots (1 - \lambda_K z^{-1})(1 - \lambda_K^* z^{-1}) = 0. \quad (13.38)$$

Therefore, our *only* problem is to find a vector **a** of linear prediction coefficients (LPC) of our signal, which is discussed in the next sections.

13.5.3 Prony Method

There are several methods used in digital filter design, for example, the Pade approximation procedure and the Prony least-squares auto-regressive model fitting, for finding coefficients $\{b_k, a_k\}$ of the digital filter difference equation (and thus the transfer function $H(z)$) for the desired/given filter impulse response h_n. In our case $h_n = x_n$ is a summation of damped sinusoids, and the calculated coefficients $\{a_k\}$ are used to estimate their frequencies and dampings, as stated above in (13.33), (13.34). Therefore, using the input–output relationship (13.30), (13.31) leads to the well-known Prony digital filter design method that is derived below.

It follows that for single damped sine component (13.1), the linear equation to be solved for a_1 and a_2 is

$$
\begin{bmatrix}
y_0 & y_1 \\
y_1 & y_2 \\
\vdots & \vdots \\
y_{N-3} & y_{N-2}
\end{bmatrix}
\begin{bmatrix}
a_2 \\
a_1
\end{bmatrix}
= -
\begin{bmatrix}
y_2 \\
y_3 \\
\vdots \\
y_{N-1}
\end{bmatrix}, \quad \mathbf{Y} \cdot \mathbf{a} = -\mathbf{y}, \quad (13.39)
$$

where **Y** is an $(N - 2) \times 2$ matrix with the Hankel structure (the lower row is the upper row shifted one position to the right). In practice estimation is based on a large number of samples (as many as thousands) with the hope that measurement noise will be averaged and cancelled this way, and thus the obtained result will be more correct.

For $N = 4$, we have four measurements y_0, y_1, y_2, y_3, and Eq. (13.39) consists of two equations with two unknowns:

$$
\begin{bmatrix}
y_0 & y_1 \\
y_1 & y_2
\end{bmatrix}
\begin{bmatrix}
a_2 \\
a_1
\end{bmatrix}
= -
\begin{bmatrix}
y_2 \\
y_3
\end{bmatrix}. \quad (13.40)
$$

In turn, for signal with K real-value damped components (13.3) we have to solve the following equation in least-squares sense in regard to vector **a**:

$$
\begin{bmatrix}
y_0 & \cdots & y_{2K-2} & y_{2K-1} \\
y_1 & \cdots & y_{2K-1} & y_{2K} \\
\vdots & \ddots & \vdots & \vdots \\
y_{N-2K} & \cdots & y_{N-2} & y_{N-1}
\end{bmatrix}
\begin{bmatrix}
a_1 \\
a_2 \\
\vdots \\
a_{2K}
\end{bmatrix}
= -
\begin{bmatrix}
y_{2K} \\
y_{2K+1} \\
\vdots \\
y_N
\end{bmatrix} \quad \mathbf{Y} \cdot \mathbf{a} = -\mathbf{y}. \quad (13.41)
$$

Let us left multiply both sides of the matrix equation $\mathbf{Ya} = -\mathbf{y}$, present in Eqs. (13.39), (13.41), by the matrix $\mathbf{Y}^T$, transposition of $\mathbf{Y}$:

$$(\mathbf{Y}^T\mathbf{Y})\mathbf{a} = -\mathbf{Y}^T\mathbf{y}. \tag{13.42}$$

Since the matrix $(\mathbf{Y}^T\mathbf{Y})$ is square, now we can calculate its inverse and write solution of (13.42) as (in Matlab)

$$\mathbf{a} = -\left[(\mathbf{Y}^T\mathbf{Y})^{-1}\right]\mathbf{Y}^T\mathbf{y}, \qquad \text{a = -pinv(Y)*y}, \tag{13.43}$$

in which:

$$(\mathbf{Y}^T\mathbf{Y})^{-1}\mathbf{Y}^T = \mathbf{Y}^I. \tag{13.44}$$

is the pseudo-inverse of $\mathbf{Y}$. In Matlab the matrix $\mathbf{Y}^I$ can be computed directly from its definition or using the built-in function pinv(Y). Knowing the vector $\mathbf{a}$, we can calculate roots λ_k of the polynomial associated with it from Eq. (13.38) and, finally, compute Ω_k and D_k using (13.33), (13.34) for each signal damped component.

In Matlab the optimal ordinary least-squares solution of (13.39), (13.41), which takes into account features of the matrix $\mathbf{Y}$, is given by a=-Y\y.

The above-described method is known as the Prony algorithm. Code of its Matlab function is presented in Listing 13.4.

Listing 13.4: Damped sinusoid parameters estimation using Prony linear prediction method (transform)

```
1   function [Om, D]=fProny(x,K)
2   % x - analyzed signal - sum of damped sinusoids
3   % K - assumed number of real damped sines (calling choose
4   N = length(x);                                % number of samples
5   P = 2*K;                                      % prediction order
6   M = N-P;                                      % max number of equations
7   FirstCol = x(1:M); LastRow = x(M:M+P-1);      % for Hankel type matrix
8   Y = hankel( FirstCol, LastRow );             % matrix MxL
9   y = x(P+1:P+M).';                            % vector Mx1
10  %a = -pinv(Y)*y;                             % classical solution
11  a = -Y\y;                                    % Matlab optimized solution
12  p = log( roots( [1 fliplr(a')] ) );          % polynomial roots
13  D = -real(p); Om = imag(p);                  % damping frequency
```

Equations (13.39), (13.41) can be solved using the following Matlab functions: arcov(), lscov(), prony(), stmb(). In the last two, transfer function coefficients of the digital linear system having the given (predefined) impulse response are to be found. The measured signal $\mathbf{y}$, input to both functions, is treated as an impulse response $\mathbf{h}$ of the system, while coefficients $\mathbf{a}$ of the transfer function denominator, returned from the functions, are the solution of (13.39), (13.41). Different methods belonging to this group are closely related to algorithms of matrix algebra and numerical analysis. The Burg estimation method (arburg()), using simultaneous forward and backward linear prediction, is not appropriate for damped sinusoids estimation and has not been tested.

Exercise 13.5 (Prony Method). Call the Prony function in the program 13.2. Compare obtained results with the `fMatPen.m` function output.

13.5.4 Steiglitz–McBride Method: Self-Filtered Signal

Our simulations have shown that the Steiglitz–McBride (STMB) method [18] (Matlab `stmb()` function), being equivalent to the iterative quadratic maximum likelihood approach, gives very good results for estimating single damped sinusoid embedded in white Gaussian noise. For the STMB method, Eq. (13.36) valid for summation of several damped sinusoids can be written in the following matrix form:

$$
\begin{bmatrix}
0 & 0 & \cdots & 0 & u_0 & 0 & \cdots & 0 \\
v_0 & 0 & \cdots & 0 & u_1 & u_0 & \cdots & 0 \\
v_1 & v_0 & \cdots & \vdots & u_2 & u_1 & \ddots & \vdots \\
v_2 & v_1 & \ddots & 0 & \vdots & \vdots & \ddots & u_0 \\
\vdots & \vdots & \ddots & v_0 & \vdots & \vdots & \ddots & \vdots \\
\vdots & \vdots & \ddots & \vdots & u_{N-1} & u_{N-2} & \ddots & u_{N-2K} \\
v_{N-1} & v_{N-2} & \cdots & v_{N-2K} & u_N & u_{N-1} & \cdots & u_{N-2K+1}
\end{bmatrix}
\begin{bmatrix}
a_1 \\
\vdots \\
a_{2K} \\
b_0 \\
\vdots \\
b_{2K-1}
\end{bmatrix}
=
\begin{bmatrix}
v_0 \\
v_1 \\
v_2 \\
\vdots \\
v_N
\end{bmatrix}
\qquad (13.45)
$$

where u_n and v_n denote filter inputs and outputs. If we write Eq. (13.45) as $\mathbf{A} \cdot \mathbf{c} = \mathbf{d}$, we should solve it with respect to vector $\mathbf{c}$ using concept of the pseudo-inverse matrix: $\mathbf{c} = (\mathbf{A}^T \mathbf{A})^{-1} \mathbf{A}^T \mathbf{d}$. Vector $\mathbf{c}$ is built in part from vector $\mathbf{a}$ of linear prediction coefficients we are looking for. Having $\{a_1, a_2, \ldots, a_{2K}\}$, we calculate $\{\lambda_1, \lambda_2, \ldots, \lambda_{2K}\}$ from Eq. (13.38) as roots of linear prediction polynomial, and then $\Omega_k, D_k, k = 1 \ldots K$ from (13.33), (13.34) for each k individually.

The idea of Steiglitz–McBride algorithm is as follows: if Eq. (13.36) and its matrix version (13.45) are valid for unitary digital impulse as an input and analyzed signal as an output, they should be also valid when both signals are passed through the same filter. Therefore, both equations are solved several times, each time for filter better adjusted to the signal components and better de-noising them. New filter coefficients are calculated using the last, improved estimation of $\mathbf{a}$. The algorithm consists of the following steps:

1. Find initial estimate of coefficients $\mathbf{a}^0 = [a_1, a_2, \ldots, a_{2K}]^T$ using the least-squares Prony solution (13.43) of (13.41),
2. Pass the signal x_n through the recursive IIR filter $H_a(z)$:

$$H_a(z) = \frac{1}{1 + a_1 z^{-1} + a_2 z^{-2} + \ldots + a_{2K} z^{-2K}}, \qquad (13.46)$$

using weights **a** recently calculated, i.e. de-noise the signal x_n; treat result as signal v_n;

3. Pass the unitary impulse $\delta(n)$ (Kronecker delta function) by the same filter and treat result as signal u_n;
4. Having v_n and u_n, build Eq. (13.45) and solve it with respect to vector **c**;
5. Extract **a** from **c** and return to step 2 or stop.

During successive iterations, estimation of linear prediction parameters is improved because in each next step the analyzed signal is better de-noised. Better is our knowledge about the signal, better filter is designed for noise reduction. Initially calculated coefficients $\mathbf{a}^{(0)}$ are used in the first filter, the signal is filtered for the first time, and the Eq. (13.45) is solved. Then new LP coefficients $\mathbf{a}^{(1)}$ are determined, and they are used in the second filter ..., and so on.

The Matlab implementation of the Steiglitz–McBride method is given in Listing 13.5.

Listing 13.5: Damped sinusoid parameters estimation using Steiglitz–McBride linear prediction method (transform)

```
 1  function [Om, D]=fSTMB(x,K,NI)
 2  % x - analyzed signal
 3  % K - number of damped components A*cos(Om*n+ph).*exp(-D*n)
 4  % NI - number of iterations, for NI=0 the Prony method is used
 5  % Om - estimated  frequency [rad]
 6  % D - estimated damping
 7
 8  N = length(x);                            % number of signal samples
 9  P = 2*K;                                  % prediction order
10  FirstCol = x(P:N-1); FirstRow = x(P:-1:1); % first row and column of Toeplitz matrix
11  X = toeplitz( FirstCol, FirstRow ); y = x(P+1:N).'; % signal matrix
12  a = X\y; % or inv(X'*X)*X'*y; or a = pinv(X)*y;    % first Prony solution
13  a1 = [1; -a];                             % first filter from Prony method
14  %a1 = [1; zeros(P,1) ];                    % or simpler filter initialization
15  delta = zeros(1,N); delta(1)=1;           % unit impulse
16  while ( NI )                              % iterative solution improvement
17      v = filter(1, a1, x);                 % input filtering
18      u = filter(1, a1, delta);             % impulse filtering
19      V = toeplitz( v, [ v(1) zeros(1,P) ] ); % input matrix
20      U = toeplitz( u, [ u(1) zeros(1,P-1) ] ); % impulse matrix
21      VU = [ V(1:N,:) U(1:N,:) ];           % combining two matrices
22      A = VU(:,2:end); d = VU(:,1);         %
23      c = A\d; % c = inv(A'*A)*A'*d; % c = pinv(A)*d; % solution, three alternatives
24      a1 = [ 1; -c(1:P) ];                  % iterative update
25      NI = NI-1;                            % decrease counter
26  end                                       %
27  p = roots( a1 );  p = log(p);             % roots, ln()
28  Om = imag(p); [Om indx] = sort( Om, 'ascend' ); Om = Om(K+1:2*K); % angular frequency
29  D = -real( p(indx(K+1:2*K)) );            % damping
```

Exercise 13.6 (Steiglitz–McBride Method). Call the `fSTMB.m` function in the program 13.2. Compare obtained results with the `fMatPen.m` function output.

13.5.5 Kumaresan–Tufts Method: Linear Prediction with SVD

In this approach, proposed by Kumaresan and Tufts (KT) [8], a principal component approximation of the pseudo-inverse matrix $\mathbf{Y}^I$ (13.44), more robust to noise, is used. The selected linear prediction order is greater than $2K$ and equals P, such that $\min(P, N-P) > 2K$, where K is assumed to be the number of sinusoids present in the analyzed signal. Additionally, the backward not forward signal self-prediction is exploited, i.e. $y_n = -b_1 y_{n+1} - b_2 y_{n+2} - \ldots - b_P y_{n+P}$ instead of $y_n = -a_1 y_{n-1} - a_2 y_{n-2} - \ldots - a_P y_{n-P}$. Therefore (13.41) is now replaced by

$$
\begin{bmatrix}
y_1 & \cdots & y_{P-1} & y_P \\
y_2 & \cdots & y_P & y_{P+1} \\
\vdots & \ddots & \vdots & \vdots \\
y_{N-P} & \cdots & y_{N-2} & y_{N-1}
\end{bmatrix}
\begin{bmatrix}
b_1 \\ b_2 \\ \vdots \\ b_P
\end{bmatrix}
= -
\begin{bmatrix}
y_0 \\ y_1 \\ \vdots \\ y_{N-P-1}
\end{bmatrix}
, \mathbf{Y}_{N-P,P}\mathbf{b}_P = -\mathbf{y}_{N-P} \qquad (13.47)
$$

and as before in (13.43):

$$
\mathbf{b} = -\left[\left(\mathbf{Y}^T \mathbf{Y} \right)^{-1} \mathbf{Y}^T \right] \mathbf{y} = -\mathbf{Y}^I \mathbf{y}. \qquad (13.48)
$$

In this approach the singular value decomposition (SVD) of the Hankel-type matrix $\mathbf{Y}$ is performed:

$$
\mathbf{Y} = \sum_{k=1}^{\min\{P, N-P\}} \sigma_k \mathbf{u}_k \mathbf{v}_k^H, \qquad (13.49)
$$

where σ_k represents singular values of $\mathbf{Y}$ ($\sigma_1 \geq \sigma_2 \geq \sigma_3 \geq \ldots$), and $\mathbf{u}_k$ and $\mathbf{v}_k$ are left and right singular vectors of $\mathbf{Y}$. The SVD of matrix $\mathbf{Y}$ produces a diagonal matrix $\mathbf{S}$ of the same dimension as $\mathbf{Y}$, with non-negative, non-increasing elements lying on the main diagonal, and unitary matrices $\mathbf{U}$ and $\mathbf{V}$ so that $\mathbf{Y} = \mathbf{U} \cdot \mathbf{S} \cdot \mathbf{V}^T$, as described in Sect. 13.3. The SVD results are then used for computation of the pseudo-inverse $\mathbf{Y}^I$ of $\mathbf{Y}$:

$$
\mathbf{Y}^I = \sum_{k=1}^{\min\{P, N-P\}} \frac{1}{\sigma_k} \mathbf{v}_k \mathbf{u}_k^H. \qquad (13.50)
$$

The exact solution of (13.47) is given by

$$\mathbf{b} = -\mathbf{Y}^I \mathbf{y} = - \sum_{k=1}^{\min\{P,N-P\}} \frac{1}{\sigma_k} \mathbf{v}_k \left[\mathbf{u}_k^H \mathbf{y} \right]. \tag{13.51}$$

In the Kumaresan–Tufts method, $P = \frac{3N}{4}$ and the summation (13.51) is limited to $2K(< P)$ terms, i.e. to the doubled number of expected real sinusoidal signals, completing the principal component approximation of the pseudo-inverse matrix $\mathbf{Y}^I$. Next, zeros (roots) λ_k of the polynomial $b_0 z^P + b_1 z^{P-1} + \ldots + b_{P-1} z^1 + b_P (b_0 = 1)$ are found (in Matlab: `roots([b0,b1,...,bP])`), and only the $2K$ that lie outside the unit circle are considered ($|\lambda_k| > 1$). Finally, as before, equations (13.33), (13.34) are used to calculate Ω and D from λ, but the value of D has to be negated. In turn, when the polynomial $b_P z^P + b_{P-1} z^{P-1} + \ldots + b_1 z^1 + b_0$ is used, we are looking for zeros inside the unit circle, and the value of D calculated from (13.33), (13.34) is not negated.

In Listing 13.6 the function `fKT()` is presented, which implements the Kumaresan–Tufts algorithm for real-value signals. It is a modified version of the `lpsvd.m` function originated from matNMR software [20] (authored by Jacco D. van Beek), working with complex-value data.

Listing 13.6: Damped sinusoid parameters estimation using Kumaresan–Tufts SVD-filtered linear prediction method

```
1  function [Om, D]=fKT(x,K)
2  % Kumaresan-Tufts method
3  % x - analyzed signal
4  % K - assumed number of real damped sine components
5
6     inside = 1;                          % 1/0=inside/outside the unit circle
7     M = 2*K;                             % number of complex components
8     N = length(x);                       % numer of signal samples
9     P = floor(N*3/4);                    % linear prediction order P = 3/4*N
10    Y = hankel( x(2:N-P+1), x(N-P+1:N) );   % backward prediction data matrix
11    y = x(1:N-P)';                       % backward prediction data vector
12    [U,S,V] = svd(Y,0);                  % singular value decomposition
13    S = diag(S);                         % take only main diagonal
14    bias = mean(S(M+1:min([N-P,P])));    % bias compensation
15    b = -V(:,1:M) * (diag(1./(S(1:M)-bias)) * (U(:,1:M)'*y));   % LP coefficients
16    if(inside==1)                        % INSIDE THE CIRCLE
17        z = roots([b(length(b):-1:1);1]);   % roots of poly
18        p=log(z); p = p( find(real(p)<0) );  % ln()
19        Om = imag(p); [Om indx] = sort( Om, 'ascend' ); Om=Om(K+1:2*K); % ang. frequency
20        D = -real(p(indx(K+1:2*K)));                      % damping
21    else                                 % OUTSIDE THE CIRCLE
22        z = roots([ 1; b]);              % roots of poly
23        p=log(z); p = p( find(real(p)>0) );  % ln()
24        Om = imag(p); [Om indx] = sort( Om, 'descend' ); Om=Om(1:K); % ang. frequency
25        D = real(p(indx(1:K)));                           % damping
26    end
27  % figure; plot(real(z),imag(z),'o'); pause, figure; plot(real(p),imag(p),'o'); pause
```

Exercise 13.7 (Kumaresan–Tufts Method). Call the `fKT.m` function in the program 13.2. Compare obtained results with the `fMatPen.m` function output.

The total least-squares (TLS) solution [12] of the discussed problem was also proposed [15]. In this method, solving (13.47) is replaced with the minimization task in respect of $\mathbf{b}_P$:

$$\|\mathbf{Y}_{N-P,P}\mathbf{b}_P + \mathbf{y}_{N-P}\|_2 \to \min, \tag{13.52}$$

in which the noisy character of not just matrix $\mathbf{Y}_{N-P,P}$ but also of vector $\mathbf{y}_{N-P}$ is taken into account. Here the SVD is performed on the augmented matrix given below (with added last column $-\mathbf{y}_{N-P}$):

$$[\mathbf{Y}_{N-P,P}| -\mathbf{y}_{N-P}] = \mathbf{U} \cdot \operatorname{diag}(\sigma_1 \dots \sigma_{P+1}) \cdot \mathbf{V}^T = \sum_{k=1}^{\min(N-P,P+1)} \sigma_k \mathbf{u}_k \mathbf{v}_k^H, \tag{13.53}$$

For $\sigma_1 \geq \dots \geq \sigma_P > \sigma_{P+1} > 0$, the solution is simply computed as:

$$\mathbf{b}_P = \frac{\mathbf{v}_{P+1}(1:P)}{\mathbf{v}_{P+1}(P+1)}, \tag{13.54}$$

using elements of the $\mathbf{v}_{P+1}$ singular vector.

Exercise 13.8 (Fast TLS Method). Program `fFastTLS.m`, given in the book repository, implements a fast Yuan–Torlak [24] total least-squares method, proposed recently for a single damped sine embedded in strong noise. Call this function in the program `lab13_ex_real_sdr.m`. Run the program for different levels of noise.

13.5.6 Matrix Pencil Method

Let us remember that the linear eigenvalue Matrix Pencil matrix problem was introduced in Sect. 13.3. The Matrix Pencil algorithm for damped sinusoids estimation consists of the following steps [6, 9, 16, 17]:

1. In the Matrix Pencil method $P = \lfloor N/3 \rfloor$, the matrix $\mathbf{Y}_{N-P,P}$ in (13.47) is denoted as $\mathbf{Y}_1$ and a new matrix $\mathbf{Y}_0$ is introduced:

$$\mathbf{Y}_0 = \begin{bmatrix} y_0 & \cdots & y_{P-2} & y_{P-1} \\ y_1 & \cdots & y_{P-1} & y_P \\ \vdots & \ddots & \vdots & \vdots \\ y_{N-P-1} & \cdots & y_{N-3} & y_{N-2} \end{bmatrix}. \tag{13.55}$$

2. Next, the matrix $\mathbf{Y}_1$ is decomposed by SVD and resultant singular vectors $\mathbf{v}_k, \mathbf{u}_k$ and values σ_k are used for calculation of the reduced $2K$-rank pseudo-inverse $\widetilde{\mathbf{Y}}_1^I$ of $\mathbf{Y}_1$ (see Eq. (13.22)):

$$\widetilde{\mathbf{Y}}_1^I = \left[\sum_{k=1}^{2K} \frac{1}{\sigma_k} \mathbf{v}_k \mathbf{u}_k^H \right]. \tag{13.56}$$

3. Now the matrix $\widetilde{\mathbf{Y}}_1^I$ is used for defining a new matrix $\mathbf{Z}$:

$$\mathbf{Z} = \widetilde{\mathbf{Y}}_1^I \cdot \mathbf{Y}_0. \tag{13.57}$$

4. Then, the matrix $\mathbf{Z}$ is decomposed by EVD (see Sect. 13.3). Let us remember that K denotes the number of signal sinusoidal components. Therefore, $2K$ the biggest eigenvalues of the matrix $\mathbf{Z}$ are found. They are equal to roots λ_k of the polynomial $b_0 z^P + b_1 z^{P-1} + \ldots + b_{P-1} z^1 + b_P$ ($b_0 = 1$), and lie outside the unit circle $|\lambda_k| > 1$).
5. Finally, as before in the Kumaresan–Tufts method, we use equations (13.33), (13.34) to estimate Ω_k and D_k from λ_k, and the value of D_k should be negated.

In Listing 13.7 a Matlab code of the Matrix Pencil algorithm is given. It is a modified version of the `itcmp.m` function originated from matNMR software [20] (author: Jacco D. van Beek), working with complex-value data. I like very much this program. A lot of advanced mathematics and only a few lines of code. Very efficient code! May be slow in execution but very robust to noise. For me it is a fantastic example that education has a sense. As a programmer, I have seen a lot of lines of code I did not understand. It is easier to write programs than to find really effective solutions.

Listing 13.7: Damped sinusoid parameters estimation using Matrix Pencil solution of the linear prediction problem (transform)

```
1  function [Om, D] = fMatPen(x,K)
2  % x - analyzed signal - sum of damped sinusoids
3  % K - assumed number of real damped sine components
4  M = 2*K;                            % number of complex components
5  N = length(x);                      % number of signal samples
6  L = floor(N/3);                     % linear prediction order L = N/3
7  X = hankel(x(1:N-L),x(N-L:N));      % building Hankel matrix
8  X0 = X(:,1:L); X1 = X(:,2:L+1);     % cutting matrices X0 and X1
9  [U,S,V] = svd(X1, 0); S = diag(S);            % SVD of X1
10 p = log( eig( diag(1./S(1:M)) * ((U(:,1:M)'*X0)*V(:,1:M)) ) );   % EVD of X0
11 Om = imag(p); [Om indx] = sort( Om, 'ascend' ); Om=Om(K+1:2*K); % ang. frequency
12 D = real(p(indx(K+1:2*K)));                   % damping
```

Exercise 13.9 (Matrix Pencil Method). Run the program 13.2 several times decreasing the SNR level: $120:-5:0$ dB. Observe when the MatPen method will stop to work properly. Compare different methods.

13.5.7 Yule–Walker Method: Linear Prediction Using Auto-Correlation Function

It is interesting to observe that equations (13.39), (13.41):

$$\left(\mathbf{Y}^T\mathbf{Y}\right)\mathbf{a} = \left(\mathbf{Y}^T\mathbf{y}\right) \tag{13.58}$$

after performing multiplication in brackets takes the following forms, respectively:

$$\begin{bmatrix} r_0 & r_1 \\ r_1 & r_0 \end{bmatrix} \begin{bmatrix} a_1 \\ a_2 \end{bmatrix} = \begin{bmatrix} r_1 \\ r_2 \end{bmatrix} \tag{13.59}$$

$$\begin{bmatrix} r_0 & r_1 & \cdots & r_{2K-1} \\ r_1 & r_0 & \cdots & r_{2K-2} \\ \vdots & \vdots & \ddots & \vdots \\ r_{2K-1} & r_{2K-2} & \cdots & r_0 \end{bmatrix} \begin{bmatrix} a_1 \\ a_2 \\ \vdots \\ a_{2K} \end{bmatrix} = \begin{bmatrix} r_1 \\ r_2 \\ \vdots \\ r_{2K} \end{bmatrix}, \tag{13.60}$$

where

$$r_k = \sum_n y(n)y(n+k), \ k = 0,1,2,3,\ldots. \tag{13.61}$$

Vector **a**, which is the solution of (13.59), (13.60):

$$\mathbf{R}_{yy}\mathbf{a} = -\mathbf{r}_{yy} \quad \rightarrow \quad \mathbf{a} = -\mathbf{R}_{yy}^{-1}\cdot\mathbf{r}_{yy}, \tag{13.62}$$

minimizes the mean squared prediction error (MSE). The generalization of (13.60) is known as the Yule–Walker equation [5, 14] and can be solved by the Matlab functions `aryule()` and `lpc()`. An estimate of the auto-correlation function r_k (13.61) of y_n needs to be found first, $\mathbf{R}_{yy}$ and $\mathbf{r}_{yy}$ are calculated next, and finally the equation $\mathbf{R}_{yy}\cdot\mathbf{a} = -\mathbf{r}_{yy}$ (13.58) is solved. One of the many existing methods is used, for example, one presented previously for solving (13.39), (13.41). However, in this case, the matrix of the linear equation ($\mathbf{R}_{yy}$) is square and symmetric in contrast to **Y**, and new possibilities exist, e.g. iteratively solving the Eq. (13.62) for increasing dimension of $\mathbf{R}_{yy}$ using the Levinson–Durbin algorithm (function `levinson()` in Matlab). For example, such method is used for fast calculations of vocal tract filter coefficients in digital speech coders. Algorithms belonging to this group represent well-known classical ARMA techniques.

Matlab implementation of auto-correlation-based estimation method (13.60) is given in the function fLPcor() presented in Listing 13.8.

Listing 13.8: Damped sinusoid parameters estimation using Yule–Walker method—linear prediction exploiting auto-correlation function

```
1   function [Om, D]=fYuleWalker(x,K,C)
2   % x - analyzed signal
3   % K - assumed number of real damped sines
4   % C - autocorrelation estimation method (choose 1,2,3 or 4)
5   N = length(x);                              % signal length
6   L = 2*K;                                    % prediction order
7   M = N-L;                                    % max numer of equations
8   if( C==1 ) r=xcorr(x,'biased');    end      %
9   if( C==2 ) r=xcorr(x,'unbiased');  end      %
10  if( C==3 ) r=xcorr(x,'coeff');     end      %
11  if( C==4 ) r=xcorr(x,'none');      end      % choose this option
12  R = r(N:end);                               % take only right part  (symmetry)
13  FirstCol = R(1:M); LastRow = R(M:M+L-1);    %
14  Y = hankel( FirstCol, LastRow );            % matrix MxL
15  y = R(L+1:L+M).';                           % vector Mx1
16  a = -Y\y; % a = -pinv(Y)*y;                 % solving linear equation
17  p = log(roots( [1 fliplr(a.')] ));          % roots, ln()
18  Om = imag(p); [Om indx] = sort( Om, 'ascend' );  Om=Om(K+1:2*K); % ang. frequency
19  D = -real(p(indx(K+1:2*K)));                % damping
```

Exercise 13.10 (Yule–Walker Method). Call the fYuleWalker.m function in the program 13.2. Compare obtained results with the fMatPen.m function output.

13.5.8 Pisarenko Method: Signal Subspace Methods

Let us remember the self-prediction equation, valid for summation of damped sinusoids:

$$x_n + \sum_{k=1}^{2K} a_k x_{n-k} = 0, \quad \mathbf{x}^T \mathbf{a} = 0, \tag{13.63}$$

where ($a_0 = 1$):

$$\mathbf{x} = [x_n, x_{n-1}, x_{n-2}, \ldots, x_{n-2K}]^T, \quad \mathbf{a} = [1, a_1, a_2, \ldots, a_{2K}]^T. \tag{13.64}$$

After setting $y_n = x_n + \varepsilon_n \rightarrow x_n = y_n - \varepsilon_n$ in (13.63), which takes into account the disturbance component adding itself to the signal x_n, we obtain

$$\mathbf{y}^T\mathbf{a} = \varepsilon^T\mathbf{a}, \quad \varepsilon = [\varepsilon_n, \varepsilon_{n-1}, \varepsilon_{n-2}, \dots, \varepsilon_{n-2K}]. \tag{13.65}$$

After setting $y_n = x_n + \varepsilon_n \rightarrow x_n = y_n - \varepsilon_n$ in (13.63), which takes into account the disturbance component adding itself to the signal x_n, we obtain:

$$\mathbf{y}^T\mathbf{a} = \varepsilon^T\mathbf{a}, \quad \varepsilon = [\varepsilon_n, \varepsilon_{n-1}, \varepsilon_{n-2}, \dots, \varepsilon_{n-2K}]. \tag{13.66}$$

After left-multiplication of Eq. (13.66) by $\mathbf{y}$ and calculation of expected values of both equation sides, we have

$$E[\mathbf{y}\mathbf{y}^T]\mathbf{a} = E[\mathbf{y}\varepsilon^T]\mathbf{a}. \tag{13.67}$$

We can re-write the right side of the above equation as follows:

$$E[\mathbf{y}\varepsilon^T]\mathbf{a} = E[(\mathbf{x}+\varepsilon)\,\varepsilon^T]\mathbf{a} = (E[\mathbf{x}\varepsilon^T]+E[\varepsilon\varepsilon^T])\mathbf{a} = E[\varepsilon\varepsilon^T]\mathbf{a} = \sigma_\varepsilon^2\mathbf{a} \tag{13.68}$$

because signal x_n is uncorrelated with noise ε_n and $E[\mathbf{x}\varepsilon^T] = \mathbf{0}$. σ_ε^2 denotes the noise variance. Finally, since $E[\mathbf{y}\mathbf{y}^T] = \mathbf{R}_{yy}$, Eq. (13.70) takes the form:

$$\mathbf{R}_{yy}\mathbf{a} = \sigma_\varepsilon^2\mathbf{a}. \tag{13.69}$$

Therefore, the vector $\mathbf{a}$ of interest is an eigenvector of the square, symmetric auto-correlation matrix $\mathbf{R}_{yy}$, associated with eigenvalue σ_ε^2 (it should be remembered that the eigenvalue problem for the given matrix $\mathbf{A}$ is solving the equation $\mathbf{A} \cdot \mathbf{v} = \lambda\mathbf{v}; \mathbf{v} \neq 0$—eigenvector of A, λ—corresponding eigenvalue of $\mathbf{A}$). Due to eigenvalue σ_ε^2, the eigenvector $\mathbf{a}$ lies in the noise subspace and is orthogonal to eigenvectors lying in the signal subspace, associated with bigger $2K$ eigenvalues. Roots of the polynomial, defined by $\mathbf{a}$, identify these eigenvectors that are signal components. Therefore, it is necessary to perform the following steps:

1. Calculation of the estimate of the auto-correlation function $\mathbf{r}_{yy}$ (13.61) of noisy measurement signal y_n as before, then building the auto-correlation matrix $\mathbf{R}_{yy}$:

$$\mathbf{R}_{yy} = \begin{bmatrix} r_0 & r_1 & \cdots & r_{2K} \\ r_1 & r_0 & \cdots & r_{2K-1} \\ \vdots & \vdots & \ddots & \vdots \\ r_{2K} & r_{2K-1} & \cdots & r_0 \end{bmatrix}, \tag{13.70}$$

2. Performing eigenvalue decomposition (EVD) of the matrix $\mathbf{R}_{yy}$, e.g. using function `eig()` in Matlab:

$$\mathbf{R}_{yy} = \sum_{k=1}^{2K+1} \gamma_k \mathbf{u}_k \mathbf{u}_k^T, \quad \gamma_1 \geq \gamma_2 \geq \dots \geq \gamma_{2K+1}, \tag{13.71}$$

3. Finding eigenvector $\mathbf{v}_{2K+1}$, associated with the smallest eigenvalue $\lambda_{2K+1} = \sigma_\varepsilon^2$, and setting $\mathbf{a} = \mathbf{v}_{2K+1}$,

4. Calculating roots λ_k of the polynomial having coefficients $[1; \mathbf{a}]$ and extracting from them frequencies of the signal components using Eq. (13.33) or (13.34).

However, due to scaling incorporated in EVD, only the signal frequency can be found from roots $\{\lambda, \lambda^*\}$ of the polynomial $\mathbf{a}$: $\Omega = |\text{imag}(\ln(\lambda))$.

The signal subspace methods (Pisarenko, EV, Min-Norm, MUSIC, ESPRIT) are very well presented in the literature [5, 12, 14, 25]. In program fPisarenko(), only the Matlab code of the Pisarenko method is given. The other methods are not discussed since using this algorithmic family is significantly inferior to the SVD-based methods, making direct use of signal samples (e.g., Steiglitz–McBride, Kumaresan–Tufts, Matrix Pencil).

Listing 13.9: Damped sinusoid parameters estimation using Pisarenko method—only frequency is found

```
1  function [ Om ]=fPisarenko(x,K,C)
2  % x - analyzed signal - sum of damped sinusoids
3  % K - assumed number of real damped sines
4  % C - autocorrelation method estimation (1, 2, 3 or 4)
5  N = length(x);                         % signal length
6  L = 2*K;                               % 2K signal components
7  if( C==1 ) R=xcorr(x,'biased'); end    % autocorrelation function
8  if( C==2 ) R=xcorr(x,'unbiased'); end  %
9  if( C==3 ) R=xcorr(x,'coeff'); end     %
10 if( C==4 ) R=xcorr(x,'none'); end      % choose this option
11 R = R(N:end);                          % only right part due to symmetry
12 RR = toeplitz( R(1:L+1) );             % Toeplitz matrix
13 [V,D] = eig(RR);                       % EVD
14 [Dmin indx] = min(diag(D));            % minimum eigen-value and its index
15 a = V(:,indx).';                       % minimum eigen-vector
16 z = roots(a); p = log(z); Om = imag(p); % roots of polynomial, then ln()
17 Om = sort( Om, 'descend' ); Om = Om(1:K); % found frequencies
```

Exercise 13.11 (Pisarenko Method). Call the fPisarenko.m function in the program 13.2. Compare obtained results with the fMatPen.m function output. Remember that only frequencies of damped sines are estimated by it.

13.6 Interpolated DFT Methods

Only for completeness of the presentation, now we shortly repeat equations of the Bertocco–Yoshida interpolated DFT algorithms [2, 23], presented already in part in chapter on FFT applications. The method addresses also the problem of damped sinusoid parameters estimation. Similar IpDFT algorithms are compared in [3]. In both methods, first N-point DFT (FFT) is calculated for signal x_n. Then the biggest magnitude value in the DFT spectrum is found, and let us denote it as X_k, and its left and right neighbors: X_{k-2}, X_{k-1} and X_{k+1}. After that both methods use different strategy summarized in Table 13.1.

Program of Bertocco–Yoshida method was presented in Listing 6.8. The program can be modified and used for finding more damped sine components in the signal, but their DFT bins (around peaks) have to be found or specified explicitly. In case of only one component, the algorithm can be further improved. When the damped sine parameters are calculated, leakage from the negative frequency component to the positive frequency one can be calculated and subtracted from DFT coefficients, used for signal estimation [21]. Equation (13.2) is used for this purpose. This leakage compensation procedure can be repeated a few times after each estimate improvement. In Listing 13.10, Matlab code of Bertocco–Yoshida function for one-component damped sine estimation is given. The program was proposed by Duda–Zielinski in [4].

Table 13.1: Algorithms of Bertocco and Yoshida DFT-based damped sinusoid estimation methods

No	Bertocco method [2]	Yoshida method [23]
1	$R = X_{k+1}/X_k$	$R = (X_{k-2} - 2X_{k-1} + X_k)/(X_{k-1} - 2X_k + X_{k+1})$
2	$\lambda = e^{j\Omega_k}(1-R)/(1-Re^{-j2\pi/N})$	—
3	$D = -\mathrm{Re}\{\ln(\lambda)\}$	$D = \frac{2\pi}{N}\mathrm{Im}\{-3/(R-1)\}$
4	$\Omega = \mathrm{Im}\{\ln(\lambda)\}$	$\Omega = \frac{2\pi}{N}\mathrm{Re}\{k - 3/(R-1)\}$

Listing 13.10: Estimation of single damped sinusoid parameters using IpDFT Bertocco–Yoshida algorithm with leakage compensation

```
1   function [we, de, Ae, pe] = fIpDFT_BY1_LC(x,NI);
2   % Leakage correction for BY1 interpolated DFT
3   % x - input signal x=A*cos(w*n+p).*exp(-d*n)
4   % NI - number of iterations, for NI=0 BY1 without LC is computed
5   % Estimated values:
6   % we - frequency (rad), de - damping, Ae - amplitude, pe - phase (rad)
7   N = length(x);
8   Xw = fft(x);                          % computation of DFT
9   [Xabs, ind] = max(abs(Xw(1:round(N/2))));
10  k = [ind-1 ind ind+1];
11  dw = 2*pi/N;                          % DFT frequency step
12  wkm1= (k(1)-1)*dw;                    % frequency of DFT bin with index k-1
13  wk = (k(2)-1)*dw;                     % frequency of DFT bin with index k
14  wkp1= (k(3)-1)*dw;                    % frequency of DFT bin with index k+1
15  [we, de, Ae, pe, lam] = BY1_in_LC(wkm1,wk,wkp1,Xw(k),N);
16
17  %% Leakage correction for negative frequencies
18  wkk = [wkm1 wk wkp1];
```

```
19   for iter=1:NI
20         for m=1:3;
21               Xw_correction(m) = ...
22               (Ae/2)*( exp(-j*pe)*(1-conj(lam)^N)/(1-conj(lam)*exp(-j*wkk(m))));
23         end
24         Xw_correction = Xw(k) - Xw_correction;
25         [we, de, Ae, pe, lam] = BY1_in_LC(wkm1, wk, wkp1, Xw_correction, N);
26   end
27
28   function [we, de, Ae, pe, lam] = BY1_in_LC(wkm1, wk, wkp1, Xw, N);
29   r = ( -exp(-j*wk)+exp(-j*wkm1) )/( -exp(-j*wkp1)+exp(-j*wk) );
30   R = ( Xw(1)-Xw(2) )/( Xw(2)-Xw(3) );
31   lam= exp(j*wk)*(r-R)/( r*exp(-j*2*pi/N)-R*exp(j*2*pi/N) );
32   we = imag(log(lam));
33   de = -real(log(lam));
34   if round(1e6*R)==-1e6  %% coherent sampling, d=0
35      Ae = 2*abs(Xw(2))/N;
36      pe = angle(Xw(2));
37   else
38      c = (1-lam^N)/(1-lam*exp(-j*wk));
39      c = 2*Xw(2)/c;
40      Ae = abs(c);
41      pe = angle(c);
42   end
```

Exercise 13.12 (Interpolated DFT Method). Call the interpolated DFT function from Listing 13.10 in the program 13.2 but only for one-component signal. Compare obtained results with remaining functions. Change signal frequency and damping and repeat a few times the experiment. Then apply the function for estimation of damped sinusoid observed in mechanical spectroscopy—call it from the program lab13_ex_real_spectroscopy.m.

13.7 Summary

What should be remembered?

1. Dumped sinusoids and their summations widely appear in surrounding world because such signals are generated by objects described by second-order differential equations that are plenty. Electrical, mechanical, and biological oscillators are the first examples. When we identify parameters of oscillations, we can deduce how an object is built that generated them or in what state it was, for example, in material spectroscopy/defectoscopy and medical magnetic resonance.

2. In case of damped oscillations, typically non-Fourier transform methods are used, mainly linear prediction ones, exploiting the fact that next signal samples are linear combination of the previous signal samples. One can find frequencies and dampings of damped sinusoids from linear prediction coefficients calculated for them.

3. Next sample of each real-value damped sinusoid can be predicted from its previous 2 samples. When signal has K such components, $2K$ next samples are required. In the presence of noise, we use more input–output relations than 2 or $2K$ and build over-determined matrix equations that are solved in mean squared error sense. We calculate LP coefficients from them.

4. Solving LP prediction equations is effective when the number of signal components, damped sines, is known. This problem was not discussed in this chapter. Final prediction error, Akaike information criterion, and minimization of description length are exemplary criteria helping us to choose appropriate K—the number of embedded damped sines.

5. Linear prediction coefficients can be calculated using matrix equation with signal samples (covariance approach) or samples of signal auto-correlation. Described Prony, Steiglitz–McBride, Kumaresan–Tufts, and Matrix Pencil methods belong to the first group, while Yule–Walker and Pisarenko methods to the second.

6. When some noise reduction technique is added to the covariance methods (for example, filtering, SVD, EVD), they offer more robust, immune-to-noise estimation of linear prediction parameters; however, they are more time consuming. The Prony method, the simplest covariance approach, is very noise sensitive.

7. Auto-correlation-based methods are used for symmetrical matrices only, are more noise sensitive than covariance ones, and do not allow damping estimation. However, they are easier to speed up, for example, using Levinson algorithm.

8. Knowing the signal model, we can also apply it to the DFT spectra of damped sinusoids and calculate their parameters from them, for example, using interpolated DFT approaches. Presented Bertocco–Yoshida method is one of them.

13.8 Private Investigations: Free-Style Bungee Jumps

Exercise 13.13 (Piano and Guitar Notes Analyzer). At webpage https://www.findsounds.com/, find sounds of single piano or guitar notes. Choose one piano and one guitar sample, desirably of the same musical note. Display signal waveforms and calculate and plot their DFT spectra (similar to the one presented in Fig. 13.2). Use the IpDFT function for finding parameters of dominant damped sines, present in both signals. Try to synthesize both signals as a summation of their oscillatory components. Listen to generated signals. How much are they similar to the original recordings? As *a starter*, you can use a program `lab13_ex_instrument_tuning.m` from the book repository. Is it possible to do the piano/guitar tuning using our programs?

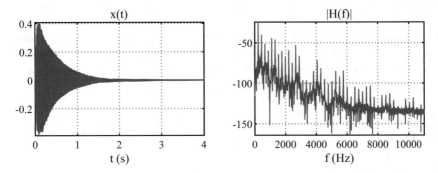

Fig. 13.2: Oscillatory damped signal of one piano note (left) and its DFT spectrum (right)

Exercise 13.14 (Damped Sines from NMR). Nuclear magnetic resonance (NMR) is used, among others, for chemical compounds recognition. In program `lab13_ex_nmr.m`, placed in book repository, one NMR signal was shown together with its DFT spectrum (see Fig. 13.3). The signal is complex-value one and has $N = 2^{16}$ samples and 22 damped sine components. Spectrometer Avance III 300 MHz of the Bruker company was used for its acquisition (sampling frequency equal 8971.291866028711 Hz). Then the file was read into the matNMR software and converted into MAT file. Try to calculate parameters of one damped sine, present in the signal, using interpolated DFT method.

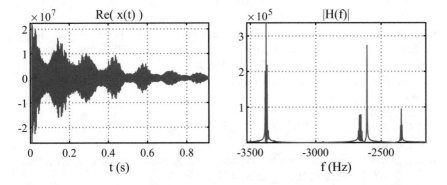

Fig. 13.3: Oscillatory signal with 22 damped sinusoids originated from nuclear magnetic resonance (left) and its DFT spectrum (right)

Exercise 13.15 (All Together Now: For Single Damped Sine). Run program `lab13_ex_test_one_component.m` from book repository. Run it for three test signals and different SNR values. Compare results. What method is a winner in your opinion? In fact, to have reliable results we should repeat experiments for random phase shifts from the range $-\pi$ to π and calculate mean values of frequency and damping estimation errors. Are you interested in program modification? Bravo!

Exercise 13.16 (All Together Now: For Multiple Damped Sines). Run program `lab13_ex_test_multi_component.m` from book repository for different SNR values. Find SNR range for proper work of each method. Compare results. Who is the winner? I do not believe!

References

1. A. Agneni, L. Balis-Crema, Damping measurements from truncated signals via Hilbert transform. Mech. Syst. Signal Process. **3**(1), 1–13 (1989)
2. M. Bertocco, C. Offeli, D. Petri, Analysis of damped sinusoidal signals via a frequency-domain interpolation algorithm. IEEE Trans. Instrum. Meas. **43**(2), 245–250 (1994)
3. K. Duda, L.B. Magalas, M. Majewski, T.P. Zieliński, DFT-based estimation of damped oscillation parameters in low-frequency mechanical spectroscopy. IEEE Trans. Instrum. Meas. **60**(11), 3608–3618 (2011)
4. K. Duda, T.P. Zieliński, Efficacy of the frequency and damping estimation of a real-value sinusoid. IEEE Mag. Instrum. Meas. 48–58 (2013)
5. M.H. Hayes, *Statistical Digital Signal Processing and Modeling* (Wiley, New York, 1996)
6. Y. Hua, T.K. Sarkar, Matrix pencil method for estimating parameters of exponentially damped/undamped sinusoid in noise. IEEE Trans. Acoust. Speech Signal Process. **38**(5), 814–824 (1990)
7. S.M. Kay, *Modern Spectral Estimation: Theory and Application* (Prentice Hall, Englewood Cliffs, 1988)
8. R. Kumaresan, D.W. Tufts, Estimating the parameters of exponentially damped sinusoids and pole-zero modeling in noise. IEEE Trans. Acoust. Speech Signal Process. **30**, 837–840 (1982)
9. Y. Li, K.J. Ray Liu, J. Razavilar, A parameter estimation scheme for damped sinusoidal signals based on low-rank Hankel approximation. IEEE Trans. Signal Process. **45**(2), 481–486 (1997)
10. S.L. Marple, *Digital Spectral Analysis with Applications* (Prentice Hall, Englewood Cliffs, 1987)
11. S.L. Marple, *Digital Spectral Analysis with Applications in C Fortran and MATLAB* (Prentice Hall, Englewood Cliffs, 1995)
12. T.K. Moon, W.C. Stirling, *Mathematical Methods and Algorithms for Signal Processing* (Prentice Hall, Upper Saddle River, 1999)

13. S.J. Orfanidis, *Optimum Signal Processing. An Introduction* (Macmillan, New York, 1988)
14. J.G. Proakis, D.G. Manolakis, *Digital Signal Processing. Principles, Algorithms, and Applications* (Macmillan, New York, 1992)
15. M.A. Rahman, K.B. Yu, Total least squares approach for frequency estimation using linear prediction. IEEE Trans. Acoust. Speech Signal Process. **35**(10), 1440–1454 (1987)
16. J. Razavilar, Y. Li, K.J. Liu, A structured low-rank matrix pencil for spectral estimation and system identification. Signal Process. **65**, 363–372 (1998)
17. T.K. Sarkar, O. Pereira, Using the matrix pencil method to estimate the parameters of a sum of complex exponentials. IEEE Antennas Propag. Mag. **37**(1), 48–55 (1995)
18. K. Steiglitz, L.E. McBride, A technique for identification of linear systems. IEEE Trans. Autom. Control **10**, 461–464 (1965)
19. P. Stoica, R. Moses, *Introduction to Spectral Analysis* (Prentice Hall, Upper Saddle River, 1997)
20. J.D. Van Beek, matNMR: a flexible toolbox for processing, analyzing and visualizing magnetic resonance data in Matlab. J. Magn. Res. **187**, 19–26 (2007). Software online: http://matnmr.sourceforge.net/
21. R.C. Wu, C.T. Chiang, Analysis of the exponential signal by the interpolated DFT algorithm. IEEE Trans. Instrum. Meas. **59**(12), 3306–3317 (2010)
22. Y. Yao, S.M. Pandit, Cramèr-Rao lower bounds for a damped sinusoidal process. IEEE Trans. Signal Process. **43**(4), 878–885 (1995)
23. I. Yoshida, T. Sugai, S. Tani, M. Motegi, K. Minamida, H. Hayakawa, Automation of internal friction measurement apparatus of inverted torsion pendulum type. J. Phys. E Sci. Instrum. **14**, 1201–1206 (1981)
24. J. Yuan, M. Torlak, Modeling and estimation of transient carrier frequency offset in wireless transceivers. IEEE Trans. Wirel. Commun. **13**(7), 4038–4049 (2014)
25. T.P. Zieliński, K. Duda, Frequency and damping estimation methods - an overview. Metrology Meas. Syst. **18**(4), 505–528 (2011). Software online: http://kt.agh.edu.pl/~tzielin/papers/M&MS-2011/

Chapter 14
Speech Compression and Recognition

Are you aware (mindful) of the fact that during your phone call a speech of your interlocutor is synthesized from a few numbers? What a dehumanized world!

14.1 Introduction

Multimedia: speech, audio, image, and video. For most people speech is the least interesting. I do not agree with this opinion. For me it is the most interesting. What does cause that we, people, understand each other? We are biologically built the same way, but we are not perfect *copies* of each other. We are smaller and bigger, for example. We have slightly different vocal folds and vocal tracts, but we generate some more-or-less standard sounds that are understood by our interlocutors. Wow! It is a real magic.

I am teaching signal processing for years. During lecture on speech compression, I am always excited. It is interesting to see how much our speech signal can be deformed in its shape but still to be very well perceived! It is fantastic to observe how smartly information is hidden in speech waveforms and how clever our mind is still understanding the speech content, even after dramatic change of speech signal shape after using very aggressive, i.e. very strong, compression.

Well. It is time to start.

Digital speech compression, used in digital telephony, is one of the best examples of DSP applications. Linear prediction coding (LPC) concept is exploited in it. Since all people have similar bodies, from the "life-functions" points of view, we can built a universal model of speech articulation, fitting relatively well to all of us. This model consists of parameters specifying two things: vocal tract filter excitation and vocal tract filter frequency response, i.e. filter frequency characteristics. Having a speech fragment, we have to find the following:

- Whether vocal folds are working; if yes, with what frequency they are opening and closing and injecting air-flow impulses to the vocal tract,

© Springer Nature Switzerland AG 2021
T. P. Zieliński, *Starting Digital Signal Processing in Telecommunication Engineering*, Textbooks in Telecommunication Engineering, https://doi.org/10.1007/978-3-030-49256-4_14

- What is the actual frequency response of the vocal tract filter, having resonance peaks, called *formants*.

Knowing answers to these two questions we should build a digital filter having exactly the same frequency response as the acoustical vocal tract filter. Then we can excite it with random noise or periodic sequence of quasi-impulses, imitating original acoustical signal coming from vocal folds, and obtain synthesized speech on the filter output. In general, excitation is speaker dependent, while filter frequency characteristic is speech content dependent. When we modify a filter excitation signal during speech synthesis, we can see that it is difficult to recognize a speaker. In contrary, while we modify the filter, we are changing a speech content, i.e. one phoneme to some other one, for example, vovel "a" to "e." Speech generated with working vocal folds is called *voiced*, while for opened vocal folds—*unvoiced*.

The digital speech synthesis filter, typically IIR filter of order 10 (see Fig. 8.5), has ten poles, which are capable of creating five peaks in the filter frequency response—please remember the zeros-and-poles method of digital IIR filter design. It is not without reason that the filter can create five spectral peaks: human speech has exactly up to five local spectral maxims, i.e. *formants*.

As already mentioned, a speech content, for example, vowels "a," "b," "c" and consonants "h," "k," "s," is hidden in a shape of speech spectrum envelope. The same curve should have frequency response magnitude of speech synthesis filter. Therefore, it is obvious that all speech recognition systems are tracking the actual speech spectrum envelope (the synthesis filter frequency response) and compare its shape with reference shapes, typical for individual phonemes or ... for phoneme doublets, or triplets, or even for whole words. In the last two cases, history of spectral changes has to be remembered. Extra knowledge about speech syntax helps additionally in decision-making. Cepstral and mel-cepstral spectral coefficients are used as spectrum shape descriptors (recognized features). Being concrete, approximately first fifteen low-frequency DCT coefficients of logarithm of speech spectrum magnitude.

In Matlab, the following functions are useful for implementation of LP-based speech coding/compression:

- `randn()`—vocal tract filter excitation for unvoiced speech,
- `xcorr()`—calculation of speech auto-correlation function,
- `lpc()`—calculation of LP coefficients for speech fragment,
- `levinson()`—fast calculation of LP coefficients,
- `filter()`—speech synthesis.

14.2 Speech Compression

Speech Production and Its Modeling As already stated in the introduction, vocal folds are muscles which are:

- permanently open for unvoiced speech like 'c," "f," "h," "s," "t,"...,
- most of the time closed but periodically opening in impulsive way during voiced speech like "a," "e," "i," "o," "u," "d," "g," "m," "n,"....

This acoustical signal of air-flow, coming from lungs, is passing through a vocal tract cavity, oral, and nasal, which creates a resonance *tube/chamber*. During speaking, we are opening mouth, moving the tongue, and permanently changing shape of the vocal tract acoustical cavity and its resonant characteristics. What is spoken is coded in natural way in speech resonance patterns: positions and heights of peaks of speech spectrum envelope. Unvoiced-speech fragments are *noisy-like*, while voiced speech fragments are oscillatory-like. A fragment of digital speech is presented in Fig. 14.1. We see mainly voiced, oscillatory speech, except samples 3500–4000, belonging to unvoiced, noisy segment. In Fig. 14.2 are presented typical Fourier spectra calculated for speech signals: one spectrum of the whole sentence/word (left) and a sequence of shorter spectra, so-called short-time Fourier transform, calculated for overlapping signal fragments (right). Both spectra belong to signal shown in Fig. 14.1.

Having samples of digital speech, we divide it into fragments and for each fragment, we find parameters describing the sound generated by the vocal folds. Mainly we are interested in resonances of the vocal tract. Typically, 30 millisecond speech fragments are used, windowed with Hamming function. They are overlapping 7.5 milliseconds (25%). Speech signal partitioning is shown in Fig. 14.3. Next, we build a digital filter having frequency response the same as the vocal tract and excite it with a signal which imitates sound generated by vocal folds. It is a random noise

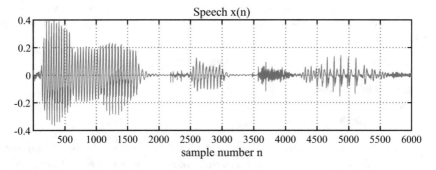

Fig. 14.1: Fragment of recorded human speech with oscillatory-like and noisy-like intervals. In oscillatory voiced parts (almost everywhere), the vocal folds are rhythmically opening and closing. In unvoiced noisy part (samples from 3500 to 4000), vocal folds are open

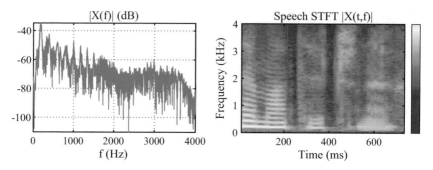

Fig. 14.2: Two types of frequency spectra calculated for speech signals: (left) magnitude of the Fourier transform of the whole signal, (right) short-time Fourier transform, a sequence of many Fourier spectra

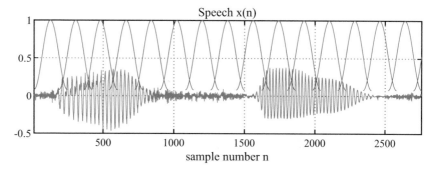

Fig. 14.3: Illustration of speech signal segmentation/partitioning: Hamming window with 240 samples is repeatedly shifted by 180 samples over the speech and multiplied with it

(for unvoiced speech) or sequence of impulses with period T (period of vocal folds opening and closing for voiced speech). A simplified scheme of speech production and modeling is presented in Fig. 14.4.

The Vocal Folds Signal Vocal folds when speaking are in two states:

- opened all the time—unvoiced case,
- opening and closing periodically with some frequency, called a fundamental one or **a pitch period**—voiced case.

In speech coding, we have to find a pitch period for each analyzed speech fragment. It is assumed that a signal part is quasi-stationary when it has $N = 240$ samples $x(n)$, $n = 0, 1, 2, \ldots, N-1$, for sampling frequency $f_s = 8000$ Hz. We calculate autocorrelation function for the speech fragment:

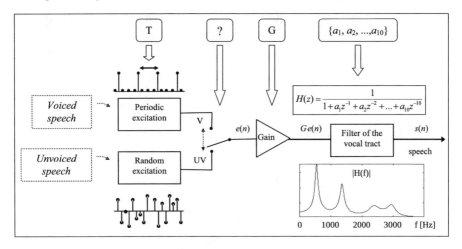

Fig. 14.4: Simplified model of speech production/synthesis [15]

$$R(k) = \sum_{n=0}^{N-1-k} (x(n) \cdot x(n+k)), \qquad \text{for} \quad k = 0,1,2,\ldots,N-1, \qquad (14.1)$$

find its maximum value for $k > k_{min}$ (for example $k > k_{min} = 20$ for $f_s = 8$ kHz), and check whether it is greater than some threshold. If so, it means the speech is periodic (*voiced*) and vocal folds are opening and closing with period equal to the argument of the found auto-correlation peak maximum. If not, the vocal folds are opened all the time and speech is *unvoiced*. We skip small values of k, because auto-correlation function for any signal has always maximum value for $k = 0$. Additionally, pitch period could be too small since the pitch frequency could not be too high, taking into account human body constraints. Voiced phoneme "a" (left) and unvoiced phoneme "sh" (right) and their auto-correlation functions (down) are presented in Fig. 14.5. For "a," the auto-correlation has side maximum for shift parameter $k \approx 70$. This is the "a," period that is confirmed by visual observation of the signal waveform. For "sh," the auto-correlation has no side maximum, which is also correct because the signal is not periodic.

The Vocal Tract Filter Digital speech synthesis filter is presented in Fig. 14.6. It is described by the following equation (10 = 2 times 5 formants to be found):

$$s(n) = G \cdot e(n) - \sum_{k=1}^{10} a_k s(n-k), \qquad (14.2)$$

valid only for one stationary speech fragment, where $e(n)$ denotes filter excitation, G constant filter gain, $s(n)$ synthesized speech, and a_k auto-regressive filter coefficients. The speech synthesis filter is a recursive IIR digital filter with transfer function:

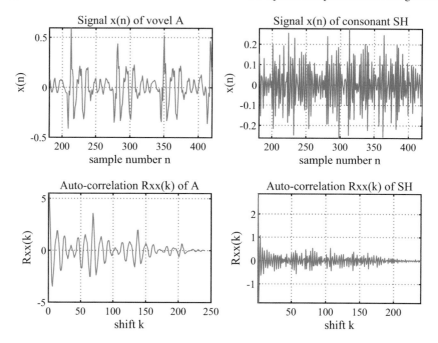

Fig. 14.5: Phonemes "a" (left) and "sh" (right): their waveforms (up) and auto-correlation functions (down)

$$H(z) = \frac{G}{A(z)} = \frac{G}{1 + a_1 z^{-1} + a_2 z^{-2} + \ldots + a_{10} z^{-10}} =$$
$$= \frac{G}{(1 - p_1 z^{-1})(1 - p_1^* z^{-1}) \ldots (1 - p_5 z^{-1})(1 - p_5^* z^{-1})}. \quad (14.3)$$

It has ten poles, five p_1, p_2, p_3, p_4, p_5 and their complex conjugates, and is capable of creating five peaks in the filter frequency response (please, remember the zeros-and-poles method of digital IIR filter design):

$$H(\Omega) = \frac{G}{1 + a_1 e^{-j\Omega} + a_2 e^{-j2\Omega} + \ldots + a_{10} e^{-j10\Omega}}, \qquad \Omega = 2\pi \frac{f}{f_s}. \quad (14.4)$$

It is not without reason that the filter can create five spectral peaks: human speech has exactly up to five local spectral maxims called *formants*. In Fig. 14.7, frequency responses of synthesis filters for voiced phonemes "a," "e," "i," and "u" are shown. We see different speech formants, resonant peaks, specific for each sound.

Calculation of Speech Synthesis Filter Coefficients Let P denote the speech linear prediction filter order. For a given speech fragment, we must find such vector of speech synthesis filter coefficients:

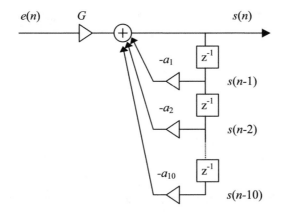

Fig. 14.6: Recursive speech synthesis filter. Denotations: $e(n)$—excitation, noise, or sequence of impulses, $s(n)$—synthesized speech, a_k—10 linear prediction filter coefficients which are modeling the frequency response of vocal tract, and G—gain [15]

$$\mathbf{a} = [1, a_1, a_2, a_3, \ldots, a_P]. \tag{14.5}$$

for which its frequency response (14.4) is the same as acoustical filter frequency response of a speaker vocal tract. How to do it? In voiced-speech excitation, $e(n)$ is impulsive, i.e. for most time we have $e(n) = 0$. In such case, Eq. (14.2) takes a form (assuming $P = 10$):

$$s(n) = -\sum_{k=1}^{P} a_k s(n-k). \tag{14.6}$$

This means that a speech signal is mostly self-predictive because each of its sample is a linear superposition of last P samples, scaled by linear prediction coefficients $\mathbf{a}$. If we assume that a fragment of coded speech signal $x(n)$ has N samples, we can

- write $N - (P-1)$ equations of the form (14.6), but for $x(n)$, not $s(n)$,
- calculate errors between original speech samples and predicted ones,
- square all errors and sum them,
- finally find vector of prediction coefficients $\mathbf{a}$ as a solution of the following least-squares error minimization:

$$J(\mathbf{a}) = \sigma^2 = \frac{1}{N-P} \sum_{n=P}^{N-1} err^2(n) = \frac{1}{N-P} \sum_{n=P}^{N-1} [x(n) - \hat{x}(n)]^2 =$$

$$= \frac{1}{N-P} \sum_{n=P}^{N-1} \left[x(n) + \sum_{j=1}^{P} a_j x(n-j) \right]^2. \tag{14.7}$$

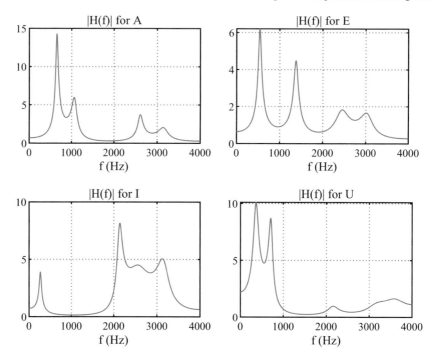

Fig. 14.7: Frequency response magnitudes of vocal tract filters calculated for voiced phonemes: "a," "e," "i," and "u" (in rows from left to right)

We calculate partial derivatives of Eq. (14.7) with respect (regard) to each coefficient a_j, set them to zeros, and solve obtained set of equations. The solution is as follows:

$$\mathbf{a} = -\mathbf{R}^{-1}\mathbf{r}, \tag{14.8}$$

where matrix $\mathbf{R}$ and vector $\mathbf{r}$ are built from samples of the signal $x(n)$ autocorrelation function $r(k)$:

$$
\mathbf{a} = \begin{bmatrix} a_1 \\ a_2 \\ \vdots \\ a_P \end{bmatrix}, \quad
\mathbf{R} = \begin{bmatrix}
r(0) & r(1) & \cdots & r(P-1) \\
r(1) & r(0) & \cdots & r(P-2) \\
\vdots & \vdots & \ddots & \vdots \\
r(P-1) & r(P-2) & \cdots & r(0)
\end{bmatrix}, \quad
\mathbf{r} = \begin{bmatrix} r(1) \\ r(2) \\ \vdots \\ r(P) \end{bmatrix}, \tag{14.9}
$$

$$r(k) = \frac{1}{N-P} \sum_{n=P}^{N-1} x(n)x(n+k). \tag{14.10}$$

Knowing vector **a** of prediction coefficients, we can calculate (from Eq. (14.7)) the minimum value of cost function $J_{\min}$, being equal to variance of the estimation error $\sigma^2_{\min}$:

$$J_{\min} = \sigma^2_{\min} = r(0) + \mathbf{a}^T \mathbf{r} = r(0) + \sum_{j=1}^{P} a_j r(j). \tag{14.11}$$

This value is used as filter gain G in our speech synthesis system, shown in Fig. 14.6. In order to synthesize a speech fragment, we are exciting the IIR filter (14.3), having already known parameters G and **a**, with a random noise or sequence of impulses with period T.

Proof. We calculate derivatives of Eq. (14.7) in regard to coefficients a_k and set them to zero (for $1 \leq k \leq P$):

$$\frac{\partial J}{\partial a_k} = \frac{\partial}{\partial a_k} \left\{ \frac{1}{N-P} \sum_{n=P}^{N-1} \left[x^2(n) + 2x(n) \sum_{j=1}^{P} a_j x(n-j) + \left(\sum_{j=1}^{P} a_j x(n-j) \right)^2 \right] \right\} = 0$$

$$\frac{1}{N-P} \sum_{n=P}^{N-1} \left[2x(n)x(n-k) + 2 \left(\sum_{j=1}^{P} a_j x(n-j) \right) x(n-k) \right] = 0$$

$$\frac{1}{N-P} \sum_{n=P}^{N-1} x(n)x(n-k) + \frac{1}{N-P} \sum_{n=P}^{N-1} \left[x(n-k) \sum_{j=1}^{P} a_j x(n-j) \right] = 0$$

$$\sum_{j=1}^{P} a_j \left[\frac{1}{N-P} \sum_{n=P}^{N-1} x(n-j)x(n-k) \right] = -\frac{1}{N-P} \sum_{n=P}^{N-1} x(n)x(n-k)$$

$$\sum_{j=1}^{P} a_j R_{xx}(k,j) = -R_{xx}(k,0), \quad k = 1, 2, \ldots, P,$$

$$\begin{bmatrix} R_{xx}(1,1) & R_{xx}(1,2) & \cdots & R_{xx}(1,P) \\ R_{xx}(2,1) & R_{xx}(2,2) & \cdots & R_{xx}(2,P) \\ \vdots & \vdots & \ddots & \vdots \\ R_{xx}(P,1) & R_{xx}(P,2) & \cdots & R_{xx}(P,P) \end{bmatrix} \cdot \begin{bmatrix} a_1 \\ a_2 \\ \vdots \\ a_P \end{bmatrix} = - \begin{bmatrix} R_{xx}(1,0) \\ R_{xx}(2,0) \\ \vdots \\ R_{xx}(P,0) \end{bmatrix}. \tag{14.12}$$

Last matrix equation is the same as Eqs. (14.8), (14.9). Why? Because $R_{xx}(k,j)$ is equal to $r(k-j)$, defined in Eq. (14.10):

$$R_{xx}(k,j) = \frac{1}{N-P} \sum_{n=P}^{N-1} x(n-j)x(n-k) = r(k-j). \tag{14.13}$$

After taking into account symmetry of the auto-correlation function, i.e. $r(-m) = r(m)$, Eq. (14.12) is the same as pair of equations (14.8), (14.9).

Now, we will derive the Eq. (14.11). To do this, we put the solution $\mathbf{a} = \mathbf{R}_{xx}^{-1}\mathbf{r}$ (14.8) into the matrix equation of cost function (14.7) and obtain Eq. (14.11)):

$$J_{\min} = \frac{1}{N-P} \sum_{n=P}^{N-1} \left[x(n) + \sum_{j=1}^{P} a_j x(n-j) \right]^2 = \frac{1}{N-P} \sum_{n=P}^{N-1} \left[x(n) + \mathbf{a}^T \mathbf{x}(n) \right]^2$$

$$J_{\min} = \frac{1}{N-P} \sum_{n=P}^{N-1} \left[x^2(n) + 2x(n)\mathbf{a}^T \mathbf{x}(n) + \mathbf{a}^T \mathbf{x}(n)\mathbf{x}^T(n)\mathbf{a} \right]$$

$$J_{\min} = \frac{1}{N-P} \sum_{n=P}^{N-1} x^2(n) + \mathbf{a}^T \frac{1}{N-P} \sum_{n=P}^{N-1} \left[2x(n)\mathbf{x}(n) + \mathbf{x}(n)\mathbf{x}^T(n)\mathbf{a} \right]$$

$$J_{\min} = r(0) + \mathbf{a}^T \left[2\mathbf{r} + \mathbf{Ra} \right] = r(0) + \mathbf{a}^T \left[2\mathbf{r} - \mathbf{RR}^{-1}\mathbf{r} \right] = r(0) + \mathbf{a}^T \mathbf{r} = r(0) + \sum_{j=1}^{P} a_j r(j).$$

Residual Excitation and Inverse Filter A digital filter inverse to $H(z)$ is defined as:

$$H^{-1}(z) = \frac{1}{H(z)} \qquad \rightarrow \qquad H^{-1}(z) \cdot H(z) = 1. \qquad (14.14)$$

Of course, passing any signal through the cascade of a filter $H(z)$ and its inverse $H^{-1}(z)$ does not change the signal (see left side of (14.14)) because *undoing* what was *done* is returning us to the initial state. Let us create an inverse of the vocal tract (speech synthesis) filter (14.3):

$$H^{-1}(z) = \frac{A(z)}{G} = \frac{1 + a_1 z^{-1} + a_2 z^{-2} + \ldots + a_{10} z^{-10}}{G}. \qquad (14.15)$$

If we pass original speech through the inverse filter and then through the direct filter, we obtain the same speech signal. Therefore output of the inverse filter, called the *residual signal*, is the best excitation of the speech synthesis filter, because it ensures perfect speech signal reconstruction. Therefore in some speech coders the residual signal is computed and compressed for each speech frame. Compressing residual signal instead of speech signal has a big sense, because for voiced speech this signal is similar to *periodic sequence of impulses* and thanks to this it is easily coded (for example, vector quantization technique can be used).

A time domain equation of the inverse filter is obtained by rewriting the speech synthesis filter formula (14.2) but treating $e(n)$ as output and $s(n)$ as input:

$$s(n) = G \cdot e(n) - \sum_{k=1}^{10} a_k s(n-k) \qquad \text{—original,}$$

$$e(n) = \frac{1}{G} \sum_{k=0}^{10} a_k s(n-k) \qquad \text{—inverse.} \qquad (14.16)$$

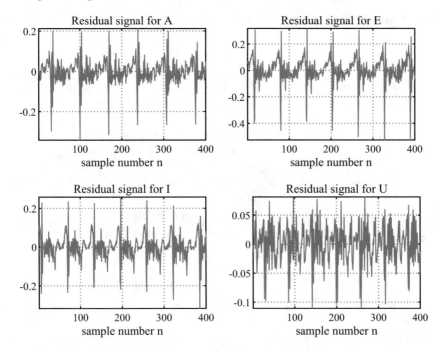

Fig. 14.8: Residual signals, output of inverse filters, for vowels (voiced phonemes) "a," "e," "i," and "u"

If we set $s(n) = x(n)$, i.e. the speech to be coded, and pass the signals $x(n)$ through the inverse filter, the perfect excitation for the speech synthesis filter is attained. EUREKA! After that, we can quantize it in a clever way and has almost perfect synthesized speech. In Fig. 14.8, residual signals for voiced phonemes "a," "e," "i," and "u" are shown. As expected, they are somehow similar to our synthetic 0/1 excitation used for voiced speech.

Speech Compression in Digital Telephony Having coefficients of speech synthesis filter and information about it excitation, we can synthesize a speech fragment. In digital phone of A SPEAKER (*content provider*), pitch period and LPC filter coefficients are: estimated, quantized, and transmitted in real time. In digital phone of A LISTENER (*content user*), the received data are de-quantized and speech is reproduced using the known synthesis model (14.2) (synthesized excitation is passed through the filter having given/sent LPC coefficients).

Signal processing during analysis (at speaker side) and synthesis (at listener side) is done in overlapping blocks of signal samples, lasting approximately 30 milliseconds. For sampling frequency $f_s = 8000$ Hz, the block has 240 samples and overlaps with 60 samples with the previous and next block (offset between blocks is equal to 180 samples).

14.3 Speech Compression Exercises

Exercise 14.1 (Synthesis with Wrong Excitation Period). Run the program `lab14_speech_compress.m`. Compare waveforms of input and output signals (different fragments). A few times listen to the input (`soundsc(x,fs)`) and to the output (`soundsc(s,fs)`). How big is the difference? Next, during speech synthesis set a constant value of speech period $T = 80, 50, 30, 0$. Simply uncomment the line:

```
% T = 80; % remove "%" and set: T = 80, 50, 30, 0
```

Now you hear what can be obtained from speech synthesis model when some "cheating" is done. First, for non-zero value of T, we hear a robot-like speech, then, for $T = 0$, a speech of a person with paralysis of vocal folds.

Exercise 14.2 (Using Different Prediction Orders). Return back to original program settings (comment the line `% T=80; % remove ...`). Repeat program execution for smaller linear prediction orders: `Np=10,8,6,4,2`. The coding should still work, worse but with speech understanding. Compare waveforms of original and synthesized speech. The difference is really very big. Set `Np=2` and `T=80`, `T=0` during speech synthesis. Compare input and output waveforms. Do you still understand the speaker? Wow! How good is our hearing system!

Exercise 14.3 (Detail Understanding Speech Coding Algorithm). Return back to original program settings. Set `ifigs=1`. Observe one-by-one four sub-figures, presenting input data and partial analysis/synthesis results for each speech segment:

- input speech fragment,
- its auto-correlation (see maximum for $k = 0$, then no peaks for noisy speech and several peaks for oscillatory speech),
- frequency response of vocal tract filter (see 4 resonant peaks, so-called formants),
- output (synthesized) speech fragment.

Compare value of speech period T, displayed on the screen, with index of the auto-correlation function maximum. Observe how T value is changing. In your opinion, is the calculated T value always correct? Well ... Try to improve the program ... in the next exercise.

Exercise 14.4 (Testing Correctness of the Voiced/Unvoiced Decision: Attempt of Improvement). Record your own speech samples for testing:

– voiced phonemes (vowels): "a," "e," "i," "o," "u," ...
– unvoiced phonemes (consonants): "c," "f," "h," "k," "s," "sh," ...

First, try to make the sound as constant as possible. Next, try to change *slightly* the intonation. Is the calculated T value constant and is it changing *slightly* as expected? Try to identify the source of the error in PITCH DETECTION part of the program ($T =$?). Modify the program if necessary.

Exercise 14.5 (Speech Parameters Quantization). Till now, we were testing qualitative limits of the LPC-based speech coding: all calculated parameters were coded in Matlab as double precision numbers. Perform quantization of vector of parameters [T, gain, a1, a2, ..., a10] to 8-bit numbers for the whole speech and store result to disc as binary numbers, for example, in WAV format, not as *.mat files. Then read the data back, display them, and play.
What is the compression ratio in this case? For 180 speech samples, 8-bit each, we require 1400 bits. For 12 parameters, 8-bit each, describing the same speech fragment, we require 96 bits. The compression ratio is about 15 times. Try to quantize 12 parameters using less than 96 bits. How does it work?

Exercise 14.6 (Improving Excitation of Vocal Tract Filter Using Inverse Filter and Residual Signal). Use your all voiced-speech recordings from task 14.4. Calculate correlation function of the whole signal for each phoneme. Use its coefficients for finding matrix **R** and vector **r**. Calculate prediction filter coefficients $[a_1, a_2, a_3, \ldots, a_{10}]$—used in Eqs. (14.2), (14.3). Build the inverse filter $H^{-1}(z) = \frac{1}{H(z)}$ from the synthesis filter $H(z)$—see Eq. (14.14).

Experiment No. 1 Pass the original speech through the inverse of synthesis filter $H^{-1}(z)$:
 y = filter([1 a1 a2 ...a10], 1, x) / G.
Observe the result—the so-called residual signal. Now pass the signal back through the synthesis filter $H(z)$:
 xs = G * filter(1, [1 a1 a2 ...a10], y).
You should obtain the original speech signal.

Experiment No. 2 Repeat experiment 1, but now try to QUANTIZE the residual signal. Observe that it is periodic and impulsive—very roughly similar to an excitation sequence [...0 0 0 1 0 0 0 1 0 0 0 1...], used in our lab14_ex_speech_compress.m program. At present, knowing the concept of residual signal, we see that our previous 0/1 excitation model was not perfect. Therefore use residual signal as better excitation for each phoneme. Try to quantize it or code it in some efficient way.

Listing 14.1: Speech compression and decompression using LPC-10 algorithm

```
1    % lab14_ex_speech_compress.m
2    % Speech compression using linear prediction coding
3    clear all; close all;
4    ifigs = 0;              %  0/1 - to show figures inside the processing loop?
5
6    % Parameters
7    Mlen=240;               % length of analyzed block of samples
8    Mstep=180;              % offset between analyzed data blocks (in samples)
9    Np=10;                  % prediction order (IIR-AR filter order)
10   where=181;             % initial position of the first voiced excitation
11   roffset=20;            % offset in auto-correlation function when find max
12   compress=[];           % table for calculated speech model coefficients
13   s=[];                  % the whole synthesized speech
14   ss=[];                 % one fragment of synthesized speech
15
16   % Read signal to compress
17   [x,fs]=audioread('speech.wav');    % read speech signal (audio/wav/read)
18   figure; plot(x); title('Speech'); % display it
19   soundsc(x,fs); pause              % play it on loudspeakers (headphones)
20   N=length(x);                      % signal length
21   bs=zeros(1,Np);                            % synthesis filter buffer
22   Nblocks=floor((N-240)/180+1);     % number of speech blocks to be analyzed
23
24   % MAIN PROCESSING LOOP
25   for  nr = 1 : Nblocks
26      % take new data block (fragment of speech samples)
27        n = 1+(nr-1)*Mstep : Mlen + (nr-1)*Mstep;
28        bx = x(n);
29      % ANALYSIS - calculate speech model parameters
30          bx = bx - mean(bx);                        % remove mean value
31        for k = 0 : Mlen-1
32          r(k+1)=sum( bx(1:Mlen-k) .* bx(1+k:Mlen) ); % calculate auto-correlation
33        end                            % try: r=xcorr(x,'unbiased')
34        if(ifigs==1)
35          subplot(411); plot(n,bx); title('Input speech x(n)');
36          subplot(412); plot(r); title('Its auto-correlation rxx(k)');
37        end
38        [rmax,imax] = max( r(roffset : Mlen) );      % find max of auto-correlation
39        imax = imax+(roffset-1);                     % its argument (position)
40        if ( rmax > 0.35*r(1) )  T=imax; else T=0; end % is the speech periodic?
```

```
41    % if (T>80) T=round(T/2); end              % second sub-harmonic found
42       T, % pause                              % display speech period T
43       rr(1:Np,1)=(r(2:Np+1))';                % create an auto-corr vector
44       for m=1:Np                              %
45          R(m,1:Np)=[r(m:-1:2) r(1:Np-(m-1))]; % build an auto-correlation matrix
46       end                                     % a = lpc(x,Np), levinson(x,Np)
47       a=-inv(R)*rr;                           % find coefficients of LPC filter
48       gain=r(1)+r(2:Np+1)*a;                  % find filter gain
49       H=freqz(1,[1;a]);                       % find filter frequency response
50       if(ifigs==1) subplot(413); plot(abs(H)); title('Filter freq response'); end
51    % compress=[compress; T; gain; a; ];       % store parameter values
52
53    % SYNTHESIS - generate speech using calculated parameters
54    % T = 80; % remove "%" and set: T = 80, 50, 30, 0
55       if (T~=0) where=where-Mstep; end         % next excitation=1 position
56       for n=1:Mstep                            % SYNTHESIS LOOP START
57          if( T==0)                             %
58             exc=2*(rand(1,1)-0.5); where=271;  % random excitation
59          % exc=0.5*randn(1,1); where=271;      % random excitation
60          else                                  %
61             if (n==where) exc=1; where=where+T; % excitation = 1
62             else exc=0; end                    % excitation = 0
63          end                                   %
64          ss(n) = gain*exc - bs*a;              % filtering excitation
65          bs = [ss(n) bs(1:Np-1) ];             % shifting the output buffer
66       end                                      % SYNTHESIS LOOP END
67       s = [s ss];                              % store the synthesized speech
68       if(ifigs==1) subplot(414); plot(ss); title('Synthesized speech s(n)'); pause, end
69    end
70
71    % Finished!
72    figure; plot(s); title('Synthesized speech'); pause
73    soundsc(s,fs)
```

14.4 Speech Recognition

The content of speech is *carried* in speech formants, i.e. resonant peaks present in envelope of speech spectrum. These peaks are changing during speaking (their positions and heights). Since speech is modeled by a filter, frequency response magnitude of the speech synthesis filter gives us also information what was said. Speech recognition is done in two steps:

- First, during training, algorithms are calculating and storing sequences of parameters describing speech spectral envelope of reference words, which is changing; typically, as descriptors of the shape, cepstral or melcepstral coefficients are used—they can be calculated directly from speech signal or from linear prediction filter coefficients which are used for speech synthesis,

- then, during recognition, sequence of parameters describing changing speech spectral envelope of an unknown word is calculated and compared with known sequences of parameters of reference words; a reference word having the most similar sequence of *spectral envelopes* is chosen as a found solution; for example, the dynamic time warping algorithm can be used for matching 2D matrices of different dimensions.

Above statements were confirmed, in part, by experiments performed by us in the previous section: we were drastically changing the speech synthesis filter excitation; after that, we could not recognize the speaker but the speech content was perceived well all the time. The filter itself was not being modified by us. We saw how the filter frequency response was tracking the formant structure of speech and observed spectral envelopes similar to the one presented in Fig. 14.7. We should remember that automatic speech recognition (ASR) systems should allow calculation of parameters describing local, peaky envelopes of speech spectrum, resulting from resonance characteristics of human vocal tract. The melcepstral coefficients are the most frequently used for this purpose.

Classical Speech Recognition Methodology In Listing 14.2, a simple Matlab program, for creating limited vocabulary of words and testing recognition of these words, is presented. First, values of our ASR system parameters are defined. Number of poles of filter transfer function Np is set to 10 (like in LPC-10 speech coders) and number of cepstral coefficient Nc, used for spectral envelope description, is set to 12. Time of recording, approximately 2–3 s, is defined by variable trec. Spectral envelope is calculated for speech fragment lasting twind=30 milliseconds with tstep=10 milliseconds offset between consecutive windows. Next, some examples of word vocabularies are given. The following main part of the program is divided into two parts: creation of vocabulary and word recognition.

In the first part, vocabulary words are recorded and processed one by one. This section starts with recording of words to be recognized (mywavrecorder() function). Then, simple voice activity detection (VAD) is performed, and silence parts, present at the beginning and at the end of the recordings, are removed (better or worse, you could improve it). Next, matrix Cref of word cepstra is calculated, which describes how resonant peaks are changing in speech spectrum in time. The matrix is stored into the archive CCref.

In the second part, one is checking recognition of any word from the vocabulary. A word is chosen, recorded once more, cleared from silence fragments, accompanied by its cepstral matrix Cx, and compared with matrices from the archive CCref. Dynamic type warping (DTW) algorithm is used for finding the shortest path in the grid of Euclidean distances between each pair of cepstra: any cepstrum from Cx and any cepstrum from Cref, being part of the CCref archive. Word from the vocabulary having the smallest accumulative distance is chosen as a recognized one.

Listing 14.2: Matlab program for speech recognition using cepstral coefficients and dynamic time warping

```matlab
 1  % lab14_ex_speech_recognition
 2  % Isolated words recognition
 3  clear all; close all;
 4
 5  global Mlen Mstep Np Nc      % global parameters described below
 6  fs = 8000;                   % sampling frequency (Hz)
 7  Np = 10;                     % number of poles in prediction filter transfer function
 8  Nc = 12;                     % number of calculated cepstral coefficients
 9  trec = 2;                    % time of recording (s), 2 or 3
10  twind = 30;                  % time of observation window (ms)
11  tstep = 10;                  % time offset between consecutive windows (ms)
12  Mlen  = (twind*0.001)*fs;    % number of window samples (in one data frame)
13  Mstep = (tstep*0.001)*fs;    % number of offset samples
14
15    words = {'zero'; 'one'; 'two'; 'three'; 'four' },
16  % words = {'start'; 'left'; 'right'; 'forward'; 'backward'; 'stop' },
17  % words = {'New York'; 'London'; 'Paris'; 'Berlin'; 'Warsaw' },
18  % words = {'John Brown'; 'Mark Smith'; 'Tom Green'; 'Betty Lewis'; 'Ann Margaret' },
19    M = length( words );       % number of words
20
21  % Creating database of reference recordings
22  Cwzr = [];
23  for k = 1 : M
24      disp( strcat('Press any key and AFTER 1 SEC say reference word:', words(k)) );
25  %   wz = wavrecord(trec*fs, fs, 1); % no of samples, sampling freq, 1 channel
26      ref = mywavrecord(trec, fs, 1); % no of seconds, sampling freq, 1 channel
27      ref = silence( ref, fs );        % remove silent parts (beginning, end)
28      [Cref, Nframes] = cepstrum( ref ); % calculate matrix of cepstral coeffs
29      [Nrow, Ncol] = size(Cref);        % read its dimensions
30      CCref(1:Nrow,1:Ncol,k)=Cref ;  Nref(k)=Nframes;  % append it to database
31  end
32
33  % Recognition of words having reference pattern in the database
34  disp('Now it is time for recognition. Press any ...'); pause
35  while( input(' Recognition (1/0) ? ') == 1 )
36      words
37      disp( 'Press any key and AFTER 1 SEC say a word to be recognized' );
38  %   x = wavrecord(trec*fs, fs, 1);  % no of samples, sampling freq, 1 channel
39      x = mywavrecord(trec, fs, 1);   % no of seconds, sampling freq, 1 channel
40      x = silence( x, fs ); pause     % remove silent parts (beginning, end)
41      Cx = cepstrum( x );             % calculate matrix of cepstral coeffs
42      num = dtw( Cx, CCref, Nref ); % dynamic type warping of matrices
43      disp( strcat( 'Recognized word: ', words(num) ) ); % recognition result
44  end
```

Voice Activity Detection: Silence Removal In Listing 14.3, an exemplary program for voice activity detection in the recorded word is presented, very simple and for sure not perfect. First, the mean value is subtracted, and signal normalization takes place, i.e. division by the signal maximum absolute value. Next, energy of each sample (E) and energy of the whole signal ($Eall$) are computed. Sample energies are sorted in descending order (sort()) and accumulated/integrated (cumsum()). Then, the first sample of accumulated sum (Ecum) bigger than the chosen energy threshold (thres*Eall) is found (Neng). This way all speech high-energy samples are found, with energy approximately equal to 100*tres=99 percentage of whole signal energy (Eall). Finally, high-energy samples are sorted according their indexes, from the lowest to the highest one, and only speech samples from ind(1) to ind(end) are taken. Of course, many other different solutions are possible. You can try to find a better one.

Exercise 14.7 (Voice Activity Detection). Propose better method for voice activity detection, more robust against impulsive disturbances occurring in the beginning and end of a recording.

Listing 14.3: Matlab function for silence removal from speech recordings

```
1  function y = silence(x, fs)
2  % removing silence from the beginning and from the end of recorded speech
3
4  N = length(x); n = 1 : N;                    % signal length
5  x = x-mean(x); xn = x/max(abs(x));           % mean value removal, normalization
6  thres = 0.99;                                % threshold (percentage of energy)
7  E = xn.^2;                                    % energy of each signal sample
8  Eall = sum( xn.^2 );                          % energy of all signal samples
9  [Esort ind] = sort(E,'descend');             % sorted energy of samples
10 Ecum = cumsum( Esort );                       % cumulative energy of samples
11 Neng = max( find(thres*Eall >= Ecum) );      % find last sample > threshold
12 [ind dummy] = sort( n(ind(1:Neng)) );        % indexes of high-energy samples
13 ny = ind(1) : ind(end);                       % indexes the lowest - the highest
14 y = x( ny );                                  % copy only silence-free speech
15 if(1) % Figure for verification
16    yz = zeros(N,1); yz( ny(1):ny(end),1) = y;                % append zeros
17    plot(n,x,'b-',ny,y+2,'r.-'); axis([1,N,min(x),max(x)+2]); % plot in one
18    grid; title('Before & After Signal Cutting'); figure(1)   % figure
19 end
```

Calculation of Speech Descriptor/Feature: Matrix of Cepstral Coefficients When we are sure that only samples of voice were selected from the recording, it is time to start to think about finding a few numbers, describing the speech spectrum envelope. However at present, in the era of deep learning, also the whole not pre-processed speech spectra could be used as feature vectors. The LPC filter coefficients **a** define a speech synthesis filter which has a peaky frequency response with a speech formant structure. However the coefficients **a** are not directly connected with the filter frequency response shape and could be used directly for this shape recognition. In ASR systems, cepstral coefficients are used as spectrum envelope descriptors. They can be calculated from coefficients **a**. Signal cepstrum is defined as follows:

$$c(n) = \frac{1}{N} \sum_{k=0}^{N-1} \ln \underbrace{\left| \sum_{m=0}^{N-1} w(m)x(m)e^{-j2\pi km/N} \right|}_{C(k)} e^{\pm j2\pi kn}. \tag{14.17}$$

The speech fragment $x(n)$ is multiplied with Hamming window function $w(n)$. Then, the discrete Fourier transform is performed, and logarithm, natural or decimal, of absolute value of each Fourier coefficient is calculated. Finally, one more Fourier transform is performed. We can shortly say that cepstrum is Fourier transform performed over the logarithm of absolute value of the Fourier transform. In other words, the second Fourier transform performs frequency analysis of the envelope of the first Fourier transform. Its first coefficients tell us about low-frequency components of this envelope, i.e. about its smooth peaks called formants. If two words have very similar 12–16 cepstral coefficients, it means that their spectral envelopes are also similar, and the words are very similar ... may be they are even the same.

It is also important to stress usage of logarithm in the cepstrum definition. At present, we know that speech can be synthesized by exciting a speech-dependent digital filter in right way. Filter output is equal to convolution of its input with filter impulse response. Therefore, spectrum of the real $(X(f))$ or synthesized $(S(f))$ speech is equal to multiplication of the excitation spectrum $E(f)$ with the filter frequency response $H(f)$:

$$X(f) = S(f) = E(f) \cdot H(f). \tag{14.18}$$

When we put Eq. (14.18) into Eq. (14.17), we get

$$c(n) = F^{-1}\left(\ln|H(\Omega)E(\Omega)|\right) = F^{-1}\left(\ln|H(\Omega)|\right) + F^{-1}\left(\ln|E(\Omega)|\right). \tag{14.19}$$

What do we see? Calculated cepstrum is summation of two cepstra. The first of them, $F^{-1}(H(\Omega))$, comes from filter frequency response which is a low-frequency one. The second of them is a cepstrum of filter excitation signal. For voiced speech, the spectrum of sequence of impulses is also a sequence of impulses, i.e. it is periodic. Therefore, the second cepstrum, as a Fourier transform of periodic spectral components, will show us the time repetition period of this components, i.e. the fundamental speech frequency (pitch frequency). Figure 14.9 presents Fourier spectrum of a 240-sample-long fragment of the phoneme "a" (up) and cepstrum of this signal (down). Over the DFT spectrum, two estimations of its envelope are overlaid. They were calculated using the linear prediction coefficients (better fit) and low-frequency cepstral coefficients (a little bit worse fit). We see that the first coefficients of the cepstrum are important for us since they give us information about spectrum envelope shape. In turn, the observed peak for $n \approx 70$ tells us about pitch frequency value. Wow!

Q cepstral coefficients can be calculated from Eq. (14.17) but also from P linear prediction coefficients, using the following equations:

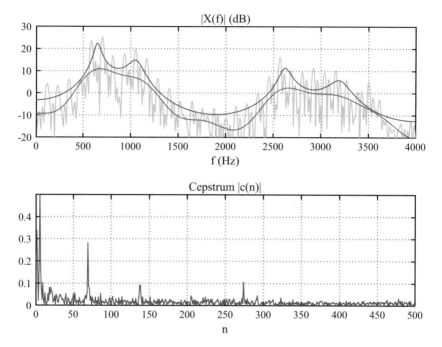

Fig. 14.9: DFT spectrum (up) and cepstrum (14.17) (down) calculated for 240 samples of phoneme "a" from Fig. 14.5, sampled at 8000 samples per second. Over the DFT spectrum, its envelopes are overlaid, estimated using LPC and cepstral coefficients

$$c_1 = a_1, \tag{14.20}$$

$$c_k = a_k + \sum_{m=1}^{k-1} \frac{m}{k} c_m a_{k-m}, \quad 2 \le k \le P, \tag{14.21}$$

$$c_k = \sum_{m=1}^{k-1} \frac{m}{k} c_m a_{k-m}, \quad P+1 \le k \le Q, \tag{14.22}$$

$$w_k = 1 + \frac{Q}{2} \sin\left(\frac{\pi k}{Q}\right), \quad 1 \le k \le Q, \tag{14.23}$$

$$cw_k = c_k \cdot w_k, \quad 1 \le k \le Q, \tag{14.24}$$

where w_k denotes weighting coefficients and cw_k cepstral coefficient after weighting.

In Listing 14.4, calculation of cepstral coefficients using different methods is presented. Parameters `Mlen=240`, `Mstep=80`, `Np=10`, `Nc=12` are global variables with values defined in the main ASR program 14.2. Signal pre-emphasis is done (amplifying higher frequencies). LPC coefficients are calculated as during speech compression. Cepstral coefficients are calculated from **a**, from Eq. (14.17) and using Matlab function `rceps()`.

Exercise 14.8 (Calculation of Cepstral Coefficients). Use program 14.4. Modify it. In cepstrum calculation loop, compute `c`, `c1`, `c2` coefficients and display them in one figure. For recordings of single voiced phonemes "a," "e," "i," ..., observe whether cepstral coefficients are similar or not, and how much they are changing. Set $N_c = 128$. Repeat observation for different phonemes and check whether the upper limit $N_c = 12$ is correct. If it is too small, increase N_c value, e.g. to 16.

Listing 14.4: Matlab function for calculation of matrix of cepstral coefficients for recorded speech

```
1   function [Cx, Nframes] = cepstrum( x )
2   % calculation of matrix of cepstral coefficients
3
4       global Mlen Mstep Np Nc                  % global parameters
5       N = length(x);                          % number of signal samples
6       Nframes = floor( (N-Mlen)/Mstep+1);     % number of frames of samples
7       m = 1:Nc; w = 1 + Np*sin(pi*m/Nc)/2;    % weighting coefficients
8       x = x - 0.9375*[0; x(1:N-1)];           % initial filtration (pre-emphasis)
9       Cx = [];                                % initialization
10  % MAIN LOOP
11      for  nr = 1 : Nframes
12          % Taking a new frame (block) of speech samples
13              n = 1+(nr-1)*Mstep : Mlen + (nr-1)*Mstep; bx = x(n);
14          % Initial pre-processing
15              bx = bx - mean(bx);             % remove mean value
16              bx = bx .* hamming(Mlen);       % multiply with window function
17          % Calculation of cepstral coefficients from prediction filter coeffs
18              for k = 0 : Np                  % # auto
19                  r(k+1) = sum( bx(1 : Mlen-k) .* bx(1+k : Mlen) ); % # correlation
20              end                             % # samples
21              rr(1:Np,1)=(r(2:Np+1))';        % auto-correlation vector
22              for m = 1 : Np                  %
23                  R(m,1:Np)=[r(m:-1:2) r(1:Np- (m-1))]; % auto-correlation matrix
24              end                             %
25              a = inv(R)*rr; a = a';          % coefficients of prediction filter
26              a = [a zeros(1,Nc-Np)];         % appending zeros
27              c(1) = a(1);                    % calculation of cepstral coeffs
28              for m = 2 : Nc                  % from LPC coefficients
29                  k = 1:m-1; c(m) = a(m) + sum(c(k).*a(m-k).*k/m); % cc=?
30              end
31          % Calculation of cepstral coefficients from speech Fourier spectrum
32          % c1 = real( ifft( log( abs(fft(bx)).^2))); % do it yourself
33          % c2 = rceps(bx);                   % Matlab function BAD
34          % c = c1; c = c'; c = c(2:Nc+1);    % choice: c1 or c2?
35          % Weighting cepstral coefficients
36              cw = c .* w;                    % multiply with weights
37              Cx = [Cx; cw];                  % storing result
38      end
```

Speech Recognition: Matching Matrices of Cepstral Coefficients At this stage, we have a collection of reference patterns of cepstral coefficients for words from the vocabulary and a CC pattern for unknown spoken word. The patterns have a form of matrices with word cepstra stored in rows. Since a word duration can be different, number of rows of each matrix can be different also. The second dimension, number of column is the same and equal to the number of cepstral coefficients $N_c = 12$, used for recognition. For each pair of matrices {*unknown, reference*} Euclidean distance $d(n_x, n_r)$ between each pair of cepstra is calculated, i.e. n_x-th unknown and n_r-th reference (r is used as reference denotation):

$$d(n_x, n_{ref}) = \sqrt{\sum_{k=1}^{N_c} \left(C_x(n_x, k) - C_{ref}(n_{ref}, k)\right)^2}, \quad n_x = 1 \ldots N_x, \quad n_{ref} = 1 \ldots N_{ref}.$$

(14.25)

The *distance* matrix created this way has N_x columns and N_r rows. In Fig. 14.10, its example is shown, in which case the cepstra of one word are treated the same time as unknown and reference cepstra. We observe the expected matrix symmetry and distance equal to zero for diagonal elements. Having such matrix for different words, the shortest accumulated path from the element d(1,1) to the element d(N_x, N_r) is calculated and is shown in Fig. 14.11. We are starting from the bottom-left corner with element d(1,1) and are moving only one position: right (10), up (10), or diagonally (11), being aware of constraints (region surrounded with dashed lines) limiting possible steps. Step is done in direction of an element having the smaller value (for diagonal move, value of an element is doubled). Matrix elements lying on the chosen

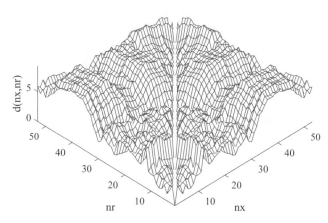

Fig. 14.10: Exemplary matrix of Euclidean errors $d(n_x, n_r)$ between cepstra of the Polish word "dwa." The same word is treated as unknown and reference word. Matrix symmetry is observed [15]

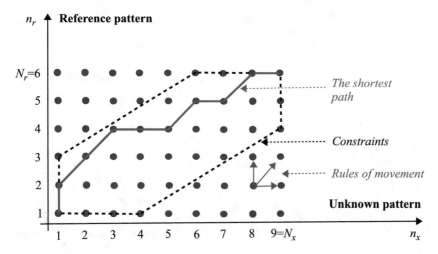

Fig. 14.11: Graphical illustration of dynamic time warping of two sequences of features. Calculation of the shortest accumulative path (from bottom-left corner to top-right corner) in the matrix of incongruity errors $d(n_x, n_r)$ (errors between cepstra of unknown word [horizontally] and reference word [vertically]) [15]

path are accumulated and treated as the shortest path measure, telling us about dissimilarity of two words. The lower this value is, the smaller is difference between words being compared. Finally, having accumulated distance measures between unknown and all reference patterns, as recognized is chosen this pattern which has the smallest accumulated distance. Equation, presented below, describes in detail the movement and accumulation strategy:

$$g(n_x, 1) = \sum_{k=1}^{n_x} d(k,\ 1), \quad n_x = 1,\ 2, \ldots, N_x, \qquad \text{(accumulation in in first row)}$$

$$g(1, n_r) = \sum_{k=1}^{n_r} d(1,\ k), \quad n_r = 1,\ 2, \ldots, N_r, \qquad \text{(accumulation in in first column)}$$

$for \quad n_x = 2, 3, \ldots, N_x :$ (vertically)

    $for \quad n_r = n_r^{(down)}(n_x), \ldots,\ n_r^{(up)}(n_x) :$ (horizontally)

$$g(n_x, n_r) = \min \begin{cases} g(n_x, n_r - 1) + d(n_x, n_r) & \text{(up)} \\ g(n_x - 1, n_r - 1) + 2 \cdot d(n_x, n_r) & \text{(diag)} \\ g(n_x - 1, n_r) + d(n_x, n_r) & \text{(right)} \end{cases}$$

where $2 \leq n_r^{(down)}(n_x)$ and $n_r^{(up)}(n_x) \leq N_r$ denote constraints of vertical movement, which depend upon n_x value. In order to obtain result independent from matrix sizes, at the end accumulated distance is divided by $\sqrt{N_x^2 + N_r^2}$.

Exercise 14.9 (Dynamic Time Warping). Use program 14.5. Modify it. Uncomment line number 16 and comment line 17. Before calculation of accumulated (accum) distance g(), display the matrix $d(n_x, n_r)$, for example, using this code:

```
Xax = ones(Nx,1)*(1:Nr); Yax = (1:Nx)'*ones(1,Nr);
mesh(Xax,Yax,Z); xlabel('nx'); ylabel('nr');
zlabel('d(nx,nr)');
axis tight; V=[-45 65]; view(V);
```

Repeat the same recognition experiment for different values of coefficient Q—width of path limiting area, for example, for 0.1, 0.2, 0.5, 0.75. Does it have influence on recognition efficiency? Can you generate a figure similar to Fig. 14.11?

Listing 14.5: Matlab function for dynamic time warping of unknown cepstral matrix with reference cepstral matrices of words being recognized

```
1   function [ number ] = dtw( Cx, Cref, Nref)
2   % word recognition via dynamic time warping (DTW) of matrices with cepstral coeffs
3
4   [ Nx, Nc ] = size(Cx);          % number of: cepstral vectors, cepstral coeffs
5
6   for nref = 1 : length(Nref)     % compare Cx with Cref of all reference patterns
7     % Calculate distance d(ns, nr) between signal (ns) and ref (nr) cepstra
8       Nr = Nref( nref );                     % number of vectors with ref cepstra
9       Q = round( 0.2 * max(Nx,Nr) );         % coefficient of path width
10      d = Inf*ones(Nx,Nr);                   % init. of distance matrix
11      tg= (Nr-Q)/(Nx-Q);                     % init. of angle tangent
12      for nx = 1:Nx                          % for each cepstrum of the word
13          down(nx) = max( 1, floor(tg*nx-Q*tg)); % lower limit
14          up(nx)   = min( Nr, ceil(tg*nx+Q));    % upper limit
15        % for nr = 1 : Nr
16          for nr = down(nx) : up(nx)         % for each cepstrum of ref word
17              d(nx,nr)=sqrt( sum( (Cx(nx, 1:Nc) - Cref(nr, 1:Nc, nref)).^2 )); % distance
18          end
19      end
20    % Calculation of accumulated (accum) distance g()
21      g = d;                                 % initialization
22      for nx = 2:Nx, g(nx,1) = g(nx-1,1) + d(nx,1); end % accum of 1st column
23      for nr = 2:Nr, g(1,nr) = g(1,nr-1) + d(1,nr); end % accum of first row
24      for nx = 2:Nx                          % accum vertically (word)
25          for nr = max( down(nx), 2 ) : up(nx)  % accum horizontally (ref)
26              dd = d(nx,nr);                 % distance: word "nx", ref "nr"
27              temp(1) = g(nx-1,nr) + dd;     % go up
28              temp(2) = g(nx-1,nr-1) + 2*dd; % go over diagonal (right up)
29              temp(3) = g(nx,nr-1) + dd;     % go right
30              g(nx,nr) = min( temp );        % choose accumulated value
```

```
31        end
32      end
33      glob(nref) = g(Nx,Nr)/sqrt(Nx^2+Nr^2)      % accum value of the "shortest" path
34    end
35    [ xxx number ] = min( glob );                % number of ref with the lowest acc value
```

14.5 Summary

Professionally, as a researcher, I am not involved in high-tech speech processing applications. But I like to deal with speech signals and teach basics of signal processing on their example. Following are the most important key points dealt in this chapter:

1. Speech signal fragments are divided into voiced ones, when vocal folds are opening and closing, and into unvoiced ones, when vocal folds are open all the time. If an auto-correlation function of speech fragment has a high local maximum for some signal shift, it means that the signal is repeating with period equal to the shift value and vocal folds are working with this period, called a *pitch period*. Otherwise, the vocal folds are opened. Signal from vocal folds is speaker dependent.

2. Vocal folds signal excites the vocal tract acoustical filter, consisting of oral and nasal cavities, being a resonating *tube*. Frequencies and strength of its resonances characterize a speech content. In speech processing language, we tell that frequency response of speech synthesis filters has 4–5 peaks called *formants*.

3. In speech compression algorithms, a speech signal is divided into segments and analyzed. For each fragment, parameters describing excitation and vocal tract filter are computed and then stored or transmitted. During decompression, speech signal is synthesized piece-by-piece; synthesized excitation is passed through a digital IIR filter imitating the vocal tract. Speech synthesis model and parameters calculation are so good that we do not hear the difference between original and synthesized speech.

4. Synthesis filter coefficients are found using linear prediction method. The speech compression is an example of linear prediction coding (LPC) of signals. In order to better track the filter change, analyzed speech fragments should overlap more. Alternatively, filter intermediate states can be found during interpolation of *line spectrum pairs (LSP) or line spectrum frequencies (LSF)*, calculated from LPC coefficients. We do not interpolate LPC coefficients, since interpolation of the filter frequency response is our goal.

5. To have higher compression ratio, calculated values of parameters of speech synthesis model should be quantized. In order to reduce influence of

this operation on the quality of synthesized signal, the synthesis filter uses typically *lattice structure* which is more robust to filter weights rounding (LPC coefficients are transformed to *reflection coefficients* which are quantized).

6. Content of a speech (*what was said*) is present in speech formant (resonance) structure which is changing in time. To recognize a word, we have to calculate its parameters describing change of its spectrum envelope and compare them with parameters of reference words. Typically, cepstral coefficients are used in this purpose, being result of discrete Fourier or cosine transform, performed over logarithm of speech spectrum magnitude. To be more precise, melcepstral ones, i.e. cepstral but calculated for mel acoustic scale.

7. In the simplest automatic speech recognition (ASR) systems, matrices of melcepstral coefficients, calculated for unknown words, are compared with matrices of melcepstral coefficients, calculated for reference words, using dynamic time warping (DTW) method. In more advanced methods, hidden Markov models (HMMs) are used for decision-making. At present, it behooves only to use *deep learning* artificial intelligence approach technology.

14.6 Private Investigations: Free-Style Bungee Jumps

Exercise 14.10 (Linear Spectrum Pairs and Speech Synthesis Filter Interpolation).** Let us assume that we have a brilliant idea of interpolating *states* of speech synthesis filter. Doing this, we expect to obtain *smoother* synthesized signal. But how to do it? Interpolate separately every LPC coefficient? No. The correct solution is as follows. We calculate coefficients of two filters $P(z)$ and $Q(z)$:

$$P(z) = 1 + (a_1 + a_{10})z^{-1} + (a_2 + a_9)z^{-2} + \ldots + (a_{10} + a_1)z^{-10} + z^{-11},$$
$$Q(z) = 1 + (a_1 - a_{10})z^{-1} + (a_2 - a_9)z^{-2} + \ldots + (a_{10} - a_1)z^{-10} - z^{-11},$$

for two consecutive speech segments. Then we find zeros of their transfer functions (always lying on unitary circle) and interpolate them, separately for $P(z)$ and $Q(z)$, between two speech *states*. Finally, we use $P_i(z)$ and $Q_i(z)$, the interpolation result, and calculate

$$A_i(z) = \frac{P_i(z) + Q_i(z)}{2},$$
$$H_i(z) = \frac{G_1 + G_2}{2A_i(z)},$$

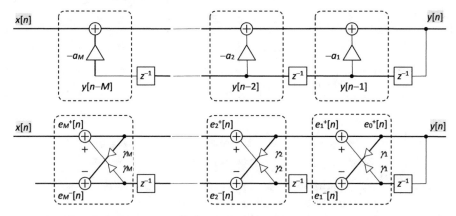

Fig. 14.12: Speech synthesis filter in classical (up) and lattice (down) version. Gain G is missing. Coefficients γ are calculated from coefficients **a** [15]

where $H_i(z)$ is a transfer function of the interpolated filter. For sure, during long winter evenings, you will find an hour or two to check efficiency of this approach.

Exercise 14.11 (** Speech Synthesis Lattice Filter). In high-tech speech compression algorithms, the speech synthesis filter is typically implemented in a lattice structure which is more robust to quantization of filter coefficients. Classical and lattice implementation of the IIR speech synthesis filter, with missing gain G, is presented in Fig. 14.12. Having the LPC coefficients **a**, one should calculate coefficients **g** (γ in Fig. 14.12) of the lattice filter, quantize them, and synthesize speech using lattice processing scheme. More practical details are given in Listing 14.6. If you are interested in this topic, try to apply this concept in our LP-based speech encoder and decoder program 14.1.

Listing 14.6: Lattice version of speech synthesis filter

```
1    % lab14_ex_lattice.m
2      clear all; clf;
3
4    % LPC filter coefficients
5      a = [ 1 -0.9 0.64 -0.576 ]; N=length(a); aa=a(2:N); P=N-1;
6    % Reflection coefficients g(i) of lattice filter
7      ax(P,1:P)=aa(1:P);
8      for i=P:-1:1
9          g(i)=-ax(i,i);
10         for j=1:i-1
11             ax(i-1,j)=(ax(i,j)+g(i)*ax(i,i-j))/(1-g(i)^2);
12         end
13     end
14     g, pause
15   % Filtered signal
16     Nx=100; n=0:Nx-1; x=sin(2*pi*n/12+pi/4);
```

```
17  % Lattice filtration
18    e1=zeros(1,N);
19    for n=1:Nx
20      e0(N)=x(n);
21      for k=N:-1:2
22        e0(k-1) = e0(k)+g(k-1)*e1(k-1);
23        e1(k)   = -g(k-1)*e0(k-1)+e1(k-1);
24      end
25      e1=[ e0(1) e1(2:N) ];
26      y(n)=e0(1);
27    end
28  % Figures - comparison with standard filtration algorithm
29    yref = filter(1,a,x);
30    figure; plot(y); title('Filter output');
31    figure; plot(y-yref); title('Difference with filter(1,a,x)');
```

Exercise 14.12 (** Melcepstrum-Based ASR). In professional ASR systems, mel-cepstrum is used instead of cepstrum for spectrum envelope description. The signal is windowed as before, but after DFT the spectral coefficients are squared and locally weighted. Since we hear higher frequencies worse, frequencies of DFT coefficients are converted into mel scale having psycho-acoustical origin (m—in mels, f—in herz):

$$m = 2595 \cdot \log\left(1 + f/700\right), \quad f = 700 \cdot \left(10^{m/2595} - 1\right). \tag{14.26}$$

In the mel scale, the DFT local weights are triangular functions having width of 200 or 300 mels that are shifted up by 100 or 150 mels, respectively, to 2100 mels (about 4000 Hz). After DFT coefficients weighting natural logarithm of the resultant spectrum is computed and DCT-II transform is performed:

$$c_k = \sqrt{\frac{2}{L}} \cdot \sum_{l=1}^{L} \ln\left(\tilde{S}(l)\right) \cos\left(\frac{\pi k}{L}(l - 1/2)\right), \quad k = 1, 2, 3, \ldots, q \tag{14.27}$$

Are you interested in extension of our ASR system to mel-cepstrum-based one? Bravo!

Further Reading

1. T.P. Barnwell, K. Nayebi, C.H. Richardson, *Speech Coding: A Computer Laboratory Textbook* (Wiley, New York, 1996)
2. W.C. Chu, *Speech Coding Algorithms. Foundation and Evolution of Standardized Coders* (Wiley-Interscience, Hoboken, 2003)
3. J. Deller, J.G. Proakis, J.H.L. Hansen, *Discrete-Time Processing of Speech Signals* (Macmillan, New York, 1993)

4. S. Furui, *Digital Speech Processing, Synthesis, and Recognition* (M. Dekker, New York, 2001)
5. B. Gold, N. Morgan, *Speech and Audio Signal Processing* (Wiley, New York, 2000)
6. L. Hanzo, F.C.A. Somerville, J.P. Woodard, *Voice Compression and Communications. Principles and Applications for Fixed and Wireless Channels* (Wiley Interscience, New York, 2001)
7. A.M. Kondoz, *Digital speech: Coding for Low Bit Rate Communication Systems* (Wiley, Chichester, 1994)
8. M.H. Kuhn, H.H. Tomaszewski, Improvements in isolated word recognition. IEEE Trans. Acoust. Speech Signal Process. **31**(1), 157–167 (1983)
9. I.V. McLoughlin, *Speech and Audio Processing. A Matlab-Based Approach* (Cambridge University Press, Cambridge, 2016)
10. P.E. Papamichalis, *Practical Approaches to Speech Coding* (Prentice Hall, Englewood Cliffs, 1987)
11. T.F. Quatieri, *Discrete-Time Speech Signal Processing* (Prentice Hall, Upper Saddle River, 2001)
12. L.R. Rabiner, R.W. Shafer, *Digital Processing of Speech Signals* (Prentice Hall, Upper Saddle River, 1978)
13. P. Vary, R. Martin, *Digital Speech Transmission. Enhancement, Coding and Error Concealment* (Wiley, Chichester, 2006)
14. S.V. Vaseghi, *Multimedia Signal Processing. Theory and Applications in Speech, Music and Communications* (Wiley, Chichester, 2007)
15. T.P. Zieliński, *Cyfrowe Przetwarzanie Sygnałów. Od Teorii do Zastosowań (Digital Signal Processing. From Theory to Applications)* (Wydawnictwa Komunikacji i Łączności (Transport and Communication Publishers), Warszawa, Poland, 2005, 2007, 2009, 2014)

Chapter 15
Audio Compression

Music has charms to soothe the savage breast, especially this
one which is streamed in the Internet :-)

15.1 Introduction

In this chapter introduction will be a little bit longer than usual. I have coded MP2 encoder and decoder in assembler on a DSP fixed-point processor. It was a long run, a lot of hard work and fantastic, new, engineering experience. So now, it is difficult for me to close the subject in one paragraph only.

Audio signal compression is completely different than speech compression. It sounds very strange but it is true. In speech compression we are modeling acoustical voice production, the vocal folds/cords excitation and the vocal tract filter. In audio compression we are modeling human hearing system. Why? Because, at first, in audio compression we can not model a sound source without knowledge about its origin, and a set of different sound sources is infinite. At second, since a listener of compressed audio is a human being, not a bat having completely different hearing system then we, people, and therefore, knowing human hearing system, we can avoid coding of sub-sounds which are not heard by us. It sounds reasonably. But what we can win knowing the psycho-acoustics of our ears and mind, *the commander center*? A lot, because single frequency tone can *mask* in our head other tones and narrow-band noises having frequency values close to it (*tone is masking tone or noise*). And vice versa, narrow-band noise can do the same: mask tones and noises lying close to it in frequency. When something is psycho-acoustically *masked*, is the sense to loose bits for its coding? No, it is not. Therefore, a psycho-acoustical model of human hearing system takes the central part in each modern audio coding algorithms. The Fourier transform (FFT) of the sound samples is performed in it and masking curves for all found tones and narrow-band noises are calculated. Then all the masking curves are combined, and added to absolute threshold of hearing. As a result a signal-to-mask-ratio (SMR) curve is obtained, having the extraordinary

© Springer Nature Switzerland AG 2021
T. P. Zieliński, *Starting Digital Signal Processing in Telecommunication Engineering*, Textbooks in Telecommunication Engineering,
https://doi.org/10.1007/978-3-030-49256-4_15

significance in audio compression. Why? Since only these frequency components which are above the final masking curve will be heard by us! Having this
information, an audio compression algorithm allocates bits only for them.

Stop! How it is possible? How we can give bits separately to some signal
frequency components having all components mixed/added together! An now,
the second very important hero of the audio compression story appears on the
stage: sub-band signal decomposition, realized as a set of parallel filters separating different frequency sub-bands, or, an orthogonal transform, namely the
modified discrete cosine transform (MDCT), doing the same but in different
manner. When the signal *big river* of *all-together-now* samples is converted
into many small *creeks/streams* of samples belonging only to some frequency
bands, signal quantization takes place. First, bits are allocated to frequency sub-
band samples proportionally to their psycho-acoustical significance. In consequence, some frequency *creeks/streams* are given many bits and some of them
(*oh, please, do not do it!*) are obtaining ZERO bits. Then: (1) normalization of
sub-band samples to the range $[-1, 1]$ is done, and (2) their quantization takes
place. When fixed-point processors and fractional binary number representation is used, in the last stage, first, samples are multiplied by some constants,
depending on the number of allocated bits, and, next, left is only the specified number of most significant bits of the result. After that some additional
loss-less compression method can be applied (for example, Huffman coding as
in the MP3 standard). Finally, the bit-stream is formed, as *a train with many
carriages*, in compression language: with many *frames*. Each frame consists of
the *header*, describing the content, and the content itself: bits describing performed normalization and quantization (here we have indexes to many tables
from standards) and quantized samples of many audio sub-band *creeks/streams*.
The bit-stream is transmitted or stored.

Ufff.

An audio decoder, first, should synchronize with data frames (*train carriages*), read header and find information how the decoding should be done (frequencies, channels, compression levels, bit-rates, etc.). Then it decodes parameters describing bit allocation and normalization, and decodes audio samples of
each frequency sub-band. Finally, all sub-band data should be up-sampled and
added: all small *creeks* are joining and *a big audio river* is restored. The original audio samples are not perfectly reconstructed because there is no sense of
such reconstruction: we do not hear the difference between the original and the
decoded sound, when all components above the masking threshold have been
fed with sufficient number of bits. It is *a magic* of psycho-acoustical audio coding! Our sense of hearing is not perfect and some audio frequency components
can be simply . . . (*oh, please, do not* . . .) REMOVED, REJECTed, CANceled,
. . . I am sorry.

From historical point of view, audio signals were first coded using the AD-
PCM method. (PCM) denotes a pulse code modulation in which audio wave-

forms are sampled and are written as a sequence of 8/16-bit samples. (D) means *differentially*: the difference between a sample and its prediction based on previous samples was quantized. (A) denotes *adaptive*: the next sample prediction was being changed in adaptive way. This compression method works well when 16-bit audio samples are coded as 4-bit ones, i.e. 4-times compression ratio is offered easily.

Next begun the era of psycho-acoustical audio coding. The second worldwide standard of audio coding, namely the MPEG-audio standard, was introduced in the beginning of 90-ties of XX century, almost 30 years ago. In MPEG, an audio stream is first processed by bank of 32 filters working in parallel, dividing a full-band sound into 32 frequency sub-bands. A prototype low-pass filter with 512 weights is modulated by 32 cosines which are shifting the filter frequency response to higher frequencies (sub-bands). After signal filtration by 32 filters, sub-band signals are sub-sampled 32 times, i.e. only each 32-th sample is left in each of 32 sub-channels. This way block of 1152 input samples is converted to 32 sub-bands having 36 samples each. In parallel 1024-point FFT is performed and psycho-acoustical analysis is done. Using its results, the adaptive bit allocation for 32 sub-band signals is performed. Then, sub-band samples are normalized, quantized according to bit allocation and put into the bit-stream. In MP3 layer each of 32 sub-bands is further divided adaptively into additional 6 or 18 sub-sub-bands (*sub-creeks*), respectively, for noisy and tonal parts. Decision, 6 either 18, is made using calculated perceptual entropy of a sound spectrum. At the last stage audio bits are losslessly coded using Huffman coder with per-calculated code-books.

Advanced audio coding (AAC) is the third the most important audio compression standard. It was proposed as a part of MPEG-2 extension in the middle of 90-ties of XX century and was improved the years later, about year 2005, with high-efficiency extensions HE-1 and HE-2. In AAC a different technique for signal splitting into sub-bands is used. A 256-point (for noisy-like audio parts) or 2048-point (for tonal-like audio parts) sine window is shifted along the signal with 50% overlapping and the MDCT transform of corresponding length is performed. Again, window is chosen adaptively according to a signal nature, *noisy* or *tonal*. 128 or 1024 DCT coefficients are coded using the perceptual model. Perceptual noise substitution is performed: decoder find area of DCT coefficients of noise, calculate their parameters, and send them to decoder which synthesize them in a block-based manner. Prediction techniques are applied in time–frequency MDCT coefficient patterns. In AAC HE-1, first high-efficiency extension, a technique of spectral band replication (SBR) is used: high-frequency DCT coefficients are synthesized in the decoder in clever way using information sent from the encoder. In AAC HE-2 extension only one audio channel is codded, the remaining ones are coded differentially in parametric way.

The latest ideas of sound compression relies of joint speech/audio coding: on recognition of sound signal type and switching between different coders. Such approach is applied in universal speech-and-audio coders (USAC).

What do you think? Will be this chapter interesting for you? If not, please, go directly to the next one.

15.2 Psycho-Acoustics of Human Hearing System

15.2.1 Fundamentals

Human inner ear is built from cochlea which is spiral tube with increasing diameter filled with some fluid. In its beginning, at the more narrow part, there are vibrating stapes generating acoustical longitudinal wave consisting of different frequencies. This wave excites the cochlea wall, called a basilar membrane which is a natural sound spectrum analyzer. The membrane is frequency selective: its different parts have different resonant frequencies and start to vibrate only when these frequencies are present in the acoustical wave propagated in the fluid. When the cochlea diameter is bigger, the resonant frequency is lower. Vibrating different parts of basilar membrane activates different neurons and this information is transmitted to human brain. But nothing is perfect. This is a mechanical system. When one membrane part is excited with one strong frequency and vibrates with big amplitude, the side membrane parts are also vibrating with the same frequency and side frequencies, if present, are *masked* and not heard. They should be stronger if they want to win the battle!

In Fig. 15.1 the frequency masking phenomena, taking place in our ears, is illustrated. Strong frequency activates the basilar membrane and deforms it creating characteristic *tent*: if excitation of side-band frequencies are below this *tent*, they are not heard. In audio coders exactly this effect is exploited: bits are allocated only to strong frequency components of a signal. A total masking threshold for frequencies is calculated and bits are allocated to frequency components of a signal proportionally to exceeding by them the masking level. The triangular masking curve, caused by some frequency excitation and describing distribution of deformation energy of basilar membrane, is presented in Fig. 15.2.

Since a strong sound is faster processed by our brain than a weak one, frequency masking has to be analyzed not only in frequency axis but in time axis also. Strong sound is masking in some extend also nearby frequencies which were BEFORE it, i.e. earlier. Additionally, when amplitude of membrane oscillation is big, the return to *no-oscillation* state takes more time, what causes that also some nearby frequencies appearing AFTER the strong sound are masked. Both time effects of frequency masking, the pre- and post-masking, are illustrated also in Fig. 15.1.

In bit allocation, general information about pressure of sound waves with difference frequencies should be taken into account also. There is no sense to loose bit for coding UN-MASKED sound waves which are too weak to be heard. Absolute threshold of hearing, evaluated for a statistically *mean* person, gives us the needed

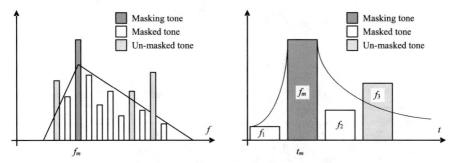

Fig. 15.1: Illustration of frequency masking effect: (left) statically in frequency domain, (right) dynamically frequency pre- and post-masking in time [12]

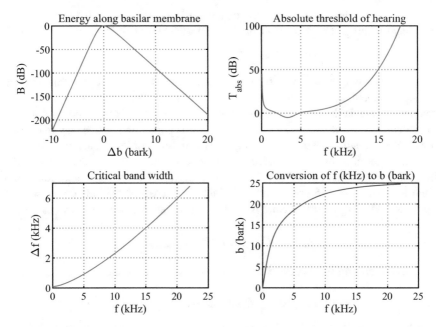

Fig. 15.2: Shapes of some mathematical functions describing psycho-acoustical effects/phenomena of human hearing system [12]

information. Its curve is presented in Fig. 15.2. Threshold for 2 kHz is treated as a reference one. We see that frequencies near 3 kHz are perceived the best, while very low frequencies (below 250 Hz) and high frequencies (above 12 kHz) are much poor heard by us. Absolute threshold of hearing is approximated by the following equation:

$$T_{abs}(f_{kHz}) = 3.64 \cdot f_{kHz}^{-0.8} - 6.5 \cdot \exp\left(-0.6 \cdot (f_{kHz} - 3.3)^2\right) + 0.001 \cdot f_{kHz}^4. \quad (15.1)$$

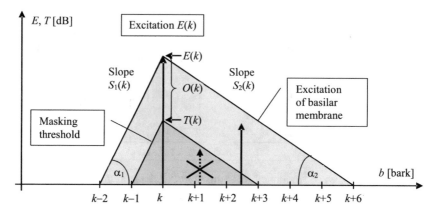

Fig. 15.3: Detail interpretation of frequency masking phenomena, helping in understanding meaning of psycho-acoustical model variables [12]

Our hearing system has not only different amplitude sensitivity for different frequencies, but also different frequency resolution. We better distinguish low-frequency sounds than high-frequency ones. This feature is described by width of the so-called critical bands. Their measurement is performed in this way that white noise is generated around an each frequency and width of its frequency band is increased in order to obtain the same sound audibility level. The critical band width increases with frequency what is shown in Fig. 15.2. When the overall audibility band is divided into 25 critical non-overlapping bands, the so-called bark scale is obtained. Width of a critical band is the following function of frequency (in kHz):

$$\Delta f_{Hz} = 25 + 75 \cdot \left(1 + 1.4 \cdot (f_{kHz})^2\right)^{0.69}, \tag{15.2}$$

while transforming frequency to bark scale is given by formula:

$$b = 13 \cdot \text{arctg}(0.76 \cdot f_{kHz}) + 3.5 \cdot \text{arctg}\left((f_{kHz}/7.5)^2\right). \tag{15.3}$$

Why we spending our precious time on bark scale discussion? Because it is exploited in psycho-acoustical models used in audio compression! Of course, first a signal FFT spectrum is calculated, but then it is transformed into bark scale when signal masking is computed.

15.2.2 Basics of Signal-to-Mask Radio Calculation

In this section we learn basics of calculation of signal-to-mask ratio, used in the MP2/MP3 standard. Equations presented below use variables which interpretation is given in Fig. 15.3.

Relative distribution of energy along basilar membrane, resulting from single tone excitation, is given by the following formula (in decibels):

$$B(\Delta b) = 15.81 + 7.5 \cdot (\Delta b + 0.474) - 17.5 \cdot \sqrt{1 + (\Delta b + 0.474)^2}, \qquad (15.4)$$

where Δb denotes the frequency distance expressed in bark scale. The function $B(\Delta b)$ is called a spreading function, spreading excitation from one critical band to neighboring ones. In Fig. 15.3 different excitations are marked with ↑. Analysis is done only for the strongest $E(k)$, which appeared in the k-th critical band. Its energy excites also neighboring part of the membrane on both sides. A characteristic tent, marked with light-gray color, is described by slopes $S_1(k)$ (left) and $S_2(k)$ (right). The strong excitation is masking weaker ones, lying below the threshold $T(k)$ (area marked with dark-gray color) obtained by shifting $E(k)$ down by $O(k)$ decibels (offset). Slopes $S_1(k)$ (left) and $S_2(k)$ are specified by

$$\begin{cases} S_1(k) = 31, \\ S_2(k) = 22 + \min(0.23/f_{kHz}, 10) - 0.2E_{dB}(k), \end{cases} \qquad (15.5)$$

where $E_{dB}(k)$ denotes the excitation in decibels and f_{kHz} its frequency in kHz. Offset $O(k)$ between excitation energy and threshold level is defined by

$$O(k) = \alpha(k) \cdot (14.5 + k) + (1 - \alpha(k)) \cdot \beta(k), \qquad (15.6)$$

where $\alpha(k)$ and $\beta(k)$ denotes, respectively, tonality index and masking index in the k-th critical band:

$$\alpha(k) = \min\{SFM(k)/SFM_{max}, 1\}, \quad SFM_{max} = -60 \text{ dB}, \qquad (15.7)$$

$$\beta(k) = 2 + 2.05 \cdot \text{arctg}(0.25f_{kHz}) + 0.75 \cdot \text{arctg}\left(\frac{(f_{kHz})^2}{2.56}\right). \qquad (15.8)$$

SFM in Eq. (15.7) denotes the spectral flatness measure, defined as ratio of geometric and arithmetic mean of squared samples of signal DFT spectrum (after windowing with Hanning window):

$$SFM(k) = 10\log_{10}\left(\frac{\left[\prod_{l=1}^{N_k} |X_k(l)|^2\right]^{1/N_k}}{\frac{1}{N_k}\sum_{l=1}^{N_k} |X_k(l)|^2}\right). \qquad (15.9)$$

Each excitation is creating its own masking tent. All of them are added and the overall masking threshold is calculated. After that, the absolute threshold of hearing is taken into account. Signals lying below the total masking threshold are not heard. They should not be coded. The signal-to-mask ratio (SMR) is defined as a difference between the signal spectrum and the total masking curve. Sub-band signals should obtain number of bits proportional to their SMR values.

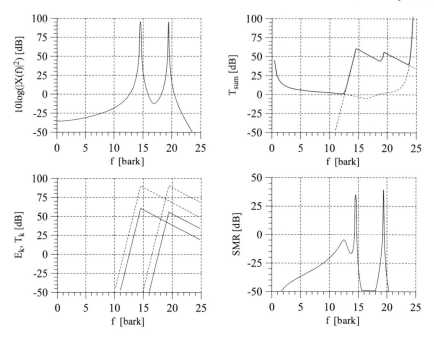

Fig. 15.4: Frequency masking example from exercise 1, in columns: (1) spectrum of two-component signal, (2) membrane energy (dashed line) and frequency masking curves (solid lines), (3) total masking threshold with absolute threshold, (4) calculated signal-to-mask ratio, used for bit allocation, difference between the signal spectrum and the total masking threshold [12]

Example of SMR calculation, for a signal consisting of two sinusoids with frequencies 2500 and 5800 Hz and amplitudes equal to 1, is presented in Fig. 15.4. We see (in columns):

- two-component signal spectrum,
- energy distribution along basilar membrane coming from each component together with resulting masking threshold,
- total masking threshold incorporating absolute masking threshold of our hearing system,
- SMR curve as a result of subtraction of the signal spectrum and total threshold; bits should be given only to frequencies having SMR higher than 0 decibels.

Exercise 15.1 (Basics of Psycho-Acoustics). Become familiar with program `lab15_ex_psycho_acoustics.m`. Run it. Carefully observe each figure. Compare program results with results presented in Fig. 15.4. Any frequency is heard if its SMR value is higher than 0 dB. Find amplitude value of the first component for which its SMR=0. Repeat this operation for the second component. Are found values the same? How do you explain this? Add a third component to the signal.

15.3 Psycho-Acoustical MP2 Model

It is assumed that overall excitation $E(k)$ of basilar membrane in any k-th critical band depends on power of all signals $S(l) = X^2(l)$ which have appeared in any critical band $l = 1, 2, 3, \ldots, K$. The signals affect the excitation in the k-th band with the strength specified by the spreading function $B(\Delta b)$ (Eq. (15.4)):

$$E(k) = \sum_{l=1}^{K} E_l(k) = \sum_{l=1}^{K} B(k-l)S(l) \tag{15.10}$$

in matrix notation:

$$
\begin{bmatrix} E(1) \\ E(2) \\ \vdots \\ E(K) \end{bmatrix} =
\begin{bmatrix}
B(0) & B(-1) & \cdots & B(K-1) \\
B(1) & B(0) & \cdots & B(K-2) \\
\vdots & \vdots & \ddots & \vdots \\
B(K-1) & B(K-2) & \cdots & B(0)
\end{bmatrix}
\cdot
\begin{bmatrix} S(1) \\ S(2) \\ \vdots \\ S(K) \end{bmatrix}
\tag{15.11}
$$

In MP1/MP2 MPEG-audio psycho-acoustical model the following operations are performed. Denotations from the standard are used.

1. Taking 1024 signal samples into the buffer: $x(i)$, $i = 1 \ldots 1024$, with delay incorporating delay of the filter bank.
2. Multiplication with Hanning window:

$$x_w(i) = x(i) \cdot h(i), \quad h(i) = 0.5 - 0.5\cos\left(\frac{2\pi(i-0.5)}{1024}\right). \tag{15.12}$$

3. Calculation of the 1024-point fast Fourier transform (FFT) of the signal, its magnitude r_ω and phase f_ω.
4. Calculation of magnitude and phase prediction, $\hat{r}_\omega$ and $\hat{f}_\omega$ on the base of their two last values from two last 1024-point data blocks, denoted as $t-1$ and $t-2$:

$$\hat{r}_\omega = 2r_\omega(t-1) - r_\omega(t-2), \quad \hat{f}_\omega = 2f_\omega(t-1) - f_\omega(t-2). \tag{15.13}$$

5. Calculation of spectral un-predictability measure c_ω:

$$c_\omega = \frac{\sqrt{\left(r_\omega \cdot \cos(f_\omega) - \hat{r}_\omega \cdot \cos(\hat{f}_\omega)\right)^2 + \left(r_\omega \cdot \sin(f_\omega) - \hat{r}_\omega \cdot \sin(\hat{f}_\omega)\right)^2}}{r_\omega + |\hat{r}_\omega|}.$$

$$(15.14)$$

6. Calculation of energy and weighted spectral un-predictability in predefined spectral bands:

$$e_b = \sum_{\omega = \omega low_b}^{\omega high_b} r_\omega^2, \quad c_b = \sum_{\omega = \omega low_b}^{\omega high_b} c_\omega r_\omega^2. \qquad (15.15)$$

7. Convolution of vectors of energy e_b and sub-band un-predictability c_b with spreading function $B(\Delta b)$ but in one-third bark scale, bark indexes are divided by 3 (*bmax*—number of last critical bark-1/3 band, *bb*—auxiliary index):

$$ecb_b = \sum_{bb=1}^{bmax} \mathbf{B}(bval_b, bval_{bb},) \cdot e_{bb}, \quad en_b = \frac{ecb_b}{\displaystyle\sum_{bb=0}^{bmax} \mathbf{B}(bval_b, bval_{bb})}, \qquad (15.16)$$

$$ct_b = \sum_{bb=1}^{bmax} \mathbf{B}(bval_b, bval_{bb}) \cdot c_{bb}, \quad cb_b = \frac{ct_b}{ecb_b}. \qquad (15.17)$$

8. Calculation of tonality index in each sub-band, taking values from 0 to 1:

$$tb_b = -0,299 - 0,43\ln(cb_b). \qquad (15.18)$$

9. Calculation of required signal-to-noise ratio in each sub-band, ensuring sound un-masking:

$$SNR_b = \max\{minval_b, \ (tb_b \cdot TMN_b + (1 - tb_b) \cdot NMT_b)\}, \quad NMT_b = 5.5 \text{ dB},$$
$$(15.19)$$

where NMT_b (Noise is Masking Tone) denotes a constant, describing tone masking by narrow-band noise in sub-band b.

10. Transforming SNR_b coefficient from decibel scale to linear scale and obtaining required power coefficient of un-masking in each sub-band:

$$bc_b = 10^{-SNR_b/10}. \qquad (15.20)$$

11. Finding energy threshold of un-masking:

$$nb_b = en_b \cdot bc_b. \qquad (15.21)$$

12. Calculation of energy threshold per one FFT coefficients:

$$nb_\omega = \frac{nb_b}{\omega high_b - \omega high_b + 1}.$$ (15.22)

13. Taking into account the absolute threshold of hearing, at present denoted by $absthr_\omega$, before as T_{abs}:

$$thr_\omega = \max\left\{nb_\omega, \; 10^{absthr_\omega/10}\right\}.$$ (15.23)

14. Calculation of signal-to-mask ratio (SMR) for each of 32 frequency channels (outputs of 32 analysis filters):

$$SMR_n = 10 \cdot \log_{10}\left(\frac{epart_n}{npart_n}\right), \quad n = 1, 2, 3, \ldots, 32,$$ (15.24)

where $epart_b$ denotes the signal energy and $npart_n$ noise level in n-th frequency channel:

$$epart_n = \sum_{\omega=\omega low_n}^{\omega high_n} r_\omega^2$$ (15.25)

$$if(width_n) == 1) \quad npart_n = \sum_{\omega=\omega low_n}^{\omega high_n} c_\omega r_\omega^2,$$

$$else \quad npart_n = \min\left\{thr_{\omega low_n} \ldots thr_{\omega high_n}\right\} \cdot (\omega high_n - \omega low_n + 1).$$
(15.26)

Exercise 15.2 (SMR Calculation in MP2 Psycho-Acoustical Model II). Become familiar with program MP2psycho.m. Compare equations given in the text with their program implementation, both in initialization and main part. The program requires $2 \cdot 1152 = 2304$ samples of analyzed signal $x(n)$, put into the buffer bx, and information about the sampling frequency f_s. Parameter show=0/1 is switching ON/OFF display of selected variables. Analyze with the MP2psycho() function two-component signal from Exercise 15.1. Call a function in the following loop:

```
N=32*36;
for iter = 1 : LastFrame
    bx1152 = x( 1+(iter-1)*N : N + (iter-1)*N );
    bx2304 = [ bx1152 bx2304(1:N) ];
    SMR = MP2psycho( bx2304, fs, show);
    plot(SMR); title('SMR (dB)'); pause
end
```

Did you obtain the same result as before? At present everything is done according to the model, before only some proof of concept was performed. Now, change test signal to any monophonic audio file sampled at 44.1 kHz, for example, a piece of music you like. Run the program. Observe SMR function values in the loop. How many sub-bands have SMR bigger than 0 dB?

Exercise 15.3 (SMR Calculation in MP2 Psycho-Acoustical Model I).
Download from Internet [8] the Matlab code of MPEG psycho-acoustical model I. Run the program `Test_MPEG.m`, observe displayed figures. A few times modify parameters of the synthesized test signal and verify visually obtained results. Then analyze the program and all functions, which are called by it. Try to find in the Matlab code equations presented is this chapter.

15.4 MP2 Filter Bank

At present we are familiar with fundamentals of human hearing systems and psycho-acoustical modeling of frequency masking effect. In order to use them in practice, we should decompose a signal being compressed into sub-bands. For this purpose in the MP2/MP3 coder $M =$32-channel filter bank is used. Its structure is presented in Fig. 15.5. Input audio signal is filtered by $M = 32$ band-pass filters working in parallel. Each of them pass only signal components belonging to different frequency bands. After bandwidth reduction, each sub-band signal is down-sampled by a compressor (reducer) which is leaving only each M-th sample and removing the rest of them. At this stage, results from psycho-acoustical signal analysis are used and available bits are allocated to sub-band signals being the most important perceptually. Then sub-bands signals are normalized and quantized, next coded into a bitstream. At the decoder part, the inverse operations are performed: re-quantization, de-normalization, zero insertion between reconstructed sub-band samples (by expanders), parallel filtration of all sub-band signals and summation of filtration results.

Input signal filtering is explained in Fig. 15.6 for $M = 3$ channel filter bank (FB) with filter impulse responses having $L_p = 8$ samples. Vector of input samples $\mathbf{x}$ is multiplied by a matrix of the FB, having in rows repeating blocks of 3 impulse responses of all filters. The blocks are shifted by 3 samples. This means that input stream of all samples is converted into 3 sub-streams of sub-band samples, but having 3 times less samples each. The total number of samples remains un-changed.

Most often the filter bank is built from one prototype low-pass filter $p(n)$, with frequency response $P(\Omega)$, which is up-shifted in frequency by real-value or

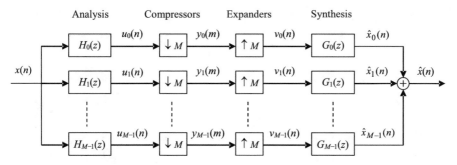

Fig. 15.5: Graphical explanation of audio compression idea: (1) audio signal is decomposed into many sub-bands by *analysis* bank of filters, (2) sub-band signals are down-sampled, (3) normalized and quantized (not denoted), (4) up-sampled and merged together using *synthesis* filter bank [12]

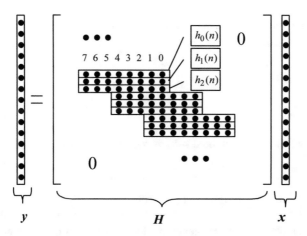

Fig. 15.6: Matrix interpretation of signal processing by an analysis bank of filters: vector of signal samples **x** is multiplied by an analysis matrix **H**, having in rows shifted (in time) blocks of impulse responses of M filters. In the figure $M = 3$ and $h_0(n), h_1(n), h_2(n)$ denote 8-samples long impulse responses of three band-pass filters of three sub-bands [12]

complex-value modulation. The original low-pass filter should ensure obtaining required features by the whole FB. For example, the following cost functions should be minimized by $P(\Omega)$:

1. linearity of the whole FB amplitude response:

$$E_1 = \int_0^{\pi/M} \left(\left| P(e^{j\Omega}) \right|^2 + \left| P(e^{j(\Omega - \pi/M)}) \right|^2 - 1 \right) d\Omega, \qquad (15.27)$$

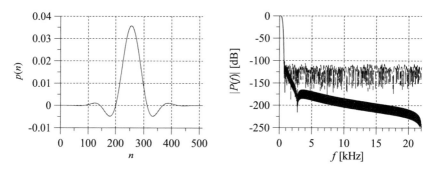

Fig. 15.7: Prototype low-pass filter of the MPEG-audio standard (left) and its frequency response (right), both use dashed line. Filter designed by us is plotted with a solid line [12]

2. high signal attenuation in the stop-band, ensuring reduction of spectral leakage between neighboring filters in the FB:

$$E_2 = \int\limits_{\Omega_s=\pi/2M+\Delta}^{\pi} \left| P(e^{j\Omega}) \right|^2 d\Omega. \tag{15.28}$$

Very often the following weighted sum of cost functions E_1 and E_2 is minimized:

$$E = \alpha E_1 + (1-\alpha)E_2. \tag{15.29}$$

Weights of low-pass prototype filter from the MP2/MP3 standard and its frequency response are shown in Fig. 15.7. They are marked with dashed lines. Solid line denotes an alternative filter, generated by our function `prototype()`. It was designed by minimizing the design objectives and assuming that prototype is a weighted sum of cosines.

In Fig. 15.8 characteristics of the whole analysis MP2/MP3 filter bank are presented. From top to bottom we see:

1. frequency responses of all 32 filters, covering the whole frequency band of 22.05 kHz for sampling frequency 44.1 kHz,
2. impulse response of the whole FB, very close (similar) to the delta Kronecker impulse, delayed by $L_p = 512$ samples, i.e. the length of the prototype filter; such response ensures near-perfect signal reconstruction introducing only signal delay at the FB output,
3. frequency response of the whole FB which is very close to 1 in the whole frequency range,
4. phase response of the whole FB being a linear one, i.e. delaying all frequencies at the FB output by the same amount of time, i.e. by $L_p = 512$ samples.

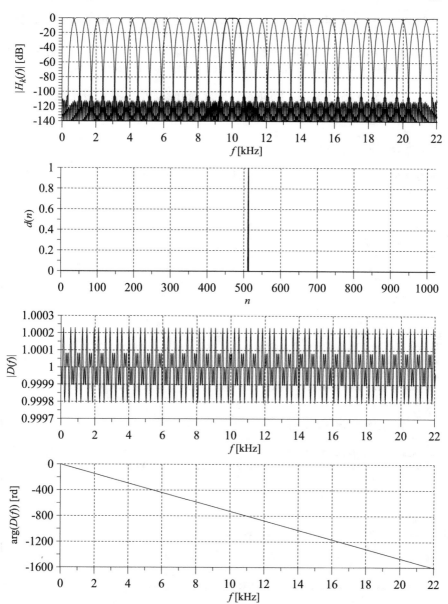

Fig. 15.8: Characterization of the MPEG-audio filter bank, from top to bottom: frequency responses of all 32 filters, the overall FB impulse response, the overall FB amplitude and phase response [12]

In program 15.1 signal decomposition into sub-bands and its synthesis is implemented in Matlab. The MP2 M =32-band analysis and synthesis filter bank is realized. Due to un-perfectness of the prototype filter, signal-to-noise ratio caused by sub-band signal splitting and reconstructing is on the level of 80 decibels.

Listing 15.1: Matlab implementation of the MP2/MP3 filter bank

```
1    % Lab15_ex_filterbank_simple
2    % MP2 filter bank
3    clear all; close all;
4
5    Nmany = 2000;                          % number of audio frames
6    N = 512;                               % filter length
7    M = 32;                                % number of sub-bands/channels
8
9    % Input signal
10   Nx = N+M*(Nmany-1);                    % number of signal samples
11   fs = 44100;                            % assumed sampling frequency
12   x = randn(1,Nx);                       % input signal
13   y = zeros(1,Nx);                       % output signal
14   %[x,fs]= audioread('bach44100.wav',[1,Nx]); x=x';
15
16   % Filter bank matrices
17   p = 2*sqrt(M)*prototype(N);            % prototype filter
18   figure; plot(p); title('p(n)'); pause %
19   pmat = ones(M,1)*p;                    % matrix with prototype in each row
20   [n,k] = meshgrid(0:(N-1),0:(M-1));     %
21   C = pmat.*cos(pi/M*(k+1/2).*(n-M/2));  % analysis  matrix with prototype
22   D = pmat.*cos(pi/M*(k+1/2).*(n+M/2));  % synthesis matrix with prototype
23   D = D';                                %
24   % Checking FB frequency response
25   figure;
26   K=5; KN=K*N; f = fs/KN*(0:KN-1);
27   figure; plot(f,20*log10(abs(fft(C',K*N)))); xlabel('f (Hz)'); grid; pause
28
29   % Analysis - synthesis, processing in M sub-bands
30   bx = zeros(1,N);
31   for m=1:Nmany
32       bx = [ x(m*M:-1:(m-1)*M+1) bx(1:N-M) ];
33       BX = C*bx';                        % analysis filter bank
34       % processing in M sub-bands
35       by = D*BX;                         % synthesis filter bank
36       n1st = 1+(m-1)*M; nlast = N + (m-1)*M;
37       n = n1st : 1 : nlast;
38       y( n ) = y( n ) + by';
39   end
40
41   % Output signal
42   Noffs=N-M+2; n1=1:(Nx-N)-Noffs+1; n2=Noffs:(Nx-N);
43   xr=x(n1); yr=y(n2);
44   figure; plot(n1,xr,'r',n1,yr,'b'); title('INPUT (red) OUTPUT (blue)'); pause
45   figure; plot(n1,xr-yr); title('Difference IN-OUT'); pause
46   error_dB = 10*log10( sum(xr.^2) / sum((xr-yr).^2) ), pause
```

Exercise 15.4 (MP2/MP3 Analysis-Synthesis Filter Bank). Become familiar with program 15.1. Note shape of the prototype filter, its frequency response and frequency responses of all filters in the analysis filter bank. Calculate and display frequency response of the synthesis filter. Excite the analysis-synthesis FB by Kronecker delta function, i.e. 1 and $N_x - 1$ zeros. Observe the impulse response. Having it, calculate frequency and phase response of the FB. Did you obtain plots similar to ones presented in Fig. 15.8? Change test signal to a fragment of your favorite song. Do you hear the difference? Modify the program: try to quantize sub-band signals. You can try to find an inspiration using the *regular*, not *simple*, version of the same program, given in the book repository.

15.5 Fast Polyphase Implementation of MP2 Filter Bank

Reading this section is only recommended for very ambitious Readers and skipped by the others. Therefore, if the audience is so much limited, why this section was written at all? Because some difficult things are extremely important, has wide applications, increase our horizons and . . . are beautiful.

Banks of many analysis and synthesis filters do their fantastic work in audio compression, splitting an audio signal into frequency sub-bands and allowing beneficial quantization of many sub-signals driven by the psycho-acoustical model. However, their usage has a very big drawback: immense computational cost of implementation. Many FIR filters with long vectors of weights, e.g. 32 filter with 512-taps in our case, require a lot of multiplications and additions. But fortunately, calculations can be significantly reduced thanks to polyphase decomposition of, both, the processed signal and the prototype filter weights. Situation is similar to relationship between DFT and FFT: calculated is exactly the same result but in different manner: significantly more efficiently by FFT. In a complex modulated filter bank a low-pass prototype filter is up-shifted in frequency by modulating it by harmonic complex-value exponents. Then, resultant impulse responses of band-pass filters are convolved with a signal to be filtered. It turns out that this operation can be significantly simplified. First, modulation signals can be grouped into a matrix and modulation simplifies to the fast, inverse, discrete Fourier or cosine transform. Additionally, thanks to polyphase decomposition, the size of the transformation can be significantly reduced. In the discussed case of the MP2/MP3 filter bank—to the DFT matrix with dimensions 64×64 or to modified DCT matrix with dimensions 32×64. $M = 32$ is a number of sub-bands and number of polyphase components of a prototype filter. Wow!

Since cosine is summation of two Fourier harmonic signals, one with positive and one with negative frequency, the real-value cosine M-channel modulation can be represented as a special form of the complex-value $2M$-channel modulation and the same savings can be done for DCT as for DFT. In practice, corresponding samples of polyphase signal and prototype filter components are multiplied and summed, and then a fast modified DCT transform of a small size is performed. As mentioned above, in the discussed case of MP2 coder $M = 32$, and $2M = 64$-channel DFT modulation is performed. After combining positive and negative frequencies, we obtain $M = 32$ real-value channels and MDCT transformation/modulation matrix with dimensions 32×64 (32 real-value filters having 64 samples in each polyphase component).

To show how this impressive reduction is get, the mathematical derivation is required. Non-interested Readers could skip the below part.

Proof. Let us perform the Z-transform of signal, writing it in polyphase version I:

$$X(z) = \sum_{m=-\infty}^{\infty} x(m)z^{-m} = \sum_{k=0}^{M-1} \left(\sum_{n=-\infty}^{\infty} x(nM+k)z^{-(nM+k)} \right) =$$

$$\sum_{k=0}^{M-1} z^{-k} \left(\sum_{n=-\infty}^{\infty} x(nM+k)\left(z^M\right)^{-n} \right) = \sum_{k=0}^{M-1} z^{-k} X_k^I \left(z^M\right). \quad (15.30)$$

Doing the same for the polyphase version II, which is defined below, we could write the signal Z-transform $X(z)$ as a summation of Z-transforms of its polyphase components $X_k^I(z)$ and $X_k^{II}(z)$, multiplied by z^{-k}:

$$X(z) = \sum_{k=0}^{M-1} z^{-k} X_k^I \left(z^M\right), \quad X_k^I(z) = \sum_{n=-\infty}^{\infty} x(nM+k)z^{-n} \quad (15.31)$$

$$X(z) = \sum_{k=0}^{M-1} z^{-(M-1-k)} X_k^{II} \left(z^M\right), \quad X_k^{II}(z) = \sum_{n=-\infty}^{\infty} x(nM+M-1-k)z^{-n}. \quad (15.32)$$

Let $p(n)$, $n = 0, 1, \ldots, L_p - 1$, denotes a low-pass prototype filter which will be used for filter bank creation. Filter impulse response of the k-th sub-band is obtained by prototype modulation:

$$h_k(n) = p(n)W_M^{-kn}, \quad W = e^{-j\frac{2\pi}{M}}. \quad (15.33)$$

The Z-transform of signal (15.33) is equal to:

$$H_k(z) = \sum_{n=-\infty}^{\infty} h_k(n)z^{-n} = \sum_{n=-\infty}^{\infty} p(n)W_M^{-kn}z^{-n} = \sum_{n=-\infty}^{\infty} p(n)\left(W_M^k z\right)^{-n} = P(zW_M^k).$$

$$(15.34)$$

Using Eq. (15.31), the Z-transform of the prototype filter can be written as summation of Z-transforms of its polyphase components:

$$P(z) = \sum_{l=0}^{M-1} z^{-l} P_l^I(z^M).$$ (15.35)

Putting Eq. (15.35) into Eq. (15.34) gives

$$H_k(z) = \sum_{l=0}^{M-1} z^{-l} W_M^{-kl} P_l^I(z^M W_M^{kM}) = \sum_{l=0}^{M-1} W_M^{-kl} \left(z^{-l} P_l^I(z^M) \right).$$ (15.36)

Last equation has the following matrix form:

$$H_k(z) = \begin{bmatrix} 1 & W_M^{-k} & W_M^{-2k} & \cdots & W_M^{-(M-1)k} \end{bmatrix} \begin{bmatrix} P_0^I(z^M) \\ z^{-1} P_1^I(z^M) \\ z^{-2} P_2^I(z^M) \\ \vdots \\ z^{-(M-1)} P_{M-1}^I(z^M) \end{bmatrix}.$$ (15.37)

When equations for all sub-band filters are combined together, one obtains

$$\begin{bmatrix} H_0(z) \\ H_1(z) \\ H_2(z) \\ \vdots \\ H_{M-1}(z) \end{bmatrix} = \begin{bmatrix} 1 & 1 & 1 & \cdots & 1 \\ 1 & W_M^{-1} & W_M^{-2} & \cdots & W_M^{-(M-1)} \\ 1 & W_M^{-2} & W_M^{-4} & \cdots & W_M^{-2(M-1)} \\ \vdots & \vdots & \vdots & \ddots & \vdots \\ 1 & W_M^{-(M-1)} & W_M^{-2(M-1)} & \cdots & W_M^{-(M-1)^2} \end{bmatrix} \begin{bmatrix} P_0^I(z^M) \\ z^{-1} P_1^I(z^M) \\ z^{-2} P_2^I(z^M) \\ \vdots \\ z^{-(M-1)} P_{M-1}^I(z^M) \end{bmatrix}$$ (15.38)

or in more condensed form:

$$\begin{bmatrix} H_0(z) \\ H_1(z) \\ H_2(z) \\ \vdots \\ H_{M-1}(z) \end{bmatrix} = \mathbf{W}^* \begin{bmatrix} P_0^I(z^M) & 0 & 0 & \cdots & 0 \\ 0 & P_1^I(z^M) & 0 & \cdots & 0 \\ 0 & 0 & P_2^I(z^M) & \cdots & 0 \\ \vdots & \vdots & \vdots & \ddots & 0 \\ 0 & 0 & 0 & \cdots & P_{M-1}^I(z^M) \end{bmatrix} \begin{bmatrix} 1 \\ z^{-1} \\ z^{-2} \\ \vdots \\ z^{-(M-1)} \end{bmatrix}.$$ (15.39)

The last equation was a goal of our mathematical *rock climbing*. What we can deduce from it? That samples of polyphase signal components should be multiplied by corresponding weights of polyphase prototype filter components and added, and then the inverse Fourier transform should be performed (multiplication by matrix $\mathbf{W}^*$).

In similar way the signal synthesis equation can be derived. Synthesis sub-band filters are defined by the equation:

$$g_k(n) = W_M^{-k} q(n) W_M^{-kn}.$$ (15.40)

When $G_k(z)$ denotes the Z-transform of the filter $g_k(n)$ and its polyphase components of type II are marked as $Q_l^{II}(z^M)$, the following equation is valid:

$$
\begin{bmatrix} G_0(z) \\ G_1(z) \\ G_2(z) \\ \vdots \\ G_{M-1}(z) \end{bmatrix}^T = \begin{bmatrix} z^{-(M-1)} \\ z^{-(M-2)} \\ z^{-(M-3)} \\ \vdots \\ 1 \end{bmatrix}^T \begin{bmatrix} Q_0^{II}(z^M) & 0 & 0 & \cdots & 0 \\ 0 & Q_1^{II}(z^M) & 0 & \cdots & 0 \\ 0 & 0 & Q_2^{II}(z^M) & \cdots & 0 \\ \vdots & \vdots & \vdots & \ddots & \vdots \\ 0 & 0 & 0 & \cdots & Q_{M-1}^{II}(z^M) \end{bmatrix} \cdot \mathbf{W}
$$
(15.41)

The overall analysis-synthesis filter bank operation is fully reversible and introduce only a delay of m_0 samples, when the following condition is fulfilled:

$$
[G_0(z) \; G_1(z) \; \ldots \; G_{M-1}(z)] \begin{bmatrix} H_0(z) \\ H_1(z) \\ \vdots \\ H_{M-1}(z) \end{bmatrix} = cz^{-m_0}.
$$
(15.42)

Using Eqs. (15.39), (15.41) in condition Eq. (15.42) one gets

$$
\begin{bmatrix} Q_0^{II}(z)P_0^I(z) & 0 & \cdots & 0 \\ 0 & Q_1^{II}(z)P_1^I(z) & \cdots & 0 \\ \vdots & \vdots & \ddots & \vdots \\ 0 & 0 & \cdots & Q_{M-1}^{II}(z)P_1^I(z) \end{bmatrix} = cz^{-m_0}\mathbf{I}
$$
(15.43)

since $\mathbf{W} \cdot \mathbf{W}^* = \mathbf{I}$. Therefore, in order to have a pure delay FB, all PP components of prototype filters $p(n)$ and $g(n)$ should fulfill the relation:

$$
Q_k^{II}(z)P_k^I(z) = cz^{-m_0}, \quad k = 0, 1, \ldots, M-1
$$
(15.44)

and they are designed using it. Thanks to this signal perfect reconstruction at the FB output is guaranteed. Ufff!

In Fig. 15.9 fast analysis-synthesis M-band filter bank, using polyphase signal decomposition and small size, complex FFT modulation, is presented. In Fig. 15.10 it is extended to $2M$ bands. After combining complex-value harmonics, used for modulation, with the same positive and negative frequency, the DCT-based filter bank is obtained with modulation realized by real-value cosines—what it presented in Fig. 15.11. Impulse responses of analysis and synthesis filters are defined in this case as follows:

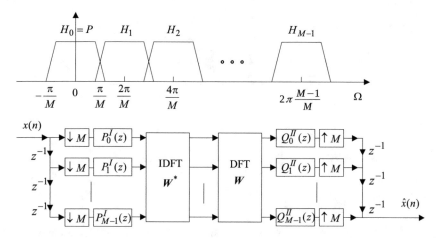

Fig. 15.9: Fast polyphase version of M-band filter bank with DFT modulation [12]

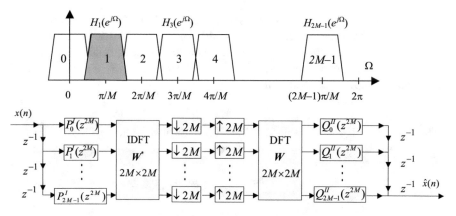

Fig. 15.10: Fast polyphase version of 2M-band filter bank with DFT modulation [12]

$$h_k(n) = 2p(n)\cos\left(\frac{\pi}{M}(k+0,5)\left(n-\frac{L-1}{2}\right)+(-1)^k\frac{\pi}{4}\right) \qquad (15.45)$$

$$g_k(n) = 2p(n)\cos\left(\frac{\pi}{M}(k+0,5)\left(n-\frac{L-1}{2}\right)-(-1)^k\frac{\pi}{4}\right). \qquad (15.46)$$

Transformation matrices used in Fig. 15.11 are defined as:

$$t_1(k,m) = 2\cos\left(\frac{\pi}{M}(k+1/2)(m-M/2)\right) \qquad (15.47)$$

$$t_2(k,m) = 2\cos\left(\frac{\pi}{M}(k+1/2)(m+M/2)\right). \qquad (15.48)$$

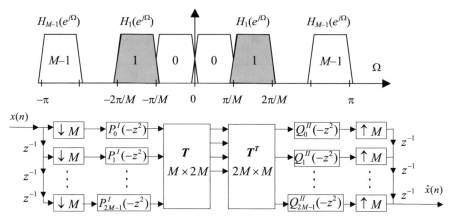

Fig. 15.11: Fast polyphase version of M-band filter bank with DCT modulation [12]

In Listing 15.2 fast polyphase Matlab implementation of M-band analysis and synthesis filter bank, originated from MP2/MP3 coding standard, is presented. Note that at present cosine transformation matrices have significantly smaller dimensions, not dimensions 32×512 and 512×32 but dimensions 32×64 and 64×32. Fast algorithms exist for these matrix transformations.

Exercise 15.5 (Fast Polyphase Filter Bank Implementation). Modify program 15.1. Use fast polyphase filter bank functions in it. Having signal decomposed into many sub-bands observe variability of sub-band samples. Try to find a sub-band quantization mechanism offering high SNR of reconstructed signal.

Listing 15.2: Fast polyphase implementation of MP2 analysis and synthesis filter banks

```
1   %################################
2   function  sb = analysisFB(x, M)
3   % M-band analysis filter bank in polyphase version
4
5   % Initialization
6     load enwindow.dat; pe=enwindow ; L=length(pe);          % read MP2 prototype filter
7     n=0:2*M-1;                                              % #
8     for k=0:M-1                                             % # polyphase analysis matrix
9         A(k+1,1:2*M)  = 2*cos( (pi/M)*(k+0.5).*(n-M/2));    % # dimensions: M x 2M
10    end                                                    % #
11
12  % Filter bank: analysis
13    sb = []; bx512 = zeros(1,L);                            % initialization
14    for k = 1 : length(x)/M
```

```
15        bx512 = [ x(k*M:-1:(k-1)*M+1) bx512(1:L-M) ];  % M new samples into buffer
16        for m = 1 : 2*M
17            u(m) = sum( bx512(m:2*M:L).*pe(m:2*M:L) );  % polyphase filtration
18        end
19        sb32 = A*u';                                    % cosine modulation
20        sb = [sb; sb32'];                               % storing the result
21    end
22
23    %###############################
24    function y = synthesisFB(sb, M)
25    % M-band synthesis filter bank in polyphase version
26
27    % Initialization
28        load dewindow.dat; pd=dewindow; L=length(pd);   % read MP2 prototype filter
29        n=0:2*M-1;                                       % #
30        for k=0:M-1                                      % # polyphase synthesis matrix
31            B(k+1,1:2*M) = 2*cos((pi/M)*(k+0.5).*(n+M/2));  % # dimensions: M x 2M
32        end                                              % #
33        MM=2*M; M2=2*M; M3=3*M; M4=4*M; Lp=L/MM; m = 0:Lp-1;  % Lp = length of PP components
34
35    % Filter bank: synthesis
36        y=[]; bv=zeros(1,2*L);                           % initialization
37        ns = length(sb(:,1));                            % number of samples in each sub-band
38        for k = 1 : ns                                   %
39            v = B'*sb(k,1:M)';                           % cosine demodulation
40            bv = [ v' bv(1:2*L-M2) ];                    % storing into buffer
41            for n = 1 : M                                % polyphase filtration
42                ys(n) = sum( bv(n+M4*m).*pd(n+M2*m) ) + sum( bv(n+M3+M4*m).*pd(n+M+M2*m) );
43            end
44            y = [ y ys ];
45        end
```

15.6 Complete MP2 Matlab Encoder and Decoder

In Fig. 15.12 the main idea of MP2 audio (en)coder is reminded: a low-pass prototype filter having 512 samples is modulated (up-converted in frequency) by 32
cosines and used as 32 band-pass filters. Impulse responses of these filters are shifted
over the audio signal with the step of 32 samples and multiplied with them. Summation of multiplication results gives us 32 samples in 32 sub-band after every 32
samples of input signal. This operation is repeated 36 times, therefore $36 \cdot 32 = 1152$
input audio samples are required and they are coded in one joint block.

In Fig. 15.13 block diagrams of MP2 encoder and decoder are presented. At the
encoder input we see 32 analysis filters and 36 samples in each sub-band, divided
into three blocks with 12 samples each. Maximum absolute value in each dozed is
find first and, then, a first scale factor (SF) from the SF table which is greater than
this maximum. This stage ends with information about SFs of all dozens and with
decision which scale factors in each sub-band will be send if the sub-band will be
coded.

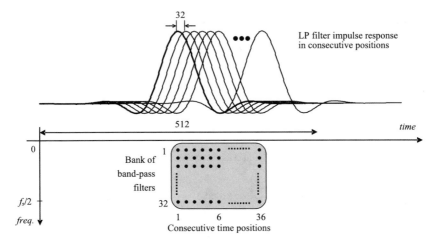

Fig. 15.12: MP2 example of sub-band signal decomposition: impulse response of a prototype low-pass filter, 512 samples long (shown on top), is first shifted-up in frequency by 32 cosines, and next is shifted 36 times over the signal with step of 32 samples. This way 36 samples in 32 sub-bands are computed (matrix of points shown in the bottom) [12]

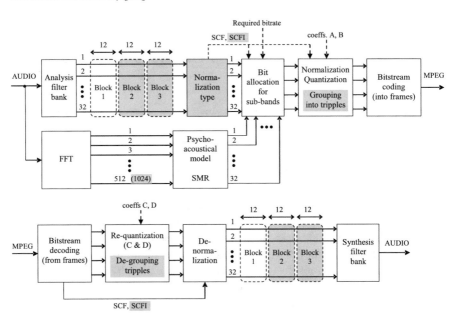

Fig. 15.13: Block diagram of MP2 encoder (up) and decoder (down) [12]

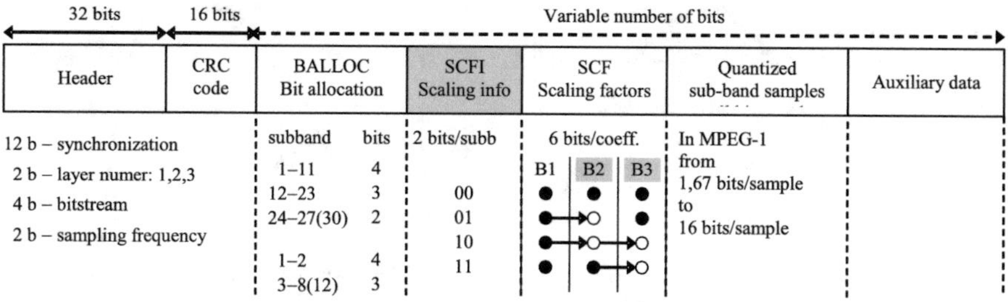

Fig. 15.14: Structure of MPEG-audio bit-stream [12]

In parallel 1024 samples are taken to the psycho-acoustical model where the 1024-point FFT is performed and 32 SMR coefficients are found, telling us about psycho-acoustical significance of each sub-band. Having this information, bits are allocated to the sub-bands in water-filling manner: (1) the sub-band with the highest SMR obtains one bit, (2) its SMR is decreased by about 6 dB, (3) a number of still available bits is calculated, and (4) again, next one bit is allocated to all samples in the sub-band with the highest SMR ... and so on, up to the moment when no bits are available, or SMRs of all sub-bands are equal or lower than 0 dB. When bits are allocated, sub-band samples, which will be transmitted, are divided by scale factors (normalized to the range $[-1, 1]$), and quantized. When any sub-band obtains small number of bits, its samples are grouped into triples and coded together.

Finally, the new data frame is appended to the bit-stream. Its organization is presented in Fig. 15.14. The frame starts with 32-bit header consisting of 12 bits of synchronization pattern and information bits, among others about the MPEG layer, bit-stream value and sampling frequency. Then the CRC code is sent allowing detection of bit errors in the header. Next, indexes to bit allocation tables are transmitted, information about scaling factors and indexes of scaling factors for each sub-band, one, two or three of them. At the end quantized sub-band samples are stored as well as possible auxiliary data.

The MP2 decoder is presented in bottom part of Fig. 15.13. All operations are performed in reverse order. First frame synchronization is done and CRC code is checked. If bit errors in the header are detected, the present frame is skipped and last audio fragment is repeated. When header is error-free, the system information is read. Knowing everything about operations performed in the coder, the decoding is done: data are de-quantized, de-normalized, removed samples are replaced with zeros and processed by synthesis filter.

Exercise 15.6 (Complete MP2 Encoder and Decoder Program). In archive supporting this laboratory, there is a program `lab15_ex_mp2.m` and two functions: `MP2write.m` and `MP2read.m`, in which complete MP2 encoder and decoder are implemented. It is possible to specify sampling frequency of audio recording and desired bit-stream value. The compressed data are stored to disc as *frame.mpg* (last audio frame) and *recording.mpg* (all data). If last file

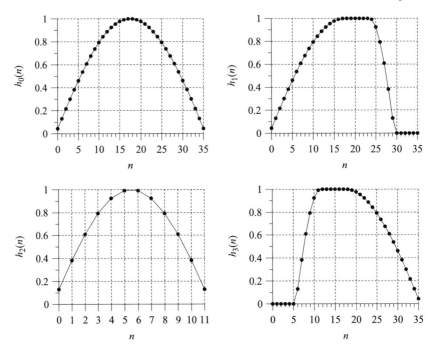

Fig. 15.15: Windows used in MDCT transform in MP3 coding, performed after the first 32-channel MP2 filter bank, in columns: long (36 points), short (12 points), long-to-short (36 points), short-to-long (36 points) [12]

already exists, new bits are appended at its end. The compressed audio can be played by different players, e.g. VLAN. Choose your favorite song. Compress it. Play it inside the lab15_ex_mp2.m program and by any player. Compare overlaid waveforms of original and decompressed signal. Note noise introduced by the compression algorithm (look at SNR value). Become familiar with the program. Set show=1. Observe data in different parts of compression and decompression programs. You can skip all operation except: (1) filter banks, (2) filter banks and normalization. Note SNR values in these two cases.

15.7 MP3 Coding Enhancements

MP2 layer of MPEG standard ensures good audio quality for compression ratio 6–8 times. The MP3 layer is better, it is offering compression ratio 10–12 times. How is it get? In MP3 each of the 32 sub-bands is decomposed further into additional 6 or 18 sub-bands, according to perceptual entropy of the signal spectrum, calculated by

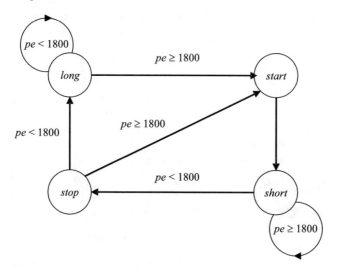

Fig. 15.16: Switching between short/long window length in MP3 adaptive filter bank, according to psycho-acoustical entropy (PE) value of the signal spectrum [12]

the psycho-acoustical model. It is, however, done not by filter bank but by sliding orthogonal modified DCT transform (MDCT). The sliding MDCT transform, used in MP3, can be interpreted also as an additional M-band cosine modulated filter bank with prototype filter equal to the sine window having $L_p = 2M$ samples:

$$p(n) = \frac{1}{\sqrt{2M}} \sin\left[(n+1/2)\frac{\pi}{2M}\right], \quad n = 0, 1, 2, \ldots, 2M-1. \qquad (15.49)$$

Such filter bank offers signal perfect reconstruction since its prototype filter fulfills the PR requirement (15.44). In MP3 parameter M takes two values: 18 and 6, bigger for tonal signals and smaller for noisy ones. In consequence, the MDCT sine windows have lengths 36 and 12 samples. In order to switch between long and short windows, two transition windows have to be used: *long-to-short* one and *short-to-long* one. All of them are presented in Fig. 15.15.

Decision what window should be used is taken by the psycho-acoustical model. It calculates perceptual entropy (PE) of the signal spectrum. When PE is smaller than 1800, long 36-point window is used, otherwise the short one. State diagram of window length switching is presented in Fig. 15.16.

The sliding MDCT transform is performed on output of each filter of the analysis filter bank. In Fig. 15.17 consecutive window positions are marked, and its transition from long window to short and back is presented. The window in each position is multiplied by sub-band samples. Obtained data are multiplied next by the MDCT matrix. Depending on its dimension, different number of additional sub-bands result: 18 for 36-point window or 6 for 12-point window. For tonal signals one obtains samples in more sub-bands but less frequently, while for noisy signals in less sub-

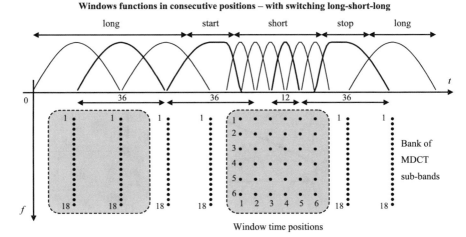

Fig. 15.17: Switching between short/long window length in MP3 adaptive filter bank according to psycho-acoustical entropy (PE) value of the signal spectrum [12]

bands but more frequently. It is correct, since for long-lasting tones we prefer higher frequency resolution of the analysis while for noisy signals better time resolution is preferred.

The final illustration of the MP3 analysis filter bank is given in Fig. 15.18 where *all bricks are put together*. We see one of 32 MP2 filters and 36 samples in its sub-band (after sub-band signal decimation). Next, these samples, initially placed in row of time–frequency plane, are exchanged with one of time–frequency matrices: the one with 18 rows and two columns (samples in 18 frequency sub-bands and in two time moments) or the one with 6 rows and 6 columns (samples in 6 frequency sub-bands and 6 time moments). Number of all samples remains the same.

Since the AAC standard, briefly described below, use the same sub-band signal decomposition technique, an exercise and program of the MDCT will be presented in the next section.

15.8 AAC Advanced Audio Coding

Merits of the advanced audio coding have been already presented in the introduction. The standard is complex and multi-thread. In this section we only concentrate on sub-band signal decomposition applied in it. The sliding MDCT approach is used, exactly the same as in the MP3 second filter bank. But significantly longer windows are used: 2048 and 256-point ones, first for tonal signals, the second for noisy signals. Transition windows and switching strategy are the same as in the MP3 standard. In MP3 for tonal signals we have 576 sub-bands, in AAC we have 1024. In MP3 192 sub-bands are used for noise, in AAC 128 sub-bands.

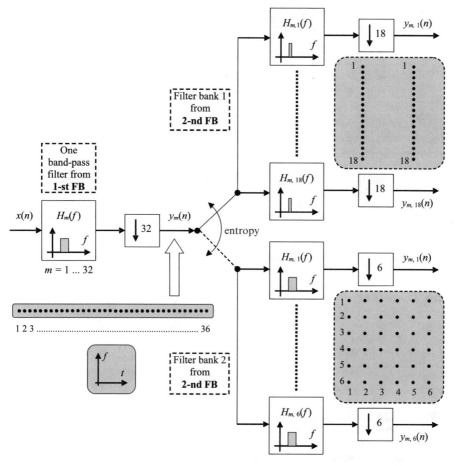

Fig. 15.18: Adaptive filter bank of MP3 encoder: each of 32-channels of MP2 filter bank is followed by the MDCT transform with variable length, offering sound division into 18 additional sub-bands (for tonal sound) either into 6 sub-bands (for noisy sound). In the first case $576 = 32 \cdot 18$ sub-bands are obtained this way. Perceptual spectrum entropy controls the MDCT switching [12]

In program 15.3 demonstration of AAC coding strategy is presented. Structure of the program is very similar to filter bank coding of the MP2 algorithm, implemented in program 15.1. It is no surprise. Both methods represent a cosine modulated filter bank but they use different parameters and modulation matrices with different sizes. The overall methodology is the same.

Exercise 15.7 (MDCT-Based AAC Audio Coding). Become familiar with program 15.3. Run it. Note perfect reconstruction of the original signal when no signal processing in sub-bands is done. Put attention to quantization of the original signal (result `xq`) and signal quantization in sub-bands (result `xqs`). Observe that the second method offers better SNR for decoded signal. Do coding of your favorite song. Apply your own quantization strategies. In the repository there is a program `lab15_ex_aac_switching.m` in which analysis and synthesis windows are switched from long to short and back. Observe that the original signal is again perfectly reconstructed.

Listing 15.3: Principles of sliding MDCT usage in AAC coding

```
1   % Lab15_ex_aac.m
2   % Principles of AAC coding using sliding MDCT transform
3   clear all; close all;
4
5   Nmany = 100;                      % number of frames
6   N = 2048;                         % window length
7   M = N/2;                          % window shift
8   Nx = N+M*(Nmany-1);               % number of signal samples
9
10  % Input signal
11  %x = 0.3*randn(Nx,1); fs=44100;
12  [ x, fs ] = audioread('bach44100.wav'); size(x), pause
13  x = x(1:Nx,1); x=x.';
14  soundsc(x,fs);
15  figure; plot(x); pause
16
17  % MDCT and IMDCT transformation matrices
18  [n,k] = meshgrid(0:(N-1),0:(N/2-1));          % indexes
19  win = sin(pi*(n+0.5)/N);                      % window
20  C = sqrt(2/M)*win.*cos(pi/M*(k+1/2).*(n+1/2+M/2)); % MDCT matrix with window
21  D = C';                                        % IMDCT matrix with window
22
23  % Bit allocation for sub-bands
24  b = [ 8*ones(M/4,1); 6*ones(M/4,1); 4*ones(M/4,1); 0*ones(M/4,1) ]; sc = 2.^b;
25  %b = 6*ones(M,1); sc = 2.^b;
26
27  % AAC analysis-synthesis with quantization in sub-bands
28  y = zeros(1,Nx); figure;                      % output signal
29  for k=1:Nmany                                 %
30      n1st = 1+(k-1)*M; nlast = N + (k-1)*M;    % next indexes
31      n = n1st : nlast;                         % samples from-to
32      bx = x( n );                              % without window
33      BX = C*bx';                               % MDCT
34  %   plot(BX); title('Samples in bands'); pause % plot of sub-band samples
35      BX = fix( sc .* BX ) ./ sc;               % quantization
36  %   BX(N/4+1:N/2,1) = zeros(N/4,1);           % some processing
37      by = D*BX;                                % IMDCT
38      y( n ) = y( n ) + by';                    % without window
39  end                                           %
40
```

```
41  n=1:Nx;
42  soundsc(y,fs);
43  figure; plot(n,x,'ro',n,y,'bx'); title('Input (o), Output (x)'); pause
44
45  m=M+1:Nx-M;
46  max_abs_error = max(abs(y(m)-x(m))), pause
47
48  % Quantization of the original signal
49  b = 6; xq = fix( 2^b * x ) / 2^b;
50
51  % Comparison
52  xqs = y;
53  [X, f]=periodogram(x, [],512,fs, 'power','onesided'); X=10*log10(X);
54  [Xq,f]=periodogram(xq,[],512,fs, 'power','onesided'); Xq=10*log10(Xq);
55  [Xqs,f]=periodogram(xqs,[],512,fs,'power','onesided'); Xqs=10*log10(Xqs);
56  figure; plot(f,X,'r-',f,Xq,'b-',f,Xqs,'g-');
57  xlabel('f (Hz)'); title('Power/frequency (dB/Hz)'); grid; pause
58
59  SNR1 = 10*log10( sum(x(m).^2) / sum( (x(m)-xq(m)).^2 ) ),
60  SNR2 = 10*log10( sum(x(m).^2) / sum( (x(m)-xqs(m)).^2 ) ),
```

15.9 Summary

Music takes important part of our life. Compression of digitized music is very interesting from *computer technology* point of views: a lot of things have to be taken into account and many tricks should be done to reduce the stereo audio bit-stream from about 1.5 megabits per second to 50 kilobits per second, i.e. about 30 times. What was the most important in this chapter?

1. Performing audio compression we exploit imperfections of human hearing system. Mechanical features of our cochlea cause that strong vibrations of some parts of basilar membrane with some frequencies do not allow vibrations of neighboring parts with slightly different frequencies. This phenomena is called the frequency masking. Since some frequencies, present in the music, are not heard by us, because they are masked by stronger ones, there is no sense to allocate bits to them during music coding. Therefore, the mathematical psycho-acoustical models play a central role in audio compression. They analyze sound and inform the remaining part of a program about psycho-acoustical significance of each frequency sub-band of music.

2. In order to exploit the frequency masking effect, audio signal has to be represented as a summation of many sub-signals, containing only frequencies of separate frequency sub-bands. This signal decomposition is done by bank of many band-pass filters, working in parallel, like in the MP2/MP3 standard, or by modified discrete cosine transform (MDCT), as

in the AAC standard. When samples of sub-band signals or MDCT co-efficients are available, they are quantized inversely proportional to their psycho-acoustical significance, i.e. less bits are given to less significant signal components.

3. In audio compression standards there are a lot tables for different sampling frequencies and compression ratios. Coefficients given in them describe different relations, resulting from frequency masking effect which was investigated in the past by big teams of researchers. It is impossible to design an effective compression algorithm without their usage.

4. Filter bank design for audio compression, like MP2/MP3, is not an easy task. Firstly, a good low-pass prototype filter has to calculated, ensuring small distortion error of the whole FB (flat magnitude response and linear phase response) as well as good frequency selectivity and small spectral leakage between sub-bands. Secondly, signal processing in the FB should be organized in computationally efficient way, namely using polyphase filter structures. Since in FBs the prototype low-pass filter is modulated by cosines, fast DCT transforms could be used, additionally of smaller sizes thanks to polyphase signal decomposition.

5. Using long 256/2048-point MDCT transforms to signal decomposition into sub-bands, instead of filter banks, turned out to be more elegant and effective way. Such approach is used in the modern advanced audio coders (AAC). They make use of many additional data compression tricks, like: prediction of MDCT coefficients, perceptual synthesis of noisy sub-band during decoding, intelligent replication of low-frequency sub-bands in high-frequency sub-bands during audio restoring, advanced differentially coding of multi-channel audio.

15.10 Private Investigations: Free-Style Bungee Jumps

Exercise 15.8 (No More Exercises!). Turn on your old gramophone and hear to some vinyl records. What pleasure!

Further Reading

1. B. Gold, N. Morgan, *Speech and Audio Signal Processing* (Wiley, New York, 2000)
2. ISO/IEC-11172, Coding of moving pictures and associated audio for digital storage media at up to about 1.5 Mbit/s. MPEG-1 International Standard, ISO/IEC, 1991

3. ISO/IEC-13818, Information technology generic coding of moving pictures and associated audio informations. MPEG-2 International Standard, ISO/IEC, 1995

4. ISO/IEC-14496, Information technology very low bitrate audio-visual coding. MPEG-4 International Standard, ISO/IEC, 1998

5. ISO/IEC-15938, Multimedia content description interface. MPEG-7 International Standard, ISO/IEC, 2002

6. M. Kahrs, K. Bandenburg (eds.), *Applications of Digital Signal Processing to Audio and Acoustics* (Kluwer, Boston, 1998)

7. I.V. McLoughlin, *Speech and Audio Processing. A Matlab-based Approach* (Cambridge University Press, Cambridge, 2016)

8. F. Petitcolas, MPEG psychoacoustic model I for MATLAB. Online: https://www.petitcolas.net/fabien/software/mpeg/index.html

9. A. Spanias, T. Painter, V. Atti, *Audio Signal Processing and Coding* (Wiley-Interscience, Hoboken, 2007)

10. K. Steiglitz, *A Digital Signal Processing Primer: With Applications to Digital Audio and Computer Music* (Pearson, Upper Saddle River, 1996)

11. S.V. Vaseghi, *Multimedia Signal Processing. Theory and Applications in Speech, Music and Communications* (Wiley, Chichester, 2007)

12. T.P. Zieliński, *Cyfrowe Przetwarzanie Sygnalów. Od Teorii do Zastosowań (Digital Signal Processing. From Theory to Applications)* (Wydawnictwa Komunikacji i Łączności (Transport and Communication Publishers), Warszawa, Poland, 2005, 2007, 2009, 2014)

13. U. Zölzer (ed.), *DAFX Digital Audio Effects* (Wiley, Chichester, 2002, 2011)

Chapter 16
Image Processing

Living in an epoch of pictures: only short message, visual icon,
funny animoi, please!

16.1 Introduction

We are living in crazy times. People start mutual relations from simple gestures and mimics, and after years of civilization progress, they return back to simple iconic interpersonal communication: :-)? :-(! In the epoch of short message communication, in the era of shortcuts of thoughts and opinions, an image creation, analysis, and processing is priceless.

What an image is? Signal is a 1D vector of numbers—samples of 1D function, while image is a 2D matrix of numbers—samples of 2D function. Signals are acquired by 1D sensors, for example, microphones, while images are acquired by 2D sensors called cameras, CCD or CMOS ones. CCD cameras are built from capacitors which are separately charged by light with different wavelengths (RGB: Red, Green, and Blue). Thanks to this, color information is not lost. In fact, triples of RGB capacitors create matries and performs spatial discretization of visual information. Both, microphones and cameras, are followed by analog-to-digital converters which perform quantization of analog quantities measured by them. At the output, we have vectors or matrices of numbers, that is, a 1D or 2D signal. In case of images, these numbers are called *pixels*. Making extension, in computer tomography (CT) or magnetic resonance imaging (MRI) devices, human body is scanned by sensors in three dimensions XYZ and 3D matrices are obtained. They consist of *voxels*. When CT is repeated a few times, dynamic CT data are obtained that have a form of 4D matrices, i.e. 3D CT repeated in time, and so on. Matrices with more and more dimensions ... But all of them are built from single numbers. When one understands 1D signal processing, she or he should have no problems with understanding basics of multi-dimensional data processing, which is shown in this chapter.

© Springer Nature Switzerland AG 2021
T. P. Zieliński, *Starting Digital Signal Processing in Telecommunication Engineering*, Textbooks in Telecommunication Engineering,
https://doi.org/10.1007/978-3-030-49256-4_16

There are special mathematical and computational design tools for process-ing 2D data, being 2D extensions of their 1D prototypes: (1) 1D orthogonal DFT and DCT transforms are replaced with their 2D versions and (2) 1D con-volution of vectors is extended to 2D convolution of matrices. If one can ride a bicycle, after short time of training, riding a motorcycle should not be a big problem for her or him. If you can drive a car, driving a truck or a bus would not be also difficult for you. The same is with 2D signal processing: it is natural extension of the 1D DSP.

During 2D DFT image transformation, first, each image row of pixels is re-placed by its DFT spectrum. Then, DFT of each column of a matrix, resulting from the first processing stage, is calculated. In summary, DFT of the DFT is computed. In 1D signal spectral analysis, a 1D signal was represented as a sum of 1D basis functions scaled by spectral coefficients. In 2D spectral analysis, an image is expressed as a sum of basis images, scaled again by spectral coeffi-cients. Basis images are generated by outer products of vertical transformation vectors $\mathbf{b}_k$ and their complex conjugation and transposition: $\mathbf{b}_k \cdot \mathbf{b}_k^H$. When we:

1. set some of the 2D DFT coefficients to zero,
2. perform inverse DFT transformations over columns of the modified spectral matrix,
3. do inverse DFT transformations of the rows,

we are coming back from the 2D spectral coefficients of an image to the image pixels. This way image content filtration is done in frequency domain. Basis im-ages, which spectral coefficients were set to zero, are lacking after the described image transformation. The situation is the same as in 1D signal processing: a signal synthesized from its modified spectrum does not have spectral compo-nents which were removed from the spectrum. In 2D processing, some images are subtracted from the original image.

Similar correspondence exits between 1D and 2D convolution. As we re-member, the FIR signal filtering had a form of convolution of two vectors: the shorter one was shifted over the longer one and for each its position correspond-ing samples of both vectors were multiplied and added. The resulting single number was representing the output filter value for given position of the filter weights. The situation is completely the same in case of convolution of images. One smaller 2D matrix of filter weights is shifted over a bigger matrix with image pixels and, for each position of the filter mask, its weights are multiplied by corresponding image pixels lying below them. Then multiplication results are summed, as before in 1D case. The calculated single value, the filter output, is put into the output image, result of filtering, into the pixel lying below the central weight of the filter. We can say that this value replaces pixel in the input image. Choice of filter weights decides about realized image processing task: image smoothing (de-noising) or contour/edges enhancement. Possible sets of weights are typically defined in long tables.

Calculated output pixels must not be obligatory a linear superposition of input pixels, taken with some weights. Some non-linear, median, or morphological image filtering is also possible. It will be mentioned in this chapter. We will also see an interesting example of multi-level, cascade image filtering using a pair of symmetrical low- and high-pass filter. This way a so-called quad-tree image decomposition is obtained: one approximation image and many multi-resolution images with vertical, horizontal, and skew details.

Similarity of one image fragment (block of pixels) to the second image fragment can be calculated using: (1) 2D cross-correlation function, or (2) mutual information, an entropy-based measure, describing also image similarity. Since image blocks can be deformed, one image block can be first pre-processed by affine transformation (doing its shifting, scaling, and rotating) and then compared with some other block.

Compression of still images is a special example of image processing. In JPEG standard, an image is decomposed into non-overlapping blocks of 8×8 pixels. Each of them is processed by 2D DCT transform and quantized. Bits of non-zero pixels are next scanned in *zig-zag* manner using variable length integer (VLI) coding and, finally, losslessly coded using entropy-based Huffman coder. The procedure will be described.

It is no surprise that video compression, compression of *moving pictures*, for example, *movies*, is more difficult than still pictures. Since consecutive film frames could be very similar, there is no sense to code them individually. Only the first picture of the block of seven pictures is coded as a whole. Next, 6 out of 7 are coded differentially. In addition, coded is not a simple difference between two images but difference between one image and the second image with compensation of block pixel movement (thanks to this, information to be coded is even smaller). A few words about movie compression will be given below.

So, it is time to start our movie! Dear Tom, turn off the light, please!

16.2 Image Representation

Nowadays digital images are permanently present in our everyday lives: in TVs, pictures from cameras, Internet content, and others. There are a lot of different applications based on image acquisition and analysis (military, industrial, medical, entertainment, etc.), and every day they are increasing in number. In Fig. 16.1, some image examples are given and applications are mentioned.

What an image is? It is a matrix of numbers as presented in Fig. 16.2. In the simplest case of monochromatic gray-scale images, the matrix consists of indexes of level of gray color. For example, for 8-bit coding we have $2^8 = 256$ integer numbers from 0 to 255, describing how much *gray* is an image pixel in certain position, i.e. in (row, column) of our matrix of numbers. "0" denotes perfectly black color, while "255" perfectly white. In right matrix in Fig. 16.2, we see exactly such integer numbers. In turn, in Fig. 16.3 a *sculpture* of letter "H" is plotted as a 3D surface/mesh

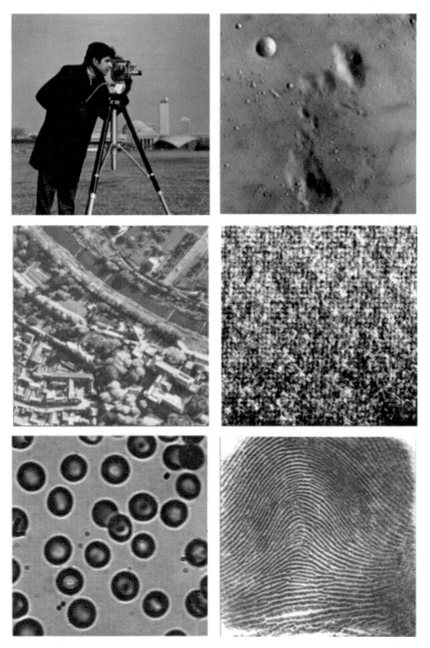

Fig. 16.1: Examples of images, in rows: (1) *Cameraman*—picture used in Matlab for testing different algorithms, (2) surface of the Moon—cosmic exploration, (3) satellite city image—remote supervision and security, (4) woolen cloth (microscopic image)—production quality monitoring, (5) blood image—medical diagnostics, and (6) fingerprint—personal authorization [17]

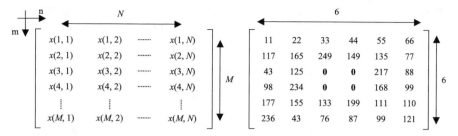

Fig. 16.2: Image as a 2D $M \times N$ matrix: (left) in general, (right) 6×6 image with an 8-bit gray-level coding (from 0 to 255), representing a black square (four central elements with 0s surrounded by a dappled pattern) [17]

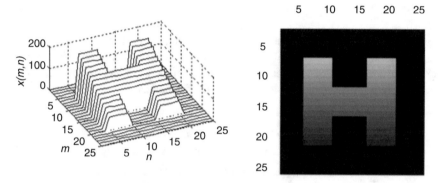

Fig. 16.3: 25×25 matrix of letter "H" *sculpture*, plotted as 3D surface/mesh (left) and as an image with gray-scale color coding (right) [17]

(left) and as an image with gray-level coding. It is important to remember that pixel marked as (0,0) or (1,1), alternatively, is located in the top-left (first row, first column) or bottom-left image corner (last row, first column).

Contemporary images are colored ones. Any computer color is superposition of three basic colors: Red, Green, and Blue (RGB):

$$color = c_R \cdot R \quad \oplus \quad c_G \cdot G \quad \oplus \quad c_B \cdot B \tag{16.1}$$

When each color is coded with 8 bits, we obtain a 24-bit number, describing one color. Therefore we have $2^{24} = 16$ millions of available colors. In order to reduce image files, combinations of *RGB* colors, occurring in an image, are tabulated, and only smaller indexes to the color table (palette) are stored or transmitted for each pixel, not color bits themselves. In Matlab, color palette has only 64 entries (i.e., Gray, Jet, HSV, Hot, Winter,...). With single table index, only one combination of RGB colors is associated. Exemplary Matlab Gray palette is presented in Table 16.1. The above-described GIF-style color system is rather an old one, but we put an attention to it since it is used in Matlab.

Table 16.1: Description of 64-element `Gray`-scale palette of Matlab—see Eq. (16.1). Color numbers (No.) come from a graphics card, and c_{RGB} values are defined by Matlab

Palette index	Color (R)ed		Color (G)reen		Color (B)lue	
	No.	c_R value	No.	c_G value	No.	c_B value
0	0	0	0	0	0	0
1	4	0.015873	4	0.015873	4	0.015873
2	8	0.031746	8	0.031746	8	0.031746
3	12	0.047619	12	0.047619	12	0.047619
...	...	...	...	...	...	...
62	251	0.984127	251	0.984127	251	0.984127
63	255	1	255	1	255	1

There are many different standards of color decomposition. In digital TV and digital media (i.e., DVD and blue-ray discs), each color is composed of: luminance (Y), measuring level of gray, as well as blue chrominance component (C_b) and red chrominance component (C_r), together YC_bC_r, defined as:

$$Y = 0.299R + 0.587G + 0.114B$$
$$C_b = 0.564(B - Y) = -0.1687R - 0.3313G + 0.5B \qquad (16.2)$$
$$C_r = 0.713(R - Y) = 0.5R - 0.4187G - 0.0813B.$$

In image analysis, when one wants to extend dynamic range of an image to the available color palette possibilities, a new color index is calculated using the equation:

$$index_{new} = \text{round}\left[\left(\frac{index_{old} - min}{max - min}\right)^p \cdot pallete\ size\right], \qquad (16.3)$$

where *min* and *max* denote found minimum and maximum color indexes in the analyzed image, and the number of available colors in the palette is equal to *size*.

How images are acquired? In case of CCDs (Charge Couple Devices), imaging sensor is a spatial matrix of capacitors which are charged by different RGB wavelengths of the light. Each pixel is built from a triple of capacitors, responsible for different colors. Capacitor voltage is quantized using a 6–8 bit analog-to-digital converter. A real-world image is analog and contains complete visual information. A digital image is discretized in space, for example, the HD image has 1920×1080 or 1280×720 pixels. In addition, color values are quantized, and not all colors are possible to obtain. In Fig. 16.4, acquisition of digital images is presented.

Each row of a digital image matrix is a 1D signal, describing change of color along one line of observed scene. In Fig. 16.5, two examples of such scanning are given. At the top, an artificial image of a house and tree, from Fig. 16.4, is scanned. In the acquisition result, we can distinguish darker pixels of the house wall and

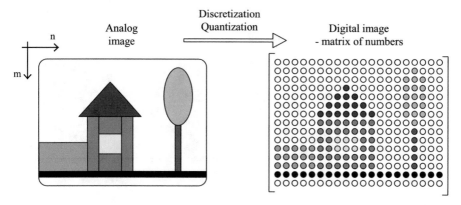

Fig. 16.4: Illustration of digital image acquisition: discretization in space and quantization in color [17]

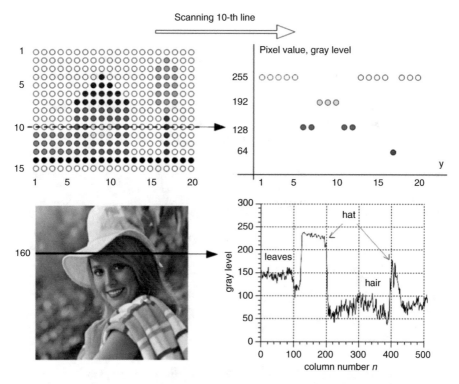

Fig. 16.5: Digital image as a collection of 1D signals: (left) artificial and real image, (right) pixel values scanned along one line, i.e. one matrix row [17]

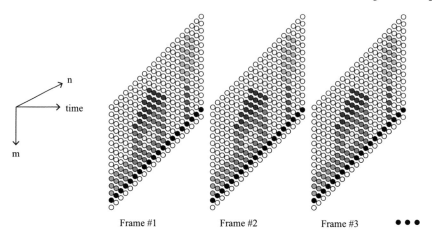

Fig. 16.6: Cinema film as a 3D data: sequence of 2D matrices (images) called film frames [17]

window, as well as tree trunk. At the bottom, scanning is presented for real-world image. In this case, one can notice brighter pixels of women hat and darker pixels of her hair.

   If 2D matrices of images are repeated, their sequence, namely a *movie*, is obtained. A cinema film is a 3D matrix of color numbers—see Fig. 16.6—with numbers in all X, Y, and Z axes. If one can analyze and process a 1D signal, working with multi-dimensional data should not represent a big problem for her or him.

Exercise 16.1 (Image Basics in Matlab). In Listing 16.1, a simple Matlab program is presented, showing how to read an image into Matlab and display it. User can select one of image lines and plot it next to the image. DFT and DCT transforms are computed for the chosen line and compared. Run the program and observe its different lines. Read different images. Make your own *selfie* with your phone and read an image into the program. Find in Matlab manual how to display separately different color image components, *RGB* or YC_bC_r. Calculate FFT or DCT of one color of one image line.

Listing 16.1: Reading and displaying images in Matlab

```
1   % lab16_ex_image_basics.m
2   % Image basics
3   clear all; close all; figure;
4
5   % Initialization - read image
6   [x,cmap] = imread('Cameraman.tif');      % read image to "x", color palette to "cmap"
7   imshow(x,cmap), title('Image'); pause  % display image using its palette
```

```
 8    [M, N] = size(x);                    % ask for number of rows and columns
 9    x = double(x);                       % pixel values to double (gray scale)
10    MN = min(M,N); N=MN; x=x(1:N,1:N);   % reduce image to square matrix
11
12    % Observation of one image line
13    Num = 150; K = 2; y = x;             % line number, marker width, copy
14    line = x(Num,1:N);                   % take one matrix row
15    y(Num-K:Num+K,1:N) = 0*ones(2*K+1,N); % mark line with black color
16    figure;
17    subplot(121); imshow(y,cmap); title('Image');        % display image
18    subplot(122); plot(line); title('One line'); pause   % show one line
19    figure;
20    subplot(211); plot( abs(fft(line))/N ); title('|DFT|');   % line DFT
21    subplot(212); plot( dct(line)/sqrt(N) ); title('DCT'); pause  % line DCT
```

16.3 2D Orthogonal Image Transformations

16.3.1 Definitions and Calculation

A pair of 2D Fourier transforms of a 2D image matrix is defined as follows:

$$X_{\text{DFT}}(k,l) = \sum_{m=0}^{M-1} \left(\sum_{n=0}^{N-1} x(m,n) e^{-j\frac{2\pi}{N}nl} \right) e^{-j\frac{2\pi}{M}mk}, \tag{16.4}$$

$$x(m,n) = \frac{1}{N} \sum_{l=0}^{N-1} \left(\frac{1}{M} \sum_{k=0}^{M-1} X_{\text{DFT}}(k,l) e^{j\frac{2\pi}{M}mk} \right) e^{j\frac{2\pi}{N}nl}, \tag{16.5}$$

where $x(m,n)$ denotes an analyzed image matrix with dimensions $M \times N$, and $X_{DFT}(k,l)$ the transformation result, matrix of Fourier coefficients, having the same size ($0 \le m,k \le M-1$ and $0 \le n,l \le N-1$). Indexes (m,n) define pixel position in an image, row and column number, respectively, while indexes k and l specify basis function frequency in vertical and horizontal dimensions. The 2D DFT consists of two series of 1D DFT presented in Fig. 16.7. First, each row of pixels is replaced with its 1D DFT coefficients. Then, DFTs are calculated for each column of the matrix, obtained in the first processing step.

As we remember, 1D DFT of a real-value signal is symmetrical in its real part and asymmetrical in imaginary part. Therefore the 2D image DFT is also symmetrical and asymmetrical (two possible equations):

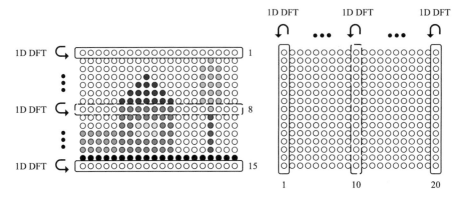

Fig. 16.7: Graphical interpretation of a 2D DFT image transformation: first image rows are replaced by their 1D DFTs, and then columns of a new matrix, resulting from the first processing level, are exchanged with their 1D DFTs [17]

$$X\left(\frac{N}{2}\pm k, \frac{M}{2}\pm l\right) = X^*\left(\frac{N}{2}\mp k, \frac{M}{2}\mp l\right), \quad 0 \le k \le \frac{N}{2}-1, \quad 0 \le l \le \frac{M}{2}-1,$$
$$(16.6)$$

$$X(k, l) = X^*(N-k, M-l), \quad 0 \le k \le N-1, \quad 0 \le l \le M-1. \tag{16.7}$$

Why symmetry of image DFT is so important? Because when we are doing image filtration in frequency domain and modifying some Fourier coefficients of an image (e.g., set them to 0), we obtain a real-value image after inverse 2D DFT only when the image (a)symmetry was not perturbed by us!

The 2D discrete cosine transform of an image is defined as:

$$X_{\text{DCT}}(k,l) = \sum_{m=0}^{M-1}\left[\sum_{n=0}^{N-1} x(m,n)\cdot\beta(l)\cos\left(\frac{\pi l}{N}(n+1/2)\right)\right]\cdot\alpha(k)\cos\left(\frac{\pi k}{M}(m+1/2)\right),$$
$$(16.8)$$

$$x(m,n) = \sum_{l=0}^{N-1}\left[\sum_{k=0}^{M-1} X_{\text{DCT}}(k,l)\cdot\alpha(k)\cos\left(\frac{\pi k}{M}(m+1/2)\right)\right]\cdot\beta(l)\cos\left(\frac{\pi l}{N}(n+1/2)\right),$$
$$(16.9)$$

where constants $\alpha(k)$ and $\beta(l)$ are equal to:

$$\alpha(k) = \begin{cases} \sqrt{1/M}, & k=0 \\ \sqrt{2/M}, & k=1\ldots M-1 \end{cases} \quad \beta(l) = \begin{cases} \sqrt{1/N}, & l=0 \\ \sqrt{2/N}, & l=1\ldots N-1. \end{cases} \tag{16.10}$$

The 2D DCT is computed exactly the same way as 2D DFT: first 1D DCT over image rows and then over columns. Since 2D DFT and 2D DCT spectra are similar

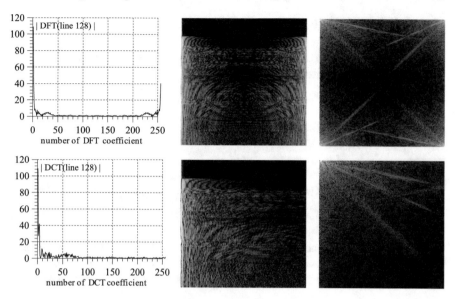

Fig. 16.8: DFT (top) and DCT (bottom) of image *Cameraman*, from left to right in columns: (1) transformation of line 128, (2) transformation of all rows, and (3) transformation of columns of matrix presented left (in figure 2) [17]

and DCT is significantly easier to compute without complex-value operations, the 2D DCT is preferred in image processing.

In Fig. 16.8, 2D DFT and 2D DCT image transformations are compared for the image *Cameramen*. The 128th line of the image is transformed by 1D DFT and 1D DCT, and then transformations of all image rows and all resultant columns are done. As we see, the 2D DFT and 2D DCT spectra are very similar. Therefore DCT ones should be preferred due to simpler calculations.

2D DCT application in image processing is beneficial also thanks to one very important 2D DCT feature. When 2D image spectra are modified and inverse 2D transforms are performed on them, images are filtered in frequency domain of transform coefficients. Multiplication in frequency domain corresponds to convolution in space domain of image pixels. Convolution realized by the sequence of operations 2D-IFFT(2D-FFT(image)*mask) is circular, while realized by 2D-IDCT(2D-DCT(image)*mask) is symmetrical. Assuming symmetrical image extension for its "unknown" part, done by convolution based on 2D-DCT, is better than assuming circular image extension used by 2D-FFT since the first operation typically creates smaller *ringing effect* (oscillations) on image borders.

16.3.2 Interpretation

Both, 2D DFT (16.4), (16.5) and 2D DCT (16.8), (16.9), as 2D orthogonal transformations, can be represented as follows:

$$Y(k,l) = \sum_{m=0}^{M-1} \left[\sum_{n=0}^{N-1} x(m,n) \cdot b_N^*(l,n) \right] \cdot b_M^*(k,m), \quad \mathbf{Y}_{MxN} = \mathbf{B}_M^* \cdot \mathbf{X}_{MxN} \cdot \mathbf{B}_N^{*T},$$

$$(16.11)$$

$$x(m,n) = \sum_{k=0}^{M-1} \left[\sum_{l=0}^{N-1} Y(k,l) \cdot b_N(l,n) \right] \cdot b_M(k,m), \quad \mathbf{X}_{MxN} = \mathbf{B}_M^T \cdot \mathbf{Y}_{MxN} \cdot \mathbf{B}_N,$$

$$(16.12)$$

where m,n are spatial and k,l are frequency indexes. $\mathbf{B}_M$ and $\mathbf{B}_N$ are orthogonal transformation matrices with dimensions $M \times M$ and $N \times N$, having in their rows basis functions of the signal decomposition, e.g. for transformation of size M:

$$b_M^{\text{DFT}}(k,m) = \sqrt{\frac{1}{M}} \exp\left(j \frac{2\pi k}{M} m \right),$$

$$(16.13)$$

$$b_M^{\text{DCT}}(k,m) = \alpha(k) \cos\left(\frac{\pi k}{M}(m+1/2) \right).$$

$$(16.14)$$

When we assume that all coefficients $X(k,l)$ are equal to 0, except the coefficient $X(k_0,l_0)$ equal to 1, and perform the inverse transformation (16.12), the following equation is derived:

$$\mathbf{X}_{k_0,l_0} = \mathbf{B}_M^T \cdot \mathbf{Y}_{k_0,l_0} \cdot \mathbf{B}_N = Y(k_0,l_0) \cdot b_M^T(k_0) \cdot b_N(l_0)$$

$$(16.15)$$

telling us that matrix $b_M^T(k_0) \cdot b_N(l_0)$ being the outer product of basis vectors $b_M(k_0)$ and $b_M(l_0)$ is associated with non-zero coefficient $X(k_0,l_0)$. Since orthogonal transformations are linear, the transform of superposition of elementary matrices $X(k_0,l_0)$ (with only one non-zero element) is equal to sum of their transforms. Therefore, we have

$$\mathbf{Y} = \sum_{k_0=0}^{M-1} \sum_{l_0=0}^{N-1} \mathbf{Y}_{k_0,l_0} \quad \Rightarrow \quad \mathbf{X} = \sum_{k_0=0}^{M-1} \sum_{l_0=0}^{N-1} \mathbf{X}_{k_0,l_0} = \sum_{k_0=0}^{M-1} \sum_{l_0=0}^{N-1} Y(k_0,l_0) \cdot b_M^T(k_0) \cdot b_N(l_0).$$

$$(16.16)$$

As we see, the $M \times N$ image $\mathbf{X}$ is a sum of scaled images $\mathbf{B}(k_0,l_0)$ with dimensions $M \times N$:

$$\mathbf{X} = \sum_{k_0=0}^{M-1} \sum_{l_0=0}^{N-1} Y(k_0,l_0) \cdot \mathbf{B}_{k_0,l_0}, \quad \mathbf{B}_{k_0,l_0} = b_M^T(k_0) \cdot b_N(l_0).$$

$$(16.17)$$

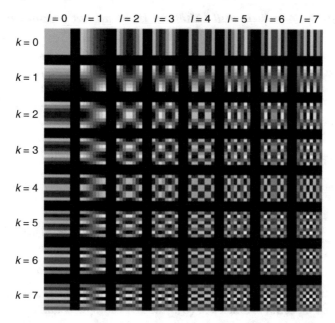

Fig. 16.9: Elementary images $\mathbf{B}_{k,l} = \mathbf{b}_M^T(k) \cdot \mathbf{b}_N(l)$ of 2D 8-point DCT transform: each 8×8 image is represented as their weighted summation [17]

Elementary images are outer products of the 1D transformation basis functions/vectors, i.e. the k_0-th row of transformation matrix $\mathbf{B}_M$ and the l_0-th row of transformation matrix $\mathbf{B}_N$. They are multiplied by 2D spectral coefficients $Y(k_0, l_0)$.

Figure 16.9 shows elementary 2D basis functions of the 2D 8-point DCT transformation. We see simple images with vertical, horizontal, and diagonal stripes, repeating with different frequencies. Each analyzed 8×8 image is represented as weighted summation of elementary images, scaled by corresponding DCT coefficients:

$$\mathbf{IMAGE} = \sum_{k_0=0}^{M-1} \sum_{l_0=0}^{N-1} c_{k_0,l_0} \cdot \mathbf{subimage}_{k_0,l_0}. \tag{16.18}$$

If some coefficient values c_{k_0,l_0} are set to zero, basis images associated with them are removed from the processed image. For example, horizontal stripes can disappear from your shirt. Wow! *Please, do not do it!*

16.3.3 Image Analysis and Filtering in Frequency Domain

In this subsection, we calculate 2D DFT and 2D DCT of an exemplary image, observe the 2D spectra, modify them, and perform inverse transformations. We note effects of image filtering realized this way. In order to better distinguish differences in spectral DFT and DCT coefficients, the following scaling of its magnitude and phase is done:

$$\frac{A(k,l) - \min_{k,l}(A)}{\max_{k,l}(A) - \min_{k,l}(A)} \cdot 255, \quad A(k,l) = \log_{10}(|Y(k,l)| + 1), \tag{16.19}$$

$$\frac{B(k,l) - \min_{k,l}(B)}{\max_{k,l}(B) - \min_{k,l}(B)} \cdot 255, \quad B(k,l) = \sphericalangle Y(k,l) = \operatorname{atan}(\operatorname{Im}(Y(k,l)), \operatorname{Re}(Y(k,l))).$$

$$\tag{16.20}$$

In Fig. 16.10, magnitude and phase of the 2D DFT of the image *Cameraman* are presented. Spectral coefficient $Y(0,0)$, located at top-left matrix corner, represents the image mean value. The 2D spectrum is symmetrical and is described by Eqs. (16.6), (16.7). If one wants to remove some 2D frequency components from the image, she or he should put zero to symmetrical and asymmetrical pairs of spectral coefficients associated with them. It is easier to do, when the (0,0) coefficient is in 2D spectrum center, not in the corner. Therefore, before 2D DFT spectrum modification, its quarter positions should be replaced, as shown in Fig. 16.11. Then, the central spectrum part is multiplied by a symmetrical mask with filter weights, and only the central spectral coefficients are left, describing low-frequency image content. In Fig. 16.11, such masking of 2D DFT spectrum of *Cameraman* image is presented in the second row. The filtering is continued in Fig. 16.12. Modified image quarters are returned to their original positions and inverse image DFT is performed. Since only low-frequency image components were left, the resultant image is *smoothed*. After zooming, some pixel value oscillations, the so-called *ringing effect*, are observed. It is a consequence of applying sharp $0/1$ filter mask in frequency domain that has oscillatory *sinc-like* impulse response in pixel-space domain. 2D filters having smoothly decaying edges in frequency domain do not create such image artifacts. A proof of this statement is presented in Fig. 16.13. In the upper row, circular Gaussian-shape low-pass filter mask is applied to *Cameraman* image, offering its very nice softening. In the lower row, the inverse Gaussian-shape high-pass filter mask is used causing edges enhancement in the image.

In Fig. 16.8, we have already seen that 2D DFT and 2D DCT spectra are very similar with this difference that the DCT spectrum is NOT redundant (symmetrical) and easier for calculation (without complex-value numbers). Therefore, the 2D DCT, not DFT, is widely applied in image processing, especially in image and video compression. In Fig. 16.14, the *Cameraman* image is filtered in domain of 2D DCT coefficients using two different approaches: *zonal* filtering/coding (top) and *thresh-*

Fig. 16.10: 2D DFT magnitude (left) and phase (right) of image *Cameraman*, shown in Fig. 16.1 as the first one [17]

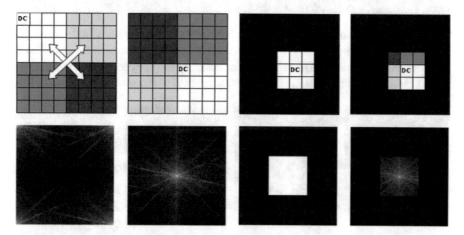

Fig. 16.11: Graphical explanation of the 2D DFT spectrum symmetry and spectrum squares re-ordering. Having (0,0) 2D DFT coefficient in the matrix center is required before applying symmetrical filter masks that modify the image 2D DFT content [17]

old filtering/coding (bottom). In the first method, DCT coefficients lying in a chosen zone are left. In our example, low-frequency ones that are located in a triangle placed at the top-left corner of the 2D DCT matrix. As we see, image smoothing is achieved but with visible *ringing effect* (oscillations of pixel values)—due to sharp 0/1 edges of the 2D filter mask. In the second approach (second row of images), only 2D DCT coefficients exceeding some threshold are left. In this case, quality of the reconstructed image is much better.

Program in Listing 16.2 presents calculation of 2D DFT and 2D DCT transforms of any image. Spectral coefficients are compared in both cases, modified, and used for backward image synthesis.

Fig. 16.12: Complete *Cameraman* image filtering in frequency domain. In rows: (1) original image, (2) its 2D DFT, (3) centered DFT, (4) 0/1 square filter mask centered at (0,0) DFT coefficient, (5) image DFT modified by the filter mask, (6) de-centered DFT, (7) filtered image (the 2D IDFT result), and (8) zoomed image fragment [17]

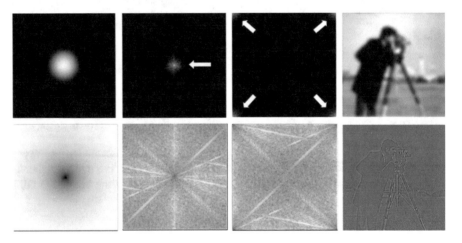

Fig. 16.13: *Cameraman* image filtering by low- (top) and high-pass (bottom) 2D circular Gaussian shape filters with softly decaying edges in frequency domain. All operations are performed in the order presented in Fig. 16.12 [17]

Fig. 16.14: Example of image filtering (compression) in domain of 2D DCT coefficients: (top) zonal filtering—only DCT coefficients from some matrix zone are left,
(bottom) threshold filtering—only DCT coefficients exceeding some threshold are
left [17]

Listing 16.2: 2D image transformations and their application to image filtering in
frequency domain

```
1    % lab16_ex_transforms.m
2    % 2D orthogonal transforms of images
3    clear all; close all; figure;
4
5    % Initialization - read image
6    [x,cmap] = imread('Cameraman.tif');      % read image to "x", color palette to "cmap"
7    imshow(x,cmap), title('Image'); pause    % display image using its palette
8    [M, N] = size(x);                        % ask for number of rows and columns
9    x = double(x);                           % pixel values to double (gray scale)
10   MN = min(M,N); N=MN; x=x(1:N,1:N);       % reduce image to square matrix
11
12   % Transformation matrices of 1D DCT (C) and 1D DFT (F)
13   c = [sqrt(1/N) sqrt(2/N)*ones(1,N-1)]; f = 1/sqrt(N); % normalizing coefficients
14   n = 0:N-1;                               % indexes of samples
15   for k=0:N-1                              % indexes of frequencies
16       C(k+1,1:N) = c(k+1) * cos( pi*k*(n+1/2) / N );   % basis functions of 1D DCT
17       F(k+1,1:N) = f       * exp(j*2*pi/N*k*n);        % basis functions of 1D DFT
18   end
19
20   % FILTERING using 2D DCT
21   K = 64; H = zeros(M,N); H(1:K,1:K) = ones(K,K);      % frequency 0/1 mask
```

```
22   X = conj(C) * x * conj(C).';                    % 2D DCT - IN
23   X1 = dct2(x);   X2=dct(dct(x')');                % 2D DCT - Matlab
24   err1 = max(max(X-X1)), err2 = max(max(X-X2)), pause  % errors
25   Y = X .* H;                                      % filtering in freq domain
26   y = C.' * Y * C;                                 % 2D IDCT - OUT
27   y1 = idct2(Y); y2 = ( idct( idct( Y )' ) )';     % 2D IDCT - Matlab
28   err1 = max(max(y-y1)), err2 = max(max(y-y2)), pause  % errors
29   XdB = scaledB( X ); YdB = scaledB( Y );          % scaling in dB
30   figure;
31   subplot(221); imshow(x, cmap);  title('Image');          % below only display
32   subplot(222); imshow(XdB, cmap); title('2D DCT');
33   subplot(223); imshow(YdB, cmap); title('2D DCT + Mask');
34   subplot(224); imshow(y(1:128,65:192), cmap); title('Filtered image'); pause
35
36   % FILTERING using 2D DFT (fftshift2D - re-ordering of 2D DFT coefficients)
37   K = 32; H = zeros(M,N);                                  % frequency mask H (MxN)
38   H(M/2+1-K : M/2+1+K, N/2+1-K : N/2+1+K) = ones(2*K+1,2*K+1); % center = (M/2+1, N/2+1)
39   h = fftshift2D( real( ifft2( fftshift2D(H) ) ) );       % impulse response
40   figure;
41   subplot(121); imshow(255*H,cmap); title('Freq Mask');       % show freq response
42   subplot(122); mesh(h); title('Filter imp. response'); pause % show imp response
43   X = conj(F) * x * conj(F).'; % 2D DFT - IN
44   X1 = fft2(x)/N;              % 2D DFT - Matlab 1
45   X2 = fft(fft(x').')/N;       % 2D DFT - Matlab 2
46   err1 = max(max(X-X1)), err2 = max(max(X-X2)), pause % errors
47   Xp = fftshift2D(X);          % 2D spectrum centering
48   Yp = Xp .* H;                % filtering = 2D DFT multiplied by filter H
49   Y = fftshift2D(Yp);          % 2D spectrum de-centering (original order)
50   y = F.' * Y * F;             % inverse 2D DFT - IN
51   y1 = ifft2(Y)*N;             % inverse 2D DFT - Matlab 1
52   y2 = (ifft(ifft(Y).'))'*N;   % inverse 2D DFT - Matlab 2
53   err1 = max(max(abs(y-y1))), err2=max(max(abs(y-y2))), pause % errors
54   y1 = real(y1);               % real part only, imaginary equal to 0
55   y1f = y1(1:128,65:192);      % image fragment for display
56   XdB = scaledB( X ); XpdB = scaledB( Xp );
57   YdB = scaledB( Y ); YpdB = scaledB( Yp );
58   figure;
59   subplot(231); imshow(x,    cmap); title('1. Image IN');
60   subplot(232); imshow(XdB,  cmap); title('2. 2D DFT');
61   subplot(233); imshow(XpdB, cmap); title('3. Centering');
62   subplot(234); imshow(YpdB, cmap); title('4. Filtering');
63   subplot(235); imshow(YdB,  cmap); title('5. De-centering');
64   subplot(236); imshow(y1f,  cmap); title('6. Image OUT'); pause
65
66   % 2D FILTERING using 2D Matlab convolution function conv2()
67   L = 32; y2 = conv2(x, h(M/2+1-L:M/2+1+L, N/2+1-L:N/2+1+L),'same');
68   figure;
69   subplot(121); imshow(y1,cmap); title('Image after FREQ filter - cyclic conv');
70   subplot(122); imshow(y2,cmap); title('Image after CONV filter - linear conv'); pause
```

Exercise 16.2 (Image as a Summation of Elementary Images). Modify program 16.2. Having matrix of the 1D DCT transformation and its basis functions lying in matrix rows, generate yourself a few elementary images of the $M = N = 512$ 2D DCT transformation, similar to ones presented in Fig. 16.9. For this purpose, you can also use the Matlab x=idct2(X) function, for example, with an argument M=512; X=zeros(M,M); X(3,3)=1.

Exercise 16.3 (Image Processing in Frequency Domain). Analyze code of the program 16.2. Note different possibilities of 2D DFT and 2D DCT computation. Carefully look at calculated 2D spectra, results of their modification and images synthesized. Next, make a *selfie* using your phone. Read this image into the program 16.2. Calculate its 2D DCT spectrum. Select a group of coefficients with the highest values and synthesize an image associated with them. Then, synthesize an image associated with low-value DCT coefficients. Finally, make a photo of your shirt with vertical or horizontal stripes and process the image.

16.4 2D Convolution and Image Filtering

Image filtering can be done by modification of 2D image spectra but also by direct modification of image pixels. The situation is exactly the same as for 1D signals: they are filtered in two ways: (1) doing direct convolution of their samples with filter weights or (2) by multiplication of DFT/DCT spectra of signals and filter weights, and calculating the inverse transform of the multiplication result. What is the conclusion for us from this observation? Instead of 2D spectra *masking* with some weights of filter frequency response, we can calculate the inverse 2D transformation of the weights, namely the 2D filter impulse response, and convolve an image with it. 2D image filtering implemented as 2D convolution is presented in Figs. 16.15 and 16.16. A pixel mask of 2D filter is moved above the image and stopped over its each pixel. Then pixels lying below the mask are multiplied by filter weighs, i.e. 2D filter impulse response, and summed. This way one pixel of the output image is calculated, the one being in the center of the filter mask. Filter weights can be chosen in some heuristic manner or as a result of inverse 2D transform of the required filter frequency response. Typically, some windowing is applied to them in image pixel domain in order to obtain their compact support (limited number of weights is non-zero) and smooth weights transition.

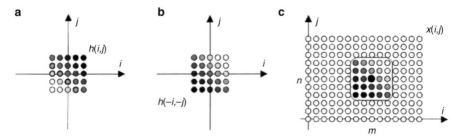

Fig. 16.15: Illustration of a 2D convolution: a small matrix of filter weights is shifted over a big matrix of image pixels and multiplied by pixel values lying below. Multiplication results are summed. Calculated value is put to an output image pixel in the position of filter weights center [17]

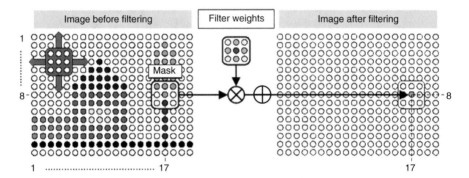

Fig. 16.16: Illustration of image filtering using 2D convolution: input image pixels, selected by a moving mask, are multiplied by corresponding filter weights and summed. Calculated value is put to the output image in the position of the moving mask center [17]

The 2D linear convolution, realized in image pixel domain, is theoretically described by the following equation:

$$y(m,n) = \sum_{i=-\infty}^{\infty} \sum_{j=-\infty}^{\infty} x(i,j)h(m-i,n-j), \tag{16.21}$$

while its frequency-domain version implements the circular (modulo-M, modulo-N) version, exactly the same as for 1D signals:

$$y(m,n) = \sum_{i=0}^{M-1} \sum_{j=0}^{N-1} x(i,j) \cdot h\big((m-i)_{\text{mod } M}, (n-j)_{\text{mod } N}\big). \tag{16.22}$$

Theoretical 2D impulse responses can be calculated analytically using 2D inverse DFT, for example, for low-pass filters with rectangular or radial support in frequency domain, respectively:

$$h_{LP}^{(f0)}(m,n) = \frac{\sin(2\pi m f_{m0}/f_{ms})}{\pi m} \cdot \frac{\sin(2\pi n f_{n0}/f_{ns})}{\pi n}, \quad h_{LP}(0,0) = \frac{4 f_{m0} f_{n0}}{f_{ms} f_{ns}},$$

$$(16.23)$$

$$h_{LP}^{(f0)}(m,n) = \frac{R}{2\pi R\sqrt{m^2+n^2}} J_1\left(R\sqrt{m^2+n^2}\right), \quad R = \frac{2\pi f_0}{f_s}, \quad (16.24)$$

where f_{ms}, f_{ns} denotes sampling frequencies in m, n directions, and $J_1(x)$ is the Bessel function of the first type and first order. Weights (impulse responses) of HP, BP, and BS filters are calculated using the following equations:

$$h_{HP}^{(f0)}(m,n) = \delta(m,n) - h_{LP}^{(f0)}(m,n), \qquad (16.25)$$

$$h_{BP}^{(f1,f2)}(m,n) = h_{LP}^{(f2)}(m,n) - h_{LP}^{(f1)}(m,n), \qquad (16.26)$$

$$h_{BS}^{(f1,f2)}(m,n) = \delta(m,n) - h_{BP}^{(f1,f2)}(m,n). \qquad (16.27)$$

2D windows, used in 2D case for tapering theoretical impulse responses, are typically outer products of 1D windows:

$$w_{2D}(m,n) = w_{1D}(m) \cdot w_{1D}(n). \qquad (16.28)$$

For the 1D Gaussian window, its 2D perfectly radial (circular) version is obtained from Eq. (16.28):

$$w_{2D}(m,n) = \left(\frac{1}{\sqrt{2\pi}\sigma} e^{-m^2/2\sigma^2}\right)\left(\frac{1}{\sqrt{2\pi}\sigma} e^{-n^2/2\sigma^2}\right) = \frac{1}{2\pi\sigma^2} e^{-(m^2+n^2)/2\sigma^2}.$$

$$(16.29)$$

In Fig. 16.17, design of 2D filter weights, using the window method, is explained. First, required filter frequency response is selected (top-left corner). Then, its theoretical impulse response (weights!) is calculated analytically (bottom-left corner). Next, a 2D window is applied to the computed filter weights (bottom-right) and, finally, the obtained filter frequency response is checked (top-right).

The 2D Gaussian function itself is a very good low-pass filter. After its differentiation in m axis and n axis separately, nice filters for detection of vertical and horizontal image edges are obtained. After calculation of second derivatives in both directions and their combination, the Laplacian of Gaussian is derived, which is an efficient filter for edge detection in all directions. Definitions of Gaussian function-based filters are presented in Table 16.2. For example, $g_{1m}(m,n)$ denotes first derivative of the 2D Gaussian function $g_0(m,n)$ in regard to variable m and $g_{2n}(m,n)$ denotes second derivative in regard to variable n.

In image processing, application of large masks of filter weights is not possible due to limited image size. Therefore, generally, standard filters of some specific types are used for: smoothing, differentiating, and enhancing image edges. The most popular filter weights are defined in Table 16.3.

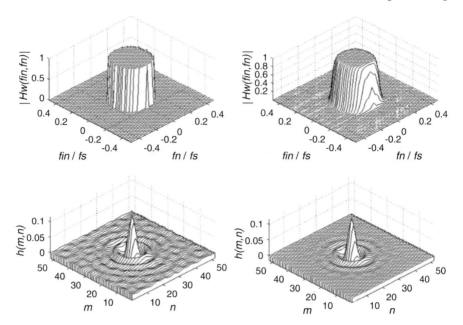

Fig. 16.17: Explanation of 2D filter design using window method: (top) required and designed filter frequency response, (bottom) calculated theoretical 2D filter impulse response and its windowed version [17]

Table 16.2: 1D and 2D filters originated from the Gaussian function ($-K \leq m \leq K$, $-L \leq n \leq L$)

Function	1D filter	2D filter
Gaussian	$g_0(m) = \frac{1}{\sqrt{2\pi}\sigma} \exp\left(-\frac{m^2}{2\sigma^2}\right)$	$g_0(m,n) = \frac{1}{2\pi\sigma^2} \exp\left(-\frac{m^2+n^2}{2\sigma^2}\right)$
First derivative	$g_{1m}(m) = -\frac{m}{\sigma^2} g_0(m)$	$g_{1m}(m,n) = -\frac{m}{\sigma^2} \cdot g_0(m,n)$, $g_{1n}(m,n) = -\frac{n}{\sigma^2} \cdot g_0(m,n)$
Second derivative	$g_{2m}(m) = \frac{m^2-\sigma^2}{\sigma^4} g_0(m)$	$g_{2m}(m,n) = \frac{m^2-\sigma^2}{\sigma^4} g_0(m,n)$, $g_{2n}(m,n) = \frac{n^2-\sigma^2}{\sigma^4} g_0(m,n)$
Laplacian		$g_2(m,n) = g_{2m}(m,n) + g_{2n}(m,n) = \frac{m^2+n^2-2\sigma^2}{\sigma^4} g_0(m,n)$

In Fig. 16.18, application of filters derived from Gaussian function to *Cameraman* image is presented. The Gaussian function parameters are as follows: $K = L = 4, \sigma = 1.4$. In first row, we see result of low-pass filtration with original 2D Gaussian function and band-pass filtration exploiting its 2D Laplacian. In the second row, first derivatives of the Gaussian function in direction m and n are used as gradient filters, causing enhancement of vertical and horizontal image edges, respectively. After taking pixels of both images to the power of two, their summation and square root calculation, the third image is obtained, having all edges enhanced.

Table 16.3: Definition of some linear and non-linear 2D filters

Filter type / Application	Definition
Low-pass / *Smoothing*	$\frac{1}{9}\begin{bmatrix}1&1&1\\1&1&1\\1&1&1\end{bmatrix}\quad \frac{1}{10}\begin{bmatrix}1&1&1\\1&2&1\\1&1&1\end{bmatrix}\quad \frac{1}{16}\begin{bmatrix}1&2&1\\2&4&2\\1&2&1\end{bmatrix}\quad \frac{1}{273}\begin{bmatrix}1&4&7&4&1\\4&16&26&16&4\\7&26&41&26&7\\4&16&26&16&4\\1&4&7&4&1\end{bmatrix}$
Differentiation / *Edges—|*	Sobel: $\begin{bmatrix}1&2&1\\0&0&0\\-1&-2&-1\end{bmatrix}\begin{bmatrix}1&0&-1\\2&0&-2\\1&0&-1\end{bmatrix}$ Prewitt: $\begin{bmatrix}1&1&1\\0&0&0\\-1&-1&-1\end{bmatrix}\begin{bmatrix}1&0&-1\\1&0&-1\\1&0&-1\end{bmatrix}$
Differentiation / *Edges* $\diagup\diagdown$	Roberts: $\begin{bmatrix}1&0\\0&-1\end{bmatrix}\begin{bmatrix}0&1\\-1&0\end{bmatrix}$ skew: $\begin{bmatrix}1&1&1\\1&-2&-1\\1&-1&-1\end{bmatrix}\begin{bmatrix}1&1&1\\-1&-2&1\\-1&-1&1\end{bmatrix}$
Double differentiation / *Any edges*	$\begin{bmatrix}0&-1&0\\-1&4&-1\\0&-1&0\end{bmatrix}\begin{bmatrix}1&-2&1\\-2&4&-2\\1&-2&1\end{bmatrix}\begin{bmatrix}-1&-1&-1\\-1&8&-1\\-1&-1&-1\end{bmatrix}\begin{bmatrix}-1&-1&-1&-1&-1\\-1&-1&-1&-1&-1\\-1&-1&24&-1&-1\\-1&-1&-1&-1&-1\\-1&-1&-1&-1&-1\end{bmatrix}$
Laplacian of Gaussian / *Any edges*	$\begin{bmatrix}0&0&1&0&0\\0&1&2&1&0\\1&2&-16&2&1\\0&1&2&1&0\\0&0&1&0&0\end{bmatrix}\quad \begin{bmatrix}0&1&1&2&2&2&1&1&0\\1&2&4&5&5&5&4&2&1\\1&4&5&3&0&3&5&4&1\\2&5&3&-12&-24&-12&3&5&2\\2&5&0&-24&-40&-24&0&5&2\\2&5&3&-12&-24&-12&3&5&2\\1&4&5&3&0&3&5&4&1\\1&2&4&5&5&5&4&2&1\\0&1&1&2&2&2&1&1&0\end{bmatrix}$
Median / *Impulse noise removal*	med $\begin{bmatrix}x_1&x_2&x_3\\x_4&x_5&x_6\\x_7&x_8&x_9\end{bmatrix}$ = central value after sorting $\{\min,\ \max\}$
Morphological / *Eroding/dilating*	min $\begin{bmatrix}x_1&x_2&x_3\\x_4&-&x_6\\x_7&x_8&x_9\end{bmatrix}$ max $\begin{bmatrix}x_1&x_2&x_3\\x_4&-&x_6\\x_7&x_8&x_9\end{bmatrix}$

In program 16.3, 2D images filtering fundamentals are put together. Filters designed using the 2D window method as well as filters derived from the 2D Gaussian function are presented and tested. Image smoothing and edge enhancement are shown.

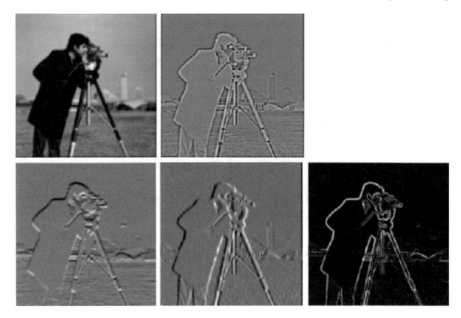

Fig. 16.18: Filtering results of *Cameraman* image using 2D filters obtained from 2D Gaussian function. First row: low-pass filtration (by Gaussian) and band-pass filtration (by its Laplacian). Second row: gradient filtering over rows and columns, and a square root of summation of first two images (after taking them to the power of two) [17]

Listing 16.3: 2D filters and their application to image filtration via 2D convolution

```
1   % lab16_ex_filter_design.m
2   % Design of 2D filters
3   clear all; close all; figure;
4
5   L = 15;   % width of filter mask (square)
6
7   K = (L-1)/2; df = 0.5/K;            % variables used for filter weights
8   m = ones(L,1)*(-K:K);              % generation and figure description
9   n = (-K:K)'*ones(1,L);            %
10  fm = ones(L,1)*(-0.5:df:0.5);     %
11  fn = (-0.5:df:0.5)'*ones(1,L);    %
12
13  % Read image to be processed
14  [x,cmap] = imread('Cameraman.tif');
15  imshow(x,cmap), title('Image'); pause
16  [N, M] = size(x);
17  x = double(x);
18
19  % Filters originated from the Gaussian function
20  sigma = 1.4; df = 0.5/K;
21  g0  = 1/(2*pi*sigma^2) .* exp( -(m.^2+n.^2)/(2*sigma^2) ); % Gaussian function
```

```
22 |  g1m = -m/(sigma^2) .* g0;                              % derivative over "m"
23 |  g1n = -n/(sigma^2) .* g0;                              % derivative over "n"
24 |  g2  = (m.^2 + n.^2 - 2*sigma^2)/(sigma^4) .* g0; % second derivative over "m", "n"
25 |  figure;
26 |  subplot(221); mesh(m,n,g0); title('Gaussian filter');
27 |  subplot(222); mesh(m,n,g2); title('Laplacian of Gaussian');
28 |  subplot(223); imshow( conv2(x,g0,'same'),cmap );
29 |  subplot(224); imshow( conv2(x,g2,'same'),cmap ); pause
30 |  figure;
31 |  subplot(231); mesh(m,n,g1m); title('Gradient "m"');
32 |  subplot(232); mesh(m,n,g1n); title('Gradient "n"');
33 |  subplot(234); imshow( conv2(x,g1m,'same'),cmap );
34 |  subplot(235); imshow( conv2(x,g1n,'same'),cmap );
35 |  subplot(236); imshow( sqrt( conv2(x, g1m,'same').^2 + conv2(x, g1n,'same').^2 ), cmap )
       ;
36 |  pause
37 |
38 |  % 2D WINDOW METHOD
39 |  chka = 1;    % 0 = rectangular mask, 1 = circular mask
40 |  w = hamming(L); w = w * w'; % 2D window
41 |  figure; mesh(m,n,w); colormap([0 0 0]); title('2D Window'); pause
42 |
43 |  if(chka==0) % Rectangular freq response - imp. responses of two low-pass filters
44 |     f0=0.25; wc = pi*f0; sinc = sin(wc*(-K:K))./(pi.*(-K:K));
45 |     sinc(K+1)=f0; lp1=sinc'*sinc;
46 |     f0=0.50; wc = pi*f0; sinc = sin(wc*(-K:K))./(pi.*(-K:K));
47 |     sinc(K+1)=f0; lp2=sinc'*sinc;
48 |  else         % Circular freq response - imp. responses of two low-pass filters
49 |     f0=0.25; wc=pi*f0; lp1=wc*besselj(1,wc*sqrt(m.^2 + n.^2))./(2*pi*sqrt(m.^2+n.^2));
50 |     lp1(K+1,K+1)= wc^2/(4*pi);
51 |     f0=0.50; wc=pi*f0; lp2=wc*besselj(1,wc*sqrt(m.^2+n.^2))./(2*pi*sqrt(m.^2+n.^2) );
52 |     lp2(K+1,K+1)= wc^2/(4*pi);
53 |  end
54 |
55 |  lp = lp1;                                 % LowPass without 2D window
56 |  lpw = lp .* w;                            % with window
57 |  hp = - lp1; hp(K+1,K+1) = 1 - lp1(K+1,K+1);  % HighPass without 2D window
58 |  hpw = hp .* w;                            % with window
59 |  bp = lp1 - lp2;                           % BandPass without 2D window
60 |  bpw = bp .* w;                            % with window
61 |  bs = - bp; bs(K+1,K+1) = 1 - bp(K+1,K+1); % BandStop without 2D window
62 |  bsw = bs .* w;                            % with window
63 |  for typ = 1 : 4                    % show imp. response, its spectrum and filtered image
64 |     switch (typ)                   % choose filter type
65 |     case 1, h = lp; hw = lpw;  % LP
66 |     case 2, h = hp; hw = hpw;  % HP
67 |     case 3, h = bp; hw = bpw;  % BP
68 |     case 4, h = bs; hw = bsw;  % BS
69 |     end
70 |     figure;
71 |     subplot(221); mesh(m,n,h);  title('Filter h(m,n)');
72 |     subplot(222); mesh(m,n,hw); title('Filter hw(m,n)');
73 |     subplot(223); mesh(fm,fn,abs( fftshift2D(fft2(h)) ) ); title(' |H(fm,fn)|');
74 |     subplot(224); mesh(fm,fn,abs( fftshift2D(fft2(hw)) ) ); colormap([0 0 0]);
```

```
75      title('|Hw(fm,fn)|'); pause
76      figure;
77      subplot(121); y = conv2(x, h,'same'); imshow(y,[min(min(y)),max(max(y))]);
78      subplot(122); y = conv2(x, hw,'same'); imshow(y,[min(min(y)),max(max(y))]); pause
79  end
80
81  % FILTER DESIGN IN FREQUENCY DOMAIN
82  % Work variables
83  N=L+1; df = 0.5/(K+1);
84  m  = ones(L+1,1)*(-(K+1):K);         n  = (-(K+1):K)'*ones(1,L+1);
85  fm = ones(L+1,1)*(-0.5:df:0.5-df); fn = (-0.5:df:0.5-df)'*ones(1,L+1);
86  % 2D window - its shape and 2D spectrum
87  w = hamming(N); w = w * w';
88  % Required frequency response - circular or rectangular
89  Q = round(K/2); % LP filter width
90  H = zeros(N,N);
91  for k = N/2+1-Q : N/2+1+Q
92      for l = N/2+1-Q : N/2+1+Q
93          if( (k-N/2-1)^2 + (l-N/2-1)^2 <= Q^2) H(k,l) = 1; else H(k,l) = 0; end %
                circular
94      end
95  end
96  % H(N/2+1-Q : N/2+1+Q, N/2+1-Q : N/2+1+Q) = ones(2*Q+1,2*Q+1); % rectangular
97  % Filter design and its testing
98  h = real( ifft2(fftshift2D(H)) ); h = fftshift2D(h);    % impulse response
99  hw = h .* w;                                            % with window
100 Hw = abs( fftshift2D( fft2( hw ) ) );                   % 2D spectrum
101 y = conv2(x, hw, 'same');                               % 2D filtration
102 % Figures
103 figure;
104 subplot(121); mesh( m,n,w ); title('2D window 2D w(m,n)');
105 subplot(122); mesh( fm,fn,abs( fftshift2D(fft2(w)) ) ); title('|W(fm,fn)|');
106 colormap([0 0 0]); pause
107 figure;
108 subplot(221); mesh(fm,fn,H);  title('Required |H(fm,fn)|');
109 subplot(222); mesh(m,n,h);    title('2D filter h(m,n)');
110 subplot(223); mesh(fm,fn,Hw); title('2D filter freq response |Hw(fm,fn)|');
111 subplot(224); mesh(m,n,hw);   title('2D filter with window hw(m,n)');
112 colormap([0 0 0]); pause
113 figure;
114 subplot(121); imshow(y, cmap); title('Image after filtering'); y = y(1:128,65:192);
115 subplot(122); imshow(y,[min(min(y)),max(max(y))]); title('Image fragment'); pause
```

Exercise 16.4 (2D Filter Design and Image Filtering). Read carefully the program 16.3. Run it and observe figures. Read your *selfie* and process it in the program instead of the image *Cameraman*. Add to the program filter weights, defined in the Table 16.3. Test them on your image.

Exercise 16.5 (Object Recognition). Using code of the program 16.3, write a new program. Make a photo of a license plate of a car. Filter the image and binarize it. Try to recognize digits written on the plate. Build a code-book of digits. Use 2D auto-correlation function for digit recognition. If needed, do normalization of the license plate size and color. Read about affine image transformations. Are they needed in this application?

Exercise 16.6 (Morphological Image Processing). Test the following Matlab commands on your own photo x:
```
% binarization
indx=find( x>100 ); y=zeros(size(x));
y(indx)=255*ones(size(indx));
z = imerode( y, [ 1 1 1; 1 0 1; 1 1 1] ); % image erosion
z = imdilate( y, [ 1 1 1; 1 0 1; 1 1 1] ); % image dilation
```
Observe resultant images.

16.5 Fundamentals of Lossless Image Compression

Huffman Coding In lossless compression, long stream of bits is segmented into smaller blocks of bits, having the same length, so-called *symbols*. Then, frequency of symbol occurrence is analyzed, and those that appear more often obtain shorted bit codes while occurring rarely—longer codes. Let us explain this idea using a simple example. Let us assume that we have the following 10-element sequence of 4 symbols:

$$\{a,a,b,a,a,c,d,d,a,a\}.$$

Because symbols are 4, 2 bits are needed for their coding. "a" repeats 6 times, "b" and "c"—1 time and "d"—2 times. In general, we can say that probabilities of symbol occurrence are equal: $p_a = 0.6, p_b = 0.1, p_c = 0.1, p_d = 0.2$. In upper part of Fig. 16.19, the symbols and their probabilities are written. Next, we are looking for two symbols, having the lowest probabilities, and join them. In our case, symbols "b" and "c" are combined, and a joint symbol "bc" with probability $p_{bc} = p_b + p_c = 0.2$ is obtained. The operation of two minimum elements selection is repeated, and the joint symbol "bcd" with probability $p_{bcd} = p_{bc} + p_d = 0.4$ is obtained. Finally, two remaining symbols "a" and "bcd" are combined, and the joint element "abcd" with probability $p_{abcd} = 1$ results. This way the *Huffman tree* for our data was constructed. Now, it is time for finding replacement codes for all sym-

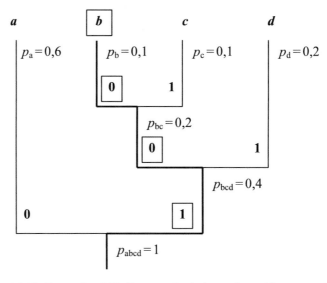

Fig. 16.19: Example of Huffman code design using a binary tree [17]

Table 16.4: Example of entropy-based lossless Huffman coding of binary symbol sequence

Symbol	Old code	Repetition number	Probability	New code
a	00	6	0.6=6/10	0
b	01	1	0.1=1/10	100
c	10	1	0.1=1/10	101
d	11	2	0.2=2/10	11

bols. We are starting from the tree trunk and going into each symbol, located at the end of tree branches. Turning left generates bit "0," while going "right" bit "1." In consequence, on the path to symbol "b" one obtains the sequence "100." This way new codes are calculated for all symbols. Our design is summarized in Table 16.4, where all old and new symbol codes are given.

What is the profit from coming from equal-length code-words to code-words of different length, but *statistically* motivated? At present, shorter code is given to most often occurring symbols.

$$
\begin{array}{lcccccccccl}
Massage: & a, & a, & b, & a, & a, & c, & d, & d, & a, & a \\
Old\ code: & 00 & 00 & 01 & 00 & 00 & 10 & 11 & 11 & 00 & 00 & = 20\ bits \\
New\ code: & 0 & 0 & 100 & 0 & 0 & 101 & 11 & 11 & 0 & 0 & = 16\ bits
\end{array}
$$

Should we be happy or not? Rather YES. Entropy of memory-less source gives as an information about minimum number of bits required for coding one symbol when probabilities of symbols occurring are known:

$$E = - \sum_{n=1}^{N_s} p_n \log_2(p_n). \tag{16.30}$$

In our case, we obtain

$$E = -0.6 \log_2(0.6) - 0.1 \log_2(0.1) - 0.1 \log_2(0.1) - 0.2 \log_2(0.2) = 1.5709506;$$

therefore, we are very close to the theoretical minimum: 10 symbols with 1.571 bits per symbol give 15.71 bits, and we used 16 bits.

In conclusion, the Huffman code of a given binary message is calculated as follows:

1. Divide a bit-stream to be coded into *symbols*, i.e. blocks of bits of equal length,
2. Find probability of symbol occurrence,
3. Build a Huffman tree and find replacements of all symbols,
4. Build a Huffman coding table with old and new code,
5. Analyze bit-stream symbol by symbol and replace bits of all symbols.

Huffman coder is not difficult. The decoder is much more complicated due to necessity of bit-stream cutting into blocks of bits with different length. Huffman tables can be calculated and added to the bit-stream or can be pre-computed for bigger data sets and be constant. In such case, they are known by encoder and decoder, and they are not embedded in the bit-stream.

Lossless Huffman coding is very popular in image compression as the last step of information packing. The same is true for audio coders, for example, MP3 and AAC. We should understand it and remember about it!

Exercise 16.7 (Huffman Coding). Write function of Huffman encoder and decoder.

Arithmetic Coding The arithmetic coding is also a very efficient lossless coding method. Since its patent has expired already, it is a sense to describe it also. Let us assume that our alphabet of symbols consists of only four letters (symbols) "a," "b," "c," and "d" occurring with probabilities 0.6, 0.1, 0.1, and 0.2, as before. Let us create for them a table of accumulated probability pa = [0, 0.6, 0.7, 0.8, 1.0]. Let us assume that each symbol has corresponding probability interval of real-value numbers: "a"—[0, 0,6), "b"—[0,6, 0,7), "c"—[0,7, 0,8), "d"—[0,8, 1). These intervals divide the whole range [0, 1) in specific proportions. If we choose any number from this range, it will belong only to one interval and *show* the symbol, the interval *owner*. If the selected interval will be divided ones more in the same proportions, and a new number will be chosen from it, this number will *hit* into a new interval and select a new symbol. Interval after interval, or symbol after symbol. Repeating this operation a few times, it is possible to define (choose) a sequence of symbols having only one, precisely chosen starting number. Of course, during coding and decoding the table of accumulated probabilities should be known since it defines the division proportions.

Let us assume that we are coding the message "aad." In the beginning, we are setting down and up pointers to the whole range: *down*=0, *up*=1, and *range*=1. Then, we are taking the first symbol from the message, "a" in our case, and read its probability bounds: [0, 0,6). Next, we use them for modifying our variables:

$$up = down + range \cdot 0.6, \qquad down = down + range \cdot 0, \qquad range = up - down. \tag{16.31}$$

At this moment, our variables have values: $up = 0.6$, $down = 0$, $range = 0.6$. Then, we are taking the second message symbol, again "a," and we are exploiting again numbers $[0, 0, 6)$ for the consecutive modification of our variable values, obtaining: $up = 0.36$, $down = 0$, $range = 0.36$. Finally, we are taking the last symbol "d," reading its probability bounds $[0, 8, 1)$ and modifying pointer values:

$$up = down + range \cdot 1 = 0.36, \qquad down = down + range \cdot 0.8 = 0.288. \tag{16.32}$$

Therefore, the result of "aad" coding is any number from the interval [0.288, 0.36). In practice, a fractional number from this range is used and their bits are sent. In our case, it is a number $0101 = \frac{0}{2^{-1}} + \frac{1}{2^{-2}} + \frac{0}{2^{-3}} + \frac{1}{2^{-4}} = \frac{5}{16} = 0.3125$. During decoding, symbols are reconstructed in similar way and in the same order like during encoding.

In Huffman coding, each symbol obtains its own unique code and is coded separately. In arithmetic coding, symbols are coded together, in groups. Binary representation of the message "aad" requires 6 bits: 3 symbols with 2 bits each. In case of Huffman coding of the same message, we obtain 4 bits: $1(a) + 1(a) + 2(d)$ (look at the code-book from the previous example), the same as for the arithmetic coder: 0101. However, for larger code-books and longer messages the arithmetic coding is more efficient.

Listing 16.4: Arithmetic encoder and decoder in Matlab

```
1  % lab16_ex_arithmetic.m
2  % Arithmetic encoder and decoder
3
4    clear all; close all;
5
6  % Initialization
7  alphabet = [ 'a'   'b'   'c'   'd' ];   % alphabet of symbols (letters)
8  p  =      [ 0.6  0.1  0.1  0.2 ];   % symbol probabilities
9  pa =      [ 0 0.6  0.7  0.8  1.0 ];   % accumulated probability to k-th symbol
10 Ns = length(alphabet);               % number of symbols
```

```
11
12    % Coding group of symbols together (one joint code-word is calculated)
13    num = [ 1, 1, 4];                        % symbol numbers: 1,1,4 =a,a,d
14    down = 0; up = 1;                         % pointers initialization
15    for k = 1 : length(num)                   % loop beginning
16        range = up-down;                      % new 'range'
17        up = down + range * pa( num(k)+1 );   % new 'up'
18        down = down + range * pa( num(k) );   % new 'down'
19    end
20
21    % Output bitstream = binary coded fractional number from the range [down, up)
22    code = 0; dv = 1/2; bits = [0];
23    while( ~((code >= down) & (code < up)) )
24        if(code < down)
25            dv = dv/2; code = code + dv; bits = [ bits 1 ];
26        else
27            if(code < up)
28                break;
29            else
30                code = code - dv; bits( end ) = 0;
31                dv = dv/2;
32                code = code + dv; bits = [ bits 1 ];
33            end
34        end
35    end
36    code, bits, pause
37
38    % Decoding message (sequence of symbols)
39    pa_rev = pa(end:-1:1); alpha = [ 0 (Ns:-1:1) ]; numdecod = []; % initialization
40    down = 0; up = 1; range = up-down;        %
41    for k = length(num) : -1 : 1              % loop beginning
42        x = (code-down)/range;                % calculate number 'x'
43        indx = find( pa_rev <= x ), indx = indx(1); % first x: pa_reverse <= x
44        up = down + range*pa_rev( indx-1 );   % new 'up'
45        down = down + range*pa_rev( indx );   % new 'down'
46        range = up-down;                      % new 'range'
47        numdecod = [numdecod alpha(indx) ];   % number of decoded symbol
48    end
49    disp('Sent symbols'); num
50    disp('Received symbols'); numdecod
```

16.6 Image and Video Compression

Image and video compression are one of the main telecommunication engineering tasks. Their methods and standards are developed for years. In this section, we learn fundamentals of JPEG standard, concerning still image coding, and MPEG standard, dealing with video coding.

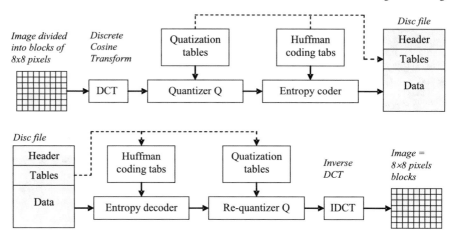

Fig. 16.20: JPEG encoder (top) and decoder (bottom) [17]

16.6.1 JPEG: Still Images

Encoder and decoder of the JPEG compression algorithm are presented in Fig. 16.20. The whole still image is divided into non-overlapping blocks of 8×8 pixels $x(m,n)$. Each such block is transformed by 2D DCT. Next, each DCT coefficient of the matrix $X_{\mathrm{DCT}}(k,l)$ is quantized using corresponding element of a quantizer matrix $Q(k,l)$, having the same 8×8 dimensions:

$$Z(k,l) = \left\lfloor \frac{X_{\mathrm{DCT}}(k,l) \pm \lfloor Q(k,l)/2 \rfloor}{Q(k,l)} \right\rfloor, \quad k,l = 0, 1, \ldots, 7, \tag{16.33}$$

where $\lfloor x \rfloor$ denotes the biggest integer number lower or equal to x, i.e. the operator $\lfloor . \rfloor$. In Eq. (16.33), sign '+' is used for $X_{\mathrm{DCT}}(k,l) \geq 0$ and sign '−' otherwise. The *RGB* components are transformed to YC_bC_r color system, and Eq. (16.33) is used separately for image luminance Y and for chrominances C_b and C_r. Each color image component has different quantization table:

$$Q_Y = \begin{bmatrix} 16 & 11 & 10 & 16 & 24 & 40 & 51 & 61 \\ 12 & 12 & 14 & 19 & 26 & 58 & 60 & 55 \\ 14 & 13 & 16 & 24 & 40 & 57 & 69 & 56 \\ 14 & 17 & 22 & 29 & 51 & 87 & 80 & 62 \\ 18 & 22 & 37 & 56 & 68 & 109 & 103 & 77 \\ 24 & 35 & 55 & 64 & 81 & 104 & 113 & 92 \\ 49 & 64 & 78 & 87 & 103 & 121 & 120 & 101 \\ 72 & 92 & 95 & 98 & 112 & 100 & 103 & 99 \end{bmatrix} \quad Q_C = \begin{bmatrix} 17 & 18 & 24 & 47 & 99 & 99 & 99 & 99 \\ 18 & 21 & 26 & 66 & 99 & 99 & 99 & 99 \\ 24 & 26 & 56 & 99 & 99 & 99 & 99 & 99 \\ 47 & 69 & 99 & 99 & 99 & 99 & 99 & 99 \\ 99 & 99 & 99 & 99 & 99 & 99 & 99 & 99 \\ 99 & 99 & 99 & 99 & 99 & 99 & 99 & 99 \\ 99 & 99 & 99 & 99 & 99 & 99 & 99 & 99 \\ 99 & 99 & 99 & 99 & 99 & 99 & 99 & 99 \end{bmatrix} .$$

$$\tag{16.34}$$

After this operation many DCT coefficients are equal to zero, for example, rather drastic, two successive blocks can look as follows:

$$
\begin{bmatrix}
520 & 50 & 0 & 0 & 0 & 0 & 0 & 0 \\
20 & 0 & 0 & 0 & 0 & 0 & 0 & 0 \\
0 & 0 & 0 & 0 & -12 & 0 & 0 & 0 \\
0 & 0 & 0 & 0 & 0 & 0 & 0 & 0 \\
0 & 0 & 0 & 0 & 0 & 0 & 0 & 0 \\
0 & 0 & 0 & 0 & 0 & 0 & 0 & 0 \\
0 & 0 & 0 & 0 & 0 & 0 & 0 & 0 \\
0 & 0 & 0 & 0 & 0 & 0 & 0 & 0
\end{bmatrix}
\begin{bmatrix}
510 & 30 & 0 & 0 & 0 & 0 & 0 & 0 \\
47 & 0 & 0 & 0 & 0 & 0 & 0 & 0 \\
0 & 0 & 0 & 0 & 0 & 0 & 0 & 0 \\
12 & 0 & 0 & 0 & 0 & 0 & 0 & 0 \\
0 & 0 & 0 & 0 & 0 & 0 & 0 & 0 \\
0 & 0 & 0 & 0 & 0 & 0 & 0 & 0 \\
0 & 0 & 0 & 0 & 0 & 0 & 0 & 0 \\
0 & 0 & 0 & 0 & 0 & 0 & 0 & 0
\end{bmatrix}. \qquad (16.35)
$$

Because sparse matrices are typically obtained with the biggest values positioned in the top-left corner, they are scanned in *Zig–Zag* manner as shown in Fig. 16.21. The detailed description of coding will be presented later; at present, we aim at qualitative description of coding and decoding process.

After scanning DCT coefficients, which survived *quantization* operation, they are losslessly coded using Variable Length Integer (VLI) method and Huffman entropy encoder. Finally, the binary file is created consisting of header, describing the applied compression details: information about quantization tables, used Huffman tables, and bits coding blocks of DCT coefficients. In the decoder, all operations are performed in reverse order: (1) sparse blocks of DCT coefficient are reconstructed, (2) zeros are put into empty spaces/positions left by non-coded DCT coefficients (equal to zero), (3) the inverse 2D DCT transform is performed, and (4) some *deblocking* filtering is eventually applied.

At present, let us look more carefully to the encoding process. DC coefficients of neighboring 8×8 blocks (one value in top-left image corner) are typically big and correlated—they represent mean values of blocks of pixels. For this reason, they are treated specially: DC coefficient of the first block and differences between DC coefficients of neighboring blocks are coded using the VLI method. Each number, let us call it D, is replaced by a pair of bit sequences (B, Y):

- B—four bits specifying the number of bits required for D magnitude coding (from 0 to 11 bits for DC differences),
- Y—magnitude of D coded binary (only significant bits); if D is negative, its bits are negated; Y is absent if $D = 0$.

Finally, the lossless Huffman coding is used (arithmetic coding is optional) but only in regard to bits of Y. Different tables for different magnitude lengths are used. In decoder, the most significant bit of Y is checked: if it is equal to 0, all bits are negated.

The remaining DCT coefficients, called AC coefficients, are coded in each block separately. Their values are scanned in *zig–zag* manner, shown in Fig. 16.21, i.e. from the lowest to the highest frequency. Due to quantization, long sequences of 0s are separated by few non-zero values. In the beginning, 8-bit objects (Z, B) are calculated:

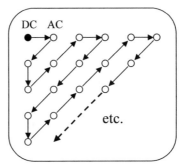

Fig. 16.21: Scanning 2D DCT coefficients of 8×8 blocks in JPEG standard [17]

- Z—(4 bits) number of 0s before a non-zero value (code RLC—*Run Length Coding*),
- B—(4 bits) the smallest number of bits required for coding magnitude of non-zero value, value changing from 1 to 10.

When sequences of zeros have more than 15 elements, they are artificially divided into sub-blocks (15,0), in order to code the pair (Z,B) on 8 bits. Thanks to this, size of used Huffman tables is limited. Record (15,0) denotes 16 zeros: 15 preceding and 1 zero as an element value. Next, value of a non-zero element Y is coded: its magnitude is written binary, most significant bits are left and all bits are negated, if Y is negative. Then, the Huffman coding is used for the group of bits obtained this way. Different tables are used for different lengths of magnitudes. As we see, a pair (B,Y) represents a VLI code of each non-zero AC DCT coefficient. Huffman codes of objects (Z,B) and Huffman codes of non-zero AC elements are placed one-by-one in the bit-stream.

In the decoder, all operations are performed in reverse order and re-quantization of each 8×8 DCT block is done:

$$\hat{X}_{\mathrm{DCT}}(k,l) = Q(k,l) \cdot \hat{Z}(k,l), \quad k,l = 0,\ 1,\ldots,\ 7, \tag{16.36}$$

where $\hat{Z}(k,l)$ denotes a result of entropy decoding. Reconstructed $\hat{X}_{\mathrm{DCT}}(k,l)$ is the image luminance or chrominance. Error introduced by image compression and decompression can be measured using the PSNR (Peak Signal-to-Noise Ratio) defined as:

$$PSNR = 10\log_{10}\left(\frac{MAX_I^2}{MSE}\right), \tag{16.37}$$

where the maximum image pixel value MAX_I and mean-square image reconstruction error MSE are equal to:

$$MAX_I = 2^b, \qquad MSE = \frac{1}{MN} \sum_{m=0}^{M-1} \sum_{n=0}^{N-1} (x_{IN}(m,n) - x_{OUT}(m,n))^2. \qquad (16.38)$$

Example: Simplified JPEG-Like Image Coding The quantization table consists of one constant: $Q(k,l) = c$. Differences of DC DCT coefficients are coded using only the VLI method: always 4 bits specifying required number of bits for difference magnitude coding and the magnitude bits themselves. For example, difference 7 written in binary is equal to 0111, and in VLI code: 0011 111. For negative numbers, the magnitude bits are negated, e.g. for -7 we have 0011 000 (analogically, $-1 \rightarrow 0001\ 0$, $6 \rightarrow 0011\ 110$). Let us assume that two first, consecutive 2D DCT blocks are given by Eq. (16.35). After *zig-zag* scanning, we obtain

Block 1 : 520, 50, 20, 0, 0, 0, 0, 0, 0, 0, 0, 0, 0, 0, 0, 0, 0, 0, 0, 0, 0, 0, 0, 0, 0, 0, 0, -12, 0, ..., 0
Block 2 : 510, 30, 47, 0, 0, 0, 0, 0, 0, 0, 12, 0, ..., 0.

Now we are creating a pair (*number of preceding zeros, non-zero value*). The pair (15,0) denotes a sequence of 16 zeros and the pair (0,0)—only zeros to the end of block. We get

$$Block\ 1: \quad DC = 520, AC = (0,50), (0,20), (15,0), (6,-12), (0,0),$$
$$Block\ 2: \quad DC = 510, AC = (0,30), (0,47), (6,12), (0,0).$$

Coding block #1:
DC = 520: 520 - 0 = 520, binary sequence: 1010 1000001000
First four bits tell us that 10 bits are required for *DC* coding, then the *DC* value is given. Because the value is positive, bits are not negated. Then *AC* coefficients are coded using many pairs (*number of preceding zeros, next AC value*). *Number of preceding zeros* is coded using 4 bits *zzzz*. If sequence of 0s has more than 15 elements, more pairs have to be used. For example, for 22 zeros two pairs are exploited: (15,0) and (6,X), where X denotes a non-zero *DC* value. The *nextDCvalue* is coded in two blocks: four bits *bbbb*, specifying the number of bits required for DC coding, and variable-length block of bits $xx...x$ with DC value (the bits are negated for negative numbers). Below the coding result is presented for our example:

	zzzz	*bbbb*	*xx......x*
$AC = (0,50)$:	0000	0110	110010
$AC = (0,20)$:	0000	0101	10100
$AC = (15,0)$:	1111	0000	
$AC = (6,-12)$:	0110	0100	0011
$AC = (0,0)$:	0000	0000	

Coding block #2:
DC = 510: 510 - 520 = -10, binary sequence: 0100 0101
First four bits tell that -10 requires 4 bits for binary coding. Then, number 10 is written in binary and all bits are negated (negative number): 1010 → 0101. DC coefficients are coded differentially, only -10 is stored since $520 - 10 = 510$—the coded value.

Exercise 16.8 (JPEG-Like Image Compression). Write your own program: (1) read any black-and-white image, for example, the *Cameraman*, (2) split it into blocks 8×8 pixels, perform 2D DCT on each block, (3) quantize DCT coefficients, dividing them by corresponding elements from the quantization table, according to Eq. (16.35), (4) perform inverse DCT of each block, (5) compare original and decompressed image, and calculate PSNR (16.37). In step 3, you can simply use single constant for quantization of all DCT coefficients. If it is interesting for you, as an extra task, try to do the *zig–zag* scanning of quantized DC coefficients of one 8×8 block and to code them into a bit-stream using the VLI method. Check whether you can recover the quantized DCT coefficients from the bit-stream with no error. If you still have *positive energy*, try to climb Kilimanjaro Peak and do the Huffman or arithmetic coding of the bit-stream.

16.6.2 MPEG: Moving Pictures

The simplest method of moving pictures compression relies on application of JPEG algorithm to each image of the sequence independently. Such solution is known as *Moving*-JPEG. However, this approach is completely inefficient when images are very similar to each other. A better idea is to compress one image as a still picture using any JPEG-like method and a few preceding images only differentially, i.e. only changes between them. Going further, higher compression of *movies* is obtained when the pure difference between two consecutive images is replaced by a difference between a new image and its prediction based on previous images, with

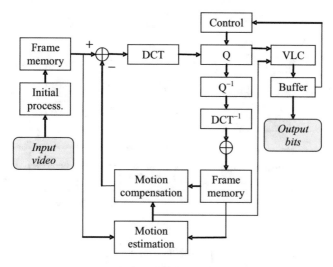

Fig. 16.22: MPEG encoder [17]

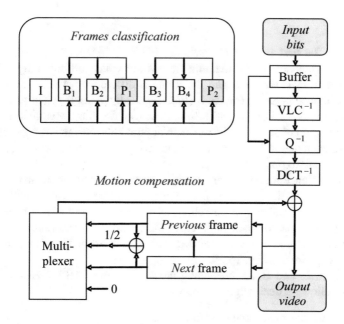

Fig. 16.23: MPEG decoder [17]

compensated motion of block of pixels. Such approach is applied in MPEG standard for moving pictures compression, used in digital TV and digital video discs.

In Figs. 16.22 and 16.23, block diagrams of MPEG encoder and decoder are presented. In MPEG standard, compression algorithm option is selected by encoder control block. Joint coding of block of 7 images (*frames*) is one possibility (see

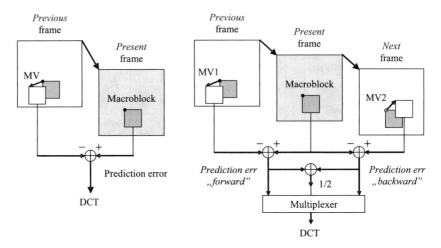

Fig. 16.24: Explanation of *motion estimation and compensation* in MPEG standard: (left) one-directional "forward," (right) two-directional "forward" and "backward" [17]

Fig. 16.23). The first frame *I* (*Intraframe*) is coded independently from the rest, in *JPEG-like* manner. The remaining frames are coded differentially: difference between frame P_1 (*Predicted*) and the frame *I*, difference between frames P_2 and P_1, and difference between frames *B* (*Bidirectionally predicted*) and two reference frames *I* and P_1 or P_2. Usage of still *I* frame aims at avoiding accumulation of errors and simplification of scrolling the *movie*. Not pure but intelligent image subtraction is done: for each macro-block of pixels (e.g., 16×16) of actual frame, coded differentially in regard to one or two frames already coded, the most similar macro-block is found in reference frame(s): only in one of previous frames or one in previous and one in next frames. Then the found similar macro-block or macro-blocks are subtracted from the actual frame. Searching of similar macro-blocks is called *motion estimation*, while macro-block subtraction, the *motion compensation*. Found macro-block shifts are stored as *motion vectors*. The motion estimation–compensation algorithm is explained in Fig. 16.24.

In Figs. 16.22 and 16.23, presenting block diagram of MPEG encoder and decoder, the following denotations are used:

- DCT—2D discrete cosine transform,
- Q—quantizer,
- VLI—lossless coding using Variable Length Integers,
- MV—motion vector,
- I—frame coded without motion compensation,
- *P*1—frame coded differentially (predicatively) in respect to frame I, with forward motion compensation,

- $P2$—frame coded differentially (predicatively) in respect to frame P1, with forward motion compensation,
- B—frames coded differentially with forward (from I) and backward (from P) motion compensation.

In simplification, in MPEG encoder the following operations are done:

1. Frames I are coded using non-overlapping macro-blocks of pixels in a manner similar to JPEG, e.g. 16×16 blocks of pixels are transformed by 2D DCT and quantized $(DCT + Q)$. Next, they are decoded $(IDCT + Q^{-1})$ and ready to be used by next frames during differential coding. Let I_r denote the reconstructed image I.
2. Motion compensation of frame P_1, in regard to frame I_r, is done, namely shifting of macro-blocks from I_r and subtracting them from P_1. The best motion vectors (MV), minimizing the difference, are found and stored. Obtained differential image is coded the same way as frame I and next decoded—the image IP_{1r} is obtained. When we have difference IP_{1r} between the frame P_1 and shifted macro-blocks of the frame I_r, and when the frame I_r and the motion vectors MV are known, we can easily reconstruct the frame P_1: $P_{1r} = IP_{1r} + MV(I_r)$. The $MV(I_r)$ denotes an image obtained from image I_r as a result of shifting of its macro-blocks using motion vectors MV.
3. Next, frame P_2 is coded, making use of reconstructed frame P_1, in the same way the frame P_1 was coded in regard to reconstructed frame I.
4. In the end, frames B_1, B_2, B_3, B_4 are coded differentially in regard to frames: decoded I and decoded P_1 either decoded P_1 and decoded P_2.

Exercise 16.9 (MPEG-Like Image Compression). Read any image into Matlab. Convert it into black-and-white colors. Let us assume that the image has dimensions 512×512. Use a 2D 256×256 observation window, shift it above the image, and cut from it a sequence of frames, *record a virtual movie*. Then, knowing details of JPEG standard and ideas of MPEG standard, try to code your *moving pictures* performing DCT on their differences and quantizing the result.

16.7 Image Watermarking

Data security and authorization/authentication is at present a very important topic. At the end of this chapter, we learn how to add a watermark to our image. The watermark is a pattern consisting of random non-overlapping blocks of squares filled with numbers -1 either 1. It has the same dimension as an image to be watermarked. The watermark is multiplied with small pseudo-random numbers and added to the

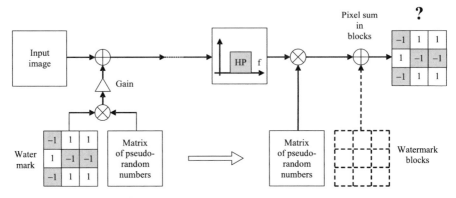

Fig. 16.25: Block diagram of implemented image watermarking algorithm [17]

Fig. 16.26: Image watermarking results: (1) original image, (2) [−1/+1] watermark, (3) watermark modulated by noise, and (4) image with added watermark [17]

image. In the watermark decoder, the image is first high-pass filtered, and its pixel values are summed separately in the area of watermark blocks. As a result, a watermark [−1, +1] pattern should be recovered. If the image has been *attacked*, some blocks could be distorted. Block diagram of the algorithm is presented in Fig. 16.25, while results from its application in Fig. 16.26.

Listing 16.5: Watermarking algorithm implemented in Matlab

```
1   % lab16_ex_watermarking.m
2     clear all; close all;
3
4   % Parameters
5     K = 32;        % block size for one watermark bit (K x K pixels)
6     gain = 1;      % watermark gain
7
8   % Read image to be watermarked
9     A = imread('lena.bmp');
10    B = double(A); [M, N] = size(B);
11
12  % Addition of watermark
13    Mb = (M/K); Nb=(N/K);                % number of blocks in row and column
14    plusminus1 = sign( randn(1,Mb*Nb) ); % random sequence of numbers +1/-1
```

```
15    Mark = zeros( size(B) );                    % watermark: chessboard with +1/-1 pattern
16    for i = 1:Mb
17        for j = 1:Nb
18            Mark( (i-1)*K+1 : i*K,  (j-1)*K+1 : j*K ) = plusminus1(i*j);
19        end
20    end
21    Noise = round( randn(size(B)) );            % noise (modulating signal)
22    MarkNoise = gain * Noise .* Mark;           % watermark = gain * noise * mark(+/-1)
23    B = uint8( B + MarkNoise );                 % image + watermark, conversion to 8 bits
24
25  % Figures
26    figure, subplot(1,2,1), imshow(Mark, []);       title('Watermark')
27            subplot(1,2,2), imshow(MarkNoise, []); title('Watermark * noise')
28    figure, subplot(1,2,1); imshow(A, []);          title('Original image')
29            subplot(1,2,2); imshow(B, []);          title('Image with watermark')
30
31  % Watermark detection
32    B = double(B);                              % conversion to double precision
33
34  % High-pass filtering
35    L = 10; L2=2*L+1;                           % 2D filter size (L x L)
36    w = hamming(L2); w = w * w';                % 2D Hamming window
37
38    f0=0.5; wc = pi*f0;  [m n] = meshgrid(-L:L,-L:L);                    % 2D
39    lp = wc * besselj(1, wc*sqrt(m.^2 + n.^2) )./(2*pi*sqrt(m.^2 + n.^2)); % low-pass
40    lp(L+1,L+1)= wc^2/(4*pi);                                            % filter
41    hp = - lp; hp(L+1,L+1) = 1 - lp(L+1,L+1);   % high-pass filter without window
42    h = hp .* w;                                % with 2D window
43    B = conv2( B, h, 'same');                   % image filtering
44
45  % Bit (+1/-1) detection in each pixel block of the watermark
46    Demod = B .* Noise;                         % noise demodulation
47    MarkDetect = zeros( size(B) );              % pixel summation in blocks
48    for i=1:Mb
49        for j=1:Nb
50            MarkDetect((i-1)*K+1:i*K, (j-1)*K+1:j*K) = \ldots
51                sign( sum( sum( Demod((i-1)*K+1:i*K, (j-1)*K+1:j*K) ) ) );
52        end
53    end
54    errors = sum(sum( abs(Mark-MarkDetect) ))
55
56  % Figures
57    figure, subplot(1,2,1); imshow(Demod, []);      title('Demodulation')
58            subplot(1,2,2); imshow(MarkDetect, []); title('Detection')
```

Exercise 16.10 (Watermarking in 2D DCT Transform Domain). Write a program in which a watermark is added to the image DCT transform coefficients. Compare visually the original and the watermarked image.

16.8 Summary

It was a very long run. Many interesting views. A lot of possibilities to own experiments. What should be remembered?

1. Image processing is not so difficult as one can think at a first moment. It is a straightforward extension of 1D signal processing into the 2D case. There are 2D orthogonal transforms, 2D FIR filters, 2D convolution, and correlation. In Matlab, there are many ready-to-use functions. But these functions can be built also with ease using their 1D versions.
2. 2D DCT is an image processing king, not the 2D DFT! It is real value and faster to compute and offers similar spectra. It is used in almost all image compression standards.
3. Classical filter design methods are not so important in image processing. Due to limited image size, all filters should be shorter. Typically, weights of basic filter types are pre-computed and given in tables. They are ready to use. The question is only in what sequence they should be applied and used for image smoothing or differentiation.
4. In telecommunication, still image compression standards (like JPEG) and video compression standards (like H.264/MPEG-4 AVC) are very widely used and understanding image/video compression basics is absolutely necessary. DCT is used as a basic tool for de-correlation of spatial information. For video, differential coding is applied.
5. At present, the most intellectually demanding is automatic image understanding, authentication, and authorization (watermarking, etc.).
6. In image processing very impressive is an immense field of its applications, which is rapidly growing due to very fast increase of computational power of contemporary computers, notebooks, tablets, and smart phones.

16.9 Private Investigations: Free-Style Bungee Jumps

Exercise 16.11 (* 2D Wavelet Transform Using Haar Filters). Separately filter the same image in rows, first using the low-pass *approximation* Haar filter $h_{LP} = \left[\frac{1}{\sqrt{(2)}}, \frac{1}{\sqrt{(2)}} \right]$, then using the high-pass *wavelet* Haar filter $h_{HP} = \left[-\frac{1}{\sqrt{(2)}}, \frac{1}{\sqrt{(2)}} \right]$. Leave every second sample. You will obtain two images two times more narrow, LP one (image approximation) and HP one (image details). Next, repeat the LP and HP filtering, separately, but on columns of two images, resulting from the first processing stage. At this moment, you should have four images: LP-LP (A-A, approximation), LP-HP (A-D, details vertical), HP-LP (D-A, details horizontal), and HP-HP (D-D, details diagonal). An example of similar decomposition is presented in Fig. 16.27. You could continue doing the same decomposition but only of the LP-

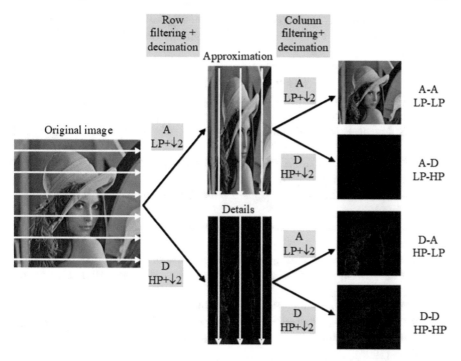

Fig. 16.27: Graphical illustration of one level 2D wavelet transform (decomposition) of an exemplary image [17]

LP (A-A) image. To turn back, you should insert one zero between each two image pixels in rows or columns, respectively, and filter images in reverse order using the following synthesis filters: $h_{LP} = \left[\frac{1}{\sqrt{(2)}}, \frac{1}{\sqrt{(2)}} \right]$ and $h_{HP} = \left[\frac{1}{\sqrt{(2)}}, -\frac{1}{\sqrt{(2)}} \right]$. Do you obtain exactly the same image as the original one? Calculate PSNR introduced by the described image processing operation. The 2D wavelet transform of any image allows its multi-resolution analysis and processing.

Further Reading

1. V. Bhaskaran, K. Konstantinides, *Image and Video Compression Standards. Algorithms and Architectures* (Kluwer Academic Publishers, Boston, 1997)
2. R.J. Clarke, *Digital Compression of Still Images and Video* (Academic Press, London, 1995)
3. R.C. Gonzales, R.E. Woods, *Digital Image Processing* (Addison-Wesley Publishing Company, Reading, 1992; Pearson, Upper Saddle River, 2017)

4. R.C. Gonzales, R.E. Woods, S.L. Eddins, *Digital Image Processing Using Matlab* (Pearson Prentice Hall, Upper Saddle River, 2004; Dorling Kindersley, 2006)

5. B.G. Haskell, A. Puri, A.N. Netravali, *Digital Video: An Introduction to MPEG-2* (Chapman & Hall, New York, 1997)

6. A.K. Jain, *Fundamentals of Digital Image Processing* (Prentice Hall, Englewood Cliffs, 1989)

7. N.S. Jayant, P. Noll, *Digital Coding of Waveforms* (Prentice Hall, Englewood Cliffs, 1984)

8. J.S. Lim, *Two-Dimensional Signal and Image Processing* (Prentice Hall, Upper Saddle River, NJ, 1990)

9. W.-S. Lu, A. Antoniou, *Two-Dimensional Digital Filters* (Marcel Dekker, New York, 1992)

10. J.L. Mitchell, *MPEG Video Compression Standard* (Chapman & Hall, New York, 1996)

11. I. Pitas, *Nonlinear Digital Filters* (Kluwer, Boston, 1990)

12. I. Pitas, *Digital Image Processing Algorithms* (Prentice Hall, Englewood Cliffs, 1993)

13. W.K. Pratt, *Digital Image Processing* (Wiley, New York, 2001)

14. B.S. Reddy, B.N. Chatterji, An FFT-based technique for translation, rotation, and scale-invariant image registration. IEEE Trans. Image Process. **5**(8), 1266–1271 (1996)

15. A.M. Tekalp, *Digital Video Processing* (Prentice Hall, Upper Saddle River, 1995)

16. J. Teuber, *Digital Image Processing* (Prentice Hall, Englewood Cliffs, 1992)

17. T.P. Zieliński, *Cyfrowe Przetwarzanie Sygnałów. Od Teorii do Zastosowań (Digital Signal Processing. From Theory to Applications)* (Wydawnictwa Komunikacji i Łączności (Transport and Communication Publishers), Warszawa, Poland, 2005, 2007, 2009, 2014)

Chapter 17
Introduction to SDR: IQ Signals and Frequency Up-Down Conversion

First steps on a new land are always dangerous and very difficult! R. Crusoe

17.1 Introduction

Digital communication systems are carrying bits. But all signals transmitted in different media (by air, twisted-wires, coaxial-cables, fiber-cables, …) are analog: their values are continuously varying in time and in space. So how the bits are transmitted? It is assumed that at one moment the analog signal is taking one state, for example 5 (101), from limited number of states, for example eight: 0, 1, 2, 3, 4, 5, 6, 7, in binary system respectively: 000, 001, 010, 011, 100, 101, 110, 111. The signal state is changing, mainly synchronously. Typically it is a sinusoid which is periodically switching values of its parameters: amplitude, frequency, and/or phase. Different sinusoid states are called its constellation points. Digital transmitter generates *continuous pulses* of one (single-carrier) or many (multiple-carrier) sinusoids in different states. Digital receiver does synchronization to the pulse beginning and measures parameters of obtained sinusoid. Knowing these values the receiver is finding an actual sinusoid/carrier state (constellation point value), e.g. 5-th, binary 101, in our example, and recovers the transmitted sequence of bits, e.g. 101. Of course, this is the simplest explanation. In order to avoid *hard* (BUNG-BUNG!) sinusoid switching, resulting in spreading the signal spectrum around the carrier and increasing its interference to neighbor transmissions, each sinusoid has to pass *softly* from one state to the other. For this reason, its parameters are smoothed by so-called pulse shaping filter.

So in digital communication mostly modulated sinusoids are used and different services exploit sinusoids having different frequencies. Each transmitter (radio and television broadcasting stations, cellular telephony base-stations, etc.) emits only its own signals which are not disturbed a lot since different services are using sinusoids (information carriers) with different frequencies.

© Springer Nature Switzerland AG 2021
T. P. Zieliński, *Starting Digital Signal Processing in Telecommunication Engineering*, Textbooks in Telecommunication Engineering,
https://doi.org/10.1007/978-3-030-49256-4_17

Each receiver should acquire only sinusoids with frequencies of its interest (using band-pass filtration) and decode their states (by signal demodulation)—recovering this way the bits transmitted.

Observation of frequency spectra of received signals should give us information about existing data *transfers*—we should see some spectral peaks associated with different services. Knowing modulation rules applied to sinusoidal frequency carrier(s) in the transmitter, one can recover information hidden in change of the sinusoidal carrier amplitude, frequency, and/or phase.

Each wireless service has allocated a separate frequency band with some frequency guard intervals on both sides (silence/zero zones), see Fig. 17.1, reducing cross-interference from/to neighboring services and allowing easier extraction of a service content. In order to decode the service data, one should know not only the service frequency band but also how it is used, e.g. number of carriers (one or many), exact carrier frequencies, types of carrier modulation used (amplitude, phase, jointly amplitude and phase, frequency, etc.), duration of one transmitted carrier state called a symbol, all allowed carrier states (carrier constellation points), sequences of symbols that are used for channel estimation and time (frame) synchronization, etc. There are many open-source programs which automatically decode service content after appropriate band-pass filtration (service extraction/isolation) of a received radio-frequency (RF) signal and its A/D sampling (A/D conversion).

In this chapter we become familiar with IQ telecommunication signals and their exemplary spectra. Since the sinusoid modulation is so important in digital communications we will digitally acquire and demodulate two analog signals: a mono FM-modulated radio broadcast signal and a mono AM-modulated speech, present in the VOR signal (VHF Omni direction Radio Range), transmitted to an airplane by the radio navigation ground station. These two examples will help us to face/focus/localize main problems related to understanding the digital transmission. We learn an IQ complex-value representation of signals returned by hardware receivers.

The SDR technology is introduced in many books, for example in the following ones: [2–4, 6, 7, 12, 14, 17]. In the end of this chapter a short list of digital communication textbooks is also recommended for further reading.

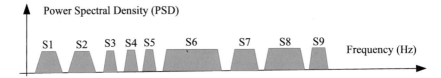

Fig. 17.1: Illustration of frequency allocation principle: each wireless service S1, S2, ..., S9 uses different frequency band which is separated from neighbor services with non-used frequency guard zones

17.2 Frequency Allocation

As already stated, wireless services use different frequency bands that are separated by not exploited frequency null zones—see Fig. 17.1. In order to extract the transmitted information, a receiver should know exactly what frequencies are used and in what manner: how they are carrying analog content (via AM, PM, and/or FM modulation) or digital content via PSK, BPSK, QPSK, DQPSK, QAM, FSK, GMSK, ...modulations.

Frequency is more precious than a gold since its *resources* are limited. International and national organizations specify general regulations for radio-frequency usage, national commissions grant rights for existence of many governmental and commercial services like public emergency services, radio and television broadcasting, telephony, etc. Only in order to have imagination about the problem complexity, the official frequency allocation table for the USA in year 2016 is shown in Fig. 17.2. It specifies frequency predestination in the band 8.3kHz–275GHz. It is interesting to note that modern instruments in year 2019 can work with frequencies in the band up to 400GHz, so with frequency about 100GHz higher then in year 2016. The technological progress in the RF technology is really impressive!

In Table 17.1 there are specified some services existing in the band from 0.1MHz to 5GHz (central frequencies and bandwidths). Some of them are interesting for us as a potential source of real-world signals to be investigated during our laboratory, the others are only listed in order to present the diversity of radio-frequency applications.

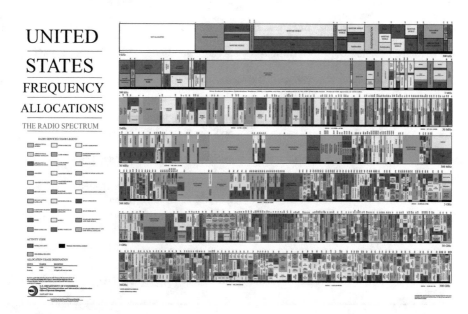

Fig. 17.2: Official frequency allocation table for the USA in year 2016, from 8.3 kHz to 275 GHz. Please, observe how many *pieces has the cake* [9]

Table 17.1: Some RF services, their frequency bands, and modulation used (DD denotes Demodulate and Decode)

Service	Description	Modulation	BW (kHz)	Band (MHz)	Algorithms
AM radio	Analog radio, long/short-waves	AM	15	0.1–30	AM demod
HAM amateur radio	Analog radio	AM-SSB	5	0.1–30	PLL, Hilbert
DRM radio	Digital radio	OFDM	10	0.1–30	Sync, channel, demod
HD radio	Digital radio	OFDM	10	0.1–30	Sync, channel, demod
RTTY	Telex	FSK	3	0.1–30	Filters, DD
HF FAX	Weather pictures	FSK	3	0.1–30	Sync
FM radio stereo, MPX	Analog radio	FM	200	60–170	PLL, FM demod
FM RDS	Digital radio text	BPSK	6	60–170	Costas, PLL, DD
NOAA APT	SAT weather pictures	FM/AM	40	137	Demod
AIS	Ship identification	GMSK	25	162	Detect, demod
DAB	Digital radio	OFDM/DQPSK	2000	170–230	Sync, channel, demod by FFT, Viterbi decoder, de-interleaving, Reed–Solomon
TETRA	Security mobile	DQPSK	25	400	Demod
Remote control	My gadgets	AM OOK, FSK	1	434	Bit timing
DVB, ISDB, DTMB	Digital TV	OFDM/QAM	8000	400–800	Sync, channel, demod
LTE	Mobile	OFDM (TDMA, FDMA, CDMA)	20000	850, 1800	Demod
GSM	Mobile	GMSK	200	900, 1800	Demod
UMTS	Mobile	CDMA	5000	900, 1800, 1900, 2100	Demod
ADS-B	Aircraft identification	PPM	1000	1090	Detect, demod
GPS	Location	BPSK	2000	1227, 1575	CDMA-like
Inmarsat Areo	SAT to aircraft	PSK	0.8, 1.6, 10500	1545, 1547	Detect
Bluetooth	Short range wireless	GFSK, DPSK, freq hops	1000	2400	Demod
Inmarsat Areo	SAT to ground station	PSK	–	3600, 3629	Detect
WiFi 802.11 a/g/n/ac	WLAN	OFDM	20000	2400, 5000	Demod

17.3 Service Frequency UP-DOWN Conversion

Typically, a signal to be transmitted is prepared around 0 Hz, in so-called frequency base-band, and then converted up to the higher target frequency f_c. The base-band (BB) signal version can be real-value one, like in AM double side-band radio or ADSL/VDSL/SDSL modems, or complex-value one, like in FM and DAB radio, DVB-T broadcast, LTE and 5G down-link. When only one sinusoid is used for bit transmission, a single-carrier communication is used. If many sinusoids are exploited in parallel, a multi-carrier/multi-tone transmission takes place. In the last case, applied sinusoids should be orthogonal to each other in order to minimize data cross-talk between them. Orthogonal Frequency Division Multiplexing (OFDM) is an official name of the multi-carrier communication. As already stated the OFDM signal can be real-value in the base-band (R-OFDM), like in telephone/cable DSL modems, or complex-value one (C-OFDM), like in DAB, DVB-T, LTE, and 5G. In order not to lose generality, here we will denote the BB signal as a complex-value signal since its imaginary part can be also equal to zero:

$$x(t) = x_{Re}(t) + jx_{Im}(t). \tag{17.1}$$

For AM suppressed carrier (SC) and large carrier (LC), FM and real-value (R) and complex-value (C) OFDM, the signal is equal, respectively:

$$\textbf{AM-SC}: \qquad x(t) = m(t) \tag{17.2}$$

$$\textbf{AM-LC}: \qquad x(t) = (1 + k_A m(t)) \tag{17.3}$$

$$\textbf{FM}: \qquad x(t) = \exp\left[j2\pi \left(0 \cdot t + k_F \int_0^t m(t)dt \right) \right] \tag{17.4}$$

$$\textbf{R-OFDM}: \qquad x(t) = a_0 + 2 \sum_{k=1}^{N/2-1} a_k \cos(k\omega_0 t + \varphi_k) \tag{17.5}$$

$$\textbf{C-OFDM}: \qquad x(t) = \sum_{k=-N/2}^{N/2-1} c_k e^{j(k\omega_0)t} = A(t)e^{j\varphi(t)} \tag{17.6}$$

In (17.5) we recognize trigonometric Fourier series of a real-value signal, having obligatory conjugate-symmetrical spectrum around DC: $X(-f) = X^*(f)$ (for example an ADSL modem case with DMT signaling), while in (17.6)—a complex Fourier series without spectral symmetry restrictions (for example a DAB radio and 4G-LTE/5G-NR case with OFDM signaling). Different states of individual carriers are specified by coefficients $\{a_k, \varphi_k\}$ and c_k.

At present the signal conversion, from the base-band to the carrier frequency f_c and back to the base-band, will be described. Spectral consequences of all operations are presented in Fig. 17.3. In the transmitter the signal $s(t)$ to be sent to an antenna is built from $x(t)$ in the following way:

$$\textbf{FREQ-UP}: \quad x_{UP}(t) = x(t) \cdot c(t) \tag{17.7}$$

$$\textbf{REAL}: \quad s(t) = \text{Re}(x_{UP}(t)), \tag{17.8}$$

where $c(t)$ is a complex-value harmonic oscillation of the service carrier:

$$c(t) = e^{j\omega_c t} = \cos(\omega_c t) + j \cdot \sin(\omega_c t). \tag{17.9}$$

Multiplication in operation FREQ-UP causes shifting the signal $x(t)$ spectrum to the carrier frequency f_c (Fig. 17.3a), while taking the REAL part of the result, which is necessary for signal transmission, causes creation of a complex-conjugated copy of the shifted spectrum in negative frequencies (Fig. 17.3b). Why mirror spectrum for negative frequencies appears? Because Fourier spectrum of any real-value signal is conjugate-symmetric in regard to 0 Hz, and our signal becomes a real-value one. Both operation can be combined into one less complex computationally, known as quadrature frequency up-shifter:

$$
\begin{aligned}
\text{Re}\left\{ x(t) \cdot c(t) \right\} &= \text{Re}\left\{ [x_{Re}(t) + jx_{Im}(t)] \cdot [\cos(\omega_c t) + j \cdot \sin(\omega_c t)] \right\} = \\
&= \text{Re}\left\{ [x_{Re}(t)\cos(\omega_c t) - x_{Im}(t)\sin(\omega_c t)] + j[x_{Re}(t)\sin(\omega_c t) + x_{Im}(t)\cos(\omega_c t)] \right\} = \\
&= x_{Re}(t)\cos(\omega_c t) - x_{Im}(t)\sin(\omega_c t)
\end{aligned}
$$

resulting in the very important equation of the quadrature frequency up-converter (modulator):

$$\textbf{TX}: \quad s(t) = x_{Re}(t)\cos(\omega_c t) - x_{Im}(t)\sin(\omega_c t) \tag{17.10}$$

In the receiver signal from the antenna is first band-pass filtered in order to separate the required service $s(t)$ from the RF background (Fig. 17.3c). Next, the result is multiplied by the complex-value carrier with negative angular frequency $-\omega_c$ scaled by 2 in amplitude:

$$\textbf{FREQ-DOWN:} \quad r_d(t) = s(t) \cdot 2e^{-j\omega_c t} = 2s(t)\cos(\omega_c t) + j \cdot [-2s(t)\sin(\omega_c t)] \tag{17.11}$$

what shifts the signal spectrum down by $-\omega_c$ (Fig. 17.3d). The scaling by 2 is justified in the proof presented below. After putting Eq. (17.10) into (17.11) the following complex-value signal is obtained (Fig. 17.3e):

$$r_d(t) = [x_{Re}(t) + j \cdot x_{Im}(t)] + [x_{Re}(t) - j \cdot x_{Im}(t)] \cdot e^{-j2\omega_c t}. \tag{17.12}$$

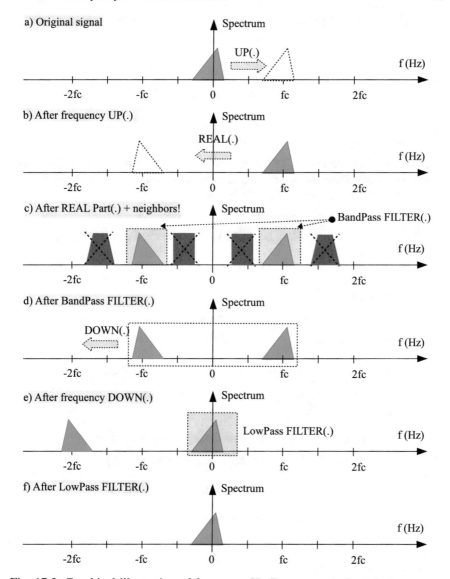

Fig. 17.3: Graphical illustration of frequency Up-Down conversion of telecommunication service

We see that the first term in (17.12) is the same as the base-band signal $x(t)$ in the transmitter, and should be remained, while the second term is high-frequency one (with doubled carrier frequency) and therefore should be removed. This is done by a low-pass filter (Fig. 17.3e):

$$\textbf{LP FILTER:} \quad \hat{x}(t) = \text{LPFilter}\,[r_d(t)]\,.) \qquad (17.13)$$

Using (17.11) in (17.13) we are obtaining the very important equation for the quadrature frequency down-converter and signal demodulator:

$$\textbf{RX:}\quad \hat{x}(t) = \text{LPFilter}\left[2s(t)\cos(\omega_c t)\right] + j \cdot \text{LPFilter}\left[-2s(t)\sin(\omega_c t)\right].$$
$$(17.14)$$

After the filter, the signal is approximately the same as in the transmitter baseband (Fig. 17.3f). Therefore we can happily conclude that signal processing TX-RX chain has finished in our example with full success.

Proof (Some Extra Math). Here Eq. (17.14) will be proved and Eq. (17.12) will be derived from Eq. (17.11):

$$
\begin{aligned}
2s(t)\left[\cos(\omega_c t)\right] &= \\
&= 2\left\{\left[x_{Re}(t)\cos(\omega_c t) - x_{Im}(t)\sin(\omega_c t)\right]\cos(\omega_c t)\right\} = \\
&= 2\left\{x_{Re}(t)\cos^2(\omega_c t) - x_{Im}(t)\sin(\omega_c t)\cos(\omega_c t)\right\} = \\
&= 2\left\{x_{Re}(t)\left[\frac{1}{2}(1+\cos(2\omega_c t))\right] - x_{Im}(t)\left[\frac{1}{2}\sin(2\omega_c t)\right]\right\} = \\
&= x_{Re}(t) + x_{Re}(t)\left[\cos(2\omega_c t)\right] - x_{Im}(t)\left[\sin(2\omega_c t)\right] \\
2s(t)\left[-\sin(\omega_c t)\right] &= \\
&= 2\left\{\left[x_{Re}(t)\cos(\omega_c t) - x_{Im}(t)\sin(\omega_c t)\right]\cdot\left[-\sin(\omega_c t)\right]\right\} = \\
&= 2\left\{-x_{Re}(t)\left[\cos(\omega_c t)\sin(\omega_c t)\right] + x_{Im}(t)\left[\sin^2(\omega_c t)\right]\right\} = \\
&= -2\left\{x_{Re}(t)\left[\frac{1}{2}\sin(2\omega_c t)\right] + x_{Im}(t)\left[\frac{1}{2}(1-\cos(2\omega_c t))\right]\right\} = \\
&= x_{Im}(t) - x_{Re}(t)\left[\sin(2\omega_c t)\right] - x_{Im}(t)\left[\cos(2\omega_c t)\right].
\end{aligned}
$$

After low-pass filters one obtains

$$
\begin{aligned}
\hat{x}_{Re}(t) &= \text{LPFilter}\left[2s(t)\cos(\omega_c t)\right] \\
\hat{x}_{Im}(t) &= \text{LPFilter}\left[-2r(t)\sin(\omega_c t)\right]
\end{aligned}
$$

which proofs the Eq. (17.14).

Now, we can put calculated equations for $2s(t)\cos(\omega_c t)$ and $-2s(t)\sin(\omega_c t)$ into Eq. (17.11) and get the Eq. (17.12):

$$2s(t)\cos(\omega_c t) - 2j \cdot s(t)\sin(\omega_c t) =$$
$$= \{x_{Re}(t) + x_{Re}(t)[\cos(2\omega_c t)] - x_{Im}(t)[\sin(2\omega_c t)]\}\ldots$$
$$+ j \cdot \{-x_{Re}(t)[\sin(2\omega_c t)] + x_{Im}(t) - x_{Im}(t)[\cos(2\omega_c t)]\} =$$
$$= [x_{Re}(t) + j \cdot x_{Im}(t)] + x_{Re}(t) \cdot \underbrace{[\cos(2\omega_c t) - j \cdot \sin(2\omega_c t)]}_{(*)\ \exp(-j2\omega_c t)}\ldots$$
$$- x_{Im}(t) \cdot \underbrace{[\sin(2\omega_c t) + j \cdot \cos(2\omega_c)t]}_{(**)\ j\cdot\exp(-j2\omega_c t)} =$$
$$= [x_{Re}(t) + j \cdot x_{Im}(t)] + [x_{Re}(t) - j \cdot x_{Im}(t)] \cdot e^{-j2\omega_c t}$$

$\square$

Frequency UP and DOWN signal conversion is demonstrated in Matlab program 17.1 but only for FM modulated signal.

Listing 17.1: Matlab program demonstrating service UP-DOWN frequency conversion

```
1   % lab17_ex_Service_UpDown_short.m
2     clear all; close all;
3
4   % FM modulated signal
5     Nx = 2000;                  % number of signal samples
6     fs = 2000;                  % sampling frequency [Hz]
7     fc = 400;                   % carrier frequency [Hz]
8     fm = 2; df = 50;            % FM modulation frequency and depth
9     dt = 1/fs; t = dt*(0:Nx-1); % time
10    xm = sin(2*pi*fm*t);        % modulating signal
11    x = exp( j * ( 2*pi*0*t + 2*pi*df/fs*cumsum(xm) ) ); % FM modulated signal
12
13  % Frequency UP and REAL part
14    c = cos( 2*pi*fc*t ); s = sin( 2*pi*fc*t );
15    xUp = x .* (c + j*s);       % Frequency UP
16    xUpReal = real( xUp );      % REAL part
17  % xUpReal = real(x).*c  - imag(x).*s;  % quadrature modulator
18
19  % Frequency DOWN - quadrature demodulator
20    xDownCos =  2*xUpReal.*c;
21    xDownSin = -2*xUpReal.*s;
22    xDown = xDownCos + j*xDownSin;
23
24  % Ideal LowPass Filter in frequency domain
25    df=fs/Nx; K = floor(fc/df);
26    XDownFilt = fft(xDown);
27    XDownFilt(K+1:Nx-K) = zeros(1,Nx-2*K);
28    xDownFilt = ifft( XDownFilt );
29
30  % ERROR after frequency UP & DOWN
31    ERROR_SIGNAL = max( abs( x - xDownFilt ) ), pause
```

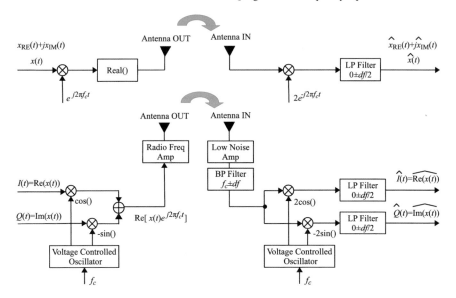

Fig. 17.4: Block diagram of the analog transmitter and receiver (frequency UP-DOWN signal conversion)

Exercise 17.1 (Frequency Up-Down Conversion of Base-Band Service). A Reader is kindly asked to modify the program 17.1. First, please, carefully analyze the program 17.1. Its longer version is given in the book repository. Add plots of spectra after each signal processing. Experimentally verify correctness of the Fig. 17.3 for the FM signal. Then, generate AM-SC, AM-LC, R-OFDM, and C-OFDM signals, defined by equations (17.2), (17.3), (17.5), and (17.6), respectively, and compare their spectra with Fig. 17.3. Some signal propositions are suggested in the longer version of the program. For example, for OFDM signals set: $N = 8, f_0 = 25, \omega_0 = 2\pi f_0, a_k = k, \varphi_k = k\pi/N, c_k = (N/2 + k + 1)e^{jk\pi/N}$.

In order to simplify the mathematical derivation, the analog TX-RX transmission scheme was discussed so far. Its functional block diagram is presented in Fig. 17.4. In modern digital communication systems, signals in the base-band are processed digitally while up-down service conversions in frequency are realized in analog or digital manner. Very often an intermediate signal processing frequency *shelf* is used: the base-band signals are first shifted to an *intermediate frequency* and later to the target one. These two frequency shifts can be: (analog+analog), (digital+analog), or (digital+digital). Two-step signal processing makes the system design easier and results in more flexible telecommunication devices. Software Design Radio (SDR) represents *more digital* designs. In the next section some practical solutions of SDR receivers will be presented. Only some because possibilities are many. In the next section we will investigate more deeply all digital, software implementation of signal processing TX-RX chain presented in this section.

17.4 RF Signal Acquisition Hardware

In order to do any practical experiments with decoding of RF transmitted data we need an antenna and signal receiver/recorder. We should prefer not expensive and flexible solutions. The best choice is usage of the so-called software defined radio (SDR) technology in which hardware is *digitally controlled* and programmable—thanks to this a program can be changed easily (on demand). Each such device should have, in simplification (see Fig. 17.5):

1. an antenna working in wide frequency band,
2. a low noise amplifier (LNA),
3. an analog band-pass (BP) filter with width Δf, allowing passing only frequencies of the interest lying around selected *central* frequency (f_0); Δf and f_0 should be tunable,
4. analog mixer/heterodyne—shifting down frequencies from the band $[f_0 - \Delta f/2, f_0 + \Delta f/2]$ to the so-called *base-band* around DC (0Hz); this operation is done by multiplying the BP filtered signal by a generated voltage controlled oscillation (VCO) having frequency f_0, precisely $-f_0$,
5. an analog low-pass filter with the cutting frequency $\pm \Delta f/2$, passing only signals around DC (0Hz),
6. analog-to-digital converter with sampling rate higher than the signal bandwidth of interest ($f_s > \Delta f$),

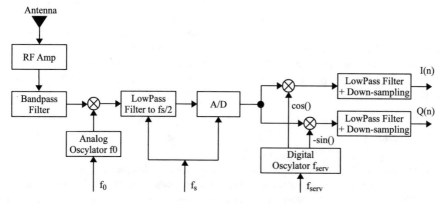

Fig. 17.5: Block diagram of a simple software digital receiver

7. DSP or/and FPGA processor for further data processing: some additional sub-band digital filtration, shifting a selected frequency sub-band to DC by digital signal mixing with numerically controlled oscillator (NCO) and sample rate reduction; NCO, called also a digital sine synthesizer (DSS), is digital version of the VCO,

8. main processor for service decoding.

The presented above RF receiver structure is an *educational* one. In fact, it can be optimized and re-designed. There are many possibilities. This introductory section is not focused on technical details—details will be presented later. Now it is important to note only that the frequency down-conversion, realized by analog and/or digital mixer, can be done in so-called *quadrature* manner. A signal to be shifted down in frequency is multiplied in parallel by signals from two oscillators, cosine and minus sine, and not one but two signals are obtained in lower frequencies, a so-called in-phase (for cos()) and in-quadrature (for −sin()) component.

In this course we will make use of signals recorded by two SDR platforms, supported by programs from Matlab Communication Toolbox:

- RTL-SDR USB stick, a commercial FM, DAB, and DVB-T receiver, having open-source Linux and Windows drivers and many signal/spectrum visualization/recording programs (Gqrx—Linux, SDRSharp—Windows);
- Analog Devices Learning Module (ADALM) called PLUTO, dedicated to self-learning or university-teaching of digital communication fundamentals on the example of wireless transmission, also having support under Linux (Gqrx, ADI IIO Oscilloscope) and Windows (SDRSharp, ADI IIO Oscilloscope).

Both devices are shown in Fig. 17.6. They send data to a PC computer using USB interface. Main information about their parameters is summarized in Table 17.2. Details concerning inner operations performed by the devices will be presented later during explanation of frequency up-down service shifting. At present we will only become familiar with their inner schematics which confirm the given above general description of RF receiver design.

Functional block diagram of the RTLSDR USB stick is presented in Fig. 17.7. The stick is using intermediate frequency $f_{IF} = 10.7$MHz and analog-to-digital converter with sampling frequency $f_{AD} = 28.8$ MHz. In the following description, to make explanation simpler, it is assumed that we are decoding an analog FM radio service having carrier frequency $f_c = 100$MHz and frequency bandwidth $\Delta f = f_{SERV} = 0.2$MHz around FM radio station. In the analog section of the RTLSDR the following operations are done:

- signal amplification using a low noise amplifier (LNA),
- selection of frequencies of interest lying in the band $[f_c - \Delta/2, f_c + \Delta f/2] = [99.9, 100.1]$ by a band-pass (BP) filter,
- shifting them down by a cos() mixer, working with frequency $f_{RF} = f_c - f_{IF} = 100$–10.7 = 89.3 MHz frequency, to the intermediate frequency $f_{IF} = 10.7$ MHz; frequency down-conversion is done in non-quadrature manner using only cosine oscillator,

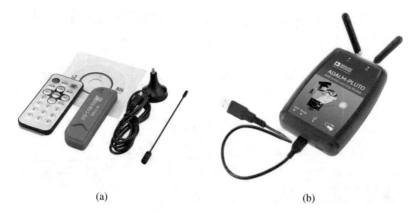

(a)   (b)

Fig. 17.6: Two very cheap software defined radio hardware platforms. (**a**) RTL-SDR USB stick (FM/DAB/DVB-T) [10]. (**b**) Analog Devices PLUTO active learning module—figure from [11]

Table 17.2: The most important parameters of two software defined radio hardware platforms: RTL-SDR (Elonics E4000 RF tuner + RTL2832U AD converter and down-sampler) and ADALM-PLUTO (AD936x RF transceiver + Xilinx Zynq 7000 SoC AD converter and down-sampler)

Feature	RTLSDR USB stick	Analog devices learning module PLUTO
Number of RX channels	1	1 (AD936x has 2)
Number of TX channels	0	1 (AD936x has 2)
Number of ADC bits	8	12
Number of DAC bits	0	12
TX band	–	47MHz–6GHz
RX band	52MHz–2200MHz (gap 1100–1250)	70MHz–6GHz
Tunable channel bandwidth	250kHz–3.2MHz	200kHz–56MHz

- removing signals with frequencies higher than $f_{AD}/2 = 14.4$ MHz by means of low-pass (LP) filtering; it is necessary since the mixer generates also unwanted high-frequency components which has to be eliminated,
- adjusting signal amplitude by means of automatic gain control (AGC) circuit.

In digital section of the RTLSDR the following tasks are performed:

- converting the signal into its digital form using an analog-to-digital converter (ADC) working with frequency $f_{AD} = 28.8$ MHz,
- shifting the digital signal down in frequency in quadrature manner using cosine() and sine() numerically controlled oscillators (NCOs) working with *intermediate* frequency $f_{IF} \approx 10.7$ MHz,

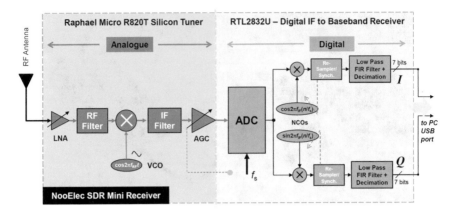

Fig. 17.7: Functional block diagram of the NooElec RTLSDR mini receiver—figure from [18]

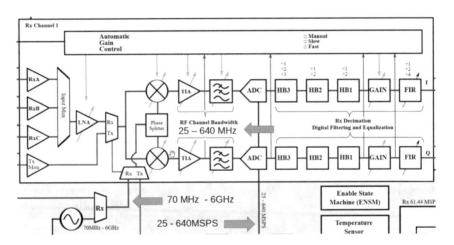

Fig. 17.8: Functional block diagram of Analog Devices AD9361 RF Agile Transceiver—figure from [1]

- fractional re-sampling of $I(n)$ and $Q(n)$ signal components and synchronization of NCOs frequency with signal intermediate carrier f_{IF},
- low-pass signal filtration in the range $[-\Delta f/2, \Delta f/2] = [-f_{SERV}/2, f_{SERV}/2] = [0.1, 0.1]$ MHz and sample rate reduction to $f_{SERV} = 200$kHz.

In turn block diagram of the ADALM-PLUTO one channel receiver module is presented in Fig. 17.8. As we can see in this case the signal conversion scenario is different:

- the system does not use intermediate frequency and directly converts the service to the base-band,

- there is no band-pass filter on the input,
- analog quadrature cos()/sin() mixer is used with f_{RF} tuned in the band 70MHz—6GHz—a Phase Splitter block is obtaining an appropriate control signal,
- signal after the mixer is passing through an analog low-pass filter which is tuned to the service frequency band in the range 200kHz–56MHz,
- next we have two analog-to-digital converters sampling in parallel analog signals $I(t)$ (the cos() path) and $Q(t)$ (the -sin() path); the converters work with sampling rate changing in the range 25–640 Msps,
- since sampling rate after ADCs is higher than the service bandwidth, i.e. the IQ signals are oversampled, their decimation takes place in the digital part of the receiver: low-pass filtering and sample rate reduction are done.

17.5 Investigating Signal Spectra

So, where to go now? Let us assume that we have an SDR signal receiver/recorder. We know a little bit how it is built. *So, it is time to ride a bicycle (!)* and observe spectra of some RF services, using available programs.

In this section we will investigate real-world spectra of some wireless services. The analyzed signals have been acquired by the RTL-SDR USB stick and ADALM-PLUTO module. Signal spectra have been calculated and displayed by SDRSharp and ADI IIO programs.

FM Analog Radio FM radio is broadcasted in frequency range 88–108 MHz. In Fig. 17.9 an FFT frequency spectrum magnitude of several FM radio stations is presented. The signal was captured by RTLSDR dongle with ratio 3.2 mega samples per second (msps) ($f_s = 3.2$ MHz) around frequency 101.6 Hz, therefore the spectrum width is equal to 3.2 Hz (1.6 MHz below and above the 101.6 Hz). In the momentum FFT spectrum (upper figure) we see more than 10 spectral peaks/stations. In the STFT spectrogram (lower figure), i.e. the time-varying FFT spectrum history plotted as an image, we can observe more than 10 time-varying frequency modulation curves associated with many FM radio stations. In the laboratory part of this chapter we will try to recover mono audio signal of one FM radio station.

Avionics VOR Omni Directional Range Signal [15] Avionics radio navigation signals are transmitted in different frequency bands, also in the range 108–138 MHz.

In Fig. 17.10 a 1 MHz-wide FFT spectrum near frequency 112.5 MHz is shown—the signal was captured by the RTLSDR dongle and its spectrum was calculated and displayed by the SDRSharp program. It is seen that some narrow-band signals are transmitted all the time: we observe relatively strong, well-visible lines in the STFT spectrogram (lower figure). Around frequency 112.8 MHz, marked with cursor, a three-component signal is observed having strong spectral peak and two weaker side-peaks, which is typical for double side-band amplitude modulation with large carrier (AM-DSB-LC), we are right: carrier of the VOR Omni Directional Range signal is not suppressed and modulated in amplitude by sum of three components:

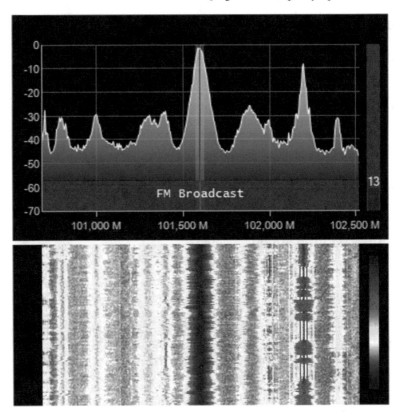

Fig. 17.9: FFT spectrum magnitude of an RF signal captured by the RTL-SDR USB stick around frequency 101.6 MHz in the frequency range of 3.2 MHz: (up) momentum, mean spectrum, and (down) short-time Fourier transform, the FFT spectral history, plotted as time–frequency image showing spectrum change in time. One can observe curves of changing instantaneous frequencies of many FM radio broadcasts. Brighter colors denote higher spectrum values

30 Hz sine (having azimuth dependent phase), 300–3300 Hz voice control signal, and 9960 Hz sine reference. The reference is modulated in frequency by 30 Hz sinusoid with modulation depth equal to 480 Hz. In the laboratory part of this chapter we will recover the voice component.

Nano-Satellite Signal in Safe, Slow Mode [16] Amateur satellites (AMSAT) have a form of very small, cheap nano-cubes, $10 \times 10 \times 10$ centimeter large, which are positioned at low earth orbit (LEO). They circle the earth during about 90 min and a ground station has a contact with them about 7–10 min. They use safe/slow-speed and high-speed telemetry transmission systems. In the first mode, for the down-link carrier frequencies around 145 MHz and frequency modulation are used. 200 bits are transmitted per second in packets lasting about 10 s. Data are transmitted under

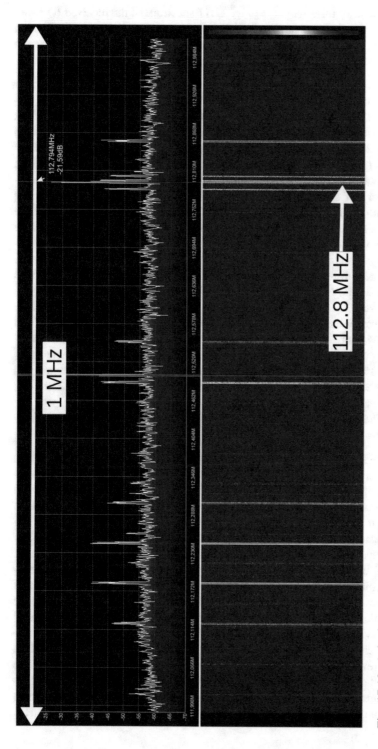

Fig. 17.10: Avionics navigation. FFT spectrum magnitude of an RF signal captured by the RTL-SDR USB stick around frequency 1125.5 MHz in the frequency range of 1 MHz: (up) momentum, local-mean spectrum, and (down) short-time Fourier transform, the FFT spectral history, plotted as time–frequency image showing spectrum change in time. One can observe peaks of avionics navigation signals

voice (DUV) control information. In Fig. 17.11 spectrum of the received IQ nano-sat signal is shown. In the central part of the spectrogram (down) we see beginning of a new 10 s data packet. In its middle part a voice description is given about a transmission mode. We can hear it during laboratory experiments.

DAB Digital Radio The momentum FFT spectrum magnitude and the time-varying STFT spectrogram of a digital audio broadcasting (DAB) radio in mode I are presented in Fig. 17.12. The signal was captured in Krakow, Poland, by the RTLSDR dongle with sampling ratio equals 2 mega samples per second (MSPS) around the frequency 229.069 MHz, therefore width of the observed spectrum is equal to 2 MHz. DAB radio is an example of multi-carrier digital data transmission which exploits Orthogonal Frequency Division Multiplexing (OFDM) signaling. In the upper momentum FFT spectrum we see *a spectral hat* of many sines/carriers *carrying* binary data in parallel, being precised—1536 carriers in the standard. They have different phase angles in consecutive data blocks (OFDM frames): their phases are changed from block to block using so-called differential quadrature phase shift keying (DQPSK) method. The phase shifts are equal to (+45, −45, 135, −135) degrees and they denote a different pair of transmitted bits (00, 01, 10, 11), i.e. one sine is carrying only 2 bits. Note that some frequencies lying on both sides of the *spectral hat* (precisely 256 in the standard) are not used by the DAB service and serve as guard frequency bands between two neighboring DAB radio emitters. In DAB blocks (OFDM frames) 2048 signal samples are synthesized at once, performing the 2048-point inverse FFT algorithm upon specially pre-set Fourier coefficients (c_k in Eq. (17.6): 1536 carrier coefficients, with phase shifts between DAB frames, and 2*256=512 coefficients of two guard side-bands, set to zero. In Fig. 17.12 one can observe in the STFT spectrogram that carriers are generated all the time, with the exception of the NULL Symbol separating the DAB frames and added for synchronization issues.

3G UMTS E-GSM Telephony In our next experiment we will observe spectrum of the up-link signal (from a phone to a base-station), in the frequency band 880–915 MHz during telephone call making use of 3G UMTS signaling named E-GSM. The signal was acquired by the ADALM-PLUTO device in the range 885–935 MHz. In Fig. 17.13 three spectra of captured signals are presented. One can observe that the signal transmitted by the phone has a spectral bandwidth equal to approximately 5 MHz. The spectrum is changing in time, during a telephone call and without it. In 3G UMTS bits are carried NOT by many orthogonal sinusoids (OFDM technique), exploiting the allocated frequency band *slice-by-slice*, BUT by a set of pseudo-random orthogonal signals, each having the spectrum spreaded along the whole available band, i.e. covering 5 MHz in the discussed case. Such orthogonal band-frequency signals are generated in the following way. First, a set of orthogonal signals is chosen, for example orthogonal Walsh sequences taking only values 1, −1. For brevity, we can assume that bit 1 corresponds to transmission of the original Walsh signal/code while bit 0 to the transmission of the negated code. Different codes are allocated to different users. Then, each code is multiplied by some spreading/scrambling sequence, e.g. Barker and Gold code specific for each user.

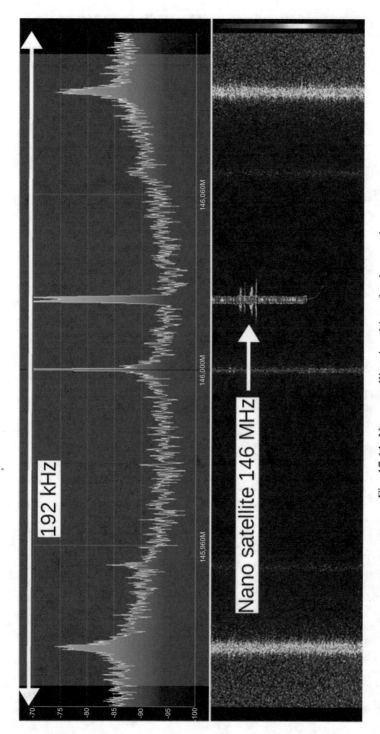

Fig. 17.11: Nano-satellite signal in safe, slow mode

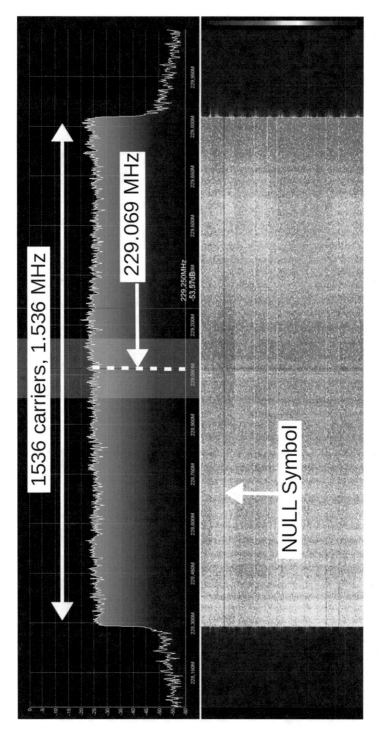

Fig. 17.12: FFT spectrum magnitude (actual mean spectrum) and short-time Fourier transform, the spectrogram, of a DAB signal captured by the RTL-SDR USB stick around frequency 229.069 MHz (central frequency of single DAB radio multiplex) with bandwidth 2MHz

Then codes of one user are collected into pairs and interpreted as in-phase $I(n)$ and in-quadrature $Q(n)$ signals. Next, these signals are used for quadrature phase shift modulation of the target service frequency using the QPSK technique. In the receiver the acquired signal is de-scrambled, correlated with orthogonal code templates and codes polarities (+/−) as well as transmitted bits are detected. Such user access to the channel and bit signaling is called CDMA (Code Division Multiple Access).

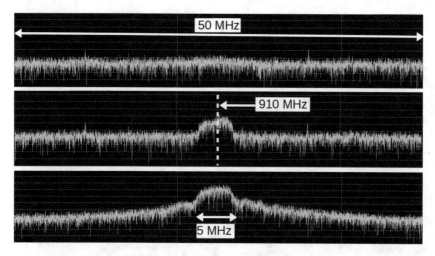

Fig. 17.13: FFT spectra magnitudes of three 3G UMTS E-GSM signals displayed by ADI IIO oscilloscope program and captured by the AD ADALM-PLUTO module: (up, middle) without telephone call and (down) during it. Observed signals have frequency bandwidth equal to about 5 MHz. Spectra are scaled in decibels

Wi-Fi Networks Finally, we will observe the Wi-Fi signal spectrum. Again, the signal has been acquired by the ADALM-PLUTO device, this time in the range 2425–2475 MHz with sampling ratio 50 megasamples per second (MSPS). An Internet video has been streamed via wireless network and watched on the PC computer screen. Three different signal spectra observed in the ADI IIO oscilloscope program are shown in Fig. 17.14. One should notice that the spectrum shape is changing in time. The observed spectrum width is equal to about 20 MHz which is in agreement with the standards:

- 802.11a/g—20 MHz (all 64 sub-carriers of OFDM), used 16.25 MHz (52 used sub-carriers of OFDM),
- 802.11b/g—22 MHz.

In the first case the OFDM signaling is used, similarly as in DAB, and bits are transferred in parallel using 52 sub-carriers with spacing 312.5 kHz (sampling rate 20 mega samples per second, 64-point FFT). In the second case CCK/DSSS (Complementary Code Keying, Direct Sequence Spread Spectrum) is used, i.e. orthogonal spectrum-spreading sequences and binary phase shift keying (BPSK) modulation, similarly to the 3G UMTS example presented above.

Summarizing we can conclude that surrounding us electro-magnetic (EM) field is a real treasure of different EM waves which are carrying different analog and digital *messages* that are hidden in changing-in-time frequencies, amplitudes, and phases of the waves.

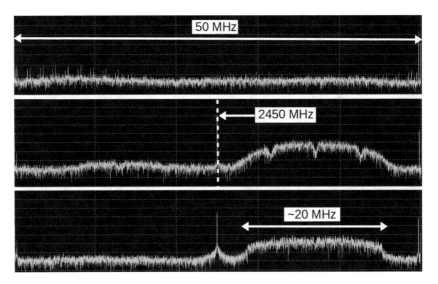

Fig. 17.14: FFT spectra magnitudes of three Wi-Fi signals displayed by ADI IIO oscilloscope and captured by the AD ADALM-PLUTO module during watching video via WLAN. Scaled in decibels

Hmm .... So we see a very *tasty cake*. But how to take a piece of it? How to decode personally (by hand) data of a simple service? How it works? Now we will try to come into the RF engineer's bakery or kitchen.

17.6 Example: Calculating Spectra of IQ Signals

Now it is time to do a next step and to calculate the signal spectrum personally. To make things easier, we will do it off-line: first record an IQ signal using existing programs (e.g. SDRSharp, Gqrx, or ADI IIO Oscilloscope), then read the recorded file into Matlab/Octave, do some computing and display the result. Before the recording we should specify: (1) central (carrier) frequency of interest (f_c), (2) the width of its neighborhood (Δf) equal to the sampling rate, (3) required signal amplification/gain or choosing automatic gain control (AGC) option, and (4) number of bits required for each I(n) and Q(n) signal sample. We will analyze four complex-value IQ signals of: (1) the FM radio, (2) the digital DAB radio, (3) the slow-mode nano-satellite FM down-link, and (4) the airplane VHF Omni direction Radio Range (VOR) airplane navigation signal, used for short and medium positioning range. On the Internet WebSDR page http://websdr.org you can decode and record personally signals from many RF receivers available on-line. After downloading them you can continue signal processing on your home computer. One of such signals is displayed in the program also.

In all cases, after recording, first we should read and display in Matlab samples of I/Q signals and then calculate and observe their power spectral density estimates, based on the DFT spectrum. Simple program performing these first initial steps is shown in the Listing 17.2.

Listing 17.2: Program for simple analysis of recorded RF signals in Matlab/Octave

```
1   % lab17_ex_IQ_DFT.m
2     clear all; close all;
3
4     m=128; cm_plasma=plasma(m); cm_inferno=inferno(m); % color maps for gray printing
5     cm = plasma;
6
7   % Read recorded IQ signal - choose one
8   % FileName = 'SDRSharp_DABRadio_229069kHz_IQ.wav'; T=1;   demod=0; % DAB Radio signal
9   % FileName = 'SDRSharp_FMRadio_101600kHz_IQ.wav';  T=5;   demod=1; % FM Radio signal
10  % FileName = 'SDRSharp_NanoSat_146000kHz_IQ.wav';  T=0;   demod=2; % Nano Satellite
11    FileName = 'SDRSharp_Airplane_112500kHz_IQ.wav'; T=5;   demod=3; % VOR airplane
12  % FileName = 'SDRWeb_Unknown_3671.9kHz.wav';       T=0;   demod=4; % speech from WebSDR
13
14    inf = audioinfo(FileName), pause          % what is ''inside''
15    fs = inf.SampleRate;                      % sampling rate
16    if(T==0) [x,fs] = audioread(FileName);    % read the whole signal
17    else     [x,fs] = audioread(FileName,[1,T*fs]); % read only T seconds
18    end                                       %
19    whos, pause                               % what is in the memory
20    Nx = length(x),                           % signal length
21
22  % Reconstruct the complex-value IQ data, if necessary add Q=0
23    [dummy,M] = size(x);
24    if(M==2) x = x(:,1) - j*x(:,2); else x = x(1:Nx,1) + j*zeros(Nx,1); end
25        nd = 1:2500;
26        figure(1); plot(nd,real(x(nd)),'bo-',nd,imag(x(nd)),'r*--'); xlabel('n'); grid;
27        title('I(n) = (o) BLUE/solid   |   Q(n)= (*) RED/dashed'); pause
28
29  % Parameters - lengths of FFT and STFT, central signal sample
30    Nc = floor( Nx/2 ); Nfft = min(2^17,2*Nc); Nstft = 512;
31
```

```
32   % Power Spectral Density (PSD) of the signal
33     n = Nc-Nfft/2+1 : Nc+Nfft/2;                    % indexes of signal samples
34     df = fs/Nfft;                                    % df - step in frequency
35     f = df * (0 : 1 : Nfft-1);                       % frequency axis [ 0, fs ]
36     fshift = df * (-Nfft/2 : 1 : Nfft/2-1);          % frequency axis [ -fs/2, fs/2 ]
37     w = kaiser(Nfft,10);                             % window function used
38     X = fft( x(n) .* w );                            % DFT of windowed signal
39     P = 2*X.*conj(X) / (fs*sum(w.^2));               % Power Spectral Density (dB/Hz)
40     Pshift = fftshift( P );                          % circularly shifted PSD
41
42   % Parameters for Short Time Fourier Transform (STFT) of the signal
43     N = Nstft; df = fs/N; ff = df*(0:1:N-1);  ffshift = df*(-N/2:1:N/2-1);
44
45     figure(2)
46     subplot(211); plot(f,10*log10(abs(P))); xlabel('f (HZ)'); ylabel(' (dB/Hz)')
47     axis tight; grid; title('PSD for frequencies [0-fs)');
48     subplot(212); spectrogram(x(n),kaiser(Nstft,10),Nstft-Nstft/4,ff,fs);
49     colormap(cm); pause
50
51     figure(3)
52     subplot(211); plot(fshift,10*log10(abs(Pshift))); xlabel('f (HZ)'); ylabel(' (dB/Hz)')
53     axis tight; grid; title('PSD for frequencies [-fs/2, fs/2)');
54     subplot(212); spectrogram(x(n),kaiser(Nstft,10),Nstft-Nstft/4,ffshift,fs);
55     colormap(cm); pause
56     subplot(111);
```

So now we are in the most risky moment of this chapter! Will the theory *work* in practice, or, as sometimes happens, something suddenly will go wrong. DFT and STFT spectra, calculated in Matlab/Octave for four recorded IQ signals, are shown in Figs. 17.15 and 17.16.

In the first figure DFT spectra of analog FM radio (left) and digital DAB radio—a multi-carrier OFDM transmission (right), are presented. In the first row results obtained from the Matlab fft() function are shown while in the second row—from the spectrogram() function. In next two rows corresponding spectra shifted circularly by the fftshift() function are presented. Why the circular spectrum rotation (right by $N/2$ spectral bins/lags) is necessary? Since the N-point FFT procedure is calculated for frequencies changing in the range $[0, \ldots, f_s)$ $(f = (f_s/N) * (0, 1, 2, \ldots, N-1)$ while we are interested in the frequency range $[-f_s/2, \ldots, f_s/2)$ $(f = (f_s/N) \cdot (-N/2, \ldots 0, \ldots, N/2-1)$. Importance of this change is especially visible in case of the DAB signal (right column in the figure) for which the spectral *hat* of the OFDM is clearly visible only after the shift. In consequence, carriers are scanned for bit recovery in proper order only after the frequency shift of the spectrum. Both signals are complex and for this reason their spectra are not symmetric around 0 Hz. In fact, after processing, the central frequency of the spectrum is lost. We should remember that for FM radio it is equal to 101.6 MHz—such was our central frequency during RF signal recording using SDR-RTL stick and program SDRSharp (with the bandwidth 3.2 MHz). In case of DAB radio the central frequency of recording was chosen 229.069 MHz and bandwidth 2.048 MHz.

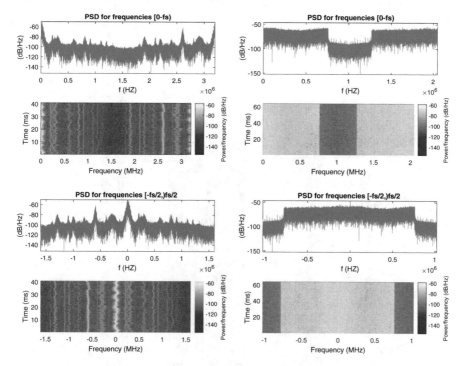

Fig. 17.15: Magnitudes of FFT and STFT spectra, scaled in decibels/Hz, calculated in Matlab/Octave for two recorded IQ signals, respectively in columns: (left) sampled analog FM radio and (right) digital DAB radio. In first two rows we see original FFT and STFT spectra while in the next two rows—their versions circularly shifted (centered at 0 Hz)

Remembering the spectrum shift necessity, in Fig. 17.16 only circularly shifted magnitudes of FFT and STFT spectra of a VOR avionic signal (left) and an slow-speed nano-satellite signal (right) are shown. In this case a Reader should notice that after rotation spectra are not symmetric around the 0 Hz frequency since both signals are complex-value ones.

The fact that obtained figures are the same as ones observed in the SDRSharp and ADI IIO oscilloscope programs should confirm us that our experiments are going in right direction. At present we know how to start RF signals analysis in Matlab, i.e. how to: (1) read signals, (2) plot them, (3) calculate their Fourier spectra, (4) scale them, and (5) display. To give a Reader an additional motivation to continue our RF journey, we will quickly demodulate the mono FM radio, the FM-modulated signal from nano-satellite and the AM-modulated VOR avionic signal. Matlab code given below in Listings 17.3, 17.4, and 17.5 is continuation of the program lab17_ex_IQ_DFT.m, presented in Listing 17.2.

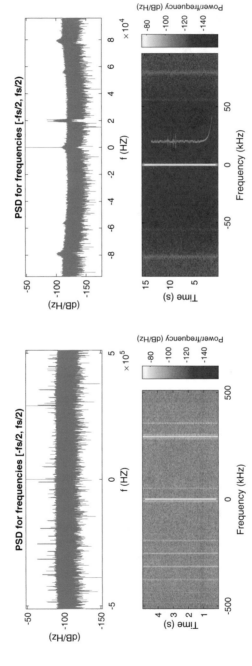

Fig. 17.16: Magnitudes of FFT and STFT spectra, circularly shifted and scaled in decibels/Hz, calculated in Matlab/Octave for two recorded IQ signals: (left) VOR avionics signal and (right) signal from nano-satellite in slow transmission mode

Two Very Important Things Have to Be Stressed Now

Generalized Sub-Band Nyquist Sampling After any DSP course student typically remember that signal sampling should be more than two times bigger than maximum signal frequency. During acquisition of an IQ signal it is not fulfilled. Why? Because in this case the generalized Nyquist sampling theorem is applied. It says that if we limit the signal frequency band with band-pass filter, and remember its parameters, the sampling can be: (1) more than two times higher than the filter band in case of real-value signal acquisition, or (2) higher than this band for complex-value signal acquisition, as in case of SDR equipment using quadrature modulators and demodulators. Of course, the acquired signal has lower frequencies than in reality. But we are not interested in absolute frequency values but in relative differences of carrier modulation frequencies in regard to the carrier frequency. And this relative values are the same.

Service Central Frequency and Bandwidth During signal acquisition two strategies are possible.

1. Precisely selecting the service frequency and service bandwidth. Thanks to this the acquired bit-stream is smaller and computational requirements lower. For example, instead of recording IQ signal of many FM radio stations it is sufficient to record only one, exactly the one to be listened to. Figure 17.17 illustrates this two strategies. On the left side, we see STFT spectrum of many FM radio stations which are present in IQ signal sampled at 3.2 MHz. On the right side, we see the STFT spectrum of only one station which is present in our IQ signal sampled at 250 kHz. In the first case significantly more bits are processed but user can choose the broadcast in digital program, i.e. in software.

2. Wider frequency bandwidth. When the first acquisition strategy is used, we have many RF services in our digital IQ data. In such case service of interest should be converted to 0 Hz, i.e. up or down shifted in frequency, and then low-pass filtered, i.e. separated from the rest. The frequency conversion, shown in Fig. 17.18, is performed exactly in the same way as during service up-conversion from the base-band to the target frequency and back, i.e. by complex-value modulation. If we assume that after DFT spectrum centering around DC, the service of interest has frequency f_0, the following frequency shift should be performed, mathematically and in Matlab:

$$x(n) = x(n) \cdot e^{-j2\pi \frac{f_0}{f_s} n}, \qquad n = 0, 1, 2, \ldots N_x - 1, \qquad (17.15)$$

```
x = x .* exp(-j*2*pi*f0/fs*(0:Nx-1)');
```

Since spectra of sampled signals, calculated via DFT, are periodic, with period f_s, the linear spectrum shift is visible is our frequency band as circular shift.

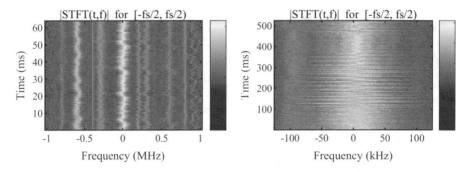

Fig. 17.17: Illustration of two IQ signal acquisition strategies: (left) wide bandwidth with many services, (right) low bandwidth with one service only

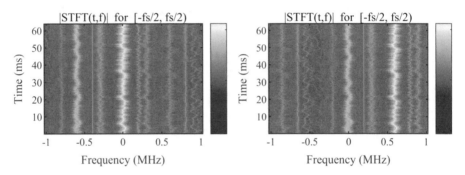

Fig. 17.18: Illustration of concrete RF service selection by means of frequency conversion (complex modulation) of acquired IQ signal: (left) STFT of original IQ signal, (right) STFT spectrum after frequency shift (conversion, modulation)

All remaining details will not be given now—they will be explained in next chapters. All Readers are kindly invited to the next part of our nuMAT/sciMAT story.

Listing 17.3: Simple program for mono FM radio decoding in Matlab/Octave

```
1   if(demod==1) % FM demodulation and mono FM radio decoding
2       bwSERV=200e+3; bwAUDIO=25e+3;                               % Parameters
3       D1 = round(fs/bwSERV); D2 = round(bwSERV/bwAUDIO);          % Downsampling ratios
4       f0 = -0.59e+6; x = x .* exp(-j*2*pi*f0/fs*(0:Nx-1)'); % Station? shift in freq
           -0.59?
5       x = resample(x,1,D1);                                % Resample down to service freq
```

```
6    x = real(x(2:end)).*imag(x(1:end-1))-real(x(1:end-1)).*imag(x(2:end)); % FM demod
7    x = resample(x,1,D2);                                     % Resample down to audio freq
8    soundsc(x,bwAUDIO);                                       % Listening
9  end
```

Listing 17.4: Simple program in Matlab/Octave for FM demodulation of speech
signal sent from nano-satellites in safe/slow transmission mode

```
1  if(demod==2) % FM demodulation and NanoSat voice control signal decoding \cite[Mare19]
2     fAudio = 48e+3; Down= round(fs/fAudio);                  % parameter values
3     f0 = 1.98e+4; x = x .* exp(-j*2*pi*f0/fs*(0:Nx-1)');     % carrier? shift spectrum
4     h = fir1(500,12500/fs,'low'); x = filter(h,1,x); x = x(1:Down:end); % filtration
5     dt = 1/fAudio; x = x(2:end).*conj(x(1:end-1)); x=angle(x)/(2*pi*dt); % FM demod
6     soundsc(x,fAudio);                                       % Listening
7  end
```

Listing 17.5: Simple program for VOR voice control decoding in Matlab/Octave

```
1   if(demod==3) % AM demodulation and VOR voice control signal decoding \cite[VOR13]
2      fc = 2.9285e+5;                         % carrier? your choice after spectrum inspection
3      fam = 10000;                            % frequency width of AM modulation around fc
4      f1 = fc-fam/2; f2 = fc+fam/2; df = 500;           % Band-Pass filter h design
5      h = cfirpm(500,[-fs/2 (f1-df) (f1+df) (f2-df) (f2+df) fs/2]/(fs/2),@bandpass);
6      x = conv(x,h,'same');                              % Band-pass filtration
7      x = sqrt( real(x).*real(x) + imag(x).*imag(x) );        % AM demodulation
8      x = decimate(x,round(fs/fam)); %[U,D] = rat(fs/fam); x = resample( x, U, D );
9      soundsc(x,fam);                                    % Listening
10  end
```

Exercise 17.2 (Observing Spectra of Recorded IQ Signals). Install and run
the program SDRSharp or Gqrs. Switch to the IQ file mode (Source panel).
Open one-by-one all IQ recordings available in the book repository. Start spectra calculation and visualization using *PLAY* ▶ button. Change service frequency to be observed, clicking on big digits displayed on the upper panel (on
their lower or upper parts—frequency decrease or increase, respectively). Set
appropriate AM or WFM demodulation (Radio panel). Listen to decoded audio. Next, analyze the program 17.2 and run it several times, selecting different
input files. Observe signal spectra. Listen to decoded audio.

17.7 Summary

This introductory chapter aims at simple explanation of RF telecommunication signals up-conversion from the base-band to the target frequency and back to the base-band. This is essential operation in wireless communications. Nowadays it is widely realized using the software defined radio technology. The SDR *basics of the basics* were presented, I hope that in a Reader-friendly manner. What should be remembered?

1. Different telecommunication services should co-exist and should not disturb themselves. Therefore they should be "orthogonal" to each other: they should use different frequency slots (FDM -frequency division multiplexing), different time slots (TDM—time division multiplexing), different time–frequency slots (TFDM, hopping frequencies), or different time/frequency/time–frequency slots in different geo-positions (SDM—spacial division multiplexing).

2. In communication system information/data are mainly carried by sine()/cosine() oscillations. The message is hidden in the oscillation parameters, it is changing them, i.e. modulating signal amplitude, phase, and frequency. In the receiver, *jumps* of the sine()/cosine() parameter values should be recovered and information coded in them should be extracted, demodulated. One or many oscillating carriers are used (single-carrier and multi-carrier transmissions).

3. Since the information is coded in *changing states* of one or many oscillations, each RF service transmitter modulates some low-frequency carriers in the base-band around the DC (0 Hz) and then converts/shifts them up to higher, target frequencies by signal mixers, but into different frequency bands. The information is not changed: higher frequencies are modulated exactly in the same manner as lower frequencies in the base-band.

4. The RF receiver is extracting signals from the frequency band of interest using a band-pass filter, and is doing the reverse frequency conversion of the signal, its down-conversion to the base-band. In the BB all service carriers are demodulated, i.e. modulating functions/numbers are recovered and transmitted information is decoded.

5. A base-band signal, synthesized in the transmitter, can have real or complex-values. Using complex-value signals simplifies frequency modulation and demodulation in the base-band, allowing the same time existence of negative frequencies. It is important to remember that after frequency up-conversion, the signal in the channel has always real values. It results from the fact that only the real part of a complex-value output of the quadrature mixer is emitted by antenna.

6. In the receiver the base-band signal is real-value (when only cos() down-frequency mixer is used) or complex-value (when quadrature cos() and –

sin() mixers are used and two signals are combined into a vector of complex numbers).

7. Therefore after up (TX) and down (RX) frequency conversions of the complex-value signal, assuming distortion free environment, one should obtain in the receiver exactly the same signal as transmitted one, for example FM or DAB radio.

17.8 Private Investigations: Free-Style Bungee Jumps

Exercise 17.3 (Up-Down Frequency Conversion of Pure Mono Speech/Music Services). Extend program from Listing 17.1: use mono speech/music recordings as signals to be transmitted, add plots of signal spectra, and check whether the signal processing presented in Fig. 17.3 really holds in each test.

Exercise 17.4 (Service Selection in Existing Wide-Band IQ Recordings). Use programs 17.2 and 17.3. Try to recover mono signals of all available FM stations present in the IQ file SDRSharp_FMRadio_101600kHz_IQ.wav. Find (read) station carrier frequencies and shift services to 0 Hz. If you have the RTLSDR stick or PLUTO module, record your own IQ file and do the same using your file.

Exercise 17.5 (Voice Control in Airplane Azimuth Detection). Try to find some other VOR signals in the IQ file SDRSharp_Airplane_112500kHz_IQ.wav and to decode their voice control audio. Use SDRSharp program. Work in IQ file mode (in source panel). In the upper bar set service frequency to 112.79285 MHz. Switch to AM decoding (in radio panel). If you have the RTLSDR stick or PLUTO module, find a new VOR signal in the range 108–138MHz (good antenna is required!), then record it and decode. Try to find airplane navigation signals on servers given in WebSDR http://websdr.org/.

Exercise 17.6 (Visiting WebSDR). Go to the page WEBSDR http://websdr.org, choose any RF signal server, choose the service type/bandwidth (CW, CW-N, LSB, LSB-N, USB, USB-N, AM, AM-N, FM), choose an RF service-by shifting the observation window using mouse, adjust the frequency edges of the window with mouse, hear the demodulated signal as audio one, record it into a file as WAV, download the file, read data into Matlab/Octave, show the signal samples, calculate and show the signal frequency spectrum.

References

1. Analog Devices, Functional block diagram of analog devices AD9361 RF agile transceiver. Available online: http://analogdevicesinc.github.io/libad9361-iio/
2. K. Barlee, D. Atkinson, R.W. Stewart, L. Crockett, *Software Defined Radio using MATLAB & Simulink and the RTLSDR*. Matlab Courseware (Strathclyde Academic Media, Glasgow, 2015). Online: http://www.desktopsdr.com/
3. T.F. Collins, R. Getz, D. Pu, A.M. Wyglinski, *Software Defined Radio for Engineers* (Analog Devices - Artech House, London, 2018)
4. B. Farhang-Boroujeny, *Signal Processing Techniques for Software Radios* (Lulu Publishing House, 2008)
5. M.E. Frerking, *Digital Signal Processing in Communication Systems* (Springer, New York, 1994)
6. C.R. Johnson Jr., W.A. Sethares, *Telecommunications Breakdown: Concepts of Communication Transmitted via Software-Defined Radio* (Prentice Hall, Upper Saddle River, 2004)
7. C.R. Johnson Jr, W.A. Sethares, A.G. Klein, *Software Receiver Design. Build Your Own Digital Communication System in Five Easy Steps* (Cambridge University Press, Cambridge, 2011)
8. J. Marek, *Programming Ground Station for Receiving Satellite Signals Using SDR Technology*. B.Sc. Thesis, AGH University of Science and Technology, Krakow, Poland, 2019
9. National Telecommunications and Information Administration, United States Frequency Allocations. The Radio Spectrum. Available online: https://www.ntia.doc.gov/files/ntia/publications/january_2016_spectrum_wall_chart.pdf, January 2016
10. OzHack, Product: RTL-SDR R820T2+RTL2832U, figure from https://ozhack.com/products/rtl-sdr-r820t2-rtl2832u
11. PLUTO Analog Devices Active Learning Module (ADALM), https://www.analog.com/en/design-center/evaluation-hardware-and-software/evaluation-boards-kits/adalm-pluto.html
12. L.C. Potter, Y. Yang, *A Digital Communication Laboratory - Implementing a Software-Defined Acoustic Modem*. Matlab Courseware (The Ohio State University, Columbus, 2015). Online: https://www.mathworks.com/academia/courseware/digital-communication-laboratory.html
13. J.G. Proakis, M. Salehi, G. Bauch, *Modern Communication Systems Using Matlab* (Cengage Learning, Stamford, 2004, 2013)
14. D. Pu, A.M. Wyglinski. *Digital Communication Systems Engineering with Software-Defined Radio* (Artech House, Boston, London, 2013)
15. Rhode-Schwarz, R & S FS-K15 VOR/ILS avionics measurements software manual, Munich, Germany 2013. Online: https://www.rohde-schwarz.com/ua/manual/r-s-fs-k15-vor-ils-avionics-measurements-manuals-gb1_78701-29018.html
16. C. Thompson, FOX-1 satellite telemetry part 2: FoxTelem. AMSAT J. **39**(1), 7–9 (2016)

17. S.A. Tretter, *Communication System Design Using DSP Algorithms* (Springer Science+Business Media, New York, 2008)

18. J. Zhao, Software defined radio with MATLAB. Available online: https://docplayer.net/31465789-Software-defined-radio-with-matlab-john-zhao -product-manager.html

Further Reading: Digital Communications

19. J.B. Anderson, *Digital Transmission Engineering* (IEEE Press/Wiley-Interscience, Piscataway NJ, 2005)

20. M. Fitz, *Fundamentals of Communications Systems* (McGraw-Hill, New York, 2007)

21. M.E. Frerking, *Digital Signal Processing in Communication Systems* (Springer, New York, 1994)

22. R.G. Gallager, *Principles of Digital Communication* (Cambridge University Press, New York, 2008)

23. S. Haykin, *Communication Systems* (Wiley, New York, 1994)

24. R.W. Heath Jr., *Introduction to Wireless Digital Communication. A Signal Processing Perspective* (Prentice Hall, Boston, 2017)

25. F. Ling, *Synchronization in Digital Communication Systems* (Cambridge University Press, Cambridge, 2017)

26. U. Madhow, *Fundamentals of Digital Communication* (Cambridge University Press, Cambridge, 2008)

27. H. Meyr, M. Moeneclaey, S.A. Fechtel, *Digital Communication Receivers* (Wiley, New York, 1998)

28. A.F. Molish, *Wireless Communications* (IEEE - Wiley, Chichester, 2005)

29. A.V. Oppenheim, G.C. Veghese, *Signals, Systems and Interference* (Pearson Education, Harlow, 2017)

30. M. Pätzold, *Mobile Fading Channels* (Wiley, Chichester, 2002)

31. J.G. Proakis, *Digital Communications* (McGraw-Hill, New York, 2001)

32. T. Rappaport, *Wireless Communication Systems: Principles and Practice* (Prentice Hall, Upper Saddle River NJ, 1996)

33. M. Rice, *Digital Communications: A Discrete-Time Approach* (Pearson Education, Upper Saddle River, 2009)

34. K. Shenoi, *Digital Signal Processing in Telecommunications* (Prentice Hall, Upper Saddle River, 1995)

35. B. Sklar, *Digital Communications: Fundamentals and Applications* (Prentice Hall, Upper Saddle River, 1988, 2001, 2017)

36. B. Sklar, F. Harris, *Digital Communications: Fundamentals and Applications*, 3rd edn. (Pearson, Upper Saddle River, 2020)

37. K. Wesolowski, *Introduction to Digital Communication Systems* (Wiley, New York, 2009)

Chapter 18
Frequency Modulation and Demodulation

*War and peace, love and hate, up and down during bicycle ride:
do not be surprised that frequency of your heartbeat is
changing!*

18.1 Introduction

We live because our hearts are beating all the time and pump the blood with oxygen to all parts of our bodies. But the heartbeat-rate is changing due to our emotional ("war and peace") and physical ("up and down," "faster or slower") stress. We can say that the frequency of heartbeat is modulated/changed by some real-world signals/events. Frequency modulation (FM), faster or slower repetition of something, is very popular in our everyday life. Wheels of our cars and bicycles are rotating with different speed before the mountain pass and after it.

In telecommunication people generate oscillatory sine signals called carriers and deliberately change their oscillation frequencies proportionally to some value—they *modulate* a carrier in frequency. When one is tracking the oscillation frequency change and recovering the value used in the transmitter for frequency shift, he is doing *carrier demodulation*. Value causing frequency change can change itself, even oscillates. In telecommunication we are transmitting functions that modulate our carriers. In analog FM radio it is a multiplex MPX signal consisting of audio and RDS binary data. In digital RTTY amateur radio, transmission bits are hidden in a carrier jumps from one frequency to the other. Frequency shift keying (FSK) methods are very popular in digital data transmission inside the human body, like in autonomous gastroscopic capsules. Carrier frequency can be softly changing from one value to the other using Gaussian function formula—such frequency modulation is known as Gaussian Minimum Shift Keying. It was widely used in the past in old 2G GSM digital telephony. Nowadays it is exploited in, for example: digital data links between mobile devices and satellites, remote control devices and Bluetooth standard. So the FM is still alive!

© Springer Nature Switzerland AG 2021   517
T. P. Zieliński, *Starting Digital Signal Processing in Telecommunication
Engineering*, Textbooks in Telecommunication Engineering,
https://doi.org/10.1007/978-3-030-49256-4_18

A crucial issue in FM modulation and demodulation is concept of instantaneous signal frequency, defined as a time derivative (divided by 2π) of an angle $\varphi(t)$ of the real-value $\cos(\varphi(t))$ or complex-value $\exp(j \cdot \varphi(t))$ oscillators. Doing an FM modulation we should ensure an appropriate change of this angle, making it a sum of two components: the first one $(\varphi_1(t) = 2\pi f_c t)$ associated with the carrier frequency and the second term $(\varphi_2(t) = 2\pi \cdot \Delta f \cdot \int_0^t x(t)dt)$, responsible for the carrier frequency change and being dependent upon modulating function $x(t)$. Frequency demodulation relies on recovering the function $x(t)$ from the carrier angle $\varphi(t)$. It is a very simple task for complex-value analytic signals of the form $\exp(j \cdot \varphi(t))$—only an angle of a complex -value number has to be computed using four quadrant $\arctan(\text{Im}(y(t)), \text{Re}(y(t)))$ function, and then its derivative (change in time) should be found. In case of real carriers with $f_c \neq 0$, the situation is only a little bit different: first the complex-value analytic version of the signal should be computed using the Hilbert transform and then the signal angle.

In this chapter we will understand the demodulation procedure of the mono FM radio broadcasting, presented in Chap. 17. We will also become familiar with some FM modulation patterns used in amateur radio and find our position in respect to airport using VOR signals transmitted from airport to airplanes.

18.2 Frequency Modulation

In this chapter the following notation is used:

$x(t), x(n)$—analog and digital signal used for *modulation*,
$y(t), y(n)$—analog and digital signal being *modulated* (modulation result),
$I(t), I(n)$—real part of $y(.)$, in-phase component,
$Q(t), Q(n)$—imaginary part of $y(.)$, quadrature component.

Instantaneous frequency of a real-value signal (1) and a complex-value *analytic* signal (2) of the form:

$$y_1(t) = \cos(\varphi(t)), \qquad y_2(t) = e^{j \cdot \varphi(t)} = \cos(\varphi(t)) + j \cdot \sin(\varphi(t)) \qquad (18.1)$$

is defined as derivative of its phase angle divided by 2π :

$$f_{inst}(t) = \frac{1}{2\pi} \frac{d\varphi(t)}{dt}. \qquad (18.2)$$

If we would like to generate a signal with instantaneous frequency changing in some predefined way, we should express $\varphi(t)$ as a function of $f_{inst}(t)$ from Eq. (18.2):

$$d\varphi(t) = 2\pi f_{inst}(t)dt \quad \Rightarrow \quad \varphi(t) = \int_0^t d\varphi(t) = 2\pi \int_0^t f_{inst}(t)dt \qquad (18.3)$$

and put the result into Eq. (18.1), which gives

$$y_1(t) = \cos\left(2\pi \int_0^t f_{inst}(t)dt\right), \qquad y_2(t) = \exp\left(j \cdot 2\pi \int_0^t f_{inst}(t)dt\right). \quad (18.4)$$

Modulation of a signal carrier frequency f_c using a function $x(t)$ with a predefined frequency modulation depth (Δf) expressed in Hz is described by the following equation:

$$f_{inst}(t) = f_c + \Delta f \cdot x(t). \quad (18.5)$$

The function $x(t)$ is typically changing its values in the range $[-1, 1]$, therefore the maximum change of signal frequency is equal to $\pm\Delta f$. After putting Eq. (18.5) into signals in Eqs. (18.4) one obtains, here only for $\exp(j \cdot \varphi(t))$:

$$y(t) = \exp\left(j \cdot 2\pi \left(f_c t + \Delta f \cdot \int_0^t x(t)dt\right)\right). \quad (18.6)$$

When $f_c = 0$ Hz, the modulation is done in the base-band. For linear (LFM) and sinusoidal (SFM) frequency modulation, Eq. (18.6) simplifies to Eqs. (18.7) and (18.8):

LFM: $x(t) = \alpha \cdot t,$ $\qquad y(t) = \exp\left(j \cdot 2\pi \left(f_c t + 0.5\alpha t^2\right)\right),$ $\qquad (18.7)$

SFM: $x(t) = \sin(2\pi f_m t),$ $\quad y(t) = \exp\left(j \cdot 2\pi \left(f_c t - \dfrac{\Delta f}{2\pi f_m} \cos(2\pi f_m t)\right)\right).$

$$(18.8)$$

In equations (18.7) and (18.8) exponent can be replaced with cosine (j should be removed). A Reader is asked to analytically check whether Eqs. (18.7) and (18.8) fulfill the phase differentiation principle (18.2).

Computer implementation of Eq. (18.6) is not difficult in Matlab:

```
y = exp( j *2*pi*(fc*t + df*cumsum(x)*dt) );
```

using the `cumsum(x)` function of cumulative summation of vector x elements:

$$\text{cumsum}(x) = [x_1, x_1 + x_2, x_1 + x_2 + x_3, \ldots].$$

While multiplied by uniform spacing dt, this is the simplest numerical approximation of a function integral in some limited interval, assuming zero-order interpolation of a function between its samples. In other words, integral in Eq. (18.6) is
calculated via additions of rectangular areas lying below the signal curve.

Function (18.6) is not linear in respect to signal $x(t)$ and difficult to analyze in
general case. Its Fourier spectrum bandwidth BW around f_c depends on the chosen
modulation depth Δf and maximum frequency of the signal $x(t)$. This relation is
approximated by the Carson's rule:

$$BW = 2f_{max} \left(\frac{\Delta f}{f_{max}} + 1 \right). \tag{18.9}$$

Knowing available BW and f_{max} of the signal $x(t)$, we can calculate required value
of Δf from the equation:

$$\Delta f = \left(\frac{BW}{2f_{max}} - 1 \right) \cdot f_{max} \tag{18.10}$$

which will be used by us in our FM radio Matlab programs.

The frequency modulation is highly non-linear. Due to this even for a very
simple case when the carrier frequency is modulated by a single pure cosine
$x(t) = \cos(2\pi f_m t)$, the resultant FM modulated signal can have many frequency
components:

$$y(t) = A_c \cos \left(2\pi f_c t + \frac{\Delta f}{f_m} \sin(2\pi f_m t) \right) = \sum_{n=-\infty}^{\infty} A_c J_n \left(\frac{\Delta f}{f_m} \right) \cos \left(2\pi (f_c + n f_m) t \right) \tag{18.11}$$

so its spectrum can be very wide. In Eq. (18.11) $J_n()$ is the Bessel function of the
1-st kind and order n, in Matlab `besselj(n,delta)`.

Proof (A Little Bit of Our Sweet Math).

At present we will discuss a special case of Eq. (18.6) for a real-value cosine carrier
$\cos(.)$ instead of the complex-value exponential one $\exp(j.)$ and for cosine modulating function $x(t) = \cos(2\pi f_m t)$. In such case one obtains

$$y(t) = A_c \cos \left(2\pi f_c t + \frac{\Delta f}{f_m} \sin(2\pi f_m t) \right) = A_c \cos \left(2\pi f_c t + \delta \sin(2\pi f_m t) \right), \tag{18.12}$$

where δ denotes $\Delta f / f_m$. Using trigonometric equality $\cos(\alpha + \beta) = \cos(\alpha)\cos(\beta) - \sin(\alpha)\sin(\beta)$ we can rewrite the above expression into the following form:

$$y(t) = A_c \cos(2\pi f_c t)\cos(\delta \sin(2\pi f_m t)) - A_c \sin(2\pi f_c t)\sin(\delta \sin(2\pi f_m t)) \quad (18.13)$$

and even transform it further:

$$y(t) \approx A_c \cos(2\pi f_c t) - A_c \delta \sin(2\pi f_c t)\sin(2\pi f_m t) \quad (18.14)$$

and further:

$$y(t) \approx A_c \cos(2\pi f_c t) + \frac{A_c \delta}{2}\cos(2\pi(f_c + f_m)t) - \frac{A_c \delta}{2}\cos(2\pi(f_c - f_m)t) \quad (18.15)$$

since for small δ values we can do in Eq. (18.13) the following substitutions:

$$\cos(\delta \sin(2\pi f_m t)) \approx 1, \quad \sin(\delta \sin(2\pi f_m t)) \approx \delta \sin(2\pi f_m t)$$

and the following replacement can be done in Eq. (18.14):

$$-\sin(\alpha)\sin(\beta) = 0.5 \cdot \cos(\alpha + \beta) - 0.5 \cdot \cos(\alpha - \beta).$$

Approximation (18.15) is only valid for small δ values. For bigger ones Eq. (18.11) should be used.

□

In Fig. 18.1 first three Bessel functions $J_n(\delta)$ are shown for $n = 0, 1, 2$: the 0-th, damped cosine-like, is the highest one for $\delta = 0$, and the others, damped sine-like, are decreasing with the increase of n. The plot was obtained using the following Matlab program:

```
d=0:0.05:25;;
J0 = besselj(0,d); J1 = besselj(1,d); J2 = besselj(2,d);;
plot(d,J0,'k',d,J1,'b',d,J2,'r'); xlabel('d'); title('J(d)');
grid; pause;
```

In turn in Fig. 18.2 two spectra of SFM signals are shown. Calculated values $J_n(\delta)$ are marked in it with red circles. In both cases carrier frequency is equal to $f_c = 1000$ Hz while modulating frequency is equal to $f_m = 100$ Hz. In the left figure Δf is equal to 0.001 Hz and in the right one—10 Hz. We can observe that increase of Δf value results in widening the spectrum which is symmetrical around f_c.

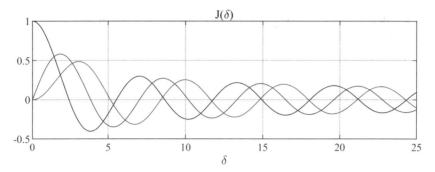

Fig. 18.1: Shape of three first ($n = 0, 1, 2$) Bessel functions for $\delta = \frac{\Delta f}{f_m}$. With the increase of n they have smaller amplitudes and they are decaying for bigger values of δ

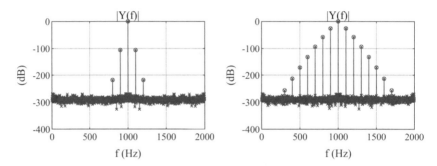

Fig. 18.2: Two Fourier spectra magnitudes of SFM modulated signals. Calculated values of $J_n(\delta)$ are marked with red circles where $\delta = \Delta f / f_m$. Values of parameters: $f_c = 1000$Hz, $f_m = 100$Hz, (left): $\Delta f = 0.001$Hz, $\delta = 0.00001$, (right): $\Delta f = 10$Hz, $\delta = 0.1$

18.3 Frequency Demodulation

Frequency demodulation of the complex-value Euler-function signals of the form (18.6) is done using Eq. (18.2) and it is not difficult because angles of the complex number are directly accessible (function `angle()` in Matlab)) and derivative of the signal phase can be easily calculated. In case of a real-value signal, $y_1(t)$ in Eq. (18.1), one should first calculate its complex-value analytic version, i.e. $y_2(t)$ in (18.1):

$$y_1^{(a)}(t) = \cos(\varphi(t)) + j \cdot \underbrace{\text{Hilbert}(\cos(\varphi(t)))}_{\sin(\varphi(t))} = e^{j \cdot \varphi(t)} = y_2(t) \qquad (18.16)$$

using the Hilbert filter/transform, being the $-\pi/2$ radians phase shifter, and then find derivative of the complex-value signal angle, as before. Therefore in this section only frequency demodulation of analytic signals is presented.

Exercise 18.1 (Angle Estimation of a Complex-Value Signal). In the beginning, we remember the problem of angle estimation of complex-value numbers (already known to us as *phase unwrapping* of filter frequency responses). In Listing 18.1 a complex-value signal is generated: $x(n) = e^{j \cdot \phi(n)}, \phi(n) = 3\pi \sin\left(\frac{2\pi}{200} * n\right), n = 0, 1, \ldots, 400$. The signal angle $\phi(n)$ is recovered from signal samples using Matlab functions `phi1=atan2(imag(x),real(x))` and `phi2=atan(x)`. Obtained results for both functions are the same— they are presented using solid line in the left plot of Fig. 18.3. However, they are wrong: the unwanted angle wrapping is observed since the arctan() function returns only values in the range $[-\pi, \pi)$. After application of the Matlab `unwrap(phi)` function, adding missing $\pm 2\pi$ jumps, the calculated angle is corrected which is shown in left plot. Do some experiments: change function describing the angle value.

Listing 18.1: Calculation of a complex-value signal angle

```
1   % lab18_ex_unwrap.m
2   % Example of signal angle wrapping and un-wrapping
3   clear all; close all;
4
5   n = 0:400;
6   phi = 3*pi*sin(2*pi/200*n);
7   x = exp( j*phi );
8
9   phi1 = atan2( imag(x), real(x) );
10  phi2 = angle( x );
11  error = max(abs( phi1 - phi2 )),
12  figure; plot(n,phi,'k--',n,phi1,'b'); grid;
13  xlabel('n'); ylabel(' [rad]'); title('\phi (n): original and calculated'); pause
14
15  phi1 = unwrap( phi1 );
16  figure; plot(n,phi,'k--',n,phi1,'b'); grid;
17  xlabel('n'); ylabel(' [rad]'); title('\phi (n): original and calculated'); pause
```

For a discrete-time signal $y(n)$ estimation of its angle derivative can be calculated using one of many existing formulae.

Arctan Methods with Phase Unwrapping In the first group of methods we first calculate the signal angle from its tangent using 4 quadrant arctan(Q, I) function:

$$\varphi_{\pm 2\pi}(n) = \arctan(Q(n), I(n)), \qquad (18.17)$$

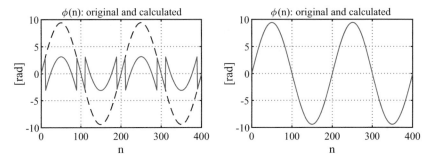

Fig. 18.3: Illustration of angle computation problem for complex-value signals: (left) original signal angle (dashed line) and its calculation using arctan() function (solid line)—the angle *wrapping* is observed, (right) estimated angle after application of Matlab *unwrapping* procedure—the angle is perfectly reconstructed

where

$$Q(n) = \text{Im}(y(n)), \quad I(n) = \text{Re}(y(n)). \tag{18.18}$$

Since the arctan(.) function returns the angle only from the interval $[-\pi, \pi)$ and the changing signal angle can also drift outside this rage, we can obtain $\pm 2\pi$ jumps in the calculated phase estimate. For this reason its unwrapping is required (i.e. jumps removal—see Fig. 7.2 in chapter on analog filters and description of the Matlab function unwrap()):

$$\varphi(n) = \text{unwrap}\left(\varphi_{\pm 2\pi}(n)\right). \tag{18.19}$$

After that we are finding derivative of the computed angle using 2-point, 3-point, or any multi-point differentiation estimator, even using a differentiation filter. Below the simplest 2-point and 3-point derivative estimator are used:

$$\widehat{f}_{inst}^{(2p)}(n) \approx \frac{1}{2\pi} \frac{\varphi(n) - \varphi(n-1)}{\Delta t}, \tag{18.20a}$$

$$\widehat{f}_{inst}^{(3p)}(n) \approx \frac{1}{2\pi} \frac{\varphi(n+1) - \varphi(n-1)}{2\Delta t}. \tag{18.20b}$$

Arctan Methods Without Phase Unwrapping In order to avoid unwanted wrapping of the angle estimate, one can use an alternative, less risky approach. Instead of calculation of an angle of a complex-value number (first) and its local derivative (then), one could do some mathematical operations on $y(n)$ samples, described below, and compute some temporal complex-value number having directly an angle equal to the signal phase difference. Such difference should be smaller than values that are subtracted and occurrence of phase wrapping effect would be less probable.

This computational trick is done in two frequency demodulation formula presented below:

$$\frac{1}{2\pi} \frac{\angle(y(n)y^*(n-1))}{\Delta t} = \frac{1}{2\pi} \frac{\angle\left(e^{j(\varphi(n)-\varphi(n-1))}\right)}{\Delta t} \approx \hat{f}_{inst}^{(2p)}(n), \qquad (18.21a)$$

$$\frac{1}{2\pi} \frac{\angle(y(n+1)y^*(n-1))}{2\Delta t} = \frac{1}{2\pi} \frac{\angle\left(e^{j(\varphi(n+1)-\varphi(n-1))}\right)}{2\Delta t} \approx \hat{f}_{inst}^{(3p)}(n), \qquad (18.21b)$$

where $\angle(c(n))$ denotes an angle of a complex number $c(n)$, $c^*(n)$—its complex conjugation and Δt—sampling period (inverse of a sampling frequency f_s). In Matlab last two equations have one line implementations:

```
finst2p=(1/(2*pi))*angle( y(2:end).*conj( y(1:end-1)) )/dt;
finst3p=(1/(2*pi))*angle( y(3:end).*conj( y(1:end-2)) )/(2*dt);
```

Fast Methods Without Arctan Computation In [3] the following fast method is described in which the necessity of arctan(.) function computation is avoided. In frequency demodulation methods we have to calculate derivative of the signal $y(t)$ angle (18.2) which is itself calculated using arctan(.) function:

$$f_{inst}(t) = \frac{1}{2\pi} \frac{d}{dt} \left[\arctan(r(t))\right], \quad r(t) = \frac{Q(t)}{I(t)} = \frac{\operatorname{Im}(y(t))}{\operatorname{Re}(y(t))}. \qquad (18.22)$$

Our sweet mathematics tells us that the following two equations hold

$$\frac{d}{dt} \left[\arctan(r(t))\right] = \frac{1}{1+r^2(t)} \cdot \frac{d}{dt} \left[r(t)\right], \quad \frac{d}{dt} \left[r(t)\right] = \frac{\frac{dQ(t)}{dt} I(t) - \frac{dI(t)}{dt} Q(t)}{I^2(t)}. \qquad (18.23)$$

Therefore Eq. (18.22) can be written as:

$$f_{inst}(t) = \frac{1}{2\pi} \cdot \frac{1}{1+Q^2(t)/I^2(t)} \cdot \frac{\frac{dQ(t)}{dt} I(t) - \frac{dI(t)}{dt} Q(t)}{I^2(t)}, \qquad (18.24)$$

which finally gives

$$f_{inst}(t) = \frac{1}{2\pi} \frac{\frac{dQ(t)}{dt} I(t) - \frac{dI(t)}{dt} Q(t)}{I^2(t) + Q^2(t)}. \qquad (18.25)$$

Function derivatives can be estimated in (18.25) using first or second order interpolation formula:

$$\frac{dF(t)}{dt} \approx \frac{F(n) - F(n-1)}{\Delta t}, \quad \frac{dF(t)}{dt} \approx \frac{F(n+1) - F(n-1)}{2\Delta t}. \qquad (18.26)$$

For the first, simpler estimator one obtains

$$f_{inst}(n) = \frac{1}{2\pi \cdot \Delta t} \frac{I(n)Q(n-1) - Q(n)I(n-1)}{I^2(n) + Q^2(n)}. \tag{18.27}$$

Application of Eq. (18.27) requires slower-variability of the $\phi(t)$.

The arctan-function-less FM demodulation methods have the following simple Matlab realizations (without normalizing division by $[I^2(n) + Q^2(n)]$):

```
finst4=(1/(2*pi))*(real(y(1:N-1)).*imag(y(2:N))-...
    imag(y(1:N-1)).*real(y(2:N)))/dt;
finst5=(1/(2*pi))*(real(y(2:N-1)).*(imag(y(3:N))
-imag(y(1:N-2)))-...
    imag(y(2:N-1)).*(real(y(3:N))-real(y(1:N-2))) ) / (2*dt);
```

In order to "demystify" the presented above equations and the program code, we will perform ourselves some FM (de)modulation examples below and observe graphically an FM modulation and demodulation magic.

18.4 FM Testing

A simple Matlab program for testing the described above frequency modulation and demodulation algorithms is presented in Listing 18.2. Integration in Eq. (18.6) is implemented by Matlab cumsum() function. Five different demodulation methods are coded. A Reader can modulate a real-value or a complex-value harmonic carrier using artificially generated sinusoid with arbitrary frequency, either using any recorded audio signal. Figure 18.4 presents results from complex-value carrier modulation using pure 2 Hz sine while Fig. 18.5 using a speech signal. Both signals are sampled at 11.025 kHz and have values in the range $[-1, 1]$. They modulate harmonic complex-value carrier 0 Hz or 4 kHz. The modulation depth in both cases is the same and equal to 1 kHz.

We should do the following remarks. The STFT spectra of modulating signals, the 2 Hz sine and speech, are not symmetrical around 0 Hz because the signals are complex-value ones. The speech spectrum is significantly wider than the spectrum of 2 Hz sine, no surprise. However, widths of the spectra of modulated carrier do not differ so much in both cases. Why? Because the modulation depth is responsible for the spectrum width and it is the same (1 kHz). Nevertheless, the spectrum of carrier modulated by speech is wider due to presence of higher frequencies in this signal— consequence of Carson rule (18.9). It is also important to observe that shapes of spectra for modulated carriers 0 Hz and 4 kHz are the same: the second is an exact copy of the first shifted up in frequency. Spectra of sampled signals are periodic in frequency which is also visible in both figures, especially for 0 Hz carrier modulated by 2 Hz sine.

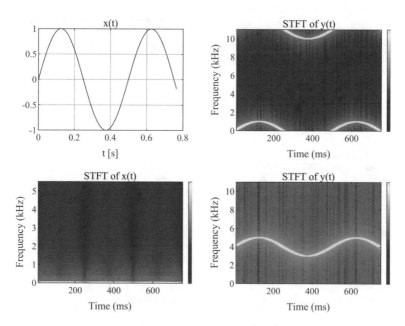

Fig. 18.4: FM modulation example: (left up) sine signal, modulating the carrier, having frequency 2 Hz and sampled with $f_s = 11.025$ kHz, (left down) its STFT time–frequency spectrum, (right up) the STFT spectrum of the FM modulated complex-value carrier 0 Hz and (right down) the STFT spectrum of the FM modulated carrier $f_c = 4$ kHz. The modulation depth in both cases is equal to $\Delta f = 1$ kHz

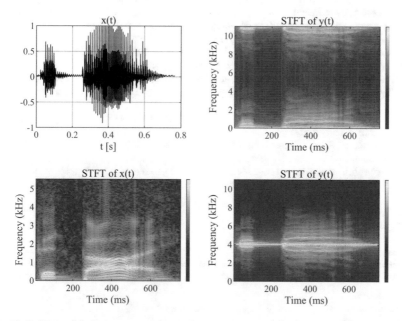

Fig. 18.5: FM modulation example: (left up) speech signal, modulating the carrier, sampled with $f_s = 11.025$ kHz, (left down) its STFT time–frequency spectrum, (right up) the STFT spectrum of the FM modulated complex-value carrier $f_c = 0$ Hz, and (right down) $f_c = 4$ kHz. The modulation depth in both cases is equal to $\Delta f = 1$ kHz

Listing 18.2: Simple program for testing in Matlab the presented frequency modulation and demodulation techniques

```
1
2  % lab18_ex_fm.m
3  % Example of FM modulation and demodulation
4  clear all; close all;
5
6  % Modulating signal m(t)
7  [x,fs] = audioread('GOODBYE.WAV');   % read speech from file
8  Nx = length(x);                      % number of signal samples
9  x = x(:,1)';                         % take only one channel, 1 or 2
10 dt=1/fs; t=dt*(0:Nx-1);              % time
11 df=1/(Nx*dt); f=df*(0:Nx-1);        % frequency
12 x = cos(2*pi*2*t);                   % alternative modulating signal
13 figure; plot(t,x); xlabel('t [s]'); grid; title('x(t)'); pause
14 figure; spectrogram(x,256,192,512,fs,'yaxis'); title('STFF of x(t)'); pause
15
16 % FM modulation
17 fc=0;                                % carrier frequency: 0 or 4000 Hz
18 BW = 0.9*fs;                         % available bandwidth
19 fmax = 3500;                         % maximum modulating frequency
20 df = (BW/(2*fmax)-1)*fmax, pause     % calculated freq modulation depth
21 df = 1000;                           % arbitrary chosen frequency modulation depth
22 y = exp( j *2*pi*(fc*t + df*cumsum(x)*dt) );
23 figure; plot(f,abs(fft(y)/Nx)); grid; xlabel('f [Hz]'); title('|Y(f)|'); pause
24 figure; spectrogram(y,256,192,512,fs,'yaxis'); title('STFT of y(t)'); pause
25
26 % FM demodulation methods
27 ang = unwrap(angle(y)); fi1=1/(2*pi)*(ang(2:end)-ang(1:end-1)) / dt;
28 fi2 = 1/(2*pi)*angle( y(2:Nx).*conj( y(1:Nx-1) ) ) / dt;
29 fi3 = 1/(2*pi)*angle( y(3:Nx).*conj( y(1:Nx-2) ) ) / (2*dt); fi3=[fi3 0];
30 fi4 = 1/(2*pi)*(real(y(2:end-1)).*(imag(y(3:end))-imag(y(1:end-2)))-...
31     imag(y(2:end-1)).*(real(y(3:end))-real(y(1:end-2))) ) / (2*dt); fi4=[fi4 0];
32 fi5 = 1/(2*pi)*(real(y(1:end-1)).*imag(y(2:end))-imag(y(1:end-1)).*real(y(2:end)))/dt;
33 figure; nn=1:length(fi1);
34 plot(nn,fi1,'r',nn,fi2,'g',nn,fi3,'b',nn,fi4,'k',nn,fi5,'m'); pause
35 xest = ( fi2 - fc ) / df ;
36 xest = xest(1:end-1);   % recovered modulating signal
37 x = x(2:end-1);
38 figure; plot(t(2:Nx-1),x,'r-',t(2:Nx-1),xest,'b-'); xlabel('t [s]');grid; pause
39 figure; spectrogram(xest,256,192,512,fs,'yaxis'); title('STFT of xest(t)'); pause
40
41 % ERROR after frequency MOD & DEMOD
42 ERROR_SIGNAL = max( abs( x - xest ) ), pause
43 soundsc(x,fs); pause
44 soundsc(xest,fs); pause
```

Exercise 18.2 (Testing FM Modulation and Demodulation Algorithms). Add more modulating signals to the program 18.2, for example, sum of sinusoids with different frequencies, sinusoid with linearly increasing frequency, different sound signals. Observe figures. Verify demodulation correctness. Compare quality of demodulation algorithms.

Exercise 18.3 (FM Bandwidth. Is the Carson's Rule Always Valid?). Use program 18.2 and check validity of formula (18.10): change value of Δf and observe the spectrum of the signal modulated by human speech, using the spectrogram() Matlab function. The spectrum width should become wider after increase of Δf.

Exercise 18.4 (How Many Trees Are in an SFM Forest? Checking Bessel Formula for FM Signals). Run program 18.2. Use signal from line 13 ($f_m = 10$, 100, 1000 Hz), set $f_c = 4000$ Hz in line 14 and $\Delta f = 1000$ in line 22. Observe Fourier spectra generated in lines 24–25. This is a case of sinusoidal frequency modulation (18.8) in which the resultant, modulated signal is described by Eq. (18.11). Check experimentally validity of this formula. Try different values of Δf.

18.5 FM Demodulation Examples

18.5.1 FM Radio Broadcasting

The FM radio, still alive despite wind and rain technological storms, is for us a very good example of frequency modulation application. It is also a very good wireless service for testing our understanding of software-based telecommunication technology. In this chapter digital quadrature complex-value IQ FM radio demodulator was discussed. We could apply with ease the described above demodulation algorithms at the mono FM receiver program (Listing 17.3), presented in Chap. 17, and test them. A Reader is encouraged to do this.

Exercise 18.5 (Swing, Swing: FM Demodulation Methods). Extend the Matlab code from Listing 17.3: please verify all described above FM demodulation methods, specified by Eqs. (18.20), (18.21a) and (18.25).

In analog times the FM radio was demodulated also using real-value signal processing and signal differentiation. In Chap. 11 on special FIR filters in Sect. 11.3

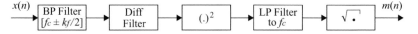

Fig. 18.6: Functional block diagram of the FM demodulator using digital differentiating filter

discrete-time version of this alternative frequency demodulation method was discussed as an example of FIR differentiation filters application. Described algorithm was implemented in Matlab program 11.8. Let us briefly repeat the method explanation. The real-value FM signal is defined as:

$$y(t) = \cos\left(2\pi\left(f_c t + \Delta f \int_0^t x(t)dt\right)\right). \tag{18.28}$$

After its differentiation we obtain

$$z(t) = \frac{dy(t)}{dt} = -[2\pi f_c + \Delta f \cdot x(t)] \cdot \sin\left(2\pi\left(f_c t + \Delta f \int_0^t x(t)dtx\right)\right). \tag{18.29}$$

In Eq. (18.29) a high-frequency signal $\sin(.)$, with frequency changing around f_c, has slowly changing amplitude/envelope $a(t) = [2\pi f_c + \Delta f \cdot x(t)]$, depending on the modulating signal $x(t)$. This envelope $a(t)$, and after that the signal $x(t)$, can be extracted from signal $z(t)$ (18.29), calculating in a cascade the following operations:

1. squaring the signal $z(t)$ (18.29), result of $y(t)$ differentiation,
2. low-pass filtering—removing $2f_c$ (2α) component of:

$$a^2(t) \cdot \sin^2(\alpha) = a^2(t)\frac{1}{2}(1 - \cos(2\alpha)) = \frac{1}{2}a^2(t) - \frac{1}{2}a^2(t)\cos(2\alpha), \tag{18.30}$$

3. calculating the square root of the result:

$$\sqrt{\frac{1}{2}a^2(t)} = \frac{1}{\sqrt{2}}a(t). \tag{18.31}$$

All together we can describe as:

$$\sqrt{\text{LPFilter}\left[z^2(t)\right]} = \frac{1}{\sqrt{2}}[2\pi f_c + \Delta f x(t)]. \tag{18.32}$$

The whole demodulator is presented in Fig. 18.6. The differentiation filter should work only around carrier frequency f_c, suppressing other frequencies. Such effect can be obtained pairing the differentiation filter with appropriate band-pass filter. The LP filter is removing the high-frequency component having frequency $2f_c$.

Exercise 18.6 (FM Radio Demodulation Using Signal Differentiation). Analyze the program 11.8. Modify it. First, adjust its parameter values for processing music signals sampled at 44100 or 48000 Hz. Then, try to use the method for demodulation of one mono FM radio signal present in complex-value IQ recording SDRSharp_FMRadio_101600kHz_IQ.wav consisting of many stations. Choose one station apart from 0 Hz, separate it using a band-pass filter, take a real value, and apply the method.

18.5.2 Amateur HAM Radio

FM modulation is widely used in amateur radio transmission. Examples of different FM modulated HAM signals and their spectra are presented in http://www.w1hkj.com/modes/index.htm. They should impress us with their diversity. In Fig. 18.7 STFT of some HAM-FM spectra are shown. Their careful inspection should give us additional confirmation about FM flexibility.

Exercise 18.7 (Beautiful FM Patters in HAM Radio Transmission). Observe frequency change in WebSDR services available at page http://websdr.org. Download some FM modulated signals from page http://www.w1hkj.com/modes/index.htm. Calculate their spectrograms, i.e. short-time Fourier transform. Observe different FM patterns. Do some literature investigations. Try to recover bits that are coded by different frequency values (states).

18.5.3 Airplane Azimuth Calculation from VOR Signal

Airplanes perform self-estimation of their azimuth angle in respect to an airport using the VOR signals emitted at the airport [5]. The VOR carrier is modulated in amplitude by a hybrid signal containing three components:

- a 30 Hz signal which phase is *azimuth* dependent and has to be found, using a signal with *reference* phase which is also transmitted,
- a speech control signal in the band [300–3500] Hz,
- a 9.96 kHz signal modulated in frequency (with 480 Hz depth) by the 30 Hz signal having *reference* phase for the *azimuth* signal.

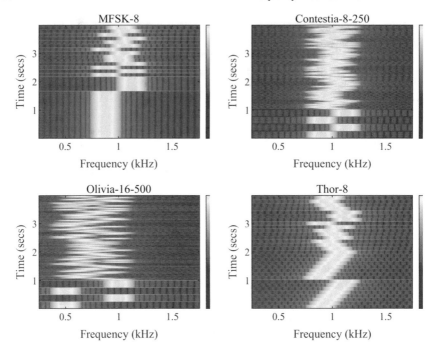

Fig. 18.7: Examples of STFT spectra of some FM modulated amateur radio signals. Modulation names are given in figure titles

In this measurement task, first, amplitude demodulation should be performed and then the frequency demodulation. Program for amplitude demodulation has been already presented in Chap. 17. Since the double side-band AM with carrier transmission is used in this case, the AM demodulation is not difficult because it does not require any carrier synchronization: only band-pass filtration of the IQ signal around the VOR service and complex number magnitude calculation are necessary. After AM demodulation the following tasks should be performed:

1. LP filtration and extraction of the 30 Hz azimuth signal,
2. BP filtration of the 9.96 kHz frequency modulated carrier,
3. doing the FM demodulation and extraction of the 30 Hz reference signal: first the signal Hilbert transform is used for the complex-value analytic signal computation and then the angle estimation is performed,
4. LP filtration (smoothing) of the reference signal,
5. estimation of the phase shift between the azimuth and the reference signal.

Program with all calculations is given in Listing 18.3. Figure 18.8 presents results obtained with its help which are wider described in the figure caption. The FM modulated 9.96 kHz component is well visible in the PSD and STFT spectra of the amplitude demodulated VOR signal (left column: second and third figure). Calculated instantaneous frequency of the 9.96 kHz signal (second column, first figure), being the reference 30 Hz sinusoid, is de-noised and scaled, and compared with the 30 Hz azimuth signal (second column, second figure). Estimated phase shift between 30Hz azimuth and reference sines is approximately equal to 1.214 radians and is fluctuating (last figure in second row). During method calibration (set 1 in if(.) for VOR test signal synthesis) very small amplitude oscillations (of order of 0.0025 radians) are observed in the calculated instantaneous phase shift. They are caused by usage of Hilbert transformer implemented in frequency domain (with rectangular window) instead of in time-domain, in the form of Hilbert filter.

Listing 18.3: Demonstrative Matlab program for computing azimuth angle of airplane position in respect to airport in which FM demodulation of 9.96 kHz carrier is performed and the reference 30Hz signal is found

```
1
2    % lab18_ex_airplane.m
3
4    % Read a recorded IQ signal - two VOR avionics signals
5      FileName = 'SDRSharp_Airplane_112500kHz_IQ.wav'; T=5; demod=3; fc = 2.9285e+5;
6    % FileName = 'SDRSharp_Airplane_112849kHz_IQ.wav'; T=5; demod=3; fc = -5.5353e+4;
7
8    % ... code from program lab16_ex_IQ_DFT.m
9
10   if(demod==3) % airplane azimuth decoding using VOR signal
11
12   M = 501; M2= (M-1)/2;            % filter length
13   fam = 25000; dt=1/fam;          % frequency width of AM modulation around fc
14   f1 = fc-fam/2; f2 = fc+fam/2;   % Band-Pass filter h design
15   h = cfirpm(M-1, [-fs/2 (f1-df) (f1+df) (f2-df) (f2+df) fs/2]/(fs/2),@bandpass);
16   x = conv(x,h); x=x(M:end-M+1);                % Band-pass filtration of VOR
17   x = sqrt( real(x).*real(x) + imag(x).*imag(x) );   % AM demodulation
18   x = decimate(x,round(fs/fam)); % [U,D] = rat(fs/fam); x = resample( x, U, D );
19   x = x - mean(x);                % mean subtraction
20
21   if(0) % only for verification test
22      N = length(x); t=dt*(0:length(x)-1);
23      x = sin(2*pi*30*t) + cos(2*pi*(9960*t-480/(2*pi*30)*cos(2*pi*30*t)));
24      x = x';
25   end
26   xc = x;
27
28   % Low-pass filtration of 30 Hz azimuth signal
29   hLP30 = fir1(M-1,50/(fam/2),'low');      % design of filter coefficients
30   x = conv(x,hLP30); x = x(M2+1:end-M2);   % filtering
31   x = x - mean(x);
32   x_azim = x(2:end-1)/max(x);              % 2,-1 due to first computation
33
34   % Extraction of 30 Hz signal with reference phase
35   hBP10k = fir1(M-1,[9000,11000]/(fam/2),'bandpass'); % BP filter design
36   x = conv(xc,hBP10k); x=x(M2+1:end-M2);   % separation of FM component
37   x = unwrap( angle ( hilbert(x) ));       % angle calculation
38   x = x(3:end)-x(1:end-2);                 % 3-point difference
39   x = (1/(2*pi))*x/(2*dt);                 % f instantaneous
```

```
40   x = conv(x,hLP30); x=x(M2+1:end-M2);      % LP filtration / denoising
41   x = x - mean(x);                          % remove mean value
42   x_ref = x/max(x);                         % normalize amplitude to 1
43
44   % Phase shift estimation
45   phi_inst = angle( hilbert(x_azim).*conj(hilbert(x_ref)) );
46   phi_estim = mean( phi_inst ), pause
47
48   end
```

Exercise 18.8 (Testing Algorithm of Airplane Azimuth Detection). Add plots to the program 18.3. You should obtain similar ones to the presented in Fig. 18.8. Possible differences can result from different filter parameters. Try to change them. Test different frequency demodulation algorithms.

18.5.4 Safe Mode Nano-Satellite Signal

The last FM demodulation example presented in this section will concern a signal transmitted from nano-satellite in safe, low bit-rate mode [6]. In the analyzed recording, the satellite signal is modulating in frequency the 146.02 MHz carrier. 200 bits are sent per one second in packets lasting 10 s and repeated once per two minutes. Binary frequency shift keying is used, i.e. the carrier frequency jumps down (-1) and up $(+1)$ 200 times per second and this way bits 0/1 are transmitted, respectively. In the middle of the packet voice control is added to the digital data signal, causing additional frequency modulation of the carrier.

In Matlab program 17.4 [4], given in Chap. 17, the signal frequency demodulation was presented. Results obtained with its use are shown in Fig. 18.9. In first column we see STFT spectra magnitudes of the following signals: (1) the original $IQ(n)$ signal sampled at 192 kHz, (2) signal down-converted in frequency to 0 Hz, (3) signal low-pass filtered around 0 Hz (with filter cut-off frequency equal to 12500/2=6250 Hz), and down-sampled to frequency 48 kHz. We can see carrier frequency *jumps* lasting about 10 s. In the middle of the bit packet we see additional carrier modulation caused by transmitted voice (DUV—data under voice technology).

In the second figure column, a recovered modulated signal is shown. In the upper figure the whole signal. In the middle figure, only signal fragment when bit transmission starts. We observe signal up-and-down (1/0) *hops* but embedded in very strong noise. In the bottom, a central part of the signal is shown where both bits and control voice are transmitted.

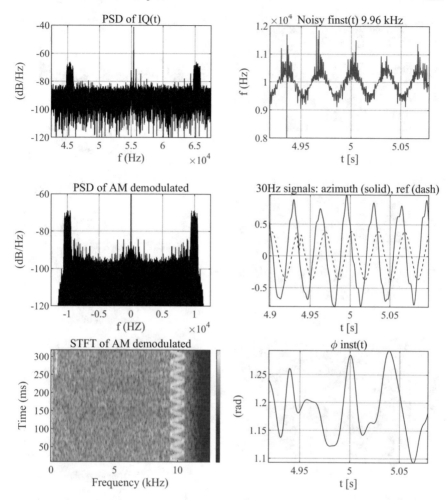

Fig. 18.8: Results obtained from the program 18.3 for airplane azimuth angle computation. First column: (1) Power spectral density (PSD) fragment of $IQ(n)$ signal—spectrum of VOR avionics signal, i.e. of AM modulated carrier, (2) PSD of a recovered signal modulating the carrier in amplitude—we can see *some* transmission around 9.96 kHz, (3) STFT of the recovered signal—we can see 30 Hz sinusoidal frequency modulation of the 9.96 kHz carrier. Second column: (1) calculated instantaneous frequency of the 9.96 kHz carrier—30Hz reference sinusoid, (2) azimuth (solid line) and reference (dashed line) 30 Hz sinusoids, de-noised and scaled in amplitudes, (3) calculated instantaneous phase shift between these two sinusoids—estimation of azimuth angle of airplane position in respect to the airport

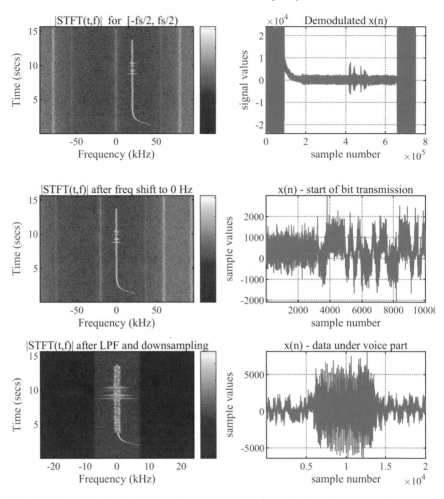

Fig. 18.9: Results obtained from the program 17.4 performing frequency demodulation of nano-satellite IQ signal sampled at 192 kHz (in slow-speed safe mode). First column: (1) STFT of the recorded $IQ(n)$ signal, (2) STFT of signal converted to 0 Hz, (3) STFT after low-pass filtration and signal down-sampling to 48 kHz. Second column: (1) whole demodulated signal $x(n)$, (2) first fragment of $x(n)$—start of bit transmission, (3) second fragment of $x(n)$—data transmitted together with voice signal (DUV—data under voice)

Exercise 18.9 (Testing Algorithm of FM Demodulation of Nano-Satellite Signal). Matlab code of FM demodulation of nano-satellite signal (in safe, slow-speed mode) was presented in Chap. 17 in the program 17.4 [4] as a start-up SDR example. Add figures to it. Run the program. You should obtain similar plots to presented in Fig. 18.9. Test different frequency demodulation algorithms. Change bandwidth of the low-pass filter. Make it smaller to better de-noise the signal and better see the transmitted bit pattern.

18.6 Summary

In this chapter frequency modulation and demodulation fundamentals were presented. What is the most important?

1. In frequency modulation we are interested in appropriate changing of instantaneous frequency of an oscillatory *carrier* according to the formula: $f_{inst}(t) = f_c + \Delta f \cdot x(t)$, where f_c is nominal carrier frequency, $x(t)$ is modulating function (signal), and Δf is modulation depth. Roughly speaking, we deliberately cause drift (fluctuation) of the carrier frequency around its nominal value.
2. To obtain this goal we have to change in proper way a phase angle $\varphi(t))$ of a real-value $\cos(\varphi(t))$ or a complex-value $\exp(j \cdot \varphi(t))$ carrier signal.
3. The required phase angle is a sum of two terms: the first one $2\pi f_c t$, associated with the carrier frequency, and the second one $2\pi \cdot \Delta f \cdot \int_0^t x(t)dt$, responsible for the carrier frequency change and being dependent upon modulating function $x(t)$.
4. During demodulation we should first recover/restore the whole angle $\varphi(t)$ of the sine oscillator (for example, an angle of a complex-value sample of the base-band $IQ(t)$ signal), then calculate its time derivative equal to $2\pi f_c + 2\pi \cdot \Delta f \cdot x(t)$, next divide the result by 2π and obtain an estimate of the instantaneous oscillator frequency $f_{inst}(t) = f_c + \Delta f \cdot x(t)$. Finally, the modulation function value is found from equation $x(t) = (f_{inst}(t) - f_c)/\Delta f$.
5. Spectral bandwidth of the FM signal can be calculated using Carson's rule (18.9) and depends on the modulation depth Δf and the maximum modulating frequency f_m.
6. In case of sinusoidal frequency modulation with single tone, the bandwidth of the resultant signal can be precisely estimated using formula (18.11) with Bessel functions for $\delta = \Delta f / f_m$.
7. Frequency modulation is not the hottest topic nowadays but it is still used in transmission of data between mobile equipment and satellites as well as short-range Bluetooth and remote control devices.

18.7 Private Investigations: Free-Style Bungee Jumps

Exercise 18.10 (Being Inquisitive Like Albert Einstein: Instantaneous Frequency). Analytically check whether Eqs. (18.7) and (18.8) fulfill the phase differentiation principle (18.2).

Exercise 18.11 (Straightening the Wheel: Unwrapping the Phase). Generate a vector of samples of the signal $x(n) = \exp(j \cdot 2\pi \cdot (0 : N - 1)/(N/10)), N = 1000$. Calculate angle of each complex-value signal sample using `angle()` or `atan2(Q,I)` Matlab function. Plot the phase change. Apply the Matlab function `unwrap()` to the vector of calculated angles and plot in one figure the function input and output. How do you comment this experiment?

Exercise 18.12 (Beautiful FM Patterns for the Tate Gallery). Modify program 18.2 and use `spectrogram()`, i.e. the Matlab function for calculation and display of time-varying signal spectra. Set $f_s = 8000$ Hz. Generate one second of a real-value (cos(.)) and a complex-value (exp(j.)) signal with:

- linear increase of frequency (18.7), for $\alpha = f_s/8, f_s/4, f_s/2$,
- sinusoidal change of frequency (18.8), for $f_m = 1Hz$ and $\Delta f = f_s/8, f_s/4$.

In both cases first use $f_c = 0$, then $f_c = f_s/4$.

Exercise 18.13 (Frequency Keying and Bit Transmission). In this chapter we were discussing carrier frequency modulation done by an arbitrary signal. Generate a signal in which frequency is switched between two values. Let the first of them denotes sending bit 0 while using the second—sending bit 1. Periodically send the same sequence of bits this way. Try to decode bits after frequency demodulation of the signal. Please, send bits of ASCII codes of your name. Calculate and observe spectrogram of the generated signal. You should see that the signal spectrum is becoming wider during abrupt frequency changes. Try to switch frequency only when the carrier has maximum amplitude (i.e. built the signal with fragments of multiple periods of cosines)—this will ensure continuity of the carrier shape. Observe the spectrogram. Try to drift smoothly between two signal frequencies using cosine-like frequency change trajectories: from 1 to −1 and back from −1 to 1. Observe the spectrogram.

References

1. K. Barlee, D. Atkinson, R.W. Stewart, L. Crockett, *Software Defined Radio using MATLAB & Simulink and the RTLSDR*. Matlab Courseware (Strathclyde Academic Media, Glasgow, 2015). Online: http://www.desktopsdr.com/
2. M.E. Frerking, *Digital Signal Processing in Communication Systems* (Springer, New York, 1994)

3. R.G. Lyons, *Understanding Digital Signal Processing* (Addison-Wesley Long-man Publishing, Boston, 1996, 2005, 2010)
4. J. Marek, *Programming Ground Station for Receiving Satellite Signals Using SDR Technology*. B.Sc. Thesis, AGH University of Science and Technology, Krakow, Poland, 2019
5. Rhode-Schwarz, R&S FS-K15 VOR/ILS avionics measurements software manual, Munich, Germany 2013. Online: https://www.rohde-schwarz. com/ua/manual/r-s-fs-k15-vor-ils-avionics-measurements-manuals-gb1_ 78701-29018.html
6. C. Thompson, FOX-1 satellite telemetry part 2: FoxTelem. AMSAT J. **39**(1), 7–9 (2016)
7. S.A. Tretter, *Communication System Design Using DSP Algorithms* (Springer Science+Business Media, New York, 2008)

Chapter 19
Amplitude Modulation, Demodulation, and Carrier Recovery

Mamma, why ocean's waves are once low and once high? Oh, Johnny, the ocean is like you: it is once whispering and once shouting.

19.1 Introduction

Each oscillation is characterized by frequency, amplitude, and phase (its starting angle). In the previous chapter we were analyzing problem of changing signal/carrier frequency in some predefined functional way (*frequency modulation*) and the problem of recovering this modulating function from the modulated signal (*frequency demodulation*). Performing this UNDO operation, we were playing a role of Sherlock Holmes and were asking the question: which modulation function was used? Which function is changing the frequency of my perfect sine?

Remark In *digital* communication the number of modulating functions is limited and finding the function and its number, for example 5 (binary 0101) out of [0, 1, 2, ..., 15], is equivalent to recovering the transmitted bits, in our example 0101. Of course, immediately a new question appears: why not to start changing other parameters of the sinusoid, for example, amplitude and/or phase, and increase this way the number of allowed carriers states, their possible numbers and number of bits transmitted by means of one sinusoid. Such idea leads us directly to concept of amplitude and phase modulation and .... to new questions. How many states should we assign for each sinusoid parameter? To use all of them together or may be only these which are less influenced (modified) by a communication channel? Answers to these and similar questions will be addressed in next chapters of this book dealing with digital transmission.

In this chapter we will learn about *analog* amplitude modulation, e.g. AM radio broadcast: in the transmitter the carrier amplitude is changed/modulated in analog way by a continuous function which shape is to be recovered in the receiver. Understanding the analog AM modulation will help us easily catch the concept of the digital AM modulation since carriers are modulated and

© Springer Nature Switzerland AG 2021

T. P. Zieliński, *Starting Digital Signal Processing in Telecommunication Engineering*, Textbooks in Telecommunication Engineering, https://doi.org/10.1007/978-3-030-49256-4_19

demodulated in the same way in both transmission techniques, only modulating functions are different. In fact, the digital AM modulation is analog in its physical realization—term *digital* means only ON/OFF switching between allowed predefined functions that are used for modulation.

The AM modulated signal has a form:

$$y(t) = a(t) \cdot c(t), \tag{19.1}$$

where $c(t)$ is a real-value or complex-value carrier with frequency f_c, equal to:

$$c(t) = e^{j2\pi f_c t} = \cos(2\pi f_c t) + j\sin(2\pi f_c t), \tag{19.2}$$
$$c(t) = \cos(2\pi f_c t), \tag{19.3}$$

and $a(t)$ is a carrier amplitude, changing in time and depending upon modulating signal $x(t)$. Therefore the AM modulation is straightforward: in the transmitter we are generating $a(t)$ using a transmitted signal $x(t)$ and then multiply $a(t)$ with the carrier $c(t)$. The $a(t)$ is equal to:

$$\begin{aligned}
\textit{Suppressed Carrier}: && a(t) &= x(t), & (19.4)\\
\textit{Large Carrier}: && a(t) &= 1 + \Delta A \cdot x(t), & (19.5)
\end{aligned}$$

where ΔA denotes the modulation depth. After putting Eqs. (19.4), (19.5) into Eq. (19.1) one obtains

$$\begin{aligned}
\textit{Suppressed Carrier}: && y(t) &= x(t) \cdot c(t), & (19.6)\\
\textit{Large Carrier}: && y(t) &= 1 \cdot c(t) + \Delta A \cdot x(n) \cdot c(t). & (19.7)
\end{aligned}$$

In the second method the carrier is *large, un-suppressed*, because it is present as a separate component in the modulated signal (first term in Eq. (19.7)), while in the first method it is not the case. Since transmission of the carrier requires extra energy, the first *suppressed carrier* method is preferred.

When carrier $c(t)$ is the complex-value signal (19.2), we can also use a complex-value signal $x(t)$ for its modulation. Otherwise, $x(t)$ is a real-value signals.

Multiplication of any signal $a(t)$ by the carrier $c(t)$ causes shift of its spectrum $A(f)$ to the carrier frequency f_c—this is the well-known Fourier transform feature. When the carrier is a complex-value harmonic signal $c(t) = e^{j2\pi f_c t}$, the spectrum $A(f)$ is only shifted to the carrier frequency f_c:

$$Y(f) = A(f - f_c). \tag{19.8}$$

For the real-value carrier $\cos(2\pi f_c t)$, the spectrum $A(f)$ is shifted to the carrier frequency fc and to its negation $-f_c$ (i.e. copied and centered at both frequencies):

$$Y(f) = 0.5A(f - f_c) + 0.5A(f + f_c), \tag{19.9}$$

since a cosine is a result of divided by 2 summation of two complex-value harmonics with frequencies f_c and $-f_c$: $\cos(2\pi f_c t) = 0.5e^{j2\pi f_c t} + 0.5e^{-j2\pi f_c t}$.

When $y(t)$ is a complex-value signal, only samples of its real part $y_{Re}(t)$:

$$y_{Re}(t) = \text{Real}(y(t)) \tag{19.10}$$

are sent to an antenna in a telecommunication transmitter. Taking only the real part of $y(t)$ causes changing shape of signal spectrum from (19.8) to (19.9): spectrum copy appears for negative frequencies.

Amplitude demodulation relies on reconstructing signal $x(t)$ from $y(t)$ given by Eqs. (19.6), (19.7). The *large carrier* modulation method has much more simpler demodulator. In the receiver:

1. the signal $y_{Re}(t)$ is band-pass filtered around f_c—the decoded service is separated from remaining services,
2. obtained signal $y_{Re}^{BP}(t)$ is transformed into its analytic complex-value version using the Hilbert transform $H(.)$ (what removes the signal spectrum for negative frequencies),
3. envelope of the complex-value analytic signal is found as magnitude of its samples:

$$\widehat{a}(t) = |y_{Re}^{BP}(t) + jH\left(y_{Re}^{BP}(t)\right)|. \tag{19.11}$$

Then $x(t)$ is recovered from $\widehat{a}(t)$. No carrier synchronization procedure is required.

In contrary, in order to demodulate the signal in the *suppressed carrier* method, we have to shift back the spectrum $A(f - f_c)$ to 0 Hz, recover signal $\widehat{a}(t)$ and reconstruct signal $x(t)$ from it using (19.4). This is done by (1) band-pass filtering, (2) calculation of a complex-value analytic version of the received signal, as before for *large carrier*, and (3) multiplying the signal by the same complex-value carrier as in the transmitter but conjugated:

$$\widehat{x}(t) = \left[y_{Re}^{BP}(t) + jH\left(y_{Re}^{BP}(t)\right)\right] \cdot e^{-j2\pi f_c t}. \tag{19.12}$$

Alternatively, the second approach can be used, derived in Chap. 17 during mathematical analysis of service up-down conversion in software defined radio technology. In it: (1) analytic version of the received signal is calculated, (2) it is down-shifted in frequency to 0 Hz by the conjugated carrier, which should be known, and (3) low-pass filtered. For suppressed carrier we have

$$\widehat{x}(t) = \text{LowPass}_{BW}\{ \quad [y_{Re}(t) + jH(y_{re}(t))]\cdot e^{-j2\pi f_c t} \quad \}. \qquad (19.13)$$

But it is easy to say: *multiply by the same cosine as in the transmitter!* How to find this cosine? How a receiver should calculate the proper carrier $c(t)$ knowing signal $y(t)$? This task is called a carrier synchronization issue and it represents the basic problem of AM demodulation of signals do not having the carrier given explicit!

In case of DSB modulation with suppressed carrier, during demodulation the carrier synchronization problem is solved by: (1) squaring $y(t)$, i.e. calculation of:

$$y^2(t) = x^2(t)\cos^2(2\pi f_c t + \varphi) = x^2(t)\cdot\frac{1}{2}(1 + \cos(2\pi(2f_c)t + 2\varphi)), \quad (19.14)$$

(2) filtering the result with very narrow band-pass filter around $(2f_c)$, and (3) using TWO adaptive phase locked loops (PLLs) for finding and tracking TWO parameters: doubled frequency $2f_c$ and doubled phase 2φ of the carrier signal $c(t)$. Knowing these parameters we can divide them by 2 and synthesize (generate) correct signal $c(t)$ and next use it for signal demodulation:

$$y(t)\cdot c(t) = x(t)\cdot\cos^2(2\pi f_c t + \varphi) = \frac{1}{2}x(t) + \frac{1}{2}x(t)\cdot\cos(2\pi(2f_c)t + 2\varphi).$$
$$(19.15)$$

In this method, first, we are doing adaptive carrier synchronization and, then, we are exploiting the recovered carrier to down-shifting the signal $y(t)$ in frequency to the base-band. In alternative approach, the adaptive Costas loop, these two things are done together: carrier recovery/synchronization and frequency downshifting.

Concluding, dealing with AM demodulation we will become familiar in this chapter with AM carrier synchronization methods which are extremely important in modern digital telecommunications. Why? Because the frequency up and down signal conversion (from the base-band to a target frequency and back), described in Chap. 17, is done exactly by the DSB-SC carrier amplitude modulation and demodulation! And we have to synchronize with not perfectly known carriers in our receivers. Since synchronization should be flexible and robust to noise and to different disturbances/interferences, adaptive solutions are preferred. And this is the second very big *Hero* of this chapter, apart from the AM modulation: practical application of adaptive system theory in telecommunication receivers.

The third *Hero* is efficiency of spectrum allocation for single services, i.e. minimization of each service frequency width aiming at allocation of more services in the available frequency band. Using only the lower or the upper frequency side-band in respect to the carrier, not both of them together, is an example of such economical resource management. Such approach is widely exploited in amateur radio. In this chapter we will learn about single side-band (SSB) modulation schemes, exploiting only lower (SSB-L, LSB) or upper (SSB-U, USB) carrier side-band.

And the last remark. In Chap. 17 we have discussed the up and down frequency conversion scheme of complex-value $I(t) + j \cdot Q(t)$ signals, which is used in software-defined radio transmission. It was making use of a complex-value carrier $e^{j \cdot 2\pi f_c t}$ and was realized by so-called quadrature, $\cos(.)$ and $-\sin(.)$, modulation and demodulation. In the present chapter we will see its connection to the other AM techniques.

So lets the AM modulation story begin!

19.2 Amplitude Modulation

In Sect. 17.3 of Chap. 17, during presentation of up and down service frequency conversion, we have described general form of quadrature *amplitude* modulation and demodulation, i.e. modulation of a complex-value carrier by a complex-value signal (in special case a real-value one). At present we will specially concentrate on amplitude modulation and demodulation of a real-value carrier by a real-value signal, with special application to FM radio software receiver (DSB-SC) and HAM amateur radio (USB, LSB). However, we will repeat also the most general AM complex-value modulation method.

In this chapter we will use the same notation as for the FM modulation: $x(t), x(n)$—real-value analog and digital signal used for *modulation* and $y(t), y(n)$—corresponding real-value *modulated* signals.

AM is defined by Eq. (19.1) as multiplication of complex-value carrier $c(t)$ (19.2) or real-value carrier (19.3) by a function $a(t)$ (19.4), (19.5) which can be interpreted as a carrier amplitude changing in time. In amplitude modulation a transmitted signal is changing amplitude of the carrier signal. We are distinguishing four common AM modulation types:

- **DSB-LC**—double side-band with large carrier,
- **DSB-SC**—double side-band with suppressed carrier,
- **SSB-U, USB**—single side-band—upper band,
- **SSB-L, LSB**—single side-band—lower band.

Additionally, we then can distinguish a special type of DSB-SC for complex-value signals and denote it as **DSB-SC-CX**. It is our name and abbreviation. Their modulators are defined as follows:

DSB-LC: $y(t) = [1 + dA \cdot x(t)] \cdot \cos(2\pi f_c t),$ (19.16)

DSB-SC: $y(t) = [x(t)] \cdot \cos(2\pi f_c t),$ (19.17)

SSB-U: $y(t) = \text{Re}\left\{ [x(t) + j \cdot \text{H}[x(t)]] \cdot e^{j \cdot 2\pi f_c t} \right\},$ (19.18)

SSB-L: $y(t) = \text{Re}\left\{ [x(t) - j \cdot \text{H}[x(t)]] \cdot e^{j \cdot 2\pi f_c t} \right\},$ (19.19)

DSB-SC-CX: $y(t) = \text{Re}\left\{ [x_{Re}(t) + j \cdot x_{Im}(t)] \cdot e^{j 2\pi f_c t} \right\}.$ (19.20)

where $x(t)$, $x_{Re}(t)$, and $x_{Im}(t)$ are real-value modulating signals and $H[.]$ denotes the Hilbert filter (-90 degree phase shifter transforming each $\cos()$ function into $\sin()$). For $x_{Im}(t) = 0$ the modulation DSB-SC-CX is equivalent to DSB-SC. After multiplication and taking the real part of the result only, the equations for SSB-U, SSB-L, and DSB-SC-CX can be written in simpler forms:

SSB-U: $y(t) = x(t)\cos(2\pi f_c t) \ -H[x(t)]\sin(2\pi f_c t),$ (19.21)

SSB-L: $y(t) = x(t)\cos(2\pi f_c t) \ +H[x(t)]\sin(2\pi f_c t),$ (19.22)

DSB-SC-CX: $y(t) = x_{Re}(t)\cos(2\pi f_c t) \ -x_{Im}(t)\sin(2\pi f_c t).$ (19.23)

As we see, the above equations have a form of Eq. (17.14). Therefore, these AM modulators are described by block diagrams presented in Figs. 17.4, 17.5 (in their transmitter part). No surprise, at present the AM carrier modulation is mainly used for frequency service up-conversion from the base-band to the target frequency.

Matlab program presented in Listing 19.1 implements the described above AM modulators. It was used for generation of waveforms (signatures) of AM modulated signals $y(t)$ presented in Fig. 19.1. In this case a carrier with frequency $f_c = 10$ kHz was used and it was modulated in amplitude by pure cosine with frequency $f_m = 2$ kHz. Sampling frequency f_s was equal to 55.125 kHz, 5-th multiplicity of frequency 11.025 kHz. In case of DSB-SC-CX, the cosine 2kHz was used for the $x_{Re}(t)$ signal, while a cosine with frequency 4 kHz—for the $x_{Im}(t)$ signal in Eqs. (19.20), (19.23).

Listing 19.1: Simple program for testing in Matlab standard amplitude modulation techniques

```
1
2   % lab19_ex_am_short.m
3   % Example of AM modulation and demodulation
4   clear all; close all;
5
```

```
 6  sig = 1;      % signal type: 1=speech, 2=cos, 3=cos+cos, 4=SFM
 7  mod = 1;      % AM type: 1=DSB-LC, 2=DSB-SC, 3=SSB-U=USB, 4=SSB-L=LSB, 5=DSB-SC-CX
 8  demod = 1;    % 1 = Hilbert transform demodulator, 2 = quadrature demodulator
 9  nosynch = 0;  % 0/1 carrier synchronization in frequency down-conversion
10  disturb = 0;  % 0/1 disturbance - second AM service using frequency (2*fc)
11  noise = 0;    % 0/1 presence of noise
12  fc = 10000;   % AM carrier frequency
13  dA = 0.5;     % AM modulation depth for DSB-LC
14  Nwin = 256; Nover = Nwin-32; Nfft = 2*Nwin;  % for STFT plot
15
16  % Modulating signal x(t)
17  [x,fss] = audioread('speech.wav');    % read mono audio from file, fs=11025 Hz
18  K=5; x = resample(x,K,1); fs=K*fss;   % upsample K-times for frequency-UP
19  N = length(x);                        % number of signal samples
20  dt = 1/fs; t = dt*(0:N-1)';           % time
21  df=fs/Nfft; f=df*(-Nfft/2:Nfft/2-1);  % frequencies for STFT display
22                                        % alternative signals for tests
23  if(sig==2) x = cos(2*pi*2000*t); end                        % 1x cos
24  if(sig==3) x = cos(2*pi*2000*t) + cos(2*pi*3000*t); end     % 2x cos
25  if(sig==4) x = cos(2*pi*(2000*t + 1000/(2*pi*5)*sin(2*pi*5*t))); end % SFM
26
27  % Create base-band signal for AM modulation of the carrier: x(t) --> a(t)
28  if(mod==1) a = (1+dA*x);                 end  % DSB-LC
29  if(mod==2) a = x;                        end  % DSB-SC
30  if(mod==3) a = x + j*imag(hilbert(x));   end  % SSB-R = USB
31  if(mod==4) a = x - j*imag(hilbert(x));   end  % SSB-L = LSB
32  if(mod==5) x = x + j*sin(2*pi*4000*t); a=x; end  % DSB-SC-CMPLX
33
34  % Carrier AM modulation - frequency-up conversion: y(t) = a(t)*c(t)
35  c = exp( j*2*pi*fc*t );
36  y = real(a).*real(c) - imag(a).*imag(c);  % y = real( a .* c );  % the same
```

In modulation the most important question is how spectrum of the modulated signal $y(t)$ depends on the spectrum of the modulating signal $x(t)$ and carrier frequency f_c. This relation follows directly from the Fourier transform modulation feature (multiplication of two signals) . A Reader is asked to remember it. As an example, we will derive spectra of signals ((19.16)–(19.20)) when carriers are modulated in amplitude by a pure cosine:

$$x(t) = x_{Re}(t) = \cos(2\pi f_m t), \qquad x_{Im}(t) = \sin(2\pi f_n t). \qquad (19.24)$$

These spectra are equal to:

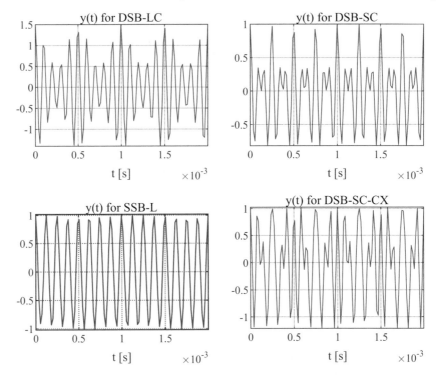

Fig. 19.1: Examples of AM signals: carriers $f_c = 10$ kHz are modulated by a pure cosine $f_m = 2$kHz: (up-left) DSB-LC, (up-right) DSB-SC, (down-left) SSB-L, (down-right) DSB-SC-CX—real part. Sampling frequency $f_s = 55.125$ kHz. For DSB-SC-CX a sine 4 kHz is used as $x_{Im}(t)$ in (19.20), (19.23)

DSB-LC: $y(t) = \dfrac{dA}{2}\cos(2\pi(f_c - f_m)t) + \cos(2\pi f_c t) +$

$$+ \frac{dA}{2}\cos(2\pi(f_c + f_m)t), \quad (19.25)$$

DSB-SC: $y(t) = \dfrac{1}{2}\cos(2\pi(f_c - f_m)t) + \dfrac{1}{2}\cos(2\pi(f_c + f_m)t), \quad (19.26)$

SSB-U: $y(t) = \mathrm{Re}\left\{e^{j2\pi f_m t} \cdot e^{j2\pi f_c t}\right\} = \cos(2\pi(f_c + f_m)t), \quad (19.27)$

SSB-L: $y(t) = \mathrm{Re}\left\{e^{-j2\pi f_m t} \cdot e^{j2\pi f_c t}\right\} = \cos(2\pi(f_c - f_m)t), \quad (19.28)$

DSB-SC-CX: $y(t) = \mathrm{Re}\left\{e^{j2\pi f_m t} \cdot e^{j2\pi f_c t}\right\} = \cos(2\pi(f_c + f_m)t). \quad (19.29)$

When coming from set of equations (19.16)–(19.20) to (19.25)–(19.29), the following identities were exploited:

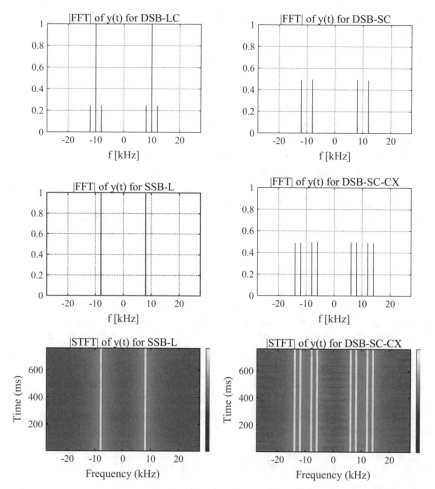

Fig. 19.2: Examples of FFT and STFT spectra magnitudes of AM signals from Fig. 19.1: carriers $f_c = 10$ kHz are modulated by a pure cosine $f_m = 2$ kHz: (up-left) FFT for DSB-LC, (up-right) FFT for DSB-SC, (middle-left) FFT for SSB-L, (middle-right) FFT for DSB-SC-CX, (bottom-left) STFT for SSB-L, (bottom-right) STFT for DSB-SC-CX. Sampling frequency $f_s = 55.125$ kHz. For DSB-SC-CX a sine 4 kHz is used as $x_{Im}(t)$ in (19.20), (19.23)

$$\cos(\alpha)\cos(\beta) = \frac{1}{2}\left[\cos(\alpha+\beta) + \cos(\alpha-\beta)\right], \tag{19.30}$$

$$-\sin(\alpha)\sin(\beta) = \frac{1}{2}\left[-\cos(\alpha-\beta) + \cos(\alpha+\beta)\right], \tag{19.31}$$

$$\cos(\alpha) \pm j \cdot \sin(\alpha) = e^{\pm j \cdot \alpha}, \tag{19.32}$$

$$e^{j\alpha}e^{j\beta} = e^{j(\alpha+\beta)}. \tag{19.33}$$

The derived equations can be concluded as follows (look at spectra presented in Fig. 19.2):

- the DSB-LC modulated signal consists of a carrier f_c and two side-band copies of the modulating signal lying symmetrically on its both sides at frequencies $f_c + f_m$ and $f_c - f_m$,
- the DSB-SC modulated signal does not have a carrier f_c, only two side-band copies of the modulating signals as in DSB-LC case (is equal to DSB-LC without carrier component),
- the SSB-U (USB) modulated signal is a copy of the modulating signal up-shifted in frequency to the frequency $f_c + f_m$ (not shown, similar to SSB-L),
- the SSB-L (LSB) modulated signal is a copy of the modulating signal up-shifted in frequency to the frequency $(f_c - f_m)$,
- signal resulting from complex quadrature DSB-SC-CX modulation has double side-band components around f_c belonging to two signals $x_{Re}(t)$ and $x_{Im}(t)$, which side-bands can overlap.

This result obtained for AM modulation by a pure cosine signal (19.24) can be generalized with ease for any real-value signal with frequency limited spectrum. This statement is verified by Fig. 19.3 presenting short-time Fourier spectra (Matlab function `spectrogram(.)`) of the carrier $f_c = 10$ kHz. It is modulated in amplitude by a cosine with frequency $f_m = 2$ kHz which is … additionally modulated sinusoidally in frequency: depth 1 kHz and modulation frequency 5 Hz. Sampling frequency is equal to 55.125 kHz (5-th multiplicity of 11.025 kHz). For DSB-SC-CX the signal $x_{Im}(t)$ is a pure cosine having frequency 4 kHz. Yes, cosine not sine, because in general case of DSB-SC-CX modulation (19.20) $x_{Re}(t)$ and $x_{Im}(t)$ can be completely different signals! All modulation schemes, apart from SSB-L, being very similar to SSB-U, are compared. The Matlab program from the Listing 19.1 was used for computations and plots generation.

Exercise 19.1 (AM Modulation). Analyze carefully the program 19.1. Run its long version `lab19_ex_am_long.m` for 4 modulating signals: speech, one cosine, summation of two cosines, and cosine with sinusoidal frequency modulation (SFM). Observe signal spectra. Change frequency values of modulating signals and note the spectra difference. Add an LFM signal as an additional modulation option. Change its starting frequency value and speed of frequency increase. Check whether the program allows carrier modulation by music sampled at 44,100 Hz. If not, do appropriate changes.

19.3 Amplitude Demodulation

Multiplication of signal $x(t)$ by a carrier with frequency f_c causes shifting the Fourier spectrum $X(f)$ of the signal to carrier frequency, i.e. f_c Hz up. In order

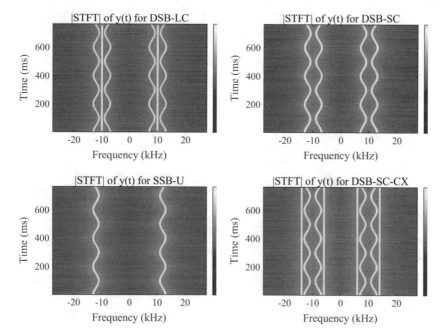

Fig. 19.3: STFT spectra of a carrier modulated by a sinusoidal FM signal, not a pure cosine: (up-left) DSB-LC, (up-right) DSB-SC, (down-left) SSB-U, (down-right) DSB-SC-CX. Description: 10 kHz carrier was modulated in amplitude by SFM signal (central frequency 2 kHz, modulation depth 1 kHz, modulation frequency 5 Hz). Sampling frequency f_s =55.125 kHz. For DSB-SC-CX the signal $x_{Im}(t)$ is a pure cosine having frequency 4 kHz

to shift back the spectrum down by $-f_c$ Hz, one of two approaches should be used. In the Hilbert filter AM demodulator sequence of operations is as follows:

(A) calculation of analytic version of the signal using the Hilbert transform $H(y(t))$: signal spectrum for negative frequency is removed,
(B) signal multiplication by the complex-value harmonic carrier but with negative frequency $-f_c$: signal spectrum of interest is shifted to 0 Hz,
(C) low-pass filtering around 0 Hz with filter bandwidth (BW) adjusted to length of the signal spectrum $X(f)$: side services lying apart from DC are removed.

Therefore the Hilbert filter-based demodulator can be summarized by the following equation:

$$\widehat{a}(t) = \text{LPF}_{BW}\{ \ \underbrace{\underbrace{[y(t) + jH(y(t))]}_{A} \cdot e^{-j2\pi f_c t}}_{B} \ \}. \qquad (19.34)$$

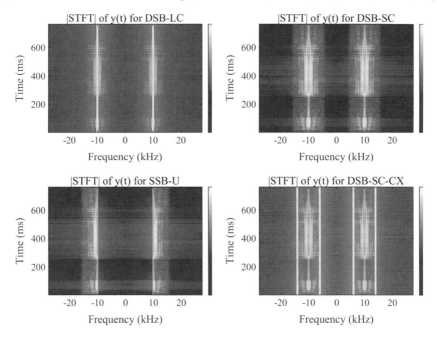

Fig. 19.4: STFT spectra of 10 kHz carrier modulated in amplitude by speech: (up-left) DSB-LC, (up-right) DSB-SC, (down-left) SSB-U, (down-right) DSB-SC-CX. Description: sampling frequency f_s =55.125 kHz, for DSB-SC-CX the signal $x_{Im}(t)$ is a pure cosine having frequency 4 kHz

The Hilbert filter-based demodulator can be used also is slightly modified version:

(A) band-pass signal filtering around the service frequency f_c, obtaining signal $y_{BPF}(t)$: other services are removed,
(B) calculation of analytic version of the signal using the Hilbert transform $H(y_{BPF}(t))$: the service spectrum for negative frequency is removed,
(C) signal multiplication by the complex-value harmonic carrier but with negative frequency $-f_c$: signal spectrum for positive frequencies is shifted back to 0 Hz.

The following set of equations describes this method:

$$\hat{a}(t) = [\underbrace{y_{BPF}(t)}_{(A)} + j\underbrace{H(y_{BPF}(t))}_{(A)}] \cdot e^{-j2\pi f_c t} \qquad (19.35)$$

$$\underbrace{\phantom{[y_{BPF}(t) + jH(y_{BPF}(t))]}}_{(B\}}$$

$$\underbrace{\phantom{[y_{BPF}(t) + jH(y_{BPF}(t))] \cdot e^{-j2\pi f_c t}}}_{(C)}$$

The AM demodulation can be alternatively done using the *quadrature* AM demodulation technique which was derived in Chap. 17:

(A) real-to-complex and frequency down signal conversion,

(B) low-pass filtering around 0 Hz with filter bandwidth (BW) adjusted to length of the signal spectrum $X(f)$: side services lying apart from DC are removed.

$$\widehat{a}(t) = LPF_{BW}\{ \underbrace{ 2y(t) \cdot \cos(2\pi f_c t) - 2j \cdot y(t) \cdot \sin(2\pi f_c t) }_{(A)} \} \quad (19.36)$$

It was described by Eq. (17.14) and graphically presented in Figs. 17.4, 17.5.

The final result of all presented above demodulators is the same: an estimation of the signal $a(t)$ is found. Having the signal $\widehat{a}(t)$ it is very easy to estimate the service signal $x(t)$ from it:

$$
\begin{array}{lll}
\textbf{DSB-LC:} & \widehat{x}(t) = \dfrac{|\widehat{a}(t)| - 1}{\Delta A}, & (19.37) \\[2mm]
\textbf{DSB-SC, SSB:} & \widehat{x}(t) = \text{Re}\,(\widehat{a}(t)), & (19.38) \\[2mm]
\textbf{DSB-SC-CMPLX:} & \widehat{x}(t) = \widehat{a}(t). & (19.39)
\end{array}
$$

A simple Matlab program doing the AM carrier demodulation is presented in Listing 19.2. It is a continuation of the Matlab code from Listing 19.1. Consecutive steps of $A(f)$ spectrum recovering using described above AM demodulation methods (19.34), (19.35), (19.36) are illustrated in Fig. 19.5. Having correct spectrum we can find $a(t)$ and then $x(t)$. The DSB-SC-CX modulation was tested. The modulating signal was complex-value: it had speech in real part and cosine 4 kHz in imaginary part—its STFT was presented in Fig. 19.4 (in left down corner). Two SFM signals were added and simulated neighboring services. Looking at the obtained results we can conclude that all approaches succeeded in reconstruction of signal $a(t)$ STFT spectrum.

Listing 19.2: Simple program for testing in Matlab the presented amplitude demodulation techniques

```
1
2   % lab19_ex_am_short.m - continuation - Example of AM demodulation
3   % ...
4
5   % Possible service using frequency (2*fc) (optional)
6   if(disturb == 1)
7     y = y + real(a).*cos(2*pi*(2*fc)*t) - imag(a).*sin(2*pi*(2*fc)*t);
8   end
9   % Additive noise
10  if(noise == 1) y = y + 0.025*randn(N,1); end
11  % Possible lack of synchronization between frequency up-shifter and down-shifter
12  if(nosynch == 1)
13    c = exp( j*(2*pi*(fc+100)*t + pi/4) ); % carrier used for freq down-conversion
14  end
15
16  % Carrier AM demodulation - frequency-down conversion
17  if(demod==1)
18    yH = hilbert( y );              % Hilbert filter - analytic signal
```

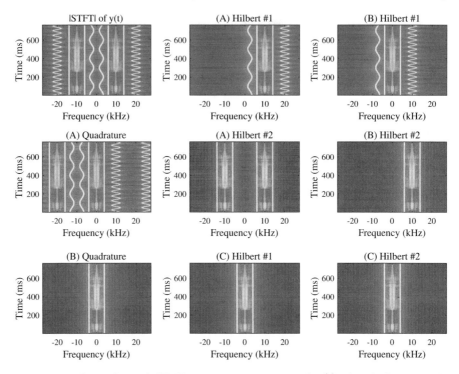

Fig. 19.5: Illustration of STFT spectrum recovery of $a(t)$ signal discussed in three AM demodulation methods: (first column) spectrum of modulated signal and quadrature demodulation (19.36), (second column) Hilbert #1 demodulation (19.34), (third column) Hilbert #2 demodulation (19.35). Magnitude of STFT spectra is shown

```
19      a1 = yH .* conj( c );            % frequency-down conversion to base-band
20   end
21   if(demod==2)
22      a1 = 2*y.*real(c) - 2*j*y.*imag(c);    % quadrature demodulator
23   end
24   M = 100; h = fir1(2*M,(fss/2)/(fs/2),kaiser(2*M+1,12)); % LP filter design
25   a2 = conv(a1,h); a2=a2(M+1:end-M);                % filtering
26
27   % Recovering x(t) from a(t)
28   if(mod==1)                   xdem = (abs(a2)-1)/dA;      end
29   if(mod==2 | mod==3 | mod==4) xdem = real(a2);            end
30   if(mod==5)                   xdem = a2;                  end
31   n = 500:N-500+1; t=t(n); x=x(n); xdem=xdem(n); % with or without it
32
33   % ERROR of demodulation
34   ERROR_Demod_SIGNAL   = max( abs( x - xdem ) ), pause
35   ERROR_Demod_SPECTRUM = max( abs( fft(x) - fft(xdem) ) )/N,
```

Exercise 19.2 (AM Demodulation). Analyze carefully the program 19.2. Run its long version `lab19_ex_am_long.m` for 4 modulating signals: speech, one cosine, summation of two cosines, and cosine with sinusoidal frequency modulation (SFM). Compare spectra of original and recovered signals. Note time and frequency errors introduced by signal processing. Propose your own modulating signals. Check whether they are recovered correctly. Check how the demodulation result will change when frequency and phase of the carrier used in the receiver differ from the transmitter carrier (use setting `nosynch=1`).

Modify the program: try to obtain figures similar to Fig. 19.5 when a few AM modulated services are present. At present you can add one extra AM service working at frequency $2f_c$ setting `disturb=1`.

Observe signal spectra and demodulation results when modulated signals are embedded in noise (use setting `noise=1`).

Modify the program: make necessary changes allowing carrier modulation by music, sampled at 44.1 kHz. Use the summation of left and right audio channel as a modulating signal for DSB-LC, DSB-SC, SSB-L, and SSB-U modulation schemes. For complex-value modulation DSB-SC-CX use left audio channel as $x_{Re}(t)$ and right audio channel as $x_{Im}(t)$.

19.4 Carrier Synchronization Importance

Now we will ask a very important question! What are the consequences of using in AM modulation and demodulation scheme a cosine with slightly different frequency and phase in the frequency up-shifter (in transmitter) and down-shifter (in receiver)? At present, as an example, we will analyze only the case of AM-DSB-SC modulation. Let us repeat definitions of the up-converted and up-down-converted versions of the transmitted signal $x(t)$, respectively, $y(t)$ and $z(t)$:

$$y(t) = x(t) \cdot \cos(2\pi f_c t), \tag{19.40}$$

$$z(t) = y(t) \cdot \cos(2\pi (f_c + \Delta f)t + \phi), \tag{19.41}$$

where Δf and ϕ denote frequency and phase shift difference between transmitter and receiver. Due to Fourier transform modulation property, frequency spectra of both signals are given by the following equations, respectively:

$$Y(f) = \frac{1}{2}[X(f+f_c)+X(f-f_c)], \tag{19.42}$$

$$Z(f) = \frac{1}{2}\left[e^{j\phi}Y(f+(f_c+\Delta f))+e^{-j\phi}Y(f-(f_c+\Delta f))\right], \tag{19.43}$$

where $X(f)$ denotes spectrum of signal $x(t)$. Putting Eq. (19.42) into (19.43) one gets

$$Z(f) = \frac{1}{2}\left\{\; \frac{1}{2}e^{j\phi}[X((f+f_c)+(f_c+\Delta f))+X((f-f_c)+(f_c+\Delta f))]+\ldots\right.$$
$$\left. +\frac{1}{2}e^{-j\phi}[X((f+f_c)-(f_c+\Delta f))+X((f-f_c)-(f_c+\Delta f))]\;\right\},$$
$$Z(f) = \frac{1}{4}\left\{\; e^{j\phi}X(f+2f_c+\Delta f)+e^{j\phi}X(f+\Delta f)+\ldots\right.$$
$$\left. +e^{-j\phi}X(f-\Delta f)+e^{-j\phi}X(f-2f_c-\Delta f)\;\right\}. \tag{19.44}$$

Spectrum $Z(f)$ consists of four copies of $X(f)$ shifted in frequency and changed in phase. After low-pass filtering (LPF) with cut-off frequency adjusted to the signal $x(t)$ bandwidth, the high-frequency components lying around frequencies $\pm 2f_c$ are removed and we obtain

$$\text{LPF}[Z(f)] = \frac{1}{4}\left\{\; e^{j\phi}X(f+\Delta f)+e^{-j\phi}X(f-\Delta f)\right\}. \tag{19.45}$$

When $\Delta f \neq 0$ two scaled copies of the spectrum $X(f)$ are slightly shifted in frequency in regard to each other and their summation does not allow exact $X(f)$ reconstruction. For $\Delta f = 0$ the spectrum (19.45) simplifies to:

$$\text{LPF}[Z(f)] = \frac{1}{4}\left\{\; e^{j\phi}X(f)+e^{-j\phi}X(f)\;\right\} = \frac{1}{2}X(f)\cos(\phi) \tag{19.46}$$

and corresponds to the signal:

$$\widehat{x}(t) = \frac{1}{2}x(t)\cos(\phi). \tag{19.47}$$

As we see, lack of carrier frequency synchronization in the receiver causes severe degradation of signal $x(t)$ recovery, while lack of the carrier phase synchronization only results in unknown attenuation of the received signal caused by $\cos(\phi)$ component in Eq. (19.47).

Exercise 19.3 (AM Demodulation with Frequency and Phase Synchronization Error). Use the programs 19.1 and 19.2. Set nosynch=0 in the first of them, while in the second modify values of $\Delta f = 100$ and $\phi = \pi/4$ in the program line c = exp(j*(2*pi*(fc+100)*t + pi/4)). For frequency down-shifting use in the demodulator a cosine with slightly different:

(A) phase only, (B) frequency only, and (C) phase and frequency together. Observe the difference between transmitted and received signal. Listen to both signal using headphones or loudspeakers. When you set $\phi = \pi/2$ in test A, $\cos(\phi) = 0$ and the $\hat{x}(t) = 0$ according to Eq. (19.47). Do you observe this effect?

19.5 Carrier Recovery in Suppressed Carrier AM Modulation

We can conclude from the above discussion that amplitude modulation and demodulation algorithms are not very difficult: in the modulator the signal spectrum $A(f)$ (or $X(f)$) is up-shifted in frequency to the frequency f_c of the carrier, while in demodulator it is back down-shifted to the base-band. The success of the whole operation depends on the perfect realization of the frequency up-down conversion. Since hardware is not perfect, frequency of the up-shifter is not perfectly known and stable. Therefore receiver should estimate the actual carrier frequency and phase for the frequency down-shifter from the high-frequency signal $y(t)$, obtained from antenna.

In this section usage of a real-value carrier is assumed:

$$c(t) = \cos(2\pi f_c t). \tag{19.48}$$

Additionally, in the below discussion we will concentrate only on AM-DSB-SC modulation technique (19.17) in which carrier is suppressed in the transmitter (TX):

$$y_{TX}(t) = x(t) \cdot c(t) = x(t) \cdot \cos(2\pi f_c t). \tag{19.49}$$

Why only on AM-DSB-SC? Because for AM-DSB-LC with large carrier (19.16) signal down-conversion is not required for finding the modulating function: simply, Eq. (19.11) can be used.

Due to channel time delay D as well as signal attenuation G and phase shift φ introduced by the channel around frequency f_c (we are assuming a narrow transmission bandwidth), a signal obtained from the receiver (RX) antenna will have the following form:

$$y_{RX}(t) = Gx(t - D) \cdot \cos(2\pi f_c(t - D) - \varphi) + e(t) =$$
$$= x_D(t) \cdot \cos(2\pi f_c t + \phi) + e(t), \tag{19.50}$$

where

$$x_D(t) = Gx(t - D), \tag{19.51}$$
$$\phi = -2\pi f_c D - \varphi, \tag{19.52}$$

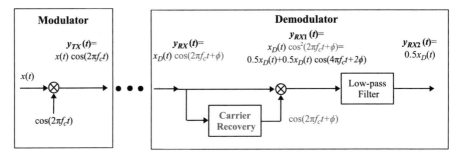

Fig. 19.6: Block diagram of real-value AM-DSB-SC modulation and demodulation

and $e(n)$ denotes noise. The phase shift ϕ is unknown. At present, we do not consider the carrier frequency change caused by the Doppler shift.

If we could generate in the receiver, exactly the same carrier as in Eq. (19.50), with the same frequency and phase, and multiply the received signal with it, we would obtain

$$y_{RX1}(t) = y_{RX}(t) \cdot \cos(2\pi f_c t + \phi) =$$
$$= x_D(t) \cdot \cos^2(2\pi f_c t + \phi) + e(n) \cdot \cos(2\pi f_c t + \phi). \quad (19.53)$$

Using trigonometric identity:

$$\cos^2(\alpha) = \frac{1}{2} + \frac{1}{2}\cos(2\alpha), \qquad (19.54)$$

Eq. (19.53) can be rewritten as:

$$y_{RX1}(t) = \frac{1}{2}x_D(t) + \frac{1}{2}x_D(t) \cdot \cos(2\pi(2f_c)t + 2\phi) + e(n) \cdot \cos(2\pi f_c t + \phi). \quad (19.55)$$

After a low-pass filtration (LPF) the second high-frequency $(2f_c)$component and the third noise component are strongly attenuated and can be neglected. Therefore after LPF we are obtaining the transmitted signal $x_D(t)$ scaled by $\frac{1}{2}$:

$$y_{RX2}(t) = \text{LPF}\left[\, y_{RX1}(t)\,\right] = \text{LPF}\left[\, y_{RX}(t) \cdot \cos(2\pi f_c t + \phi)\,\right] \approx \frac{1}{2}x_D(t). \quad (19.56)$$

Figure 19.6 illustrates described above real-value AM-DSB-SC modulation and demodulation scheme.

But how to recover the correct carrier $\cos(2\pi f_c t + \phi)$ needed by the frequency down-converter of the receiver described by Eq. (19.53)? Let us take the received signal $y_{RX}(t)$ to the power of two (neglecting the noise):

$$y_{RX}^2(t) \approx x_D^2(t) \cdot \cos^2(2\pi f_c t + \phi) \qquad (19.57)$$

and use Eq. (19.54) again:

$$y_{RX}^2(t) \approx \frac{1}{2} \cdot x_D^2(t) \cdot [1 + \cos(2\pi(2f_c)t + 2\phi)]. \qquad (19.58)$$

At present we assume that $x_D^2(t)$ consists of mean value m_{x^2} and some oscillations $v_{x^2}(t)$ around it:

$$x_D^2(t) = m_{x^2} + v_{x^2}(t). \qquad (19.59)$$

Using Eq. (19.59) in (19.58) we get

$$y_{RX}^2(t) \approx \frac{1}{2} \cdot [\, m_{x^2} + v_{x^2}(t) + m_{x^2} \cdot \cos(2\pi(2f_c)t + 2\phi) + \dots$$
$$+ v_{x^2}(t) \cdot \cos(2\pi(2f_c)t + 2\phi) \,]. \qquad (19.60)$$

Now we are using a very narrow band-pass filter (BPF) around frequency $2f_c$ to: (1) pass the third signal component with frequency $(2f_c)$, (2) remove the first components, and (3) suppress the second and the fourth component.

$$c_{(2f_c)}(t) = \text{BPF}_{(2f_c)} \left[\, y_{RX}^2(t) \right] \approx 0.5 \cdot m_{x^2} \cdot \cos(2\pi(2f_c)t + 2\phi + \psi \,). \qquad (19.61)$$

Angle ψ results from the filter delay, it is known and can be taken into account (compensated) easily.

Remark In Eq. (19.60) the second component $v_{x^2}(t)$ is a low-frequency one in comparison to the frequency $2f_c$, while the fourth component has a spectrum around frequency $2f_c$ but without it, since it represents AM-DSB-SC modulation of the $2f_c$ carrier by the signal $v_{x^2}(t)$.

When the *doubled* carrier $c_{(2f_c)}(t)$ (with $2f_c$ and 2ϕ) is extracted with success from the received signal $y_{RX}(t)$, one should synthesize a cosine well fitting to it (synchronized with it).

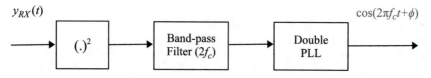

Fig. 19.7: Block diagram of possible carrier recovery module for AM-DSB-SC demodulator presented in Fig. 19.6

Typically, an adaptive double phase locked loop (PLL) algorithm, described in the next section, is used for recovering, both, the cosine frequency and phase. When working on discrete-time signals, all digital PLL is applied and finding original values of carrier parameters from its multiples is straightforward. After the carrier recovery, summarized in Fig. 19.7, it is used for signal demodulation using equations (19.48)–(19.56) and diagram presented in Fig. 19.6.

Exercise 19.4 (Carrier Recovery in AM-DSB-SC Modulation). Investigate AM-DSB-SC modulation. Use program `lab19_ex_am_long.m`, the long version of the program `lab19_ex_am_short.m` presented in Listing 19.1, and observe spectrogram of the signal $y^2(t)$ for different modulated signals $x(t)$. Is it possible to see spectra of signal components present in Eq. (19.60)? Note that the doubled carrier is very well visible in the spectrogram of single-component modulated signals, like one cosine and the SFM signal, and completely invisible for speech, a time-varying multi-component signal. Extend the program: check whether this is true also when `fft()` of the long fragment of the signal $y^2(t)$ is computed. Design a band-pass filter for separation of a doubled carrier $2f_c$. Apply the filter and observe it output signal. Try to estimate frequency $2f_c$ and phase $2\phi + \psi$ of the doubled carrier $\cos(2\pi(2f_c)t + 2\phi + \psi)$ from the calculated FFT spectrum. Knowing their values, try to synthesize the correct carrier $\cos(2\pi(f_c)t + \phi)$ and to use it for signal demodulation. Check the demodulation result. Some help: (1) find phase $2\phi + \psi$ as an angle of the `fft()` peak at frequency $2f_c$, (2) find ψ as an angle of used BP filter frequency response `fregz()` for frequency $2f_c$, (3) subtract calculated values and take modulo π of the result. An angle obtain this way is equal to ϕ since all angle calculation of a cosine are done modulo 2π and our angle has doubled value.

Remark The AM-DSB-LC demodulation could be also processed in traditional way with service down-conversion in frequency described by Eq. (19.53). Because in AM-DSB-LC the carrier is present in a received signal, it can be separated by narrow band-pass filter centered at (f_c). Since filters are not ideal, the resultant signal can be noisy and disturbed by transmitted information. Therefore the adaptive digital double phase locked loop (PLL), described below, could be used for generation of cosine well fitting to the result of filtering. The synthesized carrier should have proper values of frequency and phase. Finally, it is used for service down-conversion (19.53).

19.6 Carrier Recovery: Phase Locked Loop

19.6.1 Real-Value PLL

In this section we will design digital double PLL module, used in carrier recovery algorithms. In order to make derivation more general and easier to apply in different scenarios, we will change denotation of variables. Let an input signal to the PLL module be described by the following equation:

$$s(n) = \cos\left(2\pi \frac{f_c}{f_s} n + \alpha\right) + e(n), \qquad (19.62)$$

where $e(n)$ denotes noise. The PLL task is to synthesize output signal $c(n)$ of the form:

$$c(n) = \cos\left(2\pi \frac{f_c}{f_s} n + \beta\right) \qquad (19.63)$$

having samples of $s(n)$ and knowing values of f_c and f_s. We are assuming that amplitudes and frequencies of signals $s(n)$ and $c(n)$ are the same, different are only their phase angles at the adaptation beginning. The amplitude assumption $A = 1$ is not restrictive since the signal $s(n)$ amplitude can be always normalized to this value. The PLL algorithm should adjust value of the phase β to value of the phase α, i.e. synchronize signal $c(n)$ with $s(n)$, minimizing a cost function having minimum exactly for $\beta = \alpha$.

Let us use the squared error cost function defined as:

$$J = E\left[(s(n) - c(n))^2\right], \qquad (19.64)$$

where $E[.]$ denotes expectation value. Putting Eqs. (19.62), (19.63) into (19.64) and setting $\Omega_c = 2\pi \frac{f_c}{f_s}$ one obtains

$$J = E\left[(\,[\cos(\Omega_c n + \alpha) - \cos(\Omega_c n + \beta)] + e(n)\,)^2\right] =$$
$$= E\left[[\cos(\Omega_c n + \alpha) - \cos(\Omega_c n + \beta)]^2\right] + E\left[e^2(n)\right] + \dots$$
$$- 2 \cdot E[\,[\cos(\Omega_c n + \alpha) - \cos(\Omega_c n + \beta)] \cdot e(n)\,]. \qquad (19.65)$$

The last term in Eq. (19.65) equals zero and can be removed since mean values of noise and cos() function are equal to zero. Having this in mind and using the following trigonometric equality for the first term:

$$[\cos(\Omega n + \alpha) - \cos(\Omega n + \beta)]^2 = (1 - \cos(2\Omega n + \alpha + \beta)) \cdot (1 - \cos(\alpha - \beta)), \qquad (19.66)$$

we can rewrite Eq. (19.65):

$$J = E\left[1 - \cos(\alpha - \beta)\right] - E\left[\cos(2\Omega_c n + \alpha + \beta) \cdot (1 - \cos(\alpha - \beta))\right] +$$
$$+ E\left[e^2(n)\right]. \quad (19.67)$$

Expectation value of the middle term in last equation equals 0. Therefore, finally, we get

$$J = (1 - \cos(\alpha - \beta)) + E\left[e^2(n)\right], \quad (19.68)$$

and we can conclude that the cost function J has really minimum for $\beta = \alpha$ and the minimum is only one for $-\pi \leqslant \alpha - \beta < \pi$.

Now we remove the expectation operator in the cost function:

$$J(n) = [s(n) - c(n)]^2. \quad (19.69)$$

Minimizing adaptively momentum value of $J(n)$, in longer time horizon we are performing minimization of a least mean squared (LMS) error between two signals, similarly as in already discussed LMS adaptive filters. In order to find argument of the cost function minimum, we calculate derivative of $J(n)$ in regard to the phase β:

$$\frac{\partial J(n)}{\partial \beta} = 2 \cdot [s(n) - c(n)] \cdot \frac{\partial c(n)}{\partial \beta} =$$
$$2 \cdot [s(n) - c(n)] \cdot \sin\left(2\pi \frac{f_c}{f_s} n + \beta(n)\right), \quad (19.70)$$

and change value of $\beta(n)$ into direction of the cost function minimum, i.e. go opposite to its gradient/derivative (direction of its growth):

$$\beta(n+1) = \beta(n) - \mu \frac{\partial J(n)}{\partial \beta} =$$
$$= \beta(n) - \mu \cdot 2 \cdot [s(n) - c(n)] \cdot \sin\left(2\pi \frac{f_c}{f_s} n + \beta(n)\right), \quad (19.71)$$

where μ denotes adaptation speed coefficient. We are using the stochastic gradient minimization technique. Setting formula of $c(n)$ (19.63) to Eq. (19.71) we get

$$\beta(n+1) = \beta(n) - 2\mu \sin\left(2\pi \frac{f_c}{f_s} n + \beta(n)\right) s(n) +$$
$$+ 2\mu \sin\left(2\pi \frac{f_c}{f_s} n + \beta(n)\right) \cos\left(2\pi \frac{f_c}{f_s} n + \beta(n)\right). \quad (19.72)$$

At present, thanks to the formula $2\sin(\alpha)\cos(\alpha) = \sin(2\alpha)$, we can rewrite the above equation into the following form:

$$\beta(n+1) = \beta(n) - 2\mu \sin\left(2\pi\frac{f_c}{f_s}n + \beta(n)\right)s(n) + \mu \sin\left(2\pi\frac{2f_c}{f_s}n + 2\beta(n)\right).$$

$$(19.73)$$

Since the last term in Eq. (19.73) has mean value equal to zero, it does not influence the adaptation convergence and can be removed:

$$\beta(n+1) = \beta(n) - 2\mu \sin\left(2\pi\frac{f_c}{f_s}n + \beta(n)\right)s(n). \qquad (19.74)$$

Now we introduce an angular frequency Ω_c and entire angle $\theta(n)$ of the carrier $c(n)$:

$$\Omega_c = 2\pi\frac{f_c}{f_s}, \qquad \theta(n) = \Omega_c n + \beta(n), \qquad (19.75)$$

and add $\Omega_c(n+1)$ to both sides of the PLL adaptation Eq. (19.74). In consequence, the following final PLL adaptation form is to get

$$\theta(n+1) = \theta(n) + \Omega_c - 2\mu \sin(\theta(n))s(n), \qquad (19.76)$$

and a synthesized/recovered carrier is equal to:

$$c(n) = \cos(\theta(n)). \qquad (19.77)$$

The signal frequency f_c can be found using DFT/FFT. However, when we are interested in adaptive tracking the input signal frequency, another approach, is preferred.

When carrier frequency f_c of the input signal $s(n)$ is not precisely known, it is very probable that the synthesized signal (19.63) has wrong frequency $f_c + \Delta f$ and still the following phase difference between two signals (19.62), (19.63) is present:

$$\phi(n) = 2\pi\frac{\Delta f}{f_s}n \qquad (19.78)$$

after adjusting value of β to the value of α. This phase error increases linearly and has to be compensated also. To solve this problem, the same adaptation mechanism as before can be used but as a second adaptation step. This time, adaptation of Ω_c in Eq. (19.76) is done.

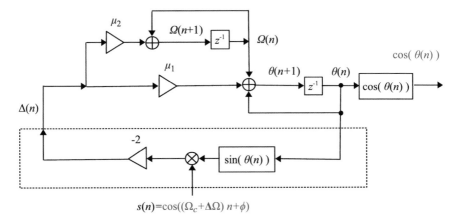

Fig. 19.8: Block diagram of the digital double PLL carrier synchronization algorithm. Having samples of signal $\cos\left(\,(\Omega_c+\Delta\Omega)n+\phi\,\right)$ the overall momentum cosine angle $\theta(n) = (\Omega_c+\Delta\Omega)n+\phi$ is tracked

The double-loop PLL algorithm, resulting from Eq. (19.76) and capable of synchronization to the carrier phase and frequency, is described by the following set of equations ($\Delta(n)$ denotes the update term):

$$\Delta(n) = -2 \cdot \sin\left(\theta(n)\right) \cdot s(n), \tag{19.79}$$
$$\theta(n+1) = \theta(n) + \Omega(n) + \mu_1\Delta(n), \tag{19.80}$$
$$\Omega(n+1) = \Omega(n) + \mu_2\Delta(n). \tag{19.81}$$

Block diagram of the PLL-based carrier synchronization algorithm is presented in Fig. 19.8. In this implementation μ_1 is multiplied by $\Delta(n)$ in Eq. (19.80) and μ_2 by $\Delta(n)$ in Eq. (19.81).

19.6.2 Complex-Value PLL

At present we would like to extend our result to the case of complex-value PLL loop when signals $s(n)$ (19.62) and $c(n)$ (19.63) are complex and defined as:

$$s(n) = \exp\left(j\cdot\Omega_cn+\alpha\right)+e(n), \tag{19.82}$$
$$c(n) = \exp\left(j\cdot\Omega_cn+\beta\right). \tag{19.83}$$

In the real-value PLL the update term $\Delta(n)$ is equal (19.79). Putting into this equation the signal model $s(n) = \cos(\Omega_c n + \alpha)$ (19.62) (without noise) and replacing $\theta(n)$ with its full version $\Omega_c n + \beta(n)$, we get

$$\Delta(n) = -2\sin(\Omega_c n + \beta(n)) \cdot \cos(\Omega_c n + \alpha). \tag{19.84}$$

Additionally using equality $2\sin(a)\cos(b) = \sin(a+b) + \sin(a-b)$ we obtain

$$\Delta(n) = -\sin(2\Omega_c n + \beta(n) + \alpha) - \sin(\beta(n) - \alpha). \tag{19.85}$$

Since the mean value of the first term is equal to zero, it has no statistical influence upon adaptation result. Important is only the second term. For small value of the difference we have $\sin(\beta(n) - \alpha) = (\beta(n) - \alpha)$. Therefore:

$$\Delta(n) \approx -(\beta(n) - \alpha). \tag{19.86}$$

When actual value of the synthesized signal angle $\beta(n)$ is greater (or smaller) than the carrier angle α, the update $\Delta(n)$ has negative (or positive) sign, what was expected.

By analogy to Eq. (19.84), in case of complex-value signals (19.82), (19.89), we can use in PLL equations (19.79)–(19.81) the following update value $\Delta\theta(n)$:

$$\Delta(n) = -2\,\mathrm{Im}\left[e^{j\cdot\theta(n)} \cdot s^*(n)\right] \tag{19.87}$$

since after assuming $\theta(n) = \Omega_c n + \beta(n)$ and using $s(n) = e^{j(\Omega_c n + \alpha)}$, the Eq. (19.87) simplifies to:

$$\Delta(n) = -2 \cdot \sin(\beta(n) - \alpha). \tag{19.88}$$

It has the same form as Eq. (19.86).

19.6.3 Using PLL Program

A Matlab program of the double PLL synchronization block, adjusting, both, carrier phase and frequency, is presented in Listing 19.3. The PLL loop values of parameters are set in it to ones corresponding to 19 kHz pilot synchronization in FM radio multiplex signal sampled at 250 kHz. The PLL adaptation process to the FM radio pilot, for real-value and complex-value signals, is presented in Fig. 19.9, respectively. A crucial issue of the program right usage is proper choice of adaptation constants μ_1 and μ_2 (see Eqs. (19.80), (19.81) and Fig. 19.8). Their values should ensure PLL adaptation convergence to the correct f_c value even despite significant initial difference Δf (known as PLL bandwidth or PLL lock frequency). Additionally, their choice decides about the convergence speed and variance around the found minimization solution. Derivation of formulas for selection of μ_1 and μ_2 values lies out of the scope of this book. Final mathematical rules are defined by the following equations [9]:

$$d = \frac{\sqrt{2}}{2}, \quad \Omega = \frac{\frac{\Delta f}{f_s}}{d + \frac{1}{4d}}, \quad \mu_1 = \frac{4 \cdot d \cdot \Omega}{1 + 2 \cdot d \cdot \Omega + \Omega^2}, \quad \mu_2 = \frac{4 \cdot \Omega^2}{1 + 2 \cdot d \cdot \Omega + \Omega^2},$$

$$(19.89)$$

where Δf denotes chosen PLL frequency bandwidth (frequency lock), i.e. allowed maximum difference between initial value of PLL frequency and input signal frequency, guaranteeing convergence of adaptation.

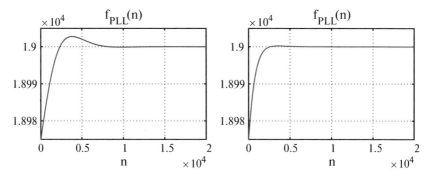

Fig. 19.9: Exemplary PLL adaptation process to FM radio 19 kHz pilot in case of real-value (left) and complex-value (right) signals. $f_s = 250$ kHz

Note that the signal $s(n)$, used in the adaptation loop for $\Delta(n)$ calculation, is divided by its maximum value, i.e. its amplitude is normalized to 1, as was assumed during algorithm derivation.

Listing 19.3: Matlab program implementing the PLL loop

```
 1
 2   % lab19_ex_PLL.m
 3   clear all; close all;
 4
 5   % PLL parameters
 6   fs = 250000;                          % sampling frequency
 7   fc = 19000; phc = pi/4;               % carrier frequency and phase
 8   fPLLstart = fc-25; dfreq = 100;       % initial PLL freq., freq. lock (synchro bandwidth)
 9   ipll = 1;                             % 1=real PLL, 2=complex PLL
10
11   % Signal generation
```

```
12   N = 50000; n=0:N-1; A = 0.1;              % number of samples, sample indexes
13   dt=1/fs; t=dt*n;                          % sampling period, sampling moments
14   if(ipll==1) s  = A * cos(2*pi*fc*t + phc);      end % real input signal
15   if(ipll==2) s  = A * exp(j*(2*pi*fc*t + phc));  end % complex input signal
16
17   % Calculation of adaptation constants
18   damp = sqrt(2)/2;                         % standard damping
19   omeg = (dfreq/fs) / (damp+1/(4*damp));    % variable
20   mi1  = (4*damp*omeg) / (1 + 2*damp*omeg + omeg*omeg); % adapt speed const #1
21   mi2  = (4*omeg*omeg) / (1 + 2*damp*omeg + omeg*omeg); % adapt speed const #2
22
23   % PLL
24   omega = zeros(1,N+1); omega(1) = 2*pi*fPLLstart / fs;
25   theta = zeros(1,N+1);
26   smax  = max(abs(s));
27   for n = 1 : N   % PLL adaptation loop
28       if( ipll==1 ) delta = -2*sin(theta(n)) * s(n)/smax;
29       else          delta = -2*imag( exp(j*theta(n)) * conj(s(n))/smax );
30       end
31       theta(n+1) = theta(n) + omega(n) + mi1*delta;
32       omega(n+1) = omega(n) + mi2*delta;
33   end
34   c = cos( theta(1:N) );   % recovered carrier
35   sr = real(s) / smax;
36   figure; plot(1:N,sr,'r-',1:N,c,'b-'); title('s(n) and c(n)'); grid; pause
37   figure; plot(1:N,sr-c,'r-'); title('s(n)-c(n)'); grid; pause
38   figure; plot(theta); title('\theta(n) [rad]');grid; pause
39   figure; plot(omega*fs/(2*pi),'b-'); xlabel('n'); title('f_{PLL}(n) [Hz]'); grid; pause
```

Exercise 19.5 (Testing PLL Loop). Test PLL loop program 19.3. You should
see plots similar to these presented in Fig. 19.9. Generate a cosine signal
(choosing `ipll=1;`) and let PLL to synchronize with it. Run the program.
Change value of the PLL frequency bandwidth Δf (`dfreq`). Note that after it
increase, the adaptation constants μ_1 and μ_2 become larger. Observe that bigger
values of adaptation coefficients change the PLL features, increasing, both, convergence speed and frequency variance of the generated cosine in steady-state
(after PLL synchronization). Set different values of the cosine carrier parameters: (1) phase, (2) frequency, (3) phase and frequency. Add different level of
white Gaussian noise (`randn()`) to the $s(n)$ signal and check the PLL convergence. Test the complex PLL setting `ipll=2`.

**Exercise 19.6 (Carrier Recovery in AM-DSB-SC Modulation Using Dou-
ble PLL Loop).** Continue Exercise 19.4. Use programs 19.2 and 19.3. Apply
program of the PLL module to recover the carrier using method presented in

Fig. 19.7. Take into account delay introduced by the band-pass filter. After synchronization with the doubled carrier divide the calculated angle $\theta(n)$ by 2 and use it for the recovery of the original carrier. After that perform signal frequency down-conversion. Compare demodulated signal with the original one. Listen to both signals.

19.7 Carrier Recovery: Costas Loop

At present we learn the second method of the AM-DSB-SC signal demodulation which is an alternative to one presented in Fig. 19.6. In the previous approach the carrier was first recovered, using method presented in Fig. 19.7 exploiting the PLL loop, and then applied to signal down-conversion (demodulation) in frequency. In the new technique, which is described now, called the Costas loop, order of these two operations is reversed: (1) first, a signal frequency down-shifting to the baseband (around 0 Hz) is done using an un-synchronized carrier, (2) then an adaptive carrier correction is performed and modulated signal is recovered.

Let us remember the problem formulation: having a signal $y(n)$, obtained by the receiver (result of band-pass filtering a signal from the antenna):

$$y(n) = x(n) \cdot \cos(\Omega_c n + \alpha), \tag{19.90}$$

we would like to recover signal $x(n)$. In the Costas method this task is performed in the following steps:

1. analytic version of signal $y(n)$ is computed using Hilbert filter/transform:

$$y_a(n) = y(n) + j \cdot H[y(n)] = x(n)\cos(\Omega_c n + \alpha) + j \cdot x(n)\sin(\Omega_c n + \alpha) = x(n) \cdot e^{j(\Omega_c n + \alpha)}; \tag{19.91}$$

2. the signal is down-converted in frequency to the base-band, using angle $\beta(n)$ which is estimated adaptively:

$$y_{bb}(n) = y_a(n) \cdot e^{-j(\Omega_c n + \beta(n))} = x(n) \cdot e^{j(\Omega_c n + \alpha)} \cdot e^{-j(\Omega_c n + \beta(n))} = x(n) \cdot e^{j(\alpha - \beta(n))};$$
$$\tag{19.92}$$

3. real part of the result is calculated:

$$y_{bb}^{Re}(n) = \mathrm{Re}\,[y_{bb}(n)] = x(n) \cdot \cos(\alpha - \beta(n)); \tag{19.93}$$

4. after adaptation of $\beta(n)$ value, the difference $\varepsilon = \alpha - \beta(n)$ is close to zero and $\cos(\varepsilon) \approx 1$, therefore we have

$$\hat{x}(n) \approx y_{bb}^{\mathrm{Re}}(n).$$
(19.94)

At present our problem is finding efficient adaptation of value of $\beta(n)$ to value of α. In this purpose let us define a cost function:

$$J(\beta(n)) = \frac{\mathrm{Re}\left[y_{bb}(n)\right] \cdot \mathrm{Im}\left[y_{bb}(n)\right]}{y_{bb}(n)y_{bb}^{*}(n)} = \frac{x(n)\cos(\alpha-\beta(n))\cdot x(n)\sin(\alpha-\beta(n))}{x^2(n)}.$$
(19.95)

Using equality $2\sin(a)\cos(b) = \sin(a+b) + \sin(a-b)$, the last equation simplifies to (in our case $a = b = \alpha - \beta(n)$):

$$J(\beta(n)) = \frac{1}{2}\sin(2(\alpha-\beta(n))).$$
(19.96)

Since $\sin(a) \approx a$ for small value of the angle a, for small value of the difference $\alpha - \beta(n)$ we obtain

$$J(\beta(n)) \approx \frac{1}{2}2(\alpha-\beta(n)) \approx (\alpha-\beta(n)).$$
(19.97)

We see that the cost function is approximately equal to the angle difference! Therefore we can directly use it as an angle error (phase error) for adaptation of $\beta(n)$ value:

$$\beta(n+1) = \beta(n) + \mu \cdot J(\beta(n)),$$
(19.98)

where μ denotes an adaptation speed constant. When $\alpha > \beta$, β value is increased, in contrary—decreased. Finally, applying Eq. (19.95) to Eq. (19.98) we obtain the searched adaptation rule:

$$\beta(n+1) = \beta(n) + \mu \cdot \frac{\mathrm{Re}\left[y_{bb}(n)\right] \cdot \mathrm{Im}\left[y_{bb}(n)\right]}{y_{bb}(n)y_{bb}^{*}(n)}.$$
(19.99)

Choice of μ is the same as in the already discussed PLL loop.

It has to be marked that the function (19.96), for one period of the sine, is equal to zero for angle difference equal to 0 and π. Therefore the adaptation can catch one of these two states. Since $e^{j\pi} = -1$ this means that the demodulated signal $x(n)$ can have a negated sign also. In analog transmission this is not a problem while in digital case special pilot signals should be sent to solve the ambiguity $0/\pi$ phase problem.

The last equation describes adaptive update of the phase shift of cosine frequency down-shifter used in Costas method: $\beta(n)$ of the down-shifter is adjusted to α of the carrier component present in the received signal $y(n)$. The equations correspond to Eq. (19.74) of the PLL adaptation. Both equations can be written in the following form (using the PLL loop notation):

$$\beta(n+1) = \beta(n) + \mu \cdot \Delta(n),$$
(19.100)

where

$$\Delta(n) = -\nabla_{PLL}(n) = -2\sin\left(\Omega_c n + \beta(n)\right)s(n) \tag{19.101}$$

$$\Delta(n) = \nabla_{Costas}(n) = \frac{\mathrm{Re}\left[y_{bb}(n)\right] \cdot \mathrm{Im}\left[y_{bb}(n)\right]}{y_{bb}(n)y_{bb}^*(n)}. \tag{19.102}$$

When we allow in the Costas loop that carrier present in signal $y(n)$ has slightly different angular frequency than Ω_c:

$$y(t) = x(n)\cos\left((\Omega_c + \Delta\Omega)n + \alpha\right), \tag{19.103}$$

we should add the second adaption mechanism for frequency tracking exactly the same way as it was introduced in case of the PLL loop. Therefore we can generalize equations (19.80), (19.81) derived for the PLL to the Costas loop case:

$$\theta(n+1) = \theta(n) + \Omega(n) + \mu_1 \cdot \Delta(n) \tag{19.104}$$

$$\Omega(n+1) = \Omega(n) + \mu_2 \cdot \Delta(n), \tag{19.105}$$

where Δ for Costas loop is specified by Eq. (19.102). Since the adapted function (carrier angle) and applied adaptation scheme is the same in PLL and Costas loops, differing only in definitions (19.104), (19.105) of the $\Delta(n)$ function used, (19.101) or (19.102), values of adaptation constant μ_1 and μ_2 are selected using the same equations (19.89).

Final Costas loop carrier synchronization and simultaneous signal demodulation algorithm are presented in Fig. 19.10. Matlab program implementing it is given in Listing 19.4.

Listing 19.4: Matlab program implementing carrier synchronization and signal demodulation using the Costas loop

```
1   % lab19_ex_dem_Costas.m
2   clear all; close all;
3
4   % Parameters
5   isignal = 1;              % 1=const, 2=fm, 3=speech, 4=bpsk, 5=qpsk
6   fs = 240000; dt=1/fs;     % sampling frequency, sampling period
7   fpilot = 19000;           % FM radio pilot frequency
8   fc = 3*fpilot; dfreq = 100;   % Costas loop: carrier frequency, frequency locking
9
10  % Choice of modulating signal
11  if(isignal==1)            % const value
12      N=100000; x = 5*ones(1,N);
```

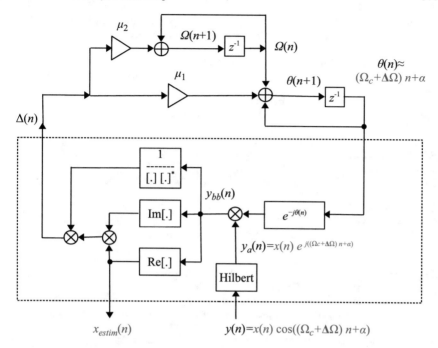

Fig. 19.10: Block diagram of the digital double Costas carrier synchronization algorithm and simultaneous signal demodulation. Having samples of the signal $x(n)\cos((\Omega+\Delta\Omega)n+\phi)$ the overall momentum cosine angle $\theta(n) = (\Omega + \Delta\Omega)n+\phi)$ is tracked

```
13  elseif(isignal==2)                    % FM signal
14      N = 100000; t = dt*(0:N-1);
15      x = cos( 2*pi*(5000*t - 4000/(2*pi*5)*cos(2*pi*5*t)) - pi/7);
16  elseif(isignal==3)                    % speech
17      [x, fx] = audioread('DANKE.WAV'); x=x.'; x=[x x]; x = interp(x,round(fs/fx));
18  elseif(isignal==4 | isignal==5)  %
19      frds = fpilot/16; Nrds = round( fs / frds );
20      if(isignal==4)     % BPSK - two symbols sin() and -sin()
21          Nsymb = 2;
22          xsymb(1,:) =  sin(2*pi*(0:Nrds-1)/Nrds);            % bit 1
23          xsymb(2,:) = -sin(2*pi*(0:Nrds-1)/Nrds);            % bit 0
24      else                % QPSK - four symbols cos(Om*n+phi),
25          Nsymb = 4;      % phi = pi/4, 3*pi/4, 5*pi/4, 7*pi/4
26          xsymb(1,:) = cos(2*pi*(0:Nrds-1)/Nrds +   pi/4);    % bits 00
27          xsymb(2,:) = cos(2*pi*(0:Nrds-1)/Nrds + 3*pi/4);    % bits 01
28          xsymb(3,:) = cos(2*pi*(0:Nrds-1)/Nrds + 5*pi/4);    % bits 10
29          xsymb(4,:) = cos(2*pi*(0:Nrds-1)/Nrds + 7*pi/4);    % bits 11
30      end
31      x = [];                                  % DIGITAL SIGNAL SYNTHESIS
32      for k = 1:100                            % building a sequence
33          x = [ x xsymb( ceil(Nsymb*(rand(1,1))),: ) ]; % of signals corresponding
```

```
34       end                                               % to symbols (one by one)
35       figure; plot(x); title('x(t) - modulating signal without PSF' ); pause
                                                           %
36       hpsf = firrcos( 6*Nrds, frds, 0.35, fs,'rollof','sqrt'); % PSF filter design
37       x = filter( hpsf, 1, x );          % low-pass filtering - bandwidth reduction
38    end
39    N = length(x); n=0:N-1; t = n*dt;     % sample indexes, sampling moments
40    figure; plot(x); title('x(t) - modulating signal'); pause
41
42    % Modulation AM-DSB-SC - frequency-up conversion
43    ph = 2*pi*(fc+25)*t + pi/4; % carrier errors: frequency  25Hz, phase pi/4 radians
44    c  = cos( ph );             % carrier samples
45    y  = x .* c;                % modulated signal, input to the Costas loop
46
47    % Disturbances
48    % y  = y + 0.01*randn(1,N);                          % additive noise
49    % D=75; n=D+1:N; y =[y(n) y(1:D)]; x=[x(n) x(1:D)]; c=[c(n) c(1:D)]; % delay
50
51    % Calculation of adatation constants
52    damp = sqrt(2)/2;                      % standard damping
53    omeg = (dfreq/fs) / (damp+1/(4*damp)); % variable
54    mi1 = (4*damp*omeg) / (1 + 2*damp*omeg + omeg*omeg),  % adapt speed const #1
55    mi2 = (4*omeg*omeg) / (1 + 2*damp*omeg + omeg*omeg),  % adapt speed const #2
56
57    % Costas loop - joint frequency-down conversion and demodulation
58    y = hilbert( y ); % real to analytic signal
59    %y = y .* exp(-j*(2*pi*(fc)*t+pi/3)); fc = 0; % test base-band adjustment only
60    omega = zeros(1,N+1); omega(1) = 2*pi*fc/fs;
61    theta = zeros(1,N+1);
62    for n = 1 : N  % Costas adaptation loop
63        bb(n) = y(n) * exp(-j*theta(n));                 % base-band signal
64        delta(n) = real( bb(n) ) .* imag( bb(n) ) / (bb(n).*conj(bb(n))); % error
65        theta(n+1) = theta(n) + omega(n) + mi1*delta(n); % 1-st update
66        omega(n+1) = omega(n) + mi2*delta(n);            % 2-nd update
67    end
68    cest = cos( theta(1:N) ); % recovered carrier
69    xest = real( bb );        % demodulated signal
70
71    % Figures
72    n=1:N;
73    figure; subplot(211); plot(real(bb)); subplot(212); plot(imag(bb)); pause
74    figure; plot(n,theta(n),'r-');              title('theta(n)');       grid; pause
75    figure; plot(n,ph(n)-theta(n),'r-');        title('ph(n)-theta(n)'); grid; pause
76    figure; plot(n,c(n)-cest(n),'b-');          title('c(n)-cest(n)');   grid; pause
77    figure; plot(n,x(n),'r-',n,xest(n),'b-'); title('x(n), xest(n)');    grid; pause
78    figure; plot(n,x(n)-xest(n),'b-');          title('x(n)-xest(n)');   grid; pause
79    figure; plot(omega*fs/(2*pi));              title('fc(n)');          grid; pause
```

Exercise 19.7 (Testing Costas Loop). Analyze Costas loop program 19.4. Compare it with the PLL loop program 19.3. At present the nominal working frequency of the loop is equal to $f_c = 19,000$ Hz and their bandwidth (locking frequency range) set to $\Delta f = 100$ Hz. Run the program with different values of frequency and phase of the carrier present in signal $y(n)$ (now their values, set in line 44, are equal to 19,025 Hz and $\pi/4$, respectively). Check in figures whether carrier synchronization is correctly performed in all cases. How fast is it done? Check correctness of signal demodulation—recovering of $x(n)$ from $y(n)$. Did you observe the case of finding minimum of $\sin(\alpha - \beta(n))$ for angle difference equal to π? In such situation recovered $x(n)$ has negated sign. Change value of the Costas loop frequency bandwidth Δf (dfreq). Note that after it increase the adaptation constants μ_1 and μ_2 become larger. Observe that bigger values of adaptation coefficients change the Costas loop features: increase, both, convergence speed and frequency variance of the generated cosine in steady-state (after synchronization).

Repeat experiments for different modulated signals (isignal=1,2, 3,4,5). Note that demodulation works also for speech and for two signals simulating digital transmission of bits (a sequence of predefined symbols, in our case short sines/cosines with the same frequency but different phase shifts— binary phase shift keying (BPSK) and quadrature phase shift keying (QPSK)). Finally check whether the Costas loop adaptation is robust to noise (uncomment line 49) and signal delay (uncomment line 50).

Exercise 19.8 (Carrier Recovery and AM-DSB-SC Demodulation Using Double Costas Loop). Using program lab19_ex_am_short.m, presented in Listings 19.1 and 19.2, generate an AM-DSB-SC signal. Add to the program the Costas loop carrier synchronization and signal demodulation, exploiting code from Listing 19.4. Frequencies and phase shifts of cosine frequency-up (transmitter) and frequency-down (receiver) converters should be slightly different but the Costas algorithm should adjust them. Compare original and demodulated signal.

19.8 Carrier Recovery Example: MPX Signal in FM Radio

As an example of application of AM modulation and demodulation with carrier recovery task we will exploit very didactic composition (generation) and decomposition of MPX (multiplex) signal used in FM radio [2]. The signal is summation of four components from which two represent AM-DSB-SC signals (stereo extension and RDS data [7]) that should be demodulated. In order to do this the carrier recovery problem should be solved efficiently in both cases. At present we deal only

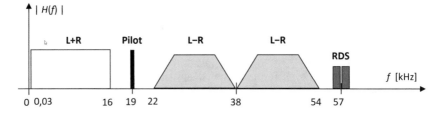

Fig. 19.11: Frequency spectrum of MPX signal used in FM radio. Denotations: L—left channel, R—right channel, RDS—digital radio data system. L-R signal is modulating in amplitude the 38 kHz carrier (doubled pilot) while RDS signal—the 57 kHz (tripled pilot). The AM-DSB-SC modulated signals L-R and RDS have to be recovered from the MPX signal, i.e. demodulated

with AM demodulation of stereo extension signal. In the next two chapters we will learn basics of digital modulation and demodulation on the example of the RDS data transmission: we will be using the same real-world data but experimenting only with the RDS signal.

In the beginning let us introduce the following notation concerning signals used in FM radio:

- $x_1(t)$—first audio channel (left L), bandwidth to 15 kHz;
- $x_2(t)$—second audio channel (right R), bandwidth to 15 kHz;
- $x_m(t) = x_1(t) + x_2(t)$—audio monophonic signal (summation L+R);
- $x_s(t) = x_1(t) - x_2(t)$—stereo extension signal (subtraction L-R);
- $x_{rds}(t)$—digital RDS signal with text (BPSK modulation, 1187.5 bits (and symbols) per second) [7];
- $p(t) = \cos(2\pi f_p t)$—cosine 19 kHz pilot signal used in receiver for synchronization.

Let additionally $\alpha(t)$ denotes an angle of the pilot signal as:

$$\alpha(t) = 2\pi f_p t, \qquad f_p = 19\,\text{kHz}. \tag{19.106}$$

The FM radio multiplex signal, used for modulating the radio station frequency carrier, is defined as:

$$x(t) = 0.9 \cdot x_m(t) + 0.1 \cdot \cos(\alpha(t)) + 0.9 \cdot x_s(t) \cdot \cos(2\alpha(t)) + 0.05 \cdot x_{rds}(t) \cdot \cos(3\alpha(t)). \tag{19.107}$$

Its spectrum is shown in Fig. 19.11. The signal consists of: monophonic signal (L+R), pilot 19 kHz, stereo extension signal (L-R) up-shifted to frequency $2f_p = 38$ kHz (by means of *doubled* pilot), and RDS bits with text up-converted in frequency to $3f_p = 57$ kHz (using *tripled* pilot). The last two components represent AM-DSB-SC modulation, analyzed with details in this chapter: low-frequency signals $x_s(t)$ and $x_{rds}(t)$ are modulating in amplitude high-frequency carriers $\cos(2\alpha(t)) = c_{38}(t)$ and $\cos(3\alpha(t)) = c_{57}(t)$. The pilot signal is added to the

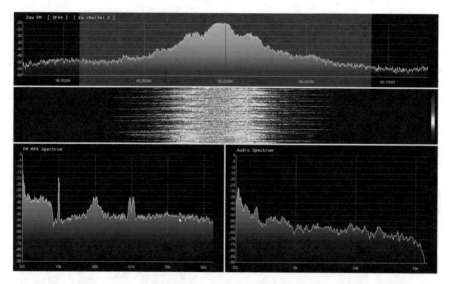

Fig. 19.12: Spectra of recorded IQ file containing only one FM radio station. From top to bottom, from left to right: (1) IQ signal accumulated FFT spectrum, (2) its time-varying STFT spectrogram, (3) decoded MPX signal spectrum (left), (4) decoded audio signal spectrum (right)

MPX since a receiver can reconstruct from its carriers $c_{38}(t)$ and $c_{57}(t)$ which are necessary for back-shifting of the up-converted signals L-R and RDS to 0 Hz.

The RDS bits $b(n)$, sent in RDS signal, are [7]: (1) coded differentially: $d(1) = b(1)$; $k = 2 : N_{rds} : d(k) = |b(k) - d(k-1)|$, (2) replaced with one period of $\sin()$ for $d(k) = 1$ and $-\sin()$ for $d(k) = 0$, both signals lasting 1/16 of the pilot period, (3) low-passed filtered (filtering is aiming at reduction of too wide signal bandwidth caused by $\sin()/-\sin()$ switching). In big simplification, neglecting differential coding and signal low-pass filtering, we can say that RDS signal is a sinusoid with frequency 1187.5 Hz in which values (samples) of some periods are negated. Such signal was generated by us in program 19.4 as `isignal==4` (see lines 23–24 and 32–35) and used for carrier modulation and demodulation using Costas loop. Generation, acquisition, and decoding of RDS data will be discussed in next chapters as an example of digital modulation and bit transmission. At present, we are becoming familiar with the complete MPX signal structure but are interested only in digital decoding of the stereo audio from sampled MPX signal.

Discretized FM radio signals can be acquired by many cheap RTL-SDR devices and ADALM-PLUTO modules using dedicated programs (SDR Sharp or GQRX SDR and ADI IIO). Frequency bandwidth of recorded IQ files is equal to sampling frequency which is used. One FM radio station occupies about 200 kHz. Using sampling frequency 2 MHz, 2.048 MHz, or 3.2 MHz we can have many stations in our recording and for further processing select only one of them as it was presented in programs 17.2 and 17.3. In Fig. 17.9 FFT and STFT spectra, of one IQ recording with many FM radio stations, were shown. Now, in Fig. 19.12 some spectra of different IQ file are presented. In this case the sampling frequency is equal to 250 kHz and only one FM broadcast is captured. In the upper figure an accumulated FFT spectrum of the IQ signal is shown, while in the middle part the IQ signal spectrogram is presented (the sequence of shorter spectra visualizing IQ signal frequency change in time). In bottom, on the left side, the MPX signal spectrum is shown, while on the right side the spectrum of the decoded audio signal. In Fig. 19.13 only spectrum of the MPX signal is shown. We clearly see, from left/lower to right/higher frequencies: (1) the L+R component around 0 Hz, (2) 19 kHz sharp peak of the pilot, (3) symmetrical spectrum of L-R component shifted to 38 kHz, and (4) spectrum of RDS data up-shifted to 57 kHz.

In this section we apply our knowledge, acquired in this chapter, to decode the FM radio stereo signal from recorded IQ data. From last two chapters we already know how to perform frequency demodulation of the IQ signal and to obtain the

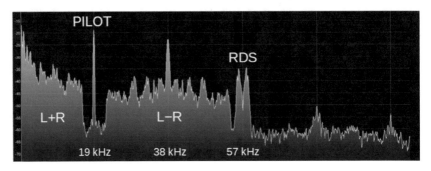

Fig. 19.13: Spectrum of an FM radio MPX signal

MPX hybrid signal. At present we are interested in frequency down-shifting of the L-R signal from 38 kHz to 0 Hz and in recovering L(t) and R(t) audio channels using these equations:

$$L(t) = 0.5 \cdot [x_m(t) + x_s(t)] = 0.5 \cdot [(L+R) + (L-R)], \qquad (19.108)$$
$$R(t) = 0.5 \cdot [x_m(t) - x_s(t)] = 0.5 \cdot [(L+R) - (L-R)]. \qquad (19.109)$$

In the next chapter we will decode the RDS bits, while in the following one-recover text coded with the bits. A nice plan, does it?

In decoding $L(t) - R(t)$ signal two topics are of crucial importance: (1) recovery of the suppressed 38 kHz carrier, (2) careful taking into account delays introduced by all used filters, since left and right audio channels will be recovered correctly from Eqs. (19.108), (19.109) only when components $L+R$ and $L-R$ are delayed by the same time. And processing path of the signal $L-R$ seems to be more difficult and longer.

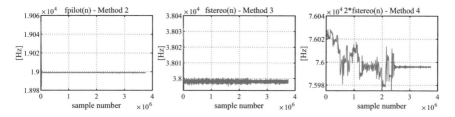

Fig. 19.14: Comparison of frequency adaptation in methods B, C and D

In present example we will perform AM demodulation of stereo extension $x_s(t) = L(t) - R(t) = x_1(t) - x_2(t)$ signal (left minus right audio channel), using four different approaches, presented below from the simplest one to the most difficult:

(A) **Raising filtered pilot 19 kHz to the square:** (1) extracting pilot signal via narrow band-pass filter around 19 kHz, (2) raising result to the square, (3) signal normalization to zero mean and amplitude 1, (4) optional wide band-pass filtration of MPX signal around 38 kHz (extracting $L - R$ signal) or only MPX signal synchronization with recovered carrier 38 kHz, (5) down-shifting the signal $L - R$ using the recovered carrier 38 kHz;

(B) **PLL synchronization with the 19 kHz pilot signal:** (1) extraction of the 19 kHz pilot signal $\cos(\alpha(n))$ via band-pass filtering, (2) synchronization with its angle $\alpha(n)$ using the PLL method, (3) generation of the doubled carrier $\cos(2\alpha(n))$ knowing the angle $\alpha(n)$, and (4) exploiting this signal for the signal $x_s(t)$ down-conversion from 38 Hz to 0 Hz;

(C) **Costas loop** working at frequency 38 kHz upon stereo extension component $x_s(t)$ extracted by the band-pass filter [22, 54] kHz;

(D) **PLL synchronization at 76 kHz with the band-passed filtered $(L(t) - R(t))^2$ signal:** (1) band-pass [22, 54] kHz filtering of stereo extension $L - R$ signal around 38 kHz, (2) raising result to the square, (3) very narrow band-pass signal filtering around 76 kHz, (4) using PLL for synchronization with obtained signal $\cos(4\alpha(n))$ being doubled 38 kHz carrier, (5) generation of the 38 kHz carrier $\cos(2\alpha(n))$ knowing the angle $4\alpha(n)$, and (6) using it for down-conversion of the stereo extension signal $x_s(t)$ to 0 Hz.

All presented above methods are implemented in Matlab program given in Listing 19.5. The program is working on recorded $(I(n), Q(n))$ files containing samples of only one analog FM radio station ($f_s = 192...256$kHz). Values of parameters are declared in the script `fmradio_params.m` included in the program beginning and not shown. Impulses responses of FIR digital filters, used in the program, are designed also in this script. They are denoted using self-descriptive names like `hLPaudio`, `hBP19`, `hBP38`, `hBP76`, . . .—low-pass (LP), band-pass (BP), around 19 kHz, 38 kHz, 76 kHz.

In Fig. 19.14 frequency adaptation in PLL/Costas loops of the methods B, C and D is shown. In the methods B and C the adaptation convergence is comparable and very fast, however in the Costas loop variance of the final frequency estimate is significantly bigger. In the method D the adaptation speed is the slowest one.

Listing 19.5: Matlab program of software decoder of stereo FM radio

```
1
2    % lab19_ex_fmradio_decoder_stereo.m
3    clear all; close all;
4
5    method = 1;        % L-R demodulation method:
6                       % 1=pilot^2, 2=BP(pilot)+PLL, 3=Costas L-R, 4=(L-R)^2 + PLL
7    BP = 1;            % 0/1 - testing necessity of using BP filter in methods #1,2
8    fmradio_params     % file with parameters: frequency values, filter weights, etc.
9
10   [y,fs] = audioread('FMRadio_IQ_250kHz_LR.wav');  % read Goodbye+Danke, L, R, LR
11   y = y(:,1) + sqrt(-1)*y(:,2);                    % [I,Q] --> complex [I+j*Q]
12
13   % DECODING MONO L+R SIGNAL
14   dy = (y(2:end).*conj(y(1:end-1)));      % calculation of instantaneous frequency
15   y = atan2(imag(dy), real(dy)); clear dy; % frequency demodulation
16   ym = filter( hLPaudio, 1, y );          % low-pass filtration of L+R (mono) signal
17   ym = ym(1:fs/faudio:end);               % leaving only every fs/faudio-th sample
18   disp('LISTENING MONO: Left + Right'); sound(ym,faudio); pause % listening to L+R
19   w = ym-mean(ym); w=w/max(abs(w)); audiowrite('FM_mono.wav',w,faudio); clear w;
20
21   % DECODING L-R SIGNAL
22   if(method==1) % pilot raised to the square and normalized
23     p = filter(hBP19,1,y);                        % band-pass filtering around 19 kHz
24     p = p.^2;                                     % raising to the square
25     p = p - mean(p(3*L+1:23*L)); T=fs/fstereo;    % mean subtraction
26     eng = sum( p(3*L+1:3*L+1+round(100*T)-1).^2 ) / round(100*T); A = 2*sqrt(eng/2);
27     c38 = p/A;                                    % normalization to -1/+1
28     if(BP==1) y38 = filter(hBP38,1,y);            % optional band-pass filtering
29     else      y38=[ zeros(L/2,1); y(1:end-L/2) ]; % only delay
30     end;      offs=L/2;                           % introduced offset
31     ys = 2*y38.*c38;                              % frequency-down conversion
32     ys = filter(hLPaudio,1,ys);                   % low-pass filtering
33   end
34   if(method==2) % pilot + PLL
35     p = filter(hBP19,1,y);                              % extracting 19 kHz pilot
36     theta = zeros(1,length(p)+1); omega = theta; omega(1) = 2*pi*(fpilot+50)/fs;
37     mi1 = 0.0011; mi2 = 5.6859e-07; p=p/max(abs(p));    %#
38     for n = 1 : length(p)                               %#
39         pherr = -p(n)*sin(theta(n));                    %# double PLL on 19 kHz
40         theta(n+1) = theta(n) + omega(n) + mi1*pherr;   %#
41         omega(n+1) = omega(n) + mi2*pherr;              %#
42     end                                                 %#
43     c38(:,1) = cos(2*theta(1:end-1));                   %# carrier 38 kHz
44     if(BP==1) y38 = filter(hBP38,1,y);                  % band-pass filtering 38 kHz
45     else      y38=[ zeros(L/2,1); y(1:end-L/2) ];       % or delay
46     end;      offs = L/2;                               % introduced offset
47     ys = 2*real(y38.*c38); clear p y38 c38 theta;       % 38 kHz --> 0 kHz
48     ys = filter( hLPaudio, 1, ys );                     % low-pass filtration of L-R
49     figure; plot(fs*omega/(2*pi)); title('fpilot(n)'); pause; clear omega;
50   end
51   if(method==3) % Costas loop working on 38 kHz
52     y38 = filter(hBP38,1,y);  offs=0;                   % band-pass filtering 38 kHz
```

```
53      y38 = hilbert( y38 );                          % analytic signal
54      theta = zeros(1,length(y38)+1); omega=theta; omega(1)=2*pi*(2*fpilot+25)/fs;
55      mi1 = 0.0011; mi2 = 5.6859e-07;                %#
56      for n = 1 : length(y)                          %# Costas loop
57          bb(n) = y38(n) * exp(-j*theta(n));         %# base-band signal
58          delta(n) = real( bb(n) ).*imag( bb(n) ) / (bb(n).*conj(bb(n))); % error
59          theta(n+1) = theta(n) + omega(n) + mi1*delta(n);   %# 1-st update
60          omega(n+1) = omega(n) + mi2*delta(n);      %# 2-nd update
61      end                                            %#
62      ys = real( bb ); ys = ys'; clear bb delta theta    %# frequency-down L-R
63      figure; plot(fs*omega/(2*pi)); title('fstereo(n)'); pause; clear omega
64   end
65   if(method==4)  % (L-R)^2 + PLL
66      y38 = filter(hBP38,1,y); offs = L/2; % extracting L-R signal, band-pass around 38kHz
67      y76 = y38.^2;                          % raising to the square
68      y76 = filter(hBP76,1,y76);             % narrow band-pass filtering around 76kHz
69      theta = zeros(1,length(y76)+1); omega = theta; pherr=theta; omega(1) = 2*pi*(4*fpilot
               +25)/fs;
70      mi1 = 5*0.0028; mi2 = 5*3.8527e-06; y76=y76/max(abs(y76));   %#
71      for n = 1 : length(y76)                        %#
72          pherr(n) = -y76(n)*sin(theta(n));          %# double PLL
73          theta(n+1) = theta(n) + omega(n) + mi1*pherr(n);   %# on 76 kHz
74          omega(n+1) = omega(n) + mi2*pherr(n);      %#
75      end                                            %#
76      c38 = cos( theta(1:end-1)/2 );                 %# 76 kHz --> 38 kHz
77      c38 = [c38(L/2+1:end)'; zeros(L/2,1)]; clear pherr theta  % delay compensation
78      ys = 2*real(y38 .* c38); % freq-down, two choices: +/-2 (sin(a)=0 for a=0, +pi)
79      ys = filter( hLPaudio, 1, ys ); clear y38 c38; % low-pass filtration of L-R signal
80      figure; plot(fs*omega/(2*pi)); title('fc(n)'); pause; clear omega
81   end
82   ys = ys(1:fs/faudio:end); % leaving every fs/faudio-th sample of (L-R)
83
84   % DECODING L,R FROM (L+R) AND (L_R)
85   % Time synchronization of L+R and L-R signals (taking into account delay of L-R)
86   delay = offs/(fs/faudio); ym = ym(1:end-delay); ys=ys(1+delay:end);
87   % Recovering L and R channel
88   n=1:min(length(ym),length(ys));
89   y1 = 0.5*( ym(n) + ys(n) ); y2 = 0.5*( ym(n) - ys(n) ); clear ym ys;
90   figure; subplot(211); plot(y1); title('y1'); subplot(212); plot(y2); title('y2'); pause
91   % De-emphasis: flat freq-response to 2.1 kHz, then decreasing 20 dB / decade
92   % y1 = filter(b_de,a_de,y1); y2 = filter(b_de,a_de,y2);
93   % Listening to L and R channel separately
94   z = zeros(length(y1),1);
95   disp('LISTENING: Left channel'); sound([y1 z ],faudio);
96   figure; plot(y1); title('L'); pause;
97   disp('LISTENING: Right channel'); sound([z y2],faudio);
98   figure; plot(y2); title('R'); pause
99   disp('LISTENING: Stereo'); sound([y1 y2],faudio); pause
00   % Writing stereo signal to disc
01   maxi = max( max(abs(y1)),max(abs(y2)) );
02   audiowrite('FM_Radio_stereo.wav',  [ y1/maxi y2/maxi ], faudio ); % clear y1 y2;
```

Exercise 19.9 (Testing Software Receiver of Stereo FM Radio). Analyze code of the program 19.5. Run it for each implemented demodulation method of the stereo extension signal $L - R$ set (`method=1,2,3,4`). Observe figures presenting trajectories of carrier adaptation if PLL/Costas loop is applied. Initial values of loop frequencies are deliberately set apart from expected ones to observe speed of adaptation process convergence. Observe left and right channel of decoded audio. Listen to it. In the beginning process WAV file `FMRadio_IQ_250kHz_LR.wav`. In left channel the word *Goodbye* is stored, while in the right the word *Danke*. Note that after decoding each word is in correct channel. Then exchange letters `LR` in the IQ file name to one letter `L` (only *Goodbye*), and next to one letter `R` (only *Danke*). After decoding we should have a very weak signal in the *empty* channel. Otherwise we have a cross-talk between channels resulting from wrong frequency down-conversion of the signal $(L - R)$ or wrong synchronization of the signals $L + R$ and $L - R$. Next, change decoded signal to `FMRadio_IQ_250kHz_Calibrate.wav`. Observe how fast PLL and Costas loops adapt for it. Finally, change the input IQ file to real radio recording `SDRSharp_FMRadio_96000kHz_IQ_one.wav`, done by RTLSDR dongle. Note how long the adaptive loops converge to the correct frequency value and observe fluctuations around this value in the steady-state.

Exercise 19.10 (Understanding DSP Operations in Software FM Radio Receiver). Modify code of the program 19.5. After each important DSP operation plot a signal fragment of the processing result and its spectrogram using the following commands:
```
n=n1st:nlast; plot(y(n)); pause
spectrogram(y(n),256,240,1024,fs,'yaxis'); pause
```
Try to understand *what is going on!*.

19.9 Summary

In this chapter the amplitude modulation and demodulation basics were presented. It looks at a first glance that the problem is very simple and solutions are well-known. However I spent relatively long time writing the chapter. Why does it happen? Because there are many modulation variations, many demodulation methods dedicated to different types of modulating signals and different carrier recovery techniques. What should be remembered?

1. In amplitude modulation a signal which is transmitted (coded) is changing amplitude of a carrier signal. Two signals are multiplied: a modulating signal $x(t)$ with information (a content) and a modulated signal $c(t)$ being a transmission medium (a container): $y(t) = x(t) \cdot c(t)$.

2. Signal $x(t)$ can be real-value or complex-value. In turn, pure cosine $\cos(2\pi f_c t)$ or complex harmonic signal $e^{j2\pi f_c t}$ can be used as a carrier. If the multiplication result is complex, its real part is only taken, simplifying.

3. If the signal $x(t)$ is real-value and the carrier is a cosine, the most popular AM with double side-bands results (AM-DSB). Why double side-bands? Because spectrum of the real-value signal has complex-conjugated symmetry around 0 Hz. Multiplication by the cosine shifts this spectrum to frequencies f_c and $-f_c$, i.e. two shifted copies of it are obtained, each with double side-bands resulting from the signal spectrum symmetry. Since the carrier component alone is absent in the signal and in the spectrum, the modulation is called as suppressed carrier one: AM-DSB-SC.

4. If the carrier $c(t)$ is multiplied not by $x(t)$ but by $(1 + \delta \cdot x(t)), 0 < \delta < 1$, in the AM modulated signal the carrier is present explicitly: $y(t) = c(t) + \delta x(t) c(t)$ (the first term). The corresponding component is present also in the modulated signal spectrum. In this case AM-DSB with large carrier is get.

5. Amplitude modulation is used, for example, in AM radio, amateur radio, multiplex signal of FM radio, up and down frequency conversion in wireless communication and digital bit transmission in the form of quadrature amplitude modulation (QAM). The QAM technique is exploited in ADSL, DVB-T, DAB, LTE, ...

6. Frequency and phase synchronization of receiver down-shifting cosine with transmitter up-shifting cosine is the most difficult part of AM technique implementation. In literature many possible approaches are proposed, most of them are strictly dedicated to the modulation signal type.

7. Typically, phase locked loops (PLL) and Costas loops are used for carrier synchronization. In case of Costas loop, the carrier synchronization is combined with frequency down-shifting (demodulation). Both loops are adaptive, they can synchronize only up and down carrier phases or, both, phases and frequencies.

8. In PLL and Costas approaches special *filter loops* are used that allow, either, faster tracking of phase/frequency changes or smaller variation of phase/frequency estimates in signal steady-states. Optimal values of loop filter parameters are derived from theoretical stability criteria of discrete-time recursive systems.

9. In this chapter we used the MPX signal of FM radio for testing different existing possibilities of carrier recovery in software receivers of discretized AM modulated signals.

19.10 Private Investigations: Free-Style Bungee Jumps

Exercise 19.11 (More MUSIC, More FUN!). Using program 17.2 you can read into Matlab a wide-band IQ file with many FM radio station and select for further processing only one them. Modify the program 19.5 and add to it this nice feature also. Test result of your work on the IQ recording `SDRSharp_FMRadio_101600kHz_IQ.wav`. In fact you could decode many FM broadcasts in parallel, individually for each of your family member. Even do their streaming in your home network.

Exercise 19.12 (My Favorite Song Coded as IQ FM Radio Broadcast). Short description of the FM radio MPX signal was given in this chapter. But if you are more interested in this topic, please, analyze code of the program `fmradio_encoder_stereo_RDS.m` given in the archive. You can skip generation of the RDS signal, we will analyze it later. At present concentrate only on the synthesis of the IQ FM radio-like signal with your favorite song. Store the IQ data to disc. Check whether program from Listing 19.5 allows it decoding. If not, because of some frequency differences, do appropriate program extensions. If you have the ADALM-PLUTO module you can even broadcast your *Ode to Joy* in real life. But, of course, using only non-restricted frequency band!

Exercise 19.13 (0,0,0,1,1,1: Does Anybody Hear Me!). In the Costas loop program 19.4 we have generated two digital signals, BPSK and QPSK, being very similar to the RDS signal used in the FM radio. Write your name as a sequence of 8-bit ASCII codes of letters used. Combine all bits together, one-by-one. Code each bit using one period of 1 kHz sine: "1" as a sine and "0" as minus sine. You will have 1000 symbols (bits) per second. Combine all $(+)(-)$sines together, create one long signal. Repeat this signal a few times. Now take the program `fmradio_encoder_stereo_RDS.m` and exchange the RDS code fragment with your own program. Use signal coding your name for modulating in amplitude the RDS 57 kHz carrier. Store result as a new IQ FM radio file. Now extend the program 19.5: add the special Costas loop module for demodulation of the 57 kHz carrier. Observe the shape of the decoded signal. Do you see flip-flopping $(+)(-)$sines. Could you decode consecutive bits? Could you combine bits into letters? Could you decode your name? If you do, the game is over: you pass the examination with very good result!

Exercise 19.14 (SDRSharp Likes GQRX and Vice Versa). Install SDRSharp (under Windows) or GQRX (under Linux) program and *run* supported IQ FM Radio WAV files by them. Change `View` options in `FFT Display plug-in`. Activate/enable `IF/MPX/Audio` options in `Zoom FFT` plug-in.

Exercise 19.15 (RTLSDR or ADALM PLUTO). If you have SDR hardware, please, do some your own recordings of FM radios and try to decode different services using our Matlab programs.

References

1. J.B. Anderson, *Digital Transmission Engineering* (IEEE Press/Wiley-Interscience, Piscataway NJ, 2005)
2. K. Barlee, D. Atkinson, R.W. Stewart, L. Crockett, *Software Defined Radio using MATLAB & Simulink and the RTLSDR*. Matlab Courseware (Strathclyde Academic Media, Glasgow, 2015). Online: http://www.desktopsdr.com/
3. T.F. Collins, R. Getz, D. Pu, A.M. Wyglinski, *Software Defined Radio for Engineers* (Analog Devices - Artech House, London, 2018)
4. B. Farhang-Boroujeny, *Signal Processing Techniques for Software Radios* (Lulu Publishing House, 2008)
5. M.E. Frerking, *Digital Signal Processing in Communication Systems* (Springer, New York, 1994)
   bibitemIEC99 IEC-62106 Standard, Specification of the radio data system (RDS) for VHF/FM sound broadcasting in the frequency range from 87,5 MHz to 108,0 MHz. International Electrotechnical Commission, 1999
6. C.R. Johnson Jr, W.A. Sethares, A.G. Klein, *Software Receiver Design. Build Your Own Digital Communication System in Five Easy Steps* (Cambridge University Press, Cambridge, 2011)
7. F. Ling, *Synchronization in Digital Communication Systems* (Cambridge University Press, Cambridge, 2017)
8. L.C. Potter, Y. Yang, *A Digital Communication Laboratory - Implementing a Software-Defined Acoustic Modem*. Matlab Courseware (The Ohio State University, Columbus, 2015). Online: https://www.mathworks.com/academia/courseware/digital-communication-laboratory.html
9. M. Rice, *Digital Communications: A Discrete-Time Approach* (Pearson Education, Upper Saddle River, 2009)
10. S.A. Tretter, *Communication System Design Using DSP Algorithms* (Springer Science+Business Media, New York, 2008)

Chapter 20
Introduction to Single-Carrier Digital Modulation Systems

2+2=4 and no more: Bits transmitted by a carrier signal in each moment are equal to the carrier state number, written binary.

20.1 Introduction

Digital information is sent using analog signals. Wow! The main difference between analog signal carrying analog content and analog signal with digital content results from the fact that, in the first case, an analog signal value, modulating an analog carrier, lies in some range but is completely arbitrary (for example, unpredictable speech or music), while in the second case the modulating analog signal can take only a finite number of states, called symbols, i.e. it is synchronously switched from one state to another state. For example, the modulating signal can be only a constant value equal to A_0 (bit 0) or A_1 (bit 1) (amplitude information, 2-PAM Pulse Amplitude Modulation), a fragment of a sine (bit 1) or minus sines (bit 0) (phase shift information, BPSK Binary Phase Shift Keying), or can have only frequency f_0 (bit 0) or f_1 (bit 1) (frequency information, BFSK Binary Frequency Shift Keying). In order to avoid abrupt signal changes, which increase width of signal spectrum, smooth transition between modulating signal states is required, for example, in GMSK (Gaussian Minimum Shift Keying) signal frequency is changed from one value to the other using the Gaussian function trajectory.

Typically, people are thinking that bits are transmitted one at a time, as in the example given above. This is not always true. When modulating signals can take more states than 2, for example, $X = 4, 8, 16, 64, 256, \ldots$ states, we are obtaining X-level-PAM/PSK/FSK modulations with X states having different amplitude, phase, or frequency value. We can increase number of states allowing a signal to change the same time two of its parameters, for example, both amplitude and phase, obtaining joint amplitude-phase modulation, known as

© Springer Nature Switzerland AG 2021
T. P. Zieliński, *Starting Digital Signal Processing in Telecommunication Engineering*, Textbooks in Telecommunication Engineering, https://doi.org/10.1007/978-3-030-49256-4_20

quadrature amplitude modulation (QAM): 4-QAM (2 bits), 16-QAM (4 bits), 64-QAM (6 bits), 256-QAM (8 bits), ... In this case one fragment of analog signal, modulated by the other analog signal, is carrying at once 2, 4, 6, 8, ... bits, respectively. In telephone and cable modems one analog carrier signal with some chosen frequency can be modulated by some other analog signal taking even $2^{15} = 32{,}768$ different states (32768-QAM) and carrying at once 15 bits. Of course, the *only* problem is in distinguishing these states in the receiver. When the states are numerous, their amplitudes, phases, and frequencies have values which are lying very close to each other and any transmission disturbance (e.g. noise, signal amplitude, and phase change, caused by the communication channel, cross-talk from neighboring frequencies/services, frequency drift due to the Doppler effect) can change them and stop our dream on transmission of gigabits per second.

In this chapter we will talk only about data transmission using a single carrier having one frequency, like in digital *old-fashioned* digital GSM telephony (Global System for Mobile Communications). But bits can be sent also using many frequency carriers transmitted in parallel (DMT—Discrete Multi-Tone, OFDM—Orthogonal Frequency Division Multiplexing) which increases achievable bit-rates. Such approaches are used nowadays in modern telecommunication standards of digital audio broadcasting (DAB), terrestrial digital video broadcasting (DVB-T), digital 4G (LTE) and 5G (NR) telephony, high-speed digital subscriber line (xDSL) cable and telephone modems, and many others. The multi-carrier transmission technique will be studied by us in last chapters of the book.

Now we learn single-carrier transmission basics. We will code an exemplary text using carrier state numbers of an arbitrary amplitude-phase modulation. We will up-convert the signal to the target frequency, simulate it passing through a transmission channel, down-convert it to the base-band and decode. After each operation we will observe signals, their eye-diagrams and phasor plots, as well as signal spectra, and carefully track their changes. We will identify sources of transmission problems—we will try to solve them in the following chapter on digital single-carrier receivers.

In this chapter digital frequency modulation, frequency shift keying, is not discussed. However ... Since instantaneous frequency is defined as a derivative of sine/cosine (or complex-value harmonic) angle divided by 2π, the presented here digital, quadrature, amplitude-and-phase signal (de)modulation can be also exploited with success for digital frequency (de)modulation. In this case, amplitude is set to one and carrier phase is changed only, for example, linearly and flipped-flopped digitally (derivative of changing signal phase corresponds to changing signal frequency).

These books present in a student-friendly way introduction to the contemporary software-based single-carrier digital transmission systems [2–8, 10, 11, 13, 19].

20.2 Basics of Single-Carrier Modulation

Let us concentrate now on a joint, digital, quadrature amplitude and phase modulation (QAM) of a single carrier. Such technique is very frequently used. A single cosine is used in it but its amplitude and phase shift are changed in regular time intervals. Number of carrier states is limited. Knowing the actual carrier state number, we can write it in binary positional system and recover bits transmitted at that moment. For example, carrier state number 9 out of 16 states corresponds to sending bits 1001. Typically, most people imagine that amplitude and phase of the cosine carrier are constant in predefined intervals (of symbol duration) and a carrier signal is composed of such smaller pieces witch are connected together:

$$y_{cut}(n) = [\dots, y_{k-1}(n), y_k(n), y_{k+1}(n), \dots].   \qquad (20.1)$$

The k-th carrier state is distinguished by unique pair of amplitude A_k and phase shift ϕ_k values ($\cos(\alpha + \beta) = \cos(\alpha)\cos(\beta) - \sin(\alpha)\sin(\beta)$):

$$y_k(n) = A_k \cos\left(2\pi \frac{f_c}{f_s} n + \phi_k\right) =   \qquad (20.2)$$

$$= A_k \cos(\phi_k)\cos\left(2\pi \frac{f_c}{f_s} n\right) - A_k \sin(\phi_k)\sin\left(2\pi \frac{f_c}{f_s} n\right) =   \qquad (20.3)$$

$$= I_k \cos\left(2\pi \frac{f_c}{f_s} n\right) + Q_k \sin\left(2\pi \frac{f_c}{f_s} n\right),   \qquad (20.4)$$

where in-phase I_k and in-quadrature Q_k components are equal to:

$$I_k = A_k \cos(\phi_k),   \qquad (20.5)$$
$$Q_k = -A_k \sin(\phi_k).   \qquad (20.6)$$

Such explanation is correct only in half. A signal generated this way has a very wide frequency spectrum around frequency f_c in the carrier *cut* positions, which disturbs services using carriers with neighboring frequencies. In order to limit this possible frequency cross-talk from one service to another, the built *cut-and-past* sequence of symbols should be next band-pass filtered by the so-called channel filter reducing the signal spectrum width around frequency f_c:

$$y(n) = \mathrm{BPF}_{(fc)}[y_{cut}(n)].   \qquad (20.7)$$

Digital carrier modulation is most often realized in alternative way by interpolation of sequence of carrier amplitudes $\{A_k\}$ and phase shifts $\{\phi_k\}$ (20.2) or, in consequence, by interpolation of sequences of $\{I_k\}, \{Q_k\}$ components (20.4). In this approach typical algorithm of an amplitude-phase digital carrier modulator has the following form:

1. carrier states, specified by amplitudes A_k and phases ϕ_k, are coded using (I_k, Q_k) values, for example:

$$\left(A_k = 1, \ \phi_k = -\frac{\pi}{4} \right): \qquad = 1e^{-j\frac{\pi}{4}} \quad = \frac{1}{\sqrt{2}} - j \cdot \frac{1}{\sqrt{2}} = I_k + jQ_k,$$

$$\left(A_{k+1} = 2, \ \phi_{k+1} = \frac{3\pi}{4} \right): \qquad = 2e^{j\frac{3\pi}{4}} \quad = \frac{-2}{\sqrt{2}} + j \cdot \frac{2}{\sqrt{2}} = I_{k+1} + jQ_{k+1};$$

2. knowing signal sampling period Δt and carrier state (*symbol*) duration T, number of samples per one state/symbol is calculated: $N = \frac{T}{\Delta t}$;
3. then $N-1$ zeros are inserted (appended) between each two values I_k and I_{k+1} of the in-phase component, in our example assuming $N = 4$:

$$I_0(n) = \left[\ldots, 0, 0, 0, \frac{1}{\sqrt{2}}, 0, 0, 0, \frac{-2}{\sqrt{2}}, 0, 0, 0, \ldots \right],$$

and between each two values Q_k and Q_{k+1} of the in-quadrature component, in our example:

$$Q_0(n) = \left[\ldots, 0, 0, 0, \frac{-1}{\sqrt{2}}, 0, 0, 0, \frac{2}{\sqrt{2}}, 0, 0, 0, \ldots \right];$$

4. next, signals $I_0(n)$ and $Q_0(n)$ are filtered separately using a low-pass, interpolating, sinc-like filter, called a pulse shaping filter (PSF); smoothed, up-sampled signals $I(n)$ and $Q(n)$ are obtained this way ($a, b, c, \ldots$ denotes interpolated values):

$$I(n) = \mathrm{PSF}[I_0(n)] = \left[\ldots, a, b, c, \frac{1}{\sqrt{2}}, d, e, f, \frac{-2}{\sqrt{2}}, g, h, i, \ldots \right],$$

$$Q(n) = \mathrm{PSF}[Q_0(n)] = \left[\ldots, k, l, m, \frac{-1}{\sqrt{2}}, n, o, p, \frac{2}{\sqrt{2}}, q, r, s, \ldots \right];$$

5. finally, interpolated signals $I(n)$ and $Q(n)$ are used for quadrature amplitude-phase carrier modulation as described in Eq. (19.20)(19.23) in the previous chapter:

$$y(n) = I(n) \cdot \cos\left(2\pi \frac{f_c}{f_s} n\right) - Q(n) \cdot \sin\left(2\pi \frac{f_c}{f_s} n\right), \tag{20.8}$$

$$= \mathrm{Re}\left[\underbrace{(I(n) + j \cdot Q(n)) \cdot e^{j2\pi \frac{f_c}{f_s} n}}_{z(n)}\right]. \tag{20.9}$$

This way information about transmitted bits is up-converted in frequency by a digital modulator.

In a digital carrier demodulator, first, a complex-value signal $z(n)$ is recovered from $y(n)$ by means of Hilbert filter ($z(n) = y(n) + jH[y(n)]$), then frequency down-conversion of information signal is done, and, finally, the same low-pass pulse shaping filter (*matched filter*) is used (see Eqs. (19.34), (19.36) and (17.14)) for removing unwanted high-frequency signal components:

$$\hat{I}(n) + j \cdot \hat{Q}(n) = \mathrm{PSF}\left[\underbrace{(y(n) + j\,\mathrm{Hilbert}[y(n)])}_{z(n)} \cdot e^{-j2\pi \frac{f_c}{f_s} n}\right] = \tag{20.10}$$

$$= \mathrm{PSF}\left[2y(n)\cos\left(2\pi \frac{f_c}{f_s} n\right)\right] - j \cdot \mathrm{PSF}\left[2y(n)\sin\left(2\pi \frac{f_c}{f_s} n\right)\right] \tag{20.11}$$

Finally, the carrier amplitude $A(n)$ and phase $\phi(n)$ are reconstructed using the following equations:

$$\hat{A}(n) = \sqrt{\hat{I}^2(n) + \hat{Q}^2(n)}, \tag{20.12}$$

$$\hat{\phi}(n) = \mathrm{atan}\left(\frac{-\hat{Q}(n)}{\hat{I}(n)}\right). \tag{20.13}$$

From instantaneous values $(\hat{A}(n), \hat{\phi}(n))$ carrier state estimates $(\hat{A}_k, \hat{\phi}_k)$ are found and carrier state numbers are deduced, for example, state number 9 out of 16 states. State numbers written binary give us transmitted bits: 9 corresponds to bits 1001 in our example.

In digital communication a term constellation diagram is used for naming a set of all possible carrier states. These states are marked as points in polar *radius-angle* $(\hat{A}(k), \hat{\phi}(k))$ coordinate system or in rectangular x-y (Cartesian) *real-imaginary* $(\hat{I}(k), \hat{Q}(k))$ coordinate system. They are like stars in the sky: we would like to see them as separated points in constellations consisting of many points. If a Reader prefers to see a graphical example of carrier states constellation diagrams *just now!*, looking at Figs. 20.5 and 20.6 is recommended.

Synthesis of signals $I(n)$ and $Q(n)$, used for generation of the cosine carrier, and carrier digital modulation by $I(n)/Q(n)$ components are not difficult. In contrary to the demodulation, in which the carrier recovery (estimation of carrier frequency and phase) should be performed using any method presented in the chapter on AM (de)modulation. For example, by means of the PLL or the Costas loop. After this, signals $\hat{I}(n)$ and $\hat{Q}(n)$ should be recovered. In this chapter we will assume perfect knowledge of carrier parameters in the receiver. We will show only consequences of wrong carrier usage. Carrier recovery methods are discussed in the next chapter.

In case of digital transmission, we are not interested in recovery of all $(I(n), Q(n))$ samples. Especially these lying between original samples of carrier states (I_k, Q_k) added by interpolation. We want to extract only the carrier states (I_k, Q_k), their numbers and their binary representations. Therefore we should synchronize with positions of the samples (I_k, Q_k). This operation in telecommunication language is called timing recovery. In this chapter we assume perfect symbol synchronization in the receiver. We will see only consequences of wrong symbol sampling. In the next chapter we learn three timing recovery methods: early-late-gate, Gardner, and Mueller–Müller.

Signal passing through a telecommunication channel is attenuated and delayed, therefore influence of channel disturbance has to be estimated and corrected in the receiver. Without these, calculated carrier states numbers and bits extracted from them could be erroneous. In order to minimize channel influence, bits can be coded not in carrier states but in their changes, for example, in phase shift difference between two consecutive carrier states. Differential coding methods help us to cope with slow unknown carrier modifications caused by the channel but at the price of higher noise sensitivity and lower bit through put, due to the fact that carrier difference depends on two noisy measurements, not on only one.

20.3 Basics of Single-Carrier Transmission Systems

In the previous section we have described concept of the digital modulation of a single signal, the carrier. Now, let look at the transmission problem from the telecommunication system perspective: a stream of bits is divided into smaller blocks, consisting of 2, 3, 4 ... bits each. Next, resulting binary numbers are interpreted as numbers of different amplitude/phase (A_k, ϕ_k) carrier states, characterizing different in-phase I_k and in-quadrature Q_k components. Then, the I_k, Q_k components are interpolated using a pulse shaping filter (PSF)—thanks to this the carrier is smoothly passing from one state to the other. The synthesized digital, modulated carrier is converted to analog signal (D/A) and transmitted through the telecommunication channel. Essential in this aspect are Eqs. (20.2)–(20.4), illustrated in the upper part of the Fig. 20.1. In the receiver, signal of our service is separated from other services by a band-pass filter, sampled by analog-to-digital (A/D) converter and demodulated. Obtained signals $I(n)$ and $Q(n)$ have two components: a low-frequency one (demodulated) and a high-frequency one (unwanted, with doubled frequency). The former have to be removed. For this purpose the same low-pass pulse shaping (*matched*) filter is used, as in the transmitter. Then, signal is down-sampled, carrier states I_k, Q_k are recovered and transmitted bits are found. The receiver part is summarized in lower part of Fig. 20.1.

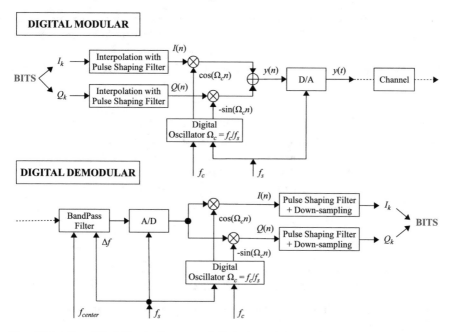

Fig. 20.1: Simplified block diagram of digital modulator (frequency up converter) and demodulator (frequency down converter) used in telecommunication systems

Well, *bits are recovered*. But how a bit-stream is formatted? General bit-stream creation is presented in Fig. 20.2. First, information to be sent should be coded and compressed (**source coding** and compression). In this place different coding&compression standards are used, depending on information type: text, speech, audio, images, video, ... Then, typically, some extra redundant bits are added (**channel coding,** forward error correction), allowing data error detection and eventual error correction in the receiver. Many different error correction codes exist, to mention only the cyclic redundancy check (CRC), low-density parity check (LDPC), polar and Reed–Solomon codes. Finally, in order to find beginning of our data in the bit-stream (**frame synchronization**), some synchronization bit pattern/stamp is put in front of our data. It is called a header or a preamble. Structure of an exemplary transmitted packet of bits is shown in Fig. 20.3. The header could be also protected against errors. In the receiver data are processed in reverse order: first the header is found, then data are extracted.

The header is used for many purposes. First, for finding the bit packet beginning in the bit-stream (frame synchronization). Carrier states of the header should represent a sequence of state numbers having very sharp auto-correlation function, typical for random signals. This feature allows robust header detection in the presence of noise. Different method of pseudo-random numbers (codes) generation can be used, only to mention Barker and Gold codes and maximum length sequences (MLS). Since the header bits are known, associated with them $I(n), Q(n)$ sequences are known also. Therefore, the receiver can use them for finding the packet/frame

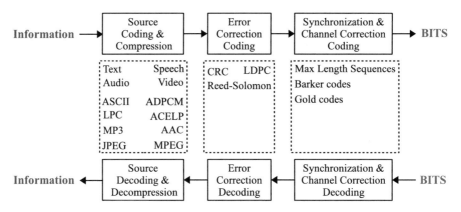

Fig. 20.2: Simplified block diagram of bit-stream generation in digital telecommunication systems

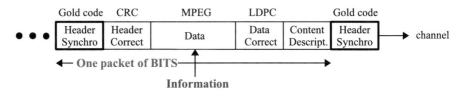

Fig. 20.3: Simplified diagram of one digital packet of transmitted bits

beginning: it calculates a cross-correlation function between the received $IQ(n)$ signal and the header $IQ(n)$ signal and looks for the function maximum. When the *sent* header signal is matched with the *received* header signal, not only the frame position is found but also positions of all header symbols are localized. Therefore, thanks to the header usage, apart from the frame synchronization, a **symbol synchronization** is realized also. Adjusting moment of carrier sampling to correct symbol position in constellation diagram of carrier states is called a symbol timing recovery.

When the header position is found, and channel input and corresponding output are known, the receiver can estimate a channel amplitude/phase influence (**channel estimation**) and remove it from the received signal (**channel equalization and correction**). Additionally, the header can be used also for estimation of frequency and phase errors of frequency down-converter, used in the receiver (**carrier synchronization, carrier recovery**), as well as for estimation of an A/D frequency sampling error (manifesting as carrier frequency offset). Ufff ... Our sweet header. So small but so important.

In the second part of the header some system information can be transmitted telling how the receiver should interpret the data content part, e.g. data type, compression standard used, etc.

20.4 Source Coding

In previous chapters on speech, audio, and images we have learned fundamental of multimedia coding. There are many different standards. (A)CELP voice coders (vocoders) are very popular for speech, MP2 and MP3 as well as AAC coders are frequently used for audio, while JPEG and MPEG standards are used for image and video compression. We will not discuss further this topic now.

One of the simplest examples of source coding is text description using ASCII character tables. We will use it in our demonstration programs in this chapter. Let us assume that we would like to transmit the following, well-known text:

Hello World!

In computer systems letters are written in ASCII code. In this code our "Hello World!" message is equal (original, then written decimally, hexadecimally, and binary):

Text	H	e	l	l	o		W	o	r	l	d	!
Dec	72	101	108	108	111	32	87	111	114	108	100	33
Hex	48	65	6C	6C	6F	20	57	6F	72	6C	64	21
Bin	0100 1000	0110 0101	0110 1100	0110 1100	0110 1111	0010 0000	0101 0111	0110 1111	0111 0010	0110 1100	0110 0100	0010 0001

In consequence, a stream of bits to be send is equal (only its beginning):

Text	H	e	l	l	o	
Bin	01001000	01100101	01101100	01101100	01101111	00100000

At present decision should be made how many bits will be transmitted together using one carrier state. It depends how many carrier states are available in the code-book. Choosing one state from the code-book having two states only, we can sent only 1 bit at any moment, e.g. choice of the first carrier state corresponds to transmitting bit 0, while choice of the second—to transmitting bit 1. If the code-book consists of 4 different states, we are choosing 1 state out of 4 possible states (0, 1, 2 or 3) and we can sent 2 bits at once (00, 01, 10 or 11). For 8 state code-book, the states have numbers {0, 1, 2, 3, 4, 5, 6, 7} and transmitting bits are equal {000, 001, 010, 011, 100, 101, 110 or 111}. Generally, more carrier states the code-book has, more bits are transmitted at once by a single carrier. Concluding, original bit-stream (b) in our example can be divided into many different ways:

(b) :   01001000011001010110110001101100011011110010000...

(8) :   01001000, 01100101, 01101100, 01101100, 01101111, 00100000,...

(6) :   010010, 000110, 010101, 101100, 011011, 000110, 111100, 100000,...

(4) :   0100, 1000, 0110, 0101, 0110, 1100, 0110, 1100, 0110, 1111, 0010,...

(3) :   010, 010, 000, 110, 010, 101, 101, 100, 011, 011, 000, 110, 111, 100,...

(2) :   01, 00, 10, 00, 01, 10, 01, 01, 01, 10, 11, 00, 01, 10, 11, 00, 01, 10, 11,...

(1) :   0, 1, 0, 0, 1, 0, 0, 0, 0, 1, 1, 0, 0, 1, 0, 1, 0, 1, 1, 0, 1, 1, 0, 0, 0, 1, 1, 0, 1, 1,...

e.g. into packets of 8 bits, 6 bits, 4 bits, 3 bits, 2 bits, and 1 bit. Having concrete packet of bits, one should go to the code-book, read parameters of associated carrier state, and synthesize the carrier using given modulation parameters, for example, having bits 0101 (binary) we should synthesize the carrier in its state number 5 (decimally) out of 16 defined in the code-book. Any sequence of bits can be divided into smaller, regular pieces of bits as described above.

Matlab is a computer language for number crunching, not for bit manipulation, therefore a low-level information processing is not very well supported in it. In Listing 20.1 two specially written functions are presented for conversion of text, written as a sequence of ASCII codes of characters, into a sequence of carrier state numbers. Carrier numbers are coded using N_{bits}, therefore their values are changing from 0 to $2^{N_{bits}} - 1$.

Listing 20.1: Matlab functions for conversion of text to carrier state numbers and back

```
1   function [numbers, bitsnum, bitschar] = text2numbers( text, Nbits )
2   % text to IQ state numbers conversion
3   bitschar = dec2bin( double(text), 8 );          % text array, letters in rows,'0'/'1'
4   [rows,cols] = size( bitschar );                 % matrix size
5   N = rows*cols;                                   % number of all bits
6   work = reshape( bitschar', [ 1, N] )';          % bits in one column
7   Nadd = Nbits-rem(N,Nbits);                       % lacking bits for the last state
8   for k=1:Nadd, work= [work;'0']; end             % appending '0' bits at the end
9   bitsnum = reshape( work', [ Nbits, (N+Nadd)/Nbits] )'; % bits of all states
10  numbers = bin2dec( bitsnum );                    % state numbers: from 0 to 2^Nbits-1
11  return
12
13  function text = numbers2text( numbers, Nbits )
14  % IQ state numbers to text conversion
15  text = dec2bin( numbers, Nbits );               % state numbers as strings of bits
16  [rows,cols] = size( text );                     % size of matrix of characters?
17  text = reshape( text', [ rows*cols, 1] )';      % one big stream of chars '0'/'1'
18  N=length(text); N=N-rem(N,8); text = text(1:N); % remove appended bits
19  text = reshape( text', [ 8, N/8 ] )';           % strings of bytes
20  text = strcat( char( bin2dec( text) )' );       % conversion to text
21  return
```

Exercise 20.1 (Conversion of Text to Carrier States Numbers and Back).
Shortly analyze both functions presented in the Listing 20.1. Simple bit operations are performed in them which require many lines of code. Write and run the
following short program (or find program `lab20_ex_text.m` in the archive):

```
text = 'Hello!', Nbits = 4;
[numbers, bitsnum, bitschar] = text2numbers( text, Nbits ),
pause
numbers2text( numbers, Nbits), pause
```

Analyze the displayed result. Change number of bits to 1, 2, 3, ... Change text
to your name.

20.5 Digital Modulation and Demodulation

There are many different digital modulation methods. Our goal is not to describe all
of them in very serious, systematic, encyclopedia-like manner. No. Our intention is
only to explain an idea of digital modulation and to present a few simple illustrating examples. When digital modulation principles are understood, learning complex
modulation is not so difficult.

20.5.1 Carrier States Definition and Generation

How different carrier states are defined? There are many possibilities. The carrier
can change its amplitude, phase, and even frequency. In this chapter we will concentrate only on amplitude, phase, and joint amplitude-phase carrier modulation using
Eqs. (20.2)–(20.4). Different carrier states are defined by different pairs (A_k, ϕ_k)
(20.2), or resulting from the pairs (I_k, Q_k) (20.4):

$$(A_k, \phi_k) \quad \rightarrow \quad (I_k, Q_k) = (A_k \cos(\phi_k), -A_k \sin(\phi_k)) \qquad (20.14)$$

Typically, possible carrier states are defined using diagrams (constellations) of available carrier parameter values, marked with points/dots. The idea is explained in
Fig. 20.4. Carrier states, available in amplitude-only 2-PAM, 4-PAM, and 8-PAM
modulations, are shown in Fig. 20.5 together with carrier constellation points for
the phase shift-only 8-PSK modulation. In turn, in figure Fig. 20.6—carrier states
for joint amplitude-and-phase 4-QAM and 16-QAM modulations are presented. One

can conclude that in the 4-QAM modulation carrier has the same amplitude in each state and it is distinguishable only by the phase shift. In fact, the 4-QAM is equivalent to the 4-PSK modulation. In telephone and cable modems, the QAM constellation can have even $2^{15} = 32{,}768$ points.

It is important to remember that neighboring carrier states in IQ constellation diagrams differ by one bit only. Code having this nice feature is called a Gray code. Thanks to it, when a neighboring state is detected in a receiver due to transmission obstacles, only one bit is lost. This is a very good news because in 4-QAM transmission all 4 bits could be lost, for example, when states "0111" (7) and "1000" (8) are located next to each other. Losing 1 bit is not a tragedy. When some redundant bits are added to the bit-stream (forward error correction, e.g. CRC, LDPC) some small amount of bit errors can be detected and eventually corrected.

When telecommunication channel is disturbing amplitude and phase of transmitted signals, and its influence changes slowly in time, it can be beneficial to send bits not in carrier states but in their changes. For example, in the $\pi/4$-DQPSK differential quadrature phase shift keying of the carrier, the carrier phase shift—between two states—by the angle: $\pi/4, 3\pi/4, 5\pi/4, 7\pi/4$ corresponds to sending bits 00, 01, 11, 10.

When phase ϕ_k in Eqs. (20.2)–(20.4) is a linear function of n (e.g. $\phi_k = 2\pi \frac{\Delta f_k}{f_c} n$), the carrier can change its frequency and different frequency switching (keying) methods can be realized, for example, Minimum Shift Keying (MSK).

Drown in color Figs. 20.5 and 20.6, showing constellation diagrams of some digital modulations, can be easily obtained using a program presented in the Listing 20.2. The program can be used also for generation of valid $\{I_k, Q_k\}$ sequences for several implemented modulation schemes. For example, we can code the text "Hello World!" with its help.

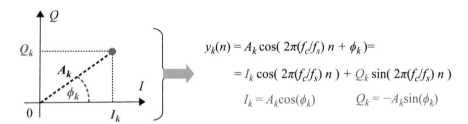

$$y_k(n) = A_k \cos(2\pi(f_c/f_s)\, n + \phi_k) =$$
$$= I_k \cos(2\pi(f_c/f_s)\, n) + Q_k \sin(2\pi(f_c/f_s)\, n)$$
$$I_k = A_k \cos(\phi_k) \qquad Q_k = -A_k \sin(\phi_k)$$

Fig. 20.4: Explanation of one carrier state $(A_k, \phi_k) = (I_k, Q_k)$ marked as a point/dot in state constellation diagrams

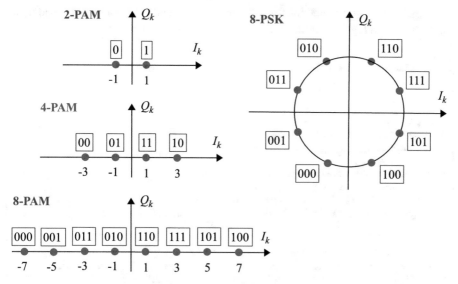

Fig. 20.5: Constellation diagrams of carrier states for amplitude modulations 2-PAM, 4-PAM, 8-PAM (left) and phase shift modulation 8-PSK (right)

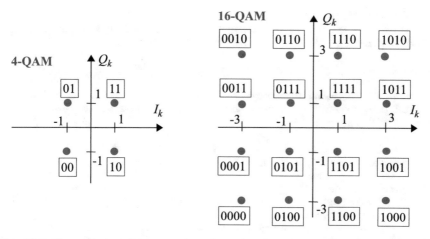

Fig. 20.6: Constellation diagrams of carrier states for joint amplitude-phase modulations 4-QAM and 16-QAM

Listing 20.2: Matlab functions for generation of I(k) and Q(k) components and their decoding for a few digital modulations with example usage

```
1   % lab20_ex_IQpoints.m
2   % Text conversion to carrier states numbers and (Ik,Qk) components and back
3   clear all; close all;
4
5   text_in = 'Hello World! 01234567890 abcdefgh',
6   modtype = '4QAM';        % 2PAM, 4PAM, 8PAM, BPSK, QPSK, DQPSK, 8PSK, 4QAM, 16QAM
7   do_texttransmit = 1;     % 0/1 optional further processing
8   do_decode = 1;           % 0/1 optional text decoding
9
10  % Definition of constellation points of carrier states
11  [IQcodes, Nstates, Nbits, R ] = IQdef( modtype );
12      phi = 0:pi/100:2*pi; c=R*cos(phi); s=R*sin(phi);
13      figure; plot(c,s,'k-',real(IQcodes), imag(IQcodes),'ro','MarkerFaceColor','red');
14      xlabel('I(k)'); ylabel('Q(k)'); title('Possible IQ(k) states'); grid; pause
15
16  % Coding our message using carrier states
17  numbers = text2numbers( text_in, Nbits ), pause      % text to IQ state numbers
18  %numbers = floor( Nstates*(rand(100,1)-10*eps) );     % random states
19  IQk = numbers2IQ( numbers, modtype, IQcodes );        % IQ state numbers to IQ values
20      figure;
21      subplot(211); stem(real(IQk),'b'); grid; xlabel('k'); title('I(k)');
22      subplot(212); stem(imag(IQk),'r'); grid; xlabel('k'); title('Q(k)'); pause
23
24  % Data transmission - our next exercise
25  % IMPORTANT: set "if(0)" in the beginning of the program lab20_ex_pulse_shaping.m
26  if( do_texttransmit ) lab20_ex_pulse_shaping; end   % our next exercise
27
28  % Text decoding - our next exercise - carrier state decoding
29  if( do_decode == 1 )
30      numbers = IQ2numbers( IQk, modtype )                 % find carrier state numbers
31      text_out = numbers2text( numbers, Nbits ), pause   % convert them to text
32  end
33
34  %###################################################################
35  function IQk = numbers2IQ( numbers, modtype, IQstates )
36  % State numbers to IQ values
37  if( isequal( modtype, 'DQPSK' ) )   % differential coding only for DQPSK
38      IQk(1) = exp(j*0);               % initial IQ state
39      for k = 1:length(numbers)-1     % loop
40          IQk(k+1) = IQk(k) * IQstates( numbers(k)+1 ); % next IQ state
41      end
42  else IQk = IQstates( numbers+1 );
43  end
44  end
```

Exercise 20.2 (IQ Constellations of Digital Modulations). Analyze code from Listing 20.2. Set do_texttransmit=0 and do_decode=0 and comment its last two lines. Run the program for all implemented modulations. Observe patterns of available carries states and shapes of generated IQ signals. Check whether number of observed signal levels is correct. Write an equation connecting number of constellation points with number of bits carried by one symbol (one carrier state). Add 64-QAM modulation scheme. Find its definition in the Internet.

20.5.2 Carrier States Interpolation: Pulse Shaping Filters

Now, we should try to realize practically the carrier changing. How to do it? In a sharp *flip-flop*, *bang-bang* manner or with smooth transitions between carrier states. Should the carrier be in one state for a while or should it be continuously and softly drifting between predefined states. In the first case, values of carrier amplitude and phase $\{A_k, \phi_k\}$ are constant for several values of n in Eqs. (20.2)–(20.4), while in the second case—they are changing permanently from sample to sample: $\{A_k(n), \phi_k(n)\}$.

In fact, we are dealing now with a problem of amplitude and phase carrier interpolation. Having samples of $\{I_k, Q_k\}$ (or $\{A_k, \phi_k\}$), we are interested in values lying in-between them. Sampling theorem specifies that analog signal can be reconstructed from its samples when sampling frequency is more than two times greater than the maximum analog signal frequency. In our situation, if we are interested only in $K-th$ times up-sampling, the following reconstruction formula should be used (10.1):

$$x(n) = \sum_{k=-\infty}^{\infty} x_z(k) h_I(n-k), \quad n = -\infty, \ldots, 0, \ldots, \infty, \qquad (20.15)$$

where $x_z(k)$ denotes original sequence of carrier states $x(k)$, with $K-1$ zeros put between each two samples, and $h_I(n)$ is an interpolation filter. Below two completely different interpolation filters are defined, having opposite features, the rectangular one $h_{RECT}(n)$ and the sinc one $h_{SINC}(n)$:

$$h_{RECT}(n) = \begin{cases} 1, & 0 \le n < K-1 \\ 0, & \text{otherwise} \end{cases}, \qquad h_{SINC}(n) = K \frac{\sin\left(\frac{\pi}{K}n\right)}{\pi n}. \qquad (20.16)$$

The rectangular filter $h_{RECT}(n)$ is the easiest zero-order, K-times interpolation filter that simply repeats the last signal sample $K-1$ times. It is simple in implementation however it offers low-quality interpolation. Additionally, due to jumps present in output signal values, spectrum of the interpolated signal is wide, what will be shown later.

In turn, an impulse response of the $h_{SINC}(n)$ filter is equal to a sinc() function, an ideal interpolation, Nyquist function. It offers perfect signal and its spectrum reconstruction. However, since $h_{SINC}(n)$ strongly oscillates and slowly decays on both sides around $n = 0$, number of required multiplications and additions are numerous and its usage is unpractical. In the below part, we will denote $h_{SINC}(n)$ as $h(n)$.

In FIR filter window design method a special window $w(n)$ was used for enforcing faster $h(n)$ decay—see Eq. (10.2). The window has a big influence upon the frequency spectrum of the resultant function $h(n)w(n)$. In our present application, the window will decide about the $I(n), Q(n)$ signal spectrum, and, in consequence, about our service spectrum after its conversion from the base-band to the target frequency. For this reason the raised cosine window is used in telecommunication applications, which spectrum edge has a shape of a raised cosine. Such spectrum is beneficial since two neighboring telecommunication services do not disturb themselves a lot and their power spectra are complementary: they are adding to 1 due to equality: $\cos^2(\alpha) + \sin^2(\alpha) = 1$. Weights of the raised cosine (RC) discrete interpolation function (pulse shaping filter) are defined as follows:

$$h_{RC}(n) = h(n)w(n) = K \frac{\sin\left(\frac{\pi}{K}\left(n - \frac{M}{2}\right)\right)}{\pi\left(n - \frac{M}{2}\right)} \cdot \frac{\cos\left(\frac{\pi}{K}r\left(n - \frac{M}{2}\right)\right)}{1 - 4r^2\left(n - \frac{M}{2}\right)^2/K^2}, \quad n = 0, 1, 2, \ldots, M$$

$$(20.17)$$

where M is an RC filter order and r denotes an roll-off parameter responsible for speed of the spectrum decay ($0 \leq r \leq 1$, 0 = maximally sharp, 1 = maximally smooth). We have $h\left(\frac{M}{2}\right) = 1$ and $h(n_0) = \frac{r}{2} \cdot \sin\left(\frac{\pi}{2r}\right)$ for values n_0 zeroing denominator of the second component in Eq. (20.17).

In digital single-carrier telecommunication systems, presented in Fig. 20.1, low-pass filter is used two times: first, in the modulator, for interpolation of carrier states $\{I_k, Q_k\}$ and obtaining *smooth* signal $\{I(n), Q(n)\}$, and then, in the demodulator, for removing high-frequency ($2f_c$) components after frequency down-shifting. **We are interested in having perfectly interpolated signal not after the first filter but after the second low-pass filter in the receiver**.

When two filters are working in a cascade, one after the other, the resultant impulse response is equal to convolution of their individual impulse responses. In consequence, the frequency response of the resultant filter is equal to multiplication of individual filter frequency responses:

$$H_{12}(\Omega) = H_1(\Omega) \cdot H_2(\Omega), \qquad \Omega = 2\pi \frac{f}{f_s}. \qquad (20.18)$$

In our application, the component filters are the same ($H_{SRRC}(\Omega) = H_1(\Omega) = H_2(\Omega)$) and the resultant filter should be a perfect raised cosine interpolation filter ($H_{RC}(\Omega) = H_{12}(\Omega)$). Therefore from Eq. (20.18) we obtain

$$H_{RC}(\Omega) = H_{SRRC}(\Omega) \cdot H_{SRRC}(\Omega) \quad \Rightarrow \quad H_{SRRC}(\Omega) = \sqrt{H_{RC}(\Omega)}. \quad (20.19)$$

We see that the filter, used in the transmitter and in the receiver, should have frequency response equal to the square root of a required one, i.e. raised cosine in our case. Such filter is called a square root raised cosine filter.

Impulse response of the square root raised cosine (SRRC) filter is defined as $(n = 0, 1, 2, \ldots, M)$:

$$h_{SRRC}(n) = \frac{4r}{\pi} \cdot \sin\left(\frac{\pi}{K}(1-r)\left(n-\frac{M}{2}\right)\right) \cdot \frac{\cos\left(\frac{\pi}{K}(1+r)\left(n-\frac{M}{2}\right)\right) + \frac{K}{4r(n-M/2)}}{1 - 16r^2\left(n-\frac{M}{2}\right)^2/K^2}.$$

$$(20.20)$$

When weights of the SRRC filter are calculated from the above equation, one should choose values of parameters do not zeroing denominator of the second component.

Remark Analog version of the raised cosine interpolation (pulse shaping) filter is defined as, similarly to (20.17):

$$h_{RC}(t) = \frac{\sin(\pi t/T)}{\pi t/T} \cdot \frac{\cos(\pi r t/T)}{1 - 4r^2 t^2/T^2} \qquad (20.21)$$

and has Fourier spectrum given by:

$$H_{RC}(f) = \begin{cases} T, & 0 \le |f| < \frac{1-r}{2T} \\ \frac{T}{2}\left[1 + \cos\left(\frac{\pi T}{r}\left(|f| - \frac{1-r}{2T}\right)\right)\right], & \frac{1-r}{2T} \le |f| \le \frac{1+r}{2T} \\ 0, & |f| > \frac{1+r}{2T} \end{cases} \qquad (20.22)$$

In digital communication systems realized in fully analog way, impulses having amplitudes equal to values of components I_k and Q_k are filtered by analog filters $h_{SRRC}(t)$ and resultant analog signals modulates analog carriers. Discrete-time impulse response $h_{RC}(n)$ (20.17) can be obtained performing IDFT upon the discretized equation (20.22), while $h_{SRRC}(n)$ (20.19)—by IDFT of a square root of Eq. (20.22). Equation (20.22) explains why discussed filters are called *raised cosine* and *square root raised cosine*—due to their Fourier transform shape.

□

In Fig. 20.7 impulse and frequency responses of a few discrete-time raised cosine filters with different value of the roll-off factor r are shown. They were calculated using program presented in Listing 20.3. For increasing value of r, impulse responses are less oscillatory and faster decaying. The same time their frequency responses have wider transition bands and higher attenuation. In the figure filters for 10-th order interpolation are presented. It should be observed that all filter impulse responses are crossing horizontal axis in exactly the same points being multiplicities of 10 (without value of 0). In turn, filter frequency responses have the same attenuation for normalized frequency equal to 0.05. Because the filter frequency characteristic is symmetric in respect to 0 Hz (only positive frequencies are shown), the filter has width equal to $1/10 = 2 \cdot 0.05$. Impulse and frequency responses of the corresponding SRRC filters are shown in Fig. 20.8.

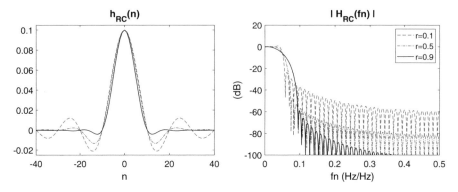

Fig. 20.7: Impulse responses of digital raised cosine filters with different roll-off factors r (left) and their frequency responses (right)

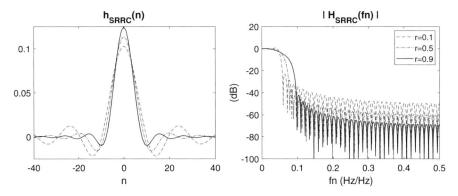

Fig. 20.8: Impulse responses of digital square root raised cosine filters with different roll-off factors r (left) and their frequency responses (right)

Listing 20.3: Calculation of raised cosine and square root raised cosine filters in Matlab

```
1   % lab20_ex_rcos.m
2   % Raised cosine (RC) and square root RC (SRRC) pulse shaping filters (PSF)
3   clear all; close all;
4
5   psf = 'sqrt-rcos';          % 'rcos', 'sqrt-rcos'
6   r = 0.5;                    % roll-off factor
7   Nsymb = 10;                 % number of samples per symbol, interpolation order
8   Ksymb = 4;                  % number of symbols per PSF filter
9   Npsf = Ksymb*Nsymb;         % filter length
10  n = 0 : Npsf; Nc = Npsf/2;  % indexes of filter samples, index of the central sample
11
12  if( isequal( psf, 'rcos') ) % definition of raised cosine
13      hpsf = sin(pi*(n-Nc)/Nsymb) ./ (pi*(n-Nc)/Nsymb) .* ...
14          cos(pi*r*(n-Nc)/Nsymb) ./ (1-(2*r*(n-Nc)/Nsymb).^2);
```

```
15    hpsf(Nc+1) = 1;
16    n0 = find( abs(1-(2*r*(n-Nc)/Nsymb).^2 ) < sqrt(eps) );
17    hpsf(n0) = r/2 * sin(pi/(2*r));
18  elseif( isequal( psf, 'sqrt-rcos') ) % definition of sqrt of raised cosine
19    n0 = find( abs(1-(4*r*(n-Nc)/Nsymb).^2) < 5*eps ), pause
20    if( prod(size(n0)) > 0 ) disp('Div by 0. Change parameters'); end
21    hpsf = 4*r/pi * ( cos(pi*(1+r)*(n-Nc)/Nsymb)+(Nsymb./(4*r*(n-Nc))) ) ...
22         .* sin(pi*(1-r)*(n-Nc)/Nsymb) ) ./ (1-(4*r*(n-Nc)/Nsymb).^2);
23    hpsf(Nc+1) = (1+r*(4/pi-1));
24  end
25  hpsf = hpsf/Nsymb; % normalization
26
27  if(1) % Comparison with Matlab
28    if( isequal( psf, 'rcos') ) type = 'normal'; else type = 'normal'; end
29    fs = 1; fcut = 1/(2*Nsymb)
30    hpsf0 = firrcos( Npsf, fcut, r, fs,'rolloff',type); % MATLAB function
31    error = max(abs(hpsf-hpsf0)), pause
32  end
```

Exercise 20.3 (RC and SRRC Filter Basics). Analyze program from Listing 20.3. Run it and its longer version (from the book repository) for the RC and SRRC filter. Choose the same filter parameters for which Figs. 20.7 and 20.8 were obtained. Add figures to the programs. Try to obtain the same figures as in the book.

Exercise 20.4 (Testing RC/SRRC Filter Length). Choose $r=0.35$; Nsymb=24; Ksymb=6. Observe shapes and frequency responses of RC/SRRC filters when number of symbols per PSF filter (Ksymb) is becoming lower (2, 4) and bigger (8, 10, 12). Plot results in one figure. From FIR filter design we should remember that increasing filter length improves sharpness of the filter transition edge.

Exercise 20.5 (RC Filter as a Convolution of Two SRRC Filters). Design RC and SRRC filters using the Matlab function firrcos() and check whether convolution of two $h_{SRRC}(n)$ filters gives the $h_{RC}(n)$ filter.

Exercise 20.6 (Design of Raised Cosine Filters in Frequency Domain). Try to calculate impulse responses of RC and SRRC filters by means of IFFT of their predefined spectra, given by equation 20.22. Remember about required spectrum symmetry in its real part and asymmetry in its imaginary part. For your inspiration: code presented in the next chapter in Listing 21.8 allows design of the SRRC filter with $r = 1$ for the Radio Data System (RDS). Copy its appropriate fragment and use for testing. Set fs=228000, Ks=8.

Exercise 20.7 (Using Pulse Shaping Filters). Apply designed filters hRC and hSRRC to interpolation of sequences Ik and Qk, generated in the program 20.2.

Use the Matlab function `upfirdown()` or personally insert zeros to both signals and filter the signals (see programs from chapter on resampling):

```
IQ0 = zeros( 1, length(IQk)*K ); IQ0(1:K:end) = IQk; IQn =
conv( IQ0, hRC);
```

Notice that fast polyphase interpolation algorithm can be used and zeros insertion and multiplication by zeros can be avoided! Plot obtained, interpolated sequences $I(n)$ and $Q(n)$ separately, e.g. `stem(real(IQn))` and `stem(imag(IQn))`. Then plot $I(n)$ as a function of $Q(n)$: `plot(real(IQn),imag(IQn));`—you will see phasor-diagrams.

And the last difficult task, which can be skipped at present: (1) try to divide each sequence into many, non-overlapping, consecutive fragments, having the same number of samples equal to the symbol length (K), and (2) display all of them together in one figure using the `reshape()` function—you should obtain eye-diagram. If you have problems, continue reading and observe figures presented below. Than go back to this exercise and finish it.

20.5.3 Carrier States Interpolation: In Action

This is the most important section in this chapter!

Well ... Having a cake nothing remains to be done, only to eat it. So, at present it is a time for demonstration of pulse shaping filters in action. This is in fact explanation of interpolation procedure, very well-known to us. If you do not agree, you should go back to Chap. 10 on re-sampling.

In Fig. 20.9 a zero-order interpolation is used for $K = 5$ times signal up-sampling. In the upper plot a sequence I_k is shown, next $K - 1 = 4$ zeros are inserted between each two I_k values, then interpolation filtering takes place and its result is presented. In this case a rectangular interpolation pulse shaping filter is used, shown in the bottom plot. As we see, each I_k value is repeated $K = 5$ times. Obtained signal is *boxy*, has abrupt changes, and one can expect that due to this its spectrum is wide, what will be shown later.

In Fig. 20.10 the up-sampling procedure is repeated for the same input signal but only different pulse shaping interpolator is used: the raised cosine one, shown in the bottom plot. What should be noticed? First, the obtained signal (on the third plot) is much smoother than before and for sure its spectrum is more narrow, what will be shown later. Next, it is important to observe that the filtering operation, realizing the data interpolation, does not change the original I_k samples marked with filled circles. Why? Because in our pulse shaping filter each $K = 5$-th sample is equal to 0 (what is valid for all K-order interpolation/up-sampling filters).

Figures 20.9 and 20.10 should learn us that after interpolation resultant signals can have different shapes. More oscillatory the interpolating filter is, more oscillatory the interpolated signal is also. Since this signal modulates the carrier, the carrier state detection becomes more difficult in the receiver. The same time oscillatory-shape filters ensure spectrum compactness and narrow service bandwidth. In Fig. 20.11 this phenomena is further demonstrated. This time two raised cosine pulse shaping filters are used with different roll-off factors $r = 0.35$ (left) and $r = 0.75$ (right). In this case 4-QAM modulation is exploited and two signals are interpolated: $I(n)$ (middle plots) and $Q(n)$ (bottom plots). As we see, for $r = 0.75$ the less oscillatory result is obtained at the cost of wider signal spectra (to be shown later). We have to choose the pulse shaping filter which is better in our specific application.

In digital communication signal its quality is very often checked by observation of some specific signal plots that are called an eye-diagram and a phasor-diagram. Both of them are demonstrated in Fig. 20.12 for signals from Fig. 20.11, e.g. 4-QAM, $K = 5$. In first two rows eye-diagrams of signals $I(n)$ and $Q(n)$ are shown for raised cosine pulse shaping with $r = 0.35$ (left) and $= 0.75$ (right). Both signals are cut into $K = 5$ samples long, consecutive fragments and displayed with overlay in one figure. Since both signals are coding carrier values "1" and "-1," in the central part of all plots we see many lines passing perfectly through these two values. We see "open eyes" of the modulating signals telling us that *the modulation pattern is perfectly visible* and there will be no problem with its detection in the receiver. When "eyes are closed", we should start to worry. In turn, sequences $Q(n)$ are displayed as functions of sequences $I(n)$ in bottom plots of the Fig. 20.12. If $Q(n)$ were a sine $\sin(\Omega n)$ and if $I(n)$ were a cosine $\cos(\Omega n)$, we would see a circle since $\sin^2(\alpha) + \cos^2(\alpha) = 1$. But signal curves are more complex and observed *phasor* shapes are more sophisticated. What is important to remember that, for high-quality transmission, lines in phasor plots are crossing exactly in points (I_k, Q_k), "1" and "-1" in our example, and they are also telling us that *everything is under control*.

Finally, in Fig. 20.13 we can admire beauty of different communication patterns: two eye-diagrams (for 2-PAM and 4-PAM) as well as four $Q(n) = f(I(n))$ phasor plots (for QPSK, DQPSK, 8PSK, and 16QAM modulations) are presented for $K = 25$ order interpolation. Now all curves are very smooth due to very high interpolation order. Such detail interpolation helps a receiver to find crossing points, carrier states, and bits sent at the cost of expensive sample throughput.

Fancy figures, presented in this section, were obtained with the help of a Matlab program given in Listing 20.4. The program can be used individually or with collaboration with the program 20.2, in which text coding is done. In the second case, the text "Hello World!" is processed by pulse shaping filters, and even converted up and down in frequency. In order to do this, a user should set do_texttransmit=1; in the program 20.2, and do_updown=1 in the program 20.4.

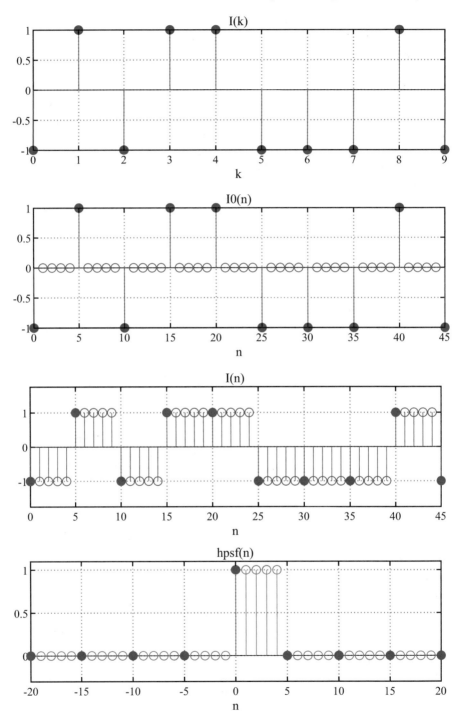

Fig. 20.9: Explanation of $I(n)$ signal generation in 4-QAM modulation (the same is valid for $Q(n)$ signal): (1) $I(k)$ values, (2) signal $I_k(n)$ after zero insertion, (3) resultant $I(n)$ signal after pulse shaping using the rectangular filter, (4) the rectangular pulse shaping filter which was used

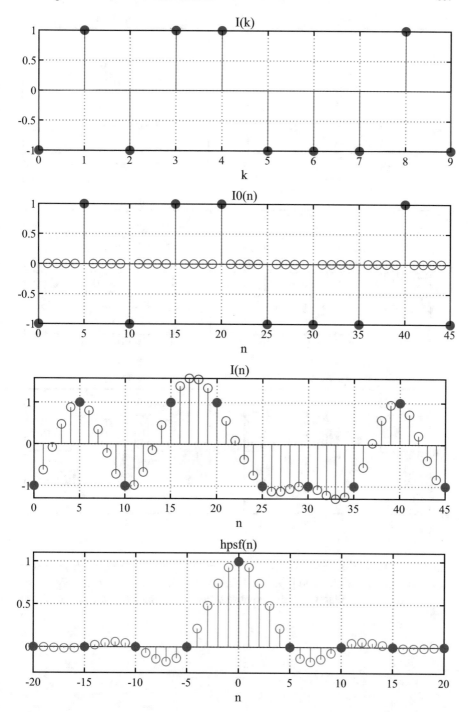

Fig. 20.10: Explanation of $I(n)$ signal generation for 4-QAM modulation (the same is valid for $Q(n)$ signal), top-down: (1) $I(k)$ values, (2) signal $I_k(n)$ after zero insertion, (3) resultant $I(n)$ signal after pulse shaping using the raised cosine filter, (4) the pulse shaping filter, raised cosine with $r = 0.35$, which was used

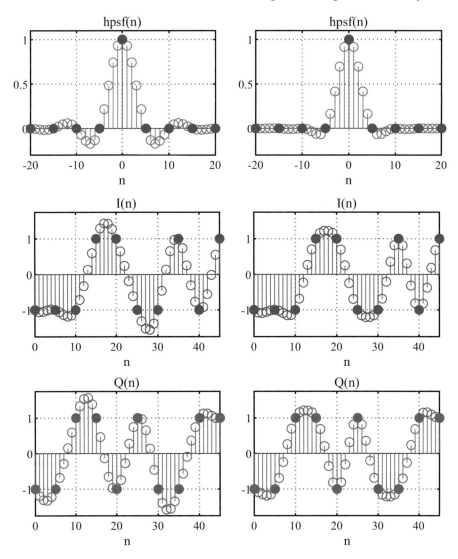

Fig. 20.11: Two raised cosine pulse shaping filters: (left) $r = 0.35$, (right) $r = 0.75$ and resultant signals $I(n)$ and $Q(n)$ for 4-QAM modulation. Observe that signals for $r = 0.75$ are less oscillatory (at the expense of less sharp spectra which is not shown here)

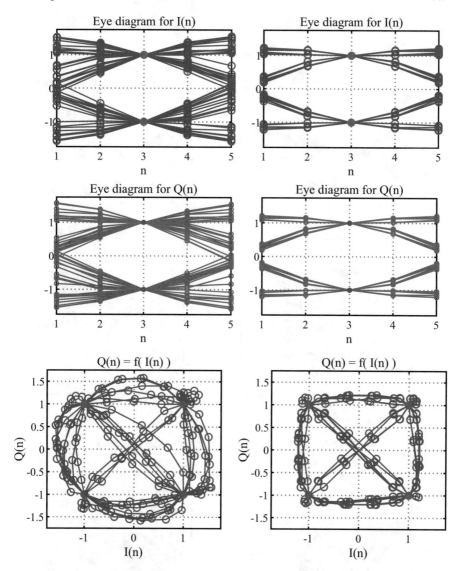

Fig. 20.12: Eye-diagrams of signals $I(n)$ and $Q(n)$ (first two rows) and theirs phasor graphs $Q(n) = f(I(n))$ (third row) for 4-QAM modulation example presented in Fig. 20.11. Left: for raised cosine pulse shaping filter with $r = 0.35$ (more oscillatory), right: for $r = 0.75$ (less oscillatory)

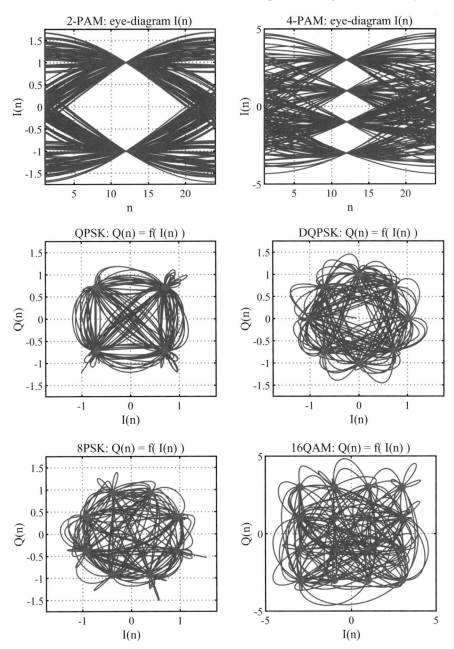

Fig. 20.13: Eye-diagrams and phasor-diagrams for different modulations, from left to right, up-down: (1) 2-PAM, (2) 4-PAM, (3) QPSK and 4-QAM, (4) DQPSK, (5) 8PSK, (6) 16-QAM. Correct states are marked with red circles

Listing 20.4: Matlab program demonstrating application of pulse shaping filters to IQ carrier states

```
1    % lab20_ex_pulse_shaping: IQk --> IQ0 --> IQ(n) --> IQk
2
3    % When THIS program IS NOT CALLED from lab20_ex_IQpoints.m, SET 1 IN IF()
4    if( 0 )  % 0/1 Optional starting program from these data
5       clear all; close all;
6       modtype = '4QAM';  % 2PAM, 4PAM, 8PAM, BPSK, QPSK, DQPSK, 8PSK, 4QAM, 16QAM
7       Nsymbols = 250;    % number of symbols to transmit
8       [IQcodes, Nstates, Nbits, R ] = IQdef( modtype );
9       numbers = floor(Nstates*(rand(Nsymbols,1)-10*eps));    % symbol generation
10      IQk = numbers2IQ( numbers, modtype, IQcodes );
11   end
12
13   do_figures = 1;             % 0/1 plotting figures
14   do_updown = 0;              % 0/1 frequency up-down conversion
15   do_disturb = 0;             % 0/1 addition of disturbances
16   rctype = 'sqrt';           % raised cosine filter type: 'normal, 'sqrt'
17   r = 0.35;                  % PS filter roll-off factor
18   K = 24; Ns = 8;            % samples per symbol, symbols per PS filter
19   fs = 240000;               % sampling frequency in Hz: 1, 250e+3, 240e+3
20   fcar = 50000;              % carrier frequency in Hz
21
22   fcut = fs/(2*K);                        % PSF cut-off frequency
23   Npsf = Ns*K+1; Mpsf=(Npsf-1)/2;         % PSF filter length and its half
24
25   numbers  = floor( Nstates*(rand(2*Ns,1)-10*eps) ); % 2*Ns random carrier states
26   dummy = numbers2IQ( numbers, modtype, IQcodes );   % prefix and postfix
27   IQdum = [ dummy IQk dummy ];                        % appending pre&postfix
28
29   IQ0 = zeros( 1, length(IQdum)*K ); IQ0(1:K:end) = IQdum; % zero insertion
30   hpsf = firrcos(Npsf-1, fcut, r, fs,'rolloff',rctype);    % 'normal' or 'sqrt'
31   IQn = conv( IQ0, K*hpsf ); IQn = IQn(Mpsf+1 : end-Mpsf); % pulse shaping in TX
32
33   if( do_updown )  % Frequency UP and DOWN, channel and disturbances are in-between
34
35    % Frequency UP conversion in TX
36       N = length(IQn); n = 0:N-1;     % signal length, sample indexes
37       y = real(IQn).*cos(2*pi*fcar/fs*n) - imag(IQn).*sin(2*pi*fcar/fs*n);
38
39    % See into the program lab20_ex_single_carrier.m for more details
40    % 1) Digital to analog conversion (DAC) and analog signal transmission.
41    % 2) Channel is filtering our signal (h = unknown channel impulse response)
42    % 3) In analog part, channel and amplifiers add noise, Signal-to-Noise-Ratio (SNR).
43    % 4) In the receiver, signal from antenna is amplified and band-pass filtered (BPF).
44    % 5) Then analog to digital conversion (ADC) takes place with sampling ratio
45    % equal to the BPF filter bandwidth.
46
47    % Frequency DOWN conversion in RX
48       df = 0; dphi = 0; % carrier frequency and phase offsets
49       IQnn = 2* y .* exp( -j*( 2*pi*(fcar/fs + df/fs)*n + dphi ) );
50   end
51
```

```
52   % Disturbances - additive noise, channel gain and phase shift, RX carrier offsets, ADC
         error
53   if( do_disturb )
54       SNR=160; chanG=1; chanPh=0; carDf=0; carDph=0; ADCppm=100*1e-6;
55       % SNR=20; chanG=0.25; chanPh=pi/7; carDf=0.001; carDph=pi/11; ADCppm=100; % Params
56       % noise
57       s = IQnn(Mpsf : end-Mpsf+1);
58       scale = sqrt( sum( s.*conj(s) ) / (2*length(s)*10^(SNR/10)) ); clear s
59       IQnn = IQnn + scale * (randn(1,N)+j*randn(1,N)); % noise addition with proper SNR
60       % IQnn = awgn( IQnn, SNR, 'measured' );                % alternative noise addition
61       % channel influence
62       IQnn = IQnn .* ( chanG * exp(j*chanPh ) );           % equivalent channel in the BB
63       % carrier frequency and phase offsets
64       carDf = carDf - ADCppm/(1+ADCppm)*fcar/fs;
65       IQnn = IQnn .* exp(j*(2*pi*carDf*(0:length(IQn)-1)-carDph)); % carrier offsets
66       if( do_figures )
67           figure, spectrogram(IQnn,Nf,Nf-64,fshift,fs); title('STFT(IQ(n))+Noise)'); pause
68       end
69   end
70
71   IQnn = conv( IQnn, hpsf ); IQnn = IQnn(Mpsf+1:end-Mpsf); % pulse shaping in TX
72   IQnn = IQnn( 2*Npsf-1: 1 : end-(2*Npsf-1)+1 );              % signal synchronization
73   IQkk = IQnn( 1: K : end );                                  % symbol extraction
74
75   IQk = IQkk; % when we are going back to the program lab20_ex_IQpoints.m
```

Exercise 20.8 (Pulse Shaping in Action: Proof of Concept). Analyze code of the pulse shaping program 20.4 which full version, with added figures, is given in the archive. Set 1 in the first if() when you prefer to run only this program or set 0 when you will call it from the program 20.2 lab20_ex_IQstates.m. In such situation in the pulse shaping program beginning data initialization is not performed. Otherwise it is done. Set do_disturb=0. Then set do_updown=0 and do_updown=1—results should be almost identical. Run the program for all supported modulations and carefully observe each figure. Change values of program parameters and try to obtain similar figures to shown in this chapter. Add the rectangular pulse shaping filter:

hpsf = [zeros(1,Ns/2*K) ones(1,K) zeros(1,(Ns/2-1)*K+1)]/K;

having the same length as RC and SRRC filters. Observe signal spectra. They should be significantly wider. Decrease number of samples per one symbol—spectra should become wider independently from the PS filter type. Change value of the roll-off parameter r of the (SR)RC filters. Observe changes in eye-diagrams and phasor-diagrams.

20.5.4 Carrier States Detection

Having in mind *fancy* plots of *eye-diagrams* and *phasor-diagrams* of $I(n)$ and $Q(n)$ signals, it is not difficult to explain the digital demodulation issue: (1) a receiver should sample signals $I(n)$ and $Q(n)$ in good moments, exactly in the positions of I_k and Q_k symbol values, i.e. at crossing points of $Q(n) = f(I(n))$ phasor functions, (2) calculate carrier state numbers having points $(I(k), Q(k))$, and (3) extract bits from the state numbers. At this moment, we are not addressing the sampling synchronization problem (so-called *symbol timing recovery*), it will be discussed later. Now we aim at carrier states finding.

This task is solved as follows. The phasor $Q(n) = f(I(n))$ plane, i.e. the carrier state plane, is divided into smaller pieces, into a grid, and for each point $(I(k), Q(k))$ the closest, reference (I_k, Q_k) point is found. Bits are extracted from its number. Listing 20.5 presents a function for carrier state decoding for some selected modulations. More examples are given in the complete version of the function, available in the book repository.

Listing 20.5: Decoding of carrier state numbers from detected constellation IQ points for exemplary digital modulations

```
1   function numbers = IQ2numbers( IQ, modtype )
2
3   % from (I,Q) values to carrier state numbers
4
5   N = length(IQ);
6
7   if( isequal(modtype,'4PAM') )                        % checking only I in predefined
8       k = 1;                                           % intervals: -2, 0, 2
9       for ns = 1 : N
10          I = real( IQ(ns) );
11          if(        I < -2      )   numbers(k) = 0;   % lower the -2
12          elseif( -2 <= I && I < 0 )  numbers(k) = 1;  % from -2 to 0
13          elseif(  0 <= I && I < 2 )  numbers(k) = 3;  % from  0 to 2
14          else                        numbers(k) = 2;  % greater then 2
15          end
16          k = k + 1;
17      end
18  end
19  if( isequal(modtype,'4QAM') || isequal(modtype,'QPSK') ) % checking I and Q
20      k = 1;                                           % positions in +/- quadrants
21      for ns = 1 : N
22          I = real( IQ(ns) );
23          Q = imag( IQ(ns) );
24          if( I>0 && Q>0 )  numbers(k) = 3; end        % +/+ right-up
25          if( I>0 && Q<0 )  numbers(k) = 2; end        % +/- right-down
26          if( I<0 && Q>0 )  numbers(k) = 1; end        % -/+ left-up
27          if( I<0 && Q<0 )  numbers(k) = 0; end        % -/- left-down
28          if( I==0 || Q==0 ) disp('ZERO'); pause; end
29          k = k + 1;
30      end
31  end
```

```
32  if( isequal(modtype,'DQPSK') ) % checking I and Q phase shifts in +/- quadrants
33      IQdiff = IQ(2:end) .* conj( IQ(1:end-1));        % finding phase shift
34      k = 1;
35      for ns = 1 : length(IQdiff)-1
36          I = real( IQdiff(ns) );
37          Q = imag( IQdiff(ns) );
38          if( I>0 && Q>0 )  numbers(k) = 0; end        % +/+ right-up
39          if( I<0 && Q>0 )  numbers(k) = 1; end        % -/+ left-up
40          if( I>0 && Q<0 )  numbers(k) = 2; end        % +/- right-down
41          if( I<0 && Q<0 )  numbers(k) = 3; end        % -/- left-down
42          if( I==0 || Q==0 ) disp('ZERO'); pause; end
43          k = k + 1;
44      end
45  end
46  if( isequal(modtype,'16QAM') )  % checking I and Q
47      k = 1;                        % first two higher (MSB) bits of I
48      for ns = 1 : N                % then  two lower  (LSB) bits of Q
49          I = real( IQ(ns) );                                   % II
50          if(        I < -2      )  numbers(k) = 0;             % 00
51          elseif( -2 <= I && I < 0 )  numbers(k) = 4;          % 01
52          elseif(  0 <= I && I < 2 )  numbers(k) = 12;         % 11
53          else                        numbers(k) = 8;          % 10
54          end
55          k = k + 1;
56      end
57      k = 1;
58      for ns = 1 : N
59          Q = imag( IQ(ns) );                                  % II + QQ
60          if(        Q < -2    )  numbers(k) = numbers(k)+0;   % ?? + 00
61          elseif( -2 <= Q && Q < 0 )  numbers(k) = numbers(k)+1;  % ?? + 01
62          elseif(  0 <= Q && Q < 2 )  numbers(k) = numbers(k)+3;  % ?? + 11
63          else                        numbers(k) = numbers(k)+2;  % ?? + 10
64          end
65          k = k + 1;
66      end
67  end
68  end
```

Exercise 20.9 (Extraction of Carrier States Numbers from (I_k, Q_k) Points).
Analyze code of the function IQ2numbers.m from the Listing 20.5, which
full version is given in the book repository. Set do_texttransmit=0 and
uncomment two last lines in the program 20.2. Run the program. Check whether
the function IQ2numbers.m allows correct text decoding for all implemented
modulation schemes. Add encoding and decoding of one extra modulation
which is not supported at present. More bits to transmit, more carrier states
are required. May be 64-QAM will satisfy your bit throughput needs?

20.6 Frequency Up-Down Conversion: Symbols and Bits per Second

If we think that everything goes right, for sure we do not notice something. And in our case? For sure we have neglected errors introduced by up and down frequency conversion of the signal and consequences of its passing through a disturbing channel. In this section we will concentrate on some important aspects of signal transmission.

First, we will investigate shape of power spectral density (PSD) of the modulating signal $I(n) + jQ(n)$. In Fig. 20.14 the PSD of a 4-QAM signal for different pulse shaping filters is presented: rectangular and three raised cosine filters with roll-off factor r equal to 0.05, 0.35, and 0.75. Sampling frequency is equal to 250 kHz, 5 samples per symbol are used and filters have 41 samples (as in our previous figures). We clearly see benefits of using the RC filter with $r = 0.35$: it represents a good compromise between narrow filter bandwidth (sharpness) and very good attenuation in the stop-band. Application of the rectangular filter gives horrible results.

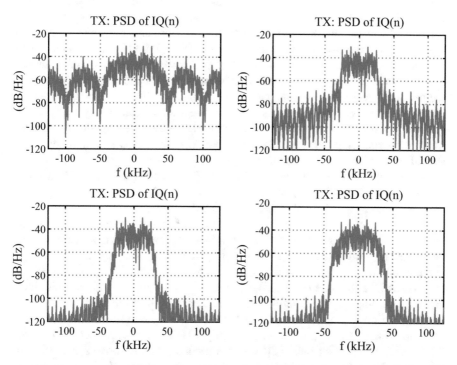

Fig. 20.14: Power spectral density of 4-QAM signal $I(n) + jQ(n)$ for different pulse shaping filters, in rows from left to right: (1) rectangular, (2) raised cosine (RC) $r = 0.05$, (3) RC $r = 0.35$, (4) RC $r = 0.75$. Sampling frequency 250 kHz, 5 samples per symbol, filter has 41 samples (as in our previous figures)

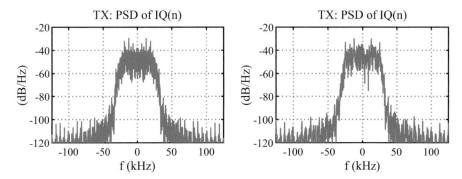

Fig. 20.15: Power spectral density of a real-value 2-PAM signal $I(n)$ (left, symmetrical with respect to 0 Hz) and a complex-value 4-QAM signal $I(n) + jQ(n)$ (right, unsymmetrical) for the same raised cosine pulse shaping filter with $r = 0.35$

If the filter is already selected, the next question typically concerns modulation choice: using only pulse amplitude modulations (PAM) or joint amplitude-phase ones (PSK or QAM). In Fig. 20.15 PSDs of two modulating signals are shown, first for 2-PAM and second for 4-QAM modulation (for the same transmitted data). The transmission parameters are the same as before ($f_s = 250$ kHz, $K = 5$ samples per symbol, $L = 41$, $f_{symb} = 250/5 = 50$ kHz). Both spectra have very similar shapes, but in the first case the spectrum is symmetric around 0 Hz and only 1 bit is transmitted per one symbol (50 kilo-bits per second (kbps) are offered, i.e. 50,000 symbols carrying 1 bit only). For 4-QAM modulation, due to signal complexity, the spectrum symmetry is not observed and two bits are transmitted by one symbol. Distance between constellation points is similar, therefore noise sensitivity in both modulations should be similar also. It looks that 4-QAM is a winner in this game offering the same symbol rate per second but two times higher bit-rate per second. With its usage we can get 100 kilobits per second (50,000 symbols carrying 2 bits each). So, what will we choose? I am choosing the 4-QAM.

At present, let us observe signal PSDs during up and down frequency shift, presented in Fig. 20.16. In this 4-QAM example the sampling and carrier frequencies are equal, respectively, $f_s = 240$ kHz and $f_c = 40$ kHz. We have 24 samples per symbol (modulating signal frequency and bandwidth are equal to $f_{symb} = 240/24 = 10$ kHz) and the RC filter ($r = 0.35$) is 8 symbols and $8 \cdot 24 + 1 = 193$ samples long. Higher oversampling offers significantly more narrow spectrum with 10 kHz bandwidth and 5 more transmission channels than before when service bandwidth was equal to 50 kHz. But at present we have only 10,000 symbols per second carrying 2 bits each, therefore bit-rate of 20 kbits per second is get. Before we had 100 kbits per second but only one service, not five. As we see, *nothing is lost in the nature*.

Let us come back to Fig. 20.16. The PSD of the base-band complex-value $I(n) + jQ(n)$ signal is not symmetrical around 0 Hz. After signal frequency-up conversion (multiplication by $e^{j\Omega_c n}$), the signal spectrum is shifted to the positive carrier frequency f_c. But because only the real part of the shift result is left, mirror copy of

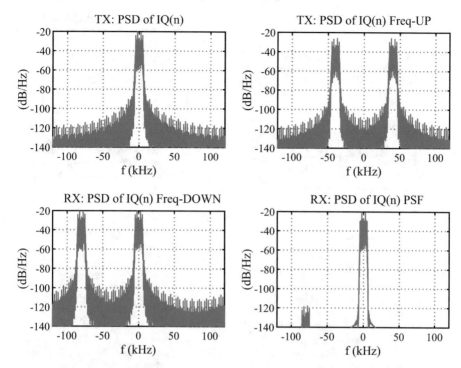

Fig. 20.16: 4-QAM carrier modulation. Power spectral densities of the following signals, in rows from left to right: (1) original $I(n) + jQ(n)$ in the base-band, (2) after frequency UP conversion, (3) after frequency DOWN conversion, (4) after low-pass pulse shaping filter in the receiver. Sampling frequency 240 kHz, carrier frequency 40 kHz, 24 samples per symbol, PSF filter length has $8 \cdot 24 + 1 = 193$ weights

the spectrum appears for negative frequencies. The PSD spectrum of any transmitted signal, the modulated carrier, looks very similarly.

In the receiver, the signal is down-converted in frequency (multiplied by $e^{-j\Omega_c n}$), therefore its spectrum is shifted left by $-\Omega_c$, what is seen in left down plot in Fig. 20.16. Now, the low-pass pulse shaping filter is used to leave only the original signal component lying the base-band around 0 Hz. As a result spectrum presented in the right down plot is get. The base-band signal is further processed in the receiver.

At this processing stage we should perfectly recover the transmitted $I(n) + jQ(n)$ signal and obtain from it the carrier states I_k and Q_k and transmitted bits. In the left part of the Fig. 20.17 eye-diagrams, phasor-diagrams, and constellation points for the transmitted (TX) and received (RX) $I(n) + jQ(n)$ signals are shown. The same RC filter was used for pulse shaping in RX and TX. Wow! A big surprise! Transmit-

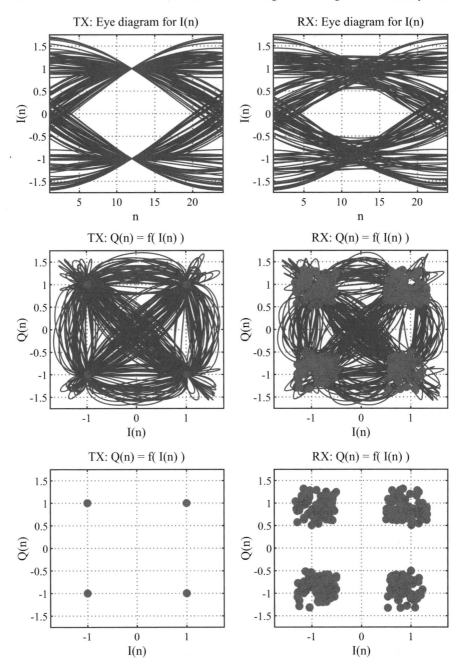

Fig. 20.17: (up-down) Eye-diagram, phasor-diagram, and constellation points only of $I(n) + jQ(n)$ signals in the transmitter (TX) (left) and in the receiver (RX) (right) for 4-QAM carrier modulation. Raised cosine shaping filter was used in RX and TX. Sampling frequency 240 kHz, 24 samples per symbol, PSF filter length has $8 \cdot 24 + 1 = 193$ weights

ted patterns are perfect (left side), the received ones—very poor (right side)! Why and how did it happen?

Why SRRC, Not RC? The answer is very simple. We used in the receiver the raised cosine (RC) pulse shaping filter. It perfectly interpolates the carrier states in the TX (left side of the figure). But after its second application in the RX (right side of the figure), the magic is breaking up, due to the fact that impulse response of a filter, resulting from convolution of two interpolation filters, is not anymore an interpolation filter! This is shown in Fig. 20.17. How to solve this problem? To use in the receiver and in the transmitter the same SRRC filter which convoluted with itself becomes the perfect interpolating RC filter. Now, in the transmitter (left side of Fig. 20.18), signal patterns are poor (badly interpolated), while in the receiver (right side of Fig. 20.18) they are perfect.

And, at the end, we will investigate two plots, which are very important in this section. In Fig. 20.19 two modulated carriers $y(n)$ for the 16-QAM modulation are presented: in the upper part—for rectangular pulse shaping filter, in the bottom part—for raised cosine filter. Typically, students, asked during examination about single-carrier digital modulations, draw a signal similar to one shown in the upper plot (or in Fig. 20.9). If the upper figure is in 99% remembered as an example of digital modulation, it means that only *bang-bang* carrier switching is well understood. In practice, the carrier smoothly comes from one state to the other reducing this way required transmission bandwidth. Therefore the second figure is correct.

Exercise 20.10 (Frequency UP and DOWN Conversion: Observing Spectra). Plots of frequency spectra, presented in this section, were obtained with the help of the pulse shaping program 20.4. Turn on the option do_updown=1 in it and run the program for the following values of sampling frequency f_s and *IQ* signal interpolation orders K: $f_s = 50, 100, 200$ kHz and $K = 5, 10, 20$. This time concentrate on spectral shapes: its widths (service bandwidth) and side-lobe attenuation (service separation). Try to obtain spectra similar to ones presented in Figs. 20.14–20.16. Notice service localization in different processing stages: (1) after the transmitter PS filter, (2) after the frequency up-conversion, (3) after the frequency down-conversion, (4) after the receiver PS filter. Add rectangular pulse shaping filter to the program:
hpsf = [zeros(1,Ns/2*K) ones(1,K) zeros(1,(Ns/2-1)*K+1)]/K;
and repeat experiments.

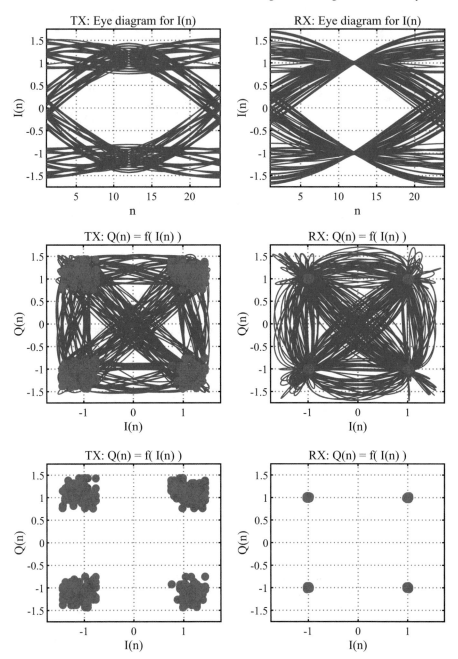

Fig. 20.18: (up-down) Eye-diagram, phasor-diagram, and constellation points only of $I(n) + jQ(n)$ signals in transmitter (TX) (left) and in the receiver (RX) (right) for 4-QAM carrier modulation. Square root raised cosine shaping filter was used in RX and TX. Sampling frequency 240 kHz, 24 samples per symbol, PSF filter length has $8 \cdot 24 + 1 = 193$ weights

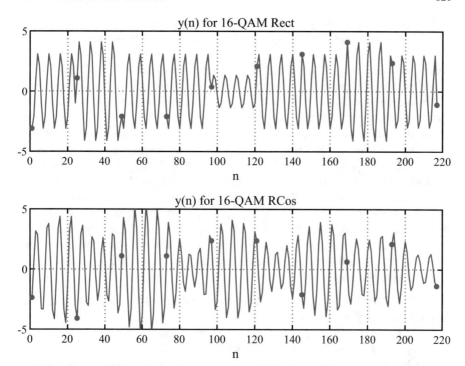

Fig. 20.19: Modulated carriers for the 16-QAM scheme for: (up) rectangular pulse shaping, and (down) raised cosine pulse shaping ($r = 0.35$, 24 samples per symbol, 8 symbols per PSF, 6 samples per carrier period). Red dots marked points of pre-defined carrier states. For rectangular filter they denote first samples of a new state. For the RC filter, the carrier is *turning left/right* in them into a next state

20.7 Disturbances and Obstacles

So, is everything going right? If it seems to us that yes, for sure we have forgotten about something. About what?

Presence of Noise The world is noisy. RF signals are disturbing themselves. Many independent noisy sources give in result a Gaussian noise. Weak signals should be amplified. Amplifier can be *low-noise* but not perfectly. During amplification noise is amplified also. In order to have reliable transmission model we have to embedded our signals in noise with correct probability density function and power. Typically additive white Gaussian noise (AWGN) generators are used and appropriate signal-to-noise ratio (SNR) is ensured (in Matlab: x=awgn(x, SNR)). In upper part of Fig. 20.20 we see constellation points of 4-QAM modulation (left), perfectly recovered in the receiver, and fuzzy constellation pattern in the presence of noise (right).

Influence of the Channel The simplest linear channel is characterized (described) by its impulse response $h(t)$ (its response to impulse on the input). Impulse consists of all frequencies. Knowing channel response to all frequencies we know what channel is doing with of each them—how attenuate and delay each of them. In fact this information is included in channel frequency response, i.e. Fourier transform of the impulse response: $H(f) = \mathrm{FT}[h(t)]$. In order to have reliable transmission model we should at least convolve our *IQ* signal, up-converted in frequency, with the sampled channel impulse response $h(n)$ (in Matlab: y=conv(y,h). Or, after assuming the very narrow bandwidth of it, at least multiply the *IQ* signal in the base-band by $|H(f_c)|e^{j\phi(f_c)}$, i.e. complex-value scaling factor incorporating channel attenuation and phase shift at frequency f_c (represented by $|H(f_c)|$ and $\phi(f_c)$, respectively). Our signal will be down-converted in frequency but attenuated and shifted in phase by the channel. In the Fig. 20.20, in the second row of plots, on the left, we see 4-QAM constellation diagram deformed by channel: attenuated and rotated. If our information is coded in carrier amplitude and phase, we have a big problem: channel influence has to be estimated (channel estimation) and corrected (channel equalization).

Imperfect Receiver Elements In the receiver we have filters separating us from near-by frequency services. How good are they?

For sure the most critical parts of the receiver are the frequency down-converter and the analog-to-digital converter. Frequency and phase of the down-converter should be exactly equal to the acquired signal carrier. If carrier of the received signal is equal to $e^{j(\Omega_c n + \phi)}$, the down-converter should multiply the received signal by $e^{-j(\Omega_c n + \phi)}$ (notice the negation sign!). But what will happen when down-converter parameters are different: $e^{-j((\Omega_c + \Delta\Omega)n + \phi + \Delta\phi)}$? Carrier frequency offset $\Delta\Omega$ will cause permanent constellation rotation (see plot in second column and row in Fig. 20.20, with constellation points lying on a circle), while carrier phase offset will rotate the constellation once. And again we have a big problem! Both carrier offsets should be estimated and eliminated.

The ADC converter can work with wrong sampling ratio giving samples to fast or two slow. If we are assuming that the sampling is correct, but it is not, we are applying to signal too low or too high frequency of the down-converter, and again we have a frequency carrier offset and observe *circles* in constellation diagrams (last plot in the Fig. 20.20. The AD converter error is specified in parts per million (ppm). Error equal $ppm = 100 \cdot 10^{-6}$ means that we can expect 100 samples more or less for 1 million samples given us by the ADC. For example, for sampling frequency $f_s = 200$ kHz we will have 20 samples more or less per one second. Therefore offset of the carrier frequency will be equal to 20 hertz. Generally if our carrier is incorrectly sampled:

$$y_1(n) = \exp\left(j2\pi \frac{f_c}{(1+\varepsilon)f_s} n\right) \qquad (20.23)$$

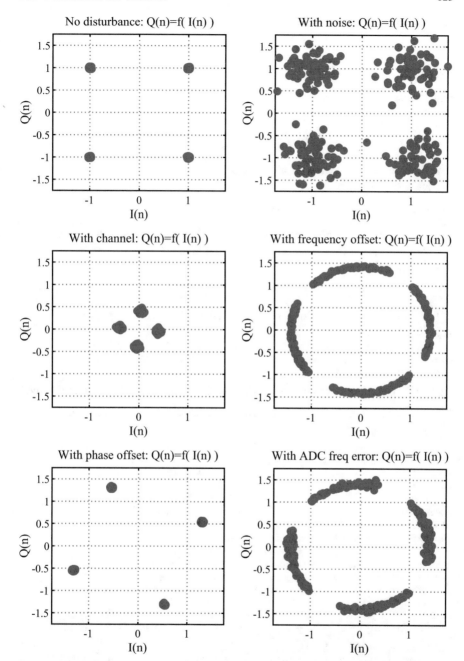

Fig. 20.20: Observed carrier constellation points for $I(n) + jQ(n)$ signal with 4-QAM modulation disturbed in different way, plots left-right up-down: (1) for the original signal in TX, (2) with channel noise SNR=0 dB, (3) after fading channel (attenuation, phase shift), (4) with carrier frequency offset in RX, (5) with carrier phase offset in RX, (6) with ADC frequency sampling error in RX

we have to solve the following equation:

$$\frac{f_c}{(1+\varepsilon)f_s} = \frac{f_c}{f_s} + x \quad \rightarrow \quad x = \frac{\left(\frac{-\varepsilon}{1+\varepsilon}\right)f_c}{f_s} \tag{20.24}$$

and correct sampled signal:

$$y_2(n) = y_1(n) \cdot \exp\left(j2\pi\frac{\varepsilon}{1+\varepsilon}\frac{f_c}{f_s}n.\right) \tag{20.25}$$

Therefore the absolute CFO error caused by ADC sampling rate error $\varepsilon = \text{ADCppm}$ is equal to:

$$\Delta f_c = \frac{-\varepsilon}{1+\varepsilon} \cdot f_c. \tag{20.26}$$

Therefore in our simplified simulations, the ADC sampling error will be represented as additional carrier frequency offset (CFO). Optionally, we can resample the signal what is more time-consuming.

Exercise 20.11 (ADC Sampling Ratio Error). Analyze program presented in Listing 20.6.

Listing 20.6: Simplified modeling in Matlab ADC sampling ratio error and its correction

```
1   % lab20_ex_adc.m
2
3   N=1000; n=0:N-1;                              % number of signal samples
4   fs=1e+6;                                      % sampling frequency
5   fc=1e+5;                                      % carrier frequency
6   ADCppm = 100                                 % ADC error in ppm
7   A = ADCppm*1e-6;                             % error scaling
8   x1 = exp(j*2*pi*fc/fs*n);                    % error-free signal
9   x2 = exp(j*2*pi*fc/((1+A)*fs)*n);           % erroneous signal with CFO
10
11  x1c = x2 .* exp(j*2*pi*A/(1+A)*fc/fs*n);    % CFO correction
12  error = max( abs( x1 - x1c ) ), pause       % correction error
```

In Fig. 20.21 we see how harmful the transmission disturbances can be, when occurring together, for 4-QAM and 16-QAM modulations. Due to length of this chapter, we will try to show solutions to all listed above transmission obstacles in the following chapter.

But now, a few exercises. Using program 20.4 we can simulate problems appearing in the single-carrier transmission: channel disturbances, noise presence, and

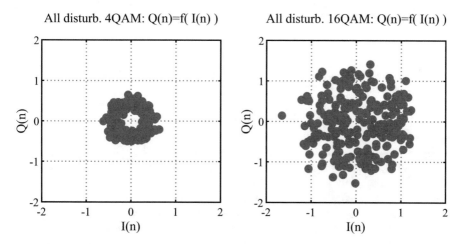

Fig. 20.21: Observed carrier constellation points for $I(n) + jQ(n)$ signals with 4-QAM (left) and 16-QAM modulation (right) disturbed by the all factors appearing together

receiver imperfections: offsets of carrier frequency and phase as well as ADC frequency sampling error using Eq. (20.23). In order to do it, it is sufficient to turn on the option do_dsturb=1 in the program. The ADCppm error is simulated by changing the carrier frequency.

Exercise 20.12 (Rainy Days: Problems in Single-Carrier Transmission). Analyze fragment of the program code in Listing 20.4 which is inside the if (do_disturb==1) xxx; end conditional field. Recognize parameters of different disturbances, find where they are used. Initially set their values to:

```
SNR=160; chanG=1; chanPh=0; carDf=0.; carDPh=0;
```

Then change each value individually to the one used in Listening 20.4. Run the program, observe figures. Compare input and output values of I_k and Q_k (last figure). You can also execute the frequency up-down pulse shaping program from the lab20_ex_IQpoints.m code and observe how well the text is decoded in the presence of different transmission imperfections. What disturbance is the most harmful? Then increase number of disturbances acting together and repeat experiments.

20.8 Summary

First round of the single-carrier marathon has ended. What should be remembered?

1. Bits are sent as numbers of predefined carrier states. Transmitted single sine/cosine is changing its frequency, amplitude, and phase, possible combinations are limited and precisely numbered. In the receiver, carrier parameters are measured (estimated), a carrier state number is found, bits are extracted from it.

2. Choosing modulation type we should take into account transmission conditions and select changing this carrier parameter which is the most robust to channel destruction. Channel features are characterized by its impulse response and resulting from its frequency response. Magnitude and phase frequency responses of a channel tell us how our carrier will be attenuated and delayed by a channel. Additionally, a channel can be noisy. And interference from side services could exist.

3. To limit the service bandwidth, switching the carrier state should not be abrupt because discontinuous signals have wide spectra. Wide service spectrum could disturb neighbor spectra of near-by services. For this reason the carrier has to pass smoothly from state to state. This effect is obtained by carrier states interpolation by a pulse shaping filter.

4. Raised cosine (RC) pulse shaping filter also should be remembered. It is a very good, low-pass interpolation filter, limiting frequency bandwidth of the single-carrier transmission. Its frequency response has a shape of raised cosine. If it is so perfect, why the square root of RC filter is used in the transmitter? Because in the receiver the second low-pass filter has to be applied, and we are interested in having perfect interpolation of carrier states at the end of processing path, not only in the transmitter. Cascade of two SRRC filters, one in the transmitter and one in the receiver, gives as a result a perfect RC interpolation filter during symbol decoding.

5. But always something is for something! The smoothly changed carrier states are more difficult to detect. This is the price paid for limiting the service bandwidth. Therefore special procedures have to be designed to sampling the carrier in the correct moment. They are called timing recovery methods. They will be discussed in the following chapter. In this one, perfect time synchronization was assumed.

6. Apart from signal deformations caused by a transmission channel and problems with carrier sampling in proper moments, there are also hardware imperfections which make the transmitted bits recovery difficult. For example, during signal frequency down-converting in the receiver, wrong carrier frequency and phase can be used which results in deformation of the carrier state constellation grid. The A/D converter can work also with wrong sam-

pling ratio causing the phase error also. Methods solving these problems are presented in the next chapter.

7. Finally, in single-carrier communications different definitions/quantities are used, and we have to be familiar with them, to mention only the most important: symbols as carrier states, constellations/grids of carrier states, and different measures of transmission quality: bits per second, symbols per second, eye-diagrams, phasor-diagrams.

8. In good transmission systems the $I(n) + jQ(n)$ signals are crossing in the same points in eye-diagrams and phasor-diagrams. In such situation we are telling that these eye/phasor-diagrams are open, because we see free space between carrier line trajectories.

20.9 Private Investigations: Free-Style Bungee Jumps

Exercise 20.13 (Speech/Audio Transmission). In this chapter we have transmitted text as exemplary data. Record short fragment of your own speech. Modify the programs and transmit the speech in place of the "Hello World!" message. Check the speech quality in the receiver when no disturbance is present. After that *inject* every disturbance separately, increase its level and listen to the decoded signal. Test different digital modulation schemes, i.e. 2-PAM ... 16-QAM. Exchange speech with an audio file of your favorite song and repeat the experiment.

Exercise 20.14 (Image Transmission). At present try to transmit any image. May be your own photo. Modify the programs. Check the received image quality in disturbance-free conditions. Next add every disturbance separately, increase its level and evaluate the decoded image quality visually.

Exercise 20.15 (64-QAM and 256-QAM Encoder and Decoder).** Add to the functions, presented in this chapter, possibility of encoding and decoding 64-QAM and 256-QAM modulations. Apply them to transmission of a longer text message. Observe eye and phasor-diagrams when disturbances are absent and present. Check number of transmission errors in both cases. Display the decoded text.

Exercise 20.16 (* Frequency Shift Keying).** Instantaneous frequency is defined as a derivative of a signal phase. Set carrier amplitude to one and code sent bits in linearly increasing or decreasing carrier phase. Each carrier symbol should have different phase change. In order to limit the signal spectrum, filter the sequence of carrier symbols, with abrupt changes in connection points, by a low-pass pulse shaping filter (channel filter). Try to decode the binary message.

References

1. J.B. Anderson, *Digital Transmission Engineering* (IEEE Press/Wiley-Interscience, Piscataway NJ, 2005)
2. K. Barlee, D. Atkinson, R.W. Stewart, L. Crockett, *Software Defined Radio using MATLAB & Simulink and the RTLSDR*. Matlab Courseware (Strathclyde Academic Media, Glasgow, 2015). Online: http://www.desktopsdr.com/
3. T.F. Collins, R. Getz, D. Pu, A.M. Wyglinski, *Software Defined Radio for Engineers* (Analog Devices - Artech House, London, 2018)
4. B. Farhang-Boroujeny, *Signal Processing Techniques for Software Radios* (Lulu Publishing House, 2008)
5. M.E. Frerking, *Digital Signal Processing in Communication Systems* (Springer, New York, 1994)
6. C.R. Johnson Jr., W.A. Sethares, *Telecommunications Breakdown: Concepts of Communication Transmitted via Software-Defined Radio* (Prentice Hall, Upper Saddle River, 2004)
7. C.R. Johnson Jr, W.A. Sethares, A.G. Klein, *Software Receiver Design. Build Your Own Digital Communication System in Five Easy Steps* (Cambridge University Press, Cambridge, 2011)
8. L.C. Potter, Y. Yang, *A Digital Communication Laboratory - Implementing a Software-Defined Acoustic Modem*. Matlab Courseware (The Ohio State University, Columbus, 2015). Online: https://www.mathworks.com/academia/courseware/digital-communication-laboratory.html
9. J.G. Proakis, M. Salehi, G. Bauch, *Modern Communication Systems Using Matlab* (Cengage Learning, Stamford, 2004, 2013)
10. M. Rice, *Digital Communications: A Discrete-Time Approach* (Pearson Education, Upper Saddle River, 2009)
11. B. Sklar, *Digital Communications: Fundamentals and Applications* (Prentice Hall, Upper Saddle River, 1988, 2001, 2017)
12. S.A. Tretter, *Communication System Design Using DSP Algorithms* (Springer Science+Business Media, New York, 2008)

Chapter 21
Digital Single-Carrier Receiver

The telecommunication receiver drama: putting together all
pieces of a broken crystal vase.

21.1 Introduction

A digital telecommunication receiver is much more complicated than a digital transmitter. The transmitter works in sterile conditions up to the D/A converter. Then different obstacles begins:

1. transmitter hardware imperfections: wrong D/A sampling ratio, wrong up-converter frequency in the transmitter,
2. channel disturbances: signal attenuation, delay, Doppler frequency shift and interference from some other signals/services,
3. receiver hardware imperfections (wrong down-converter frequency and phase in the receiver, wrong A/D sampling ratio, hardware thermal noise).

Seeing clear constellations of carrier states in the receiver is only a dream in the beginning. A lot of hard work has to be done to obtain an acceptable picture.

Receiver synchronization and channel equalization are crucial tasks in all telecommunication receivers, in the discrete-time ones also. I hope that at present, after reading the previous, introductory chapter on single-carrier transmission, all of us are totally convinced of that. These topics are very difficult and well explained only by the greatest telecommunications guru. Unfortunately, it is not me. I would like to apologize for that! Nevertheless, I will try to do my best in a few following pages.

In digital telecommunication standards, generation of bit-streams and corresponding discrete-time $I(n), Q(n)$ signals, sent to AD converters, is perfectly specified. Receiver algorithms, dealing with bit recovery from the $I(n), Q(n)$ signals, are very well documented also, but mainly for the noise&disturbance-free scenarios. In such case, all transmitter blocks are un-do step-by-step in the

© Springer Nature Switzerland AG 2021
T. P. Zieliński, *Starting Digital Signal Processing in Telecommunication Engineering*, Textbooks in Telecommunication Engineering,
https://doi.org/10.1007/978-3-030-49256-4_21

receiver in reverse order. Synchronization procedures, required for *real-world* transmission conditions, typically are not part of a standard. Telecommunication equipment vendors compete in the market in solving problems better. I remember very well my first attempts in decoding DAB radio IQ signals. Everything went well for *clear* transmitter IQ streams. In contrary to real-world IQ DAB recordings with carrier offsets and wrong sampling frequency of AD converters for which software, used by me, collapsed. Recently, I had similar situation with synthetic and recorded LTE signals: for the first of them, I saw very well sharp grids of QAM constellation points, while for the second only *thousands of stars in the sky*.

Dr Marek Sikora from Telecommunication Department, AGH Technical University of Science and Technology, Krakow, Poland, for His priceless help with preparing programs for this chapter!

My next comment: universal solutions are not optimal. In digital telecommunication systems numerous modulation types as well as numerous synchronization and error correction codes are used. It is not a problem to modulate and code in the transmitter the original, *clean* data. The problem is to perform efficiently in the receiver: synchronization, demodulation, and error correction of the disturbed, *dirty* data. There are many possible *un-do* operations for each *do*. It is impossible in this book to discuss all existing solutions which can be used in the receiver. No. Our goal is to explain only source of the problem, its consequences and first intuitive *software emergency exits* from the signal processing point of view. When the issue and its first, simple solution are well understood, it is much easier to analyze more difficult state-of-the-art algorithms. Unfortunately, the advanced algorithms are not discussed in this book—this is not our intention. There are many good books addressing professional high-tech telecommunication solutions. But they are very difficult to understand for the beginners.

In this chapter we analyze an example of a Wi-Fi-like bit packet transmission, in which regular bit-stream is proceeded by a synchronization pattern, e.g. Barker or Gold code, which is used for complete identification of existing problems and their solution: frame/symbol synchronization, frequency and phase carrier frequency offset estimation and correction as well as channel estimation and equalization. We will use Gardner and Mueller–Müller timing recovery methods with optional Farrow filter interpolator and test influence of signal oversampling (number of samples per symbol) for bit detection quality.

The slow-mode AMSAT nano-satellite down-link signal and RDS FM radio signal will serve us in this chapter as examples for testing different DSP algorithms used in digital single-carrier receivers. We will decode bits of both signals which are using the BPSK modulation. The RDS signal demodulation makes use of the PLL or Costas loop and one of the three symbol timing recovery methods: early-late-gate, Gardner, and Mueller–Müller.

21.2 Receiver Synchronization and Channel Equalization Program

We will try to attack difficult topics mentioned in the introduction using simple examples and simple, intuitive solutions. What do you think: is it a good idea to discuss difficult topics using a simple example? Yes? Therefore in the beginning of this section I will present a complete, exemplary, working program of a single-carrier receiver and will refer to it in the following subsections during describing specific issues. A full version of the program, with some extra figures, is available in the book repository. Since the program is well commented and self-presenting, it will be not described now line by line. Its analysis is a Reader's task. In comparison with our previous stand-alone programs 20.2 and 20.4, program presented in the Listing 21.1 uses simple functions in the transmitter part, which should be well recognized at present, and concentrate on the receiver synchronization code. Here algorithms of exemplary solutions are explicit given. The program is coding and decoding carrier state numbers for implemented digital modulation types and is calculating number of transmission errors.

In the program the following denotations are used:

- `IQk`—complex-value carrier states $I(k) + jQ(k)$,
- `IQn`—samples of interpolated complex-value signals $I(n) + jQ(n)$,
- `numX`—numbers of carrier states coding a variable X,

In the program user can choose:

- different modulation type (`modtype()`),
- RC or SRRC pulse shaping filter used in TX and RX (`rctype()`),
- doing frequency UP/DOWN conversion or not, together with channel simulation at target frequency (`do_updownchan()`),
- adding disturbances in the base-band or not (`do_disturb()`), allowing also channel modeling as one complex-value tap,
- doing synchronization in the RX or not (`do_synchro()`), with synchronization algorithm selection—none, using pure pilot signals or their differential versions—and carrier frequency offset algorithm selection—simple or polynomial fitting (`do_cfoequ`),
- doing `IQ` signal decimation in the RX or not (`do_decim()`), with decimation order selection (`Mdecim`).

The following functions are used in order to reduce the code length:

- `IQdef()`—generation of possible carrier states (`IQcodes`) for a chosen modulation type,

- `modtype2header()`—generation of carrier state numbers of the header for a chosen modulation,
- `numbers2IQ()`—conversion of carrier state numbers to carrier states IQk, taking only values specified by IQcodes,
- `IQ2numbers`—conversion of IQk states to carrier state numbers,
- `IQ2psf()`—passing IQk carrier states through a pulse shaping filter and obtaining smooth, transmitted IQn signals,
- `IQ2disturb()`—addition of disturbances and receiver imperfections in the base-band.

Exercise 21.1 (Understanding the Single-Carrier Transmission Program).
Analyze code of program presented in Listing 21.1. Try to understand: (1) transmitter part, (2) frequency up-down converter part, (3) disturbance addition part, (4) signal decimation in the receiver. Run the full version of the program available in the book repository. Observe figures, notice encountered numbers of transmission error. Change modulation type, switch ON/OFF different DO options.

Listing 21.1: Exemplary Matlab program of a single-carrier receiver

```
1   % lab21_ex_receiver.m
2   % Testing synchronization procedures in single carrier receiver
3   clear all; close all;
4
5   modtype = '4QAM';   % 2PAM, 4PAM, 8PAM, BPSK, QPSK, DQPSK, 8PSK, 4QAM, 16QAM
6   rctype = 'sqrt';    % 'normal, 'sqrt' : raised cosine filter type for TX and RX
7   Ndata = 250;        % number of carrier states (symbols) to be transmitted
8   K = 24; Ns = 8;     % PSF: samples per symbol, symbols per pulse shaping filter (PSF)
9   r = 0.35;           % PSF: filter roll-off factor
10  fs = 240000;        % sampling frequency in Hz: 1, 250e+3, 240e+3
11  fcar = 40000;       % carrier frequency in Hz
12
13  do_updownchan = 0;  % 0/1 frequency up-down conversion plus channel simulation
14  do_disturb = 0;     % 0/1 addition of disturbances in the baseband
15  do_synchro = 0;     % 0/1/2: 0 = none, 1 = using x(n),s(n), 2=using xD(n),sD(n)
16  do_decim = 0;       % 0/1 signal decimation in the receiver
17  do_cfoequ = 2;      % 0/1/2 freq carrier offset estimation: 0=none, 1=simple, 2=polyfit
18  do_chanequ = 3;     % 0/1/2/3 channel correction methods: 0=none, 1=ZF, 2=LS/MSE, 3=NLMS
           filter
19
20  %chan = [ 0.5, -0.25, 0.1, -0.1 ]; % channel impulse response in baseband symbol-spaced
21  chan = [ 1 ];                     % perfect channel
22  SNR=160; chanG=1; chanPh=0; carDF=0.0000; carDPh=0; ADCdt=0;        % No disturb
23  %SNR=40; chanG=0.25; chanPh=pi/7; carDF=0.0002; carDPh=pi/11; ADCdt=0.5; % Disturb
24  Mdecim = 1;         % decimation order, Mdecim = 1, 2, 3, 4, 6, 8, 12, 24
25  Mdelay = 0;         % decimation delay: in samples before decimation
26
```

```
27   % Old transmission part - as before
28   Npsf = K*Ns+1; Mpsf = (Npsf-1)/2;
29   [IQcodes, Nstates, Nbits, R ] = IQdef( modtype );              % take carrier IQ codes
30   % IQk of Header
31   [numHead, Nhead ] = modtype2header( modtype );                % take header IQ numbers
32   IQkHead= numbers2IQ( numHead, modtype, IQcodes );             % calculate IQ states
33   % IQk of Data
34   numData = floor( Nstates*(rand(Ndata,1)-10*eps) );            % generate random IQ numbers
35   IQkData = numbers2IQ( numData, modtype, IQcodes );            % calculate IQ states
36   % Numbers ALL, IQk ALL
37   num = [ numData' numHead' numData' numHead' numData' ];       % ALL transmitted IQ numbers
38   IQk = [ IQkData IQkHead IQkData IQkHead IQkData ];            % ALL transmitted IQ states
39   % IQn of Header only (pulse shaping)
40   IQnHead = IQ2psf( IQkHead, K, Ns, r, 'normal' );              % IQn of Header
41   % IQn of everything (pulse shaping)
42   [IQn, hpsf ] = IQ2psf( IQk, K, Ns, r, rctype );              % IQn of ALL
43
44   if(0) % 0/1 Testing symbol timing recovery methods
45       alg = 1; % 1/2/3/4/5 timing recovery algorithm
46       for Mdelay = 1 : 1 : K
47           Mdelay
48           IQnn = IQn( 1+Mdelay : Mdecim : end ); n1st = 1;
49           [ ns, IQs ] = timing_recovery( IQnn, alg, K/Mdecim, n1st );
50       end
51   end
52
53   % Optional frequency UP and DOWN conversion in TX plus channel simulation
54   if( do_updownchan )
55       if( length( chan ) > 1 )            % when more than one channel/filter weight
56           chan = resample(chan,K,1);      % upsampling channel impulse response
57           figure; plot(chan); title('h(n)'); pause
58       end
59       df = 0; dphi = 0; % CFO added in the base-band OR df = carDF*fs; dphi = carDPh;
60       IQn = IQupchandown( IQn, fcar, fs, chan, df, dphi );
61   end
62
63   % Addition of disturbances in the base-band
64   if( do_disturb )
65       IQn = IQdisturb( IQn, SNR, chanG, chanPh, carDF, carDPh, ADCdt, Npsf );
66   end
67
68   % Possible signal down-sampling in the receiver
69
70   if( do_decim )
71      if(M>1)
72       M = Mdecim;                         % copy initial program parameter
73       IQn = [ zeros(1,Mdelay) IQn ];      % signal delay
74       IQn = resample(IQn, 1, M);          % low-pass filtering and decimation
75       IQnHead = resample(IQnHead, 1, M);  % low-pass filtering and decimation
76     % IQnHead = IQnHead(1:M:end);         % only decimation
77       hpsf = M*hpsf(1:M:Npsf); K=K/M; Npsf=(Npsf-1)/M+1; Mpsf=(Npsf-1)/2;
78       N = length(IQn); N = floor(N/K)*K; n = 1:N; IQn = IQn(n);
79      end
80   else M=1;
```

```
81   end
82
83   % Low-pass pulse shaping filter in the receiver
84   IQn = conv( IQn, hpsf );
85
86   s = IQnHead;    % IQn of synchronization header
87   x = IQn;        % IQn of received signal in the base-band
88
89   % Auto correlation function
90   sD = s(2:end) .* conj(s(1:end-1));                        % sD(n)
91   ms = mean(numHead); Css1 = conv( numHead-ms, numHead(end:-1:1)-ms );  % auto corr
         NumHead
92   ms = mean(s);       Css2 = conv( s-ms, conj(s(end:-1:1)-ms) ); % auto corr s(n)
93   ms = mean(sD);      Css3 = conv( sD-ms, sD(end:-1:1)-ms );     % auto corr sD(n)
94
95   % Cross correlation function
96   Cxs = conv( x-mean(x), conj(s(end:-1:1)-mean(s)) );    % cross corr of s(n) with x(n)
97   xD = x(2:end)  .* conj( x(1:end-1));                   % signal xD(n)
98   CxDsD = conv( xD-mean(xD), sD(end:-1:1)-mean(sD) );    % cross corr of sD(n) with xD(n)
99
100  if( do_synchro == 0 )   % NO SYNCHRONIZATION
101
102     IQn = IQn( 2*Npsf-1 : end - (2*Npsf-1)+1 );                % remove transients
103     received = IQ2numbers( IQn(1:K:end), modtype );           % decimate
104     errors = sum( received ~= num( Ns + 1 : end-Ns ) ), pause % errors
105
106  else                       % SYNCHRONIZATION
107
108     Nsync = length( IQnHead );
109     if( do_synchro==1 ) % less robust to noise, using x(n), s(n)
110        [dummy, nmax ] = max( abs( Cxs ) );   % maximum position
111        n1st = nmax - Nsync + (Npsf-1)/2 + 1,  % 1st header symbol
112     else                    % more robust to noise, using xD(n),sD(n)
113        [dummy, nmax ] = max( abs( CxDsD ) ); % maximum position
114        n1st = nmax - Nsync + (Npsf-1)/2 + 2,  % 1st header symbol
115     end
116
117     % Use only header symbols for frequency offset and phase shift estimation
118     nsynch = n1st : K : n1st+(Nhead-1)*K; nhead = Mpsf+1 : K : Mpsf+1+(Nhead-1)*K;
119     work = IQn( nsynch ) .* conj( IQnHead( nhead ) ); % # the same
120     % work = IQn( nsynch ) .* conj( IQkHead );        % #
121     if( do_cfoequ == 0 ) df=0; dph=0; end
122     if( do_cfoequ == 1 )       % simple frequency carrier offset estimator
123        df = mean( angle( conj( work(1:end-1) ) .* work(2:end) ) ); % simple version
124        df = df / (2*pi*K); dph = 0;
125     end
126     if( do_cfoequ == 2 )       % phase polynomial fitting method
127        phi0 = angle(work(round(Nhead/2))); work=work.*exp(-j*phi0);
128        % phi0=0;
129        ang = unwrap( angle(work) ); nn = 0 : K : (Nhead-1)*K;
130        temp = polyfit( nn, ang, 1); df = temp(1)/(2*pi); dph = temp(2)+phi0;
131           figure;
132           plot( nn,ang,'b-',nn,temp(2)+temp(1)*nn,'r-'); grid; title('Angle'); pause
133        if( carDF == 0) allPhase_estim = dph, expected = (chanPh+carDPh), end
```

```
34        chanPh = 0; carDPh = 0;
35      end
36    % Do frequency offset and phase correction - from channel and freq-down converter
37      IQn = IQn( n1st : length(IQn) ) .* exp(-j*(2*pi*df*(0:length(IQn)-n1st)+dph));
38      N=length(IQn); n=1:N; ns=1:K:N;
39    % Results
40      carOffs_estim = df/M,  expected = carDF, pause
41
42    % Amplitude and phase correction - using header for channel estimation & correction
43    % Knowing input and output, we can estimate a channel and correct it
44      IQkHeadEstim = IQn( 1 : K : 1+(Nhead-1)*K );              % detected header states
45      if( do_chanequ==0 ) gain=1;dph = 0; end                  % no corrector
46      if( do_chanequ==1 ) % one-tap corrector
47        gains = IQkHeadEstim .* conj(IQkHead) ./ abs(IQkHead).^2; % compared with known
48        gain = mean(real(gains)) + j*mean(imag(gains));         % mean channel "gain"
49        IQn = IQn / gain;                                       % signal correction
50      end
51      % ... methods do_chanequ==2 and do_chanequ==3
52    % Results
53      if( carDF == 0) allPhase_estim = angle(gain), expected = chanPh+carDPh, end
54      chanGain_estim = abs(gain), expected = chanG, pause
55    % Errors
56      received = IQ2numbers( IQn( 1 : K : (Nhead+Ndata)*K ), modtype );
57    % errors = sum( received ~= [ numHead; numData ] ), pause
58      errors = sum( received(Nhead+1:Nhead+Ndata)' ~= [ numData ] ), pause
59    end % end of synchronization using header
```

21.3 Preambles Detection and Frame/Symbol Synchronization

Transmission correction technology is nothing new. To trust in a measurement equipment we have to calibrate it. Calibration relies on measuring some *reference quantity* which is known, checking what the device is showing and changing the device setup (like gain and delay in oscilloscopes), in order to see on the display what is expected. The same is true for telecommunication apparatus: some signals known for receivers (i.e. *headers, pilots*) should be transmitted and the receivers, having received data and knowing the transmitted ones, should estimate values of disturbing parameters. These values should be used next, for a given period of time, for correction of all received data. Therefore, in front of unknown data bits some known preamble/header bits should be sent, giving receiver a chance for disturbance estimation. Usage of such headers has an additional benefit. When we detect them, they can be used not only for adjustment of receiver parameters but also for finding the data frame beginning (frame recovery)—since data are after the header—and even for finding symbol position in received signals (symbol timing recovery)—since a header consists of symbols also. Wow, wow, wow! Like in the movie "Little Big Man." All thanks to our small, very useful header!

A good preamble/header is such a sequence of carrier states numbers which is converted by a pulse shaping filter into a signal $I(n) + jQ(n)$ having very *sharp* auto-correlation function, i.e. with high, distinctive maximum for zeroth shift and small values for others shift arguments. Having this feature, a header well *fits/matches* to itself and its position in the received signal can be found precisely. Signals with *good/sharp* auto-correlation functions are essential in all echography techniques, radars, and sonars, where we correlate reflected signals with transmitted ones in order to find copies of transmitted signals. Then we calculate distance and velocity of an object from which transmitted signals were reflected.

Sequence of numbers describing preambles also well correlates with itself allowing precise finding preamble position in the detected sequence of carrier states. This way beginning of our own data can be found, i.e. data frame recovery/synchronization is performed.

Numerous good synchronization sequences (headers/preables) are used in telecommunication receivers, among them codes of Barker and Gold, complementary/maximum length sequences and Zadoff–Chu sequences. Values of exemplary preambles, proposed for use in our tests, are given in Listing 21.2 as a part of our target single-carrier transmission program given in the repository (lab20_ex_single_carrier.m).

Listing 21.2: Examples of some synchronization headers

```
 1  function [ synch, Nsynch ] = modtype2header( modtype )
 2  % Generation of synchronization headers for different modulations
 3
 4  switch modtype
 5  case {'2PAM','BPSK' }
 6    synch = [ 1 1 1 0 0 0 1 0 0 1 1 0 1 0 1 1 1 0 0 0 1 0 0 1 1 0 1 0 1 1 ...
 7              1 1 0 0 0 1 0 0 1 1 0 1 0 1 ]; % pseudo random binary sequence
 8  %synch=[ 0 1 0 0 0 0 0 0 0 1 1 1 0 0 0 0 0 1 0 0 1 1 1 0 1 1 0 0 0 1 1 0 ];   % Gold 32
 9  %synch=[ 0 0 0 0 0 0 1 0 0 0 1 1 0 1 1 0 0 0 0 1 1 0 0 1 1 1 0 0 1 1 ];   % Gold #2, 30
10  %synch=[ 1 0 0 0 1 1 1 1 1 1 0 0 0 0 1 0 0 0 1 1 1 1 0 0 0 1 0 0 1 0 0 ];   % Gold #3, 30
11  %synch=[ 1 1 1 1 1 0 0 1 1 0 1 0 1 ]; synch = [ synch synch synch ]        % 3x Barker13
12  case {'4PAM','QPSK','DQPSK','4QAM' }
13  %synch=[ 3 0 0 1 2 1 3 0 3 2 2 1 3 0 0 1 2 1 3 ]; synch=[synch synch(end:-1:1)]; % STS
14    synch=[ 0 3 0 0 0 0 0 0 0 3 3 3 0 0 0 0 0 3 0 0 3 3 3 0 3 3 0 0 0 3 3 0 ]; % Gold 32
15  % synch=[ 3 3 3 0 0 0 3 0 0 3 3 0 3 0 3 3 3 3 0 0 0 3 0 0 3 3 0 3 0 3 3 ...
16  %         3 3 0 0 0 3 0 0 3 3 0 3 0 3 ]; % pseudo random binary sequence
17  % Nsynch=99; synch = floor( Nstates*(rand(1,Nsynch) - 10*eps) );
18  case {'8PAM','8PSK' }
19    synch = [ 1, 5, 1, 3, 6, 1, 3, 1, 1, 6, 3, 7, 7, 3, 5, 4, 3, 6, 6, 4, 5, 4, 0, ...
20              2, 2, 2, 6, 0, 7, 5, 7, 4, 0, 7, 5, 7, 1, 6, 1, 0, 5, 2, 2, 6, 2, 3, ...
21              6, 0, 0, 5, 1, 4, 2, 2, 2, 3, 4, 0, 6, 2, 7, 4, 3, 3, 7, 2, 0, 2, 6, ...
22              4, 4, 1, 7, 6, 2, 0, 6, 2, 3, 6, 7, 4, 3, 6, 1, 3, 7, 4, 6, 5, 7, 2, ...
23              0, 1, 1, 1, 4, 4, 0, 0, 5, 7, 7, 4, 7, 3, 5, 4, 1, 6, 5, 6, 6, 4, 6, ...
24              3, 4, 3, 0, 7, 1, 3, 4, 7, 0, 1, 4, 3, 3, 3, 5, 1, 1, 1, 4, 6, 1, 0, ...
25              6, 0, 1, 3, 1, 4, 1, 7, 7, 6, 3, 0, 0, 7, 2, 7, 2, 0, 2, 6, 1, 1, 1, ...
26              2, 7, 7, 5, 3, 3, 6, 0, 5, 3, 3, 1, 0, 7, 1, 1, 0, 3, 0, 4, 0, 7, 3 ];
```

```
27    case{'16QAM'}
28       synch = [ 8 8 8 2 2 2 8 2 2 8 8 2 8 2 8 8 8 2 2 2 8 2 2 8 8 2 8 2 8 8 ...
29                 8 8 2 2 2 8 2 2 8 8 2 8 2 8 ]; % pseudo random binary sequence
30    otherwise disp('Unknown modulation type');
31       Nsynch=99; synch = floor( Nstates* (rand(1,Nsynch) - 10*eps) );
32    end
33    synch = synch' ; Nsynch=length(synch) ;
```

At present we will look more carefully to one arbitrary chosen synchronization header. *The Wheel of Fortune* is rotating and . . . *Bonk!* . . . the 32-element binary Gold code is selected. Since we will exploit the 4-QAM modulation with 4 states (00, 01,10, 11), we are changing 0/1 elements of the code to two opposite modulation states 0 (00) and 3 (11). Results of our *inquiry* are presented in Fig. 21.1. Covariance functions (correlation functions of sequences with subtracted mean values) were calculated using Matlab conv() function and not scaled. In the first plot the header sequence values (0/3) are shown (left up). In the second plot (right up), the auto-covariance function of the sequence is depicted. Next, in the left-center plot, absolute values of the auto-covariance function of the $K = 24$-times up-sampled and SRRC filtered Gold sequence are shown (8 symbols per PSF filter, $r = 0.35$). Let us denote this interpolated synchronization sequence as $s(n)$. In turn, in right-center plot, similar auto-covariance function is presented but calculated for artificially created signal $s_\Delta(n) = s(n)s^*(n-1)$ which allows proper detection of the header position even in the presence of bigger carrier frequency offsets, which will be shown later. Finally, in two bottom plots, results from synchronization preamble application are shown in noise/disturbance-free scenario. In the first case, in left-bottom plot, cross covariance between the synchronization signal $s(n)$ and the received signal $r(n)$ is shown, while in the second right-bottom plot—cross covariance between the synchronization signal $s_\Delta(n)$ and the differential received signal $r_\Delta(n) = r(n)r^*(n-1)$, calculated similarly as $s_\Delta(n)$.

As we see, for un-correlated random sequences, all auto covariance functions have well distinguished peaks for shift $k = 0$. Thanks to this feature, one can localize precisely the header position in the received signal which is seen in two bottom plots.

Figure 21.1 shows that, in low-noise and without carrier frequency offset, both synchronizations techniques can be used for header detection and data frame synchronization. However, the situation changes completely when strong carrier frequency offset occurs. Such case is presented in Fig. 21.2 for offset equal to 0.1% of the carrier value. Now, using the signal $s(n)$ for synchronization failed (left plot) while exploiting the signal $s_\Delta(n)$ ended with full success (right plot).

At present, fully convinced in efficiency of some header detection methods, we derive mathematical explanation of the observed experimentally phenomena. Let us denote the $IQ(n)$ synchronization signal as $s(n)$:

$$s(n) = A(n)e^{j\phi(n)}. \tag{21.1}$$

Let us assume additionally that our transmission is narrow-band, around frequency f_c, and the channel disturbance is described by one complex-value number, i.e. the

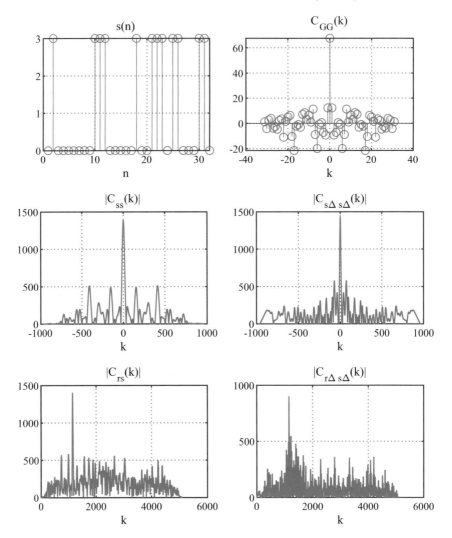

Fig. 21.1: Example of synchronization header usage in noise/disturbance-free situation: (left up): 32-element Gold binary synchronization sequence, (right up): its covariance function, (right-middle) auto-covariance function of PSF-interpolated header $s(n)$, (right-middle) auto-covariance function of sequence $s_\Delta(n) = s(n)s^*(n-1)$, (left-bottom) header position detection using signal $s(n)$—cross covariance with the received signal $r(n)$, (right-bottom) header position detection using signal $s_\Delta(n)$—cross covariance with the received signal $r_\Delta(n) = r(n)r^*(n-1)$

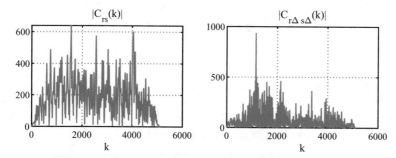

Fig. 21.2: Example of synchronization header usage in a presence of strong carrier offset (0.1% of the frequency carrier value): (left) header position is not detected using the reference signal $s(n)$ and cross covariance with the received signal $r(n)$, (right) header position is detected using the reference signal $s_\Delta(n)$ and cross covariance with the received signal $r_\Delta(n) = r(n)r^*(n-1)$

channel scales only transmitted data in amplitude (gain G) and shift them in phase (angle α):

$$h(n) = Ge^{j\alpha}. \tag{21.2}$$

Imperfect receiver adds to the received signal two offsets: a carrier frequency offset ($\Delta\Omega_c$) and a carrier phase offset (β):

$$c(n) = e^{j(\Delta\Omega_c n + \beta)}. \tag{21.3}$$

In consequence, the received signal is mathematically described as follows (first the transmitted signal $s(n)$, then the one-tap channel impulse response $h(n)$, finally the receiver down-converter error $c(n)$):

$$r(n) = c(n) \cdot h(n) \cdot s(n) = GA(n-M) \cdot e^{j(\Delta\Omega_c(n-M)+\phi(n-M)+\alpha+\beta)}, \tag{21.4}$$

where M denotes the signal delay expressed in signal samples.

Frame Synchronization Method #1: Sensitive to Carrier Frequency Offset We know what synchronization signal $s(n)$ was sent and should find its position in the received signal $r(n)$. Let us calculate cross-correlation function of these two signals:

$$\hat{R}_{sr}(k) = \sum_{n=0}^{N-1} r(n)s^*(n-k) =$$

$$= \sum_{n=0}^{N-1} \left(GA(n-M)e^{j(\Delta\Omega_c(n-M)+\phi(n-M)+\alpha+\beta)} \cdot A(n-k)e^{-j\phi(n-k)} \right) =$$

$$= Ge^{j(\alpha+\beta)}e^{-j\Delta\Omega_c M} \sum_{n=0}^{N-1} \left(e^{j\Delta\Omega_c(n)}A(n-M)A(n-k)e^{j(\phi(n-M)-\phi(n-k))} \right). \tag{21.5}$$

When carrier frequency offset is missing or it is very small ($e^{j\Delta\Omega_c n} \approx 1$), we have

$$\hat{R}_{sr}(k) \approx Ge^{j(\alpha+\beta)}e^{-j\Delta\Omega_c M}\sum_{n=0}^{N-1}\left(A(n-M)A(n-k)e^{j(\phi(n-M)-\phi(n-k))}\right) =$$

$$= Ge^{j(\alpha+\beta)}e^{-j\Delta\Omega_c M}R_{ss}(k-M). \quad (21.6)$$

As we see, absolute value of the cross-correlation function $R_{sr}(k)$ will have maximum for $k = M$ (index of $R_{ss}(k-M)$ maximum), when the synchronization signal $s(n)$ *finds itself* in the signal $r(n)$. The method can be used for very small carrier frequency offset and frequent repetition of the header.

Frame Synchronization Method #2: Robust to Carrier Frequency Offset Alternative solution for bigger carrier offsets is as follows. A differential version $s_\Delta(n)$ of the sent header (synchronization) signal $s(n)$ (21.1) is defined:

$$s_\Delta(n) = s(n)s^*(n-1) = A(n)A(n-1)e^{j(\phi(n)-\phi(n-1))} \quad (21.7)$$

together with the differential version $r_\Delta(n)$ of the received signal $r(n)$ (21.4):

$$r_\Delta(n) = r(n)r^*(n-1) =$$

$$= G^2 A(n)A(n-1)\cdot e^{j(\Delta\Omega_c n+\phi(n)+\alpha+\beta)}e^{-j(\Delta\Omega_c(n-1)+\phi(n-1)+\alpha+\beta)} =$$

$$= G^2 A(n)A(n-1)\cdot e^{j(\Delta\Omega_c+\phi(n)-\phi(n-1))}. \quad (21.8)$$

Auto-correlation of the signal $s_\Delta(n)$ is equal:

$$\hat{R}_{s_\Delta s_\Delta}(k) = \sum_{n=0}^{N-1} s_\Delta(n)s_\Delta^*(n-k) =$$

$$= \sum_{n=0}^{N-1}\left(A(n)A(n-1)e^{j(\phi(n)-\phi(n-1))}A(n-k)A(n-k-1)e^{j(\phi(n-k)-\phi(n-k-1))}\right)$$

$$(21.9)$$

while cross-correlation of the signals $s_\Delta(n)$ and $r_\Delta(n)$ can be expressed as:

$$\hat{R}_{r_\Delta s_\Delta}(k) = \sum_{n=0}^{N-1} r_\Delta(n)s_\Delta^*(n-k) =$$

$$= \sum_{n=0}^{N-1}\left(G^2 A(n-M)A(n-M-1)\cdot e^{j(\Delta\Omega_c+\phi(n-M)-\phi(n-M-1))}A(n-k)A(n-k-1)e^{j(\phi(n-k)-\phi(n-k-1))}\right) =$$

$$= G^2 e^{j\Delta\Omega_c}\sum_{n=0}^{N-1}\left(A(n-M)A(n-M-1)\cdot e^{j(\phi(n-M)-\phi(n-M-1))}A(n-k)A(n-k-1)e^{j(\phi(n-k)-\phi(n-k-1))}\right) =$$

$$= G^2 e^{j\Delta\Omega_c}R_{sDsD}(k-M). \quad (21.10)$$

Absolute value of the $R_{r_\Delta s_\Delta}(k)$ function does not depend on $\Delta\Omega_c$ and has maximum when signal $s_\Delta(n)$ *finds itself* in the signal $r_\Delta(n)$, i.e. for $k = M$, index of the $R_{sDsD}(k-M)$ maximum.

Exercise 21.2 (Where Does a Synchronization Header Start?). More carefully analyze code of the program presented in Listing 21.1, in the part concerning the header definition, its addition to the bit-stream and finding its occurrence in the received signal. Initially run program in disturbance-free scenario:

```
do_updownchan = 0; do_disturb = 0; do_synchro = 1; do_decim=1;
```

Observe shapes of auto and cross-correlation functions calculated for different headers, proposed for different modulations inside the function modtype2synchro. Choose different sequences if more are available. Generate and test longer headers using the program line (add it if necessary):

```
Nsynch=99; synch = floor ( Nstates*(rand(1,Nsynch) - 10*eps) );.
```

The longer the header is, the larger peaks should have the correlation functions and more robust to noise the header detection could be possible. Turn on the option do_disturb = 1. Then set different levels of disturbances in the line:

```
SNR = 40; chanG = 0.25; chanPh = pi/7;
carDF = 0.0001; carDPh = pi/11; ADCdt=0.5;
```

and run the program several times, checking visually shapes of the correlation functions and the calculated value nlst—found first header sample. It should be the same for headers having the same length.

21.4 Carrier Offsets Detection and Correction

After matching position of the synchronization signal $s(n)$ (21.1) to the received signal $r(n)$ (21.4) in the receiver, multiplication of the received and the sent header, conjugated and appropriately delayed, should us give:

$$d(n) = r(n)s^*(n - M) = GA^2(n - M) \cdot e^{j(\Delta\Omega_c(n-M)+\alpha+\beta)}, \qquad (21.11)$$

allowing identification of Ω_c and $\alpha + \beta$ from the angle: $\theta(n) = \Delta\Omega_c n + \alpha + \beta$ of the complex value (21.11).

Frequency Carrier Offset Estimation Method In this method only $\Delta\Omega_c$ value is estimated. Angles α and β are eliminated from Eq. (21.11) by calculating a temporary variable $z(n)$:

$$z(n) = d(n)d^*(n - 1) = G^2A^2(n - M)A^2(n - M - 1) \cdot e^{j\Delta\Omega_c}. \qquad (21.12)$$

Next, we compute angles of $z(n)$ for many indexes n and average results, obtaining relatively good estimator of the carrier frequency offset:

$$\Delta \Omega_c = \text{mean}\left(\angle z(n) \right). \tag{21.13}$$

The method can be further improved when only $z(n)$ values corresponding to IQ_k symbols of the synchronization sequence are used in Eq. (21.13):

$$\Delta \Omega_c = \text{mean}\left(\angle z(n) \big|_{n=k} \right). \tag{21.14}$$

This way we avoid problems of complex numbers with small real or imaginary part.

Joint Frequency Carrier Offset and Phase Offsets Estimation Method We can calculate many samples of $d(n)$ (21.11), than its angles $\theta(n)$, and finally perform least-squares polynomial line fitting:

$$\theta(n) = \Delta \Omega_c(n - M) + \alpha + \beta = \Delta \Omega_c n + \underbrace{\alpha + \beta - \Delta \Omega_c M}_{\gamma}, \tag{21.15}$$

to the known points $\{n, \theta(n)\}$. This way we can find optimal values of $\Delta \Omega_c$ and γ. In this method the phase offset problem is solved at the same time.

Both carrier frequency offset estimation methods, presented above, are implemented in the program 21.1. But they can be used ONLY when signals $s(n)$ and $r(n)$ are already synchronized! Results presented in Fig. 21.3 confirm their usefulness.

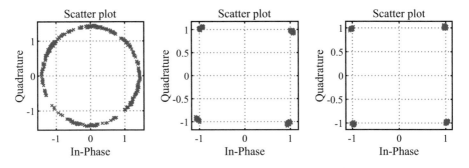

Fig. 21.3: Confirmation of CFO estimation methods efficiency. IQ scatter plots of detected 4-QAM carrier states for three cases, from left to right: no CFO correction, CFO correction using the simple method (mean value of the angle), CFO correction using polynomial fitting of the angle curve. CFO is the only imperfection, $K = 7$ samples per symbol, pulse shaping filter SRRC with $r = 0.35$

Exercise 21.3 (Carrier Frequency Offset Detection). Continue Exercise 21.2. Choose your favorite modulation, use the proposed header or generate yourself a random sequence of arbitrary selected length, run the program systematically increasing level of disturbances (noise, carrier offsets), choose different header synchronization methods ($r(n)s^*(n)$ and $r_\Delta(n)s^*_\Delta(n)$) and carrier frequency offset estimation methods (simple or (polyfit())). Observe final eye and phasor-diagrams of the enhanced received signal. Check how many transmission errors are found in each case.

21.5 Channel Estimation and Equalization

Repetition of Channel Basics Signal phase shift corresponds to signal time delay:

$$\cos(\omega_0 t - \phi_0) = \cos\left(\omega_0\left(t - \frac{\phi_0}{\omega_0}\right)\right), \quad e^{j(\omega_0 t - \phi_0)} = e^{j\omega_0\left(t - \frac{\phi_0}{\omega_0}\right)}. \qquad (21.16)$$

Signal acquired in the receiver is a result of convolution of the transmitted signal $x(t)$ with channel impulse response $h(t)$. Convolution of signals in time domain is equivalent to signal spectra multiplication in frequency domain:

$$y(t) = \int_{-\infty}^{\infty} x(\tau)h(t-\tau)d\tau = \int_{-\infty}^{\infty} h(\tau)x(t-\tau)d\tau \quad \Leftrightarrow \quad Y(f) = X(f)H(f). \quad (21.17)$$

For single frequency complex-value input signal:

$$x(t) = e^{j\omega t}, \qquad (21.18)$$

the channel output is equal:

$$y(t) = \int_{-\infty}^{\infty} h(\tau)x(t-\tau)d\tau = \int_{-\infty}^{\infty} h(\tau)e^{j\omega(t-\tau)}d\tau = \left[\int_{-\infty}^{\infty} h(\tau)e^{-j\omega\tau}d\tau\right]e^{j\omega t} =$$

$$H(\omega)e^{j\omega t} = |H(\omega)|e^{j(\omega t + \angle H(\omega))} \quad (21.19)$$

i.e. the signal is changed in amplitude by channel gain/attenuation and shifted in phase (delayed), according to Eq. (21.16). Different signal frequency components are processed (attenuated and delayed) by channel in different way. When channel has linear-phase response $\angle H(\omega) = -\alpha t$ all signal frequencies are delayed by the same time α:

$$y(t) = |H(\omega)|e^{j\omega(t-\alpha)}. \qquad (21.20)$$

Because $\cos(\omega t) = 0.5(e^{j\omega t} + e^{-j\omega t})$, the last equation holds also for real-value signals:

$$y(t) = |H(\omega)|\cos(\omega(t-\alpha)). \qquad (21.21)$$

Channel Estimation and Equalization: Zero Forcing Method Since the channel changes amplitudes and phases of passed sinusoidal signal components, which are passing through it, information contained in these quantities is modified/destroyed. In order to recover it, we have to estimate channel frequency response (Fourier transform of the channel impulse response $h(t)$):

$$\widehat{H}(f) = |\widehat{H}(f)|e^{j\angle\widehat{H}(f)}, \tag{21.22}$$

and equalize the channel, i.e. remove (minimize, reduce) its influence upon the channel output $y(n)$:

$$\frac{Y(f)}{\widehat{H}(f)} = \frac{X(f)H(f)}{\widehat{H}(f)} \approx X(f). \tag{21.23}$$

Having $X(f)$ we can recover bits contained in the signal.

In example presented in this chapter, after header synchronization and carrier recovery (carrier frequency offset estimation and removal), the transmission channel has to be estimated and equalized. How is it done in our example? At this processing stage we know sent carrier states $(I_k^s, Q_k^s) = A_k^s e^{j\phi_k^s}$ of the synchronization header and have received carrier states $(I_k^r, Q_k^r) = A_k^r e^{j\phi_k^r}$ of the header. If the bandwidth of the transmitted signal is very small (around f_c), we can assume that all frequency components in this bandwidth are attenuated by $G = |H(f_c)|$ and delayed by $\alpha = \angle H(f_c)$. Therefore:

$$IQ_k^r = G \cdot e^{j\alpha} \cdot IQ_k^s. \tag{21.24}$$

For each transmission symbol we can calculate

$$\frac{IQ_k^{(r)} \cdot \left(IQ_k^{(s)}\right)^*}{IQ_k^{(s)} \cdot \left(IQ_k^{(s)}\right)^*} = \frac{G_k A_k e^{j(\phi_k+\alpha_k)} A_k e^{-j\phi_k}}{A_k e^{j\phi_k} A_k e^{-j\phi_k}} = G_k e^{j\alpha_k}, \quad k = 1, 2, 3, \ldots, N_{header}. \tag{21.25}$$

The method is forcing to zero the detection error for each $IQ_k^{(s)}$ reference pilot state—what gives the method name. As a channel estimation result, a mean value of found G_k and α_k is taken:

$$G = \text{mean}(G_k), \qquad \alpha = \text{mean}(\alpha_k). \tag{21.26}$$

Channel influence is removed from the received $IQ(n)$ signal samples by a simple correction:

$$\frac{IQ_n^{(r)}}{Ge^{j\alpha}} = \frac{GA_n e^{j(\phi_n+\alpha_k)}}{Ge^{j\alpha}} = A_n e^{j\phi_n} = IQ_n^{(s)} = I(n) + jQ(n). \tag{21.27}$$

As we see, values of the sent $I(n) + jQ(n)$ signal can be theoretically perfectly recovered. Hmm…

In Fig. 21.4 IQ scatter plot of the received signal is shown for noisy channel (SNR=10 dB) with gain equal to 0.25 and phase shift $-\frac{\pi}{4}$. Four different cases are considered: no channel estimation and correction as well as channel estimation and correction using three methods: ZF, presented above, as well as LS/MSE and adaptive NLMS filter, discussed below. As we see all methods succeeded to remove the channel distortion from the received signal. User can also apply the channel to an up-converted version of the transmitted signal, using the function IQupchandown.m. In this case a multi-tap channel can be simulated. What is left for Reader for home exercise.

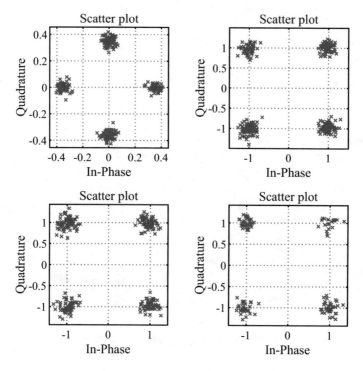

Fig. 21.4: Results of channel corrections for different methods, in rows from left to right: (1) no correction, (2) ZF corrector, (3) LS/MSE corrector, (4) NLMS corrector ($\mu = 0.05$). Simulation conditions: 4-QAM, $K = 8$ samples per symbol, $N_s = 8$ symbols per PSF filter (square root raised cosine, $r=0.35$), SNR=10 dB, 1-tap channel: gain=0.25, phase shift=$-\frac{\pi}{4}$

Exercise 21.4 (Channel Estimation and Equalization—ZF Method). Continue Exercise 21.3. Set do_updownchan=0. At present concentrate on short program fragment performing channel estimation and equalization, chosen by setting: do_chanequ=1. It implements channel correction using Eq. (21.27). Change a few times values of channel gain and phase shift (for example, set chanG=0.1; chanPh=pi/9; Run the program. Observe consequences of channel distortion on eye and phasor-diagrams of the uncorrected received signals. Compare them with diagrams obtained for the signals when channel was equalized.

Now set do_updownchan=1, chan=[0.5,−0.25,0.1,−0.1] and simulate channel at target high frequency. Check how the ZF method works in this case.

Exercise 21.5 (Channel Simulation Methods).

Analyze fragment of a program presented in the Listing 21.3. Its full version is given in the book repository. Three channel simulation methods are implemented and compared:

1. complete, with frequency up-down conversion and signal convolution with channel impulse response at high, target frequency,
2. simplified, but giving exactly the same results, without signal frequency up-down conversion but with down-shifting the channel impulse response to the base-band and performing channel convolution in the base-band,
3. the simplest one, as a one-tap complex-value signal modifier.

Test different available channel impulse responses. Change number of samples per symbol and observe modification of exploited transmission bandwidth and different errors given by the simplest channel simulation method.

Listing 21.3: Testing channel simulation methods in Matlab

```
1   % lab21_ex_channel.m
2   % Testing channel simulation methods
3   clear all; close all; figure;
4
5   % TESTING CHANNEL SIMULATION #############################################
6   % Frequency UP conversion in TX
7   N = length(IQn); n = 0:N-1;                    % signal length, sample indexes
8   y = real(IQn).*cos(2*pi*fcar/fs*n) - imag(IQn).*sin(2*pi*fcar/fs*n);
9   %y = 0.5 * IQn .* exp(j*2*pi*fcar/fs*n);   % alternative
10  Y = freqz( y, 1, f, fs );
11  plot(f,abs(H),'b',f,abs(Y),'r'); grid; xlabel('f (Hz)'); title('H(f), Y(f)'); pause
12  % Channel simulation in high frequency
13  y = filter( h,1,y);
14  % Frequency DOWN conversion in RX
15  IQnn = 2 * y .* exp( -j*( 2*pi*fcar/fs*n ) );
16  % Pulse shaping in RX
17  IQdown1 = conv( IQnn, hpsf ); IQdown1 = IQdown1(Mpsf+1:end-Mpsf); % pulse shaping in TX
18
```

```
19   % Channel simulation in the base-band
20   hdown = h .* exp( -j*( 2*pi*fcar/fs*(0:length(h)-1) ) );
21   hdown = conv( hdown, hpsf );
22   IQdown2 = conv( IQn,hdown ); IQdown2 = IQdown2(Mpsf+1:end-Mpsf);
23
24   % Other simplified method - one point deformation, good for small signal bandwidth
25   IQdown3 = IQn .* chanGain * exp(j*chanPhase); % equivalent channel in the BB
26   IQdown3 = conv( IQdown3, hpsf ); IQdown3 = IQdown3(Mpsf+1:end-Mpsf);
27
28   figure;
29   plot(1:length(IQdown1),real(IQdown1),'ro-',...
30        1:length(IQdown2),real(IQdown2),'bx-',...
31        1:length(IQdown3),real(IQdown3),'gs-'); grid; title('Compare I(n)'); pause
```

Channel Estimation and Equalization: Least-Squares Method At present we will interpret the channel estimation and correction problem from the linear algebra point of view, i.e. we will solve in least squares (LS) sense a matrix equation, describing data transmission through a linear time-invariant channel. Let $h_n, n = 0 \ldots M - 1$, denote M taps of a channel impulse response, u_n—input signal, and v_n—output signal. The input–output matrix relation, describing the transmission, is as follows:

$$
\begin{bmatrix}
u_{M-1} & \cdots & u_1 & u_0 \\
u_M & \cdots & u_2 & u_1 \\
\vdots & \ddots & \vdots & \vdots \\
u_{N-1} & u_{N-2} & \cdots & u_{N-M}
\end{bmatrix}
\begin{bmatrix}
h_0 \\
h_1 \\
\vdots \\
h_{M-1}
\end{bmatrix}
=
\begin{bmatrix}
v_{M-1} \\
v_M \\
\vdots \\
v_{N-1}
\end{bmatrix}, \qquad \mathbf{Uh} = \mathbf{v}. \qquad (21.28)
$$

When we are synchronized and when we know which signal samples in the receiver are connected with the header (pilot) states and which states where transmitted, we can use the above equation as a template and apply it for header (pilot) samples only. Therefore, we solve this equation, taking into account that data has complex-values now:

$$
\mathbf{Uh}^* = \mathbf{v} \quad \rightarrow \quad \mathbf{h}^* = (\mathbf{U}^H \mathbf{U})^{-1} \mathbf{U}^H \cdot \mathbf{v}. \qquad (21.29)
$$

In such case the vector $\mathbf{h}$ represents decimated channel impulse response (with number of samples per symbol as the decimation order). To simplify correction of unknown transmitted data, which are sent after the header, we can inverse the Eq. (21.28):

$$
\begin{bmatrix}
v_{M-1} & \cdots & v_1 & v_0 \\
v_M & \cdots & v_2 & v_1 \\
\vdots & \ddots & \vdots & \vdots \\
v_{N-1} & v_{N-2} & \cdots & v_{N-M}
\end{bmatrix}
\begin{bmatrix}
g_0^* \\
g_1^* \\
\vdots \\
g_{M-1}^*
\end{bmatrix}
=
\begin{bmatrix}
u_{M-1} \\
u_M \\
\vdots \\
u_{N-1}
\end{bmatrix}, \qquad \mathbf{Vg}^* = \mathbf{u}, \qquad (21.30)
$$

and directly calculate from it the correction filter $\mathbf{g}$:

$$
\mathbf{g}^* = (\mathbf{V}^H \mathbf{V})^{-1} \mathbf{V}^H \cdot \mathbf{u}. \qquad (21.31)
$$

Having weights **g** calculated from Eq. (21.31) for the header (from input–output relation), next we can use them for estimation of transmitted data **u** using received data **v** (from new input–output relation) using Eq. (21.30): $\mathbf{u} = \mathbf{V}\mathbf{g}^*$. Many different LS methods can be used for solving linear equation (21.30). Details can be found in Chap. 13. In Listing 21.4 Matlab code, implementing the LS/MSE channel corrector, is given. In Fig. 21.4 result of its application is presented.

Listing 21.4: Channel estimation and correction using LS solution of matrix input–output equation written for pilot states

```
1   % lab21_ex_receiver.m - fragment
2       if( do_chanequ==2 ) % solving linear input-output equation in LS (MSE) sense
3           M = 9;                              % channel length (in number of symbols)
4           L = Nhead;                          % number of input symbols
5           v = IQkHeadEstim;                   % values of output symbols
6           V = toeplitz(v(M:L),v(M:-1:1));     % matrix with output signal
7           u = IQkHead(M:L).';                 % input symbols
8           g = (V\u),                          % channel inverse filter
9           v = IQn( 1+(Nhead-1-(M-2))*K : K : 1 + (Nhead+Ndata-1)*K); % data to correct
10          L = length(v);                      % number of input data
11          V = toeplitz(v(M:L),v(M:-1:1));     % matrix with input data
12          IQkDataEstim = V*g;                 % data correction
13          [ IQkHeadEstim(1:10).' IQkHead(1:10).' ], pause      % compare IQ header
14          [ IQkDataEstim(1:10)  IQkData(1:10).' ], pause       % compare IQ data
15          rxData = IQ2numbers( IQkDataEstim, modtype );        % IQ to state numbers
16          [ rxData(1:15).' numData(1:15) ],                    % numbers in and out
17          errors = sum( rxData.' - numData ), pause            % error
18          figure;
19          plot( IQkDataEstim,'r*'); grid; title('Q(n) = f( I(n) )'); pause
20          return
21      end
```

Channel Estimation and Equalization: Adaptive Least Mean Squares (LMS) Filter Instead of using the matrix block-based solution (21.31), one can apply also the LMS complex-value adaptive filters for channel inverse filter calculation:

$$err_n = u_n - \begin{bmatrix} v_n & v_{n-1} & \cdots & v_{n-(M-1)} \end{bmatrix} \cdot \begin{bmatrix} g_0^{(n)} \\ g_1^{(n)} \\ \vdots \\ g_{M-1}^{(n)} \end{bmatrix}^* \tag{21.32}$$

$$\begin{bmatrix} g_0^{(n+1)} \\ g_1^{(n+1)} \\ \vdots \\ g_{M-1}^{(n+1)} \end{bmatrix} = \begin{bmatrix} g_0^{(n)} \\ g_1^{(n)} \\ \vdots \\ g_{M-1}^{(n)} \end{bmatrix} + \mu_{cmplx} \cdot err_n^* \cdot \begin{bmatrix} v_n \\ v_{n-1} \\ \vdots \\ v_{n-(M-1)} \end{bmatrix}. \tag{21.33}$$

The LMS algorithm can be replaced by: (1) normalized LMS algorithm (second term in Eq. (21.33) is divided by energy of last M samples of signal v_n, i.e. by the $E_n = \sum_{k=0}^{M-1} v_{n-k} v_{n-k}^*$) or (2) more computationally extensive but faster converging RLS algorithm. Details can be found in Chap. 12 on adaptive filters.

The adaptive channel correction filters can be also applied during regular data transmission, not only for pilots. When disturbances are not big or when the system is already *roughly* equalized and our signal v_n is close to a correct constellation point, we can use the difference between observed and correct carrier state as an error for adaptive filter weights improvement. In such case u_n in Eq. (21.33) is replaced by recognized (predicted) constellation point $\lfloor v_n \rfloor$. Algorithms using this approach are called decision-directed.

In Listing 21.5 Matlab code, implementing the NLMS adaptive filter as the channel corrector, is given. Since the header used in our program is short, we allow the filter to adapt longer time using the transmitted data also (the are known to it). In Fig. 21.4 result of the program application is presented. In turn, in Fig. 21.5 adaptation of the first filter tap is presented, for the same simulation.

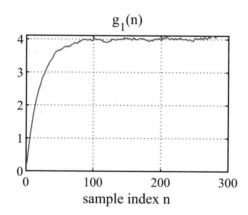

$$g_1(n)$$

Fig. 21.5: Adaptation of the first weight of NLMS adaptive filter ($\mu = 0.05$) during simulation which results are presented in Fig. 21.4

Listing 21.5: Channel estimation and correction using LMS adaptive filter

```
1  % lab21_ex_receiver.m - fragment
2      if( do_chanequ==3 ) % NLMS adaptive filter of symbols - weights adaptation using
                header
3          u = [ IQkHead IQkData ];                      % sent
4          v = IQn( 1 : K : 1+(Nhead+Ndata-1)*K );       % received
5          bv = zeros(1,M); g = zeros(1,M); mi=0.1; ghist = []; % initialization
6          for n = 1 : length(u)                         % filter loop
7              bv = [ v(n) bv(1:M-1) ];                  % input to buffer
8              uest(n) = sum( bv .* conj(g) );           % estimated value
9              err(n) = u(n) - uest(n);                  % error
```

```
10        g = g + mi * conj(err(n)) * bv / (bv*bv');      % filter weights update
11        ghist = [ ghist g(1) ];                          % history
12      end                                                %
13      figure; plot( abs(ghist) ); title('ghist(n)'); pause % figure
14      rx = IQ2numbers( uest, modtype ).';                % received state numbers
15      tx = [ numHead; numData ];                         % transmitted state numbers
16      errors = sum( rx(end-Ndata/2:end) - tx(end-Ndata/2:end) ), pause  % error
17      figure;
18      plot( uest(end-Ndata/2:end),'r*'); grid; title('Q(n) = f( I(n) )'); pause
19      return
20    end
```

Exercise 21.6 (Channel Estimation and Equalization: LS and NLMS Adaptive Filter Methods). Continue Exercise 21.3. Set do_updownchan=1. At present concentrate on program fragment performing channel estimation and equalization using: (1) LS (matrix-based), and (2) NLMS (adaptive filter-based) methods. They are, respectively, chosen by settings: do_chanequ=2 and do_chanequ=3. Run the program for different values of chan coefficients. Observe consequences of channel distortion on eye and phasor-diagrams of uncorrected received signals. Compare them with diagrams obtained for signals when channel is equalized. Compare efficiency of equalizer do_chanequ=1 (ZF), do_chanequ=2 (LS), and do_chanequ=3 (NLMS). Use different values of μ. Change the NLMS filter to LMS and test it. You can also try to use the RLS adaptive filter which is converging fast and its application is beneficial for short headers.
Now set do_updownchan=1, chan=[0.5,-0.25,0.1,-0.1] and simulate channel at target high frequency. Check how LS and NLMS methods work in this case.

Phase Delay and Group Delay Group delay is an important parameter characterizing signals in analog communication channels. Let us assume that we are interested in the following modulation scheme:

$$x(t) = x_m(t) \cdot e^{j2\pi f_c t}, \qquad X(f) = X_m(f - f_c), \qquad (21.34)$$

where $x_m(t)$ denotes a modulation signal, which has very narrow frequency spectrum (bandwidth) in comparison with carrier frequency f_c. The signal is passed by a channel having frequency response $H(f) = G(f)e^{j\varphi(f)}$. Since we are interested in channel *activity* around carrier frequency f_c, we express value of $\varphi(f)$ around f_c using the Taylor series:

$$H(f) = G(f) \cdot e^{j(\varphi(f_c)+(f-f_c)\Phi(f_c))}, \qquad \Phi(f_c) = \left.\frac{d\varphi(f)}{df}\right|_{f=f_c}, \qquad (21.35)$$

assuming that the channel gain $G(f)$ is constant in the neighborhood of f_c. Fourier transform of the channel output is equal to:

$$
\begin{aligned}
Y(f) = X(f)H(f) &= X_m(f - f_c) \cdot G(f)e^{j(\varphi(f_c)+(f-f_c)\Phi(f_c))} = \\
&= \left[X_m(f - f_c)e^{j(f-f_c)\Phi(f_c)} \right] \cdot G(f)e^{j\varphi(f_c)} = \\
&= \left[X_m(f - f_c)e^{-j2\pi(f-f_c)t_g} \right] \cdot G(f)e^{-j2\pi f_c t_p} = \\
&= \left[X_m(f - f_c)e^{-j2\pi f t_g} \right] \cdot G(f)e^{j2\pi f_c(t_g-t_p)}, \quad (21.36)
\end{aligned}
$$

where t_p and t_g, signal phase delay and group delay, are defined as:

$$
t_g = -\frac{\Phi(f_c)}{2\pi} = -\frac{1}{2\pi} \cdot \frac{d\varphi(f)}{df}\bigg|_{f=f_c}, \qquad t_p = -\frac{\varphi(f_c)}{2\pi f_c}. \qquad (21.37)
$$

Spectral relation (21.36) corresponds to the following time relation:

$$
y(t) = x_m(t - t_g)e^{j2\pi f_c(t-t_g)} \cdot e^{j2\pi f_c(t_g-t_p)} \star g(t) = x_m(t - t_g)e^{j2\pi f_c(t-t_p)} \star g(t), \qquad (21.38)
$$

where $\star$ denotes a convolution and $g(t)$ is an inverse Fourier transform of $G(f)$. We have assumed that $G(f)$ has constant value G around f_c. Since spectrum of our modulating signal is very narrow around f_c, the rest of channel frequency response, outside the f_c neighborhood, is not significant for the calculation result. Therefore we can assume gain G in the whole frequency range. In such situation $g(t) = G\delta(t)$ (the Dirac delta function) and Eq. (21.38) simplifies to:

$$
y(t) = G \cdot x_m(t - t_g) \cdot e^{j2\pi f_c(t-t_p)}. \qquad (21.39)
$$

For real-value cosine-modulation we have

$$
y(t) = G \cdot x_m(t - t_g) \cdot \cos(2\pi f_c(t - t_p)). \qquad (21.40)
$$

At present we can do physical interpretation of times t_g and t_p. The t_g is called a group delay, i.e. a delay of signal $x_m(t)$ (carrier envelope in (21.34)). In turn, the t_p is called a phase delay, i.e. a delay of the complex signal carrier in (21.34). In fact, during simulation of a telecommunication channel in the base-band by means of one-tap data modifier, we were only taking into account the carrier delay, neglecting the group delay what is correct for flat channel phase response in the neighborhood of frequency f_c.

Exercise 21.7 (Phase Delay and Group Delay). It is impossible to include all programs inside the book. In the webpage of the book one can find a program lab21_ex_gdelay.m for simulation of analog transmission channels with very narrow-band carrier modulation. Become familiar with this program. Run it for Gaussian window and raised cosine window as a modulation function. Input signals $x_m(t)$ and $x(t)$ are plotted in black. The expected signal after channel passing and its expected envelope are plotted in blue. Real signal of the channel output is drown in red and should overlap with blue curves. Notice values of shifts ng and np, corresponding to t_g and t_p. You should see delay of the carrier envelope and delay of the carrier itself when these values are bigger than 1. Change filter taps. Change value of carrier frequency: choose different point of channel phase response.

21.6 Decreasing Signal Sampling Ratio

Our single-carrier transmission program, used in this chapter, is an educational one. Despite its simplicity, it should give Reader a chance to learn several things at once: fundamental transmission techniques, appearing problems and their solutions. From this point of view, it meets our needs. However, up to now, its implementation was unpractical and unrealistic in one aspect: all developed solutions worked well, more or less, but had been tested only for very oversampled signals, i.e. having many samples per one carrier symbol. Now we can reduce the signal sampling rate in the receiver, setting do_decim=1 and Mdecim=2,3,4,6,8,12 in the program, and analyze exemplary results of the program usage for different signal decimation ratios. In Fig. 21.6 recovered $I(k)$ and $Q(k)$ carrier states are presented for 4-QAM modulation (in one plot since only values 1 and −1 are expected). If two distinctive lines of values 1 and −1 are visible, it means that the applied algorithms succeeded to recover correctly all carrier states. In turn, when we see widening clouds of points, problems with carrier states decoding are becoming bigger and bigger. In Fig. 21.6 we see that after decimation of the received signal the obtained results are still satisfactory. Being precise: they are the same, what is very strange! May be something was forgotten by us? Will we use a Phone-A-Friend? Hmmm . . . Yes, yes. We have not taken into account in our simulation a time offset of the ADC sampling (in fraction of the sampling period) as well as a time offset in respect to the symbol maximum during its down-sampling. After considering these two occurring phenomena and repeating the simulation, results presented in Fig. 21.7 were obtained. Originally, we had 24 samples per symbol, after down-sampling only 3 samples per period. We have assumed 50% time offset in ADC sampling and signal delay by 1, 2 and 4 samples before its re-sampling. Since we were taking every 8-th sample, now we had 12.5%, 25%, and 50% delay in respect to the symbol maximum. As we see, the obtained results are significantly worse now. We can conclude that it is much better when we are synchronized with the symbol positions.

We should remember that signals in efficient receivers are drastically down-sampled, and only a few samples per symbol are used in them, from 1 to 3–5 during the symbol recognition. If necessary, missing samples are obtained by means of interpolation. But this is beginning of completely new story. We will later tackle the problem briefly during presentation of real-world examples.

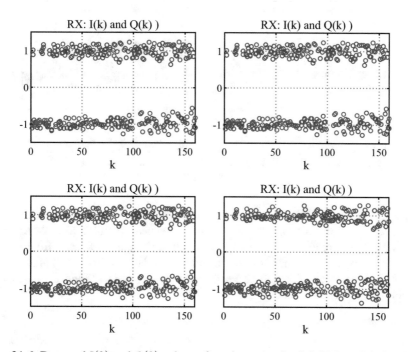

Fig. 21.6: Detected $I(k)$ and $Q(k)$ values of carrier states for 4-QAM modulation and different signal decimation orders Mdecim=2,4,6,8 in the receiver. From left to right and up-down, number of samples per one symbol period is equal: 12, 6, 4, 3. Simulated transmission conditions were as follows: SNR = 10 dB, chanG = 0.25, chanPh = $\pi/7$, carDF = 0.0001, carDPh = 0, without sampling time offset (!). Noise sequence was the same in all simulations

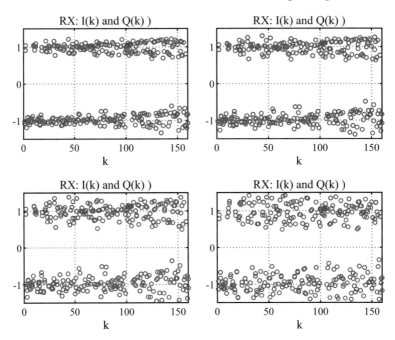

Fig. 21.7: Detected $I(k)$ and $Q(k)$ values of carrier states for 4-QAM modulation and Mdecim=8 times signal decimation in the receiver (coming from 24 samples per symbol to 3 samples per symbol). From left to right and up-down, for different values of sampling time offsets in respect to the symbol maximum: (1) precisely in the symbol maximum, (2) with 12.5% delay (1 sample of 8), (3) with 25% delay (2 samples of 8), (4) 50% delay (4 samples of 8)

Exercise 21.8 (Decreasing Number of Samples per Symbol in the Receiver). Use program from Listing 21.1 and try to obtain plots similar to ones presented in the Fig. 21.6. Set do_decim=1, Mdecim=2,3,4,6,8,12. Test algorithm robustness to single disturbance, adding one disturbance after the other. Increase the disturbance level. Catch the value for which first transmission errors occur. Are these thresholds the same for all decimation schemes? Check whether your conclusions are valid for other modulation techniques, for example, 8PSK or 16QAM. Do program modification, allowing simulation of sampling time offsets, discussed in the text:

```
ADdt=0.5; n=1:N; IQ(1,n)=interp1(n-1,IQ(1,n),n+ADdt,'spline');
SYMdt=0.5; IQ = resample(IQ(1+round(SYMdt*Mdecim):N),1,Mdecim);
```

After the modification, try to generate plots similar to shown in the Fig. 21.7.

21.7 Timing Recovery Methods

For demonstration of different problems, arising in digital single-carrier communication, and for presentation of different templates of possible solutions, we have used a transmission example similar to a Wi-Fi packet sending. In such application the header (preamble) is used for many different purposes: carrier recovery, channel estimation and equalization, frame and symbol synchronization, i.e. timing recovery. In telecommunication technology there are many different methods dedicated for different modulations that are used, different channels types, ...etc. It is impossible to present all existing solutions. In next chapters on multi-carrier digital transmission, some methods used in multi-carrier scenario will be presented.

At present, to complete the view of single-carrier systems, we learn some additional timing recovery methods which are used for finding states of one carrier. Why are they important? Because together with carrier recovery methods, presented in the chapter on amplitude modulation (PLL and Costas loops), they can be applied to a signal transmitted without a header/preamble.

The early-late-gate, Gardner and Mueller–Müller timing recovery methods are very popular in case of BPSK, QPSK (4-QAM), and DQPSK modulations.

Early-Late-Gate Method The early-late-gate method will be presented with more details in the following section in which demodulation of nano-satellite signal is described. At present, we can shortly summarize it as a method in which slope of the $IQ(n)$ signal is tested in its real part and imaginary part. After taking a sample near the $IQ(n)$ signal maximum (peak (+1)) and its two, left and right neighbors, we should observe: (1) a positive signal slope (left), (2) a maximum peak (in the center), or (3) a negative signal slope (right). Degree of slope of positive and negative slopes inform us that we are taking signal samples too early, correctly or too late than required, and that the next symbol sampling should be performed later, as before or earlier. When two neighbors of the central sample have similar values, the sampling step is correct and should be repeated. Of course, for the $IQ(n)$ signal minimum (peak (−1)) interpretation of slope signs is reversed (overall *positive* means "too late" while overall *negative*—"too early"). We should have at least 5 samples per symbol in order to perform the slope detection.

Gardner Method In the Gardner timing recovery method at least two samples per symbol are required. In this approach we are looking for such value of n for which the following cost functions (four possibilities) are equal to 0:

$$C_1(n) = [\text{Re}\{x(n+N)\} - \text{Re}\{x(n)\}] \cdot \text{Re}\{x(n+N/2)\}, \tag{21.41}$$

$$C_2(n) = [\text{Im}\{x(n+N)\} - \text{Im}\{x(n)\}] \cdot \text{Im}\{x(n+N/2)\}, \tag{21.42}$$

$$C_3(n) = [\text{Re}\{x(n+N)\} - \text{Re}\{x(n)\}] \cdot \text{Re}\{x(n+N/2)\} + ...$$
$$.... + [\text{Im}\{x(n+N)\} - \text{Im}\{x(n)\}] \cdot \text{Im}\{x(n+N/2)\}, \tag{21.43}$$

$$C_4(n) = \text{Re}\{[x^*(n+N) - x^*(n)] \cdot x(n+N/2)\}, \tag{21.44}$$

where N denotes the number of samples per symbol. An idea leading to proposition of these cost function is as follows:

1. when two neighbor symbols, separated by N samples, have the same values (both are equal to $+1$ or to -1), their difference is equal to zero—therefore they are subtracted in all cost functions causing their zeroing,
2. when two neighbor symbols have opposite values (-1 and $+1$) there is a small value (desirable 0) which is lying in-between them—therefore their difference is multiplied in all cost functions by this value causing their zeroing.

The last cost function is not sensitive to carrier frequency offset when the offset is small in comparison with symbol frequency and approximately constant during symbol period ($e^{j(\Delta\Omega_c n + \Delta\phi_c)} = e^{j\theta} = \text{const}$), what can be proofed:

$$[(x_R(n+N) - jx_I(n+N)) - (x_R(n) - jx_I(n))]e^{-j\theta} \cdot (x_R(n+\frac{N}{2}) + jx_I(n+\frac{N}{2}))e^{j\theta} =$$
$$= x_R(n+\frac{N}{2})[x_R(n+N) - x_R(n)] + x_I(n+\frac{N}{2})[x_I(n+N) - x_I(n)] + \dots$$
$$\dots + j\left\{ x_R(n+\frac{N}{2})[x_I(n) - x_I(n+N)] + x_I(n+\frac{N}{2})[x_R(n+N) - x_R(n)] \right\}. \quad (21.45)$$

The real part of Eq. (21.45) gives Eq. (21.43).

Mueller–Müller Method In the Mueller–Müller method zero of the following cost function is searched:

$$C_5(n) = \text{Re}[x(n+N)]\,\text{sign}\{\text{Re}[x(n)]\} - \text{Re}[x(n)]\,\text{sign}\{\text{Re}[x(n+N)]\} + \dots$$
$$\dots + \text{Im}[x(n+N)]\,\text{sign}\{\text{Im}[x(n)]\} - \text{Im}[x(n)]\,\text{sign}\{\text{Im}[x(n+N)]\}. \quad (21.46)$$

The method requires minimum one sample per symbol. Its justification is as follows:

1. if two neighbor symbols are the same (both $+1$ or both -1), their values and signs are the same and their subtraction is equal to zero,
2. if two neighbor symbols are different, the cost function is equal to 0 also, for example, for a=1 and b=-1 we have

$$a \cdot \text{sign}(b) - \text{sign}(a) \cdot b = 1(-1) - 1(-1) = 0. \quad (21.47)$$

In order to check shapes of the defined above cost functions and to verify whether they are equal to zero when index n of $IQ(n)$ is equal to the symbol position k of $IQ(k)$, we have generated $IQ(n)$ sequence for 4-QAM modulation and calculated mean values of two cost functions (21.44) (Gardner) and (21.46) (Mueller-Müller) for different values of the index n, in respect to symbol maximum position $k = 0, N, 2N, \dots$:

$$C_x^{mean}(n) = \sum_m C_x(n+mN), \quad n = 0, 1, 2, \dots, N-1, \quad (21.48)$$

where N denotes the number of samples per symbol. Obtained results are presented in Fig. 21.8. As we see, in both methods cost functions are equal to zero for $n = 0$

and $\frac{N}{2}$. Therefore, one should find initially the approximate symbol position using some other criterion, for example, sign change of $I(n)$ or $Q(n)$ signals $((1) \rightarrow (-1)$ or $(-1) \rightarrow (1))$ and then locally, adaptively search for the exact symbol position looking for cost function zeroing. Figure 21.8 was generated by the program 21.6, presented below.

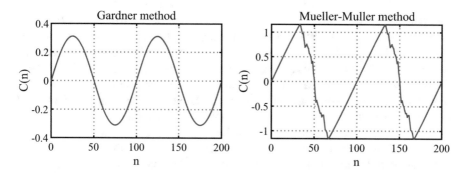

Fig. 21.8: Mean values (21.48) of Gardner (21.44) (left) and Mueller-Muller (21.46) (right) cost functions for symbol position detection. $N = 100$ samples per symbol

One can use the above presented cost functions for adaptive selection of these $IQ(n)$ values, which are expected to be carrier states $IQ(k)$. As an adaptive tracker one can exploit the same adaptive filter which was used for the carrier recovery in the PLL and Costas loops, described in chapter on AM (de)modulation. In Figs. 21.9 and 21.10 results of such adaptation, performed for the Gardner method, are shown. Both figures were obtained in the program 21.6. In our experiment, one symbol had 24 samples and symbol sampling offset was equal to 8 samples. In the first figure, we can validate the adaptation procedure, observing how values of program variables were changing:

- `partition(n)` is adapting to value bigger than 0.3, what is correct since $\frac{1}{3}24 = 8$,
- `error(n)` is converging to 0 while sample `offset(n)` is moving to 8 (Bravo!)

In Fig. 21.10 we can compare results of initial, wrong sampling of $IQ(n)$ with offset 8, on the left side, and correct, after symbol timing recovery, on the right side.

Of course, the above presented synchronization methods to symbol position, without usage of the synchronization header, are very useful. However, they were implemented by us in inefficient way with high signal oversampling. This inconvenience can be omitted by reducing the sampling rate and applying any interpolation procedure, for example, by using an efficient Lagrange interpolator using Farrow filters, described in the chapter on signal re-sampling. Let us try to implement this technique in our case. Now we can have only two samples per symbol in the Gardner method and estimate true position $IQ(k)$ of any symbol using surrounding it

$IQ(n)$ samples, for example, quadratic interpolation with 3 samples. Prediction procedure of the next symbol position should be adaptive and robust to high variance of an input information. For this purpose we can apply the same robust adaptive tracker (filter) which was used in the PLL and Costas loops, described in the chapter on amplitude modulation. Both above mentioned techniques, the Farrow interpolator and the adaptive tracker have been implemented in the function presented in Listing 21.6. When the $IQ(n)$ signal has more than two samples per symbols, interpolation is not performed: samples $IQ(n)$ lying close to carrier states $IQ(k)$ are adaptively selected only. When the $IQ(n)$ signal has 2 or 1 sample per symbol, the Gardner and the Mueller–Müller methods are combined with Farrow interpolators. Results from the program usage are presented in Figs. 21.11 and 21.12 for Gardner+Farrow solution. As we can see, even for such small number of samples per modulation symbol results are satisfactory: carrier states have been well recovered and the adaptation process has converged to the correct solution.

One should remember that the detection curves of the Garner and Mueller–Müller methods are equal to zero in two points (see Fig. 21.8) therefore some good initial guess of the approximate symbol position is required. I am very sorry. I will not continue further this story.

Exercise 21.9 (Timing Recovery University). Analyze code of the function presented in Listing 21.6. IQ denotes a signal sent to it, alg=1,2,3,4,5 is selected algorithm number, Nsps denotes number of samples per symbol and n1st is a number of the first IQ sample to be processed. In first part of the function, a user can plot detection curves for five cost functions $C_1(n) - C_5(n)$— Eqs. (21.41)–(21.44) and (21.46). In order to do this, if(1) should be set. In the remaining part of the function only algorithms 3–5 are available only. When Nsps>2, the Gardner and Mueller–Müller algorithms without signal interpolation are used. For Nsps=2 the Gardner procedure with the Farrow interpolator is used, while for NSps=1 the Mueller–Müller method and the Farrow filters. Add the following code to the program lab21_ex_receiver.m just after calculation of the IQn signal:

```
if(1)              % 0/1 Testing symbol timing recovery methods
    alg = 1;       % 1/2/3/4/5 timing recovery algorithm
    for Mdelay = 1 : 1 : K
        IQnn = IQn(1+Mdelay : Mdecim : end); n1st = 1;
        [ns,IQs] = timing_recovery(IQnn,alg,K/Mdecim,n1st);
    end
end
```

Run the program. Observe figures. Change values of the following parameters: modtype = 'BPSK', '4QAM', rctype='normal' and Mdecim=1,2,4,8,12,24.

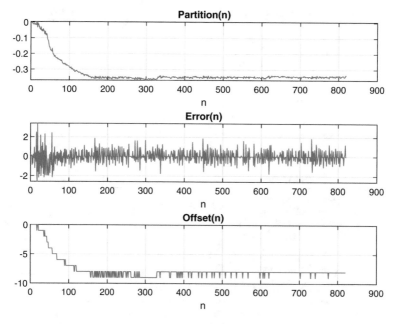

Fig. 21.9: Exemplary results from adaptive timing recovery using the Gardner method. $N = 24$ samples per symbol. Correct symbol sampling offset equal to 8 is found: partition=$\frac{1}{3} \cdot 24 = 8$, error=0 and offset=8

Listing 21.6: Matlab function for testing Gardner and Mueller-Muller symbol timing recovery algorithms

```matlab
function [ns, IQs] = timing_recovery( IQ, alg, Nsps, n1st )
% Calculation of symbol positions (ns) and carrier states values (IQs)

  N = length(IQ); I = real(IQ); Q = (imag(IQ));
% Adaptation parameters
  damp = sqrt(2)/2;                                    % adaptation loop damping
  band = (0.5*pi/500) / (damp+1/(4*damp));             % adaptation loop bandwidth
  mi1  = (4*damp*band) / (1 + 2*damp*band + band*band); % adaptation coeff 1
  mi2  = (4*band*band) / (1 + 2*damp*band + band*band); % adaptation coeff 2

% Checking detection characteristics of all timing recovery methods
if(1) % select 0/1 NO/YES
  N = floor(N/Nsps)*Nsps; mN = 1 : Nsps : N-4*Nsps+1;
  for n=0:2*Nsps-1
    if(alg==1) % Gardner - frequency offset sensitive
      a = (I(n+Nsps+mN) - I(n+mN)) .* I(n+Nsps/2+mN);
      cost(n+1) = mean(a);
    end
    if(alg==2) % Gardner - frequency offset sensitive
      a = (Q(n+Nsps+mN) - Q(n+mN)) .* Q(n+Nsps/2+mN);
      cost(n+1) = mean(a);
    end
    if(alg==3) % Gardner - frequency offset sensitive
```

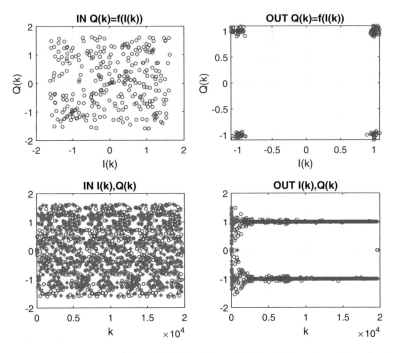

Fig. 21.10: Phasor plots $Q(k) = f(I(k))$ (up) and $I(k), Q(k)$ values (down) of estimated carrier states for wrong symbol sampling with 8 sample offset (left) and for correct sampling after adaptive timing recovery using the Gardner method (right). $N = 24$ samples per symbol

```
24          a = (I(n+Nsps+mN) - I(n+mN)) .* I(n+Nsps/2+mN);
25          b = (Q(n+Nsps+mN) - Q(n+mN)) .* Q(n+Nsps/2+mN);
26          cost(n+1) = mean(a + b);
27        end
28        if(alg==4) % Gardner - frequency offset not sensitive
29          cost(n+1) = mean(real((conj(IQ(n+Nsps+mN))-conj(IQ(n+mN)) .* IQ(n+Nsps/2+mN)))
                );
30        end
31        if(alg==5) % Mueller & Muller
32          a = I(n+mN).*sign(I(n+Nsps+mN)) - I(n+Nsps+mN).*sign(I(n+mN));
33          b = Q(n+mN).*sign(Q(n+Nsps+mN)) - Q(n+Nsps+mN).*sign(Q(n+mN));
34          cost(n+1) = mean(a + b);
35        end
36      end
37      figure; plot(0:2*Nsps-1,cost,'b.-'); xlabel('n'); ylabel('C(n)');
38      title('Detection curve'); grid; pause
39    end
40
41    % Adaptive timing recovery for highly over-sampled signals
42    if( Nsps > 2 ) % big Nsps: choosing best sample
43    k=1; ns(1) = n1st; offs = 0; adap1 = 0; adap2 = 0;
```

```
44    for n = n1st : Nsps : length( IQ )-2*Nsps
45       if(alg==3) % Gardner
46          a = (I(n+Nsps+offs) - I(n+offs)) .* I(n+ (Nsps)/2+offs);
47          b = (Q(n+Nsps+offs) - Q(n+offs)) .* Q(n+ (Nsps)/2+offs);
48          err = -(a + b);
49       end
50       if(alg==4) % Gardner
51          err = -real((conj(IQ(n+Nsps+offs)) - conj(IQ(n+offs))) .* IQ(n+ (Nsps)/2+offs));
52       end
53       if(alg==5) % Muller & Muller
54          a = I(n+offs).*sign(I(n+Nsps+offs)) - I(n+Nsps+offs).*sign(I(n+offs));
55          b = Q(n+offs).*sign(Q(n+Nsps+offs)) - Q(n+Nsps+offs).*sign(Q(n+offs));
56          err = -(a + b);
57       end
58       adap2 = adap2 + mi2 * err;                % 1-nd update
59       adap1 = adap1 + adap2 + mi1 * err;        % 2-st update
60       while(adap1 >  1) adap1 = adap1 - 1; end  % # wrapping to interval (-1,1)
61       while(adap1 < -1) adap1 = adap1 + 1; end  % # (mapping)
62       offs = round( adap1 * Nsps );             % offset calculation as % of Nsps
63          partition(k) = adap1;                  % for figure
64          error(k)  = err;                       % for figure
65          offset(k) = offs;                      % for figure
66       k = k+1;                                  % index update
67       ns(k) = n+Nsps+offs;                      % storing symbol position
68    end
69    IQs = IQ(ns);                                % estimated carrier states
70    k = n1st : Nsps : length(IQ)-1;              % for figures
71
72    end % Nsps > 2
73
74    % Adaptive timing recovery for critical sampling - signal interpolation
75    if( Nsps == 1 || Nsps == 2 ) % Nsps = 1 or 2: Muller (1) or Gardner (2) method
76    % Farrow filtration for Lagrange quadratic polynomial interpolation
77       x2 = filter( [ 1/2  -1   1/2], 1, IQ ); x2 = x2(3:end); % sample before
78       x1 = filter( [ 1/2   0  -1/2], 1, IQ ); x1 = x1(3:end); % sample central
79       x0 = filter( [  0    1    0 ], 1, IQ ); x0 = x0(3:end); % sample after
80       adap1 = 0; adap2 = 0; k = 1;
81       for n = n1st+1 : Nsps : length(x0)-1
82          xm1 = x2(n-1) * adap1*adap1 + x1(n-1) * adap1 + x0(n-1);
83          xc0 = x2(n)   * adap1*adap1 + x1(n)   * adap1 + x0(n);
84          xp1 = x2(n+1) * adap1*adap1 + x1(n+1) * adap1 + x0(n+1);
85          if(Nsps==1) % Muller
86             a = real(xc0).*sign(real(xp1)) - real(xp1).*sign(real(xc0));
87             b = imag(xc0).*sign(imag(xp1)) - imag(xp1).*sign(imag(xc0));
88             err = -(a + b);
89          end
90          if(Nsps==2) % Gardner
91             err = (real(xp1)-real(xm1))*real(xc0) + ...
92                   (imag(xp1)-imag(xm1))*imag(xc0);
93          end
94          adap2 = adap2 + mi2 * err;             % 1-nd update
95          adap1 = adap1 + adap2 + mi1 * err;     % 2-st update
96          IQs(k) = xc0;                          % carrier state value
97             partition(k) = adap1;               % for figure
```

```
98              error(k)  = err;              % for figure
99              offset(k) = 0;               % for figure
100         ns(k)  = n+adap1;                % storing symbol position
101         k = k+1;                         % index update
102     end
103 end % Gardner & Muller
```

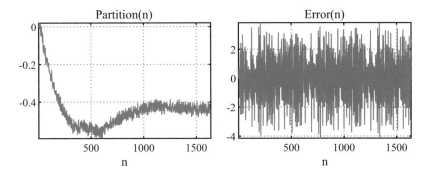

Fig. 21.11: Exemplary results from adaptive, fractional timing recovery using the Gardner method and the Farrow filter interpolator. $N = 2$ samples per symbol

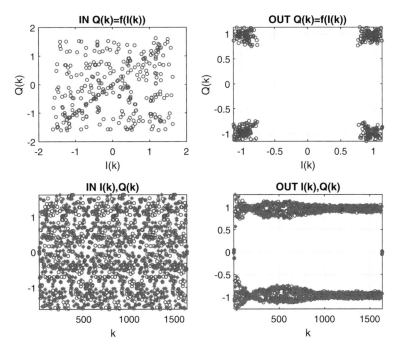

Fig. 21.12: Phasor plots $Q(k) = f(I(k))$ (up) and $I(k), Q(k)$ values (down) of esti-mated carrier states for wrong sampling (left) and for correct sampling (right) after adaptive timing recovery using the Gardner method and the Farrow filters interpo-lator. $N = 2$ samples per symbol)

21.8 Real-World Examples

At present we will apply our knowledge, acquired in this chapter, to demodulation of three real-world digital signals:

- a safe mode nano-satellite signal using digital BPSK modulation, transmitted together with voice (DUV—data under voice technique),
- an RDS (Radio Data System) signal using digital bi-phase BPSK modulation with differential coding,
- synchronization continuous down-link burst signal of the TETRA standard using $\pi/4$-DQPSK modulation.

Two first IQ signals were already FM-decoded by us in Chap. 17 on software defined radio (see listings: 17.2, 17.3 and program 17.4). Additionally, in Chap. 18 on FM modulation, we dealt with the multiplex (MPX) FM radio signal demodulation in Sect. 18.5.1 and learn nano-satellite signal morphology in Sect. 18.5.4. Finally, in Chap. 18 on AM demodulation, we did carrier synchronization (via PLL and Costas loops) and decoded stereo FM radio signal in program 19.5. In this section we will extract the RDS 57 kHz component from the MPX FM radio signal, shift it to the 0 Hz (to the *base-band*), decode RDS bits, and sent them to RDS bit parser.

21.8.1 Decoding Bits from Nano-Satellites

In this subsection we will try to write a simple Matlab bit decoder for the Fox-Telem AMSAT (Radio Amateur Satellite Corporation) ground station, receiving save-mode signals from Fox-I satellites [17].

As mentioned above, the save-mode nano-satellite signal was already tackled by us. A Reader is asked to see Fig. 18.9 and its short description in the text of Sect. 18.5.4. At present we will concentrate on real-value BPSK signal demodulation and bit recovery only. Let us recall: the sampling frequency is equal to 48 kHz, one BPSK symbol has 240 samples and 200 bits are sent per second. We analyze one voice beacon with embedded data. It consists of the following parts, transmitted one-by-one: (1) 10 bits of a synchronization header, (2) 960 data bits, (3) second 10 bits of synchronization pattern, (4) 960 data bits, (5) one more 10 bit synchronization pattern. In the beacon beginning, one starting bit is sent (0 or 1, it depends what synchronization header is used). In the middle part of the digital signal, voice is overlaid, describing the transmission type. In Fig. 21.13, in the first column of plots we see: (1) the whole signal, (2) it first part with starting bit transmission (notice very strong noise!), (3) central signal part were low-frequency digital signal and voice information overlaps. In plots of the second column, there are shown: (1) detected carrier states for square root raised cosine pulse shaping filter with roll-off parameter $r = 0.5$ and having length equal to two BPSK symbols (positive values corresponds to bit 1, while negative to bit 0), (2) beginning of the filtered signal with marked detected carrier states (notice a peaky curve of the BPSK modulation—long sequences of the same value 1 or -1 do not occur because data are scrambled), (3) central part of the filtered signal with marked detected carrier states—here voice works as disturbance).

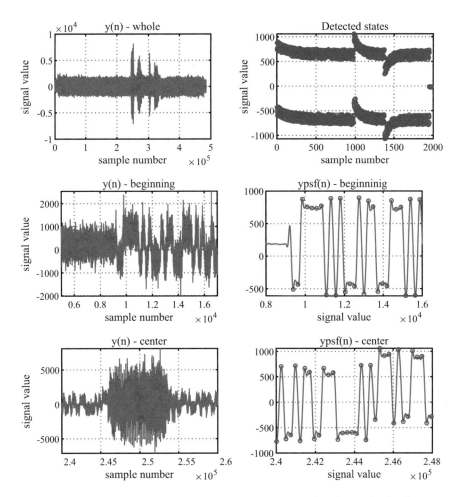

Fig. 21.13: Detection of bits transmitted in a nano-satellite signal, in the first column: (1) the whole signal, (2) its first part with starting bit transmission, (3) its central part with starting voice message; in second column: (1) detected carrier states in the whole signal, (2) beginning of the low-pass filtered signal with detected carrier states, (3) central part of the signal when voice transmission starts with detected carrier states

How bit positions were detected? First, we found visually the first positive $(+1)$ peak of the starting symbol. Next, knowing that one BPSK symbol has 240 samples, we jump forward 240 samples and check the signal value: the positive value means occurrence of bit 1 while negative one of bit 0. After decision taking, we are checking two neighborhood samples which are lying on both sides of the carrier state sample. If the central sample is positive and bigger than its neighbors, our symbol timing is correct (we are in the peak maximum). Therefore our next jump is again equal to regular 240 samples. But if the next sample is bigger than the central one, it means that our central sample estimate is before the peak (*we are too early*, we are sampling too fast) and jump value should be increased by 1 to 241 samples. In contrary, when the left neighbor is bigger than the central carrier sample estimate, it means that we are sampling too slow (*we are too late*) and the jump value should be decreased by 1. The same idea is used when "negative" peak is found: at present the central point should be lower than its left and right neighbors. Therefore we are checking this relation: minimum (OK), negative slope (too early) or positive slope (too late). No surprise that the method is known as early-late-gate symbol timing recovery. It is illustrated in Fig. 21.14 and implemented in the program 21.7. Results presented in Fig. 21.13 positively validate the algorithm.

When bit positions are detected, we start searching synchronization bit patterns 0011111010 and 1100000101—look at the program 21.7. In Fig. 21.13, in the second and third plot of the second column, we can recognize both synchronization words (look for sequence of five zeros 00000 and five ones 11111). When the synchronization words are found, we have to check whether they are separated by 960 bits: 96 words with 10 bits each. We have exactly this number of bits what confirms correctness of header detection. In 10 bit words are coded 8 bit bytes (code 8b10b). We should recover the bytes using a look-up table. If a received 10 bit word is missing in the table, it means that bits were corrupted. Then, recovered 96 bytes together with side information about encountered errors are passed to the Reed–Solomon decoder which extracts 64 bytes from 96 bytes and tries to perform error correction. Finally, header is decoded and digital information sent from satellite is extracted. Our program 21.7 stops after detection of synchronization words and after checking numbers of received bits. The program is presented without any cutoffs because it demonstrates all important steps in IQ signal processing: data reading, plotting, service selection and separation (down-shifting in frequency and band-pass filtering) and signal demodulation (pulse shaping filtering and symbol timing recovery).

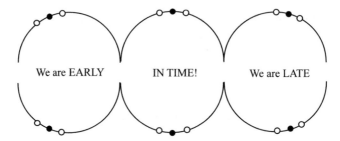

Fig. 21.14: Illustration of the early-late-gate symbol sampling synchronization method: (up) for symbols with positive values, (down) for symbols with negative values. (left) too early symbol sampling—increasing of the sampling step is required, (middle) correct sampling—the sampling step remains unchanged, (right) too late symbol sampling—decreasing of sampling step is necessary

Listing 21.7: Early-late-gate symbol timing recovery in analysis of nano-satellite signal

```
1  % lab21_ex_nanosat.m
2    clear all; close all;
3
4    m=128; cm_plasma=plasma(m); cm = plasma;          % color maps for gray printing
5
6  % Read recorded IQ signal - choose one
7    FileName = 'SDRSharp_NanoSat_146000kHz_IQ.wav';   % file to be read
8    [x,fs] = audioread(FileName);                     % reading, fs=192 kHz
9    Nx = length(x),                                   % signal length
10   x = x(:,1) - j*x(:,2);                            % IQ signal formation
11 % FM demodulation and NanoSat voice control signal decoding
12   faudio = 48000; Down= round(fs/faudio);           % parameter values
13   f0 = 1.98e+4; x = x .* exp(-j*2*pi*f0/fs*(0:Nx-1)'); % carrier? spectrum shifting
14       figure;
15       N = 512; df = fs/N; ffshift = df*0.125*(-N/2:1:N/2-1);
16       spectrogram(x,kaiser(N,10),N-N/4,ffshift,fs);
17       colormap(cm); pause
18   h = fir1(500,12500/fs,'low',chebwin(501,120));    % low-pass filter design
19   x = filter(h,1,x); x = x(1:Down:end);             % filtration
20   dt = 1/faudio; x = x(2:end).*conj(x(1:end-1)); y=angle(x)/(2*pi*dt); % FM demod
21   %soundsc(y,faudio);                               % listening
22       figure; plot(y); grid;
23       title('Digital signal in time domain - all samples');
24       xlabel('sample number'); ylabel('Signal values'); pause
25
26 % Bits decoding ##############################################################
27 % fsymb = 200Hz = 200 symbols per second, Nsps = 240 samples per symbol
28   fsymb = 200;                % symbol rate [Hz], 200 symbols per second
29   r = 0.5;                    % PSF: 0.5 - roll-off factor of RC filter
30   Nspan = 2;                  % PSF: 2 - number of symbols per RC filter
31   Nsps = faudio/fsymb;        % PSF: nominal number of samples per symbol, equal 240
32 % Taking signal fragment, first or second voice beacon
```

```
33   n = 170000:655000;            % selected manually voice beacon, visual inspection
34   y = y(n); y = y-mean(y);      % cutting signal fragment
35 % Low-pass square root raised cosine pulse shaping filter (PSF) design
36   h = firrcos(Nspan*Nsps, fsymb, r, faudio,'rollof','sqrt'); % PS filter
37 % Low-pass filtering
38   ypsf = conv(y,h);
39   ypsf = ypsf - mean(ypsf);
40 % Removing filter delay - transients
41   Nh = length(h); Nskip = (Nh-1)/2;
42   ypsf = ypsf(Nskip+1:end-Nskip);
43
44 % Manual selection of the position of the starting symbol (-1 or +1)
45 % It is present just before the first synchronization word of the beacon
46   Nmax = length(ypsf); n = 1:Nmax;
47   indstart = 9383;  % 9383/9384 starting position of the voice beacon
48
49 % Adaptive bits detection - changing step +/-1, approx. 240 samples long
50   bstring = '';       % message (bits) received, written to string - at present nothing
51   bpos = [];          % indexes of bit positions, at present nothing
52   pos = indstart;     % position of the starting bit "0"(-1) or "1"(1), found manually
53   m = 1;              % working variable for indexing vector elements
54   ybits = [];         % vector of signal values at bit positions, to be filled
55   while( pos + 1 <= Nmax )
56       ybits(m) = ypsf(pos);
57       bpos = [ bpos pos ];
58       if( ypsf(pos) > 0) % bit "1=+1", should be maximum
59           bstring=strcat( bstring, '1');
60           if ( ypsf(pos-1)<ypsf(pos) && ypsf(pos)<ypsf(pos+1) )
61               pos = pos+1; % increase position by one, you are early, positive slope
62           elseif (ypsf(pos-1)>ypsf(pos) && ypsf(pos)>ypsf(pos+1) )
63               pos = pos-1; % decrease position by one, you are late, negative slope
64           end
65       else                % bit "0=-1", should be minimum
66           bstring=strcat( bstring, '0');
67           if ( ypsf(pos-1)<ypsf(pos) && ypsf(pos)<ypsf(pos+1) )
68               pos = pos-1; % decrease position by one, you are late, positive slope
69           elseif (ypsf(pos-1)>ypsf(pos) && ypsf(pos)>ypsf(pos+1) )
70             pos = pos+1; % increase position by one, you are early, negative slope
71           end
72       end
73       pos = pos + Nsps; % do next step, after position correction
74       m = m + 1;          % increase index of detected bits
75   end
76
77 % Searching for synchronization words
78   synchro1 = '1100000101';
79   synchro2 = '0011111010';
80 % bstring(100:109)=synchro1; bstring(200:209)=synchro2; % for test
81   for n = 1 : length(bstring)-9
82       result1(n) = strcmp( bstring(n:n+9), synchro1 );
83       result2(n) = strcmp( bstring(n:n+9), synchro2 );
84   end
85     figure;
86     n = 1 : length(bstring)-9;
```

```
87      plot(n,result1,'ro',n,result2,'bo'); grid; xlabel('n');
88      axis([ n(1), n(end), 0, 1.1 ] );
89      title('Detection of Synchronization Words');
90      pause
91
92   % Finding bit positions and number of bits between synchronization words
93      ind1 = find( result1 == 1 );   % 1st bit of sync word #1
94      ind2 = find( result2 == 1 );   % 1st bit of sync word #2
95      ind = sort( [ ind1, ind2 ] );  % sorting them
96      i1 = ind(1)+10; i2 = ind(2)-1; % indexes of 1st and last bit in 1st data block
97      len1 = i2-i1+1,                % number of bits in the first data block
98      i3 = ind(2)+10; i4 = ind(3)-1; % indexes of 1st and last bit in second data block
99      len2 = i4-i3+1,                % number of bits in the second data block
```

Exercise 21.10 (Decoding Data from AMSAT Nano-Satellites). Analyze code of the program presented in Listing 21.7. Play a while with it. Run it for different pulse shaping filters. May be we could perform bit decoding more efficiently using a down-sampled signal, at present we have 240 samples per symbol. If you are a fun of nano-satellite transmission, you might interest in further bit processing (10 bits to 8 bits conversion, Reed–Solomon decoder, . . .).

21.8.2 Decoding RDS Data in FM Radio

Introduction Radio Data System (RDS) [7] used in FM radio, which is almost everywhere available in the world at present, will be our next digital transmission example in this chapter. It is old but still working. We will test on it our practical understanding of bit transmission using a single carrier TX-RX knowledge acquired till now. Principles of creating MPX hybrid signal, which is used for frequency carrier modulation in FM radio, were presented in Sect. 19.8. Performing decoding real-world stereo FM broadcast was tested by us in the program 19.5. In this section we will concentrate on:

1. creation of MPX signals with left and right audio channels and with pseudo-RDS 0/1 data,
2. generation in the base-band of an FM radio $IQ(n)$ signal, i.e. modulating in frequency 0 Hz complex-value carrier by the MPX samples,
3. FM demodulation of an artificially created or practically recorded FM radio $IQ(n)$ signals and extracting from them transmitted RDS bits without their interpretation/parsing (however, there are available programs which allow doing it and, of course, we will use them).

In this section the simplest solutions will be presented. More complex are left for home exercises. Because here we summarize some part of our *telecommunication* knowledge, decoding of stereo signal also will be done by the accompanying programs but not presented in listings in the book.

FM Radio Parameters In order to FM radio signal processing, its synthesis and analysis, coding and decoding, we should choose values of some important parameters and prepare required filters. It is done in program fmradio_params.m, presented in Listing 21.8. We assume sampling ratio equal to 250 kHz because RTLSDR USB stick, which we are using, offers this value as the lowest available sampling frequency—we would like to use our programs to decode real-world recordings also. In consequence, in order to avoid difficult signal re-sampling in the receiver, we choose sampling ratio of audio signal in the RX equal to 25 kHz, ten times lower value.

In FM radio high frequencies of transmitted audio are pre-emphasized in a transmitter and de-emphasized in a receiver. We design only pre/de-emphasis filters for audio frequency 25000 Hz, thinking about decoding of read-words recordings. During tests of $IQ(n)$ FM radio signals generated by us, we can use the de- and pre-emphasis filters or do not. Reader could make encoder and decoder program more flexible in this aspect.

In RDS square root raised cosine filter with roll-off factor $r = 1$ is used. For sampling frequency 250000 Hz and RDS symbol frequency 19000/16=1187.5 Hz, we have 250000/1187.5=210.5263 samples for symbol. Choosing $f_s = 228$ kHz, we would obtain 192 samples per symbol which would offer simpler signal generation. In Listing 21.8 the PSF filter is designed by us in frequency domain and compared with Matlab function design—the difference is very small. Variable phasePSF denote the filter delay which should be taken into account during signals synchronization in the receiver.

Finally, in the program four band-pass filters are designed for central frequencies: 19, 38, 57, and 76 kHz, multiplicity of 19 kHz, the pilot frequency.

Listing 21.8: Initialization Matlab program of software FM radio transmitter and receiver

```
1   % FM radio_params.m
2   % FM Radio - initialization
3   % Parameters
4     fs = 250000;           % sampling frequency of one FM radio station (to be changed)
5     fpilot = 19000;        % frequency of the pilot, 19000 Hz
6     fsymb = fpilot/16;     % frequency of RDS symbols 1187.5 Hz, 19000/16
7     fstereo = 2*fpilot;    % frequency of L-R signal carrier, 38000 Hz
8     frds = 3*fpilot;       % frequency of RDS signal carrier, 57000 Hz
9     faudio = 25000;        % frequency of audio signal (assumed, can be changed)
10    L = 500;               % length of used FIR filters
11    Ks = 6;                % number of symbols in PSF filter
12    dt = 1/fs;             % sampling period
13    K = L/2;               % half of the filter length
14  % Pre-emphasis and de-emphasis filters (for TX and RX) - for frequency faudio
15    f1 = 2120; tau1 = 1/(2*pi*f1); w1 = tan(1/(2*faudio*tau1));
```

```
16   b_de = [w1/(1+w1), w1/(1+w1)]; a_de = [1, (w1-1)/(w1+1)];
17   b_pre = a_de; a_pre = b_de;
18 % Pulse shaping filter (PSF) for RDS symbols
19   Tsymb = 1/fsymb;                % time duration of one RDS symbol
20   Nsymb = round(fs/fsymb);        % number of samples per RDS symbol
21   Npsf = Ks*Nsymb;                % number of samples of PSF filter
22   if(rem(Npsf,2)==1) Npsf=Npsf+1; end
23   df = fs/Npsf; f = 0 : df : 2/Tsymb; Nf=length(f);        % spectrum
24   H = zeros(1, Npsf); H(1:Nf) = cos(pi*f*Tsymb/4);
25   H(end:-1:end-(Nf-2)) = H(2:Nf);
26   hpsf1 = fftshift(ifft(H)); hpsf1=hpsf1/max(hpsf1);        % imp. response #1
27   hpsf2 = firrcos( Npsf, fsymb, 1.0, fs,'rolloff','sqrt'); % imp. response #2
28   hpsf2 = hpsf2/max(hpsf2); hpsf2 = hpsf2(1:end-1);
29   n1=1:length(hpsf1); n2=1:length(hpsf2);
30 % figure; plot(n1,hpsf1,'ro-',n2,hpsf2,'bx-'); grid; pause
31   hpsf = hpsf2;
32   phasePSF = angle( exp(-j*2*pi*fsymb/fs*[0:Npsf-1]) .* hpsf' ); % phase shift
33 % Low-pass (LP) filter for recovery of L+R signal
34   hLPaudio = fir1(L, (faudio/2)/(fs/2),kaiser(L+1,7));
35 % Narrow band-pass (BP) filter for separation of pilot signal (around 19 kHz)
36   fcentr = fpilot;  df1 = 1000; df2 = 2000;
37   ff = [ 0 fcentr-df2  fcentr-df1 fcentr+df1 fcentr+df2 fs/2 ]/(fs/2);
38   fa = [ 0 0.01        1          1          0.01       0 ];
39   hBP19 = firpm(L,ff,fa); % in Octave firls()
40 % Narrow band-pass filter for separation of RDS signal (around 57 kHz)
41   fcentr = frds; df1 = 2000; df2 = 4000;
42   ff = [ 0 fcentr-df2  fcentr-df1 fcentr+df1 fcentr+df2 fs/2 ]/(fs/2);
43   fa = [ 0 0.01        1          1          0.01       0  ];
44   hBP57 = firpm(L,ff,fa); % in Octave firls()
45 % Wide band-pass (BP) filter for separation of L-R signal (around 38 kHz) ...
46 % Narrow band-pass filter for separation of 2*fstereo component (around 76 kHz) ...
```

Exercise 21.11 (Digital Filter Design for Software FM Radio Receiver). Plot frequency responses, magnitude, and phase, of all filter used in software FM radio receiver, including pre/de-emphasis and pulse shaping filters.

FM Radio Encoder Matlab program for generation of $IQ(n)$ signal of one FM radio station is presented in Listing 21.9. The signal can be transmitted into the air by any talented SDR TX-RX hardware, for example, by the ADALM PLUTO, in any free frequency band. After that we could receive the signal by less capable SDR RX-only equipment, for example, RTLSDR dongle. At present we analyze code of the program 21.9.

1. We start with including the file fmradio_params.m with chosen parameter values and designed filter coefficients.
2. Reading audio signals. Then we read two monophonic recordings x1, x2, one for left and one for right channel. They can have different lengths and different

sampling frequencies fx1, fx2. Of course, we can use one stereo signal. If you prefer such option, modify the program.

3. Processing audio signals. Next, two monophonic signals are re-sampled to frequency faudio, default 25000 Hz, and are aligned in length. Then each of them is pre-emphasized with short IIR filter and both are up-sampled to the target FM radio service frequency. We have chosen 250000 Hz, however, frequencies 256, 228, 192 kHz are also acceptable. Warning: activation of the signal pre-emphasis option cause generation of spurious frequency around 16 kHz in the MPX signal. What is the reason?

4. Processing RDS bits. This section of the program is for us the most important in present chapter [7]. In part A, knowing service and symbol frequencies, we calculate number of samples per one RDS symbol (Nsps=fs/fsymb) and then number of RDS symbols which can be send with an audio signal having Nx samples: Nrds=ceil(Nx/Nsps). We decide to transmit an RDS-like bit pattern which is easy to check: four 1s, three 0s, two 1s, and one 0: [1 1 1 1 0 0 0 1 1 0. Finally, we repeat this pattern many times in loop. In part B we encode bit-stream differentially, simply subtracting next rds(k) bit from the bit r(k−1) already coded. In part C each bit r(k) equal to 0 is replaced by a bi-phase sequence [−1, 1], and bit r(k) equal to 1 by a sequence [1, −1]. Next, in part D, the most difficult, appropriate number of zeros are inserted between each two values of the vector resulting from the previous step: vector brds is replaced with sparse vector uprds. First, a long vector uprds having only zeros is created, then values of brds are copied into appropriate positions of uprds, i.e. when the function sin(2*pi/Tsymb*t) changes its sign. Finally, in part E, the signal is processed by a pulse shaping filter hpsf and synchronized with pilot (i.e. filter delay is removed). In this program section, real-value 0/1 carrier states $I(k)$, corresponding to RDS data, are transformed into a smooth $I(n)$ signal.

5. Generation of MPX signal. Now an MPX signal is synthesized, which consists of: mono x1+x2 component, 19 kHz pilot cos(alpha), stereo x1-x2 component multiplied by the doubled pilot cos(2alpha) (38 kHz), and RDS signal prds, multiplied by the tripled pilot cos(3alpha) (57 kHz). All signals are appropriately scaled.

6. FM modulation. At the end, frequency modulation depth df is calculated using the Carson rule and complex-value carrier exp(j2*pi...) with frequency 0 Hz (base-band) is modulated in frequency using the Matlab cumsum() function. Result is stored to a file as a target I,Q signal of the FM radio broadcast, to be transmitted by an SDR hardware.

In Fig. 21.15 creation of RDS-like $I(n)$ signal is presented: r(k) —RDS bits differentially encoded, brds(k) —corresponding bi-phase sequence, uprds(n) — bi-phase sequence with inserted zeros, prds(n) —final $I(n)$ signal after pulse shaping with square root raised cosine filter $r = 1$.

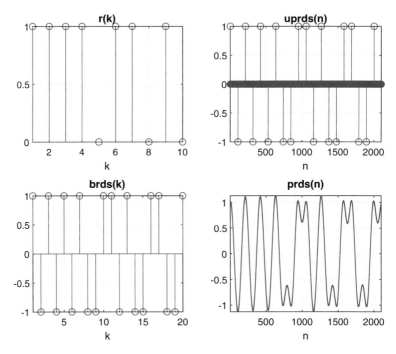

Fig. 21.15: Illustration of RDS signal generation. The following signals are presented, in columns: `r(k)`, `brds(k)`, `uprds(n)`, and `prds(n)`, generated in the program 21.9

Exercise 21.12 (Investigating Signals in FM Radio Encoder). Present version of the program, offered in the book repository, displays only the MPX signal and the $IQ(n)$ signal of the modulated carrier together with their spectra. Add plots in the initial part of the program: (1) investigate how are changing spectra of audio signals after two up-sampling steps, from `fx` through `faudio` to `fs`, (2) using `stem()` function, observe temporary signals after each processing step during creation of the RDS signal `prds`—you should obtain plots presented in Fig. 21.15. Observe that switching signal pre-emphasis ON, causes generation of some spurious 16 kHz component in the MPX signal. Investigate this problem.

Listing 21.9: Matlab program for preparation of $IQ(n)$ base-band signals of one FM radio station

```
1   % lab21_ex_FMradio_encoder_stereo_RDS.m
2     clear all; close all;
3
4     FMradio_params    % read parameters from the included file
5
6   % Read two monophonic audio signals - audioread or wavread
7     [x1, fx1 ] = audioread('GOODBYE.WAV'); x1=x1.'; % plot(x1);
8     [x2, fx2 ] = audioread('DANKE.WAV');   x2=x2.'; % soundsc(x1,fx1);
9   % Re-sampling to FM radio audio frequency - faudio
10    [N1,D1] = rat(faudio/fx1,1e-6); x1 = resample(x1,N1,D1);
11    [N2,D2] = rat(faudio/fx2,1e-6); x2 = resample(x2,N2,D2);
12    Nx1=length(x1); Nx2=length(x2); Nx = max( Nx1, Nx2 );
13    x1 = [ x1 zeros(1,Nx-Nx1) ]; x2 = [ x2 zeros(1,Nx-Nx2) ];  % append zeros
14  % x2 = zeros(1,Nx); % for testing cross-talk between channels in the receiver
15  % Filters were designed for faudio=25000 Hz, our signals can have different frequency
16  % Pre-emphasis, flat freq response to 2.1 kHz, than increasing 20 dB per decade
17    x1 = filter(b_pre,a_pre,x1); x2 = filter(b_pre,a_pre,x2);
18  % Up-sampling to FM radio service frequency - fs
19    [N,D] = rat(fs/faudio,1e-6); x1 = resample(x1,N,D); x2 = resample(x2,N,D);
20  % RDS BITS ###########################################
21  % A. Generation of dummy RDS bits for encoding
22    Nx = length(x1); Nsps=fs/fsymb; Nrds = ceil(Nx/Nsps); rds = [];
23    for k=1:ceil(Nrds/10), rds = [ rds 1 1 1 1 0 0 0 1 1 0 ];
24    end
25  % rds = round(rand(1,Nrds)); % for test
26  % B. Differential encoding of RDS bits
27    Nrds = length(rds); r = 0;
28    for k=2:Nrds, r(k)=abs(rds(k)-r(k-1));
29    end
30  % C. Generation of bi-phase impulses: -1 -> +1, +1 -> -1,
31    bip = [-1 1; 1 -1 ]; brds = bip( r+1, : ); brds=brds'; brds=brds(:);
32  % D. Zero insertion (appending) between -1/+1 impulses
33    Nr = ceil(fs/fsymb *length(r)); uprds = zeros(1,Nr); t = dt* (0:Nr-1);
34    clk = abs(diff(sign(sin(2*pi/Tsymb*t))))/2; clk = [clk 0];
35    indx=1;
36    for n=1:Nr
37        if(clk(n)==1) uprds(n)=brds(indx); indx=indx+1;
38        end
39    end
40  % E. Signal smoothing using pulse shaping filter (PSF), synchro with pilot
41    prds = conv( uprds, hpsf); prds = prds(Npsf:Npsf+Nx-1);
42    clear t clk r brds uprds;
43  % RDS BITS ###########################################
44  % Generation of multiplex (MPX) signal of FM radio
45    n=0:Nx-1; alpha = 2*pi*fpilot/fs*n;
46    x = 0.9*(x1+x2)+0.1*cos(alpha)+0.9*(x1-x2).*cos(2*alpha)+0.05*prds.*cos(3*alpha);
47  % Final frequency modulation of the complex carrier in the base-band
48    BW = 160000;                % overall FM radio bandwidth > 2*fmax=2*60000 Hz
49    fmax = 60000;               % maximum modulating frequency
50    df = (BW/(2*fmax)-1)*fmax,   % from Carson's bandwidth rule
51    beta = df/fmax,             % modulation index
```

```
52    fc = 0;                           % carrier frequency of FM radio in the base-band
53    x = exp( j*2*pi*( fc/fs*n + df*cumsum(x)/fs ) ); x=x.';   % FM modulation
54    I = real(x); Q = imag(x);                   % I, Q components
55    audiowrite('FMRadio_IQ_250kHz_LR.wav',[I Q],fs);    % store result
56  % audiowrite('FMRadio_IQ_250kHz_L.wav',[I Q],fs);    % store result
```

FM Radio Decoder Matlab program for decoding $IQ(n)$ signal of one FM radio station is presented in Listing 21.10. IQ file can be generated by us or taken from any SDR receiver, for example, from RTLSDR USB stick or ADALM PLUTO. At present we analyze the program code.

1. Reading IQ file. Program starts with selection of an IQ file to be processed and reading it. The IQ data should be synthesized or recorded (by SDR hardware) earlier by us. The only problem frequently occurring here is that different Matlab versions use different audio reading functions: wavread() in the past and audioread() at present.
2. Restoring IQ signal. Frequency demodulation. After reading the data y, the two-column matrix of numbers [I; Q] is converted into a complex-value signal $IQ(n) = I(n) + jQ(n)$. Then, frequency demodulation is performed, i.e. instantaneous signal frequency is calculated as the signal phase derivative. No extra signal scaling is done.
3. Decoding monophonic L+R signal. This is the easiest task, *a starter* in our *menu*. We should only leave the low-frequency part of our signal, i.e. remove the remaining stuff: pilot, stereo L-R, and RDS. Audio of FM radio signal is typically sampled at 32 ksps (32 kHz) therefore the filter cut-off frequency should be smaller than 16 kHz. We are using 25000 sps sampling ratio and cut-off frequency equal to 90% of the faudio/2=12500 kHz.
4. Carrier/pilot recovery. In order to go further, to decode stereo and RDS signals, we should recover their carriers having frequencies 38 kHz and 57 kHz, respectively, i.e. the second and third multiplicity of the 19 kHz pilot frequency. The pilot is transmitted. We are lucky: not always in telecommunication transmission this is the case. Therefore, we will use a well-known to us the double PLL loop, investigated in chapter on AM, and synchronize it with pilot signal, extracted from the MPX signal by means of very narrow, 2 kHz wide, band-pass filter. At this stage the most important is selection of PLL loop adaptation constants mi1 and mi2, deciding about its dynamic features (convergence and its speed as well as variance in steady-state solution). Appropriate equations were given in chapter on AM. After computation of the pilot angle theta, we calculate cosines with the following angles: theta/16 (for RDS 1187.5 Hz timing recovery), 2theta (for frequency down-conversion of the 38 kHz L-R signal), and 3theta (for frequency down conversion of the 57 kHz RDS signal). Now we can check the PLL loop convergence and estimate time when all carriers have correct frequencies.
5. Decoding L−R stereo extension signal. To enjoy the music, we are down-converting to 0 Hz the L−R signal, multiplying it with the recovered 38 kHz carrier and then low-pass filtering the result with the same LP filter which was

used for the L+R signal. Having signals (L+R) and (L−R) we are adding and subtracting them, obtaining signals 2L and 2R. Right. But this is true only when signals (L+R) and (L−R) are not shifted in time in respect to each other due to different delays in their processing paths. When number of filters and their delays in both paths are the same, reconstruction of audio channels is perfect. For this reason usage of band-pass filter for extraction of L-R signal from the MPX signal is optional. When it is done, clearness of the signal (L−R) is better but one more filter is used in the (L−R) signal processing path and the signal (L+R) should be delayed by the same amount also before addition and subtraction of (L+R) and (L−R). We can avoid using the `hBP38` filter at the cost of application of better and longer the `hLPaudio` filter. We can test both options.

6. Decoding RDS bits. This is our main story [7]. Beginning is typical, without problems—*only blue sky* (**part A**): optional band-pass filter around RDS frequency 57 kHz. Why optional? Because we have in reserve a low-pass pulse shaping filter `hpsf` playing now a role of `hLPaudio` filter. After that we perform signal down-conversion using the `c57` carrier, recovered by PLL, and remove the doubled carrier component $2 \cdot 57 = 114$ kHz with use of a low-pass pulse shaping filter. Please, add spectra computation to the program to observe these *star wars*. In **part B** *first clouds appears*: at present we have the $I(n)$ RDS signal but we have to sample it in proper places to recover carrier states and transmitted bits. Therefore, the symbol timing recovery should be done. The easiest but not very robust way for doing this is finding in the signal two *camel-like humps*, both positive or both negative and notice that this is a place of change of transmitted bit value. After positive *cammel*, bit 1 starts, after negative bit 0. Having correct position of one bit, we could jump forward one symbol length, read signal value, decode the next bit, and adjust length of next jump using the early-late-gate method, already used by us in case of nano-satellite signal. A Reader is encouraged to implement this idea. We will not risk and apply a robust coherent bit detection. Carrier `c1`, synthesized after the PLL loop, is synchronized with the RDS bit transmission. However, it is 16 times slower than the pilot and we do not know for which of pilot cycle the RDS symbol starts. To find this, we correlate signal `y` with `c1` with a time shift equal to a multiplicity of the pilot period. Correlation maximum tells us when the RDS symbol begins. Now, in **part C** we use a concept of coherent detection: modulated carrier (signal `y`) is multiplied by recovered carrier (signal `c1`). When 1 (+1) is transmitted, both signals are synchronized sines, when 0 is sent (−1)—the signal `y` is equal to minus sine. In the first multiplication, result is positive (+1), in the second negative (−1). Having this in mind, in **part D** we apply to the signal `y` a moving filter with 1s only, having length equal to the symbol period. And we filter the signal `y`. In **part E**, we are finding zero-crossing locations of the carrier `c1` for its positive (`maxi2`) and negative (`maxi1`) slopes. Then, we check what is standard variation of signal `y` in these points, and select this set of points, `maxi1` or `maxi2`, for which variation is bigger. *Here we are!* We have found positions of our RDS symbols. *Clouds disappeared! Blue sky won.* In **parts F, G, and H** we are finishing the game. First, we are recovering bits 0/1 from bi-phase sequences $(\pm1) \rightarrow (\mp1)$.

Next, we remove differential encoding. Finally, we store detected bits for further processing by other programs. *Lucy in the sky with diamonds!*

Some signals, illustrating idea of coherent symbol detection, are presented in Fig. 21.16. All plots were generated by the program 21.10 for data acquired by the RTLSDR USB stick. In the upper plot we see the received signal y after pulse shaping and the carrier c1 synchronized with it after step B. In the middle plot, signal y after multiplication with c1 and integration is shown (look at the program) together the carrier c1 and points of its zero crossing. Finally, in the bottom plot, the detected carrier states are presented for longer signal fragment.

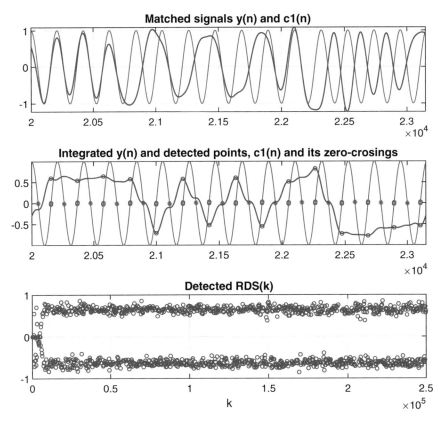

Fig. 21.16: Illustration of RDS bits detection. The following signals are presented: (up) signal $y(n)$ after the pulse shaping low-pass filter and synchronized with it carrier $c1(n)$, (middle) (1) carrier $c1(n)$ and points of its positive and negative zero-crossings, (2) locally integrated multiplication of signals $y(n)$ and $c1(n)$ in the interval of RDS symbol duration (see program), (bottom) detected carrier states of RDS signal

Exercise 21.13 (Investigating Signals and Spectra in the FM Radio Decoder). Analyze in detail code of the program 21.10. Run it for different available IQ files. Switch ON/OFF using of band-pass filters hBP38 and hBP57. Add spectra plots to observe spectral results of all signal modifications. Observe how fast is the PLL loop convergence (variable omega). Change values of mil and observe results of it.

Exercise 21.14 (RDS Bits Parser). Use the program 21.10 for processing FM/RDS IQ files (with one station only), recorded by the SDRSharp program. The decoded RDS bits are stored to the file FM_Radio_RDS.txt. Apply the program RDS.m from RDS_parser.zip to recover text from the RDS bits. Have a look to the RDS standard [7] and try to add a new functionality to the program RDS.m, i.e. decode and display more information transmitted by the RDS service.

Exercise 21.15 (My Own FM Broadcast). Prepare you own FM/RDS IQ file using the program FMradio_encoder_stereo_RDS.m. As RDS bits sent periodically ASCII codes of "Hello World!" or any text of your choice. Decode the text in the end of the program 21.10.

Listing 21.10: Complete Matlab program for decoding stereo FM radio and RDS bits (without their interpretation)

```
1   % lab21_ex_FMradio_decoder_stereo_RDS.m
2     clear all; close all;
3
4     FMradio_params    % reading parameters from the included file
5
6   % READ IQ SIGNAL - synthesized or recorded
7     [y,fs] = audioread('FMRadio_IQ_250kHz_LR.wav');                          % synth
8   % [y,fs] = audioread('SDRSharp_FMRadio_96000kHz_IQ_one.wav',[1,1*250000]); % record
9
10  % FM DEMODULATION
11    y = y(:,1) - sqrt(-1)*y(:,2);           % IQ --> complex
12    dy = (y(2:end).*conj(y(1:end-1)));      % calculation of instantaneous frequency
13    y = atan2(imag(dy), real(dy)); clear dy; % frequency demodulation
14
15  % DECODING MONO L+R SIGNAL
16    ym = filter( hLPaudio, 1, y );          % low-pass filtration of L+R (mono) signal
17    ym = ym(1:fs/faudio:end);               % leaving only every fs/faudio-th sample
18    disp('LISTENING MONO: Left + Right'); sound(ym,faudio); pause
19    w = ym-mean(ym); w=w/max(abs(w)); audiowrite('FM_mono.wav',w,faudio); clear w;
20
21  % CARRIER RECOVERY
```

```
22    p = filter(hBP19,1,y);                        % extracting 19 kHz pilot
23    theta = zeros(1,length(p)+1);                 %#
24    omega = theta; omega(1) = 2*pi*fpilot/fs;     %# double PLL on 19 kHz
25    mi1 = 0.0025; mi2 = mi2^2/4; p=p/max(abs(p));  %# see chapter on AM
26    for n = 1 : length(p)                         %#
27        pherr = -p(n)*sin(theta(n));              %#
28        theta(n+1) = theta(n) + omega(n) + mi1*pherr;  %#
29        omega(n+1) = omega(n) + mi2*pherr;        %#
30    end                                           %#
31    c1(:,1)    = cos(theta(1:end-1)/16);
32    c1PSF(:,1) = cos(theta(1:end-1)/16+phasePSF);  % RDS 19 kHz / 16 = 1187.5 Hz
33    c38(:,1) = cos(2*theta(1:end-1));              % L-R carrier 38 kHz
34    c57(:,1) = cos(3*theta(1:end-1)); clear p; clear theta; clear freq;  % RDS carrier 57
          kHz

35
36  % DECODING STEREO
37    if(1) ys = filter(hBP38,1,y); delay = (L/2)/(fs/faudio);  % extraction of L-R signal
38    else ys=y; delay=0; end                       % (optional BP filtration around 38 kHz)
39    ys = real(ys .* c38); clear c38;              % L-R signal: 38kHz --> 0kHz
40    ys = filter( hLPaudio, 1, ys );               % low-pass filtration
41    ys = ys(1:fs/faudio:end);                     % leaving every fs/faudio-th sample
42    ym = ym(1:end-delay); ys=2*ys(1+delay:end);   % synchronization of L+R and L-R
43    clear ymm yss;
44    y1 = 0.5*( ym + ys ); y2 = 0.5*( ym - ys ); clear ym ys;  % recovering L and R
45    y1 = filter(b_de,a_de,y1); y2 = filter(b_de,a_de,y2);     % de-emphasis
46  % disp('LISTENING TO STEREO'); soundsc([y1' ; y2'],faudio); pause
47    maxi = max( max(abs(y1)),max(abs(y2)) );
48    audiowrite('FM_Radio_stereo.wav', [ y1/maxi y2/maxi ],faudio); clear y1 y2;
49
50  % DECODING RDS \cite{IEC99}
51  % A. Initial operations
52    if(1) y = filter(hBP57,1,y); c1 = c1PSF; end  % extraction of RDS (optional BP filter
          )
53    y = y .* c57; clear c57;                      % frequency conversion: 57kHz-->0kHz
54    y = filter( hpsf, 1, y );                     % low-pass filer - SRRC pulse shaping
55  % B. Signal correlation with clock c1 shifted 16x by 1/fpilot (exact timing recovery)
56    nstart = 20000; Nmany = 50*210;               % skipping time of PLL loop adaptation
57    for n = 1 : round(fs/fpilot) : round(fs/fsymb)+1;
58        synchro(n)=sum((y(nstart+n-1:nstart+n-1+Nmany-1).* c1(nstart:nstart+Nmany-1)));
59    end
60    [v, indx] = max(synchro);                     % finding maximum
61    y = y(indx:end); c1=c1(1:end-indx+1);         % final signal synchronization
62  % C. Signal multiplication with synchronized carrier - coherent detection
63    y = y .* c1; y = y/max(y);
64  % D. Moving average of last fs/Tsymb samples
65    h = ones(1,round(fs/fsymb))/round(fs/fsymb); y = filter(h,1,y);
66  % E. Finding negative and positive slopes of synchronized clock
67    N = length(c1); M=5; M2=(M-1)/2; maxi1 = []; maxi2 = [];
68    for n = M2+1:N-M2
69        if( (c1(n-2)>0) && (c1(n-1)>0) && (c1(n+1)<0) && (c1(n+2)<0) )  % negative slope
70            maxi1 = [ maxi1 n ];
71        end
72        if( (c1(n-2)<0) && (c1(n-1)<0) && (c1(n+1)>0) && (c1(n+2)>0) )  % positive slope
73            maxi2 = [ maxi2 n ];
```

```
74    end
75   end
76   if( std(y(maxi1)) > std(y(maxi2)) ) maxi=maxi1; else maxi=maxi2; end % best fitting?
77   % F. Changing signal levels: {-1, +1} --> {0,1}
78   SCRX = (-sign( y( maxi ) )+1)/2; clear c1 maxi maxi1 maxi2;
79   % G. Differential decoding with taking into account bi-phase signal nature
80   SCRX = abs(SCRX(2:end) - SCRX(1:end-1)); SCRX = SCRX(2:2:end)
81   % H. Storing detected bits to disk for further analysis by outside program
82   save FM_Radio_RDS.txt SCRX -ascii
```

21.8.3 Decoding Carrier States in TETRA Digital Telephony

TETRA (TErrestrial Trunked Radio) TETRA (1995) [4] is a special-purpose digital telephony system designed for government, emergency, safety, transport, and military applications. It uses $\pi/4$-DQPSK (Differential Quaternary Phase Shift Keying) modulation already known to us. The symbol frequency is equal to 18 kHz. The square root raised cosine PSF filter with roll-off factor $r = 0.35$ is applied. In this section we will recover carrier states of synthetic IQ TETRA signals, simulating different real-world transmission conditions:

- increasing level of noise (files: XX_-22.mat,...,XX_20.mat),
- different fading channels: FLAT, TU50 Typical Urban channel at 50 kph, HT200 Hilly Terrain channel at 200 kph.

In our case we are interested in finding 19-elements STS synchronization header which is present in synchronization continuous down-link burst TETRA signal. We would like to decode 129 carrier states which are after it. For noisy signals we know values of the first data block.

For TETRA signals decoding, we will use the code presented in Listing 21.1— we will only slightly modify it. Since core of the program lab21_ex_tetra.m is the same, it is not presented here—Reader can take it from the book repository. In Fig. 21.17 results from analysis of the file TETRA_423.4125MHz_0.mat are presented: power spectral density of the signal, cross-correlation function between the signal and the STS-19 header, and detected carrier states. In Fig. 21.18 the same characteristics are presented for the file TETRA_423.4125MHz_tu50.mat. As we see in both cases the program 21.1 allowed us to recover correctly, more or less, the IQ constellation points, without and with very big problems, despite the fact that no timing recovery procedure is implemented in it. It should convinced us that always it has a sense to make an effort and write a little bit more universal program because it can be used later, after minor modifications, to variety of *special cases*.

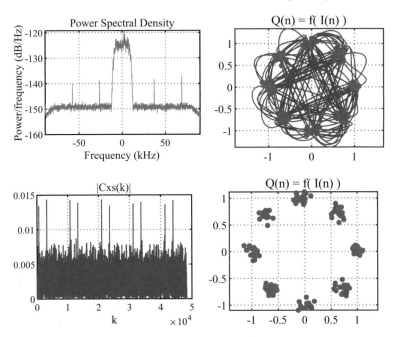

Fig. 21.17: Analysis of noisy TETRA signal, from left to right in columns: (left up) signal PSD, (left down) signal and header cross-correlation function, (up-right) decoded phasor-diagram, (down-right) decoded constellation points

Exercise 21.16 (Testing Demodulation of TETRA Signal). Analyze code of the program `lab21_ex_tetra.m`. Run it for all supported files. Choose different CFO and channel correction method (with exception of the NLMS filter). Observe that the last signal, for *hilly terrain at high speed*, is not correctly decoded. Change value of K, i.e. the number of samples per symbol. Observe that result of decoding is changing. Why? Try to solve this problem.

For signals embedded in noise only, data block after the first STS-19 synchronization header is known and given in the program, therefore we can find number of transmission errors for it, i.e. number of wrong decoded carrier states. 129 symbols after the STS-19 header has precisely defined role. They are grouped into packets consisting of: 15, 108, 1, 5 symbols. Observe carriers states after different STS headers and find which symbols are changing their values.

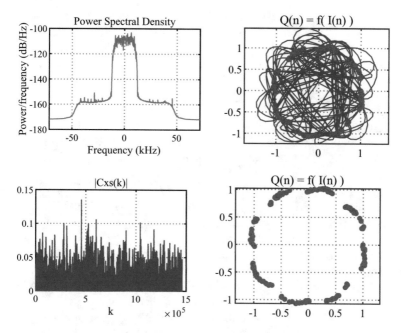

Fig. 21.18: Analysis of typical urban TETRA signal, from left to right in columns: (left up) signal PSD, (left down) signal and header cross-correlation function, (up-right) decoded phasor-diagram, (down-right) decoded constellation points

21.9 Summary

The second round of the single-carrier marathon has ended. What should be remembered?

1. Since our information is sent in amplitude and phase of a carrier and channel is modifying them, we have to estimate the channel parameters and correct their influence in the receiver. If channel is fast varying, channel estimation and equalization have to be repeated permanently. Pilot signals, the pseudo-random signals known to receivers, are used for channel de-conspiracy. Having input and output, channel impulse and frequency response can be calculated. For narrow-band single-carrier transmission channel influence can be simplified to signal attenuation and delay.

2. Even without channel disturbance signal demodulation is not a simple *undo* of the modulation. First, the carrier parameters are not changed in the transmitter in abrupt way because such ON/OFF carrier switching will make the carrier spectrum wider and disturbing neighbor services. Carriers are changing their states smoothly. To be precise, signal between the carrier

states is interpolated using the so-called pulse shaping filters. When carrier shape is so smooth, how to distinguish that carrier is already in its *bit-symbol* state, important for us, not in a transition part. This is a task for timing recovery methods. In this chapter we have used Barker and Gold codes as synchronization patterns (preambles/headers) for symbol and frame synchronization in packet-based transmission. We have also become familiar with three timing recover methods which do not require a special synchronization templates: the early-late-gate, Gardner, and Mueller–Müller techniques. They attempt to adaptively *sample-and-hold* the symbol position in the $IQ(n)$ signal by tracking the zero value of special cost functions.

3. During frequency up-conversion in the transmitter and frequency down-conversion in the receiver, the same carrier should be synthesized. But it is impossible. Some frequency and phase shift errors are present. The receiver should recognize the transmitter carrier in the received data and to synchronize with it. Without this, all transmission magic will collapse. This is a task for different carrier recovery methods. To remind only the PLL and Costas loops. We investigated them in the previous chapter on amplitude modulation but they were used also in this chapter.

4. As we see in digital telecommunication we have a lot of *recoveries*, to mention only the channel, timing, and carrier ones. In order to perform everything in a smart and efficient way, periodically are transmitted synchronization headers, preambles, pilots: special, unique signals having sharp auto-correlation functions. Thanks to this, they are easy to found and difficult to mask. Therefore they are used for identification and correction of all possible transmission disturbances. We have used Gold and Barker sequences and outputs from pseudo-random generators.

5. Digital single-carrier receivers have to be computationally efficient also. The winner should do the job with the lowest computational effort. Therefore $I(n) + jQ(n)$ signals should be highly decimated and we should have a few samples per symbol only for the carrier state detection. Missing sampled should be interpolated. For this purpose the Farrow filter interpolators are used. Nevertheless ... the bit detection robustness has to maintained.

21.10 Private Investigations: Free-Style Bungee Jumps

Exercise 21.17 (* **Channel Estimation: Robust Solutions of the Transmission Input–Output Equation**). Use program `lab21_ex_equ_mse_nlms.m` from the book repository. Analyze its code, run it, observe figures. Try to apply different algorithmic solutions for solving matrix input–output transmission equation more robust to noise. Some of existing algorithms are presented in Chap. 13 on damped sinusoid estimation, together with programs which can be easily modified to our new task. In Matlab there are also functions ready to use. Try to apply them.

Exercise 21.18 (* Channel Estimation: Usage of the RLS Adaptive Filter). Use program `lab21_ex_equ_mse_nlms.m` from the book repository. Try also to implement RLS adaptive filter as a channel corrector. You can find its code in Chap. 12. Compare its convergence with the NLMS filter. The RLS filter should be faster.

Exercise 21.19 (** Confidential RDS Transmission). Replace standard bi-phase BPSK RDS modulation scheme with some other, multi-level, phase shift modulation method (described in the previous chapter—see functions `IQdef()`, `numbers2IQ()`, `IQ2psf()`, `IQ2numbers()`, for example, by the QPSK (4-QAM). You can implement generation of an $I(n) + jQ(n)$ RDS signal similarly as in the program 21.1 and then replace in the FM/RDS encoder the following command (in line 46):

```
x = 0.9*(x1+x2)+ ... + 0.05*prds.*cos(3*alpha)
```

with

```
x = 0.9*(x1+x2)+ ... + 0.05*(I.*cos(3*alpha)-Q.*sin(3*alpha))
```

Add your code to programs of FM radio encoder and decoder. Modify the coherent RDS symbol detector for the QPSK case. Test the program. Try to improve it.

Exercise 21.20 (** RDS Decoding with Costas Loop Demodulation and Early-Late-Gate Symbol Tracker). In the FM radio decoder program, we have used the standard RDS decoding procedure. The pilot was used and coherent detection. Try to write a program in which Costas loop will do the RDS signal down-conversion (blind carrier recovery) and early-late-gate method will be applied for symbol detection (blind timing recovery). You can think about signal down-sampling from 210 samples per symbol to significantly lower value. If this task is too difficult for you, analyze code of two programs, presented in the book repository:

- `rdsbits_decoder_costas_earlylate.m`.
- `rdsbits_decoder_decim_costas_earlylate.m`.

In the second program 14-times signal decimation is done but, as a consequence, the RDS symbol detection is less robust. Compare codes of both programs. Run them. Use `RDS.m` program from `RDS_parser.zip` to decode text from RDS bits stored to the file `FM_Radio_RDS.txt`.

Exercise 21.21 (*** RDS Decoding with Costas Loop and Gardner/ Muller Method). Try to solve the RDS bit detection problem, combining the Costas loop and Gardner or Mueller–Müller method for finding symbol timing. For very small number of samples per symbol you can think about signal interpolation using Farrow filters. Try to use Matlab code fragments from the program `timing_recovery.m`.

Exercise 21.22 (** Timing Recovery in TETRA). Add Gardner or Gardner and Mueller–Müller timing recovery procedure to the program `lab21_ex_tetra.m`. For small number of samples per symbol implement the Farrow filter interpolator. We test all modules in timing recovery subsection of this chapter.

Acknowledgements I would like to express my sincere thanks to **Dr Marek Sikora** from Telecommunication Department, AGH Technical University of Science and Technology, Krakow, Poland, for His priceless help with preparing programs for this chapter!

References

1. J.B. Anderson, *Digital Transmission Engineering* (IEEE Press/Wiley-Interscience, Piscataway NJ, 2005)
2. K. Barlee, D. Atkinson, R.W. Stewart, L. Crockett, *Software Defined Radio using MATLAB & Simulink and the RTLSDR*. Matlab Courseware (Strathclyde Academic Media, Glasgow, 2015). Online: http://www.desktopsdr.com/
3. T.F. Collins, R. Getz, D. Pu, A.M. Wyglinski, *Software Defined Radio for Engineers* (Analog Devices - Artech House, London, 2018)
4. ETSI EN-300-396-2, Terrestrial trunked radio (TETRA); Technical requirements for direct mode operation (DMO); Part 2: Radio aspects, V1.4.1, Dec 2011. Online: https://www.etsi.org/deliver/etsi_en/; https://www.etsi.org/deliver/etsi_en/300300_300399/30039602/01.04.01_60/en_30039602v010401p.pdf
5. B. Farhang-Boroujeny, *Signal Processing Techniques for Software Radios* (Lulu Publishing House, 2008)
6. M.E. Frerking, *Digital Signal Processing in Communication Systems* (Springer, New York, 1994)
7. R.W. Heath Jr., *Introduction to Wireless Digital Communication. A Signal Processing Perspective* (Prentice Hall, Boston, 2017)
8. IEC-62106 Standard, Specification of the radio data system (RDS) for VHF/FM sound broadcasting in the frequency range from 87,5 MHz to 108,0 MHz. International Electrotechnical Commission, 1999
9. C.R. Johnson Jr., W.A. Sethares, *Telecommunications Breakdown: Concepts of Communication Transmitted via Software-Defined Radio* (Prentice Hall, Upper Saddle River, 2004)
10. C.R. Johnson Jr, W.A. Sethares, A.G. Klein, *Software Receiver Design. Build Your Own Digital Communication System in Five Easy Steps* (Cambridge University Press, Cambridge, 2011)
11. F. Ling, *Synchronization in Digital Communication Systems* (Cambridge University Press, Cambridge, 2017)
12. H. Meyr, M. Moeneclaey, S.A. Fechtel, *Digital Communication Receivers* (Wiley, New York, 1998)
13. L.C. Potter, Y. Yang, *A Digital Communication Laboratory - Implementing a Software-Defined Acoustic Modem*. Matlab Courseware (The Ohio State University, Columbus, 2015). Online: https://www.mathworks.com/academia/courseware/digital-communication-laboratory.html
14. J.G. Proakis, M. Salehi, G. Bauch, *Modern Communication Systems Using Matlab* (Cengage Learning, Stamford, 2004, 2013)

15. M. Rice, *Digital Communications: A Discrete-Time Approach* (Pearson Education, Upper Saddle River, 2009)
16. B. Sklar, *Digital Communications: Fundamentals and Applications* (Prentice Hall, Upper Saddle River, 1988, 2001, 2017)
17. C. Thompson, FOX-1 satellite telemetry part 2: FoxTelem. AMSAT J. **39**(1), 7–9 (2016)
18. S.A. Tretter, *Communication System Design Using DSP Algorithms* (Springer Science+Business Media, New York, 2008)

Chapter 22
Introduction to Digital Multi-Carrier Transmission: With DSL Modem Example

Old is still alive: a simple story about times of first big telecommunication breakthrough when DMT won by knockout with a 56k telephone modem

22.1 Introduction

In this chapter we will be talking about modems: joint modulators-and-demodulators. Do you remember the late nineties of the twentieth century? I did. I was using slow 56 kilo-bit per second (kbps) telephone modem and asymmetric digital subscriber line (ADSL) modems appeared. They offered 256, 512, 1024, 2048, 4096, and 8192 kbps! Wow. At present the same technology with new telephone and cable lines and short distances can offer even 80 Mbps. The same multi-carrier technique is used at present in power line communication (PLC) home modems, with HomePlug technology, offering approximately 100 Mbps. As we see, the wired, copper line transmission is still alive. It is true that nowadays fiber optics solutions are preferred due to its significantly higher bit-rates (giga and tera bits per second), but when the backbone copper cable infrastructure is present, why not to use it?

We are starting presentation of the multi-carrier telecommunication technology from ADSL modems because they were first using multi-carrier technology and solutions used in them represent a golden standard in some sense. In the next chapters of the book, we will switch to multi-carrier wireless solutions which are extremely important because they offer mobility for user and *things* (IoT—Internet of Things).

In single-carrier transmission, one cosine carrier is choosing one state from predefined table of possible states, differing in frequency, amplitude, and phase. Tracking carrier states in the receiver, we are finding their numbers and extract bits coded in them. It is straightforward that using more carriers with different frequencies, transmitted the same time (in parallel), leads to increase of the total transmission bit-rate. This is an idea. How to synthesize many *parallel* carriers? By using the inverse discrete Fourier transform: carrier amplitudes and phase shifts are coded in Fourier transform coefficients, one coefficient decides about

© Springer Nature Switzerland AG 2021
T. P. Zieliński, *Starting Digital Signal Processing in Telecommunication Engineering*, Textbooks in Telecommunication Engineering, https://doi.org/10.1007/978-3-030-49256-4_22

one carrier (or two of them as in ADSL modems), then the inverse DFT (FFT) is performed, and, *here we are*, we have a multi-tone (multi-carrier) signal, i.e. a summation of many orthogonal components with different frequencies. And each of them is carrying bits. The bit-rate grows with number of carriers. In first ADSL modems we had 512 complex carriers, 2048 in PLC modems, 4096—in VDSL modems. All of them around 0 Hz (base-band), excluding bandwidth 0–4000 kHz for analog speech/fax transmission.

Since a DSL modem transmit data in the base-band, no frequency up-conversion is performed. Therefore the signal can be only a real-value one. If such, its spectrum is symmetrical around 0 Hz ($X(-f) = X^*(f)$) and therefore bits can be allocated only to signal components with positive frequencies. If 512-point IDFT is used, maximally 256 sub-carriers are available. And this is the main difference between multi-carrier wired and wireless transmission: in wireless one all complex-value carriers are used, 512 in our example, then signal is up-converted in frequency and real part of the result is taken. The same was in the single-carrier communication (see Exercise 17.1 and program 17.1). The second main difference, which should be remembered, is that the single carrier smoothly passes from state to state due to carrier states interpolation done by the pulse shaping filter. In multi-carrier transmission the situation is different: all carriers, parallel in frequency, are synthesized at once by means of one IFFT procedure: we obtain one signal, block of samples, being summation of all carriers. Each carrier in this block is in one state only, all the time. Next IFFT gives us a next block of samples of a real-value signal to be transmitted. Block after block after block ... Yes, multi-carrier transmission is block-based, carriers are changing their states in *bung-bung* manner, not smoothly as a single carrier with the PSF-based states interpolation.

Channel impulse response is convolving with transmitted signal, which consists of separate blocks. The impulse response is *looking back*, to the past signal samples. Therefore, on the block edges/borders, channel impulse response is multiplied with samples of the previous data block and an inter-symbol (block) interference (ISI) occurs! In order to mitigate this effect some samples from the block end are copied into its beginning: a cyclic prefix (CP) is created. The CP causes that channel influence is limited to one data block only, when correct block beginning is found. For this purpose a special block synchronization algorithms are required. CP simplifies also performing frequency channel equalization.

When number of carriers grows, the signal bandwidth becomes wider. For ADSL, sampling frequency is equal to $f_s = 2.208$ MHz, signal bandwidth is equal to $2.208/2 = 1.104$ MHz, and carrier spacing is equal to $1104000/256 = 4312.5$ Hz. Lower carriers are not used. Some carriers are used for down-link (more) and some for up-link (less). Some carriers are used in duplex mode for both, up- and down-link. When copper wire is long, higher frequencies are more attenuated. If the signal is weak, noise is more disturbing and number of available carrier states decreases. During modem initialization central office (CO) and user modems are choosing optimal number of states for each sub-carrier on the base of estimated SNR for it. Number of states varies from to 2 to 2^{15}, i.e. modulation is changed from 4-QAM up to 32768-QAM. Bit allocation is performed in the beginning and remains unchanged during modem work. And this is the main difference between wired and wireless multi-carrier transmission: in wire-line transmission carriers deliver different number of bits, while in wireless one, due to different, reflection-like channel characteristics all carriers deliver the same number of bits (the same modulation is used for all carriers). In wireless case, bit allocation is not performed for a single carrier but in a mean sense for all carriers together: 4-QAM or 256-QAM for all of them, depending on transmission conditions. Wrong bits from carriers more attenuated by a channel are corrected then.

During ADSL modem start-up, situation is similar to our example from the chapter on single-carrier transmission. In the beginning, sequence required for channel estimation and design of its corrector is transmitted. Receiver uses it for the design of time (TEQ) and frequency (FEQ) channel equalizers. First of them is used for shortening the impulse response of the overall transmission path, while the second for calculation of channel attenuation and phase shift of each carrier. Then a random synchronization pattern (preamble) is sent, just before a regular bit transmission, and next bits themselves. In DSL transmission signal is not up- and down-converted in frequency; therefore, there is no frequency and phase carrier offset problem. However, when analog-to-digital converter in the receiver is working with wrong sampling frequency, amplitudes and phases of all carriers are wrongly interpreted. Additionally cross-talk between them occurs since their orthogonality is lost. This is a severe problem which has to be solved to ensure the high bit-rate.

The goal of this chapter is to learn basic problems and solutions existing in multi-carrier transmission, called discrete-multi-tone (DMT) in copper line DSL world and orthogonal frequency division multiplexing (OFDM) in wireless services. ADSL is only an example [4, 5]. In fact, the chapter is a general introduction to DMT/OFDM technology.

The following books present in detail the DSL technology: [3, 7, 10, 11]. In turn, in [13, 19] a Reader can find not only a brief description but also exemplary laboratory experiments dealing with multi-carrier transmission, in particular with ADSL modems [13].

So let us start our flight. "Here captain is speaking. Please, fasten your seat belts."

22.2 Concept of Discrete Multi-Tone Transmission

Typically, *multi* is obtained as repetition of *single*. This is also true in transition from single-carrier communication systems to multi-carrier ones. Before, one sine was periodically changing its parameters, i.e. it was modulated in amplitude, phase, and frequency. It was taken different *states*. The states were numbered. Bits were transmitted in these numbers. At present we have many sines. They cannot change frequencies lest they overlap in frequency and become indistinguishable. They can change their amplitudes and phases only, i.e. change *simpler* carrier states described by the QAM modulation. We transmit them simultaneously through the channel and receive simultaneously. Thanks to *multi*-technology, number of transmitted bits is increased in a receiver.

Typically, after such introduction, all listeners imagine direct parallel application of the single-carrier transmission with: (1) carrier states, (2) their pulse shaping in the base-band in TX, (3) up and down signal conversion in frequency, (4) pulse shaping in RX, and (5) carrier states detection. But in such scenario, the up-down high-frequency conversion, resource-consuming and expensive, is repeated many times for each sub-carrier! So, it is impractical. Therefore, let us generate the summation signal in the base-band: let us up-convert each $IQ(n)$ component to a different, low frequency lying around 0 Hz (in the so-called *base-band*), then sum them all and up-convert the combined signal to a target, high frequency. Then, in the receiver, down-shift the summation signal back from high frequency to the base-band, and demodulate each sub-carrier. In the base-band we can use sub-carrier spacing equal to 1 kHz, can have 1000 sub-carriers, and exploit 1 MHz bandwidth (1 kHz times 1000). After frequency up-conversion of the base-band, combined signal to the target frequency, let us say 1 GHz, we still use the same bandwidth equal to 1 MHz. The method is more efficient than the first one in which single carriers are individually converted up-and-down in frequency, but creation of the summation signal in the base-band and its backward base-band decomposition is still very confusing: a lot of low-frequency up-down shifts are required!

So, how is it done! The answer is simple: via FFT—fast Fourier transform. The combined base-band signal is generated using inverse FFT, while its decomposition using direct FFT. In the transmitter, sent bits decide about $X(k)$, i.e. about amplitude and phase of each k-th Fourier harmonics $x(n) = e^{j\frac{2\pi}{N}kn}$, $n, k = 0, 1, ..., N - 1$. Performing IFFT in the transmitter, we synthesize a signal being summation of many Fourier harmonics taking different states. In turn, performing FFT of the summation signal in the receiver, we find Fourier $X(k)$ coefficients of signal harmonic components, i.e. their *states* and bits. Since Fourier harmonics are orthogonal to each other, this multi-carrier FFT-based transmission technique has two names in two different areas of its application: digital multi-tone (DMT) in copper-wire modems (modulators–demodulators) and orthogonal frequency division multiplexing (OFDM) in wireless radio communication. Names are different, but the method is the same.

In Fig. 22.1 very simplified comparison of single-carrier and multi-carrier digital transmission is done. Instead of repeating many individual single-carrier schemes, a joint multi-carrier signal is generated in the base-band using the inverse fast Fourier transform (IFFT). It is next optionally up-converted in frequency, passed through the communication channel, wired or wireless, optionally down-converted in frequency, and then decomposed into single carriers using the FFT.

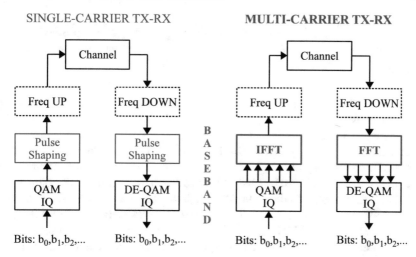

Fig. 22.1: Simplified comparison of single-carrier and multi-carrier digital transmission, with optional base-band signal up- and down-conversion in frequency. For single carrier a quadrature amplitude modulation (QAM) was selected only as an example. IQ denotes in-phase and quadrature carrier components/states

We start in this chapter our discussion from digital subscriber line (DSL) telephone or cable transmission because it is a simpler one as a precursor of OFDM which was developed later. Why simpler?

Firstly, because signal is transmitted in the base-band: no frequency up-down conversion is performed since each user has its own pair of twisted-wires. As a consequence, signal realness is required. It causes that the signal spectrum is conjugate-symmetric around the 0 Hz and bits can be allocated only to half of the complex-value Fourier harmonics (carriers), typically to positive frequencies only.

Secondly, transmission conditions in DSL are easier: channel impulse response is constant during transmission as the telephone line is at it is. In wireless communication situation is completely different: user can move all the time, even very fast in difficult environment with many reflections, channel characteristics are changing permanently and have to be estimated and corrected all the time, for example, per 1 millisecond. In home asymmetric DSL modems (asymmetric because we are more taking from the net then giving to it), the channel is estimated only once, during the modem start-up, and equalized all the time but using parameters calculated at the beginning. When something has changed and transmission *crashed*, modem restarting is required, as usual in programmable electronics devices when one of more if() instructions is missing.

What should be remembered? The multi-carrier communication is block-oriented. We do not have in it many carriers smoothly passing from one state to the other with the help of pulse shaping filter. No. Carriers are in one state in time duration of one DMT/OFDM symbol, i.e. one IFFT output. Since signal switching is abrupt, an extra analog channel filtering of the summation signal is required to reduce the transmission bandwidth.

In Fig. 22.2 concept of multi-tone transmission is presented graphically: many sub-carriers with different frequencies are used simultaneously. Typically, distance between sub-carriers is equal to a few, over a dozen or several dozen kilohertz (sampling frequency divided by FFT size): 4.3125 kHz in 512-point ADSL (2.208 MHz/512), 24.4140625 kHz in HomePlug Green power line communication (PLC) systems (50 MHz/2048, 75 MHz/3072), 1 kHz in 2048-point digital radio DAB (2.048 MHz/2048), 3.90625 kHz in digital terrestrial television DVB-T (8 MHz/2048), and 15 kHz in LTE (1.92 MHz/128, 3.84 MHz/256, 7.68 MHz/512, 15.36 MHz/1024, 23.04 MHz/1566, 30.72 MHz/2048). Many modern services exploit multi-carrier transmission. This chapter stands for introduction to all of them.

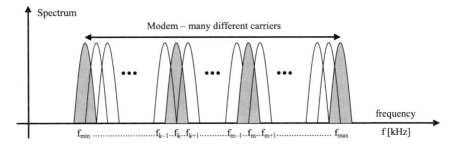

Fig. 22.2: Graphical illustration of discrete multi-tone (DMT) transmission concept—data are transmitted simultaneously using many carriers which are modulated in amplitude and phase (QAM) [14]

In Fig. 22.3 signal generation in DMT/OFDM systems is explained:

1. stream of bits is divided into smaller bit packets,
2. bit packets are allocated to different carriers and interpreted as numbers of allowed carrier states; corresponding $IQ(k)$ values in constellation points are found,
3. many sub-carriers Ω_k are generated, each in its specific $(I(k), Q(k))$ state,
4. all sub-carriers are summed.

There are different strategies for bit allocation to individual sub-carriers depending on channel type. In DSL and PLC wire-line modems, signal-to-noise ratio (SNR) is estimated for each sub-carrier and sub-carriers having bigger SNR obtain more bits (exact equation will be given later). This strategy results from the fact that frequency response of a wire-line is very fast decaying in frequency, i.e. it is significantly lower for higher frequencies. Therefore, typically, each low-frequency carrier

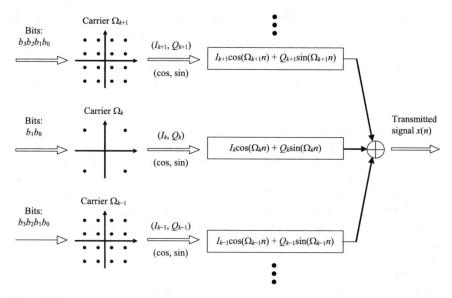

Fig. 22.3: Discrete multi-tone (DMT) in action: bit-stream is cut into portions of bits, bits denote numbers of $IQ(k)$ states taken by different carriers, QAM modulated carriers are synthesized and summed [14]

can code 10–15 bits (2^{10}–2^{15}-QAM), while high-frequency one only a few bits (2^2–2^5-QAM). On the contrary, wireless radio channels have relatively flat frequency response (FR) in the whole frequency range. The FR is as a whole better or worse, with many notches, and the same number of bits is allocated to each carrier depending on the overall received signal quality. Errors occur for sub-carriers strongly attenuated by the channel but they are corrected using special coding.

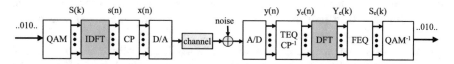

Fig. 22.4: Discrete multi-tone (DMT) transmission formally as a pipeline of the following operations: QAM—quadrature amplitude modulation, IDFT—inverse discrete Fourier transform, CP—addition of cyclic prefix, D/A—digital-to-analog converter, A/D—analog-to-digital converter, TEQ—channel time equalizer (channel shortening), CP^{-1}—CP removal, DFT—discrete Fourier transform, FEQ—channel frequency equalizer, QAM^{-1}—QAM demodulation [14]

Figure 22.4 concludes the discussion presented in this section. A simplified block diagram of multi-carrier base-band modem is shown in it (ADSL, HomePlug PLC, and coaxial cable). Incoming bits are divided into groups, allocated to sub-carriers, and interpreted as their $(I(k), Q(k))$ QAM states. Then inverse N-point DFT (FFT) is performed and the so-called cyclic prefix (CP) is added to each DSL/OFDM group of samples (some samples from each block end are copied to the block beginning—purpose of this operation will be explained later). Then signal samples are arranged serially, one-by-one, and sent to D/A converter, amplifier, and antenna. Transmitted signal is passing through disturbing channel with some constant or time-varying impulse response, noise, and disturbances (cross-talk, narrow-band, and impulsive interference). In the receiver all operations are reversed. Signal from antenna and low-noise amplifier is given to the input of A/D converter. In order to correctly decode transmitted data, the channel impulse response has to be estimated—which allows frequency equalization (FEQ) of the channel influence. Additionally, a special digital FIR filter, called a TEQ filter, can be designed optionally and applied to the input signal—it causes shortening of the overall impulse response of the transmission path to the length of the cyclic prefix. After synchronization to the DMT/OFDM symbol beginning, the cyclic prefix is removed and the N-point direct FFT is performed. Next, the obtained Fourier spectrum is divided by estimated frequency response of the overall signal transmitter-receiver path, including TEQ, and this way the transmission frequency equalization (FEQ) is done. Finally, carrier state numbers are extracted from FFT coefficients and transmitted bits are recovered from them. Hurrah!

Diagram from Fig. 22.4 can be easily made more general. After addition of frequency up-conversion module before or after the D/A converter in the transmitter and frequency down-conversion module before or after the A/D converter in the receiver, one obtains a multi-carrier scheme used for wireless radio communications. Why *before* or *after*? It depends on technology used. In software defined radio solution: before the D/A and after the A/D.

Let us describe a TEQ role in a few words. In case of wire-line links, telephone or cable, a channel impulse response can be very long. We intent to extract carrier states from one DMT/OFDM symbol but impulse response *is looking back* to the past and it *sees* also samples of the previous symbol. A DMT/OFDM inter-symbol interference (ISI) occurs! In order to remove it, we copy an ending fragment of any frame and put it into its beginning, *cheating* this way the impulse response: now it *sees* only samples of the present frame and there is no ISI. But doing this, repeating the signal, we are transmitting less bits. Therefore, we are interested to make the copy as short as possible. And this is a task of the TEQ filter. Since the incoming data are filtered by it in the receiver, the overall impulse response of the whole transmission path is changed. The task of the added filter is to make it shorter than the length of the cyclic prefix (number of copied samples). For this reason, the filter is called a TEQ one, since it causes time equalization (shortening) of the channel. The TEQ filter is missing in multi-carrier OFDM wireless transmission because impulse response of the wireless channel is short and length of the used cyclic prefix is sufficiently adjusted to it. Additionally, wireless channel is changing all the time and permanent TEQ designing could be cumbersome in such application scenario.

Exercise 22.1 (Generation of Multi-carrier Signal and Its Up-Down Frequency Conversion). Use long version of program 17.1 from the book repository. Set `isignal=2;`, i.e. the complex-value OFDM. Observe signal spectrum after each processing step. Note that the multi-sub-carrier signal can be recovered after the up-down frequency conversion. Now change `exp(.)` to `cos(.)` in the signal generation loop. This is the case of **base-band** multi-carrier transmission: we can also convert the signal up and down in frequency, losing however the half of the bandwidth used, because the spectrum is conjugate-symmetric. Finally, set `Nx=2048` and generate the summation signal $x(n), n = 1, ..., 2048$, performing the 2048-point IFFT of some random complex values $X(k), k = 1, ..., 2048$, with exception of $X(1) = 0$ and $X(1025) = 0$:

```
X = zeros(1,Nx);
X([2:Nx/2, Nx/2+2:Nx]) = randn(1,Nx-2) + j*randn(1,Nx-2);
```

Convert the signal $x(n)$ up-down in frequency. Observe spectra. Perform FFT upon the received signal. Compare received and transmitted Fourier coefficients. How big the errors are? Then, change values of Fourier coefficients $X(k)$ for negative frequencies: they should be conjugate versions of coefficients for positive frequencies:

```
X( Nx:-1:Nx/2+2 ) = conj( X(2:Nx/2) );
```

Perform IFFT upon $X(k)$: the generated signal $x(n)$ should have only real values. Check it. Convert the signal in frequency up and down. Perform FFT of the received signal. Check errors between transmitted and received signals, $x(n)$ and $y(n)$, as well as between their spectra, $X(k)$ and $Y(k)$. In the last experiment, you have synthesized one DMT/OFDM symbol, transmitted it, and received.

Magnitude DFT spectra presented in Fig. 22.5 demonstrate idea of wireless transmission of multi-carrier signal. First, the multi-component complex signal is created in the base-band (a), then up-shifted in frequency (b), next only real part of the result is left (c). Then the signal is down-shifted in frequency by multiplication with the cosine (giving $I(n)$, the real part of the result) and by sine (giving $Q(n)$, the imaginary part). Spectrum of the complex signal $(I(n) + jQ(n)$ is shown in plot (d), while the spectrum after low-pass signal filtration in plot (e). Presented spectra were obtained for option `isignal=2` from program `lab17_ex_Service_UpDown_long.m` given in the book repository. It is a long version of the program from Listing 17.1, in which the second signal is generated in the base-band as summation of 5 complex-value Fourier harmonics. In all plots only magnitudes of DFT signal spectra are presented.

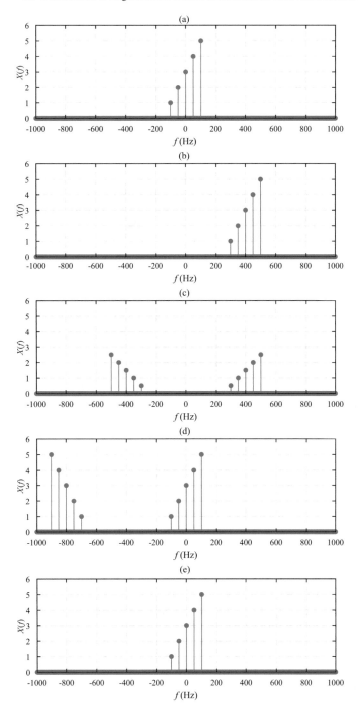

Fig. 22.5: Illustration of the concept of multi-carrier signals and their wireless transmission. Description in the text. (**a**) Base-band. (**b**) Freq. up. (**c**) Real. (**d**) Freq. down cos+sin. (**e**) LP filter

22.3 Examples of Multiple Carriers Technology

In Table 22.1 some important technical parameters of digital subscriber line (DSL) and power line communication (PLC) technology are summarized and compared with two existing alternative broadband Internet access solutions, like LTE 4G telephony (multi-carrier based also) and fiber optics links. Given values are approximate (illustrative) only and aim at showing existing similarities and differences between technologies. All of them, including new fiber optics links, exploit frequency division multiplexing (FDM), in particular orthogonal FDM (OFDM)—data are transmitted on carriers having different frequencies, i.e. many tones or waves.

Table 22.1: Short summary of DSL (typical channel spacing 4.3125 kHz) with rough comparison with other available multi-carrier technologies: PLC, LTE 4G, and fiber optics

Standard	Downstream (Mbit/s)	Upstream (Mbit/s)	FFT/Prefix length	Max sampling (MHz)	Max distance (km)
ADSL	6.144	0.640	512 / 32	1.104	5.5
ADSL2	8	0.800	512 / 32	1.104	5.5
ADSL2+	16	0.800	1024 / 64	2.208	4.0
VDSL	52	2.3	2783 / 174	12	1.2
VDSL2	100	100	$1783 / 112 - 8192 / 512$	8.5–35	0.2–1.3
G.fast	500	500	2048 / 320, 4096 / 320	106, 212	0.25
PLC Indoor	100	100	128 / 32, 3072 / 417	50, 75	0.10
LTE 4G	100	30	$128 / var - 2048 / var$	1.92–30.72	3–6
Fiber	1000	1000	–	–	10–60

Telephone DSL Modems As already mentioned, multi-carrier transmission in telephone lines will serve in this chapter as an illustration example. In digital subscriber line (DSL) modem channel is assumed as constant but its parameters are estimated and both, time and frequency, channel equalization take place (TEQ and FEQ). Additionally, bit allocation is performed according to the estimated SNR for each subcarrier. A wrong sampling ratio of the A/D converter in the receiver causes loss of the signal orthogonality and appearance of inter-channel (inter sub-carrier) interference (ICI) which should be eliminated or at least significantly reduced. Inter-symbol interference (ISI) can be present in DSL transmission also, when channel impulse response, even after shortening, is longer than a cyclic prefix. In fact, only channel time-varying and up-down carrier frequency offset problems (resulting from pure electronics precision and stability (ppm) as well as existing Doppler effect) are missing in this transmission scenario. But they will be discussed by us in the following chapters.

Because the simplest DSL modem will serve for us at present as the main example of multi-carrier transmission, it will be briefly introduced now. DSL connection can be symmetric (SDSL), with similar bit-rate in down-link (*to me*) and up-link (*from me*), and asymmetric (ADSL) when the down-link is bigger. The second possibility is offered for private home users who are mainly digital content consumers, while the first one—for companies requiring fast links in both directions. Both modems are almost identical differing only in details (different values of parameters). DSL is dedicated for *last mile* connection between traditional telephone central office and our office or home. Copper wires are used as a transmission medium. DSL methodology is exploited also in cable TV infrastructure making use of copper coaxial cables. The idea of DSL modem operation was described above: (1) transmission in the base-band, (2) many orthogonal carriers which are summed using the IFFT, (3) in each signal block, obtained from IFFT, last samples are copied to the block beginning, giving a cyclic prefix aimed at reduction of inter-block (inter-symbol) (ISI) interference.

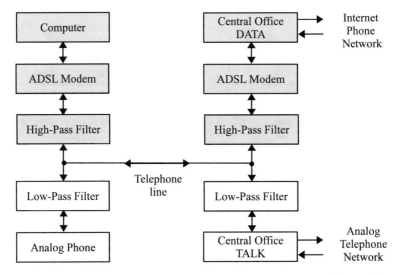

Fig. 22.6: ADSL copper wire modem connection with central office: old analog phone (low frequencies) and new data link (high frequencies) use special analog low-pass and high-pass filters in order to co-exist and do not disturb one another [14]

In Fig. 22.6 connection between home ADSL modem and telephone central office (CO) of plain old telephone service (POTS) is shown. Analog telephone copper-wire line connects our home and CO. There is a low- and high-frequency POTS splitter on both sides, separating analog speech signal (low-pass filter to 3400 Hz) and digital data (high-pass filter above 4800 kHz). In the ADSL standard sampling frequency is equal to 2.208 MHz. Since $N = 512$ carriers are used, the sub-carrier spacing is equal to 2.208 MHz/512=4.3125 kHz—see Fig. 22.7. Because transmission is in the base-band, without up-down frequency conversion, the signal has to be real-value, not complex-value. Its spectrum is conjugate-symmetric around 0 Hz and bits can be allocated only to one half of it—the negative part of the spectrum is determined by the positive part. In the positive part, the DC and first 5 sub-carriers are not used: here there is an analog phone connection to 4 kHz and separation zone to 26 kHz. In the upper part of the spectrum two scenarios can be applied: (1) using separate carriers for transmitting (carriers 6–28) and receiving (36–256) with separation zone between them (29–35), or (2) using duplex mode in which all available carriers (6–256) are used for receiving data, while carriers (6–32) are additionally used for transmitting. In the first scenario, extra filters are used for separation of down-link and up-link. In the second scenario, the separation zone is missing and echo canceling (EC) techniques have to be applied for carriers working in frequency zone with duplex. The second technique is more challenging from DSP point of view but it is preferred because it offers bigger bit-streams in the down-link: for lower frequencies channel attenuation is lower, SNR is bigger and more bits are carried by low-frequency sub-carriers.

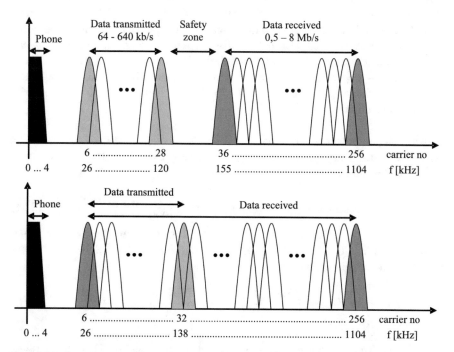

Fig. 22.7: Frequency division/usage in ADSL modems: (up) separate bands for transmitting and receiving with safety zone, (down) partial overlapping of transmission and receiving bands—necessity of echo canceling algorithms [14]

During initialization both linked modems inform themselves about transmission will. Modem asking for a connection is sending a 32 ms signal using one of the four frequencies: 189.75 kHZ, 207 kHz, 224.25 kHz, or 258.75 kHz. During the first 16 ms the signal is transmitted with power level −4 dBm and during the next

16 ms—with power level -28 dBm. Acceptation is sent back using one of the three frequencies: 34.5, 43.125, or 60.375 kHz. Then a specially formed preamble/header signal is transmitted which allows both modems to identify the channel (its impulse and frequency response) and to design TEQ filter, shortening channel impulse response in time, as well as FEQ frequency equalizer. At this stage A/D sampling ratio is estimated and corrected, most often in hardware. Then, the second training signal is transmitted which is used for ADSL synchronization, i.e. finding beginning of N-point DMT/OFDM symbols. FFT is periodically computed for precisely selected blocks of N-samples and SNR is calculated for each sub-carrier. Having this knowledge both modems, after *short discussion over coffee*, make decision how many states can take each carrier, i.e. how many bits it can *carry*. And regular transmission starts: carrier states are found and bits are extracted from them.

What problems are existing in DSL data transmission? Channel is strongly attenuated and delaying transmitted signal (for higher frequencies and longer lines). Physical side taps, present in DSL lines, give strong notches in channel frequency responses. Near-band and impulsive disturbances exist. Near-end (NEXT) and far-end (FEXT) cross-talk appear between modems using copper twisted-pair wires in the same shielded cable. Long channel impulse response causes interference between transmitted consecutive DMT/OFDM symbols. Sampling ratio of a cheap A/D converter can be wrong, causing loss of sub-carrier orthogonality and very strong inter-carrier (inter-channel) interference (ICI). Everything happens in the simplest DSL modem connection.

Remark Signal power is measured in telecommunication in dBm, i.e. in decibels referred to 1 milliwatt. If P denotes signal power expressed in watts, its power level in dBm is defined as:

$$S_{\text{dBm}} = 10\log_{10}(1000 \cdot P), \qquad P = \frac{1}{1000} \cdot 10^{S_{\text{dBm}}/10}. \qquad (22.1)$$

Signal 1 mW has a power level equal to 0 dBm. Stronger signals have more, weaker—less than 0 dBm. In turn, dBm/Hz denotes power level referred to 1 mW for 1 hertz of signal bandwidth:

$$S_{\text{dBm/Hz}} = 10\log_{10}\left(\frac{1000 \cdot P}{\Delta f}\right), \qquad P = \frac{\Delta f}{1000} \cdot 10^{S_{\text{dBm/Hz}}/10}. \qquad (22.2)$$

The G.Fast The G.Fast is an incoming new, advanced, broadband DSL technology. Typically sub-carrier spacing in ADSL is equal to 4.3125 kHz or 8.6250 kHz. In G.fast it is equal to 51.75 kHz due to significantly higher sampling ratio (106 or 212 MHz) which results in almost flat channel for each tone, easier to equalize. High bit-rate is obtained, thanks to usage of advances signal processing techniques but service distance is limited to a few hundreds of meters since sampling frequency is very high and signal is stronger attenuated for longer lines. It is not *last mile* but *street-to-home* technology.

Power Line Communication (PLC) PLC can be outdoor or indoor, but in some countries only indoor transmission is permitted and its called HomePlug in such case. It is divided into AV standard for fast home Audio-Video networks or GP standard for general-purpose usage. In comparison with ADSL, the time equalization of the channel is not performed in PLC, only longer symbol prefix is used. Additionally, due to turning on/off of different home electronic devices and changing channel features, channel estimation and equalization have to be performed in PLC more often, not only during modem restart. But not permanently like in wireless transmission.

22.4 Transmission Channels

In each telecommunication system, features of a channel used for transmission are extremely important. In the simplest approach the channel is modeled as a linear time-invariant system, characterized by its impulse response $h(t)$ (system output for delta Dirac function excitation) and corresponding frequency response $H(f)$, equal to Fourier transform of $h(t)$:

$$H(j\omega) = \int_{-\infty}^{\infty} h(t)e^{-j\omega t}\,dt, \quad \omega = 2\pi f. \tag{22.3}$$

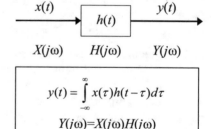

Fig. 22.8 Model of ADSL channel: linear-time invariant system—an output signal $y(t)$ is result of convolution of an input signal $x(t)$ and a channel impulse response $h(t)$ [14]

The simplest convolutional transmission model is presented in Fig. 22.8: output signal $y(t)$ is a result of convolution of input signal $x(t)$ with the channel impulse response $h(t)$:

$$y(t) = \int_{-\infty}^{\infty} x(\tau)h(t-\tau)\,d\tau. \tag{22.4}$$

If two signals are convoluted, their spectra are multiplied:

$$Y(j\omega) = X(j\omega) \cdot H(j\omega). \tag{22.5}$$

Fourier spectrum of each transmitted signal with arbitrary frequency f_0 is modified by the channel in the following way:

$$Y(f_0) = |X(f_0)| \cdot |H(f_0)| \cdot e^{j(\angle X(f_0) + \angle H(f))}. \tag{22.6}$$

In DSL each received sub-carrier will have different amplitude and phase than the transmitted one. We have to know $|H(f)|$ and $\angle(H(f))$ in the receiver in order to correct $Y(f)$ and obtain estimation of $X(f)$:

$$\hat{X}(f) = \frac{Y(f)}{H(f)}. \tag{22.7}$$

In software realization, when N-point DFT is exploited for Fourier spectrum calculation, DFT coefficients $X(k)$ of the transmitted signal $x(n)$ are recovered from the following equation:

$$X(k) = X(k \cdot f_0) = \frac{\text{DFT}[y(n)]}{\text{DFT}[h(n)]}, \quad f_0 = \frac{f_s}{N}, \quad k = 0, 1, 2, ..., N-1. \tag{22.8}$$

Summarizing Copper-wire telephone line transmission channel can be modeled as a linear time-invariant filter. If the channel impulse response is changing in time, as in wireless mobile channels, the filter becomes time-variant and channel estimation and equalization become much more difficult. Pilot signals known to receiver are used, for estimation of actual channel frequency response $H(f)$. Knowing $H(f)$ of a channel and having $Y(f)$ of a received signal, one can find $X(f)$ of a transmitted signal: $X(f) = Y(f)/H(f)$.

If communication channels are so important, let us have a look at their exemplary impulse responses and frequency responses. In Fig. 22.9 characteristics of a few twisted-pair copper-line telephone channels are shown, on the left side for lines without the high-pass POTS splitter and filter, and on the right side—with the high-pass filter. We can observe a significant change in shape of the impulse responses and almost invisible change in their spectra (only in their very beginning). We should notice also the spectrum smoothness (with some notches) and the fast spectrum decaying in frequency, causing a decrease of number of bits which can be allocated for high-frequency sub-carriers.

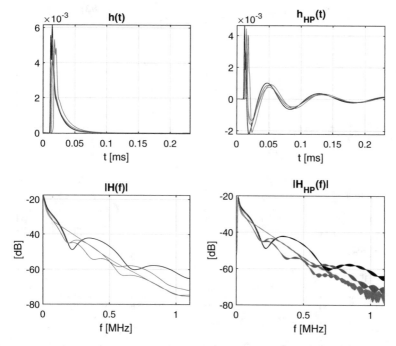

Fig. 22.9: Characteristics of several telephone lines without (left) and with (right) high-pass filter (POTS splitter): (up) impulse responses $h(t)$, (down) corresponding frequency responses $H(f)$. Notice time delay of $h(t)$ and notches present in $H(f)$

In Fig. 22.10 two power line channels are presented, both medium quality, first long, 350 m, outdoor (top) and second short, indoor (bottom). This is a moment for big reflection: how different are observed impulse responses! The first is long-lasting and smooth. The second is short lasting and highly impulsive. Looking at their frequency responses, presented in bottom plots, we see that the first is fast decaying, allowing bit transmission only in a very narrow low-frequency band and requiring precise bit allocation. In turn, the second frequency response has many notches but it is not decaying. This transmission PLC channel is very similar to wireless channels, for example, to the channel EVA from LTE 4G standard, presented in Fig. 22.11. Impulse responses of both channel are impulsive, both have not-decaying frequency responses with notches. In wireless transmission all sub-carriers use the same number of states, i.e. the same number of bits is allocated to each of them, based on overall signal strength. For this reason, since errors occur, transmitted are not only the original bits but also some additional, redundant bits, protecting the original ones against errors (forward error correction (FEC) code "3/4" means that the original bits represent 75% of all sent bits). It looks that this should be also a recommended scenario of bit allocation for the PLC indoor transmission.

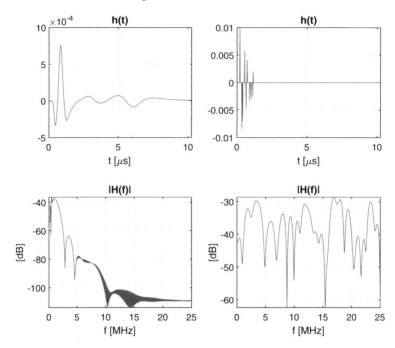

Fig. 22.10: Characteristics of two medium quality power lines: (left) long 350m outdoor, (right) shorter 100m indoor: (up) impulse responses, (down) corresponding frequency responses [14]

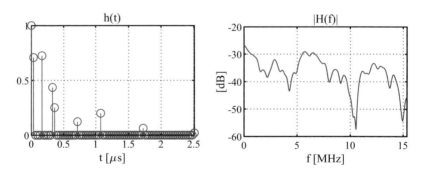

Fig. 22.11: Impulse response (exactly: statistical power density profile of channel filter taps) of an exemplary wireless channel EVA from the LTE 4G standard 3GPP TS 36.104 (left) and its frequency response (right) [14]

22.5 DMT/OFDM Modulator and Demodulator

In digital communication *game* we fight against channel, interfering signals, and hardware imperfection of transmitter and receiver elements: should be A but unfortunately B is set. In this section we learn the first multi-carrier transmission problem arising from channel disturbing, i.e. the DMT/OFDM inter-symbol interference (ISI). And we become familiar with its solution: addition of the so-called cyclic prefix (CP), the repeated part of each DMT/OFDM signal/symbol.

In Fig. 22.12 the ISI problem is explained graphically and the solution is intuitively derived. The figure consists of four plots. In the first one, at the top, we see blocks of signal samples (marked as a diamond, left triangle, rectangle, and right triangle), synthesized by IFFT from specified carrier states for each DMT/OFDM symbol. The DMT/OFDM symbol is summation of many individual sub-carriers being in different states. Above the symbol chain, we see an impulse response of a channel which is convoluting with our signal, *looking* to its past. We can observe that impulse response beginning can lie above one symbol but its tail—above the previous symbol. Received signal $y_k(t)$ (for the $k-th$ symbol) depends not only upon the transmitted symbol $x_k(t)$ but also on its predecessor $x_{k-1}(t)$. Wow!

In order to eliminate this unwanted inter-symbol interference (ISI) we could transmit every symbol twice and perform FFT in the receiver only for the second symbol, after symbol synchronization. Such situation is shown in the second figure. Symbol coping is done and impulse response *standing* at the beginning of the second transmitted symbol *looks back* and see ... the same symbol. DFT and FFT are discrete versions of the Fourier series. In Fourier series signal periodicity is assumed. Copying does not change the result: our signal after the channel is artificially made periodic and FFT of the second symbol gives us the correct value $Y_k(f)$. When the impulse response is shorter than the symbol length, and when we are not synchronized perfectly and take, for the FFT calculation, the end of the first (copied) symbol also, the FFT result in magnitude will be the same—only phase shift of spectral coefficients will occur but this effect can be easily corrected, which we will investigate later.

So, are we happy? Not completely. Repeating each DMT/OFDM symbol twice, we have reduced two times the bit-rate! How counteract this? To make a shorter signal copy which is shown in the third figure: at present only some limited number of last samples of the DMT/OFDM symbol is copied to the symbol beginning which is presented in the third plot of Fig. 22.12. It is possible, because in the receiver a special digital filter was designed, the so-called TEQ channel equalizer (Time EQualizer), and incoming samples $y(n)$ were filtered by it. Therefore, transmitted data were convoluted with two filters:

1. the physical channel impulse response $h(t)$,
2. a designed impulse response $e(n)$ of the TEQ FIR digital filter—to be discussed later in Sects. 22.8.3 and 22.8.8.

Effective impulse response $h_e(n)$ of the whole ADSL channel is, in such case, equal to convolution of both impulse responses, $h(t)$ and $e(n)$. The TEQ $e(n)$ is specially designed in ADSL to make the $h_e(n)$ short: having maximally 32 plus 1 sample (first sample of $h_e(n)$ is overlapping with the first sample of a new DMT/OFDM symbol). Thanks to the TEQ usage, only the last 32 samples of the DMT/OFDM symbol have to be copied to the symbol beginning, not 512. The repeated DMT/OFDM symbol part, in ADSL 32-samples long, is called the cyclic prefix (CP). Connection between length of the shortened channel impulse response $h_e(n)$ and the cyclic prefix length is shown in last, fourth plot in Fig. 22.12.

Why TEQ equalizer is not always used in multi-carrier transmission? Because design of a TEQ filter is not simple. Some TEQ design examples are presented in Sects. 22.8.3 and 22.8.8. The filter depends on $h(t)$ and when the channel is changing very fast it is no sense, and time, to design new TEQ filters permanently. Telephones lines are not-varying and TEQ is designed only during modem initialization. In PLC indoor communication and in wireless channels, impulse responses are time-varying, very impulsive (look at figures of the previous section), and difficult to make shorter. But there is no need for doing this because they are not long. Therefore only prefix with sufficient length is added to each DMT/OFDM symbol.

Till now, very intuitive explanation of frequency channel equalization (FEQ) and time channel equalization (TEQ) in base-band DMT/OFDM systems was presented as well as the main concept of signal generation in their transmitters and signal analysis in their receivers. At present, we can re-draw block diagram of the DMT/OFDM signal processing path, presented in Fig. 22.4, with more understanding in an enlarged form, connecting in pairs corresponding modules of the transmitter and receiver. This is done in Fig. 22.13. Spectra and signals are denoted in it as in the program which will be used later by us for simulation of a working ADSL modem.

Going one step further, in Fig. 22.14 only ADSL modem transmitter is shown with precise description of:

1. QAM bit allocation to positive-frequency sub-carriers (1–255) only,
2. conjugate-symmetric, carrier states copying to negative frequencies (257–511),
3. performing $N = 512$-point FFT algorithm and addition of a 32-point cyclic prefix (544 samples result),
4. data serialization,
5. and their converting to analog signal.

In turn, in Fig. 22.15 a receiver of the base-band ADSL modem is presented. We have in a cascade:

1. an A/D converter,
2. effective impulse response channel shortening by the TEQ filter to 33 samples or less,
3. synchronization with DMT symbol beginning,
4. serial-to-parallel samples re-organization,
5. removing the cyclic prefix (first 32 samples),
6. performing 512-point FFT,
7. leaving only positive sub-carriers (1–255),
8. performing their correction (FEQ channel equalization),
9. recovering sub-carrier states (de-QAM), their numbers and bits transmitted.

In the remaining part of the chapter we will discuss selected technical issues connected with modem implementation.

In the program presented in Listing 22.1, the idea of DSL modem can be tested. Number of `Kiter` DMT symbols are transmitted, having N samples each with cyclic prefix P samples long. The symbols can be defined directly in time domain (`inputType=1`) or in frequency domain (`inputType=2`) as in DSL modems.

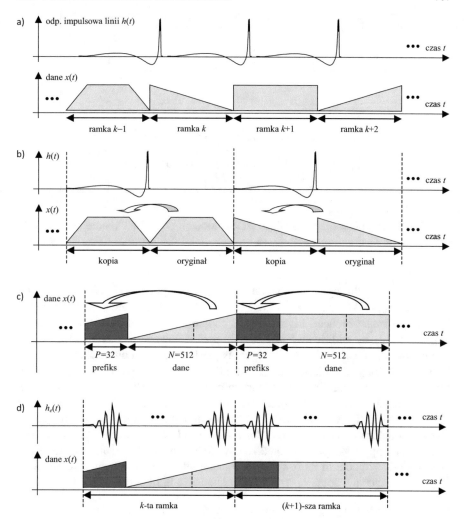

Fig. 22.12: Explanation of a cyclic prefix addition to each data block and necessity of this operation, from top to bottom: (1) transmission without a prefix, (2) maximum prefix—doubling the transmitted symbol for long channel impulse response, (3) short prefix—cyclic copying a signal fragment from the data block end to the beginning, (4) short prefix—cyclic convolution of the transmitted data with the channel impulse response shortened by the TEQ equalizer [14]

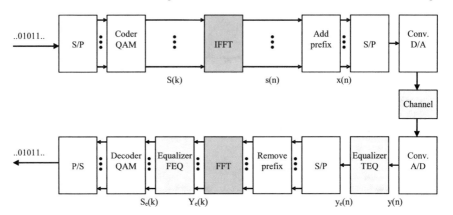

Fig. 22.13: Complete diagram of the ADSL modem with IFFT modulator and FFT demodulator. Denotations: S/P—serial-to-parallel, P/S—parallel-to-serial [14]

Signal is passing through a channel, user can choose taps of its impulse response (at present there are 4 taps and P=4, i.e. length of the CP is sufficient to eliminate the inter-symbol interference). Signal can be contaminated by different disturbances, at present only by additive white Gaussian noise. Now, perfect synchronization is assumed in the receiver. Received signal is partitioned into consecutive fragments having P+N samples each, which are put into columns of a matrix y1p. After that, first P rows of the matrix, with cyclic prefixes, are removed. Then fft() upon columns of y1 matrix is performed, result is divided by channel frequency response H, calculated earlier, and this way the channel influence is removed, i.e. frequency channel equalization (correction) is done. At the end, we are going back to time domain and compare samples of equalized, received signal y1e (matrix having samples of DSL symbols in its rows) with transmitted samples. When DSL signal is defined in frequency domain, error is calculated also in frequency domain.

Listing 22.1: Matlab program for testing idea of base-band multi-carrier DSL transmission

```
1   % lab22_ex_dsl_idea.m
2   clear all; close all;
3
4   inputType = 1;                              % 1=signal, 2=spectrum
5   Kiter = 10;                                 % number of DMT symbols
6   N = 16; P = 4;                              % symbol and cyclic prefix lengths
7   nstd = 0;                                   % channel noise standard deviation
8   h = [ 1, 0.5, 0, -0.25 ]; Nh = length(h);  % channel impulse response
9   H = fft( [h zeros(1,N-Nh)] )/N; H = H.';    % channel frequency response
10  plot( 20*log10(abs(H)) ); title('|H(k)|'); % its figure
11  Nx = Kiter*(N+P);                           % transmitted signal length
12  x = []; x1ref = []; S1ref = [];             % initialization
13  for k = 1 : Kiter                           % DMT signal generation
14      if(inputType==1) x1 = randn(1,N);       % # real time-domain signal
```

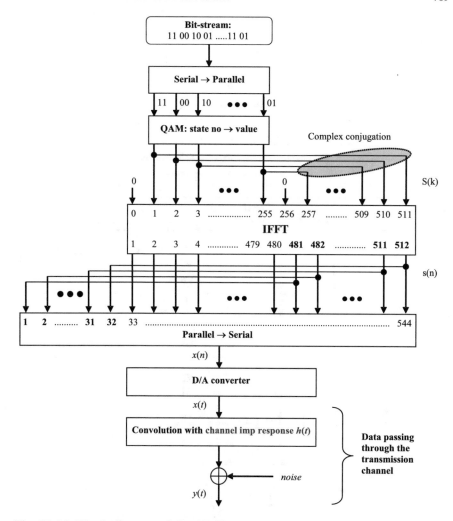

Fig. 22.14: Block diagram of the ADSL transmitter and data passing through the channel [14]

```
15 ||    else                                    % @
16 ||      S0 = randn(1,N/2-1) + j*randn(1,N/2-1);% @ half of the DFT spectrum
17 ||      S1  = [ 0 S0 0 conj( S0(end:-1:1) ) ]; % @ its Hermitian symmetry
18 ||      x1 = ifft( S1 );                       % @ real signal from IDFT
19 ||      S1ref = [ S1ref, S1.'];                % @ storing carrier states
20 ||    end
21 ||    x1p = [ x1(1,N-P+1:N), x1 ];             % one symbol plus prefix
22 ||    x = [ x, x1p ];                          % chain of DMT symbols
23 ||    x1ref = [ x1ref,  x1' ];                 % storing transmitted signals
24 || end                                         % end of signal generation loop
25 || y = conv(x ,h); y = y(1:Nx);               % signal is coming through the channel
```

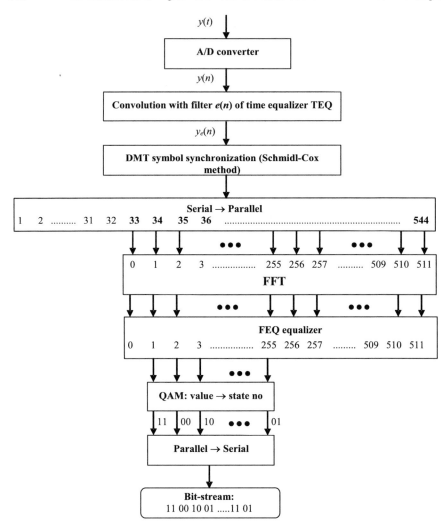

Fig. 22.15: Block diagram of the ADSL receiver [14]

```
26  y = y + nstd*randn(1,Nx);              % the channel is noisy, change nstd value
27  y1p = reshape(y,N+P,Kiter);            % in columns RX signals with prefixes
28  y1 = y1p( P+1:N+P,1:Kiter);            % removing prefixes
29  Y1e = fft(y1)/N ./ repmat(H,1,Kiter);  % channel equalization
30  if(inputType==1)                       % FOR INPUT SIGNAL
31      y1e = ifft( Y1e );                 %    going back to time domain
32      error1 = max( max( abs( x1ref - y1e ))),   %    error in the receiver
33  else                                   % FOR INPUT SPECTRUM
34      error2 = max( max( abs( S1ref - Y1e ))),   %    error in freq domain
35  end
```

Exercise 22.2 (Testing Idea of the Base-Band Multi-Carrier DSL Transmission).

(1) Analyze program 22.1. Run it for two types of input signals. Observe negligible errors. Change length of the cyclic prefix. Notice that for P ≥ 3, the channel influence is perfectly removed because channel has 4 taps and cyclic prefix with length 3 is sufficient. Observe that for P=1,2 errors occur. After noise addition (e.g. `nstd=0.1`) the received signal is different than the transmitted signal.

(2) Modify the program. Define carrier states in frequency domain as 4-QAM ($\pm 1 \mp j$):

```
S0=(2*round(rand(1,N/2-1))-1)+j*(2*round(rand(1,N/2-1))-1);
```

and detect them in the receiver. Observe sent and received carrier constellation points with sufficient and too short cyclic prefix length. For this purpose add appropriate figure:

```
Y1e = Y1e([2:N/2, N/2+2:N],:); Y1e=Y1e(:);
plot(real(Y1e(:)),imag(Y1e(:)),'bo'); grid; title('I=f(Q)');
```

Repeat experiment for different levels of noise.

(3) Now, set long prefix and absent noise (simply comment the noise addition instruction). After channel convolution add to the signal sinusoid having frequency different than transmitted carriers, i.e. a narrow-band interference:

```
y = y + 0.05*sin(2*pi/N*10.5*(0:Nx-1));
```

At present, the disturbing sine has frequency exactly in the middle between the 10-th and 11-th carrier. Observe result. It should be significantly worse. Change amplitude and frequency of the sine.

(4) At present, simulate wrong sampling ratio of the receiver A/D converter:

```
step=(Nx-1)/Nx; % step=0.99999;
y(1:Nx)=interp1([0:Nx-1],y(1,1:Nx),[0:step:(Nx-1)*
step],'spline');
```

The ADC is working with faster sampling ratio and it is giving us Nx samples in time of transmitted $Nx - 1$ samples—one transmitted sample is lost! Observe constellation points of the carriers. Than set also `step=0.99999, 0.9999, 0.999, 0.99`. Wow! *Where are our constellation points?!*

(5) Finally, de-synchronize receiver delaying the signal after channel noise addition by one sample: `y=[ 0 y(1:end-1) ]`. Observe in the receiver carrier constellation points which should rotate: the phase shift is caused by the time shift of the signal and should be different for different carriers. Check it. Take a break. Switch off the computer. Go for a walk to the nearest forest. Enjoy its beauty.

Plots which should be observed during realization of Exercise 22.2 are presented in Fig. 22.16. States of all carriers are observed in one plot. Because 4-QAM modulation is used, four points: $(1+j)$, $(1-j)$, $(-1+j)$, and $(-1-j)$ should be visible. Should be. But are not.

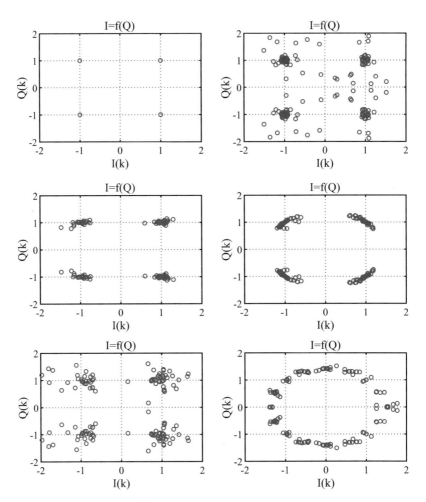

Fig. 22.16: Exemplary figures of $I = f(Q)$ 4-QAM constellation points which should be seen by a Reader during realization of Exercise 22.2. All carrier states are presented in one plot. In columns for: (1) sufficiently long CP P=3, (2) too short CP P=2—ISI occurs, (3) noise with nstd=0.1, (4) disturbing sinusoid ($A = 0.1, k = 10.5$), (5) for wrong ADC sampling rate step=0.999—ICI occurs, (6) for synchronization error/delay equal to 1 sample

22.6 Carrier Orthogonality Importance

Results of doing Exercise 22.2 from the last section should make us think more carefully about orthogonality concept exploited in multi-carrier transmission. When vectors of two signals are orthogonal, their inner product is equal to 0. DFT consists of calculating many inner products *between* a received signal and predefined set of Fourier harmonics. If the received signal is a superposition of these harmonics, scaled in amplitude and shifted in phase, there are no detection problems in the receiver: each signal component, scaled and shifted, is orthogonal to other harmonics of the DFT detector and it is not seen/measured by them. All inner products are equal to zero except the one, in which the received Fourier harmonic is compared with its own DFT reference. The result tells us about the amplitude and phase of the concrete received sub-carrier. This ideal carrier state detection scenario fails in case of losing signal orthogonality.

Let us perform Exercise 22.3, checking signal detection in DMT/OFDM transmission scenarios with many orthogonal carriers.

Exercise 22.3 (Testing Orthogonal Signal Detection Realized in Multi-Carrier Transmission). Let us assume the following experiment. In a DMT/OFDM receiver we calculate DFT of a received signal. Now, in our experiment, we are taking from DFT only one Fourier harmonic and calculate its inner product with the input signal. We would like to check which input signals will influence the DFT coefficient associated with the tested DFT harmonic reference and how much. Therefore, in a loop we are generating cosines with different frequencies and calculate their inner products with the selected DFT harmonic. Finally, we collect results of all inner products in one plot. Program doing this is given in Listing 22.2. Analyze it. Run it. Choose complex-value or real-value signal components inside the loop. Change frequency of the harmonic of interest. Are you surprised or not? How do you conclude this exercise?

Listing 22.2: Matlab program for checking quality of different signal components detection realized by the DFT frequency analyzer

```
1  % lab22_ex_ortho.m
2  clear all; close all;
3
4  N  = 20;         % number of orthogonal complex-value carriers, their lengths
5  fs = 20;         % sampling frequency
6  k  = 5;          % number of the carrier which value is checked
7  df = fs/1000;    % frequency resolution
8
9  dt = 1/fs; t = 0 : dt : (N-1)*dt;   % sampling period, sampling instants
10 f0 = 1/(N*dt); fk = k*f0;           % fundamental and checked carrier frequency
11 b = exp( -j*2*pi*fk*t );            % reference Fourier harmonic of the carrier
12 w = boxcar(N)';                     % "no window"
```

```
13   %w = hanning(N)';                      % window used optionally
14   f = 0 : df : fs;                       % frequencies of received signal components
15   for m = 1 : length(f)                  % analysis / detection loop
16      x = exp( j*2*pi*f(m)*t );           % component in a received signal
17      % x = cos(   2*pi*f(m)*t );          % component in a received signal
18      X(m) = sum( b .* (w.* x) )/N;       % its inner product with the reference harmonic
19   end
20   figure; plot( f, abs(X),'b-',f(1:f0/df:end),abs(X(1:f0/df:end)),'ro');
21   grid; xlabel('f (Hz)'); title(' |X_{k} (m) |'); pause
```

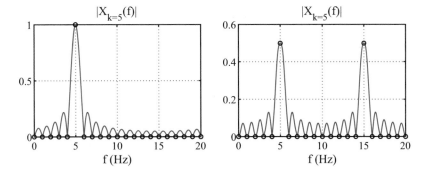

Fig. 22.17: Plots generated in example 22.3, showing values of inner products between the Fourier harmonic number 5 taken from 20-point DFT and mono-component signals with frequency changing linearly from 0 Hz to $f_s = 20$ Hz: (left) for complex-value input, (right) for real-value one

Figure 22.17 presents plots obtained in Exercise 22.3 when the received signal has complex values (left) and real values (right). We see that the inner product operation *does not see* signals which are orthogonal to the searched DFT harmonic exploited in the inner product, i.e. the remaining DFT harmonics; however, it *sees* all components having frequencies in-between. And this is a bad news! Noise from the whole frequency band and any non-orthogonal disturbing sine/cosine will influence all DFT coefficients, in our case—all the carriers! Now, we can understand the results presented in Fig. 22.16, showing how harmful the noise and narrow-band signals are for clearness of carrier state constellations. At present, the importance of correct ADC sampling ratio is understood also: while the sampling frequency is wrong, frequencies of received signal components do not overlap with frequencies of the DFT analyzer and inner product of any DFT harmonic with the signal is sensitive to all signal components, because these components are no longer orthogonal to the Fourier harmonic signal whose presence is verified! When orthogonality is lost, the detection performance is also lost. This effect is illustrated in Fig. 22.18.

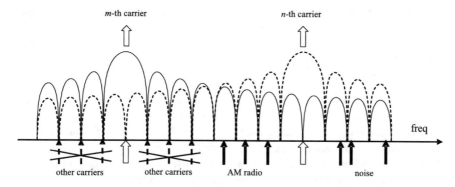

Fig. 22.18: Graphical illustration how other carriers and disturbances influence the detection of constellation state of any carrier [14]

22.7 Transmission Disturbances and Distortions

The aim of this section is to convince us that each transmission, even looking to us so simple like ADSL, is faced with many problems. Everything is easy at the stage of drawing diagrams on sheet of paper, especially when the diagrams are to be implemented by somebody else. In real word we are living in a jungle of many co-existing physico-chemical phenomena, fighting against each other, which are independent from us, and there is no easy answer or one solution only. Signal analysis and processing are performed usually in combat conditions.

Attenuation Attenuation in wire-lines depends on signal frequency and wire resistance. Since higher signal frequencies are much stronger attenuated (as in RC, resistance-capacitance circuit), maximum signal frequency used in DSL cannot be extremely high, typically it is equal to 2.208 MHz. Situation is additionally complicated by using inductive coils in some telephone lines. In turn, bigger wire resistance causes bigger signal attenuation. Resistance decreases with wire diameter (wider wires are better) but increases fast with cable length (shorter wires are better). For this reason, the DSL technology is typically used as a *last mile connection*, about 1.6 km; however, longer links to 10 km happen also. Copper is a good conductor. Wire technology production is important also.

Change of Twisted-Pair Diameter Typical telephone connection is built from a few sections (fragments) with different wire diameters. Thin wires are used closer to CO since cables have more twisted-pairs then. Because wave impedance depends on wire thickness, in places of wire connections impedance mismatch can occur, causing signal reflections.

Branches Unfinished branches of old subscriber lines, not used at present, occur very often. They cause also signal reflections: some energy of the transmitted signal can pass into the branch, reflects from its end and go back to branch beginning.

From this point, the reflection can go back to the transmitter as an echo and go forward to the receiver as a disturbance. Reflection adds to the signal. When its wave delay is equal to 180 degrees, some signal frequency components may be strongly attenuated, even completely canceled. It happens when branch length is equal to odd multiple of wavelength quarter.

White Noise Additive noise, present in receivers of ADSL modems, has typically Gaussian probability density function and flat power spectrum (AWGN additive white Gaussian noise). It is adding evenly to a signal in the whole frequency range. AWGN origins from quantization noise, thermal phenomena existing in analog parts of transmitter and receiver and electromagnetic disturbances coming from environment. Typically its power level is assumed as -140 dBm/Hz.

Impulsive Noise It sources are difficult to specify precisely. It is generated by commutative and signaling devices, also by electrostatic charge. It is characterized by short, impulsive signal shape (duration from 30 to 150 μs) and big amplitude (from 5 to 20 millivolts). Impulses appear typically 1–5 times per minute. Due to its big amplitude, the impulsive noise is very harmful to DSL modems.

FEXT and NEXT Cross-Talk Cross-talk between working DSL modems of different types, but using the same frequency band, even only in small part, is the most dangerous disturbance in wire-line base-band transmission. *Cross-talk* means that transmission in one copper twisted-pair disturbs transmission in some other twisted-pair in the same cable. The cross-talk takes place in distribution part of the connection when wires of big, infrastructure CO cables are changed (crossed) to individual subscriber links. Cross-talk are divided into far-end (FEXT) and near-end (NEXT).

FEXT During FEXT, transmitter and receiver having cross-talk are on opposite ends of the same cable. Typically, FEXT occurs between modems of the same type, e.g. ADSL. FEXT increases with line length. However, in comparison with NEXT, it is less harmful because far-end cross-talk has to pass through the line, so it is also attenuated. FEXT is a dimensionless value, estimated using the following equation:

$$FEXT_N = \left(\frac{N}{49}\right)^{0.6} k d f^2 |H(f)|^2, \tag{22.9}$$

where $k = 8 \cdot 10^{-20}$ is an empirical constant, d denotes loop length expressed in feets (1 ft=0.305 m), f is frequency in hertz, and $H(f)$—loop frequency response.

NEXT NEXT cross-talk between modems is significantly more harmful. It occurs only when transmitter and receiver are working on the same end of the infrastructure cable and are using different twisted-pair wires. They are close to each other: the cross-talk is not passing through the line, is not attenuated and therefore is strong. NEXT, a dimensionless value again, is estimated by the following equation:

$$NEXT_N = \left(\frac{N}{49}\right)^{0.6} \frac{1}{1.134 \cdot 10^{13}} \cdot f^{\frac{3}{2}}, \tag{22.10}$$

where N denotes number of disturbing lines (maximum 49), and f is frequency of disturbing signal in hertz.

NEXT cross-talk increases when distance between twisted-pairs becomes lower and distance of their interaction increases. NEXT is dual: both twisted-pairs are *infected*. In multi-pair cable all wires can disturb each other. Pairs having lower twist are more robust to cross-talk.

When ADSL modem is working in FDM mode (with frequency division between transmitter and receiver), the NEXT does not occur (since different frequencies are transmitted and received by two modems being on the same side of a cable) but FEXT does (since the same frequencies are transmitted by different modems on one side of the cable). In ADSL modems with echo cancellation, both NEXT and FEXT are present, due to the fact that the same frequencies are used, in some band, for TX and RX. The situation is more complicated when twisted-wires of the same cable are used also by different technologies, like ISDN or HDSL.

22.8 DSL Modem Implementation Issues

Having in mind a complete program of ADSL transmission model, which will be presented at the end of this chapter, we will discuss now step-by-step, mathematically, all DSP algorithms which are used in it, in order to do their software implementation in the modem program. The short outline of this section is as follows:

1. channel estimation will take place: channel is disturbing transmitted data, we have to know it in order to perform correction of the received signal;
2. ADC sampling ratio is estimated and corrected (in hardware) since DMT/OFDM systems are very sensitive to losing carrier orthogonality—due to this an inter-carrier interference (ICI) is minimized;
3. channel time equalizer (TEQ) is designed in order to shorten the impulse response of the overall transmission path to the length of cyclic prefix and to avoid inter-symbol interference (ISI);
4. knowing the channel and having TEQ filter designed, the final channel frequency equalizer (FEQ) is calculated;
5. at present, DMT/OFDM symbol synchronization takes place, using, for example, the Schmidl–Cox algorithm;
6. next, we can observe constellations of carrier states and test influence of different disturbances upon carrier states detection;
7. at the moment when we are *feeling the blues*, signal-to-noise (SNR) ratio for each sub-carrier is calculated and maximum available number of bits is allocated to the sub-carrier;
8. finally, having everything ready, regular transmission can start.

We end this section with short discussion about available strategies of TEQ filter design: what its length should be preferred, short or long?

22.8.1 Channel Estimation/Identification

Telecommunication channel constant in time, like DSL, is modeled as a linear time-invariant (LTI) filter. When channel is varying not very fast, we can always assume local validness of the LTI model and identify it in some specified time slots. What does the phrase channel identification mean? To find it impulse response $h(t)$ or frequency response $H(f)$, both connected by Fourier transform (FT): $H(f) = FT(h(t)), h(t) = FT^{-1}(H(f))$. Frequency response of any processing module/block/object tells us how this module modifies amplitude and phase of each input frequency component. To check it, we should excite the module with signal having all frequencies and observe the module output: Fourier transform of the output (state of each frequency component at the module output) divided by Fourier transform of the input (state of each frequency component at the module input) specifies change of each frequency component done by the module and it is called the module frequency response. In order to succeed, the input signal should contain all frequencies. Only the ideal Dirac delta function and ideal white noise have such feature. The Dirac function is very difficult to realize in practice; therefore, channel identification is realized technically by exciting with pseudo-random digital sequences, e.g. maximum length sequences. In Fig. 22.19 such 512-samples long sequence is presented in the left plot, while in the right plot its auto-correlation function, which is almost perfectly impulsive. Since signal power spectral density (PSD) is defined as the Fourier transform of the signal auto-correlation function, we can expect almost perfectly flat PSD of our signal, which was our goal.

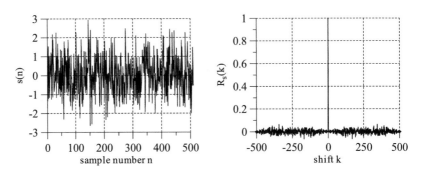

Fig. 22.19: Channel identification sequence (left) and its auto-correlation function (right) [14]

The excitation sequence is known to the receiver: knowing channel input and output, the channel impulse response is found using Eqs. (22.7), (22.8). In ADSL $N =$ 512-point DFT is used. Spectral coefficients $X(0)$ and $X(256)$ are set to zero and $X(1)...X(255)$ to random states of 4-QAM modulation. Coefficients $X(511)...X(257)$ are equal to complex conjugation of coefficients $X(1)...X(255)$, respectively. Then IFFT is calculated and resultant signal $x(n)$, presented in Fig. 22.19, is transmitted many times, without a cyclic prefix, which is shown in Fig. 22.20.

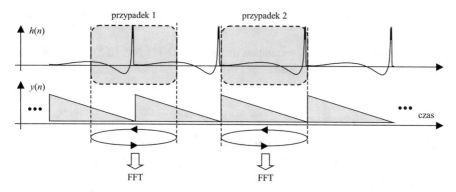

Fig. 22.20: Illustration of the circular shift of the impulse response $h(n)$, calculated from Eq. (22.11) (presented on gray background), depending on acquisition time of the signal $y(n)$ fragment sent to the FFT analyzer (in the second case the shift is missing) [14]

Receiver, initially, is not synchronized with the beginning of repeated signal samples. It is taking consecutive blocks of N signal samples and calculating DFT of each of them. Next, it adds results of all N-point FFTs and, according to Eq. (22.8), divides it by N-point FFT of the transmitted signal ($k = 0, 1, ..., N-1$, $f_0 = \frac{f_s}{N}$:

$$\hat{H}(kf_0) = \frac{\frac{1}{M} \sum_{m=1}^{M} \text{FFT}_N[y_m(n)]}{X_N(k)}, \qquad \hat{h}(n) = \text{FFT}_N^{-1}\left[\hat{H}(kf_0)\right]. \qquad (22.11)$$

This way estimation of channel frequency response $H(kf_0)$ is obtained and its inverse FFT gives us estimation of the channel impulse response $\hat{h}(n) = h_{est}(n)$. The $h_{est}(n)$ is presented in Fig. 22.21, on the left side for subscriber line without high-pass POTS splitter and on the right side with this high-pass filter. In both plots we see two impulse responses. The ones located in figure central parts (marked with solid lines) are obtained from the first FFT in Fig. 22.19, which is not synchronized with received signal blocks. Impulse responses present in figures beginning (marked with dashed lines) come from the second FFT in Fig. 22.19, which is synchronized with signal blocks.

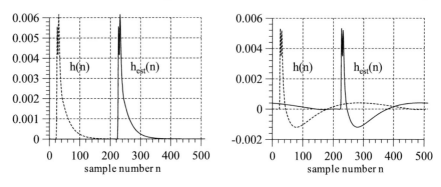

Fig. 22.21: Results of $h(n)$ estimation from Eq. (22.11) for: (left) telephone line alone, (right) line with high-pass filter [14]

Exercise 22.4 (DSL Channel Estimation and Correction). In program 22.1 ideal DSL transmission has been simulated. Signal was passing the channel and it was perfectly corrected in the receiver because RX knew the channel frequency response. Modify the program and implement in it channel estimation procedure based on Eq. (22.11). In case of difficulties, try to find inspiration in program 22.7, presented at the end of this chapter.

Exercise 22.5 (Reverberation Signal Generation for Channel Identification). In programs 22.1 and 22.7 Matlab pseudo-random number generator is used for synthesis of a signal used for channel identification. Replace it with the reverberation signal generated by code, presented in Listing 22.3. Check shape of the signal auto-correlation function. Notice that vector num is a random sequence of numbers {1,2,3,4}.

Listing 22.3: Matlab program for generation of reverberation signal used for channel identification in ADSL modems

```
1  % lab22_ex_reverb.m
2  clear all; close all;
3
4  N = 512;
5  b = zeros( N, 1 ); b( 1:9 ) = 1;  % initialization
6  for i = 10:length(b)
7      b(i) = xor( b(i-4), b(i-9) );  % recursive calculations
8  end
9  for i = 1:N/2-1
```

```
10      num( i ) = ( 2 * b( 2*(i-1)+1 ) + b( 2*(i-1)+2 ) ) + 1;   % random numbers 1,2,3,4
11   end
12   coder = [ -1-j, -1+j, +1-j, +1+j ];      % QPSK/4-QAM constellation points
13   S0 = coder( num );                       % their random selection
14   S1  = [ 0 S0 0 conj( S0(end:-1:1) ) ];   % spectrum
15   x1 = ifft( S1 );                         % real value signal
16
17   figure; plot( real(S1), imag(S1),'bo'); grid; title('I=f(Q)'); pause
18   figure;
19   subplot(211); plot( real(x1) ); grid; title('Re(x(n))')
20   subplot(212); plot( imag(x1) ); grid; title('Im(x(n))'); pause
21   figure; plot( real(xcorr( x1 )) ); grid; title('Auto-correlation of x1(n)'); pause
```

22.8.2 ADC Sampling Rate Estimation and Correction

In the fifth plot of Fig. 22.16 we can see how harmful is losing carrier orthogonality due to wrong sampling ratio of the ADC converter in the DMT/OFDM receiver. Total drama, not romantic comedy! Therefore, the sooner the better, we have to check sampling frequency value and sent the command to the hardware (*This is captain speaking!*) to correct it.

Let us assume situation presented in Fig. 22.19 when the same frequency-reach signal blocks are transmitted many times and we expect to have N samples per block. Let us assume also ADC sampling ratio error ε_s and frequency offset error ε_c, typical for OFDM systems with frequency up-down conversion. Frequency carrier offset, resulting from service frequency up-down conversion, is absent in DMT systems but presented here the Sliskovic method [9], developed for OFDM systems, is more general and allows also its estimation.

Let $X(k)$ denote sent carrier states. When transmission is ideal and channel is perfect, not deforming the signal, two consecutive DMT/OFDM symbols are represented in time domain by the following samples:

$$y(n) = \frac{1}{N} \sum_{k=0}^{N-1} X_k e^{j\frac{2\pi}{N} kn}, \quad n = 0, 1, ..., 2N - 1. \tag{22.12}$$

Even with the imperfect channel, samples $y(n)$ are deformed the same way and should be the same in N-samples long blocks. In the presence of ADC sampling and carrier frequency errors, respectively, ε_s and ε_c, DFT of the first N-sample block of $y(n)$ is equal to:

$$Y_1(k) = \sum_{n=0}^{N-1} y(n) e^{-j\frac{2\pi}{N} n \, (k(1+\varepsilon_s)+\varepsilon_c)}, \quad k = 0, 1, ..., N - 1. \tag{22.13}$$

For DFT of the second block of $y(n)$ we have

$$Y_2(k) = \sum_{n=N}^{2N-1} y(n)e^{-j\frac{2\pi}{N}n\,(k(1+\varepsilon_s)+\varepsilon_c)} = \sum_{n=0}^{N-1} y(n+N)e^{-j\frac{2\pi}{N}(n+N)\,(k(1+\varepsilon_s)+\varepsilon_c)}.$$

(22.14)

It can be easily shown that both DFT are connected by the following equation:

$$Y_2(k) = Y_1(k) \cdot e^{-j2\pi(k(1+\varepsilon_s)+\varepsilon_c)} = Y_1(k) \cdot e^{-j2\pi k}e^{-j(k\varepsilon_s+\varepsilon_c)} = Y_1(k) \cdot e^{-j(k\varepsilon_s+\varepsilon_c)}.$$

(22.15)

We see that having values $Y_1(k), Y_2(k)$ for two indexes k, we obtain a set of two equations with two unknowns ε_s and ε_c which can be solved. In fact two pilot signals are sufficient during normal modem work. However, during initialization, we have many block of samples and many k values available, and we can find sampling frequency error precisely solving over-determined set of many equations in the mean square sense. Knowing the error, we have to do signal re-sampling in software or adjust hardware sampling.

Exercise 22.6 (Simulation of Wrong Signal Sampling and Correcting This During Signal Spectrum Calculation). We would like to check validity of Eqs. (22.13–22.15). Analyze program 22.4. We generate in it signal x correctly sampled and signal xe incorrectly sampled by ADC and with carrier frequency offset (i.e. with errors). After that we calculate DFT of both signals. Next, we would like to obtain the spectrum of the second signal modifying DFT of the first signal. With success. Finally, we intend to modify DFT of the second signal and obtain spectrum of the first signal. With success. Run the program for different values of k0, ppm and cfo. Notice that wrong signal sampling can be corrected during signal spectrum computation.

Listing 22.4: Matlab program demonstrating how to simulate ADC sampling rate error expressed in ppm for arbitrary signal

```
1   % lab22_ex_ADCppm1.m
2   clear all; close all;
3
4   N=64;                               % number of signal samples
5   k0 = 25;                            % frequency index of tested harmonic
6   ppm = 10000e-6; cfo = 0.25;         % assumed ADC ppm error and CFO error
7   x  = exp( j*2*pi/N*k0*(0:N-1));                % signal 1 well sampled
8   xe = exp( j*2*pi/N*( k0*(1+ppm)+cfo )* (0:N-1) );   % signal 2 wrong sampled
9   Xref = fft(x)/N;                    % spectrum of reference signal 1
10  Xe1  = fft(xe)/N;                   % spectrum of reference signal 2
11
12  for k=0:N-1                         % spectrum of signal 1 with DFT error
13      Xe2(k+1) = sum( x .* exp( -j*2*pi/N*(k*(1-ppm)-cfo)*(0:N-1) ) )/N;
14  end
15  figure; plot(1:N,abs(Xref),'ro',1:N,abs(Xe1),'bo',1:N,abs(Xe2),'b*'); grid; pause
16
```

```
17   for k=0:N-1                          % spectrum of signal 2 with DFT correction
18       Xcor(k+1) = sum( xe .* exp(-j*2*pi/N*(k/(1-ppm)+cfo)*(0:N-1) ) ) / N;
19   end
20   figure; plot(1:N,abs(Xref),'ro',1:N,abs(Xcor),'g*'); grid; pause
```

Exercise 22.7 (Estimation of ADC Sampling Rate Error). In Exercise 22.2 we have modified program 22.1 and simulated wrong sampling rate of AD converter in the receiver. At present, add to program 22.1 the Matlab code from Listing 22.5, in which the ADC error is specified in ppm (parts per million). Try to implement the method of sampling rate error estimation, described in this section. In case of difficulties, see how it is done in program 22.7. Note that in program 22.5 signal has not 10000 samples but 10001 because for ADCppm=100 we have: $100 \cdot 10^{-6} \cdot 10000 = 1$.

Listing 22.5: Matlab program demonstrating how to simulate ADC sampling rate error expressed in ppm for arbitrary signal

```
1    % lab22_ex_ADCppm2.m
2    clear all; close all;
3
4    y = sin(2*pi/100*(0:10000));        % input signal
5    Ny = length(y),                     % its length
6    ADCppm = -100;                      % ADC ppm error (+/-) in the receiver
7    Nppm = Ny + (ADCppm*1e-6)*Ny;       % more signal samples after faster/slower sampling
8    yi = interp1( [0:Ny-1], y(1,1:Ny), [0 : Ny/Nppm : (Ny-1)],'spline');
9    Nyi = length(yi),                   % signal length after interpolation
10   figure; plot(1:Ny,y,'ro',1:length(yi),yi,'bx'); grid;
11   title('y(n) (o)  and  yADC(n) (x)'); pause
```

22.8.3 Time Equalization of the Channel

Choosing in this chapter the DSL modem as an example of multi-carrier transmission was motivated by two facts: (1) in this technology channel is equalized in time also (TEQ) and effective impulse response of the overall signal transmission path is made shorter and (2) bit allocation is performed according to SNR of each subcarrier. At present, we will focus on TEQ design.

Time duration of the DSL impulse response can be very long. Coping all DMT/OFDM blocks as remedy against the inter-block (inter-symbol) interference (ISI), as shown in Fig. 22.12, is not an effective solution since bit-rate is significantly reduced in it. In order to allow usage of a shorter signal copy, $P = 32$ samples instead of $N = 512$, energy of the *shortened-by-TEQ* channel impulse should concentrate in $P + 1 = 33$ consecutive samples only (i.e. in the most *effective* part of it). For this purpose signal in the receiver is passed by additional digital FIR filter $e(n)$, specially designed and called the TEQ, which causes that the resultant impulse response of serial connection of the channel $h(t)$ (after discretization $h(n)$) and the designed TEQ filter $e(n)$ has a significant part of its energy concentrated in small number of consecutive samples, e.g. 33 in the discussed case. Maybe it looks strange at a first glance that an impulse response of a cascade of two filters can be *effectively* shorter than the longer impulse response of the filter pair, but it is possible. Let us consider a slowly varying and slowly decaying impulse response and the filter with weight $[1,-1]$: subtracting neighbor weights of the first filter gives very small value for similar filter coefficients. This is the trick!

Our optimization problem is explained in Fig. 22.22. We aim at *pushing* almost all energy of the shortened impulse response $h_e(n)$:

$$h_e(n) = \sum_{k=-\infty}^{+\infty} h(k)e(n-k) \tag{22.16}$$

into a rectangular window which is non-zero only for $P+1$ samples.

There are many methods for TEQ design, for example, [2]. Matlab code of their program implementation is available here [1]. Below the Tkacenko–Vaidyanathan eigen-filter TEQ design method [12] is presented. The following cost function is minimized in it:

$$J = \frac{\sum\limits_{n \notin window} h_e^2(n)}{\sum\limits_{n \in window} h_e^2(n)} = \frac{\sum \left[h_e^{IF}(n)\right]^2}{\sum \left[h_e^{signal}(n)\right]^2} \xrightarrow[e(n)=?,\ \mathbf{ee}^T=1]{} \min. \tag{22.17}$$

Energy of $h_e(n)$ samples, lying outside the rectangular window, is minimized in nominator of Eq. (22.17), i.e. $h_e^{IF}(n)$ in Fig. 22.22, the interference part. The same time, energy of $h_e(n)$ samples lying inside the window is maximized, i.e. $h_e^{signal}(n)$ in Fig. 22.22. The following derivation is an important lesson showing that without mathematics there is no sense to dream about impressive bit-streams in DSL transmission.

Let us introduce the following denotations ($L_{he} = L_h + L_e - 1$):

- $h(n)$—original channel impulse response, $n = 0, 1, ..., L_h - 1$,
- $e(n)$—TEQ filter, $n = 0, 1, ..., L_e - 1$,
- $h_e(n)$—shortened channel impulse response, $n = 0, 1, ..., L_{he} - 1$, having energy concentrated in $P+1$ consecutive samples,
- $\mathbf{H}$—channel matrix with dimensions $L_{he} \times L_{he}$:

$$\mathbf{H} = \begin{bmatrix} h(0)\ h(1) & \cdots & h(L_h-1) & 0 & \cdots & 0 \\ 0\ h(0)\ h(1) & & \cdots & h(L_h-1) & \cdots & \vdots \\ \vdots & \ddots & \ddots & \ddots & \ddots & 0 \\ 0 & \cdots & 0 & h(0) & h(1) & \cdots & h(L_h-1) \end{bmatrix}_{L_{he} \times L_{he}} \tag{22.18}$$

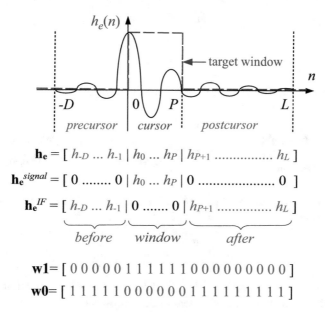

$$\mathbf{h_e} = [\, h_{-D} \,...\, h_{-1} \mid h_0 \,...\, h_P \mid h_{P+1} \,................\, h_L \,]$$

$$\mathbf{h_e}^{signal} = [\, 0 \,........\, 0 \mid h_0 \,...\, h_P \mid 0 \,....................\, 0 \,]$$

$$\mathbf{h_e}^{IF} = [\, h_{-D} \,...\, h_{-1} \mid 0 \,.......\, 0 \mid h_{P+1} \,................\, h_L \,]$$

$$\underbrace{\qquad\qquad}_{before} \quad \underbrace{\qquad\qquad}_{window} \quad \underbrace{\qquad\qquad}_{after}$$

$$\mathbf{w1} = [\,0\,0\,0\,0\,0\,1\,1\,1\,1\,1\,0\,0\,0\,0\,0\,0\,0\,0\,]$$

$$\mathbf{w0} = [\,1\,1\,1\,1\,1\,0\,0\,0\,0\,0\,0\,1\,1\,1\,1\,1\,1\,1\,1\,]$$

Fig. 22.22: Illustration of channel impulse response partition into three parts: precursor, cursor (window), and post-cursor. Signal energy in cursor (window) part is maximized during optimization [14]

- $\mathbf{W}_1$—matrix with dimensions $L_{he} \times L_{he}$, having on its main diagonal rectangular window with length $P + 1$, shifted down D elements, which cut fragment of $h_e(n)$ for its energy maximization ($\mathbf{0}_D$—zeros matrix with dimensions $D \times D$, $\mathbf{I}_{P+1}$—square unitary matrix with dimensions $(P + 1) \times (P + 1)$, etc.):

$$\mathbf{W}_1 = \begin{bmatrix} \mathbf{0}_D & \mathbf{0} & \mathbf{0} \\ \mathbf{0} & \mathbf{I}_{P+1} & \mathbf{0} \\ \mathbf{0} & \mathbf{0} & \mathbf{0}_{L_{he}-D-(P+1)} \end{bmatrix}_{L_{he} \times L_{he}} \tag{22.19}$$

- $\mathbf{W}_0$—matrix with dimensions $L_{he} \times L_{he}$, having on its main diagonal elements reversed in regard to matrix $\mathbf{W}_0$, i.e. ones are exchanged with zeros and vice versa:

$$\mathbf{W}_0 = \begin{bmatrix} \mathbf{I}_D & \mathbf{0} & \mathbf{0} \\ \mathbf{0} & \mathbf{0}_{P+1} & \mathbf{0} \\ \mathbf{0} & \mathbf{0} & \mathbf{I}_{L_{he}-D-(P+1)} \end{bmatrix}_{L_{he} \times L_{he}} \tag{22.20}$$

Assuming that the vector $e(n)$ is horizontal, the vectors $h_e(n)$, $h_e^{signal}(n)$, and $h_e^{IF}(n)$ can be calculated from the following matrix equations:

$$\mathbf{h}_e = \mathbf{eH}, \tag{22.21}$$

$$\mathbf{h}_e^{signal} = \mathbf{h}_e \mathbf{W}_1, \tag{22.22}$$

$$\mathbf{h}_e^{IF} = \mathbf{h}_e \mathbf{W}_0. \tag{22.23}$$

In consequence, the cost function (22.17) can be rewritten as:

$$J = \frac{\mathbf{h}_e^{IF}\left(\mathbf{h}_e^{IF}\right)^T}{\mathbf{h}_e^{sig}\left(\mathbf{h}_e^{sig}\right)^T} = \frac{\mathbf{h}_e \mathbf{W}_0 (\mathbf{h}_e \mathbf{W}_0)^T}{\mathbf{h}_e \mathbf{W}_1 (\mathbf{h}_e \mathbf{W}_1)^T} = \frac{\mathbf{h}_e \mathbf{W}_0 \mathbf{W}_0^T \mathbf{h}_e^T}{\mathbf{h}_e \mathbf{W}_1 \mathbf{W}_1^T \mathbf{h}_e^T}. \tag{22.24}$$

Since $\mathbf{W}_0 \mathbf{W}_0^T = \mathbf{W}_0$ and $\mathbf{W}_1 \mathbf{W}_1^T = \mathbf{W}_1$, after using Eq. (22.21), the last equation can be rewritten as:

$$J = \frac{\mathbf{h}_e \mathbf{W}_0 \mathbf{h}_e^T}{\mathbf{h}_e \mathbf{W}_1 \mathbf{h}_e^T} = \frac{\mathbf{eH}\mathbf{W}_0\mathbf{H}^T \mathbf{e}^T}{\mathbf{eH}\mathbf{W}_1\mathbf{H}^T \mathbf{e}^T}. \tag{22.25}$$

If we assume that matrix $\mathbf{HW}_1\mathbf{H}^T$ is of full rank and positive defined, it is equal to multiplication of two matrices (Cholesky decomposition):

$$\mathbf{HW}_1\mathbf{H}^T = \mathbf{G}_1^T \mathbf{G}_1, \tag{22.26}$$

where matrix $\mathbf{G}_1$ is an upper triangular matrix. Therefore:

$$J = \frac{\mathbf{eH}\mathbf{W}_0\mathbf{H}^T \mathbf{e}^T}{\mathbf{e}\mathbf{G}_1^T (\mathbf{G}_1 \mathbf{e}^T)}. \tag{22.27}$$

After setting:

$$\mathbf{v} = \mathbf{G}_1 \mathbf{e}^T, \tag{22.28}$$

we can express $\mathbf{e}$ as a function of vector $\mathbf{v}$ and matrix $\mathbf{G}$:

$$\mathbf{e} = \mathbf{v}^T \left(\mathbf{G}_1^{-1}\right)^T \tag{22.29}$$

and use Eqs. (22.28) and (22.29) in Eq. (22.27):

$$J = \frac{\mathbf{v}^T \left(\mathbf{G}_1^{-1}\right)^T \mathbf{HW}_0\mathbf{H}^T \mathbf{G}_1^{-1} \mathbf{v}}{\mathbf{v}^T \mathbf{v}} = \frac{\mathbf{v}^T \mathbf{Tv}}{\mathbf{v}^T \mathbf{v}}. \tag{22.30}$$

From lectures on numerical analysis we remember (I do not doubt!) that vector $\mathbf{v}$ minimizing Eq. (22.30) is an eigenvector $\mathbf{v}_{min}$ of matrix $\mathbf{T}$:

$$\mathbf{T} = \left(\mathbf{G}_1^{-1}\right)^T \mathbf{HW}_0\mathbf{H}^T \left(\mathbf{G}_1^{-1}\right), \tag{22.31}$$

associated with the smallest eigenvalue λ_{min}. Therefore *the game is over*: we should calculate the matrix $\mathbf{T}$, perform its eigenvalue decomposition, find λ_{min} and $\mathbf{v}_{min}$, and use it in our ADSL modem joy, *after so long climbing!*

In Fig. 22.23 results from described algorithm usage are shown. On the left side we see long, estimated channel impulse response $\hat{h}(n)$ and its shortened version $\hat{h}_e(n)$ for DSL line without high-pass filter, and on the right side—with the high-pass filter. As we see, after channel time equalization (TEQ), both shortened impulse responses have energy concentrated in a smaller number of samples than the ADSL cyclic prefix plus 1 ($P+1 = 33$).

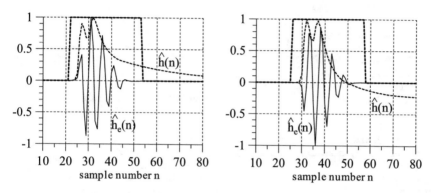

Fig. 22.23: Graphical illustration of the overall channel (telephone line) impulse response shortening by means of designed digital TEQ filter (time equalizer). Denotations: $\hat{h}(n)$—impulse response to be shortened, which was estimated in the previous section, $\hat{h}_e(n)$—shortened impulse response (with energy concentrated in $P+1$ samples). Rectangular window is shown in which energy of the shortened impulse response $h_e(n)$ was maximized. Optimal rectangular window delay is a parameter to be chosen [14]

Exercise 22.8 (Time Equalization of the DSL Channel). Analyze the function presented in Listing 22.6. Find in it all equations given in the text. Design coefficients $\{b, a\}$ of a digital low-pass IIR Butterworth filter having two poles only, cut-off frequency equal to 100 kHz and working with sampling frequency $f_s = 2.208$ MHz. Filter with its help the Kronecker impulse, i.e. 1 in the beginning and many 0s after it. Try to shorten the impulse response to $P = 32$ samples using the TEQ filter. Next take from the book repository any reference `csaloopX.time` impulse response (or from http://users.ece.utexas.edu/~bevans/projects/adsl/) and repeat the experiment. Call the function with different values of the D parameter. Compare results.

Listing 22.6: Matlab function for TEQ filter design using Tkacenko–Vaidyanathan method [12]

```
1   function [e] = AdslTEQ(h, Le, Lp, D)
2   % h  - DSL (channel) impulse response to be shortened
3   % Le - chosen length of the shortening filter - TEQ equalizer
4   % Lp - chosen length of the impulse response after shortening (P-prefix)
5   % D  - chosen delay of the shortened impulse response (in samples)
6   % e  - calculated weights of the TEQ equalizer
7
8   disp('Designing TEQ equalizer ...');
9
10  Lh = length(h); Lz = Lh+Le-1-D-Lp;
11  H = zeros(Le, Lh+Le-1);
12  for i=1:Le
13      H(i,i:i+Lh-1) = h;                      % convolutional channel matrix
14  end
15  W1 = zeros(Lh+Le-1, Lh+Le-1);               % matrix with zeros
16  W0 = zeros(Lh+Le-1, Lh+Le-1);               % matrix with zeros
17  rect= [zeros(1,D),ones(1,Lp),zeros(1,Lz)];  % vector [ ...0001111000... ]
18  W1 = diag(rect);                            % 1s in the center
19  W0 = diag(1-rect);                          % 0s in the center, 1s outside
20  E1=H*W1*H';                                 % h(n) energy in the window
21  G1 = chol(E1);                              %
22  E0=H*W0*H';                                 % h(n) energy outside the window
23  T = inv(G1)'*E0*inv(G1);                    %
24  [V,D] = eig(T);                             % eigenvalue decomposition of T
25  LambdaMin = min(diag(D));                   % minimum eigenvalue
26  LambdaMinIdx = find(diag(D)==LambdaMin);    % its position
27  e = V(:,LambdaMinIdx)' * inv(G1)';          % final solution, weights of TEQ filter
```

22.8.4 Frequency Equalization of the Channel

TEQ digital filter (corrector) $e(n)$ is designed considering only time channel features. But as a filter, it has a frequency response $E(k) = \text{FFT}[e(n)]$ (k is frequency index) which changes the overall frequency response of the whole transmission path:

$$\hat{H}_e(k) = \hat{H}(k)E(k), \quad k = 0, 1, ..., N-1, \tag{22.32}$$

where $\hat{H}(k)$ is frequency response of the already estimated original channel impulse response $\hat{h}(n)$:

$$\hat{H}(k) = \text{FFT}\left[\hat{h}(n)\right]. \tag{22.33}$$

After TEQ equalization (filtering) - ADD of $y(n)$ by $e(n)$, we calculate in the receiver the FFT of the resultant signal $y_e(n)$ obtaining:

$$Y_e(k) = \text{FFT}\left[y_e(n)\right]. \tag{22.34}$$

For any k, values of $Y_e(k)$ are equal to states $S(k)$ of the k-th carrier which are modified in amplitude and phase by corresponding value $\hat{H}_e(k)$:

$$\underbrace{Y_e(k)}_{\text{output}} \approx \underbrace{\hat{H}_e(k)}_{\text{line}} \cdot \underbrace{S(k)}_{\text{input}} + \underbrace{\Xi(k)}_{\text{noise}}. \tag{22.35}$$

States $S(k)$ can be recovered from $Y_e(k)$ by FEQ equalization of $Y_e(k)$:

$$\hat{S}(k) \approx \frac{1}{\hat{H}_e(k)} Y_e(k) = \text{FEQ}(k) \cdot Y_e(k). \tag{22.36}$$

Figure 22.24 presents exemplary shapes of frequency responses $|E(k)|$, $|\hat{H}(k)|$, and $|\hat{H}_e(k)|$ (left) and a shape of $|\text{FEQ}(k)|$ as an inverse of $|\hat{H}_e(k)|$ (right). Multiplication of $|\text{FEQ}(k)|$ and $|\hat{H}_e(k)|$ gives us 1 in linear scale (0 in decibels) for all frequencies, i.e. distortion-less overall signal processing path. We observe some notches (frequencies of very strong signal attenuation) in $|\hat{H}_e(k)|$. It is important to ask the question of their origin: do they come from the channel either from the TEQ. If they are caused by the TEQ filter, some additional constraints should be applied during TEQ design in order to avoid very strong signal attenuation by it. It will be shown later that signal attenuation leads to significant reduction of ADSL modem throughput.

We can use the TEQ filter design example to stress that digital filters do not always have 0/1 frequency responses. In the same applications, very often in measurement systems, they should reverse the frequency response of a non-ideal device/channel/object, i.e. to correct non-ideal device frequency response.

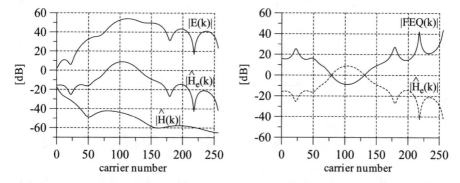

Fig. 22.24: Illustration of successive computational steps leading to design of FEQ equalizer. Frequency responses (amplitude-frequency characteristics) of: $H(k)$—telephone line, $E(k)$—TEQ equalizer, $H_e(k)$—telephone line with TEQ equalizer, FEQ—FEQ equalizer. $\hat{a}$ denotes estimated value of a [14]

22.8.5 DMT/OFDM Symbol Synchronization

So, we have a channel already equalized. In time and in frequency. At present, we want to recover carrier states and extract bits from their numbers. In order to do this we have to synchronize with the DMT/OFDM frame beginning. Below two methods are presented, the first originated from the DSL technology, and the second, the Schmidl–Cox technique—coming from the wireless world.

Synchronization Using Cyclic Prefix Cyclic prefix is an all-in-one solution. It has helped us already to remove ISI between DMT/OFDM symbols and allowed easy FEQ channel equalization, thanks to artificially created signal periodicity: linear convolution has taken a form of circular one and the channel was equalized doing simple DFT/IDFT operations. But ... the CP can be also additionally used for symbol synchronization. CP is a repeating part of any DMT/OFDM symbol. When we correlate P samples of the received, equalized signal $y_e(n)$ with itself, but with the shift equal to symbol length N, the correlation function maximum tells us about the CP beginning. In Fig. 22.25 concept of the CP usage for symbol detection is explained, while in Fig. 22.26—values of the calculated correlation function are shown. As we see, the maximum is present for the shift $d = 32$, i.e. exactly for the CP length.

To be precise, in the CP symbol synchronization method maximum of the following cost function is searched for $d = 1, 2, 3, ..., N + P$:

$$J(d) = \frac{Q(d)}{R(d)}, \tag{22.37}$$

where functions $Q(d)$ and $R(d)$ are defined as:

$$Q(d) = \sum_{m=0}^{P-1} y_e(d+m)y_e(d+m+N) \tag{22.38}$$

$$R(d) = \sum_{m=0}^{P-1} |y_e(d+m)|^2 + \sum_{m=0}^{P-1} |y_e(d+m+N)|^2. \tag{22.39}$$

Division by $R(d)$ is only a normalization. Values of $Q(d)$ and $R(d)$ can be calculated iteratively, e.g.:

$$Q(d+1) = Q(d) - y_e(d)y_e(d+N) + y_e(d+P)y_e(d+P+N). \tag{22.40}$$

Synchronization Using Schmidl-Cox Method The second symbol synchronization method comes from OFDM wireless transmission and was proposed by Schmidl and Cox [8]. Its concept is explained in Fig. 22.27. A special DMT/OFDM synchronization symbol is formed which consists of a cyclic prefix and two identical parts (sub-signals) having $N/2$ samples each. At present, we calculate correlation function of $N/2$ signal samples with the $N/2$ shift and its maximum tells us about the

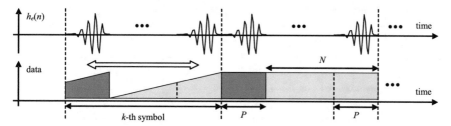

Fig. 22.25: Illustration of the cyclic prefix usage for the DMT/OFDM symbol synchronization: end of each symbol is correlated with its cyclic prefix which should be identical! [14]

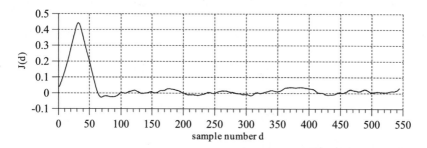

Fig. 22.26: Exemplary shape of the averaged synchronization cost function (22.37)–(22.39) (for 100 realizations) in case of the lack and presence of additive white Gaussian noise with power level -110 dBm/Hz (both functions overlap) [14]

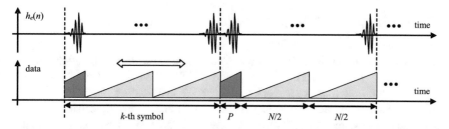

Fig. 22.27: Illustration of the Schmidl-Cox method of DMT/OFDM symbol synchronization: in the first and in the second half of each symbol, with cyclic prefix, the same signal is transmitted which is correlated with itself [14]

beginning of the first signal block, i.e. about the DMT/OFDM symbol beginning. Results of the method usage are presented in Fig. 22.28.

In the Schmidl-Cox method in cost criteria (22.37) the following functions are used for $d = 1...N$:

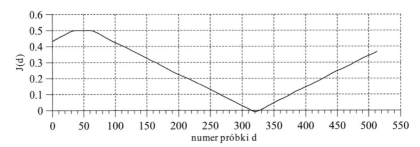

Fig. 22.28: Exemplary shape of the averaged synchronization cost function (22.37), (22.41), (22.42) (after 100 realization) in case of the lack and presence of additive white Gaussian noise with power level -110 dBm/Hz (both functions overlap) [14]

$$Q(d) = \sum_{m=0}^{N/2-1} y_e\,(d+m)y_e\,(d+m+N/2), \qquad (22.41)$$

$$R(d) = \sum_{m=0}^{N/2-1} |y_e\,(d+m+N/2)|^2, \qquad (22.42)$$

which can be calculated recursively:

$$Q(d+1) = Q(d) + y(d+N/2)y(d+N) - y(d)y(d+N/2). \qquad (22.43)$$

22.8.6 Influence of Disturbances

At present, a little bit tired after a few *mathematical expeditions*, and after completing TEQ and FEQ channel equalization and DMT/OFDM symbol synchronization, we can check *how that is all work together?* We are sending bits using all available sub-carriers by means of 4-QAM modulation and observing carrier constellation points in the receiver for different levels of noise. We see plots presented in Fig. 22.29. With the noise increase, we observe *clouds* of detected states that becoming bigger and bigger. It is obvious that we can make the constellation grid less dense (more spare) for stronger noise in order to avoid incorrect states interpretation. However, less points mean lower number of transmitted bits.

In the second experiment, very short, we assume noise-less transmission conditions, perfect synchronization but FEQ corrector missing. We see in Fig. 22.30 that constellation points are significantly rotated (left, states are specially, didactically enlarged in amplitude) and heavily, both, rotated and strongly attenuated (right). We see how dramatic consequences are caused by the FEQ absence.

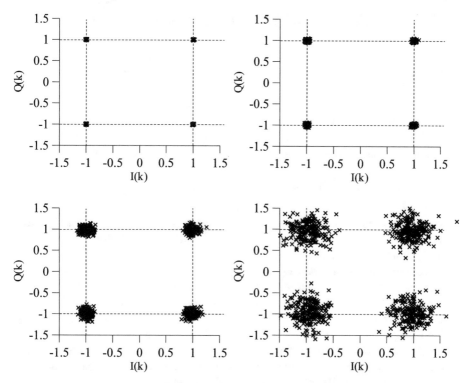

Fig. 22.29: Detection of 4-QAM constellation points of one frequency carrier in case of FEQ usage. Additive white Gaussian noise is present having the following power level: $-120, -110, -100, -90$ dBm/Hz (in turn horizontally) [14]

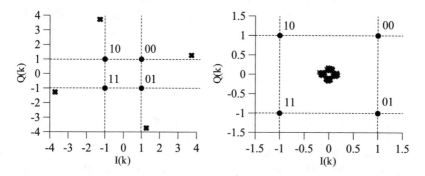

Fig. 22.30: Detection of 4-QAM constellation points of one frequency carrier in case of FEQ missing. Additive white Gaussian noise has power level -120 and -90 dBm/Hz (in turn horizontally) [14]

22.8.7 Bit Allocation and Channel Information Capacity

Are you ready, Teddy? Yes, I am. And we are ready also, after channel equalization, symbol synchronization and some *test drives*, to transmit pilot data upon each
sub-carrier, known to the receiver, and calculate SNR for each sub-carrier. Why is
it needed? Because, as we see in the last section, number of bits, which can be
allocated to any sub-carrier, depends on the noise level in concrete transmission
sub-channel. The exact formula specifying total number of bits b, transmitted by the
DSL modem in one DMT symbol, is defined by the following formula:

$$b = \sum_k b_k = \sum_k \log_2 \left(1 + \frac{SNR_k}{\Gamma} \right), \quad SNR_k = \frac{P_{x,k}}{P_{n,k} + P_{ISI,k}}, \tag{22.44}$$

where k denotes sub-carrier number, b_k number of bits transmitted by the k-th sub-
carrier (in DSL modem limited to 2...15 bits), and SNR_k signal-to-noise ratio for
the k-th sub-carrier, i.e. transmitted pilot power $P_{x,k}$ divided by the power of disturbances (summation of power of noise $P_{n,k}$ and power of inter-symbol (ISI) interference $P_{ISI,k}$). Because transmitted bits can be additionally coded using some forward
error correction (FEC) methods, some transmission errors can be found and corrected in the receiver. This is taken into account in Eq. (22.44) by using Γ scaling
factor, depending on: (1) a *coding gain G* of the FEC used, (2) assumed margin
of safety M and assumed bit error rate (BER). In ADSL case, convolutional codes
and Viterbi decoder are exploited, and the following values of parameters are used
($G = 4.2$ dB, $M = 6$ dB, BER $= 10^{-7}$) for Γ calculation:

$$\Gamma = 10^{(9,8+M-G)/10} = 14,4544. \tag{22.45}$$

Because pilot signals are transmitted, known to the receiver, the transmitted state
$S_m(k)$ of the k-th sub-carrier in the m-th DMT symbol is known. Let $\hat{S}_m(k)$ denote
the received value. Therefore SNR_k for the carrier k-th is equal to:

$$SNR_k = \frac{\sum_m [S_m(k)]^2}{\sum_m [S_m(k) - \hat{S}_m(k)]^2}. \tag{22.46}$$

Usage of a bit allocation procedure is explained in Fig. 22.31. In the upper part,
we see calculated SNR_k for each sub-carrier for DSL line without a high-pass POTS
splitter (left) and with it (right). Different noise levels are considered, varying from
-300 dBm/Hz to -100 dBm/Hz. One can observe how strong is the noise influence
upon the SNR. In the lower part of the figure, we see results from bit allocation for
both DSL lines and for all levels of noise. For weak noise even 15 bits are allocated
for single carriers. For strong noise, number of bits allocated to high-frequency sub-
carriers is decreasing very fast, even they are not used at all.

Of course, choice of the bit allocation algorithm represents a topic for special
discussion. Maybe it is a *water-filling* one, as in the MPEG audio standard where
bits are allocated to sub-band signals after the $M = 32$ channel filter bank.

It is interesting, how long we have walked forward and forward without asking and answering the question about the channel information capacity. What does the Shannon communication theory say?

Information capacity, measured in bits per second, of continuous, memory-less, ideal channel with bandwidth B hertz, disturbed by additive Gaussian noise with power spectral density $\frac{N_0}{2}$ is equal to:

$$C = B \log_2 \left(1 + \frac{P}{N_0 B} \right), \qquad (22.47)$$

where P denotes mean power of the transmitted signal. What does it mean? We can increase C, increasing P and B and decreasing N_0. Using Eq. (22.47) one can derive bit-rates R_b for different real modulation systems. Dividing this value by B, one obtains their spectral efficiency. The bigger, the better. This allows a comparison of different modulation schemes in regard to spectrum width usage. Different transmission systems are also compared using energy per bit

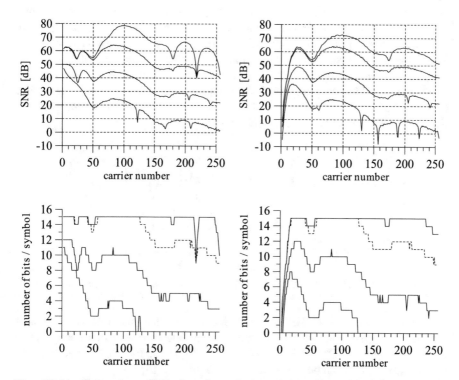

Fig. 22.31: Exemplary SNR functions (up) and bit allocations (down) for a telephone line without a high-pass POTS splitter (left) and with it (right). Additive white Gaussian noise is present having the following power levels: $-300, -140, -120, -100$ dBm/Hz [14]

to noise power spectral density ratio E_b/N_0, required for ensuring requested bit-error-rate (BER). Link budgets allow to choose telecommunication system components and values of their parameters, for example, size of carrier constellation states, ensuring required BER. There are many good Telecomm Books covering these topics in detail. We are DSP-oriented!

22.8.8 Choice of TEQ Equalizer

In this section, we will go back once more to the TEQ filter design problem. The TEQ filter, shortening channel impulse response, modifies also the channel frequency response. It is improving something, worsening something else. In Fig. 22.32 TEQ with different lengths ($L_e = 2, 8, 16$) is designed in noise-less conditions. On the left plot we see residual signal which remained after shortening operation. The worst is short TEQ $L_e = 2$, leaving more channel impulse response energy out of the required rectangular window. It is also the worst in the right plot offering the lowest SNR. However the situation is changing in Fig. 22.33, where TEQ was designed in the presence of strong -110 dBm/Hz noise. Here only two TEQ lengths are considered: $L_e = 2$ (solid line) and $L_e = 8$ (dashed line). The short TEQ has *less-notchy* frequency response (left) and it offers also less-notchy SNR curve, in general overlapping with SNR characteristics for TEQ $L_e = 8$. As we see, typically, many circumstances should be considered. There are no simple answers! No free lunches. Always carefully look around.

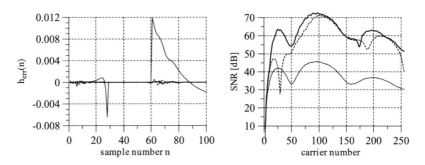

Fig. 22.32: Application of TEQ filters having different lengths in case of telephone line with high-pass filter when noise is not present. (left) error of shortening, (right) corresponding SNR. Denotations: TEQ $L_e = 2$ (thin solid line), 8 (dashed line), 16 (thick solid line)—from bottom to top in the right plot [14]

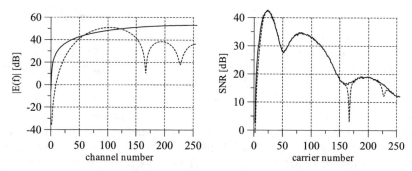

Fig. 22.33: Application of TEQ filters having different lengths in case of telephone line with high-pass filter for noise −110 dBm/Hz. (left) frequency response of the TEQ equalizer alone, (right) SNR of the whole transmission path. Denotations: TEQ $L_e = 2$ (solid line), 8 (dashed line)—frequency response with notches [14]

22.9 Program of DSL Modem

We are ending one of the last chapters of this book. We were climbing for months, week by week, higher and higher, seeing more and more. It is time for something spectacular. In this section we will analyze a program written by group of my former co-workers, now friends, more than 15 years ago. The work lasted several months, with block delay and synchronization problems, and missing many *small details* which *made the mission impossible* for a long time. Thanks a lot Paweł Turcza, Jarek Bułat, Tomek Twardowski, and Krzysiek Duda!

General Comments The program is presented in Listing 22.7. It is widely commented and self-explaining/presenting. However, some of its initial general description is recommended. In the beginning, values of all program parameters are set and all vector/matrix initialization are done. The program consists of a main loop repeating `Niter` times, in which one-by-one are executed operations performed by two co-working ADSL modems, one transmitting (signal x) and one receiving (signal y). In one iteration one DMT/OFDM frame is transmitted and received. In order to consider the channel impulse response, convoluting with the transmitted signals, and to synchronize all blocks, the transmitter output and the receiver input are buffered (variables bx and by). Last three DMT symbols are stored. In fact valid is the fourth symbol in the receiver.

In order to distinguish what should be executed during actual iteration of the main loop, special flags are turned ON/OFF. In the beginning, their values are as follows:

`ADCCorrect=1; ChanEstim=0; Synchronization=0; SNREstim=0.`

The flag value is checked inside the loop, and when 1 is set, the specified operation is executed, in the beginning `ADC Correction` only. When one task has finished its work, it is setting its own flag to 0 and a flag of the next task to 1, i.e. when the ADC correction is done, we have

```
ADCCorrect=0; ChanEstim=1; Synchronization=0; SNREstim=0;
```

Because one modem task can require execution of several iterations, each task has a counter which is initially set to some value: `CounterADC=50`, `CounterChan =50`, `CounterSynchro=50`, and `CounterSNR=50`, telling how many main loop iterations are required for this task. During loop executions counters are decreased. When 0 is reached, some final decision is taken by the task (estimation of: ADC frequency error, channel impulse response, symbol synchronization, and SNR for each sub-carrier) and modem is switched to the next operation. First, ADC frequency sampling error is estimated, then channel estimation task is performed, ending with TEQ and FEQ design, next DMT symbol synchronization is done, and, finally, SNR for each sub-carrier is estimated and number of bits is allocated to each sub-carrier, according to Eqs. (22.44 and 22.46).

Each task requires different transmitted data. During ADC and channel testing, in the beginning, special reverberation DMT symbols are transmitted, synthesized from random 4-QAM carrier states. The cyclic prefix is not added. During DMT symbol synchronization, a pseudo-random signals are synthesized directly in time domain and transmitted as DMT symbols with the cyclic prefix. Pseudo-random noise takes uniformly values from the range $\sqrt{(3)} \cdot [-1, 1]$. During SNR estimation standard DMT symbols with cyclic prefix are sent, but they are synthesized from 4-QAM carrier states only. Reader can extend the program, adding to it: (1) regular transmission of data with higher-level QAM modulations (exploiting the calculated bit allocation), (2) decoding the received bits, and (3) calculation of symbol error rate and bit error rate.

The following disturbances are simulated: wrong ADC sampling ratio, additive noise, and narrow-band interference.

ADC Error Estimation and Correction The method described in Sect. 22.8.2 is implemented. Since the same DMT symbols are transmitted one-by-one without cyclic prefix, having two spectra we can calculate the phase shift (22.15) between them:

$$\phi(k) = k \cdot \varepsilon_s + \varepsilon_c, \tag{22.48}$$

for each carrier k, and find parameters $(\varepsilon_s, \varepsilon_c)$ of the line (22.48) performing line fitting to the set of points $(k, \phi(k))$. Figure presenting the line fitting result can be uncommented in the program. This operation is repeated many times and calculated values ε_s and ε_c, in the program ppm and cfo, are averaged, after removing of extreme values (outliers). When error of ADC sampling rate is known, it is assumed in the program that it is corrected in hardware. Therefore, in the beginning we should: (1) set, for example, `ADCppm=10`, (2) check whether the correct value was estimated, (3) observe terrific plots of carrier state constellation points and 0 bits allocated, (4) set `ADCppm=0` and perform next exercises. Simpler program `lab21_ex_ADCppm3.m`, given in the book repository, allows to check correctness of the described ADC sampling ratio estimation. Analyze it and run, if necessary.

Channel Estimation and Correction Channel estimation and correction is the next task of the receiver in the main loop. Equation (22.11) is implemented: accumulated are many N-point FFT spectra (due to this operation noise influence is suppressed) and the result is divided by the known spectrum of transmitted reverberation signal. Frame synchronization is not required at the moment: its lack causes *only* that the channel impulse response, calculated by IFFT from the estimated channel frequency response, is circularly rotated. This effect can be easily eliminated by de-rotation of `hest` samples. When the long channel impulse response is known, the TEQ filter e can be designed for shortening it. Next, the shortened channel impulse response is computed `he=conv(e,hest)` and its FFT is calculated, i.e. HE—the frequency response of the shortened channel. The channel estimation is finished, its flag is turned off and synchronization flag is turned on. Plots of `hest`, he, and HE are displayed in the program.

DMT Symbol Beginning Synchronization DMT symbol beginning synchronization is in the program realized using the cyclic prefix method (22.37)–(22.39)— uniform pseudo-random noise is sent as DMT symbols equipped with the prefix. The detection function is calculated for many symbols and averaged. It maximum informs us about the cyclic prefix position. Knowing it we can calculate `SynchDelay` of the DMT symbol beginning and correct phase of HE taking into account the time delay information. Inverse of HE gives us a FEQ which will be used for correction of received spectra of DMT symbols during regular bits/data transmission. Plot of the frame beginning detection function is shown in the program.

SNR Estimation and Bit Allocation Procedure SNR estimation and bit allocation procedure starts then. Just before the regular transmission, number of bits should be allocated to each sub-carrier, depending on its SNR value. During the modem startup only known 4-QAM carrier states are transmitted and SNR for each carrier is calculated. Transmitted signal is obtained from the IFFT of the spectral requirements. The cyclic prefix is added to each DMT symbol. In the receiver, we are already synchronized and have calculated channel influence correction. Therefore, only FFTs are performed on selected samples of the received signal and FFT results are divided by FEQ. Knowing transmitted Fourier coefficients S0 and received ones R, the SNR for each sub-carrier is computed and allowed number of its bits is found from Eq. (22.44). In this part of the program, constellation points of 4 carriers are plotted and final results of bit allocation are shown.

Regular Transmission Regular transmission follows. It is not implemented— remains as an exercise.

Program 22.7 is not short. Somebody can say that it is much too long. Most of us started our *game of life* having approximately 3–3.5 kg and 50 cm. At present, our weight is bigger than 50 kg and height larger than 150 cm (I do not want to offend my wife, I am over 80 kg/180 cm). We have grown up. As our knowledge during this course, I hope of the manuscript. After 700 pages of the manuscript, I propose you a DSP *star wars*.

Listing 22.7: Matlab program for demonstration of work of DSL modems

```
1   % lab22_ex_adsl.m
2   % Simulation of multi-carrier transmission used in ADSL modems
3   clear all; close all;
4
5   % PARAMETERS
6     Le = 16;                  % length of the TEQ equalizer
7     Lp = 32;                  % length of the cyclic prefix (CP)
8     N = 512;                  % length of the DMT/OFDM frame, FFT size
9     Np = N + Lp;              % length of the DMT/OFDM frame with cyclic prefix
10    fs = 2.208e6;             % sampling frequency [Hz]
11    NoiseDbmHz = -160;        % power level of noise   [dBm/Hz]
12    SignalDbm = 23;           % power level of input signal in the receiver [dBm]
13    f0 = 100.5*fs/N;          % frequency of narrow-band disturbance
14    A0 = 0;                   % amplitude of narrow-band disturbance: 0 or 50
15    CodingGain = 4.2;         % coding gain [dB]; e.g. without Viterbi enc/dec=0, with it=4.2
16    CodingMargin = 6;         % SNR margin [dB]
17    FilterTel = 1;            % 1 = high-pass filter, 0 = without it
18    ADCppm = 0;               % ADC ppm error in receiver, e.g. -5,0,5
19    CounterADC = 50;          % number of loops for ADC sampling error estimation
20    CounterChan = 50;         % number of loops for impulse response estimation
21    CounterSynchro = 50;      % number of loops for synchronization
22    CounterSNR = 50;          % number of loops for SNR calculation and bit allocation
23    TransmitDelay = 100;      % number of samples of the transmission delay
24    NR1 = 20;  NR2 = 80;      % indexes of pilot carriers, may be used in the receiver
25    NR3 = 60;  NR4 = 120;     % indexes of data carriers, observed in the receiver
26    rand('state',0); randn('state',0);  % initialization of random generators
27    SignalPower = 0.001 * 10^( SignalDbm/10 );          % to linear scale
28    NoisePower  = 0.001 * fs/2 * 10^( NoiseDbmHz/10 );  % to linear scale
29  % Reading impulse response of the transmission line
30    if(FilterTel == 0) h = load('DSLine.dat')';    % csaloop2 from [Arslan]
31    else               h = load('DSLineHP.dat')';  % csaloop2 with high-pass filter
32    end
33  % Generation of training sequence for channel impulse response identification
34    QAM4 = [ 1+j, -1+j, 1-j, -1-j ];          % 4-QAM constellation: 00, 01, 10, 11
35    number = floor((4-eps)*rand(1,N/2))+1;    % random sequence of numbers { 1, 2, 3, 4 }
36    X = QAM4( number );                        % generation of the spectrum half
37    X( 1 ) = sqrt(2); X( 257 ) = sqrt(2);      % DC and fs/2 are not used
38    X( 258:512 ) = conj( fliplr( X(2:256) ) ); X = X.';  % conjugate symmetry
39    x = real( sqrt(N/2)* ifft(X) ); st = x/std(x);  % transmitted time signal
40    Swe = X;                                    % its spectrum
41    Swy = zeros(N,1);                           % accum. for received spectra
42
43  % ################################################################################
44  % MAIN LOOP - BEGINNING   ########################################################
45  % ################################################################################
46
47    Niter = CounterADC+CounterChan+CounterSynchro+CounterSNR; % all iterations
48    nr = 2 : N/2;                  % numbers of used frequency carriers with data
49    Nb = N;                        % DSL/OFDM frame length
50    bx  = zeros(3*Nb,1); by = bx;  % input and output buffer for TX and RX signal
51    bye = zeros(2*Nb,1);           % buffer for signal after TEQ equalizer
52    S  = zeros(N,1); S1 = S;       % the last sent frame and one frame before
```

```
53    sig = zeros(N/2-1,1); err = sig; % accumulators for signal and error energy
54    scSUM = zeros(1,N+Lp);            % accumulator for the synchronization function
55    QamHistory = [];                  % for constellation states for the carrier NR-th
56    CopyCounterChan = CounterChan; TxDelay = TransmitDelay;    % set initial values
57    ADCCorrect=1; ChanEstim=0; Synchronization=0; SNREstim=0;  % control parameters
58    ppm = zeros(1,CounterChan); cfo = zeros(1,CounterChan); cnt = 0;
59    n0 = 1;                           % first sample of narrow-band disturbance
60    Nmax = Niter*Np;                  % maximum number of signal samples
61    nnorm = 0:Nmax-1; nln = 1;        % signal sample indexes without ADCppm
62    %Nppm=Nmax+(ADCppm*1e-6)*Nmax; nppm=0 :Nppm/Nmax: (Nmax-1)*Nppm/Nmax; nlp=1; % ADCppm
63    Nppm=Nmax+(ADCppm*1e-6)*Nmax;  nppm=0 :Nmax/Nppm: (Nmax-1)*Nmax/Nppm; nlp=1; % ADCppm
64    dfref = ADCppm*1e-6*fs,
65
66    for iter = 1:Niter  % ----------------------------------------------------------
67      % Generation of data to be transmitted
68        if( ADCCorrect==1 || ChanEstim==1) s = st; end  % ADC & Channel estimation
69        if( Synchronization == 1 )                % Synchronization
70            s = sqrt(3)*(2*rand(N, 1)-1);         % uniform noise in [-1, 1]
71        end
72        if( SNREstim == 1 )                       % Data transmission:
73        %  disp('###'); iter, pause
74            S0 = S1; S1 = S;                      % remember two last sent spectra
75            number = floor((4-eps)*rand(1,N/2-1))+1; % random sequence of: 00, 01, 10, 11
76            S = QAM4( number );                   % generation of the first spectrum
                   half
77            S(NR1-1) = 1+j; S(NR2-1) = -1-j;      % optional states of pilot signals
78            S = [0 S 0 conj(fliplr(S))]; S =S.';  % if conjugate symmetry: ifft(S) is
                   real!
79            s = real( sqrt(N/2)* ifft(S) );       % inverse FFT, scaling
80        end
81        s = s * sqrt( SignalPower ) * sqrt( 10^( CodingGain/10 ) );
82        % Adding cyclic prefix
83        if( ADCCorrect || ChanEstim == 1 ) x = s; % Channel estimation: without the
               prefix
84        else x = [ (s(N-Lp+1:N)); s ];            % Synchronization and TX: add prefix
85        end
86        % Convolution with the original (un-shortened) channel impulse response
87        bx = [ bx(Nb+1 : 3*Nb); x ];                      % put data into the buffer
88        y = conv( bx(Nb+1-TxDelay : 3*Nb- TxDelay ),h );  % convolution
89        y = y(Nb+1 : Nb+Nb);                              % cut the result
90        % ADC sampling error in the receiver
91        y(1:Nb,1)=interp1(nnorm(nln : nln+Nb-1),y(1:Nb,1),nppm(nlp : nlp+Nb-1),'spline');
92        nln = nln + Nb; nlp = nlp + Nb;
93        % Adding noise (dBm/Hz) and narrow-band disturbance
94        noise = sqrt( NoisePower ) * randn(Nb,1);
95        y = y + noise;
96        nband = A0*sin(2*pi*f0/fs*(n0:n0+Nb-1)'); n0 = n0+Nb;
97        y = y + nband;
98        % Receiver: data scaling, taking into the buffer
99        y = y / ( sqrt( SignalPower ) * sqrt( 10^( CodingGain/10 ) ) );
00        by = [ by(Nb+1 : 3*Nb); y ];
01        % TEQ - channel impulse response shortening
02        if( ADCCorrect == 0 && ChanEstim == 0 )                    % after channel
               estim
```

```
103          ye  = conv(by(1+ReceiveDelay : 2*Nb+ ReceiveDelay),e);   % conv with TEQ
104          ye = ye(Nb+1 : Nb+Nb);                                    % cut the result
105          bye = [ bye(Nb+1 : 2*Nb); ye ];                           % store in the
                  buffer
106       end
107
108       % ADC SAMPLING RATE ESTIMATION AND CORRECTION  #########################
109       if( (ADCCorrect == 1 ) && (iter > 4) )                       % check condition
110          CounterADC = CounterADC-1                                 % decrement the counter
111          Y1 = fft( by(1:N) ); Y2 = fft( by(N+1:2*N) );             % spectra of two consecutive
                  ...
112          Y21 = Y2 .* conj(Y1);                                     % symbols, should be the
                  same
113          ang = fftshift( angle( Y21 ) );                           % check phase increase
114          Ksize = 30; kf = (-Ksize : Ksize)';                       % neighborhood around 0Hz
115          angi = ang( (N/2+1)+kf );                                 % angles around 0Hz
116          coef = polyfit(kf, angi, 1); ppm1=coef(1)/(2*pi); cfo1=coef(2); % pause
117          cnt = cnt+1; ppm(cnt)=ppm1; cfo(cnt)=cfo1;                % store ppm and cfo
118        % figure; plot(kf,angi,'b-',kf,coef(2)+coef(1)*kf,'r-'); title('Angle(n)');
                  pause
119        % Y1=sum(by(1:N).*exp(-j*2*pi/N*(k/(1-ppm1)+cfo1)*(0:N-1)))/N; % correction
120          if(CounterADC==0)                                         % end of ADCCorrect
121             Koffs = round(0.25*cnt);                               % number of skipped ppms
122             ppm=sort(ppm); ppm=-mean(ppm(Koffs:end-Koffs+1)),      % ppm estimation
123             cfo=sort(cfo); cfo=mean(cfo(Koffs:end-Koffs+1)),       % cfo estimation
124             pause                                                  % do hardware ppm correction
125             ADCCorrect=0; ChanEstim=1; iterADCCorrect = iter;      % next operation
126          end
127       end
128
129    % ESTIMATION OF THE CHANNEL IMPULSE RESPONSE AND SYSTEM DELAY ############
130       if( (ChanEstim == 1) && (iter > iterADCCorrect + 4 )) % check condition
131          CounterChan = CounterChan-1                               % decrement the counter
132          Swy = Swy + fft( by(1:N) ) /sqrt(N/2);                    % accumulate the spectrum
133          if(CounterChan==0)                                        % end of ChanEstim
134             hest = real( ifft( Swy ./ Swe ) )/CopyCounterChan; hest=hest'; % h=?
135             [ hmax, nrmax ] = max(abs(hest));                      % find maximum
136             K=15; n1st = nrmax-K; ReceiveDelay = n1st-1; %
137             if(n1st<=0) n1st = n1st + N; end                       %
138             hest = [ hest(n1st:end) hest(1:n1st-1)];               % rotate circularly
139             figure; plot(hest); title('hest(n)'); pause            % figure
140             [ e ] = AdslTEQ(hest, Le, Lp, 0);                      % design TEQ filter
141             he = conv(e, hest);                                    % shorten hest(n)
142             figure; plot(he); title('he(n)'); pause                % figure
143             he = he(1:N);                                          % leave N samples
144             HE = fft(he).';                                        % FFT of he(n)
145             figure; plot(fs/N*(0:N-1),20*log10(abs(HE))); title('|H(f)|'); pause
146             Nb = Np; bx=zeros(3*Nb,1); by=zeros(3*Nb,1); bye=zeros(2*Nb,1); % set
147             ChanEstim = 0; Synchronization = 1; iterChanEstim = iter;       % next
148          end % end of CounterChan = = 0
149       end % ChanEstim == 1
150
151    % SYNCHRONIZATION #######################################################
152       if( (Synchronization == 1) && (iter > iterChanEstim + 4) ) % condition
```

```
53          CounterSynchro = CounterSynchro-1                    % decrement counter
54          for n = 0 : N+Lp-1                                   % synchro function
55              q = sum(bye(n+1:n+Lp) .*bye(n+N+1:n+N+Lp));      %
56              r = sum(bye(n+1:n+Lp) .^2) + sum(bye(n+1+N:n+Lp+N) .^2);
57              sc(n+1)=q/r;
58          end
59          scSUM = scSUM + sc;                                  % accum synchro function
60          if (CounterSynchro == 0)                             % end of synchronization
61             figure; plot(scSUM); title('SYNCHRO function');   % optional figure
62             SynchDelay = find(scSUM == max(scSUM));           % find argument of max
63             SynchDelay = SynchDelay + Lp/2, pause             % take prefix into account
64             HEC = HE .* exp(j*2*pi/N * (0:N-1)' * (SynchDelay-Lp));  % correction
65             FEQ = 1./HEC;                                     % FEQ
66             Nb = Np; bx=zeros(3*Nb,1); by=zeros(3*Nb,1); bye=zeros(2*Nb,1); % set
67             Synchronization = 0; SNREstim = 1; iterSynchronization = iter;  % next
68          end % CounterSynchro = = 0
69       end % Synchronization = = 1
70
71   %   SNR ESTIMATION  ##################################################
72       if( (SNREstim == 1) & (iter > iterSynchronization + 4) )     % condition
73          r = bye(1+SynchDelay : N+SynchDelay);     % synchronize, remove prefix, received
74          R = sqrt(2/N) * fft(r);                   % FFT
75          R = R .* FEQ;                             % FEQ correction
76          % figure;
77          % subplot(211); plot(nr,real(R(nr)),'ro',nr,real(S0(nr)),'bx'); grid;
78          % subplot(212); plot(nr,imag(R(nr)),'ro',nr,imag(S0(nr)),'bx'); grid; pause
79          SRerr = S0 - R;                           % reconstruction error
80          sig = sig + S0(nr).*conj(S0(nr));         % accumulate signal energy in sub-
                   bands
81          err = err + SRerr(nr).*conj(SRerr(nr)); % accumulate error energy in sub-bands
82          QamHistory = [ QamHistory; R(NR1) R(NR2) R(NR3) R(NR4) ];% NR-th sub-carriers
83       end % SNREstim == 1
84
85    end % main loop -----------------------------------------------------------------
86   % #################################################################################
87   % MAIN LOOP - END  ################################################################
88   % #################################################################################
89   % Figures: SNR in sub-bands, received values for the NR-th carrier (4-QAM is sent)
90     SNR = (sig + eps) ./ (err + eps); SNRdB = 10*log10( SNR );
91     figure; plot(nr, SNRdB,'b'); title('SNR [dB]'); xlabel('Channel number'); pause
92     figure;
93     subplot(221); plot(real(QamHistory(:,1)),imag(QamHistory(:,1)),'bx'); title('4QAM-1')
              ;
94     subplot(222); plot(real(QamHistory(:,1)),imag(QamHistory(:,2)),'bx'); title('4QAM-2')
              ;
95     subplot(223); plot(real(QamHistory(:,3)),imag(QamHistory(:,3)),'bx'); title('4QAM-3')
              ;
96     subplot(224); plot(real(QamHistory(:,4)),imag(QamHistory(:,4)),'bx'); title('4QAM-4')
              ;
97     pause
98   % Do yourself detection of constellation state (number), i.e. do the de-QAM
99   % ...............................................................................
00   % Figure: ALLOCATION OF BITS for sub-carriers
01     gamma = 10 ^ ( (9.8 + CodingMargin - CodingGain)/10 );
```

```
202    ab = log2( SNR/gamma+1 ); ab = floor( ab );        % log2( snr/gamma +1 )
203    idx = find( ab > 15 ); ab( idx ) = 15;             % no more than 15 bits per channel
204    idx = find( ab < 2 ); ab( idx ) = 0;               % no less than 2 bits per channel
205    figure; plot(nr, ab,'b'); xlabel('channel number'); ylabel('number of bits'); pause
206    BitsPerSymbol = sum( ab ), BitsPerSecond = BitsPerSymbol * fs/Np, pause
```

Exercise 22.9 (Testing ADSL Modem Program). Analyze program 22.7. First, carefully read section in which values of all important modem parameters are initialized. Then, concentrate on the main TX-RX loop. Find the section where different transmitted signals are set according to actual modem initialization level. Notice, when cyclic prefix is added and when is not. Then find where signal is convoluted with the channel impulse response and disturbances are *injected*. Observe that after channel estimation and TEQ design, the TEQ is used during DMT symbol synchronization and, next, during data transmission. Analyze program fragments were: (1) ADC sampling error is estimated (we assume that it is corrected by hardware), (2) channel is estimated and TEQ&FEQ are designed, (3) DMT symbol synchronization is done, (4) data transmission starts for bit allocation. Run the program. Observe all figures—uncomment them. Set ADCppm = 10. Notice severe degradation of carrier constellations and allocation of 0 bits. Check correctness of ppm estimation. Set again ADCppm = 0. Increase noise level NoiseDbmHz=$-140, -130, \ldots$. Observe growing *clouds of points* in received constellation grids. Exchange the line he = conv(e, hest); with he=hest; and see how important is time channel equalization. Return to the original version. Now comment the line %R = R .* FEQ; and see how important is frequency channel equalization. Test some other DSL channels using their impulse responses from book repository (csaloopX.time).

22.10 Summary

Multi-carrier transmission is in attack. Almost all new high-speed Internet access and multi-media delivery standards used many carriers. The New Radio 5G technology also. This chapter is an introductory only. It aims not at detailed technical presentation but at more intuitive explanation of FFT-based multiple carrier transmission. What should be remembered?

1. In multi-carrier transmission technology, a signal consisting of many amplitude-and-phase QAM-modulated sub-carriers is synthesized by the inverse fast Fourier transform. Sent bits decide about the Fourier coefficients, i.e. all carrier states. For example, transmitting any carrier in its state number 7 from 16 states available in 16-QAM modulation means that bits 0111 are sent since this is a binary representation of the decimal number 7.

2. The multi-carrier transmission is block-based. The IFFT output is a summation of long-lasting Fourier harmonics which are all the time in one state. This is a different situation than in the single-carrier transmission, already known to us, where carrier states are smoothly interpolated by the pulse shaping filter and each carrier is permanently *on the run* from one state to the other. We must have very precise timing recovery methods to sample the single carrier exactly when it *passes* through its state.

3. While each multi-carrier transmitter uses IFFT for signal generation, each multi-carrier receiver exploits FFT for signal analysis and demodulation. After synchronization with each block beginning: the FFT is performed, states of all carriers are found and bits extracted from them (as numbers of carrier states). Slight synchronization errors result in rotation of QAM constellations. How many IFFTs and FFTs are executed at this moment all around the world!

4. Channel impulse response (CIR) convolves with transmitted signal across the DMT/OFDM blocks/symbols and the inter-symbol interference (ISI) occurs. In order to mitigate this effect, part of each block, longer than CIR, is transmitted twice: last block samples are copied to block beginning in the form of cyclic prefix. After this operation, despite channel filtering, result of this filtering depends only on the samples of one DMT/OFDM block: the inter-symbol interference is canceled. The cyclic prefix is very important!

5. Cyclic prefix makes a signal ... cyclic. Signal convolution realized in frequency domain by means of DFT/IDFT is also cyclic. Thanks to the CP usage, not only ISI is canceled, but channel frequency equalization (FEQ) becomes very easy also: FFT of a synchronized block of received signal samples is simply divided by FFT of the estimated channel impulse response!

6. Usage of cyclic prefix causes bit-rate reduction because portion of copied samples should be longer than the impulse response which in DSL systems is ... long. Transmitting something twice is not a big fun. Therefore, in DSL systems the channel impulse response is artificially shorten by the special, digital, channel time equalization (TEQ) filter which is put into the signal processing path in the receiver.

7. Only known channels can be equalized. We cannot *un-do* unknown *do*. So, channel impulse response and frequency response have to be estimated. Special pilot signals are sent, known to receiver. Knowing input (sent) and

having output (received) signals, the receiver can estimate signal change introduced by the channel and use it for correction of unknown transmitted data.

8. In multi-carrier transmission orthogonal signals are used. When their orthogonality is lost in the receiver, for example, due to incorrect ADC sampling ratio, all carriers interfere with each other and the inter-channel interference (ICI) appears. At all cost orthogonality has to be recovered, by hardware (clock change) or software solution (signal interpolation). In wireless systems the carrier frequency offset and the Doppler effect cause orthogonality loss and ICI existence.

9. When everything what can be estimated and adjusted is already done, only noise remains. We have to estimate signal-to-noise ratio for each sub-carrier and allocate the number of bits proportionally to the SNR: sub-carrier less disturbed by the noise obtains more bits to ... carry. In DSL channels high frequencies are more attenuated and typically ... carry less bits.

22.11 Private Investigations: Free-Style Bungee Jumps

Exercise 22.10 (SNR and Bit-Rate Curves for Different Noise Levels). Run program of complete ADSL modem with perfect ADC sampling. Increase the noise level NoiseDbmHz=−160, −150,..., −80. Store SNR and bit-allocation curves for all sub-carriers and then plot them all in two separate figures. Calculate allowed bit-rates and plot them in one figure also. Check different channels.

Exercise 22.11 (Noise Level, Channel Estimation, and TEQ/FEQ Design). For stronger noise, it is more difficult to estimate reliably the channel frequency response and correct TEQ and FEQ. Set strong noise NoiseDbmHz=−80, −90. Run the modem program many times, increasing the number of main loop iterations for channel estimation, i.e. CounterChan=1,5,10,20,50. Calculate, store, and plot in one figure all: (1) estimated channel impulse responses hest (Fig. 22.1), (2) all designed TEQ filters e (Fig. 22.2), (3) all shortened channel impulse responses he (Fig. 22.3), (4) all frequency responses HE of the shortened channels (Fig. 22.4).

Exercise 22.12 (Schmidl-Cox Synchronization Method). Implement the Schmidl-Cox method [8] and use it for DMT symbol synchronization in the complete ADSL modem program.

Exercise 22.13 (* Text Transmission by ADSL Modem). At present the ADSL modem program does not transmit bits, only performs the modem initialization. Add bit transmission and detection to it. Send any text of arbitrary length using only 4-QAM modulation. Having the program, calculate number of errors. Check how this number grows with the increase of noise level.

Exercise 22.14 (* 16-QAM Modulation in ADSL Modem). Perform bit allocation for the ADSL modem. If 2 or 3 bits are allocated, use 4QAM, for bigger number of bits use 16QAM. Send any bit-stream, detect the bits, and find number of errors for each sub-carrier. Repeat experiment for different levels of noise and plot all error curves in one plot (i.e. number of errors for each sub-carrier for different noise levels).

References

1. G. Arslan, M. Ding, B. Lu, M. Milosevic, Z. Shen, B.L. Evans, Matlab DMT-TEQ toolbox. Online: http://users.ece.utexas.edu/~bevans/projects/adsl/
2. G. Arslan, B.L. Evans, S. Kiaei, Equalization for discrete multitone transceivers to maximize bit rate. IEEE Trans. Signal Process. **49**(12), 3123–3135 (2001)
3. J.A.C. Bingham, *ADSL, VDSL, and Multicarrier Modulation* (Wiley, New York, 1999)
4. ITU-T, Transmissions systems and media, digital systems and networks - Asymmetric digital subscriber line (ADSL2) transceivers, Recommendation G.992.3, April 2009. Online: https://www.itu.int/rec/T-REC-G.992.3-200904-I
5. ITU-T, Transmissions systems and media, digital systems and networks - Asymmetric digital subscriber line (ADSL) transceivers, Recommendation G.992.1, June 1999 (available in the book repository)
6. J.G. Proakis, M. Salehi, G. Bauch, *Modern Communication Systems Using Matlab* (Cengage Learning, Stamford, 2004, 2013)
7. D. Rauschmayer, *Adsl/Vdsl Principles: A Practical and Precise Study of Asymmetric Digital Subscriber Lines and Very High Speed Digital Subscriber Lines* (Macmillan Technical Publishing, New York, 1998)
8. T.M. Schmidl, D.C. Cox, Robust frequency and timing synchronization for OFDM. IEEE Trans. Commun. **45**(12), 1613–1621 (1997)
9. M. Sliskovic, Carrier and sampling frequency offset estimation and correction in multicarrier systems, in *Proc. IEEE Global Communication Conf. GLOBE-COM'01*, San Antonio, 25–29 Nov. 2001
10. T. Starr, J.M. Cioffi, P.J. Silverman, *Understanding Digital Subscriber Line Technology* (Prentice Hall, Upper Saddle River, 1999)
11. T. Starr, M. Sorbara, J.M. Cioffi, P.J. Silverman, *DSL Advances* (Prentice Hall, Upper Saddle River, 2002)

12. A. Tkacenko, P.P. Vaidyanathan, A low-complexity eigenfilter design method for channel shortening equalizers for DMT systems. IEEE Trans. Commun. **51**(7), 1069–1072 (2003)
13. S.A. Tretter, *Communication System Design Using DSP Algorithms* (Springer Science+Business Media, New York, 2008)
14. T.P. Zieliński, *Cyfrowe Przetwarzanie Sygnałów. Od Teorii do Zastosowań (Digital Signal Processing. From Theory to Applications)* (Wydawnictwa Komunikacji i Łączności (Transport and Communication Publishers), Warszawa, Poland, 2005, 2007, 2009, 2014)

Chapter 23
Wireless Digital Multi-Carrier Transmission: With DAB Example

The best way to understand how something work is to
disassemble it into pieces and ... assemble it ... hmm ... back

23.1 Introduction

Digital radio broadcasting (DAB/DAB+) standard is result of European project Eureka started in 1986. Its first version was finished in 1995 and it was using MP2 layer of MPEG-1 audio standard from 1992 (only with frequency 48 kHz). Not all first DAB installations in Europe were successful due to high building cost of new infrastructure and listeners non-interest due to similar audio quality to the analog FM radio. In response, new technological solutions were added to the standard and in 2006 an enhanced DAB+ version appears. It allowed using the newest extension of the modern high-efficiency advanced audio coder (AAC-HE2) and packet transmission with multimedia services, e.g. *Slide Show*. The DAB+ is described in the ETSI EN 300 401 standard being permanently upgraded. Despite these efforts acceptance of DAB technology is not high. It results from decrease of general radio popularity among young people and rapid increase of usage of other multimedia services, especially Internet multimedia streaming, i.e. Internet radio and television, audio and TV podcasts, music and video servers. Nevertheless, in many countries digital radio broadcast is available and we can use its signal for practical learning different DSP algorithms which are widely allied in telecommunication systems, not only the wireless ones. In this book we will decode exemplary recorded IQ files of DAB multiplex signal.

The DAB multi-carrier transmission and signal processing is similar to the DSL one, described in the previous chapter. The biggest difference is that the synthesized $IQ(n)$ signal is in the transmitter up-converted in frequency and emitted in a wireless RF way, and in the receiver—down-converted to the baseband. This up-down operation is a potential source of carrier frequency offset

© Springer Nature Switzerland AG 2021
T. P. Zieliński, *Starting Digital Signal Processing in Telecommunication Engineering*, Textbooks in Telecommunication Engineering,
https://doi.org/10.1007/978-3-030-49256-4_23

(CFO) in the receiver, which is absent in the DSL modem. The CFO has to be obligatory estimated and corrected. The second important difference is that channel in DAB can be time-varying due to allowed receiver mobility (for example, when DAB is listened in a car moving with a high speed or in urban environment). Despite these new problems, the reception of DAB transmission is relatively robust, due to small bit-rate with high error correction and simple but effective solutions implemented in it:

1. the DAB bit-stream is not high, equal to about 1.7 megabits per second, and secured by strong 3/4 forward error correction (FEC) coding,
2. very robust, low-level differential phase shift keying modulation is used in it (DQPSK) with only two bits per sub-carrier,
3. channel estimation is not necessary, even in the presence of Doppler effect, due to specially chosen values of transmission parameters (frequency band around 200 MHz, sampling frequency 2.048 MHz, OFDM symbol having 2048 samples equipped with additional long prefix with 504 samples, differential modulation).

The applied differential modulation helps us to reduce the channel influence upon the transmitted data, even in the case of relatively fast channel variations. In turn, long cyclic prefix, longer than the channel impulse response, eliminates inter-symbol interference (ISI) and helps not to lose sub-carrier orthogonality.

Error detection and correction is a very important part of each telecommunication systems. In this chapter this topic will be covered on the example of the DAB standard. The cyclic redundancy check coding (CRC) as well as convolution coding with Viterbi decoding will be shortly described. Details of the Reed–Solomon encoder and decoder application in the DAB radio will be presented also.

We will learn DAB+ digital radio using text of the ETSI EN 300 401 standard entitled "Radio Broadcasting Systems: Digital Audio Broadcasting (DAB) to mobile, portable and fixed receivers" (2006). It will be a big lesson for us to see how telecommunication standards look like: how many details they contain and how difficult to read they are.

This chapter will have a different structure from the rest of the book. Since the Matlab program implementing DAB decoder is not short, it is cut into pieces and presented part-by-part with minimum required explanation. The *bottom-up* approach will be used: we will analyze the data and methods in order of their appearing, and then do the generalization in form of block diagrams.

In this chapter we will use a Matlab program of DAB radio receiver written in 2011 by Michael Häner-Höin as his M.Sc. Thesis "SW-Realisierung eines DAB-Empfängers mit GNU Radio" [5] at School of Engineering (the Center for Signal

Processing and Communications Engineering) of the ZHAW Zurich University of Applied Science https://www.zhaw.ch/en/university/, under supervision of Prof. Dr Marcel Rupf. The program was made available at GitHub repository and used by my students in 2014-16 in real-time SDR DAB radio project https://sdr.kt.agh.edu.pl/ sdrdab-decoder, under permission of ZHAW. The Matlab DAB code was extended by myself from mode I to modes II, III, and IV, as well as from DAB to working DAB+ decoder. However, the most important was addition of missing time and frequency synchronization procedures, i.e. carrier and timing recovery methods. I would like to thank a lot the ZHAW for possibility of using its programs in the past and starting my DAB students projects from so high level as well as for possibility of using the Matlab DAB code, with my extensions, in this book.

DAB standard is very well described in [4, 6]. Its norms are available online [2, 3]. OFDM and digital wireless transmission are simulated in Matlab in [19].

23.2 Reading Recorded DAB Files

IQ signals of the DAB radio, processed by us in this chapter, were recorded by RTLSDR USB stick with the use of SDRSharp program. Next, recorded files, long two-column IQ matrices are read into Matlab, combined into one complex-value $I(n) + jQ(n)$ vector and decoded. Since the DAB sampling frequency is equal to 2.048 MHz we should record the IQ file exactly with this frequency. It is available in the USB stick. Otherwise, after reading into Matlab the signal has to be re-sampled to the DAB sampling frequency. In Fig. 23.1 a 150 ms fragment of $I(n)$ and $Q(n)$ signals is shown. We clearly see one DAB frame, about 100 ms long, separated from the rest by two NULL zones.

In the first part of this chapter we will use the program `lab23_ex_dab_fic.m`. It is a simpler version of the program `lab23_ex_dab_all.m` which will be used by us later. It allows only decoding information about available services from one DAB frame, it is not decoding an audio broadcast. The program will be cut into pieces and presented *slice-by-slice* as an illustration to the DAB presentation performed in the text. First part of the program is given in the Listing 23.1. In the beginning values of control program parameters are set, DAB parameters are initialized, and required tables are generated. At present meaning of the parameters will not be described, it will be understand by us later. Then one of two available DAB recordings can be chosen: a clear DAB IQ signal, prepared for a transmitter but not transmitted, and recorded real-world DAB signal with severe disturbances. When sterile DAB signal is used, we can artificially add to it by a function `disturbDAB()` some typical anomalies: wrong ADC sampling ratio, carrier frequency offset, channel simulation, and noise.

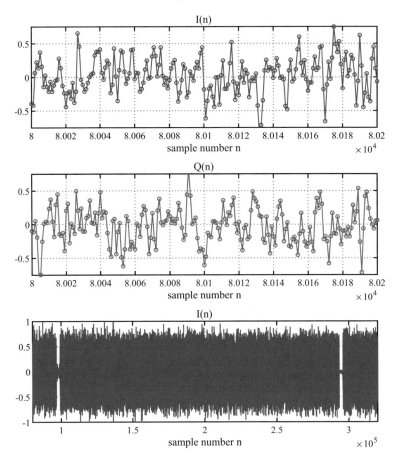

Fig. 23.1: In-phase $I(n)$ and in-quadrature $Q(n)$ components of DAB+ digital radio: (top) short fragment of $I(n)$, (center) short fragments of $Q(n)$, (bottom) long fragment of $I(n)$—one DAB frame, lasting about 100 ms, is visible in the middle, separated on both sides by two NULL zones

Listing 23.1: Reading IQ file of DAB+ radio signal

```
1   % lab23_ex_DAB_fic.m
2   % Simple Matlab program for decoding service information from one DAB frame
3
4   % Masterthesis
5   % Zurcher Hochschule fur Angewandte Wissenschaften
6   % Zentrum fur Signalverarbeitung und Nachrichtentechnik
7   % (c) Michael Haner-Hoin, 12.4.2011 ZSN, info.zsn@zhaw.ch
8   % Extended:
9   % AGH University of Science and Technology, Krakow, Poland
10  % (c) Tomasz Zielinski, 31.12.2019, tzielin@agh.edu.pl
11
12    clear all; close all; fclose('all');
13
```

```
14 | % --------------------------
15 | % PROGRAM CONTROL PARAMETERS
16 | % --------------------------
17 |   CorrectADC_ON         = 1; % 0/1 CORRECT ADC sampling frequency by re-sampling
18 |   CorrectFreqOffset_ON  = 1; % 0/1 CORRECT carrier frequency offset (CFO)
19 |   CorrectTimeSynchro_ON = 1; % 0/1 CORRECT time synchro using known DF (delta freq)
20 |   Disturbances_ON       = 0; % 0/1 ADD disturbances
21 |
22 | % --------------------------
23 | % GENERAL DAB PARAMETERS
24 | % --------------------------
25 |   fs = 2.048e6;              % sampling frequency used in the DAB algorithm
26 |   NSymbPerFrame = 76;        % L, number of OFDM symbols per one DAB frame
27 |   NCarrPerSymb = 1536;       % K, number of transmitted carriers
28 |   NSampPerFrame = 196608;    % Tf, duration of the transmission frame (in samples)
29 |   NSampPerNull = 2656;       % TNULL, duration of the Null symbol (in samples)
30 |   NSampPerSymb = 2552;       % Ts, duration of OFDM symbols
31 |   Nfft         = 2048;       % Tu, the inverse of the carrier spacing
32 |   NSampPerPrefix = 504;      % delta, duration of the guard interval
33 |
34 |   NFIBsPerFrame = 12;        % (FIC) number of FIBs per one DAB frame (non-interleaved)
35 |   NCIFsPerFrame =  4;        % (MSC) number of CIFs per one DAB frame (time-interleaved)
36 |   NSymbPerFIC   =  3;        % (FIC) number of OFDM frames per one FIC
37 |   NFIBsPerCIF   =  3;        % (FIC) number of FIBs per one CIF
38 |
39 |   NSampPerData = NSymbPerFrame*NSampPerSymb; % DAB frame length without NULL
40 |
41 | % Generation of Phase Reference Symbol (used for frame synchro) (ETSI pp. 147-149)
42 |   [specPhaseRef, sigPhaseRef] = PhaseRefGen( 0 );
43 | % Generation of a Frequency Interleaving and De-interleaving Tables (ETSI pp. 157-161)
44 |   [FreqInterleaverTab]   = FreqInterleaverTabGen( Nfft );
45 |   [FreqDeinterleaverTab] = FreqDeinterleaverTabGen( FreqInterleaverTab, Nfft );
46 | % Generation of trellis for polynomial (ETSI, page 129-130, below fig. 72)
47 |   VitTrellis = poly2trellis(7, [133 171 145 133]); % 7 = delay, then octal numbers
48 |
49 | % Read an IQ file for FIC decoding
50 |   Nframes = 3;                               % number of read DAB frames
51 |   if(1)                                      % synthesized, disturbance-free
52 |     ReadFile = fopen('DAB_PolishRadio_TX_IQ.dat', 'rb');
53 |     x = fread( ReadFile, [2, Nframes * NSampPerFrame], 'float' );
54 |     x = x(1,:) + j*x(2,:); x = x.' / 2^15;
55 |     fclose( ReadFile );
56 |   else                                       % recorded, with strong disturbances
57 |     ReadFile = 'DAB_PolishRadio_RX_IQ.wav';       % file name
58 |     n1st = 1; nlast = n1st + Nframes * NSampPerFrame-1; % samples from-to
59 |     [x, fs] = audioread( ReadFile, [n1st, nlast] ); % read
60 |     x = x(:,1) - j*x(:,2);                     % complex IQ
61 |   end
62 |   figure; plot(real(x)); xlabel('n'); title('Real(x(n))'); grid; pause
63 |
64 | % ####################################################################################
65 | % Optional addition of disturbances
66 | if( Disturbances_ON ) x = disturbDAB( x ); end
```

Exercise 23.1 (Reading DAB+ IQ Files, Observing Signal Samples). Add figures to the program 23.1 displaying IQ samples of the DAB signal. Find zones where signal samples have low energy. How long (in samples) are these zones and how many samples are between them. Notice difference between clear IQ data in transmitter (TX) and receiver (RX), i.e. presence of noise. Observe also that low-energy zones are of two types in the IQ RX file: *silent* ones, where only noise is present, and *oscillatory* ones—with deterministic tones. Calculate 2048-point FFT and find frequencies of these tones. Calculate and display FFT spectrum of the signal in high-energy intervals.

23.3 DAB Physical Layer: Samples and Carriers

Frequency Domain Initially DAB had 4 different application modes, at present only mode I remained, working in the band III (175–240 MHz). Each DAB multiplex signal is sampled with frequency 2,048 MHz, has the 2.048 MHz bandwidth, and consists of 2048 sub-carriers distant 1 kHz apart (2.048 MHz/2048). Complex-value time signal $I(n) + jQ(n)$ of single DAB multiplex, used in the base-band, is obtained by means of the $N = 2048$ (1536+512)-point inverse FFT from the pre-set $IQ(k)$ sub-carrier states. Differential Quadrature Phase Shift Keying (DQPSK) modulation is exploited. Only central 1536 sub-carriers are used, symmetrical around 0 kHz: 768 left (negative frequencies) and 768 right (positive frequencies). Remaining 512 side sub-carriers, 256 for negative frequencies and 256 for positive frequencies, are set to zero. They are used as separating zones between neighbor DAB multiplexes. As a result, in practice, only 1.536 MHz bandwidth is exploited. Of course, bits are allocated to used sub-carriers only. In Fig. 23.2 spectrum of a multiplex DAB signal is shown. In fact, it is typical to all wireless OFDM services.

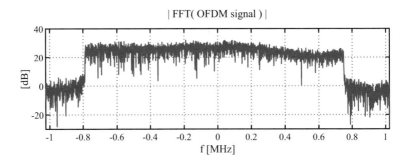

Fig. 23.2: Frequency spectrum of recorded DAB+ signal: the OFDM *hat*, consisting of 1536 sub-carriers lying 1 kHz apart, is clearly visible. Sampling frequency is equal to $f_s = 2.048$ MHz and NFFT length—2048 samples

For example, for terrestrial digital video broadcasting (DVB-T2) the OFDM *hat* has different widths and spacing, depending on the sampling frequency (1.7, 5, 6, 7, 8, 10 MHz) and FFT size used (1k, 2k, 4k, 8k, 16k, 32k, e.g. 1k = 1024). In this case the following modulation types are available: QPSK, 16-QAM, 64-QAM, and 256-QAM. In LTE 4G digital telephony, for down-link signal, sampling frequency varies from 1.30 MHz to 30.92 MHz, FFT size from 128 to 2048, and modulation from QPSK, 16-QAM, 64-QAM, and 256-QAM (recently).

Exercise 23.2 (FFT Spectrum of the DAB+ Signal). Calculate and compare 2048-point FFT spectra of the TX and RX DAB IQ files. Cut signal fragments in the middle of high-energy signal parts. Display the spectra in two ways: (1) in separate two figures and (2) in one figure with overlay. Observe noise components and frequency shift of the RX spectrum.

Time Domain In Fig. 23.1 a time-domain DAB signal was presented, precisely its real $I(n)$ and imaginary $Q(n)$ components. At present, we can describe it with more details. The DAB frame starts with a pre-amble having a form of zeroth (no-data) symbol, the so-called NULL symbol. The NULL symbol has 2656 samples (1297 μs). From time to time, typically every second DAB frame, a weak tonal signal is transmitted inside the NULL zone, allowing transmitter identification (TII—transmitter identification information). After the NULL, there are 76 time OFDM symbols, each having 2552 samples (1246 μs): this is an IFFT result (2048 samples) of actual sub-carrier states, with 504 last samples copied to the beginning as a cyclic prefix (CP) (guard interval (GI)). The CP protects us against inter-symbol interference, as explained in the previous chapter, and simplifies carrier frequency offset estimation. The overall DAB frame has $2656 + 76 \cdot 2552 = 196608$ samples and lasts about 100 μs. Because the differential QPSK modulation is exploited, only phase shifts of each sub-carrier from OFDM symbol-to-symbol are important. Due to the DQPSK usage, the first transmitted OFDM symbol stands for starting Phase Reference (this is its name!) for the second symbol. There are four phase shifts allowed per sub-carrier ($\pm\frac{1}{4}\pi$ and $\pm\frac{3}{4}\pi$) and each sub-carrier *carries* two bits only ($4 = 2^2$). The PR symbol, with precisely defined time and frequency pattern, is used in DAB receiver for time and frequency synchronization also. The Phase Reference symbol, the first OFDM symbol of the DAB frame, is followed by 75 regular OFDM symbols carrying bits: first 3 belong to the fast information channel (FIC), describing the multiplex content, and next 72 symbols to the main service channel (MSC) with multimedia data. Logical structure of the DAB frame is shown in Fig. 23.3.

Thanks to differential coding, the channel is not to be obligatory estimated and corrected in DAB. It is possible due to carefully chosen sampling frequency value as well as appropriate FFT and CP lengths, taking into account the expected channel impulse response length (CP) and a receiver mobility—Doppler effect (FFT).

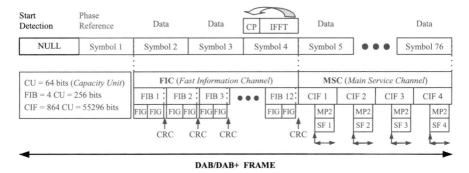

Fig. 23.3: Structure of the DAB/DAB+ frame. Denotations: FIB—*Fast Information Block*, FIG—*Fast Information Group*, CIF—*Common Interleaved Frame*, MP2—layer 2 of MPEG-2 audio coder, SF—*Super Frame* of DAB+ with AAC audio coding. Each FIB block is followed by 16-bit CRC protection

Exercise 23.3 (Sequence of DAB+ Signal FFT Spectra). Calculate a sequence of 2048-point FFT spectra of the TX DAB+ signal: (1) find the first sample after the NULL zone, skip first 504 samples and calculate FFT of the next 2048 samples, (2) repeat this operation (504 samples skipping and FFT computing) 76 times. Plot all spectra (abs(), real(), imag() in a loop, one after the other with delay of 0.5 s (pause(0.5)). How do you conclude this experiment? Notice that the first spectrum after the NULL zone is all the time the same—this is a spectrum of Phase Reference signal.

23.4 Synchronization

Synchronization techniques applied in DAB for synchronization purposes are described in many papers, for example, in [1, 4, 8, 12, 13, 15].

Initial DAB Frame and OFDM Symbol Synchronization The NULL symbol is used in DAB receiver for coarse finding of 100 μs DAB frame beginning. We are looking for a minimum of summation of signal samples absolute values inside a time window 2656 samples long:

$$\sum_{n=0}^{N_{NULL}-1} |x(n_0 - n)| \xrightarrow[n_0=?]{} \min, \qquad (23.1)$$

i.e. tracking envelope of this cost function is performed. In Fig. 23.4 two types of NULL symbols are presented, with and without TII transmitter identification, their

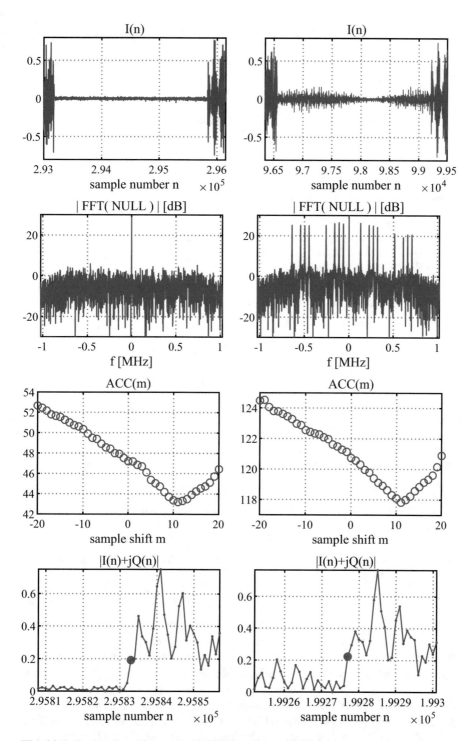

Fig. 23.4: Processing of recorded NULL signals, (left) without transmitter identification information (TII) and (right) with TII. First row: $I(n)$ signal samples, second row: signal spectra, third row: detection curves $Acc(m)$, fourth row: beginning of Phase Reference signals with marked the first PR sample (with circle)

FFT spectra and calculated DAB frame beginning detection curves. In both cases
it is not difficult to find the curve minimum. Detection curves and first samples of
DAB frames were calculated by the program 23.2.

Listing 23.2: Finding NULL symbols in DAB IQ signal

```
1   % lab23_ex_DAB_fic.m - continuation, x denotes a signal being decoded
2
3   % Detection of first two NULL symbols: sum(abs(samples(...)))
4     xabs = abs(x);
5     NSampPerSynchro = NSampPerFrame + NSampPerNull;
6
7   % First DAB frame
8     acc = zeros(1,NSampPerSynchro); acc(1) = sum( xabs(1:NSampPerNull) );
9     for n = 1 : NSampPerSynchro;
10        acc(n+1) = acc(n) - xabs(n) + xabs(n+NSampPerNull);
11    end
12    [~, imin] = min(acc);
13    i1st = imin, % first sample of DAB #1 frame w/ NULL
14
15  % Second DAB frame
16    ADCppmMAX = 400; % assumed maximum ADC ppm error
17    M = round( 0.5 * ADCppmMAX * NSampPerFrame/fs );   % maximum possible offset
18    i2nd = imin + NSampPerFrame - M;  % first sample index to test
19    acc = zeros(1,2*M); acc(1) = sum( xabs( i2nd : i2nd + NSampPerNull-1 ) );
20    for n = 1 : 2*M-1
21        acc(n+1) = acc(n) - xabs(i2nd+n) + xabs(i2nd+n+NSampPerNull);
22    end
23    [dummy imin] = min(acc); clear xabs;
24    i2nd = i2nd + imin,  % first sample of DAB #2 frame /w NULL
```

**Exercise 23.4 (Using NULL Symbol for Finding DAB Frame Begin-
ning).** Try to replace absolute value detector (xabs(n)) with energy one
(x(n)*conj(x(n))). Add different impairments to the clear TX DAB signal using function x=disturbDAB(x), and observe the detection function
shape.

ADC Sampling Ratio Estimation and Correction Yes. I was personally programming the Matlab software decoder. And everything worked perfectly well for clear
IQ multiplex files, prepared for RF emission. But attempts of running programs
upon real SDR recordings have failed immediately. Why? Because of the carrier frequency offset. And it was corrected. For a while everything looked fine ... until a next
crash occurred. This time problems were caused by a wrong ADC sampling ratio.
Instead of 196608 samples per DAB frame, I had 7 to 15 samples more: error from
70 to 150 ppm. And it was enough to stop the game. The ADC sampling rate can be
corrected in hardware and we did it in our C++ real-time DAB+ student project
(google sdrdab or go directly to https://sdr.kt.agh.edu.pl/sdrdab-decoder/). Or in

software, performing time-consuming interpolation of a very long $I(n) + jQ(n)$ signal. In Matlab program the spline interpolation is done.

Listing 23.3: Software correction of wrong ADC sampling ratio by signal interpolation

```
1  % lab23_ex_DAB_fic.m - continuation
2
3  % ADC correction of the first DAB frame starting from NULL symbol
4  if( CorrectADC_ON )
5      NSampPerFrameADC = i2nd - i1st, % found frame length
6      if( NSampPerFrameADC ~= NSampPerFrame )  % comparison with expected length
7          step = NSampPerFrameADC / NSampPerFrame;
8              x(i1st : i1st+NSampPerFrameADC+NSampPerNull-1), ...
9              [0:step:NSampPerFrameADC+NSampPerNull-1]','spline'); % signal resampling
10     end
11 else
12     x = x( i1st : i2nd -1 );
13 end
```

Exercise 23.5 (Checking Influence of the Wrong ADC Sampling Rate). Use the function x=disturbDAB(x) for modification of TX IQ signal, simulating wrong ADC sampling rate (expressed in ppm). Compare 2048-point FFT spectrum of the Phase Reference symbol (first 2552 samples of the DAB frame, skip first 504 of them) before and after this modification.

Phase Reference Symbol Definition As we remember, carrier and timing recoveries require special synchronization patterns (preambles, headers). The first OFDM symbol, transmitted in DAB frame, is a header-type one. It is called a Phase Reference (PR) and is precisely defined (see Fig. 23.5). Its name results from its main application in DAB radio: since differential phase modulation is used (namely, the DQPSK), the PR symbol is used as a phase reference for the first data OFDM symbol transmitted after it. The Phase Reference signal has sharp auto-correlation function, which can be exploited for time synchronization, as well as impulsive auto-correlation of its FFT—useful for integer frequency shift (offset) estimation.

Exercise 23.6 (Phase Reference Symbol). Compare in one plot samples of PR signals which are present in two DAB+ IQ files, TX and RX. Compare them also to reference IQ samples, generated by the function PhaseRefGen(). Do the same for 2048-point FFT spectra. Correlate samples of both IQ signals, TX and RX, with PR samples returned by the function PhaseRefGen(). Use function conv(). Is it possible to find the DAB frame beginning using the correlation results?

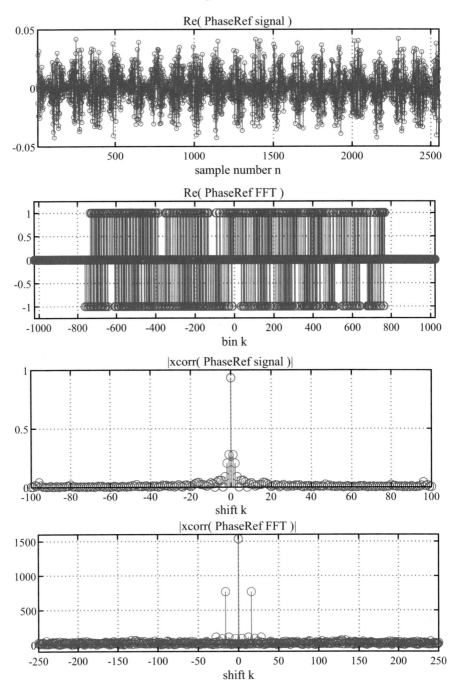

Fig. 23.5: Phase reference OFDM symbol, up-down: (1) real part of 2552 (504+2048) signal samples, (2) real part of FFT coefficients, (3) magnitude of signal auto-correlation, (4) magnitude of FFT spectrum auto-correlation

Carrier Frequency Offset Estimation and Correction The Phase Reference symbol, the first, true OFDM symbol in each DAB frame, having strictly defined structure, is for us priceless. Apart from being the starting phase reference for all used sub-carriers, switched in phase, it can be also used for additional time OFDM symbol synchronization as well as for fractional carrier frequency offset estimation (CFO), i.e. value of carrier de-tuning from nominal value measured in a fraction of the FFT lag (e.g. $0.12345 \cdot f_0, f_0 = \frac{f_s}{N_{FFT}}$).

After simplifying assumptions that: (1) the channel impulse response is constant and short in comparison to the CP length, (2) noise is absent, and (3) all operations performed in the receiver are ideal inverses of operations done in the transmitter, then the last samples of the received signal $x(t)$ and the cyclic prefix should be almost the same. Because the same were transmitted in signal $s(t)$, this is the CP concept. However, when the frequency f_{down} of signal down-conversion to the baseband, done in the receiver, is different than the frequency f_{up} of signal up-conversion from the base-band to intermediate or target frequency, done in the transmitter, we have a problem. In such case, the received signal is equal to:

$$x(n) = s(n)e^{j2\pi \frac{f_{up}-f_{down}}{f_s} n} = s(n)e^{j2\pi \frac{\Delta f}{f_s} n}. \tag{23.2}$$

Let us assume that we multiply last samples of an OFDM symbol by corresponding samples of the CP, but with complex conjugation. In ideal case, the same samples are multiplied, only delayed by the OFDM symbol length N equal to the FFT length. If f_s denotes sampling frequency, $f_0 = f_s/N$ is a fundamental FFT frequency and a frequency shift error is equal to $\Delta f = \Delta k \cdot f_0$, where $\Delta k \in \mathbb{R}$, we should obtain

$$z(n) = x(n) \cdot x^*(n-N) = |s(n)|^2 \cdot e^{j2\pi \frac{\Delta f}{f_s} n} e^{-j2\pi \frac{\Delta f}{f_s}(n-N)} = |s(n)|^2 \cdot e^{j2\pi \frac{\Delta f}{f_s} N} =$$
$$= |s(n)|^2 e^{j2\pi \frac{\Delta f}{f_0}} = |s(n)|^2 \cdot e^{j2\pi \frac{\Delta k f_0}{f_0}} = |s(n)|^2 \cdot e^{j2\pi(\Delta k)}. \tag{23.3}$$

Δk, being a real-value number, denotes the overall carrier frequency offset errors expressed in multiplicity of fundamental DFT frequency f_0. For example, $\Delta k = 3.456$ tells us that carrier frequency is shifted up $3.456 \cdot f_0$ Hz, i.e. 3 DFT bins (integer shift) plus 0.456 of the DFT bin (fractional shift). In general, when Δk has integer part k_Δ and fractional part ε_Δ:

$$\Delta k = k_\Delta + \varepsilon_\Delta, \tag{23.4}$$

the integer and fractional CFO errors in hertz are equal to:

$$\Delta f_{int} = k_\Delta \cdot f_0, \quad k_\Delta \in \mathbb{Z}, \tag{23.5}$$
$$\Delta f_{fract} = \varepsilon_\Delta \cdot f_0, \quad 0 < \Delta \varepsilon_\Delta < 1. \tag{23.6}$$

We can calculate ε_Δ (23.4) from Eq. (23.3):

$$\frac{\sphericalangle z(t)}{2\pi} = \frac{2\pi\varepsilon_\Delta}{2\pi} = \varepsilon_\Delta \tag{23.7}$$

and then use it for calculation of Δf_{fract} from Eq. (23.6). In Eq. (23.7), during calculation of $\sphericalangle z(t)$, we cannot distinguish integer multiplicities of 2π caused by Δk, and only fractional CFO drift ε_Δ can be found.

The described above auto-correlation-based CFO estimation method was proposed in [10] for repeated OFDM symbol. In [14] the method was extended for cyclic prefix application. It was proofed in both works that the maximum likelihood version of Eqs. (23.3), (23.7), minimizing noise influence upon the estimation error of the angle in Eq. (23.3), has the following form (N—FFT length, P—CP length, $L = P + N$, $K < P$):

$$\varepsilon_\Delta = \frac{\sphericalangle \sum\limits_{n=L-K}^{L} x(n)x^*(n-N)}{2\pi}. \tag{23.8}$$

Knowing fractional CFO we can remove it, multiplying the received IQ signal (23.2) with the following complex correction signal (notice the minus sign in its exponent):

$$x^{(1)}(n) = x(n) \cdot e^{-j2\pi\frac{\varepsilon_\Delta}{f_s}n} \approx \left(s(n) \cdot e^{j2\pi\frac{k_\Delta+\varepsilon_\Delta}{f_s}n}\right) \cdot e^{-j2\pi\frac{\varepsilon_\Delta}{f_s}n} \approx s(n) \cdot e^{j2\pi\frac{k_\Delta}{f_s}n}. \tag{23.9}$$

The correction eliminates very severe blur (smear) of the received OFDM symbols caused by fractional CFO, but the integer CFO, giving the circular spectrum shift, can still be present. Typically, it is estimated now by: (1) FFT calculation of the received PR symbol, (2) circular correlating this spectrum with the theoretical (expected) PR spectrum for several left/right shifts, and (3) looking for the maximum and its argument. Alternatively, fast cross-correlation algorithm is used using FFT: (1) instead of cyclic spectra correlation, signal $y^{(1)}(n)$ (23.9) of the received, partially corrected PR is multiplied with the complex conjugation of its reference template, (2) then FFT of the multiplication result is performed, and (3) its maximum is found: the maximum argument is equal to k_Δ and informs us about the integer spectrum shift. The required correction Δf_{int} (23.5) is realized upon the signal $y^{(1)}(n)$ obtained from Eq. (23.9):

$$x^{(2)}(n) = x^{(1)}(n) \cdot e^{-j2\pi\frac{k_\Delta}{f_s}n} \approx \approx s(n) \cdot e^{j2\pi\frac{k_\Delta}{f_s}n} \cdot e^{-j2\pi\frac{k_\Delta}{f_s}n} \approx s(n). \tag{23.10}$$

Results of carrier frequency offset estimation, obtained for the RX DAB signal, are presented in Fig. 23.6. The calculated instantaneous fractional CFO error, presented in left figure, is noisy, but it is averaged during computation of Eq. (23.8). The integer CFO error detection, presented in right figure, is not difficult: the signal spectrum is shifted 20 FFT bins in direction of negative frequencies. Here three peaks dominate. In presence of impairments not always the central one is the highest, nevertheless, its position should be selected as a correct result.

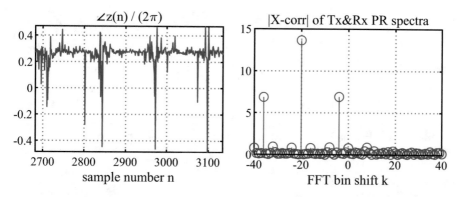

Fig. 23.6: Carrier frequency offset (CFO) estimation results: (left) instantaneous fractional CFO estimated from cyclic prefix of PR symbol—see Eq. (23.7), (right) integer CFO found from cross-correlation of theoretical and received PR symbol

Listing 23.4: Carrier frequency offset estimation

```
1   % lab23_ex_DAB_fic.m - continuation
2
3   % ESTIMATION of CFO fractional part: df_fract
4     Ncut = 25;                % Assumed max time offset error +/-
5     icp  = (i1st+Ncut     : i1st+NSampPerPrefix+Ncut-1)';      % CP indexes
6     isymb = (i1st+Ncut+Nfft : i1st+NSampPerPrefix+Ncut-1+Nfft)';  % SYMBOL indexes
7     z = sum( conj(x( icp )) .* x( isymb ) );  % correction using cyclic prefix
8     df_fract = angle(z)/(2*pi),
9   % df_fract COMPENSATION of the PHASE REFERENCE symbol
10    icpsymb  = (i1st : i1st + NSampPerPrefix+Nfft-1)';         % CP+SYMBOL indexes
11    s = x(icpsymb)  .* exp(-j*2*pi/Nfft*df_fract* (icpsymb));  % CFO correction
12
13  % ESTIMATION of CFO integer part: df_int
14  % Multiplication in time domain = cyclic convolution (corr) in frequency domain
15  % Cyclic correlation of signal spectrum with the PhaseRef spectrum --> find maximum
16    M = 20;   % M = 10? 20? max search region = [-M, +M] around 0 spectrum shift
17    isymb = (NSampPerPrefix+1 : NSampPerPrefix+Nfft)';
18    XC = fft( s( isymb ) .* conj( sigPhaseRef( isymb ) ) );  % fft( multiplication in
           time )
19    XC = [ XC(end-M+1:end); XC(1:1+M) ];
20    [ val, indx ] = max( XC );
21    df_int = indx - (M+1), % pause
22
23  % SUMMATION
24    df = df_int + df_fract;
```

Exercise 23.7 (Carrier Frequency Offset Estimation). Calculating the fractional carrier frequency offset (CFO) we do not explicitly use the information about the PR symbol values. The method can be used for any OFDM symbol. Therefore, apply it for all 75 OFDM DAB symbols with data, for TX and RX signals. Compare individual values and average them for one DAB frame. You can disturb the given TX DAB signal using the `disturbDAB()` function, in respect to CFO, and try to estimate the fractional CFO error from the signal. Is it possible to estimate the integer CFO using any OFDM symbol with unknown data?

Improvement of OFDM Symbol Synchronization After the ADC sampling ratio correction and carrier frequency offset correction, the Phase Reference time signal can be used additionally for enhancement of the DAB frame beginning estimate, initially calculated using the NULL symbol only. We can calculate cross-correlation between the known, sent PR signal, and its received copy. The operation can be performed in time domain for limited number of signal shifts, or in frequency domain using *fast convolution/correlation* FFT-based procedure. In the second case, sequence of the following operations is performed:

1. cut the $IQ(n)$ signal fragment, considered at present as the Phase Reference,
2. calculate its FFT,
3. multiply the spectrum with conjugation of the expected PR spectrum,
4. calculate IFFT of the multiplication result,
5. look for the maximum of cross-correlation function magnitude, telling us about a time shift of our PR estimate.

Matlab code of the PR-based DAB frame beginning synchronization is presented in Listing 23.5.

Listing 23.5: Improvement of synchronization with Phase Reference—better timing recovery

```
1   % lab23_ex_DAB_fic.m - continuation
2
3   % START: Extra time synchronization using PHASE REFERENCE SYMBOL
4   % Frequency offset correction of the Phase Reference Symbol
5     s = x( icpsymb ) .* exp(-j*2*pi/Nfft*df*( icpsymb ));
6   % TIME OFFSET CALCULATION IN FREQ DOMAIN - mult in freq --> conv/corr in time
7     S = fft( s( isymb ) );
8     work = ifft( S .* conj( specPhaseRef ) ) / NSampPerSymb;  % ifft( mult in freq )
9     M=7; work=[ work(end-M+1:end); work(1:M+1) ]; % Assumed max +/- 7
10    [ dummy indx ] = max(abs(work(1:2*M+1)));
11    ioffsA = indx- (M+1),
12  % TIME OFFSET CALCULATION IN TIME DOMAIN - direct signal correlation in time domain
13    M = 7; XC = []; Nsymb = NSampPerSymb;  % Assumed max +/-7
14    XC(M+1) = sum( s .* conj(sigPhaseRef) )/Nsymb;
```

```
15   for m = 1 : M
16       XC( M+1+m ) = sum( s(1+m:Nsymb) .* conj(sigPhaseRef(1:Nsymb-m)) )/(Nsymb-m);
17   end
18   for m = 1 : M
19       XC( M+1-m ) = sum( s(1:Nsymb-m) .* conj(sigPhaseRef(1+m:Nsymb)) )/(Nsymb-m);
20   end
21   [ dummy ioffsB ] = max(abs(XC));
22   ioffsB = ioffsB - (M+1), pause
23   % CORRECTION of DAB frame beginning (without NULL)
24   ioffs = round( (ioffsA + ioffsB)/2 ); % use mean value
25   i1st = i1st + ioffs;
26   % FINAL CFO CORRECTION OF THE WHOLE LONG DAB FRAME WITHOUT NULL WITH PHASEREF
27   x = x(i1st : i1st + NSampPerData-1) .* exp(-j*2*pi/Nfft*df*(0 : NSampPerData-1)).' ;
```

The method was applied to the RX DAB signal, obtained results are presented in
Fig. 23.7: in left plot for direct time-domain cross-correlation implementation, and
in right plot for fast frequency-domain cross-correlation implementation. The results
are almost identical: only (almost) due to different scaling applied.

Exercise 23.8 (Time Synchronization Using Phase Reference Symbol).
Check efficiency of the PR-based DAB frame synchronization method using
TX DAB signal disturbed by the function disturbDAB().

Final Verification of Applied Time/Frequency Synchronization Methods There
are many references dealing with time and frequency synchronization methods dedicated exactly for DAB+ radio, simply google *DAB+ synchronization*, for example, BBC DAB White Paper. However, general methods designed for OFDM systems can be applied also, like described CFO estimation using the cyclic prefix or

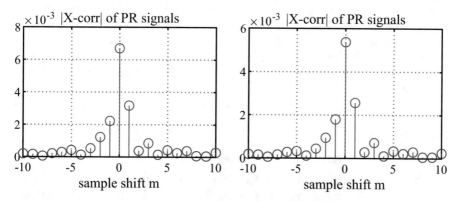

Fig. 23.7: Verification of the Phase Reference signal beginning by means of cross-
correlation of the received and transmitted PR signals. Results for: (left) direct time-
domain and (right) fast frequency-domain calculations

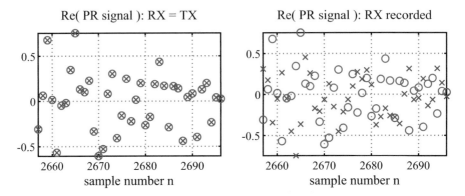

Fig. 23.8: Verification of time & frequency synchronization signal processing algorithms applied in the receiver. Comparison of the Phase Reference signal recovered in the receiver for two different input IQ signals: (left) non-distorted transmitter signal, (right) recorded, distorted signal. Denotation: circles—transmitted values, crosses—received values

OFDM symbol synchronization exploiting special synchronization patterns like the Phase Reference. Final synchronization results obtained in our case are presented in Fig. 23.8. For recorded real-world signal (right) strong channel influence is observed. It has hower minor significance for data decoding due to differential phase modulation which is used in DAB.

23.5 DQPSK-OFDM Demodulation

After taking decision about the position of the first Phase Reference sample, we skip 504 samples of the cyclic prefix and calculate FFT of the next 2048 samples, belonging to the PR. This operation is repeated 76 times for 76 OFDM symbols, i.e. skipping the 504 samples and calculating 2048-point FFT. As a result we obtain 76 spectra, take FFT coefficients of the used 1536 sub-carriers around 0 kHz only and compute their phase shifts, i.e. 75 values for each sub-carrier ($k = 1 \dots 75$, $n = 1 \dots 1536$):

$$\Delta X(k,n) = X(k+1,n) \cdot X^*(k,n). \tag{23.11}$$

From $\Delta X(k,n)$ values we could immediately decode the sent bits:

$$\text{Real}(\Delta X) > 0 \ \& \ \text{Imag}(\Delta X) > 0 \quad \rightarrow 00, \tag{23.12}$$
$$\text{Real}(\Delta X) > 0 \ \& \ \text{Imag}(\Delta X) < 0 \quad \rightarrow 01, \tag{23.13}$$
$$\text{Real}(\Delta X) < 0 \ \& \ \text{Imag}(\Delta X) < 0 \quad \rightarrow 10, \tag{23.14}$$
$$\text{Real}(\Delta X) < 0 \ \& \ \text{Imag}(\Delta X) > 0 \quad \rightarrow 11, \tag{23.15}$$

but we are not doing this. We left calculated numbers ΔX because we intend to use a better, *soft* not *hard*, version of the Viterbi forward error correction (FEC) decoder. In the transmitter some additional bits are added using convolutional FEC encoder. In the decoder we can exploit the added data redundancy and correct some limited amount of transmission errors. What will be explained later.

Listing 23.6: Signal demodulation—recovery of carrier IQ states

```
1  % lab23_ex_DAB_fic.m - continuation
2
3  % FFT
4    Ncp = NSampPerPrefix; Ncar = NCarrPerSymb;   % shorter names
5    offs = 0;                                     % take a few last CP samples
6    X = zeros( [NSymbPerFrame, Ncar] );           % creating empty array for all FFTs
7    for nsymb = 1 : NSymbPerFrame                  % processing all OFDM symbols
8      i = (nsymb-1) * NSampPerSymb;               % index of the first sample
9      s = x( 1+i : Ncp+Nfft+i );                  % samples of one whole OFDM symbol
10     s = s( Ncp+1-offs : Ncp+Nfft-offs );        % take only Nfft part minus offs
11     S = fft( s );                               % FFT of the selected samples
12     S = fftshift( S );                          % negative frequencies to the left
13     X( nsymb, 1 : Ncar/2 )   = S( Nfft/2-Ncar/2+1 : Nfft/2 );   % store to FFT
                                                                     array
14     X( nsymb, Ncar/2+1 : Ncar ) = S( Nfft/2+2 : Nfft/2+Ncar/2+1 );  % used carriers
                                                                         only
15   end
16   DX = X( 2:NSymbPerFrame, : ) .* conj( X( 1:NSymbPerFrame-1, : ) ); % phase shift
17
18   figure; plot(angle(DX(:,:)),'b*');
19   xlabel('symbol k'); title('Phase Shifts');
20   grid; pause
21   figure; plot(real(DX(:,100)),imag(DX(:,100)),'b*');
22   xlabel('I'); title('Q(k)=f(I(k))'); grid; pause
```

In Fig. 23.9 calculated phase shift angles $\angle \Delta X(k,n)$ (23.11) for all DAB+ sub-carriers in one frame are shown (left)($\pm\frac{1}{4}\pi$ and $\pm\frac{3}{4}\pi$ are expected) as well as complex-values of $\Delta X(k,n)$ for one sub-carrier only in one DAB+ frame (right).

Exercise 23.9 (Observing States of DAB Carriers). Run the FIC decoder program for disturbance-free TX DAB signal. Switch off all synchronization methods. Observe all constellation points of all carriers in one DAB frame. Then use function `disturbDAB` to *inject* all available impairments individually (one-by-one) and in mixtures. Observe changing constellation shapes. What disturbances are the most harmful in your opinion? Finally, turn on, one-by-one, all the synchronization procedures and observe improvement of carrier state visibility—you should obtain plots similar to ones presented in Fig. 23.10.

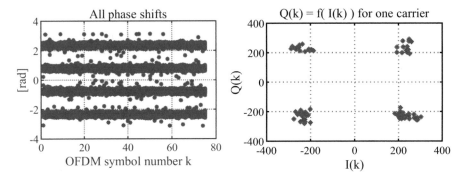

Fig. 23.9: Phase shift angles of all DAB+ sub-carriers in one frame (left) as well as complex-values $I(k) + jQ(k)$ of ΔX (23.11) for one sub-carrier only in one DAB+ frame (right)

Exercise 23.10 (Testing Efficiency of Synchronization Procedures for Recorded DAB Signal). Run the FIC decoder program for the recorded disturbance-full RX DAB signal. Initially, switch off all synchronization procedures. Observe all constellation points of all carriers in one DAB frame. Then, turn on, one-by-one, all the synchronization procedures and observe improvement of carrier state visibility—you should obtain plots similar to ones presented in Fig. 23.10.

23.6 Removing Frequency Interleaving

Signal passing through a telecommunication channel is disturbed: attenuated, delayed, and interfered. Apart the wide-band noise, disturbing all sub-carriers all the time, two types of disturbances are typical for DAB: (1) very strong, narrow-band, notch-like attenuation of selected frequencies and (2) short impulses very harmful for all sub-carriers but only from time to time. In DAB+ there is no channel identification and correction. All carriers obtain 2 bits and take one from available four states. There are added some redundant bits allowing correction of limited amount of errors (to be discussed later) but occurring information *gaps (loss)* in frequency and time domain should not be very big. In order to ensure this, transmitted bits are interleaved in frequency and time. Frequency interleaving, performed in DAB+ for all sent data, means that consecutive bits are not transmitted linearly, sub-carrier after sub-carrier, but are randomly mixed (scrambled) between sub-carriers. A special hashing table is used. Thanks to this, if a channel frequency response has a deep notch attenuating several sub-carrier, the error is uniformly distributed along

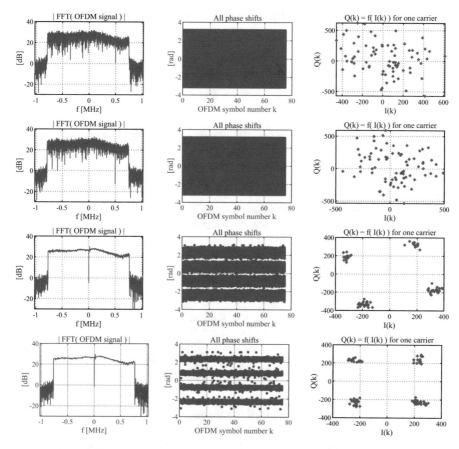

Fig. 23.10: OFDM signal spectrum (left), phase shifts of all carriers (center) and $Q(k) = f(I(k))$ plot for one carrier (right) for (up-down in rows): (1) original input IQ signal, (2) IQ signal after ADC sampling rate correction only, (3) carrier frequency offset correction only, (4) ADC and CFO correction together

the frequency axis after hashing and it is easily eliminated using error correction methods.

Being concrete, values of $\Delta X(k,n)$, calculated from Eq. (23.11) for each k (OFDM symbol number), are exchanged in regard to n (sub-carrier number) in a way described in the standard. Frequency interleaving (hashing), done in the transmitter, is now removed in the receiver. Next the data-stream is formatted in the following way:

$$\text{Real}(\Delta X(k,n)) \quad \text{for } k=1,\, n=1\ldots1536 \text{ (first)} \tag{23.16}$$

$$\text{Imag}(\Delta X(k,n)) \quad \text{for } k=1,\, n=1\ldots1536 \text{ (next)} \tag{23.17}$$

$$\ldots$$

$$\text{Real}(\Delta X(k,n)) \quad \text{for } k=75,\, n=1\ldots1536 \text{ (first)} \tag{23.18}$$

$$\text{Imag}(\Delta X(k,n)) \quad \text{for } k=75,\, n=1\ldots1536 \text{ (next)} \tag{23.19}$$

Listing 23.7: Frequency de-interleaving and DQPSK de-mapping

```
1   % lab23_ex_DAB_fic.m - continuation
2
3   % Frequency de-interleaver  - using FreqDeinterleaverTab table
4     DeintDX = zeros( size( DX ) );
5     for r = 1 : length( FreqDeinterleaverTab(:,4) )
6         DeintDX( :, FreqDeinterleaverTab(r,4) ) = DX(:,r);
7     end
8
9   % D-QPSK demodulation: complex --> real, imag
10  % D-QPSK symbol de-mapped - from 2D matrix to 1D vector, ETSI pp. 157 &&&&&&
11    DataFIC = zeros(1, NSymbPerFIC*NCarrPerSymb*2);            % 3*1536*2
12    for f = 1 : NSymbPerFIC % FIC DATA from OFDM symbols, first real, then imag
13        DataFIC( 1+(f-1)*NCarrPerSymb*2 : (f)*NCarrPerSymb*2 ) = ...
14                            [ real(DeintDX(f,:)) imag(DeintDX(f,:)) ];
15    end
```

23.7 Structure of FIC and MSC Block

Detail logical structure of DAB+ frame has been shown already in Fig. 23.3. Data of the first 3 out of 75 OFDM symbols (all together $3\cdot1536\ sub-carriers\cdot2\,\text{Re/Im}=9216$ numbers/bits) contains so-called fast information channel (FIC) about available services. Next 72 OFDM symbols contain the so-called main service channel (MSC) with service data. Access to the FIC is fast, because above 5 DAB frames long time-interleaving is only applied to the MSC. The MSC data are therefore better secured against impulsive disturbances than the FIC data. When un-recoverable errors are found in actual FIC, the data should be neglected and previously decoded service settings are used.

The FIC data are divided into 12 blocks called FIBs (Fast Information Blocks), having 768 numbers (bits) each. 12 FIBs are grouped into 4 parts having three successive FIBs ($3\cdot768=2304$ numbers (bits)). Each part is associated with one from four CIF blocks (Common Interleaved Frame), into which the MSC multimedia data are put, and describes the CIF content. The overall MSC data section has 221184 numbers (bits) (72 symbols × 1536 sub-carriers × 2 Re/Im numbers), and one CIF is equal to $\frac{1}{4}$ of the MSC, i.e. 55296 numbers (bits).

23.8 FIC Decoding

De-puncturing Bits sent in the transmitter are subject to the so-called channel coding in which some extra, redundant bits are added to an original bit message. There are many methods for generation of the redundant bits. In DAB 1/4 convolutional coding is used, increasing 4 times number of transmitted bits, i.e. in a group of 4 bits only 1 bit is original and 3 are defending it against errors. The coding principle is explained in Fig. 23.11. Input bits b_i are stored, delayed, and shifted in a buffer, and convolved with 4 binary patterns:

- `1 011 011`, octally 133,
- `1 111 001`, octally 171,
- `1 100 101`, octally 145,
- and once more `1 011 011`, octally 133.

Additions present in Fig. 23.11 are realized as binary exclusive OR operation, named as XOR (the addition result is equal to 1 only when two added bits are different). In the DAB convolutional coder the input bit-stream is replaced by 4 synthesized bit-streams, in which the original bits are *hidden*. If the $4\times$ redundancy is too high, the 4 generated bit-streams are punctured in a controlled way, i.e. some bits are removed from them in specially defined way, known for encoder and decoder. This operation is presented in Fig. 23.12: not transmitted bits are marked with $\times$. Then left bits are scanned vertically in columns (in direction of dashed lines ended with arrows) and transmitted. Thanks to this redundancy level can be changed and user has a choice: to send less data better secured or more data worse secured. Therefore, in DAB number of services and their quality can vary.

DAB decoder is putting zeros into the received bit-stream at positions where bits were removed (punctured) by encoder. The operation is called de-puncturing. As a result, after appropriate zero insertion in the DAB receiver, each 1/4 part of the received FIC increases its length from 2304 to 3096 numbers (bits).

Viterbi Decoder of Convolutional Channel Coding and Error Correction Next, from each FIC quarter having 3096 numbers (bits), sent bits are recovered using "soft" or "hard" Viterbi decoder. In the first case, Real() and Imag() real-value numbers are the decoder input while in the second case sequence of bits 0 and 1. Due to 4-times oversampling, 3096/4=774 bits are obtained. 6 last bits are removed—these are transient bits. Their number (6) is equal to the number of delay elements in convolutional encoder used in the transmitter (see Fig. 23.11). After this operation we obtain from FIC 4 blocks, each having 768 bits.

Bit De-scambler, Bit Energy De-dispersal Next, each block is de-scrambled `b=XOR(b,prbs)`: its bits are XOR-ed with corresponding bits of the same pseudo-random binary sequence (PRBS) which was exploited during data scrambling in the transmitter. XOR, logical exclusive OR operation performed upon two bits, gives 1 as a result, only when two elements/operands are different, i.e. for {0,1} or {0,1}.

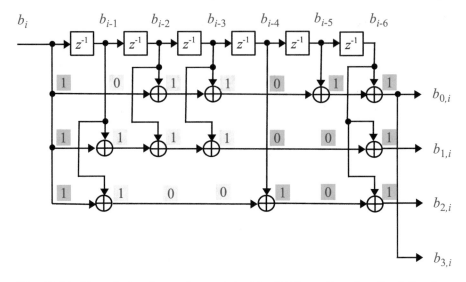

Fig. 23.11: Convolutional encoder exploited in DAB+ radio using the following, binary patterns: 1 011 011, 1 111 001, 1 100 101, and, one again, 1 011 011, respectively, octally 133, 171, 145, and 133

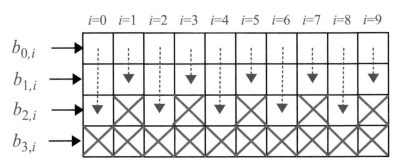

Fig. 23.12: Explanation of bit-stream puncturing and de-puncturing applied in DAB+ radio: some bits are removed in the encoder (marked with ×) and *restored* with 0s in the decoder. Bits are scanned vertically along dashed lines and transmitted

When two binary sequences are XOR-ed, their bits are added modulo-2. XOR performed after XOR results in coming back to the original bit sequence. XOR-ing bits of the sequence in the encoder causes dispersal of the transmitted bits energy, i.e. removal of long sequences of the same bit. Repeating this operation in the decoder results in restoring possible long sequences of the same bits. PRBS sequence, used in DAB encoder and decoder, is generated by a shift register with a loop back: next bit is calculated using the polynomial $x^9 + x^5 + 1$, see Fig. 23.13.

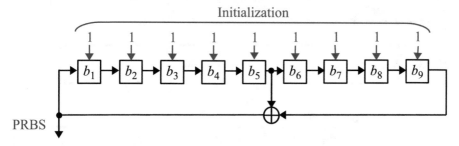

Fig. 23.13: Diagram for generation of pseudo-random binary sequence (PRBS) used for energy dispersal in DAB radio

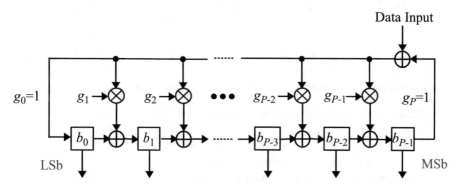

Fig. 23.14: Diagram for generation of cyclic redundancy check (CRC) code used in DAB+ radio

CRC Verification At present, we have 4 FIC blocks with 768 bits after bit energy de-dispersal. Each block is divided into 3 equal parts and 12 blocks with 256 bits result. They are called Fast Information Blocks (FIBs). Last 16 bits of each 256-bit block are checked: this is a CRC (Cyclic Redundancy Check) code, added to block of first 240 bits in the encoder. The added 16-bit code represents remainder of two polynomial division: the first represented by the codded 240 bits, and the second equals to $g(x) = x^{16} + x^{12} + x^5 + 1$. When we will calculate in the receiver the CRC code for the first 240 bits of each 256-bit block and when we will obtain the last 16 bits of the block, it means that there is no error. In such case, we can process the 240 bits further, otherwise we skip them. In the DAB standard, diagram presented in Fig. 23.14 is recommended for the CRC algorithm implementation. One example of CRC coding and decoding is presented in the end of this chapter.

FIG Decoding Bits obtained this way are divided into blocks called Fast Information Group (FIG), with variable length. First 3 bits define block type (from 0 to 7), next 5 bits block length in bytes, then the main data are placed. For FIG type "0" (see Fig. 23.15) we have in a data field: 1 bit C/N (Current/Next configuration 0/1), 1 bit OE (this/Other Ensemble 0/1), and 1 bit P/D (Programme/Data service 0/1).

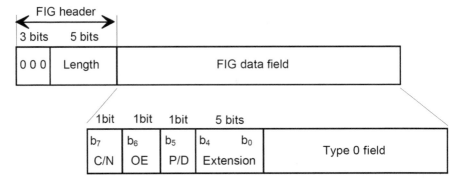

Fig. 23.15: Description of FIG type number 0 from the standard [2]

For P flag, the radio programme identifier is written on 16 bits, then 5 bits define data type (Extension, from 0 to 31), and finally the regular data are put. In particular, for Extension=1, there are 24-bit or 32-bit long bit packets in the data field:

- 6 bits with a number identifying a service (SubChId),
- 10 bits with starting address of the service in MSC (SubChStartAddr, binary number without sign in the range 0–863, pointing to the first 64-bit block CU (Capacity Unit) with audio bits),
- 1 bit specifying remaining number of bits (Short/Long 0/1, i.e. 7/15 bits),
- if 7 bits follows, the DAB service is sent and MP2 audio coding is used:

  - 1 bit for switching tables in the standard (0/1 =table 6/reserved),
  - 6 bits denote selected option in the table (from 0 to 63: SubChSize in CU, Protection Level, Audio Bit-rate),

- if 15 bits follows, the DAB+ service is sent and AAC audio coding is applied:

  - 3 bits for choosing the Option of audio coding,
  - 2 bits specifying the error Protection Level,
  - 10 bits defining SubChSize (see tables 7 or 8 in the standard).

In similar way other information is decoded. All details are given in the ETSI EN 300 401 standard. We are especially interested in values of parameters which are necessary for decoding concrete audio broadcasting, transmitted in system DAB+: SubChID, SubChStartAddr, SubChSize, and Protection. Parameter values are sent in the FIC block while audio samples in the MSC block, transmitted just after the FIC (see Fig. 23.3).

Listing 23.8: FIC decoding

```
1   % lab23_ex_DAB_fic.m - continuation
2
3   % Constants for FIC
4     NSampAfterTimeDepunct = 3096; NSampAfterVit = NSampAfterTimeDepunct/4-6; % 768;
5
6   % Put bits of FIBs corresponding to one CIF in separate row ------------------
7     NficbitsPer1CIF = NSymbPerFIC*NCarrPerSymb*2/NCIFsPerFrame; % 2304/2304/3072/2304
8     FIC = zeros(NCIFsPerFrame, NficbitsPer1CIF);
9     for r = 0:NCIFsPerFrame-1
10        FIC(r+1,:) = DataFIC( r*NficbitsPer1CIF+1 : (r+1)*NficbitsPer1CIF );
11    end
12
13  % De-puncturing using special function ------------------------------------
14    DataDep = [ depuncturing( FIC(:,1:2016),16 ) depuncturing( FIC(:,2017:2292),15 ) ...
15                depuncturing( FIC(:,2293:2304),8 )];
16
17   % Viterbi decoder using Matlab function vitdec() ------------------
18     DataVit = zeros(NCIFsPerFrame,NFIBsPerCIF*256+6); % Bitvector + Tail
19     for f=1:NCIFsPerFrame                             % Matlab. +/-real or {-1,+1}
20        DataVit(f,:) = vitdec( DataDep(f,:), VitTrellis, 1, 'trunc', 'unquant');
21     end
22     DataVit = DataVit(:,1:end-6);                     % Tail removing
23
24  % Energy dispersal using special function EnergyDispGen()------------
25    DataEnerg = zeros(size(DataVit));
26    for m=1:NCIFsPerFrame
27        DataEnerg(m,:) = xor( DataVit(m,:), EnergyDispGen( NFIBsPerCIF*256 ) );
28    end
29
30  % FIBs building --------------------------------------------------------------
31    FIB = reshape( DataEnerg, 256,NFIBsPerFrame )';
32
33  % CRC checking of FIBS -------------------------------------------------------
34    DABFrameNr = 1;                          % only one frame is analyzed
35    FIBCRCCheck = zeros(1,NFIBsPerFrame);
36    for k = 1 : NFIBsPerFrame
37        FIBCRCCheck(k) = CRC16( FIB(k,:) );  % special function CRC16()
38    end
39    if sum(FIBCRCCheck) == NFIBsPerFrame     % 12?
40       disp('CRC OK');                        % FIGs building from FIBS
41       for FIBNr = 1  : NFIBsPerFrame         % to 12
42          pos = 1;
43          while pos < 241
44              if FIB(FIBNr,pos:pos+7) == [1 1 1 1 1 1 1 1], break, end
45              % FIG type and length finding
46              Type = BinToDec( FIB(FIBNr,pos:pos+2), 3 );
47              FIGDataLength = BinToDec( FIB(FIBNr,pos+3:pos+7), 5 ); % in bytes
48              % FIG building (reconstruction) using special function FIGType()
49              FIGType( DABFrameNr, FIBNr, Type, FIGDataLength, ...
50                       FIB(FIBNr, pos+8:pos+8+8*FIGDataLength-1) )
51              pos = pos + (1+FIGDataLength)*8;
52          end
```

```
53      end
54      disp('############################### Ready !');
55      % disp('PRESS ANY KEY !'); % pause
56    else
57      disp(['CRC Fail! Frame: ',num2str( DABFrameNr )]);
58    end
```

23.9 MSC Decoding

Program of MSC Decoder Decoding the audio broadcast, i.e. the MSC information of the DAB and DAB+ frame, is very similar to the FIC decoding. The only
big difference relies on application of time-interleaving, which is not used for FIC,
aiming at strong protection against impulsive disturbances. For this reason a Matlab program of the MSC decoder is not presented here, inside the book, however,
it is given in the book repository. First, DAB and DAB+ decoding, up to time de-
interleaving operation, are performed together. After that, based on UEP flag value,
DAB and DAB+ are processed in similar way but with different parameter values
(for de-puncturing and Viterbi decoder). In DAB decoder, after energy de-dispersal,
samples of MP2 are obtained and stored, while in DAB+ the resultant bit-stream
should be further processed as a so-called Super Frame.

> **Exercise 23.11 (Main Service Channel Decoding).** In our DAB+
> decoder, the function decodeMSC(), called from the program
> lab23_ex_dab_fic_msc.m and given explicit in the program
> lab23_ex_dab_all.m, is responsible for Main Service Channel han-
> dling. A Reader should continue reading this section, looking for the
> corresponding Matlab code and analyzing it: (see Figs 23.16 and 23.17)

Time De-interleaving After reading from FIC the information about radio programs available in DAB+ signal, we are choosing one service and start reading their
bits from the Main Service Channel (MSC) block, built from 4 Common Interleaved
Frames (CIFs). The bits are located exactly in the same position in each block, starting from SubChStartAddr address and having SubChSize size, expressed in Capacity Units (CUs), having 64 bits. First, we are removing time-interleaving of these
bits, which was done in the transmitter in order to defend the data against impulsive
disturbances. In DAB 19 last CIFs from 5 last 5 DAB frames (lasting 100 µs each)
are interleaved, giving a delay of about 0.5 s.

De-puncturing After that, similarly like in the FIC block, *data puncturing* performed in the transmitter is removed. The convolution encoder used in TX for channel coding is the same: it increases the data size 4 times. But the data puncturing
method is different than in FIC, additionally different for DAB standard (Unequal

Error Protection UEP with step changes, tables 30 and 31) and DAB+ standard (Equal Error Protection EEP with exactly linear changes, tables 33–36). Zero values are inserted into positions of the *punctured* data removed in the TX.

Viterbi Decoding Next soft Viterbi decoder is used for optimal recovery of sent bits, separately for data obtained from 4 CIF blocks. Size of the input data is different. It depends on compression level of audio signal and level of applied error protection.

De-scrambling Finally, similarly as for FIC, resultant bits are logically XOR-ed with bits of the PRBS sequence: this way bits energy dispersal, done in TX, is removed, i.e. long sequences of the same bit are restored. In case of DAB radio, this is the last processing step: the bit sequence is sent now to the MP2 audio decompression algorithm. However, for DAB+ standard, bits do not represent a pure audio track and have to be decomposed further. For DAB+, our *onion* has more layers. The most upper one is called a Super Frame.

23.10 DAB+ Super Frame Decoding

Super Frame Structure As mentioned above, for the DAB+ standard the bit processing continues. Bits read from 5 consecutive CIF blocks give a Super Frame (SF), created in the transmitter (see Fig. 23.16). In receiver we have to find the super frame beginning. Super Frame has a header consisting of:

- 16 bits of the Fire code, calculated in the TX using the polynomial $(x^{11} + 1)(x^5 + x^3 + x^2 + x + 1)$ and protecting 9 bytes lying after it,
- 8 bits specifying AAC coding parameters, among others sampling ratio, spectral replication, parametric stereo, surround mode configuration,
- num_aus equal to 2, 3, 4, or 6, specifying number of 12-bit blocks, defining starting addresses au_start[n] of blocks with audio samples (their number results from AAC settings),
- eventual 4 padding zeros; their presence can be predicted from preceding values.

After the header, AAC bit packets are located, starting from addresses au_start[n]. Each of them has 16-bit CRC code, computed with the use of the same polynomial as in FIC: $x^{16} + x^{12} + x^5 + 1$. CRC code of the first AAC bit packet can be used for finding Super Frame beginning, i.e. the first of its CIFs.

Exercise 23.12 (Super Frame Decoding). In our DAB+ decoder, the function decodeSuperFrame() is responsible for Super Frame synchronization, its error correction, re-packing AAC bit-stream into ADTS container, and storing decoded radio program into a file. A Reader should continue reading the below

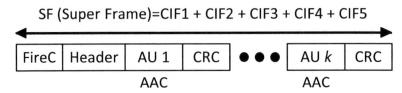

Fig. 23.16: Structure of the DAB+ Super Frame with AAC audio collected from five CIF blocks. Denotations: FireC—Fire Code, AU—Audio Unit, CRC—Cyclic Redundancy Check

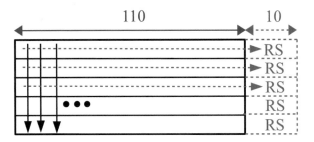

Fig. 23.17: Graphical explanation of Reed–Solomon error protection coding in DAB+ digital radio: (1) data bytes are put column-wise into a matrix having 110 elements in each row, (2) then RS algorithms encode individually each row and adds 10 protecting bytes to it

part of this section, looking for the corresponding code in the function and analyzing it (see Figs. 23.16 and 23.17).

Super Frame Synchronization Super Frame synchronization is described in [3]. It can be realized also in the following way [5]. We are assuming that a chosen CIF is the first one in a Super Frame. We correct its beginning using the Fire code. Then, we decode values of `au_start[n]`. If they do not increase linearly, we jump to the next CIF. Otherwise, we take the first audio block and check it CRC. If it is correct, we have just found the first CIF of the Super Frame. Otherwise, we go to the next CIF.

Reed–Solomon Decoder At the Super Frame end, there are some bits added in the transmitter for the overall SF error protection. The Red-Solomon (RS) $(120,110,t = 5)$ encoder is used for this purpose: each sent 110 bytes of the super frame are protected by 10 bytes (110+10=120). Generator used in the encoder is equal to:

$$G(x) = \prod_{i=0}^{9} (x + \alpha^i). \tag{23.20}$$

Galois field GF(2^8) is exploited with $\alpha^1 = 2$ and polynomial $P(x) = x^8 + x^4 + x^3 + x^2 + 1$. This is the most difficult part of the book: number-theoretic math known as number group theory. I am sorry. Not this time. And not me.

If we want to make use of protecting bits in the receiver, we should write consecutively all SF bytes into a matrix having 120 elements (bytes) in each row, but putting them up-down into the consecutive matrix columns. Then the Reed–Solomon decoder program must be called for each matrix row. Each input row should be exchanged with the RS output. Next, the repaired Super Frame is get by scanning the resultant matrix again by columns. Only first 110 columns should be taken, redundant part is removed. As RS decoders we can use programs for very popular RS(255,245,$t = 10$) option, e.g. used in satellite transmission. In such case, we should add 135 zero bytes in the beginning of each matrix row (135+120=255), then call the decoder program and take 110 last elements from each row of the resultant matrix. In Matlab there is a special RS encoding and decoding function. In the end of this chapter, in special subsection, its usage in DAB+ receiver is presented.

Packing Audio Bits to a Chosen Container, e.g. ADTS At this stage, we know starting addresses `au_start[n]` of corrected, AAC compressed, audio blocks and have to re-pack them into any popular AAC transport stream, the so-called *container*, supported by used multimedia decompressing library. We have chosen ADTS standard. Re-packing is done in block-to-block mode. We are checking CRC, last 16 bits, of each block. AAC is packed into the ADTS transport stream which is played by different media players.

Playing Audio VideoLAN program can be used, for example, for listening to recovered AAC bit-streams.

A function `decodeSuperFrame()`, given in the book repository, is responsible for DAB+ super frame processing (synchronization and error checking), ACC audio samples extraction, re-packing them into ADTS container and storing to disc. A Reader is encouraged to analyze its code personally.

23.11 Final DAB Block Diagrams

In this chapter we have started from the bottom, we have been carefully collecting together many details and just now ... we have reached a top. It is time to see the final panorama of DAB+ digital radio and to make general conclusions.

In Figs. 23.18 and 23.19 creation of DAB multiplex signal is summarized. Individual audio services: data, DAB MP2, and DAB+ AAC, are processed separately and combined together into the Main Service Channel (MSC) which is time-interleaved and accessed with delay of about 0.5 s. In parallel information about digital content is collected in Fast Information Channel, not time-interleaved and faster accessed, protected always by strong forward error protection (FEC 1/3) and CRC16 (FIBs). Then IQ carrier states of MSC and FIC are joint with Phase Reference IQs and multiplexed. Finally, the following operations take place: frequency

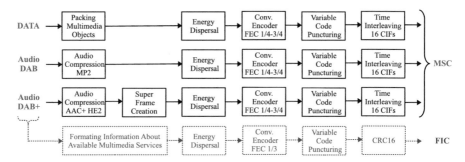

Fig. 23.18: DAB+ encoder: FIC (service information), and MSC (service content) creation

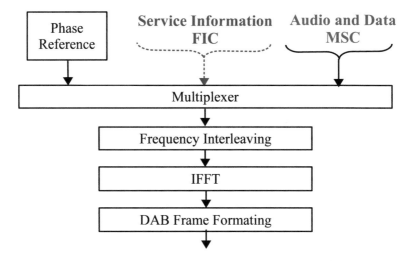

Fig. 23.19: DAB+ encoder: IQ signal generation of DAB+ multiplex—combining carrier states of Phase Reference (synchronization pattern), FIC (service information) and MSC (service content)

interleaving, series of many FFTs, DAB frame formatting, NULL symbol insertion and signal transmission.

Algorithm of DAB/DAB+ decoder is summarized in Fig. 23.19. It starts with NULL symbol detection, ADC sampling rate estimation and correction, carrier frequency offset estimation, and time synchronization using the Phase Reference symbol, FFT calculation, frequency de-interleaving and soft DQPSK carrier states demodulation. Then FIC (service information) and MSC (digital content) are processed separately. In case of FIC, data are de-punctured, processed by Viterbi decoder, energy dispersal is removed, FIB blocks—protected by CRC—are un-packed and FIGs are decoded. In case of MSC, first time-interleaving is undone, then data are: de-punctured, Viterbi decoded, and de-dispersed. At this moment the DAB

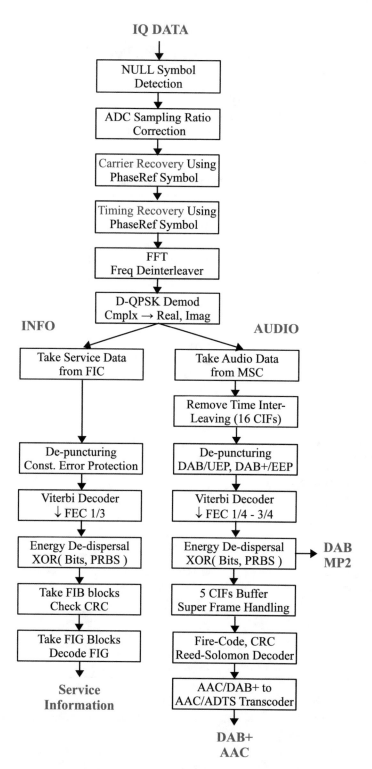

Fig. 23.20: Block diagram of DAB+ decoding algorithm

algorithm stops: the MP2 stream is ready to play. However the DAB+ algorithm continues further. The Super Frame, composed of 5 CIFs, is synchronized, corrected by Fire code and Reed–Solomon decoder, CRC of audio units are checked and AAC audio data are re-packed from DAB+ container into any other format if necessary, for example, to ADTS. It is important to note that, despite the same processing path, de-puncturing and Viterbi decoder have different parameters for DAB and DAB+ (see Fig. 23.20).

The last but not least is Table 23.1 presenting DAB/DAB+ main parameters values and offering the last encyclopedia view of this chapter.

Table 23.1: DAB+ Specification from [6]

Parameter	Value	Time duration
DAB frame	196608 samples	96 ms
NULL symbol	2656 samples	1297 µs
OFDM symbol w/o CP	2048 samples	1000 µs
FFT	2048	–
Sub-carrier used	1536	–
Sub-carrier spacing	1 kHz	–
Cyclic prefix (CP)	504 samples	246 µs
OFDM symbol w/ CP	2552 samples	1246 µs
OFDM symbols:	76	–
– Phase Reference	1	–
– with FIC (info)	3	–
– with MSC (media)	72	–
FIC (Fast Info Chan):		
– FIBs in DAB frame	12	–
– FIBs for 24 ms	3	–
MSC (Main Serv Chan):		
– CIFs in DAB frame	4	–
– CIFs for 24 ms	1	–
Bits per:		
– OFDM symbol	3.072 kbit	–
– DAB frame	230.4 kbit	–
Bit-rate (w/o PR):		
– FIC (FEC 1/3)	96 kbit/s	–
– MSC (overall)	2.304 Mbit/s	–
– MSC (max per serv)	1.824 Mbit/s	–
Max echo delay	–	300 µs
Max propag. path diff.	90 km	(1.2 CP)
Max RX velocity:		
– urban	260 km/h	–
– countryside	390 km/h	–

23.12 Error Correction in DAB

As we see, error detection and correction methods represent a very important part of the DAB standard. Let us recapitulate:

- convolutional error protection and correction of FIC and MSC bit-streams,
- CRC error detection of FIBs in FIC,
- CRC error detection of Audio Units in DAB+/AAC Super Frame,
- Fire code protection of first 88 bits of DAB+/AAC Super Frame allowing error detection and correction of the Super Frame beginning (capable of detecting and correcting most single error burst of up to 6 bits),
- Reed–Solomon error detection and correction of the whole Super Frame.

Error correction coding is not a topic of this book. We are interested in fundamental digital signal processing methods applied in modern digital telecommunication systems, mainly in source coding, digital modulation, channel estimation and correction as well as synchronization techniques. Error protection and data encryption/security lie outside our interests and scope of the book.

However, when error coding methods are so important and widely used, we should understand their merits. A goal of this short section is to introduce error protection fundamentals and practical application from a user point of view. For further reading the book [9] is recommended as a very valuable source of practical information.

23.12.1 Cyclic Redundancy Check Encoder and Decoder

Addition of a parity bit, causing that the overall number of 1s in our message is even, is the simplest method of error detection. For example, 0 should be appended to the end of a message 01010101 (four 1s) and 1 to a message 01010111 (five 1s). When in the received message the number of bits is odd, it means that error occurred and information should be re-transmitted.

Cyclic redundancy check is not so simple as parity error addition, nevertheless it is one of the simplest and practically efficient in realization method. The idea is not very difficult also. A transmitted sequence of bits (a binary number) has to be divided in the transmitter by a properly selected binary divisor and a binary division reminder is appended to the transmitted bits. In the receiver, obtained bits are divided by the same divisor, know to the transmitter. If the division result is the same as bits appended to the message, sent and obtained in the RX, it is assumed that obtained data are error-free. Binary numbers (objects) and their division (operation upon them) can be described by mathematical theory of groups/fields of polynomials which helps to find *good* divisors. For example, the following divisors and polynomials represent CRC *pairs*:

1 0 0 0 0 0 1 0 1 $x^8 + x^2 + 1$

1 0 0 0 1 1 1 0 1 $x^8 + x^4 + x^3 + x^2 + 1$

1 1 0 0 0 0 0 0 0 0 0 0 0 0 1 0 1 $x^{16} + x^{15} + x^2 + 1$

1 0 0 0 1 0 0 0 0 0 0 1 0 0 0 0 1 $x^{16} + x^{12} + x^5 + 1$

and the last one is used in DAB for FIC, MSC, and AU in Super Frame protection. From the point of view of a user (as we in DAB standard at present), when the divisor is already chosen, application of the CRC technology is not difficult.

Exercise 23.13 (CRC by hand). Let us look at Fig. 23.21. In general, the CRC encoder and decoder have a form presented in Fig. 23.14 but for polynomial $x^3 + x^2 + 1$ (divisor 1101) it simplifies to the diagram shown in Fig. 23.21. The binary adder $\oplus$ is modulo 2 and implemented as xor() function. Let us assume that we code the data input equal to 1111. Initially bits $b_0 b_1 b_2$, stored in a shift register, are set to zero. In the first table presented on the right side, the coding operation is tracked: we see consecutive shift register contents. After the last input bit coming in, the shift register is filled with three 1s: it is the division reminder. These bits are appended to the end of the input message: 1111 111 and transmitted. In the receiver the same algorithm is used for CRC decoding, only input is different, at present three bit longer. In the second table presented on the right side, consecutive contents of the shift register are shown. After four clock cycles, content of the shift buffer is exactly the same as during coding (what is marked with a dashed line rectangle): the divisor is the same, there is no transmission error. After processing the last, the seventh input bit, the shift buffer ($b_0 b_1 b_2$) is filled with zeros, indicating the lack of errors. In Fig. 23.21 bottom, two manual binary number divisions are performed, both in CRC coder and decoder. They confirm correctness of CRC=111 and the overall CRC methodology. First, in encoder, the divided sequence 1111 is appended with $P - 1$ zeros (number of polynomial taps minus one) and a series of divisions takes place. Then, in decoder, the calculated CRC is appended to the error-protected bit sequence and, again, a series of divisions is performed.

As an exercise, please, verify correctness of all calculations performed in Fig. 23.21.

Exercise 23.14 (CRC by computer). In Listing 23.9 two implementations of CRC encoder and decoder are presented. The first one, CRCeasy(), attempts to realize manual binary polynomial division, as in calculations presented in bottom part of Fig. 23.21. The second implementation, CRCuniversal(), programs diagrams from Figs. 23.14 and 23.21. It is possible in it to complement last $P - 1$ input bits, as it is done in DAB radio, as well as set initial values

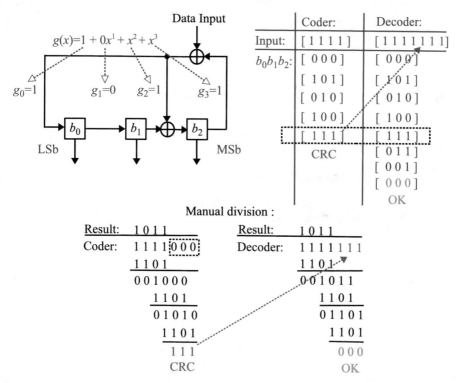

Fig. 23.21: Graphical illustration of CRC encoding and decoding

of the shift buffer—in DAB with 1s. Analyze the code. Run the program. Notice final 0s in CRC variable informing us about error absence. Introduce 1 or 2 errors in error protected bit-streams. Check whether they are detected by the CRC decoder. Change coded bits and the polynomial to ones used in Fig. 23.21. Using the program check whether data presented in figure tables are correct.

Listing 23.9: Matlab program for testing CRC coding and decoding

```
1  % lab23_ex_crc
2  % Testing CRC error protection
3
4  % Bit sequence to be coded
5  N = 32; Bits = round( rand(1,N) );      % e.g.  [ 1 1 0 ... 1 1 0 1];
6  Bits = [ 1 1 1 1 ]; N = length( Bits ); % in our example
7
8  % Generating polynomials
9  %            x^16 + x^12   +   x^5  +   1    % CRC-16
10 % Polynomial = [ 1 1 0 1 ]; % Test - our example
```

```
11   Polynomial = [1 0 0 0 1 0 0 0 0 0 0 1 0 0 0 0 1]; %
12   P = length( Polynomial );
13
14   % Encoder and decoder demo with binary number division
15   disp('Easy');
16   CRC = CRCeasy( [ Bits zeros(1,P-1)], Polynomial),
17   CRC = CRCeasy( [ Bits CRC ], Polynomial ), pause  % should be [ 0 0 ... 0]
18
19   % Encoder and decoder with shift buffer
20   disp('Universal encoder');
21   CRC = CRCuniversal( [ Bits ], Polynomial, zeros(1,P-1), 0 ),
22   disp('Universal decoder');
23   CRC = CRCuniversal( [ Bits CRC ], Polynomial, zeros(1,P-1), 0 ), pause
24
25   %#######################################
26   function CRC = CRCeasy( Data, Polynomial )
27   D = length( Data );
28   P = length( Polynomial );
29   N = D- (P-1);
30
31   n1st = 1;
32   while( Data(n1st) == 0 ) n1st = n1st+1; end
33   CRC = Data( n1st : n1st+P-1);
34   for n = n1st : N
35      % n
36      if( CRC(1)==1 ) CRC = xor( CRC, Polynomial ), % xor-ing
37      else            CRC = xor( CRC, zeros(1,P) );
38      end, % pause
39      if(n < N)       CRC = [ CRC(2:P) Data(n+P) ], % shifting
40      else            CRC =    CRC(2:P);
41      end, % pause
42   end
43
44   %#################################################################
45   function [ CRC ] = CRCuniversal( Data, Polynomial, CRC, cFlag )
46   % Data       - input bit message
47   % Polynomial - used, with 1s on the most/least significant bits
48   % CRC        - initial CRC values in the shift buffer, all 1s or all 0s
49   % cFlag      - 0/1 complement flag: optional complement of last (P-1) input bits
50
51   P = length( Polynomial );                % polynomial length
52   Polynomial = Polynomial( 2 : P-1 );      % remove MSB and LSB 1s
53
54   % Optional complement of (P-1) last input bits
55   if( cFlag == 1 )
56      Data(end- (P-2):end) = xor( Data(end- (P-2):end), ones(1,P-1) );
57   end
58
59   % CRC calculation - buffer shifting and data xor-ing
60   % CRC, % initial value
61   for i = 1 : length(Data)
62      if( xor( CRC(1), Data(i) ) == 1)
63         CRC = [ xor( Polynomial, CRC(2:end) ) 1 ];
64      else
```

```
65          CRC = [ CRC(2:end) 0 ];
66      end
67    % CRC, pause % in fig.23.21 order of bits is reversed
68  end
```

23.12.2 Application of Reed–Solomon Coding

Reed–Solomon coding concept is similar to CRC: input data are divided by a polynomial and a division remainder is appended to the original data. However, byte-based, 8-bit Galois field theory and arithmetic is exploited, not a bit-based one. The method allows not only the error detection, but also their correction, up to some error level. Mathematical derivation of RS encoder and decoder principles and algorithms are as climbing the Mount Everest for city dwellers. I have done it with students once or two times. But each effort, after doing something else, is a starting from the beginning, like from the sea level. Therefore, here we will concentrate only on RS decoder application in DAB radio, as an example of practical RS coding usage and demonstration of its very big efficiency. Our intention is to interest a Reader in answering the question: *how on Earth is it done?*.

In program in Listing 23.10 one Super Frame, extracted from the TX IQ DAB file, is decoded by the RS encoder (see Fig. 23.17). The data are error-free but we are artificially adding some number of errors (ErrPerRS) to each RS-coded 120 bytes. And then checking whether the *injected* error were detected and corrected. Input bits of a Super Frame are re-packed into bytes. Then, the number of obtained bytes (Nlen) is divided by 120 and number of rows (Nrows) of the matrix, presented in Fig. 23.17, is computed. Then each Nrows-th element of the byte-stream is taken (i.e. each matrix row) and decoded. Bytes returned by the RS decoder replace corresponding bytes in the input bytes-stream. Finally, output bit-stream is formed.

Listing 23.10: Matlab program for testing Reed–Solomon decoder in DAB+ application scenario

```
1   % lab23_ex_rs.m
2   % Reed-Solomon in DAB
3   clear all; close all;
4
5   ErrPerRS = 5;                    % injected errors per 120 bytes, per one RS
6
7   load SuperFrameIN.dat;           % data coming from the function decodeMSC()
8   BITS = SuperFrameIN;             % new, shorter name
9   NBitsPerCIF = 2688;              % number of bits per one CIF
10  RSpoly = rsgenpoly(255,245,285,0);  % 285 = 100011101 = x^8+x^4+x^3+x^2+1
11
12  % Reformating bit-stream to byte-stream
13  BYTES = ( reshape(BITS(1:5*NBitsPerCIF)',8,[])' * [128 64 32 16 8 4 2 1]' )';
14
15  % Main Reed-Solomon loop - processing Nrows-th polyphase sequences, i.e. matrix rows
```

```
16   Nlen = length(BYTES); Ncols=120; Nrows=Nlen/Ncols;
17   for k = 1 : Nrows        % for each polyphase sequence having 120 bytes each
18     % Take data
19       DataIN   = BYTES(k:Nrows:end);   % take 120 bytes, k-th polyphase sequence
20     % Add errors
21       DataINCopy = DataIN;
22       for m = 1 : ErrPerRS                          % consecutive errors
23           pos = round( rand(1,1)*119 )+1;           % random error position
24           val = gf( round(rand(1,1)*255), 8 );      % Galois field error
25           oldval = gf( DataIN(pos),8 );             % Galois field old value
26           newval = oldval + val;                    % Galois field new value
27           DataIN(pos) = double( newval.x );         % modification of the input
28       end
29     % RS decoder
30       if(0) % RS: Original code, Matlab only
31           [DataOUT, num_errs] = rsdec( gf( DataIN, 8), 120, 110, RSpoly);
32       else     % RS: Longer code that is shortened, Matlab and Octave
33           [DataOUT, num_errs] = rsdec( gf([ zeros(1,135) DataIN], 8), 255, 245, RSpoly);
34           DataOUT = DataOUT(136:245);
35       end
36       DataOUT = double( DataOUT.x );
37     % Checking correction results
38       RS_ERRORS_BEFORE = num_errs, % in bytes
39       ERRORS_AFTER_RS = sum( sum( abs( DataINCopy(1:110) - DataOUT ) ) ),
40     % Put corrected bits back into the bit-stream
41       start = 1+(k-1)*8;
42       step = Nrows*8;
43       for m = 1:length(DataOUT)
44           BITS(start+(m-1)*step + 0 : 1 : start+(m-1)*step + 7) = DecToBin( DataOUT(m), 8
                 );
45       end
46   end % of Main RS loop
```

Exercise 23.15 (Testing Reed–Solomon Decoder in DAB Scenario). Analyze
Matlab code presented in the Listing 23.10. Run the program. Step-by-step increase number of errors: find the moment when errors are not corrected. Use
long and shortened version of the RS decoder. Then add displaying some extra information to the function decodeSuperFrame(), i.e. the number of
RS-found errors (num_errs) and the number of RS-changed bytes (sum(
abs(DatIN(1:110) - DatOUT))). Choose TX IQ DAB input in the
program lab23_ex_dab_all.m. Disturb the IQ data using the function
disturbDAB(). Run the program and observe number of errors found by
the RS decoder. Finally, switch to the RX IQ DAB input file. Do not disturb the
data. Again, observe the number of reported errors.

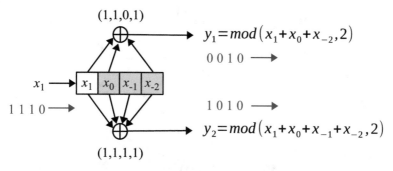

Fig. 23.22: Analyzed examples of convolutional coder

23.12.3 Convolutional Coding and Viterbi Decoder

In this subsection we briefly present a concept of convolutional error protection coder and Viterbi decoder using a simple example. Our coder of interest is presented in Fig. 23.22. Answer of the coder to the input sequence of bits `0 1 1 1` is equal to `00 11 00 01`. Let us assume that in a receiver error occurred on the first position: `10 11 00 01`.

In the beginning we should build a transition diagram between our coder states which is presented in Fig. 23.23. We are assuming that in the previous coder state bits $x_0x_{-1}x_{-2}$ of the shift register are equal to [0 0 0], [0 0 1], ..., [1 1 1], i.e. to the values specified on the left side. Bit x_1, coming in, can have a value 0 or 1, and bits $x_0x_{-1}x_{-2}$ are changing their values to new ones shown on the diagram right side. When $x_1 = 0$, a solid line to a new state goes up, for $x_1 = 1$—a dashed line goes down. In square boxes connected with lines are given values of bits y_1y_2 being a coder output for all possible transitions.

At present we analyze step-by-step the Viterbi decoder work for the discussed example. Let us start from Fig. 23.24. In its upper part, received pairs of bits are shown in square boxes with gray background. Because shift buffer in the transmitter was filled initially with zeros, in the receiver decoder we are starting from the state `000` and goes *up* (solid line) and *down* (dashed line) to the next allowed state. In square boxes, connected with lines, we have expected values of bits y_1y_2, at present 00 and 11. We compare them with obtained bits y_1y_2 (bold numbers upon gray background, at present **10**) and calculate accumulated metrics of bit patterns similarity (1=the same, 0=different) for each available path. We write result on the right side of the plot, individually for each path. At present we have values (1) and (1).

Now we switch to Fig. 23.25. New pair of bits y_1y_2 is equal to **11**. Four possible transitions are possible, marked with lines, expecting input pairs of bits: 00, 11, 11, 00 (up-down). We again calculate similarity measures and accumulate metrics for all possible coder states paths, obtaining (1)(3)(3) and (1) (please, calculate them on hand).

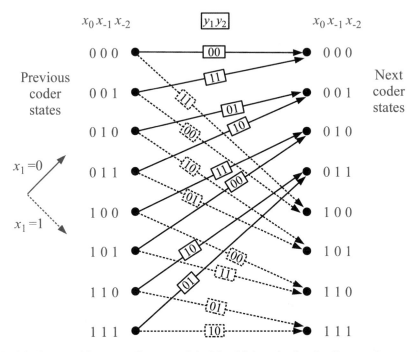

Fig. 23.23: Transition state diagram of the Viterbi decoder for the discussed example

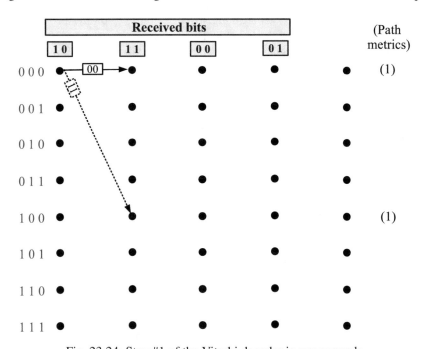

Fig. 23.24: Step #1 of the Viterbi decoder in our example

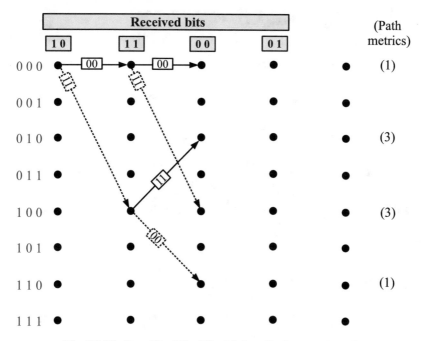

Fig. 23.25: Step #2 of the Viterbi decoder in our example

We continue the same operations in Figs. 23.26 and 23.27, for new bit pairs **00** and **01**, respectively. We plot available state connections and accumulate metrics of all paths. In Fig. 23.26 we have 8 different paths ending in 8 different coder states having the following metrics (up-down): (3)(4)(3)(2)(1)(4)(5)(2). In turn, in Fig. 23.27 we have 16 possible paths, two to each state (with some regular up-down pattern)—the best one is marked with yellow/gray background. When we use this path, go back and associate bit 1 with each state higher then 011 and bit 0—otherwise, we obtain the following sequence of decoded bits: 0111, which is correct.

When we continue decoding, the last transition lattice will be repeating. There-fore, in some algorithms, the exact starting in neglected: we directly start from the last lattice diagram, being in fact the diagram presented in Fig. 23.23, perform more steps than required (about $5 \cdot (N_{buffer} - 1)$), and use the best path to go back to one bit of interest and to decide about its values.

When bits are on input, the so-called hard decoding takes place. When real values are on input (bits probable values), the soft decoding is used.

In Listing 23.11 the presented convolutional coding example and the hard Viterbi decoder are implemented.

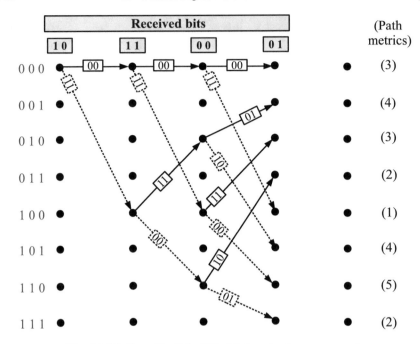

Fig. 23.26: Step #3 of the Viterbi decoder in our example

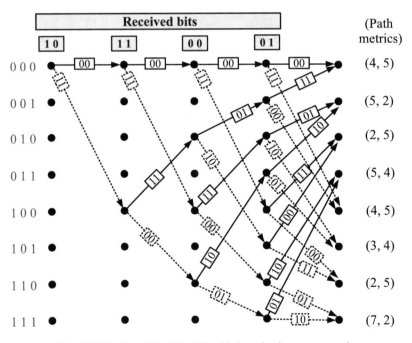

Fig. 23.27: Step #4 of the Viterbi decoder in our example

Listing 23.11: Matlab program for testing convolutional coding and Viterbi decoding

```
1  %lab23_ex_viterbi.m;
2  clear all; close all;
3
4  x = [ 0 1 1 1 ];              % input
5  g = [ 1 1 0 1;               % filter #1
6        1 1 1 1 ];             % filter #2
7
8  Mtaps = length(g(1,:));      % number of coefficients/delays
9  Kfilt = length(g(:,1));      % number of filters
10 MaxIndx = 2^(Mtaps-1)-1;     %
11 mult = 2.^[(Mtaps-2):-1:0];  % used for binary to decimal conversion
12
13 % Lattice diagram of analyzed convolutional encoder
14 y0ref = zeros(2^(Mtaps-1),Kfilt); y1ref = y0ref;          % reference states
15 transit = zeros(2^(Mtaps-1),3);                           % transition table
16 for indx = 0: MaxIndx
17     xbuf = str2num( reshape(dec2bin(indx,Mtaps-1), Mtaps-1, 1 )); xbuf=xbuf'; % pause
18     for k = 1 : Kfilt
19         y0ref(indx+1,k) = mod( [ 0 xbuf ] * g(k,:)', 2); % up
20         y1ref(indx+1,k) = mod( [ 1 xbuf ] * g(k,:)', 2); % down
21     end
22     transit(indx+1,1:3) = [ indx, [0 xbuf(1:Mtaps-2)]*mult', [1 xbuf(1:Mtaps-2)]*mult
           '];
23 end
24 y0ref, y1ref, transit, pause
25
26 % Encoder
27 % x(n) --> y12(n)
28 xbuf = zeros(1,Mtaps);
29 yout = [];
30 for n = 1 : length(x)
31     xbuf = [ x(n) xbuf(1:Mtaps-1) ];
32     for k = 1 : Kfilt
33         y(1,k) = mod( xbuf * g(k,:)', 2);                % convolution modulo 2
34     end
35     yout = [ yout; y(1,1:Kfilt) ];
36 end
37 yout
38
39 % Optional addition of errors
40 yin = yout;
41 if( yin(1)==0 ) yin(1)=1; else yin(1)=0; end              % errors
42 %if( yin(end)==0 ) yin(end)=1; else yin(end)=0; end       % errors
43 yin, pause
44
45 % Decoder
46 npoints = 1; step = 2^(Mtaps-1);
47 metric = zeros(2^(Mtaps-1),2);                            % accumulated metrics
48 path = zeros(2^(Mtaps-1),length(x)+1);                    % path history, initialized
49 path(1,1) = 1;                                            % starting point
50 for ndata = 1 : Mtaps-1                                   % first steps, incomplete tree
```

```
51        for indx = 1 : 2^(Mtaps-1)
52            if( path(indx,ndata) ~= 0 )
53                path( transit(indx,2)+1, ndata+1 ) = indx;
54                path( transit(indx,3)+1, ndata+1 ) = indx;
55                val0 = sum( y0ref(indx, :)==yin(ndata, :) );
56                val1 = sum( y1ref(indx, :)==yin(ndata, :) );
57                metric( transit(indx,2)+1,2 ) = val0 + metric( indx,1 );
58                metric( transit(indx,3)+1,2 ) = val1 + metric( indx,1 );
59            end
60        end
61        [metric(:,2), path] ,pause
62        metric(:,1) = metric(:,2);
63    end
64    for ndata = Mtaps : length(x)                     % next steps, regular tree
65        for indx = 1 : 2 : 2^(Mtaps-1)               % two branches DOWN and UP
66            val0A = sum( y0ref(indx, :)  ==yin(ndata, :) ) + metric( indx,1 );
67            val0B = sum( y0ref(indx+1, :)==yin(ndata, :) ) + metric( indx+1,1 );
68            val1A = sum( y1ref(indx, :)  ==yin(ndata, :) ) + metric( indx,1 );
69            val1B = sum( y1ref(indx+1, :)==yin(ndata, :) ) + metric( indx+1,1 );
70            % [ val0A, val0B], [ val1A, val1B], pause
71            idown = (indx+1)/2; iup = idown + 2^(Mtaps-1)/2;
72            if( val0A > val0B )                         % down
73                metric( idown, 2 ) = val0A;
74                path(  idown, ndata+1 )  = indx;
75            else
76                metric( idown, 2 ) = val0B;
77                path(  idown, ndata+1 ) =  indx+1;
78            end
79            if( val1A > val1B )                         % up
80                metric( iup, 2 ) = val1A;
81                path(  iup, ndata+1) = indx;
82            else
83                metric( iup, 2 ) = val1B;
84                path(  iup, ndata+1) =  indx+1;
85            end
86        end
87        [metric(:,2), path] , pause
88        metric(:,1) = metric(:,2);
89    end
90
91    % Decision taking
92    [dummy,pos] = max(metric(:,1));
93    for p = 0:ndata-1
94        if( pos  >= 2^(Mtaps-2) ) bits(p+1) = 1; else bits(p+1)=0; end
95        pos = path(pos,end-p),
96    end
97    bits = bits(end:-1:1), pause
```

Exercise 23.16 (Convolutional Coding and Viterbi Decoder). In Listing 23.11 Matlab program for convolutional coding and hard Viterbi decoder are implemented. Analyze the code. Run the program. Change input bits and filter coefficients. Modify the program and implement soft Viterbi decoding.

23.13 Summary

In this chapter we explore details of DAB+ digital radio standard. DAB signal is easily accessible in many countries and therefore it was used in our course for further mastering and testing our DSP telecommunication skills. What should be remembered?

1. Telecommunication standardization documents, including DAB norm and many others, are as thick cookery books. They are very condensed and difficult to read. It is interesting to notice that on hundreds of pages, filed with closely related equations, diagrams and tables, only one *recipe* of only one *technical meal* is presented. In this chapter we become familiar with one telecommunication standard from the beginning to the end.
2. DAB radio makes use of OFDM technology. It exploits 1536 carriers with 1 kHz spacing (1.536 MHz bandwidth) in Band III (174–240 MHz). 2.048 MHz sampling frequency and 2048-point (I)FFT are used. The DAB symbol has 2048+504=2552 samples, including the cyclic prefix. Since differential modulation is used (4-state DQPSK) a transmission channel is not estimated and corrected. Special NULL and Phase Reference symbols are used for synchronization purposes.
3. The DAB multiplex frame has a length of about 100 miliseconds and consists of 75 informative OFDM symbols: 3 first of them describe the digital content (fast information channel), next 72 carry digital data. MP2 audio coder is used in DAB radio while AAC and AAC+ in DAB+.
4. DAB+ is equipped with typical defense tools against channel impairments: time and frequency interleaving, CRC error correction, Fire code as well as punctured convolutional and Reed–Solomon channel coding. In fact, brief presentation of these method application, using the DAB example, was the most distinctive feature of this chapter.
5. Error correction coding is not easy to explain. Group theory and advanced polynomial algebra are applied for development of good codes and efficient coders and decoders. Since we are DSP-oriented, only fundamental issues of error correction usage were shown in this chapter: CRC encoding and

decoding, Viterbi decoder of convolutional codes, and practical application of Reed–Solomon encoder and decoder.

23.14 Private Investigations: Free-Style Bungee Jumps

Exercise 23.17 (Testing CRC16 in DAB Scenario). In Exercise 23.14 and program 23.9 we were testing concepts of CRC error detection. At present test CRC exactly in DAB scenario. Generate random sequence of 240 values 0/1, use CRC16 DAB polynomial, calculate 16 bits of the CRC code, append the code to be protected message, introduce some errors, and check whether the CRC decoder is informing you about errors.

Exercise 23.18 (Testing CRC in RDS FM Radio Scenario). In Chap. 21 we have written programs for detection of RDS bits, coded by bi-phase BPSK. But how to interpret them? RDS bit-stream consists of frames which are built from 4 blocks A, B, C, and D with 26 bits each: 16 original bits plus 10 CRC error protection bits (polynomial $x^{10} + x^8 + x^7 + x^5 + x^4 + x^3 + 1$ is used). The whole RDS frame has 104 bits. In DAB+ we do synchronization with Super Frame checking CRC code of audio blocks. The same technique can be used for finding beginning of RDS ABCD blocks. After having them, we are decoding bits of a packet of interest, for example, name of a radio station. Apply CRC programs/functions, presented in this chapter, for finding beginnings of ABCD blocks in RDS bit-streams, calculated in the program 21.10. After that you could start to think about decoding the radio station name. And answer how to do it, you will find in the RDS standard [7] You can also try to find help in the file RDS_Parser.zip, supporting chapter 21 in the book repository.

Exercise 23.19 (Testing Viterbi Decoder in DAB Scenario).** In Exercise 23.16 and program 23.11 we were testing concepts of convolutional coding and Viterbi decoding. Now perform tests exactly in DAB scenario. Generate a random stream of 3072 bits, use DAB + convolutional coding filters, code your message, puncture obtained four bit-stream in arbitrary way. Introduce some transmission errors into four bit-streams. Then, put zeros in place of punctured values, perform Viterbi decoding, and compare the output bit-stream with the input one. Looking for help, see how it is done in functions DABViterbiTZ() and DABViterbiTZinit() called by the program lab23_ex_dab_all.m.

Exercise 23.20 (Fire Code Implementation and Testing).** First 16 bits of the DAB+ Super Frame represent a Fire code, added in the encoder, which protects 72 bits $(= 9 \cdot 8)$ located after it. We have not discussed in this chapter the problem of Fire code generation and usage for error protection. Personally, I did some experiments in the past but I was not very satisfied with obtain results. May be somebody would like to continue my work. In the book repository there is a program `lab23_ex_firecode`, a function `CRC16FireCodeCorrect()` and `FireCodeData.dat` for tests. After success, the Fire code *functionality* could be added to our DAB software.

References

1. J. Cho, N. Cho, K. Bang, M. Park, H. Jun, H. Park, D. Hong, PC-based receiver for Eureka-147 digital audio broadcasting. IEEE Trans. Broadcasting **47**(2), 95–102 (2001)

2. ETSI EN 300 401, Radio broadcasting systems: Digital audio broadcasting (DAB) to mobile, portable and fixed receivers,' 2006. Online: https://www.etsi.org/deliver/etsi_en/300400_300499/300401/02.01.01_20/en_300401v020101a.pdf

3. ETSI TS 102 563, Digital audio broadcasting (DAB): Transport of advanced audio coding (AAC) audio. Online: https://www.etsi.org/deliver/etsi_ts/102500_102599/102563/01.02.01_60/ts_102563v010201p.pdf

4. C. Gandy, *DAB: an introduction to the Eureka DAB System and a guide to how it works*. BBC R&D White Paper, June 2013. Online: https://www.bbc.co.uk/rd/publications/whitepaper061

5. M. Häner-Höin, SW-Realisierung eines DAB-Empfängers mit GNU Radio, M.Sc. Thesis, ZHAW Zurich University of Applied Science, School of Engineering, Center for Signal Processing and Communications Engineering, 2011

6. W. Hoeg, T. Lauterbach, *Digital Audio Broadcasting. Principles and Applications of Digital Radio* (Wiley, Chichester, 2003)

7. IEC-62106 Standard, Specification of the radio data system (RDS) for VHF/FM sound broadcasting in the frequency range from 87.5 MHz to 108.0 MHz, International Electrotechnical Commission, 1999

8. H. Lu, Z. Dong, Carrier frequency offset estimation of DAB receiver based on phase reference symbol. IEEE Trans. Consum. Electron. **46**(1), 127–130 (2000)

9. T.K. Moon, *Error Correction Coding. Mathematical Methods and Algorithms* (Wiley Interscience, Hoboken, 2005)

10. P. Moose, A technique for orthogonal frequency division multiplexing frequency offset correction. IEEE Trans. Commun. **42**(10), 2908–2914 (1994)

11. J.G. Proakis, M. Salehi, G. Bauch, *Modern Communication Systems Using Matlab* (Cengage Learning, Stamford, 2004, 2013)

12. Ch.-R. Sheu, Y.-L. Huang, Ch.Ch. Huang, Joint symbol, frame, and carrier synchronization for Eureka 147 DAB system, in *Proc. 6th Int. Conf. on Universal Personal Communications*, San Diego, 12–16 Oct. 1997

13. K. Taura, M. Tsujishita, M. Takeda, H. Kato, M. Ishida, Y. Ishida, A digital audio broadcasting (DAB) receiver. IEEE Trans. Consum. Electron. **42**(3), 322–327 (1996)

14. J.J. van de Beek, M. Sandell, P.O. Börjesson, ML estimation of time and frequency offset in OFDM system. IEEE Trans. Signal Process. **45**(7), 1800–1805 (1997)
15. F. van de Laar, N. Philips, J. Huisken, Towards the next generation of DAB receivers. *Digital Audio Broadcasting - EBU Technical Review*, pp. 46–59, Summer 1997

Chapter 24
Modern Wireless Digital Communications: 4G and 5G Mobile Internet Access (with Grzegorz Cisek as a co-author)

Where are my binoculars? Living in a labyrinth of thousands of options and settings.

24.1 Introduction

I am sure that this chapter will be very interesting for many Readers. New times are coming with powerful 5G gadgets. Gadgets equipped with very advanced technologies and offering completely new functionalities. But how do they work from the signal processing point of view? This chapter aims at answering this question in the simplest possible way: by decoding physical broadcast channel (PBCH) of the 4G LTE mobile digital telephony signal. Having a Matlab program of the PBCH signal receiver, we will test it on synthetic and recorded (real-world) LTE signals. Finally, at the chapter end, after some generalization, a short introduction to the 5G wireless communication standard will be presented.

Legacy digital telephony technologies, like GSM and TETRA, made use of one carrier only. In GSM (Global System for Mobile Communication) bit 1 was represented by one period of sine lasting T microseconds, while bit zero by two periods of sine, but also lasting together T microseconds (2 times higher frequency). Since the signal was changed during zero-crossing, its phase continuity was ensured this way. Finally, the signal was filtered by Gaussian weights and minimum shift keying (MSK) modulation was being changed to Gaussian MSK (GMSK). In TETRA (Terrestrial Trunked Radio), used by government agencies and public emergency services (police forces, fire departments, ambulance), the differential quadrature phase shift keying $\pi/4$-DQPSK of one carrier

© Springer Nature Switzerland AG 2021
T. P. Zieliński, *Starting Digital Signal Processing in Telecommunication Engineering*, Textbooks in Telecommunication Engineering, https://doi.org/10.1007/978-3-030-49256-4_24

was used. Since only one carrier was used by one user in GSM and TETRA, the available bit-stream was limited in these two services.

In response to growing need for significantly higher throughput of mobile digital links, the multi-carrier technology, initially developed for DSL modems and DAB digital radio broadcasting, was applied with success in mobile telephony in the form of 4G Long Term Evolution (LTE) standard. In many aspects the multi-tone transmission used nowadays in modern mobile communications, called Orthogonal Frequency Division Multiplexing (OFDM) here, is very similar to DSL and DAB, already known to us. Main differences come from: (1) user mobility, (2) very fast channel variability (in regard to DSL), (3) different user access to the channel (dynamic sharing frequency resources between many users, in contrary to DSL and DAB)—in other words OFDMA (orthogonal frequency division multiple access), and (4) much more available application scenarios (for example, Internet of Things (IoT)).

How does the digital telephony work? In this chapter we will recover the broadcast PBCH signal of the LTE telephony standard. The PBCH contains essential information about a physical radio cell, for example, frequency bandwidth and system time (frame number), which is absolutely necessary for user equipment (UE) to connect to the network. The PBCH is the first signal which is decoded by a phone after power on.

As in DSL and DAB, a time-domain signal is a sum of many carriers and it is obtained by means of inverse FFT of complex-value Fourier coefficients. Their magnitudes and angles are defining carrier states, i.e. amplitudes and phase shifts. Each carrier can take only limited number of states and transmitted bits are equal to the carrier state number written binary, for example, state number 7 out of 16 means that bits 0111 are sent. Neighbor states should have given numbers which differ by one bit only (Gray code) in order to minimize bit errors caused during carrier state decoding. Time duration of each IFFT output is precisely defined. The synthesized signal is called an OFDM symbol. Precisely defined number of its last samples is copied to the signal beginning as a cyclic prefix (CP), already known to us. Typical carrier spacing in LTE is equal to 15 kHz. Due to different bandwidths used (i.e. 5, 10, 20 MHz), different number of carriers is exploited (300, 600, 1200, respectively), the IFFT has different lengths (512, 1024, 2048) and the OFDM symbol also. To observe the IQ signal spectrum we should calculated it via FFT, center it around 0 Hz and display

```
X=fft(x(1:Nfft))/Nfft;
X=fftshift(X);
f=(fs/Nfft)*(-Nfft/2:Nfft/2-1);
plot(f,20*log10( abs(X) );
```

The digital telephony frame is a sequence of OFDM symbols (IFFT outputs with inserted cyclic prefixes). One 4G LTE frame lasts 10 milliseconds and consists of ten 1 millisecond long sub-frames, each composed of two time slots. In consequence, one LTE frame has 20 slots lasting 0.5 milliseconds each. Since each slot is built from 7 OFDM symbols, the LTE frame has 140 symbols. Each OFDM symbol exploits 300, 600, or 1200 carriers. Brr... Too many numbers per a sentence! Let start from the beginning.

In LTE we are using a precisely defined time–frequency grid. Each carrier carries symbols delivering bits in number of their states. But the frequency bandwidth is shared between many users and they obtain a right to use some number of carriers in specified time, both in down-link (*to me*) and in up-link (*from me*). The smallest allocation amount of resources is called a Resource Block (RB), in LTE 12 consecutive carriers in 7 consecutive OFDM symbols. A user smart-phone has to obtain information from the base-station about resources allocated to it. In the beginning a smart-phone read an information from a SIM card where LTE frequency bands are located, scan them, acquire an IQ signal and look for some system/control information transmitted in broadcast mode. After reading the base-station physical cell identification, it reads the broadcast physical channel (PBCH) information and start dialog with the base-station *office* in order to obtain resource blocks for personal data up-down transmission. In this chapter, we will recover states of PBCH carriers.

To do this we have to perform a set of *time/frequency synchronizations* and system information decoding. First, during PBCH decoding, we assume that signal frequency bandwidth is equal to 1.4 MHz, i.e. estimate and correct the integer carrier frequency offset, i.e. integer multiplicity of FFT carrier spacing (15 kHz in LTE). Then, find primary synchronization signals (PSS), the Zadoff–Chu sequences, transmitted in the 6-th (0...6) OFDM symbol of the 0-th and 10-th slot of the LTE frame. Next, find the secondary synchronization signal (SSS) transmitted in the OFDM symbols just before the PSS symbols. Both searches exploits the cross-correlation function as a computational engine. When numbers of detected PSS and SSS signals are known, it is possible to calculate the physical radio cell identification (PCI) number and use it for generation of cell-specific reference signals (CRS) sent from different transmitter antennas. Correlating these antenna-specific pilot signals with the received signal we can find a number of antennas. Knowing this number, as well as transmitted and received CRS signals, we can:

1. estimate a channel frequency response in time–frequency grid points occupied by the CRS sub-carriers,
2. interpolate this channel estimate for remaining TF points,
3. correct all data (equalize the channel) for different antenna ports,
4. perform appropriate space–time signal decoding, the Alamouti one in our case.

In case of Alamouti space–time diversity coding, in the transmitter the same two signals are sent one-by-one by two antennas—but in different order and with sign negation. Then, in the receiver, they should be recovered from signals acquired by 1 or 2 receiver antennas. This space–time information diversity makes the transmission more immune to noise.

Since the broadcast PBCH information, sent to the smart-phone, is located in 4 OFDM symbols transmitted (located) just after the PSS symbol in the first LTE sub-frame, we perform the above-described CRS-based channel equalization and decoding in regard to the PBCH information. This will be the last step of our practical LTE signal decoding lesson: we will see a clear, more or less, scatter plot of PBCH carrier constellation points of the QPSK modulation. Bits can be recovered from them. And our dialog with a base-station starts.

Many modern *TeleDSP* topics are explained through Matlab exercises in [19]. Flexible contemporary telecommunication systems allow many options and settings. Too many to describe even 0.1% of them in an introductory book like this. But we hope that description how our smart-phones do initial synchronization with down-link LTE data and how they recover first bits of the very important broadcast information from it, presented in this chapter, helps us to understand philosophy of modern mobile digital telephony and, wider, modern wireless communication. After laboratory on DAB, my students were faced with LTE transmission standard in their professional work and I heard their opinion: "Wow, it is so similar!" I am sure that after reading this short description of the first steps performed by the 4G wireless receiver, the 5G one will be also very familiar to us. This chapter will end short DSP characterization of the incoming 5G New Radio standard. We will become 5G Ready!

When I am traveling in Krakow by a crowded bus and everybody around me are using their phones, I am thinking: "how on earth is it possible?!" But I see with my own eyes that it is possible. Even if students have problems with decoding an LTE signal acquired by ADALM-PLUTO hardware in university laboratories.

24.2 LTE Basics

Detail information concerning the 4G LTE telephony can be found in ETSI standards [9, 11]. Essential LTE characterization is also given in the book [27].

In the 4G LTE digital telephony, the down-link uses standard OFDMA modulation like in DSL and DAB, the discrete multi-tone one, while up-link—the DFT-spread OFDMA, called also a linearly-precoded OFDMA (LP-OFDMA) and a single-carrier frequency division multiple access (SC-FDMA). The abbreviation OFDMA (Orthogonal Frequency Division Multiple Access) in contrary to OFDM (Orthogonal Frequency Division Multiplexing), used in DSL and DAB, allows dynamic carrier allocation to different users: OFDMA = OFDM + Access. Parameters of the LTE down-link and up-link are summarized in Table 24.1.

First important information concerns the service bandwidth: 5, 10, and 20 MHz are the most frequently used. The carrier spacing Δf is assumed to be equal to 15 kHz and we are using 300, 600, or 1200 carriers, respectively. When we, for example, multiply 1200 carriers times 15 kHz, an approximate 18 MHz service bandwidth is obtained. The 2 MHz difference between declared 20 MHz and practically used 18 MHz is utilized for band-pass signal filtering and as frequency separation zone between different wireless services. The total number of carriers is increased to 512, 1024, or 2048, the nearest power of 2. After multiplication of the DFT length by 15 kHz, we obtain sampling frequency equals 7.68, 15.36, and 30.72 MHz. Therefore, during IQ signal synthesis in the base-band, we are:

1. specifying values of 512, 1024, or 2048 Fourier coefficients (carrier values/states),
2. performing inverse fast Fourier transform (IFFT) with appropriate length, obtaining 512, 1024, or 2048 samples of complex-value IQ signal,
3. inserting a cyclic prefix to the signal,
4. sending signal samples to quadrature modulator and D/A converter, working with frequencies 7.68, 15.36, or 30.72 MHz.

The cyclic prefix length, normal or extended, is specified in Table 24.2. 12 consecutive carriers lying in 7 consecutive IQ OFDM symbols constitute one Resource Block (RB) that is individually allocated to different users. It is the smallest allocation resource in LTE. Channels with different bandwidths use different number of RBs.

Table 24.1: 4G LTE down-link and up-link specification

LTE Mode	1	2	3	4	5	6
Channel bandwidth BW [MHz]	1.4	3	5	10	15	20
Sub-carrier spacing [kHz]	15	15	15	15	15	15
Number of sub-carriers used	72	180	300	600	900	1200
FFT/IFFT size	128	256	512	1024	1536	2048
Sampling rate [MHz]	1.92	3.84	7.68	15.36	23.04	30.72
Number of RBs[a] in frequency	6	15	25	50	75	100
Signal samples per slot	960	1920	3840	7680	11520	15360

[a]Resource Block = 12 carriers in 7 OFDM symbols

Table 24.2: Cyclic prefix length in LTE down-link in microseconds

Scenario	First symbol in RB[a]	Next symbols in RB
Normal CP, $\Delta f = 15$ kHz, 7 symbol/slot	5.21 µs	4.69 µs
Extended CP, $\Delta f = 15$ kHz, 6 symbol/slot	16.67 µs	16.67 µs
Extended CP, $\Delta f = 7.5$ kHz, 6 symbol/slot	33.33 µs	33.33 µs

[a]Resource Block

The 10 millisecond long LTE frame is presented in Fig. 24.1. It consists of ten 1 millisecond long sub-frames, each having two time slots. The total number of time

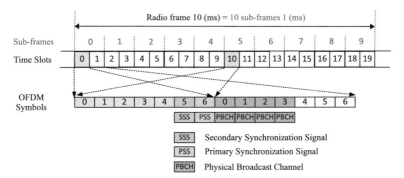

Fig. 24.1: Frame structure used in 4G LTE digital telephony

slots is, therefore, equal to 20 and the slots are numbered from 0 to 19. Each slot is built typically from 7 OFDM IQ symbols/signals (6 in case of extended cyclic prefix), numbered from 0 to 6. In the 5-th and 6-th OFDM symbols of time slots number 0 and 10, Secondary (SSS) (first) and Primary (PSS) (then) Synchronization Signals are sent. In first four OFDM symbols of time slot number one, the Physical time slot number one, the Physical Broadcast Channel (PBCH) information is transmitted to all phones: it carries the Master Information Block (MIB) message, which encapsulates the most essential cell configuration parameters and must be decoded by the mobile before the network connection can be established. Our goal in this chapter is to see clear QPSK constellation points of the PBCH data for an LTE signal recorded by ADALM-PLUTO SDR receiver.

In Fig. 24.2 a more detail view of the first two time slots of the LTE frame is presented. Data are changing in time (horizontally, time slots and OFDM symbol numbers) and in frequency (vertically, carrier numbers in Resource Blocks and numbers of RBs). What is important to observe? In the upper figure only 8 central RBs (around 0 Hz in the base-band) are shown. Dots "•••" means that there are RBs up and down, for higher and lower frequencies. In fact, we are interested at present only in 6 central RBs, i.e. in 72 central sub-carriers, lying 15 kHz apart from each other, 36 for positive and 36 for negative FFT frequencies. On the bottom figure only 12 of them are shown, i.e. from -36 to -25. In the 5-th time slot, the SSS signal is sent, while in the 6-th—the PSS signal. The PSS signal allows initial LTE frame synchronization and initial channel equalization, while the SSS signal is used, in conjunction with PSS, for identification of the actual physical radio cell number. In the following 4 time slots, the PBCH data are transmitted using QPSK modulation. Unused time–frequency cells (slots) are marked with "X." In TF points "A1," "A2," "A3," and "A4" the LTE base-station Cell-Specific Reference Signals (pilots)(CRS) are sent by different transmitter antenna ports. They allow: (1) identification of number of antenna ports, used by the base-station (CRS signal in "Ak" point is present or not), (2) channel identification and channel equalization for each antenna port. Reference CRS signals are sent permanently by the base-station. They should be generated in the mobile receiver for cross-correlation purposes but it is

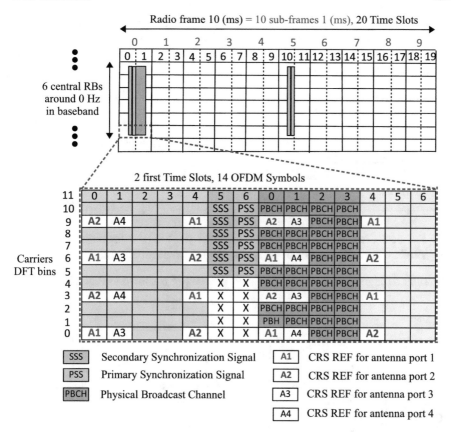

	SSS	Secondary Synchronization Signal		A1	CRS REF for antenna port 1
	PSS	Primary Synchronization Signal		A2	CRS REF for antenna port 2
	PBCH	Physical Broadcast Channel		A3	CRS REF for antenna port 3
				A4	CRS REF for antenna port 4

Fig. 24.2: Position of PSS (primary) and SSS (secondary) synchronization signals in the LTE radio frame as well as positions of physical cell-specific reference signals (CRS) from transmitter antennas A1, A2, A3, and A4 (antenna ports 0–3 in the standard)

possible only after identification of the physical radio cell identifier (from PSS and SSS). Our goal in this chapter is to decode the PBCH carrier states.

The bottom diagram of Fig. 24.2 is a very good resume of modern telecommunication technology: repeat it 100 times vertically (100 RBs for $N_{FFT} = 2048$) and *look for a needle in a haystack!*

At present, let us look at the 4G LTE telephony from the system point of view, top-down perspective. LTE coverage is divided into physical cells, supervised by base-stations. Control signals are physical cell-dependent, including the Zadoff–Chu primary synchronization PSS sequences, secondary synchronization SSS sequences, and CRS pilots of different antenna port. Data are transmitted to many users at the same time but different users exploit different carriers (frequencies) in different time slots. The frequency allocation is dynamic and a user phone/equipment is permanently instructed by a down-link data control information (DCI).

Carrier states are *carrying* bits. These bits belong to different *streams* called channels, having different destination. Transmission is bi-directional, down (to the user smart-phone) and up (to the base-station). In this chapter we are interested in down-link mainly. Here we have:

1. the Physical Broadcast Channel (PBCH), constant in time, which allows a phone configuration according to the base-station instructions,
2. the Physical Downlink Control Channel (PDCCH) dynamically informing a phone about resources allocated to it (for example, about time–frequency resource blocks to be used),
3. the Physical Downlink Shared Channel (PDSCH) containing data transmitted to the user,
4. Physical HARQ Indicator Channel (PHICH) carrying ACK/NACK acknowledge information for Up-link transmission (1 bit),
5. Physical Control Format Indicator Channel (PCFICH) informing phone about PDCCH format (number of OFDM symbols), transmitted in each sub-frame.

Simplified diagram of 4G LTE down-link transmitter is shown in Fig. 24.3.

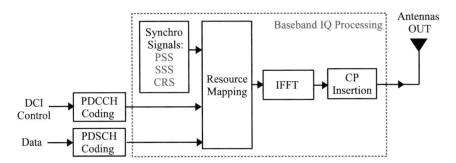

Fig. 24.3: Simplified down-link LTE transmitter diagram: PSS—primary synchronization signal, SSS—secondary SS, CRS—cell-specific reference signal

The control PDCCH and data PDSCH channels are coded in different way (see Fig. 24.4). The PDCCH data are attached with CRC, convolutionally encoded (then for sure Viterbi decoded in the receiver), scrambled, time-interleaved, and used for only two-bit QPSK modulation. In turn, the PDSCH data are attached with CRC, turbo coded, interleaved and they are exploiting higher order QAM modulations (16, 64, 256). In PDSCH rate matching is more complex than in PDCCH. The AMC (Adaptive Modulation and Coding) technique is used: number of redundancy bits, added to the original ones and defending them against errors, is dynamically changed and depends on transmission conditions. Additionally the HARQ (Hybrid Automatic Repeat reQuest) protocol is exploited: during transmission only a part of a code-word is sent. When a receiver has failed to decode the code-word, its another part is sent. During decoding, the receiver makes use of all previous unsuccessful decoding attempts. This method, called IR (Incremental Redundancy), was used in 3G-HSPA also.

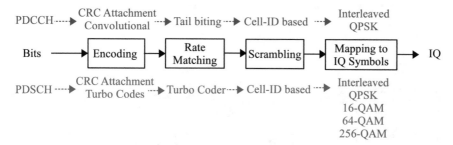

Fig. 24.4: Physical Downlink Control Channel (PDCCH) and Shared Channel (PDSCH) encoding in 4G LTE standard (pre-coding and layer mapping are also performed but not shown)

The 4G LTE receiver diagram, the simplest one, is presented in Fig. 24.5. First, we should find the LTE frame beginning in the input IQ base-band signal. For this purpose we should calculate the cross-correlation function between the received signal and the reference three Zadoff–Chu sequences, known to the receiver (one of them was used by transmitter for the primary synchronization signal (PSS) generation). When we synchronize with the OFDM symbol of the PSS signal, we can use its cyclic prefix (CP) for initial carrier frequency offset estimation, exactly the same way as we did in DAB radio. After that, knowing the index of the transmitted and received PSS ($N_{ID}^{PSS} = 0, 1, 2$), we can perform initial estimation of a channel frequency response. Then, knowing already the channel, we can go to the OFDM symbol before the PSS, to the secondary synchronization signal SSS, equalize it using the PSS-based channel estimation (optionally) and decode. Since the SSS can take *only* one of 168 different forms, we find its number $N_{ID}^{PSS} = 0...167$, and use it for calculation of physical radio cell identification:

$$N_{ID}^{CELL} = 3 * N_{ID}^{SSS} + N_{ID}^{PSS} \tag{24.1}$$

Knowing N_{ID}^{CELL} of the physical wireless cell we can generate the cell-specific reference synchronization signals (CRS) (marked as A1, A2, A3, and A4 in Fig. 24.2). And after this, use them for: (1) determination of number of transmitting base-station antenna ports, and (2) local, in time and frequency, channel equalization during decoding the control (PDCCH) and data (PDSCH) bit-streams. Implementation of channel interpolation for non-CRS TF bins can be linear over frequency axis and zero-hold-order over time axis (using last estimate)—the standard does not restrict it. The regular work of the down-link receiver is presented in Fig. 24.5. After initial synchronization, the PSS and SSS signals are tracked all the time in order to allow a phone performing handover to another base-station with stronger signal. The cyclic prefix of each OFDM symbol can be used for permanent adaptation of the carrier frequency offset. When a base-station registers a phone as an active player in the game and at least 4×4 MIMO antenna system is used, it can send to the phone user-equipment specific reference signals (UERS), which is not marked in Figs. 24.3 and 24.5 since it is optional in the standard.

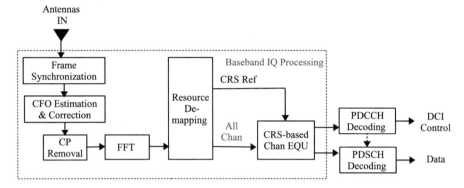

Fig. 24.5: Simplified down-link LTE receiver diagram: CRS—cell-specific reference signals, PDCCH—physical downlink control channel, PDSCH—physical downlink shared channel

The main difference between 4G LTE down-link (from base-station) and up-link (from our phone) transmitter relies on the fact that the base-station IFFT takes into account data transmitted to all users over all available carriers—data for different users are transmitted using different resource blocks RBs (12 consecutive carriers in 7 OFDM symbols). In case of up-link transmitter, each phone performs its own IFFT, which should be synchronized with other phones IFFTs, and sends "up" only data of one user exploiting allocated RBs (specified carriers in specified time slots). These data are cut into blocks, extra processed by DFT, allocated to target Fourier coefficients, and finally performed by IFFT. Due to the DFT usage and special spectral data allocation, peak transmission power is reduced. Difference between down-link (CP-OFDMA) and up-link (Single-Carrier FDMA) FFT usage is presented in Fig. 24.6.

Physical Antenna vs. Antenna Port Typically distinguishing between a physical antenna and an antenna port concept, used in 3G/4G/5G standards, is one of the biggest problems for a novice in digital telephony world. For this reason, the earlier we clear the difference between these two terms, the better for further reading. The antenna port, defined in the 4G-LTE and 5G-NR standards, does not directly correspond to the physical antenna on which the radio signal is transmitted or received. It is a logical concept that puts together (combines) different data streams and reference signals. More strictly speaking, the mobile terminal can assume that all the signals conveyed over the same antenna port

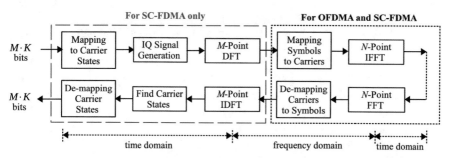

Fig. 24.6: Block diagram explaining the difference between CP-OFDMA used in 4G LTE down-link transmitter and SC-FDMA used in 4G LTE up-link transmitter. $N > M$ and K denotes modulation order. Each input data symbol in SC-FDMA makes use of $M \cdot \Delta f = M \cdot 15$ kHz

undergone the same radio propagation conditions. Based on this, the mobile can determine which data stream can be equalized using a particular reference signal. In case of LTE down-link, PBCH, PDCCH, and PDSCH channels are sent over the same set of antenna ports as the CRS signal. Thanks to this the channel impulse response estimate based on the CRS can be used to equalize the aforementioned channels. In practice, signals mapped to the same logical antenna port are transmitted using the same physical antenna (or multiple antennas). In some specific cases, signals mapped to different antenna ports may be still transmitted on the same physical antennas, but the mobile terminal is not allowed to make such assumption.

More on SC-FDMA and PAPR At this point, the reader is probably asking the questions: Why there are different waveforms used in LTE down-link and up-link? What is the benefit of using SC-FDMA in up-link direction? I am not convinced after above one sentence long explanation!

To answer the questions, let us start from the beginning. In OFDMA systems, modulation IQ symbols are mapped to corresponding sub-carriers in frequency domain prior to IFFT. The IFFT is in fact a linear transformation, so the resulting time-domain signal is likely to have non-constant amplitude with occasional higher peaks. To characterize a signal in term of its relative peak amplitude, the Peak-to-Average Power Ratio (PAPR) metric is commonly used, expressing the ratio between the highest peak value x_{peak} to an average Root-Mean-Square (RMS) signal level x_{rms} over an arbitrary period:

$$\text{PAPR} = \frac{|x_{\text{peak}}|^2}{x_{\text{rms}}^2}$$

Occurrence of high peaks is very undesirable, as it leads to signal distortion occurring in the analog circuitry of the transmitter. One of the components that is most vulnerable to high-energy peaks in the signal is the power amplifier due to its high non-linearity characteristics. Therefore, the higher the PAPR is (or in other words: the higher energy peaks are occurring in the signal), the higher is the signal distortion due to circuit non-linearity, which in turn lowers the probability of successful information decoding at the receiver.

Thanks to the additional DFT spreading stage at the transmitter, SC-FDMA is characterized by a much lower PAPR than OFDMA scheme. This is achieved at a cost of negligible performance degradation [16]. In turn, utilization of SC-FDMA technique in LTE up-link allows the mobile phone manufacturers to equip less expensive power amplifiers into the devices, which reduces the total price that have to be paid for a new mobile phone. Secondly, lower PAPR allows the mobile phone to transmit with reduced power, so it can operate longer on a single battery charge.

OFDM vs. OFDMA And important remark at the end of this section. In OFDM technique channel is divided in frequency for many users *once forever*, i.e. one user exploits the same frequency. In turn, in OFDM Access (OFDMA) technology, used in 4G LTE and 5G NR, the carrier allocation to users is changed dynamically.

24.3 Decoding 4G LTE Physical Broadcast Channel

In this chapter we write a Matlab program for DSP processing an base-band IQ LTE signal, recorded by ADALM PLUTO, and displaying constellation points of the Physical Broadcast Channel (PBCH) signal. Its location in time–frequency grid of the LTE frame is presented in Fig. 24.2. In order to obtain this goal we will have to extract PSS and SSS, primary and secondary synchronization signals, from the LTE signal and find a physical radio cell ID as intermediate tasks.

There are several references telling us in synthetic way how this should be done [6–8, 13, 22–25, 27]. Of course, we can also try to extract directly the required information from the ETSI standards [9, 11].

24.3.1 LTE Signal Recording and Its Spectrum Observation

To perform this, we should first choose an LTE provider. In this web page we have listed all European countries and different telephony vendors. We have chosen Poland, Py Telecom and its 5 MHz LTE service in B20 800 MHz frequency band (on the green background). In table 5.5-1 "E-UTRA operating bands" of the 3GPP TS 36.101 standard (ver. 13.4) and on this web page we have found exact frequency specification of the B20 band:

B20, FDD, 800 MHz, up-link 832–862 MHz, down-link 791-821 MHz, channels 5, 10, 15, 20 MHz.

Then we have used ADALM PLUTO, observe the signal spectrum in the down-link frequency range, find the local spectrum minimum, which should be visible for the not used DC LTE carrier, and took decision to record an IQ signal using center frequency 804 MHz and sampling frequency equal to 30.72 MHz. Then after spectrum calculation and its more careful observation in Matlab, we repeat recording for the center frequency equal to 803.505 MHz. For this frequency our goal was reached: the LTE spectral "hat" was centered around 0 Hz in the base-band, additionally the spectrum had local minimum at the 0 Hz spectral bin which is the case of LTE since it is not using the DC carrier. This file is processed in the next subsections.

The analyzed IQ LTE signal, sampled at 30.72 MHz, has 2,457,600 samples and consists of 8 LTE frames of the 5 MHz LTE service, lasting together 80 milliseconds. Its spectra are presented in Fig. 24.7. In the upper-left figure we see an FFT result of the whole signal as a function of frequency. In the bottom-right plot one can observe a spectrum resulting from averaging sixteen 2048-point FFTs—only its central part is shown with the LTE service. The signal spectrogram (STFT) and Welch PSD complement the figure.

Listing 24.1: Matlab program for PBCH decoding in LTE signal

```
1  % lab24_ex_4g_lte.m
2    clear all; close all;
3
4    do_LPfilter = 0;        % 0/1 low-pass filtration, passing only used carriers PBCH
                               carriers
5    do_disturbing = 0;      % 0/1 disturbing the signal
6    do_ADCcorrect = 0;      % 0/1 correcting ADC sampling rate
7    do_plots = 1;           % 0/1 displaying figures
8
9  % Load input signal: real-world recorded or synthesized
10    load('LTE_803.505MHz_30.72MHz_Short.mat'); y = samples; clear samples;
11  % load('LTE20.mat'); y = new.'; clear new;
12  fs=30.72e6; fs_lte=7.68e6; THR=15; REPEAT=0;
13
14  % Observe high-resolution signal spectrum
15    N = length(y);                                          % signal length
16    %y = y .* exp( -j*2*pi*(0 : N-1) * ( 5*df_lte )/ fs );  % TEST integer CFO
17    plot_freq(y, fs); title('Input signal spectrum'); pause % function call
18    figure; pwelch(y,2048,2048-1024,2048,fs,'centered'); pause % PSD Welch
19
20  % Observe low-resolution signal spectrum
21    Nfftup = Nfft*fs/fs_lte; Ncut = 16*Nfftup; df = fs/Nfftup;
22    plot_freqbins(y(1:Ncut), Nfftup); title('|Y(k)| before shift, 15kHz step'); pause
```

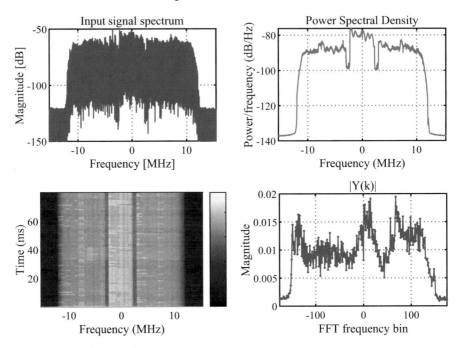

Fig. 24.7: Spectrum of IQ LTE signal sampled at $f_s = 30.72$ MHz: (up-left) one FFT of all 2,457,600 IQ signal samples, around 0 Hz we see a 5 MHz LTE service, (up-right) Welch power spectral density, (down-left) short-time Fourier transform, (down-right) 2048-point FFT, averaged 16 spectra, only central part of the spectrum with the LTE service is shown

```
23
24  function plot_freq(x, fs)
25    N = length(x);                        % signal length
26    X = abs( fftshift (fft(x)) ) / N;     % FFT calc., centering, magnitude, scaling
27    f = (fs/N * (-N/2 : N/2-1) ) / 1e+6;  % frequency axis in MHz
28    figure;
29    plot(f, 20 * log10(X), 'b'); xlim( [f(1), f(end)] ); grid;
30    xlabel('Frequency [MHz]'); ylabel('Magnitude [dB]');
31  end
```

Exercise 24.1 (Observing   LTE   Spectra).   Run   program lab24_ex_4g_lte.m for all supported MAT and TXT files with LTE signals, real-world and synthesized. Observe four types of spectra: one long FFT, averaged sixteen 2048-point FFTs, Welch PSD, and STFT spectrogram. Analyze code of the function plot_freqbins(x, Nfft).

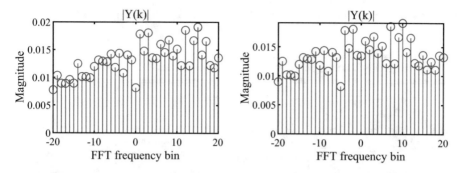

Fig. 24.8: Zooming LTE signal spectrum from Fig. 24.7: (left) spectrum of the recorded IQ signal with no integer carrier frequency offset (CFO), (right) spectrum of the signal with integer CFO equal 5 FFT bins left, added artificially

24.3.2 Signal Pre-processing and Integer CFO Correction

Integer CFO Estimation and Correction Observing spectrum of a recorded signal allows us to find and correct the integer carrier frequency offset (CFO), if it is present. It is demonstrated in Fig. 24.8, presenting a zoomed spectrum from the Fig. 24.7. In the left plot the integer CFO is not observed since the spectrum has local minimum at frequency 0 Hz. In turn, in the right plot we see the integer CFO equals minus 5 FFT bins, which was artificially added to the IQ signal by means of its multiplication with the signal:

$$e^{-j2\pi \frac{\Delta f}{f_s} \cdot (0:N-1)} = e^{-j2\pi \frac{k f_0}{f_s} \cdot (0:N-1)} = e^{-j\frac{2\pi}{N_{fft}} k \cdot (0:N-1)}, \qquad (24.2)$$

where $f_s = 30.72$ MHz, $N_{fft} = 2048$, $f_0 = \frac{f_s}{N_{fft}} = 15$ kHz, $k = 5$, $\Delta f = k \cdot f_0 = 75$ kHz, N—number of signal samples. Of course, the *injected* integer CFO can be corrected by the *sign-reversed* signal multiplication:

$$e^{j2\pi \frac{\Delta f}{f_s} \cdot (0:N-1)} = e^{j2\pi \frac{k f_0}{f_s} \cdot (0:N-1)}. \qquad (24.3)$$

Listing 24.2: Matlab program for PBCH decoding in LTE signal (cont.)—integer CFO

```
1  % lab24_ex_4g_lte.m - continuation
2
3  % Calculate or choose required integer spectrum shift, then perform it upon the signal
4    Kmax = 20;                                       % select search size
5    s = reshape( y(1:Ncut), Nfftup, Ncut/Nfftup );   % create a signal matrix
6    S = fftshift( fft(s) / Nfftup );                 % do FFTs of its columns
7    S = mean( abs(S.') );                            % accumulate all spectra
8    kcentr = Nfftup/2+1;                             % index of spectrum center
9    [val, indx ] = min( S(kcentr-Kmax:kcentr+Kmax) );% find minimum and its position
10   SHIFT = indx-Kmax-1, % 1234?                     % choose/calculate shift value
```

```
11     y = y .* exp( -j*2*pi*(0 : N-1) * (SHIFT*df) / fs );    % do integer shift
12     % y = y .* exp( -j*2*pi*(0 : N-1) * (-1250)/ fs );       % do fractional shift?
13
14     % Observe spectrum of the resultant signal
15     plot_freqbins(y(1:Ncut), Nfftup); title(' |Y(k)| after shift, 15kHz step'); pause
```

Exercise 24.2 (Integer CFO Observation and Correction). Run program
`lab24_ex_4g_lte.m` for the recorded LTE signal. Uncomment line TEST
integer CFO. Set different spectrum shift value, i.e. integer multiples of df.
Observe results. Write better function for automatic finding of the spectrum
shift that makes use of the overall width of the LTE "hat."

Signal Re-sampling and Low-Pass Filtration In our example in the IQ signal
three LTE services are present: one 5 MHz (in the middle, around 0 Hz) and
two 10 MHz (one in negative frequencies and one in positive frequencies). We
have to extract the service of interest from the signal. For this purpose the Matlab
`resample()` function is used. It performs two things in our case: (1) the low-pass
filtration reducing the signal bandwidth 4 times from 30.72 MHz to 7.68 MHz, and
(2) 4:1 signal decimation from sampling frequency 30.72 MHz to 7.68 MHz. Then,
we can further reduce the signal bandwidth from 7.68 MHz to 5 MHz—in our service only 300 central carriers are used out of 512 (the (I)FFT length). In Fig. 24.9,
in two upper plots the following signal PSD spectra are shown: (1) after 4-times
bandwidth reduction and down-sampling, (2) after additional low-pass filtering. In
the bottom plot a mean 512-point FFT spectrum of the signal is presented—16 FFT
spectra are computed and averaged.

Listing 24.3: Matlab program for PBCH decoding in LTE signal (cont.)—signal
re-sampling and low-pass filtration

```
1     % lab24_ex_4g_lte.m - continuation
2
3     % Signal re-sampling - when actual sampling frequency is not a standard one
4     if( fs ~= fs_lte )
5         y = resample( y, 1, fs/fs_lte ); fs = fs_lte;    % LP filtering and down-sampling
6         N = length(y);                                    % new signal length
7         plot_freq(y, fs); title('Down-sampled'); pause
8     end
9
10    % Optional low-pass filtration - passing only carriers used by LTE or PBCH
11    if( do_LPfilter )
12        Mtaps = 125;                                       % number of filter weights
13        % freq = [0, Nsc * df_lte,  BW_lte, fs] / fs;      % only used LTE sub-carriers
14        freq = [0, 72 * df_lte, 4*72*df_lte, fs] / fs;     % only PBCH sub-carriers
15        gain   = [ 1 1 0 0 ];                              % gains
16        weights = [ 1    0.8 ];                            % bands significance
17        h = firls( Mtaps-1, freq, gain, weights);          % least-squares design
18        y = conv(y, h, 'same');                            % filtering
```

```
19  plot_freq(y,fs); title('Filtered'); pause        % result
20  plot_freqbins(y(1:Ncut), Nfft); title('Filtered'); pause
21  end
```

Exercise 24.3 (Signal Low-Pass Filtration). Design different taps of low-pass filters using Matlab functions `fir1()`, `fir2()`, `firls()`, `firpm()`. Add to the program plotting filter frequency response. Design filters passing only central 72 FFT frequency bins, not 300 bins, i.e. only the PBCH part. Observe how remaining part of the program accepts more narrow signal bandwidth.

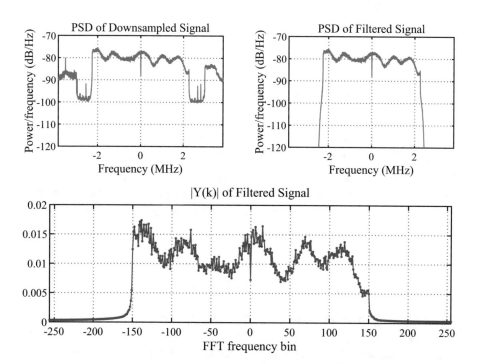

Fig. 24.9: Spectra of IQ LTE signal: (up-left) PSD after down-sampling from 32.72 MHz to 7.68 MHz, (up-right) PSD after low-pass filtration with BW=5 MHz—leaving only 300 carriers used, (down) averaging sixteen 512-point FFT spectra

24.3.3 Signal Auto-Correlation: Fractional Carrier Frequency Offset Estimation and Correction

The fractional carrier frequency offset is the most harmful for OFDM-based signal transmission because it leads to losing carrier orthogonality. Additionally, time synchronization procedures making use of special sent synchronization patterns usually work better when the CFO is compensated. We are assuming that the integer CFO has been already removed by us during signal acquisition—using spectrum inspection and correction. At present we should take care of the fractional CFO only. Since Cyclic Prefix is used in LTE OFDM symbols, we can exploit it for: (1) OFDM symbols localization, as described in the Chap. 22 on DSL, and (2) fractional CFO estimation, as described in the Chap. 23 on DAB. In the first task, we are looking for maxims of the following auto-correlation-based cost function (see Eqs. (22.37), (22.38), (22.39) in Chap. 22 on DSL):

$$J(n) = \frac{Q(n)}{R(n)}, \tag{24.4}$$

$$Q(n) = \sum_{m=0}^{N_{CP}-1} y(n+m)y^*\left(n+m+N_{fft}\right), \tag{24.5}$$

$$R(n) = \sum_{m=0}^{N_{CP}-1} |y(n+m)|^2 + \sum_{m=0}^{N_{CP}-1} \left|y\left(n+m+N_{fft}\right)\right|^2. \tag{24.6}$$

Their arguments indicate first samples n_{1cp} of the OFDM symbol cyclic prefixes. In the second task, we are calculating the fractional CFO for each OFDM symbol using its CP, which the first sample n_{1cp} has been already found (see Eq. (23.8) in DAB chapter):

$$\Delta f = f_0 \cdot \frac{\text{angle}\left(\frac{1}{N_{cp}} \sum_{n=n_{1cp}}^{n_{1cp}+N_{cp}-1} y^*(n)y(n+N_{fft})\right)}{2\pi}, \tag{24.7}$$

where $f_0 = \frac{f_s}{N_{fft}}$. In order to obtain a noise-robust fractional CFO estimate, initially we should calculate CFO for many OFDM symbols, find the mean value, and slowly adapt it using new CFO estimates. The fractional CFO correction is realized by multiplication of the IQ signal $y(n)$ with:

$$e^{-j2\pi\frac{\Delta f}{f_s}\cdot(0:N-1)}. \tag{24.8}$$

In upper plot of Fig. 24.10 the calculated cost function $R_{CP}(n)$ is shown. Its maxims tell us about cyclic prefix positions in the IQ signal. Having these knowledge, we can calculate fractional CFO for different OFDM frames (see bottom plot in Fig. 24.10) and average obtained results. In our case frequencies -1189.1 Hz and

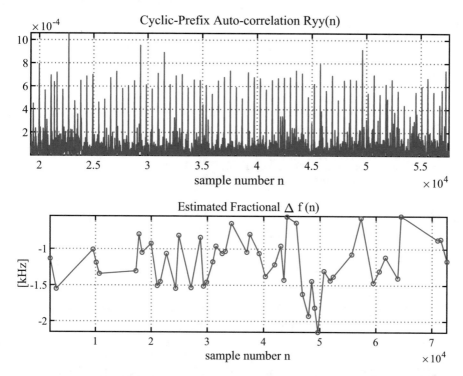

Fig. 24.10: Fractional CFO estimation using CP-based auto-correlation of the LTE signal defined by Eq. (24.6): (up) auto-correlation, we observe repeating peaks of cyclic prefix self-matching: we can use them for time synchronization with OFDM symbols and for estimation of fractional carrier frequency offset, (down) estimated fractional carrier frequency offset

−1221.6 Hz were found this way, the first when the additional low-pass filter was used, the second—without the filter.

Listing 24.4: Matlab program for PBCH decoding in LTE signal (cont.)—signal auto-correlation function and fractional CFO estimation and correction

```
1  % lab24_ex_4g_lte.m - continuation
2
3  % 3. Signal Auto Correlation - fractional carrier frequency offset estimation
4
5     Ryy=[];
6     for n = 1 : Nframe % find correlation coefficients
7         n1 = n:n+Ncp2-1; n2 = n1 + Nfft;
8         a1 =  abs( sum( y(n1) .* conj( y(n2) ) ) );
9         b1 =  abs( sum( y(n1) .* conj( y(n1) ) + sum( y(n2) .* conj( y(n2) ) ) ));
10        Ryy(n) = (a1*a1) / (b1*b1);
11    end
12    figure; plot(Ryy); grid; title('Original Ryy(k)'); pause
```

```
13    for k = 1 : 20 % find 20 max peaks
14        [val,indx(k)] = max(Ryy);                                    % find max
15        Ryy( indx(k)-Nsymb2 : indx(k)+Nsymb2 ) = zeros(1,2*Nsymb2+1); % zero around
16    end
17    indx = sort(indx);
18    for k = 1 : 20 % use CP for fractional CFO estimation
19        n1 = indx(k):indx(k)+Ncp1-1;
20        CFO_cp(k) = fs/Nfft*mean( angle( conj( y(n1) ) .* y( n1+Nfft ) ) / (2*pi));
21    end
22
23      if( do_plots ) % Fractional CFO
24          figure; plot( indx(1:20), CFO_cp(1:20),'bo-' ); grid;
25          xlabel('sample index n'); ylabel(' [Hz]'); title('\Delta f (n)'); pause
26      end
27
28    CFO = mean( CFO_cp ), pause                   % fractional CFO estimation
29    y = y .* exp(-j*2*pi*(0:length(y)-1)*(CFO/fs)); % fractional CFO correction
30    CFO = 0;                                       % already corrected
```

Exercise 24.4 (Signal Auto-Correlation Function and Fractional CFO Estimation and Initial Correction). Set do_disturbing=0. Run the program lab24_ex_4g_lte.m for different synthesized LTE signals and observe auto-correlation function shapes as well as plotted fractional CFO detection curves. Set do_disturbing=1, ADCppm=0, G0=0, G1=0, G2=0, npwr=−160 (see Listing 24.11). Run the program changing cfo=0, 100, 500, 1000, 2000, 5000. Then systematically increase the noise power npwr=−100,−80,−60,−40,−20 and repeat experiment with growing fractional cfo.

24.3.4 PSS Signals: Frame Synchronization, CFO Correction, and Channel Estimation

The primary synchronization signal (PSS) is sent two times per LTE frame, in the beginning and in the middle, in time slot number 0 and 10, both time in OFDM symbol number 6. Only 62 carriers around DC are exploited which is shown in Fig. 24.2. The PSS signal is used for LTE frame synchronization, necessary channel estimation and equalization for SSS signal decoding, and physical LTE radio cell identification from Eq. (24.1). Knowing the physical wireless cell identity (PCI) allows for physical broadcast channel (PBCH) decoding.

Introduction to Zadoff–Chu Sequences The ZC sequences [4, 12] were successfully adapted to 4G LTE due to their unique properties, succeeding the Walsh–Hadamard codes used in the legacy 3G communication. They are applied in several places in the LTE standard, mostly for synchronization purposes, including PSS sequence in down-link, Physical Random Access Channel (PRACH) used for asynchronous initial cell access in up-link direction, as well as user-specific reference signals for PUSCH channel. Moreover, the Zadoff–Chu sequences are also used in other fields of study, e.g. radar signal processing.

A Zadoff–Chu sequence of length N_{ZC} is given as a series of complex numbers with constant, unitary magnitude, so the values differ only by the phase. Subsequent element in the sequence is determined by the following equation:

$$x_u(n) = e^{-j\frac{\pi un(n+1+2q)}{N_{ZC}}}, \tag{24.9}$$

where $u \in \{0, 1, \ldots, N_{ZC} - 1\}$ is a base sequence index. Remark that LTE standard assumes $q = 0$ for all its applications.

The Zadoff–Chu sequences exploit four important properties:

1. If N_{ZC} is a prime number, the sequence is cyclic with period N_{ZC}:

$$x_u(n + N_{ZC}) = x_u(n). \tag{24.10}$$

2. Auto-correlation of a Zadoff–Chu sequence with its shifted version is equal to 0 if the cyclic shift is not a multiplication of the sequence length N_{ZC}:

$$x_u(n)x_u^*(n+c) = \begin{cases} 0 & \text{if } \mathrm{mod}(c, N_{ZC}) \neq 0 \\ 1 & \text{otherwise} \end{cases}. \tag{24.11}$$

3. If N_{ZC} is a prime number, Fourier transform of the Zadoff–Chu sequence is also a Zadoff–Chu sequence with different root index u, but conjugated and scaled.
4. Cross-correlation between two Zadoff–Chu sequences of prime length N_{ZC} is constant and equal to:

$$x_{u_1}(n)x_{u_2}^*(n+c) = \sqrt{\frac{1}{N_{ZC}}}, \tag{24.12}$$

if the absolute difference between the root indices $|u_1 - u_2|$ is relatively prime to N_{ZC}.

Zadoff–Chu PSS Sequences and Their Generation After initial fractional CFO correction of the recorded IQ LTE signal, three Zadoff–Chu sequences are generated in the LTE receiver, which are used for primary synchronization. First their 64-point DFT spectra $S(k)$ are specified for a *root parameter* $u = 25, 29, 34$, according to the following formula [11]:

$$S_u(k - 31) = \exp^{-j\pi uk(k+1)/63}, \quad k = 0, 1, \ldots, 30, 32, 33, \ldots, 62. \tag{24.13}$$

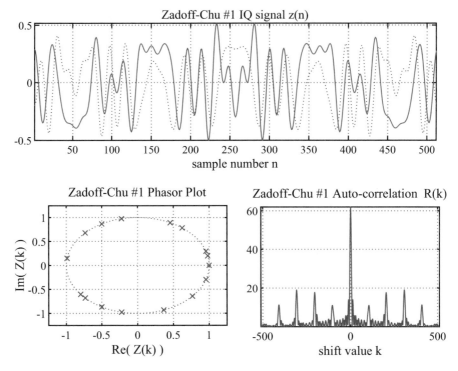

Fig. 24.11: Characterization of Zadoff–Chu PSS synchronization sequence number 1: (up) 512-samples long IQ signal $z_1(n)$, synthesized from 62 Fourier coefficients $Z_1(k)$, (down-left) phasor plot of all 62 Fourier coefficients $Z_1(k)$, i.e. imaginary part as a function of real part, (down-right) auto-correlation function of the time sequence

$S_u(k), k = -32...31$, represent values of DFT coefficients, from the lowest negative to the highest positive frequency. $S_u(-32)$ and $S_u(0)$ are equal to 0. Then, the spectrum is appended on both sides with zeros to the length N_{fft}, 512 in our example, and the inverse FFT is performed.

In Fig. 24.11 Zadoff–Chu sequence number one is characterized. In the upper plot we see real part (solid line) and imaginary part (dotted line) of the time sequence ($N_{fft} = 512$ samples). In the bottom-left plot values of the ZC spectrum coefficients are presented as a phasor-diagram, i.e. imaginary part as a function of real part. In turn, in the bottom-right plot magnitude of the time-domain auto-correlation is drawn, demonstrating good time localization properties of the sequence (high and sharp main-peak, low side-peaks).

Listing 24.5: Matlab program for PBCH decoding in LTE signal (cont.)—PSS signal generation

```
1   % lab24_ex_4g_lte.m - continuation
2
3   % PSS generation in frequency domain - Zadoff-Chu sequences
4     ZC = zeros(Nfft,3);     % NFFT=512  (226-288), NFFT=1024 (482-544)
5     Nc = Nfft/2+1;          % NFFT=1536 (738-800), NFFT=2048 (994-1056)
6     n = (0:62)';
7     ZC(Nc-31:Nc+31,1) = exp( -j*pi*25*n.*(n+1)/63 ); % from ZC definition
8     ZC(Nc-31:Nc+31,2) = exp( -j*pi*29*n.*(n+1)/63 );
9     ZC(Nc-31:Nc+31,3) = exp( -j*pi*34*n.*(n+1)/63 );
10    ZC(Nc, 1:3)       = zeros(1,3);
11
12  % Going to time domain
13    for nn = 0 : 2
14        ZCC(:,nn+1) = ifftshift( ZC(:,nn+1) );
15    end
16    zc = ifft( ZCC )*sqrt(Nfft);
```

Exercise 24.5 (PSS Signal Generation and Observation). Run the program `lab24_ex_4g_lte.m`. Observe plots from Fig. 24.11 for all Zadoff–Chu sequences, number 1, 2, and 3.

Cross-Correlation of the PSS Sequences with LTE Signal: LTE Frame Synchronization Calculated three PSS synchronization sequences, for $u = 25, 29, 34$, are then cross-correlated with the analyzed LTE signal:

$$R_{ys_u}(k) = \sum_n y(n) \cdot s_u^*(n+k). \tag{24.14}$$

One of the calculated cross-correlation functions should have bigger magnitude maxims. Since their arguments inform us about the PSS signal positions in the IQ signal, they allow timing recovery of the LTE frame. In Fig. 24.12 signal cross-correlations with the first and the third PSS synchronization sequences are shown. We see that the third PSS sequence is present. Due to our observation, we can choose the first physical radio cell identification number N_{ID}^{PSS}, making use of the following rule (u—ZC Zadoff–Chu sequence root):

$$
\begin{aligned}
PSS = 1,\, u = 25 &\quad \rightarrow \quad N_{ID}^{PSS} = 0, \\
PSS = 2,\, u = 29 &\quad \rightarrow \quad N_{ID}^{PSS} = 1, \\
PSS = 3,\, u = 34 &\quad \rightarrow \quad N_{ID}^{PSS} = 2.
\end{aligned}
\tag{24.15}
$$

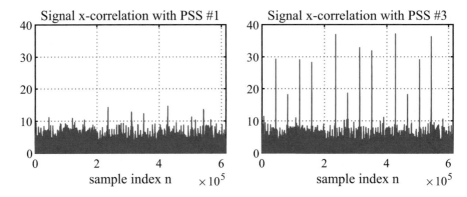

Fig. 24.12: LTE signal cross-correlation with PSS synchronization sequences Zadoff–Chu: (left) with PSS #1: decision=absent, (right) with PSS #3: decision=present

Remark The PSS sequence can be used also for integer CFO estimation. Having this in mind, one should synthesize several ZC time-domain sequences from their spectra shifted a few DFT bins left and right, and correlate all of them with the LTE signal. The best correlating sequence will inform us about the integer CFO also in such case. Additionally, the PSS signals have a mirror symmetry $(s_u(n) = s_u(N_{fft} - n), n = 1, 2, ..., N_{fft} - 1$, see Fig. 24.11) which can be exploited for timing synchronization. Last not least, knowing positions of OFDM symbols with PSS, we can calculate number of samples between them and find whether the ADC sampling rate is correct. In our program we exploit this knowledge doing, if necessary, a signal interpolation. In our smart-phones the ADC sampling rate is corrected in hardware.

Listing 24.6: Matlab program for PBCH decoding in LTE signal (cont.)—cross-correlation of PSS signal with LTE signal

```
1  % lab24_ex_4g_lte.m - continuation
2
3  % PSS: signal cross-correlation with Zadoff-Chu (zc) sequences
4  for k = 1:3
5      Rpss(:,k) = abs( conv( y.', conj( zc(end:-1:1,k)) ) );
6  end
7
8  % Choosing the best sequence (with the highest Rpss), finding arguments of its peaks
9  % Detection N_PSS_ID
10     [dummy, PSS_ID ]  = max( max( Rpss ) );
11     Rpss = Rpss(:,PSS_ID); zc = zc(:,PSS_ID); ZC = ZC(:,PSS_ID).';
12     PSS_ID = PSS_ID-1,
13     threshold = THR;               % 0.85*max(Rpss); % <--- CHOOSE THRESHOLD
14     counter = 1;
```

```
15    for m=2:length(Rpss)-1
16        if( (Rpss(m) > threshold) & (Rpss(m) > Rpss(m-1)) & (Rpss(m) > Rpss(m+1)) )
17            imax(counter) = m; counter = counter+1;
18        end
19    end
20
21    % Final calculations
22    PSS_idxs = imax - Nsymb2;
23    SSS_idxs = PSS_idxs - Nsymb2;
24    display('PSS beginnings:'); PSS_idxs'
25    pause
26    display('Frame/HalfFrame lengths:'); (PSS_idxs(2:end)-PSS_idxs(1:end-1))',
```

Exercise 24.6 (PSS Signal Cross-Correlation with LTE Signal). Set `do_disturbing=0`. Run the program `lab24_ex_4g_lte.m` for different synthesized LTE signals and observe the PSS-vs-LTE cross-correlation function shapes. Set `do_disturbing=1`, `ADCppm=0`, `G0=0`, `G1=0`, `G2=0`, `npwr=-160` (see Listing 24.11). Run the program changing `cfo=0`, `100`, `500`, `1000`, `2000`, `5000`. Then systematically increase the noise power `npwr=-100,-80,-60,-40,-20` and repeat experiment with growing fractional cfo. Then add channel influence: `G0=0.5`, `G1=0.1`, `G2=0.02`.

Using PSS for CFO Estimation, Its Update and Local Signal Correction PSS is robust to moderate level of the fractional CFO. Once the PSS is found, we can use it for the fractional CFO estimation (24.7), update of the stored fractional CFO value, which is shown in left plot of Fig. 24.13, and eventually do local CFO correction.

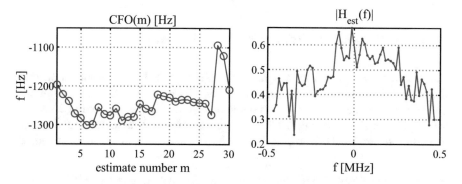

Fig. 24.13: Two examples of additional PSS sequence usage for: (left) adaptive carrier frequency offset estimation, (right) channel estimation for 62 FFT bins around 0 Hz

Using PSS for Channel Estimation Knowing transmitted and received PSS signal we can find local channel estimate valid for central 62 carriers around DC:

$$\hat{H}_{PSS}(k) = \frac{Y(k)}{S_u(k)} = Y(k)S_u^*(k), \quad k = -31, \ldots -1, 1, \ldots 31. \tag{24.16}$$

Exemplary channel estimate calculated this way is shown in Fig. 24.13 (right).

Listing 24.7: Matlab program for PBCH decoding in LTE signal (cont.)—cross-correlation of PSS signal with LTE signal

```
1   % lab24_ex_4g_lte.m - continuation
2
3   % Next iteration
4     PSS_idx = PSS_idxs( iter ),            % position of the next PSS
5     nlsss = SSS_idxs( iter );              % position of the next SSS
6
7   % CFO estimation using PSS
8     n = PSS_idx+Ncut1 : PSS_idx+Ncp2-1-Ncut2;        % sample numbers
9     CFO_pss = fs/Nfft * angle( sum( conj(y(n)) .* y(n+Nfft) ) ) / (2*pi);
10    CFO = alpha*CFO + (1-alpha)*CFO_pss,             % estimation update
11    display( sprintf( 'Estimated CFO: %f Hz', CFO ) );
12    CFO_history(iter+1) = CFO;                        % store it
13
14  % CFO correction of the LTE frame half (using calculated CFO) - starting from SSS
15    n = nlsss : nlsss + Nframe2-1;                    % sample numbers
16    y( n ) = y( n ).* exp(-j*2*pi*(0:Nframe2-1)*(CFO/fs));  % CFO correction
17
18  % Channel estimation using PSS
19    PSS_td = y( PSS_idx+Ncp2 : PSS_idx+Ncp2+Nfft-1 );    % OFDM symbol samples
20    PSS_fd = fftshift( fft( PSS_td ) ) / sqrt(Nfft);      % FFT
21    Hest = PSS_fd( k_pss ) .* conj( ZC( k_pss ) );        % LS channel estimate
```

24.3.5 SSS Signals: Physical Radio Cell Identification

The secondary synchronization signal (SSS) is sent in the beginning and in the middle of the LTE frame in sub-frames 0 and 5: exactly in the 5-th OFDM symbol, just before the PSS sequence, of time slots number 0 and 10. It uses only 62 carriers around DC as shown in Fig. 24.2. It is used for physical radio cell identification according to Eq. (24.1). Knowing it, the phone can decode the physical broadcast channel (PBCH).

SSS Sequences and Their Generation The SSS signal can have 168 different forms numbered from 0 to 167. Together with possible 3 PSS signals it gives 504 different combinations, 504 possible physical cell ID identifiers (PCIs). The SSS sequence is a concatenation of two binary sequences $D(2k)$ and $D(2k+1)$ which are defined in different way for frame 0 (time slot 0) and frame 5 (time slot 10) [11]:

$$\text{subframe } 0: \quad \begin{cases} D_0(2k) = s_{m_0}(k) \cdot c_0(k) \\ D_0(2k+1) = s_{m_1}(k) \cdot c_1(k) \cdot z_{m_0}(k) \end{cases} \qquad (24.17)$$

$$\text{subframe } 5: \quad \begin{cases} D_1(2k) = s_{m_1}(k) \cdot c_0(k) \\ D_1(2k+1) = s_{m_0}(k) \cdot c_1(k) \cdot z_{m_1}(k) \end{cases} \qquad (24.18)$$

$$m_0, m_1, k = 0, 1, 2, ..., 30$$

In the above equations, given in the LTE standard, the vectors $D_0(.)$ and $D_1(.)$ represent Fourier coefficients of 62 SSS sub-carriers that are next mapped to DFT bins $-31, ..., -1, 1, ...31$, exactly the same way as it was done for PSS sequences (see Eq. (24.13)). All sequences, i.e. $s_{m_0}(k)$ and $s_{m_1}(k)$, z_{m_0} and $z_{m_1}(k)$ as well as $c_0(k)$ and $c_1(k)$, have 31 elements and they are built from maximum length sequences (MLS). They are generated in similar way as pseudo-random binary sequences (PRBS) $(-1, 1)$ in DAB program (see Fig. 23.13): first base sequences $s(k), z(k)$ and $c(k)$ are synthesized using different polynomials ($x^2 + 1$, $x^3 + 1$ and $x^4 + x^2 + x^1 + 1$, respectively), and then they are cyclically shifted:

$$s_{m_i}(k) = s((k + m_i) \quad \mathrm{mod}\ 31), \quad i = 0, 1,$$
$$z_{m_i}(k) = z((k + (m_i \quad \mathrm{mod}\ 8) \quad \mathrm{mod}\ 31), \quad i = 0, 1,$$
$$c_0(k) = c((k + N_{ID}^{PSS}) \quad \mathrm{mod}\ 31),$$
$$c_1(k) = c((k + N_{ID}^{PSS} + 3) \quad \mathrm{mod}\ 31).$$

Note, that sequences $c_0(k)$ and $c_1(k)$ depend on already found N_{ID}^{PSS}, equal to 0, 1, or 2.

Listing 24.8: Matlab program for PBCH decoding in LTE signal (cont.)—SSS signal generation

```matlab
1  function [d_sf0, d_sf5] = lte_sss_gen(N_1_id, N_2_id)
2
3      assert( ismember(N_1_id, 0:167 ), 'invalid N_1_id: allowed range is 0 to 167');
4      assert( ismember(N_2_id, 0:2 ),   'invalid N_2_id: allowed values are 0, 1 or 2');
5
6      % generate m0 and m1
7      % (NOTE: can use Table 6.11.2.1-1 directly as LUT instead of below calculations)
8      q_p = floor(N_1_id / 30);
9      q   = floor( (N_1_id + q_p*(q_p+1)/2) / 30);
10     m_p = N_1_id + q*(q+1)/2;
11     m0  = mod(m_p, 31);
12     m1  = mod(m0 + floor(m_p/31) + 1, 31);
13
14     % generate s0 and s1 sequences
15     [s0, s1] = sss_seq_scrambler([0,2], m0, m1);
16
17     % generate c0 and c1 sequences
18     [c0, c1] = sss_seq_scrambler([0,3], N_2_id, N_2_id+3);
19
20     % generate z0 and z1 sequences
```

```
21    [z0, z1] = sss_seq_scrambler([0,1,2,4], mod(m0,8), mod(m1,8));
22
23    % generate SSS sequences for subframe 0 and 5
24    d_sf0 = zeros(62,1);
25    d_sf5 = zeros(62,1);
26    n = 0:30;
27
28    d_sf0(1+2*n)   = s0(n+1) .* c0(n+1);
29    d_sf0(1+2*n+1) = s1(n+1) .* c1(n+1) .* z0(n+1);
30
31    d_sf5(1+2*n)   = s1(n+1) .* c0(n+1);
32    d_sf5(1+2*n+1) = s0(n+1) .* c1(n+1) .* z1(n+1);
33  end
34
35  % Helper function - generator for s, c and z sequences
36  function [seq0, seq1] = sss_seq_scrambler(polynomial, shift0, shift1)
37    x = zeros(31,1); x(4+1) = 1;
38    for i = (0 : 25)+1
39      for p = polynomial
40          x(i + 5) = mod(x(i + 5) + x(i+p), 2);
41      end
42    end
43    y = 1 - 2*x;
44    seq0 = circshift(y, -shift0);
45    seq1 = circshift(y, -shift1);
46  end
```

Signal Cross-Correlation with SSS Sequences and Physical Cell Identification
After LTE timing recovery using PSS sequences, we can find and cut an LTE signal
fragments, corresponding to OFDM symbols with SSS signals. Next, correlation
coefficients between appropriate IQ samples and all SSS signals are calculated (two
sequences $D_{0,j}$ and $D_{1,j}$ in 168 different versions for $j = 0...167$):

$$C(i,j) = \sqrt{\left| \left[\sum_{k \in SSS, k \neq 0} \frac{Y_{SSS}(k)}{\hat{H}_{PSS}(k)} \cdot D_{i,j}^*(k)) \right]^2 \right|}, \quad i = 0,1, \ j = 0,...,167. \quad (24.19)$$

Note that in Eq. (24.19) the correlation coefficient is calculated in frequency domain
and that the LTE OFDM symbol spectrum is equalized by $\hat{H}_{PSS}(k)$ in part belonging
to the SSS carriers. Next, the biggest correlation coefficient is found. Its number
N_{ID}^{SSS} allows to find the physical radio cell ID from Eq. (24.1) and helps to distinguish
LTE frame 0 from frame 5 and active frame synchronization. Exemplary calculated
correlation coefficients for SSS sequences for frame 0 (left) and frame 5 (right) are
presented in Fig. 24.14. We observe that the SSS signal number 111 is detected for
frame 0. Therefore, the $N_{ID}^{CELL} = 335$.

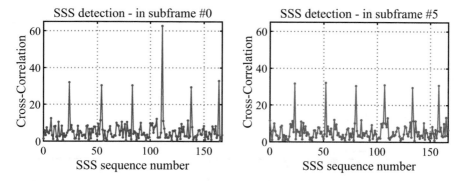

Fig. 24.14: Results of LTE signal cross-correlation with 168 SSS secondary synchronization sequences (0...167) for LTE sub-frame number 0 (left) and number 5 (right). Calculated correlation coefficients are presented

Listing 24.9: Matlab program for PBCH decoding in LTE signal (cont.)—crosscorrelation of SSS signal with LTE signal

```matlab
1  % lab24_ex_4g_lte.m - continuation
2
3  % Extraction of SSS sequence and it equalization using the estimated channel
4    SSS_td = y(n1sss+Ncp2 : n1sss+Ncp2+Nfft-1);        % OFDM symbol samples
5    SSS_fd = fftshift( fft( SSS_td ) ) / sqrt(Nfft);   % FFT
6    SSS_fd = SSS_fd( k_pss  ) ./ Hest;                 % correction
7
8  % Detection of N_SSS_ID and subframe number using the SSS
9    [SSS_ID, subframe, SSS_corr] = lte_sss_detect( SSS_fd.', PSS_ID );
10   if( do_plots )
11       figure;
12       plot(0:167, SSS_corr(:,1), 'r', 0:167, SSS_corr(:,2), 'b');
13       title('SSS detection'); xlabel('N^{(1)}_{ID}'); ylabel('Correlation');
14       legend('Subframe 0', 'Subframe 5'); pause
15   end
16   % Calculating the physical radio cell indentifier
17     Ncell_ID = 3 * SSS_ID + PSS_ID;
18     display(sprintf('N_cell_ID: %0d; Subframe: %0d; Sample: %0d', ...
19     Ncell_ID, subframe, PSS_idx)),
20
21  %##################################################################
22  function [N_SSS_ID, subframe, R] = lte_sss_detect(rx_seq, N_PSS_ID)
23    R = zeros(168, 2);
24
25    % calculate correlation with each of 168*2 sequences
26    for nid1 = 0 : 167
27        [d_sf0, d_sf5] = lte_sss_gen(nid1, N_PSS_ID);
28
29        R(nid1+1,1) = sqrt( abs( sum( rx_seq .* conj(d_sf0) ) .^2 ) );
30        R(nid1+1,2) = sqrt( abs( sum( rx_seq .* conj(d_sf5) ) .^2 ) );
31    end
```

```
32
33      [v,i] = max(R.^2);
34
35      if v(1) > v(2)
36         subframe = 0;
37         N_SSS_ID = i(1)-1;
38      else
39         subframe = 5;
40         N_SSS_ID = i(2)-1;
41      end
42   end
```

24.3.6 CRS Signals: PBCH Decoding

Knowing the PCI, physical cell ID, we can calculate the cell-specific reference signals (pilots) (CRS) and use them for antenna port number detection and channel identification. In Fig. 24.2 these special-task carriers are marked as A1, A2, A3, and A4. Having the channel estimate and knowing number of the transmitter antenna ports we can think about decoding the physical broadcast channel data, sent in 4 OFDM symbols after the PSS and exploiting central 72 sub-carriers (once again see Fig. 24.2).

PBCH Signal Demodulation Aiming at PBCH decoding, we start with finding 4 OFDM symbols possessing it. Next we synchronize with symbol beginnings, perform 4 FFTs, and store selected spectral coefficients, i.e. 72 PBCH carrier states values.

CRS Signal Generation Next we generate the CRS signals. They are QPSK-modulated sub-carriers with pseudo-random IQ states which depend on the physical radio cell number N_{ID}^{CELL} and are generated using Gold sequence. Details can be found in the program.

Using CRS Signals for Number of Antenna Ports Detection The exact number of antenna ports used by the base-station transmitter is coded in PBCH. In order to avoid blind data decoding for different antenna configuration, we use CRS signals for detection of the antenna port number. We correlate transmitted and received states of CRS carriers for each antenna port A_k, $k = 1, 2, 3, 4$, individually:

$$C_{A_k} = \frac{\left| \sum_{k \in CRS} Y(k) X_{A_k}^*(k) \right|}{\sqrt{\left| \sum_{k \in CRS} Y(k) Y^*(k) \right|} + \sqrt{\left| \sum_{k \in CRS} X_{A_k}(k) X_{A_k}^*(k) \right|}}, \quad k = 1, 2, 3, 4. \quad (24.20)$$

In our example the following values of correlation coefficients were obtained:

$$C_{A_1} = 1.0773, \quad C_{A_2} = 1.0405, \quad C_{A_3} = 0.2520, \quad C_{A_4} = 0.1887,$$

indicating that transmitter was using two antenna ports to transmit CRS signal.

Using CRS Signals for Channel Equalization Knowing transmitted and received states of CRS signals (i.e. DFT coefficients), we can calculate channel estimate in CRS signal positions:

$$\hat{H}(k_{CRS}) = \frac{Y(k_{CRS})}{S_{CRS}(k_{CRS})} = Y(k_{CRS}) \cdot S^*_{CRS}(k_{CRS}), \tag{24.21}$$

and then interpolate this estimate in frequency (linearly) and in time (repeating the last value). Channel estimation described by Eq. (24.21) is known as least squares (LS), i.e. least-square error between transmitted and received values is minimized when this formula is used. This is the simplest and the least efficient/precise method of channel estimation but sufficient for BPCH decoding due to low-order modulation used (QPSK) and high data redundancy.

Listing 24.10: Matlab program for PBCH decoding in LTE signal (cont.)—final operations

```
1   % lab24_ex_4g_lte.m - continuation
2
3   % PBCH is in the first sub-frame only
4     if subframe == 0
5
6       % starting positions of 4 PBCH symbols
7       PBCH_idxs = PSS_idx + Ncp2 + cumsum( [Nfft+Ncp1, Nfft+Ncp2, Nfft+Ncp2, Nfft+Ncp2] );
8       % OFDM demodulation of 4 PBCH symbols, extraction of center 72 REs
9       % (resource elements) (TS36.211 sec. 6.6.4, 1 RE = 1 subcarrier)
10      % (TS36.211 sec. 6.6.4)
11      PBCH_grid = zeros(72,4);                              % initialization
12      for n_sym = 1 : numel( PBCH_idxs )                   % for 4 PBCH
13          PBCH_td = y( PBCH_idxs(n_sym) : PBCH_idxs(n_sym)+Nfft-1); % samples
14          PBCH_fd = fftshift( fft(PBCH_td) ) / sqrt(Nfft); % FFT
15          PBCH_grid(:,n_sym) = PBCH_fd( k_pbch );          % store
16      end
17      % figure; plot(real(PBCH_grid),imag(PBCH_grid),'o'); grid; title('PBCH_grid'); pause
        '
18
19      % symbol numbers containing CRSs (Cell-Specific Ref Signals) for a given antenna
            port
20      l_crs_ant = [0, 0, 1, 1];
21
22      % constants
23      N_PBCH_symb = 4;    % number of PBCH symbols
24      N_PBCH_RB   = 6;    % number of Resource Blocks used by PBCH
25      N_ap_max    = 4;    % maximum number of antenna ports
26
27      % pre-allocate memory
28      x_crs       = zeros( N_PBCH_RB *2, N_ap_max ); % CRS transmitted (ref)
29      y_crs       = zeros( N_PBCH_RB *2, N_ap_max ); % CRS received
```

```
30      k_crs      = zeros( N_PBCH_RB *2, N_ap_max);  % CRS indexes
31      H_est_crs  = zeros( N_PBCH_RB *2, N_ap_max);  % H estimated for CRS
32      H_est_PBCH = zeros( N_PBCH_RB*12, N_ap_max);  % H estimated for PBCH
33      CRS_corr   = zeros( N_ap_max, 1 );            % correlation of TX and RX CRS
34
35    % Iterate for all four antenna ports
36    for ant = 1 : N_ap_max
37      % Generate ideal CRS for slot 1 and current antenna port
38      [x_crs(:,ant), k_crs(:,ant)] = lte_crs_gen( N_PBCH_RB, Ncell_ID, 1, ...
39                                      l_crs_ant(ant), ant-1, cp_type);
40      % De-map CRS from the received signal
41      y_crs(:,ant) = PBCH_grid( k_crs(:,ant)+1, l_crs_ant(ant)+1 );
42
43      % Find correlation coefficient between CRS TX and RX
44      % CRS_corr(ant) = abs( corr( x_crs(:,ant), y_crs(:,ant) ) );
45      CRS_corr(ant) = abs( sum( conj(x_crs(:,ant)) .* y_crs(:,ant) ) ) / ...
46            (  sqrt( abs( sum( conj(x_crs(:,ant)) .* x_crs(:,ant)) ) * ...
47                sqrt( abs( sum( conj(y_crs(:,ant)) .* y_crs(:,ant)) ) ) ));
48
49      % Estimate channel (least squares) on pilot sub-carriers
50      H_est_crs(:,ant) = conj(x_crs(:,ant)) .* y_crs(:,ant);
51
52      % Interpolate to non-pilot frequencies
53      H_est_PBCH(:,ant) = lte_interp_crs_hest( H_est_crs(:,ant) ); % better
54
55    end
56
57    display(sprintf('CRS xcor for PBCH: AP0=%1.4f, AP1=%1.4f, AP2=%1.4f, AP3=%1.4f',...
58                  CRS_corr(1), CRS_corr(2), CRS_corr(3), CRS_corr(4)));
59
60  % Determine REs that are occupied by PBCH by removing the CRS sub-carriers
61    PBCH_grid_RE_idx = ones( N_PBCH_RB*12, N_PBCH_symb );      % init: all 1s
62    for ant = 1 : N_ap_max                                    % for all ant ports
63        PBCH_grid_RE_idx( k_crs(:,ant)+1, l_crs_ant(ant)+1 ) = 0; % 0s for CRS
64    end                                                       %
65
66  % Un-map PBCH modulation symbols and channel estimates from the RE grid
67    y_pbch = PBCH_grid( PBCH_grid_RE_idx ~= 0 );        % take non-zero PBCH
68    h_pbch = zeros( length(y_pbch), N_ap );             % initialize h_pbch
69    for ant = 1 : N_ap                                  % for antenna ports
70        H_est_PBCH_ant = repmat( H_est_PBCH(:,ant), [1,N_PBCH_symb] ); % copy to N_ap
                  cols
71        h_pbch(:,ant)  = H_est_PBCH_ant( PBCH_grid_RE_idx ~= 0 );     % only for RE
72    end
73    size(y_pbch), size(h_pbch), pause
74
75  % Equalize TX diversity and do layer de-mapping to get constellation symbols
76    PBCH_IQ = lte_eq_and_demap_tx_div( y_pbch, h_pbch(:,1:N_ap) );
77
78    figure; scatterplot( PBCH_IQ [], [],'bx' ); title('Scatter for PBCH'); pause
79
80  end % of subframe==0
```

Exercise 24.7 (PBCH in LTE Subframe 0). A proverb says that "the devil is in the details." Look inside all functions that are called in the Listing 24.10 and analyze their code. Find where they are described, if at all, in the book. If you are brave enough, propose your own solutions, modify the code, and verify its correctness running the program.

Shortly About MIMO and Alamouti TX Diversity Scheme The multi-antenna techniques in modern wireless communications can be used in different ways depending on the needs and system capabilities. The most common MIMO (multiple input-multiple output data streams) modes of operations are:

- *Spatial Multiplexing*—where a signal with high data rate is split into a multiple lower-rate streams and then each of them is transmitted using a separate antenna. This allows for increasing of peak throughput of the system by a factor equal to a number of transmit antennas.
- *Diversity coding*—where a signal is coded using space–time or space–frequency redundancy-adding code and then emitted from each antenna. The coding paired with a very simple post-processing in the receiver allows for increase of the SNR without a need for explicit channel knowledge at the transmitter.
- *Beam-forming*—where the same signal is emitted from each of the transmit antennas with prior phase and amplitude adjustment. The beam-forming allows for maximization of the antenna gain in a given direction, providing spatial separation between different users and increasing the range of the transmission. However, unlike the diversity coding technique, beam-forming requires explicit knowledge of the channel state at the transmitter.

In LTE standard, PBCH channel is transmitted on multiple antennas using space–frequency diversity coding technique, which is a form of transmit diversity scheme proposed by Siavash Alamouti in 1998 [2]. By applying a simple pre-coding at the transmitter in configuration with two transmit and a single receiving antennas, a similar SNR gain may be achieved as in case of a single transmit and two receiving antennas. Due to this fact, the scheme allows to increase the probability of successful information decoding at the mobile terminal without increasing its complexity (price), just by exploiting an additional transmit antenna at the base-station.

Equalization of TX Space Diversity: Alamouti Coding and De-coding In Release 8 the LTE transmitter can use 1, 2, or 4 antennas at the same time. If there are 2 or 4 antennas, the same signal is encoded to produce multiple streams transmitted simultaneously from the antennas. They use the same FFT sub-carriers but sent them in different states. Let us assume the following situation, presented in

Fig. 24.15: a transmitter has two antennas, A_1 and A_2, and a receiver, our phone or ADALM-PLUTO module, only one, marked as RX. Both transmitter antennas use two neighboring sub-carriers in one OFDM symbol, i.e. with frequencies f_1 and $f_2 = f_1 + 15$ kHz.

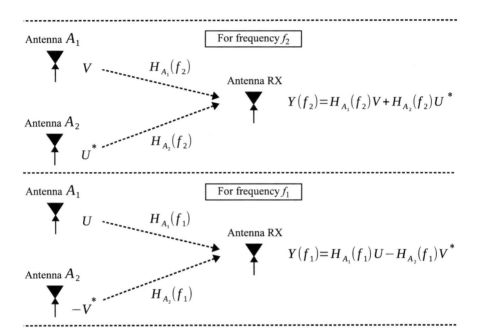

Fig. 24.15: Graphical illustration of space data coding using two transmitting antennas A_1 and A_2 and one receiving antenna RX: (up) for frequency f_2. (down) for frequency f_1. Complex-value numbers V and U are transmitted as states of sub-carriers used (f_1 and f_2) but with possible negation and complex conjugation

The channel frequency responses for each antenna ($n = 1, 2$) and each frequency ($k = 1, 2$) is a complex-value number:

$$H_{A_n}(f_k) = |H_{A_n}(f_k)|e^{j\angle H_{A_n}(f_k)}. \tag{24.22}$$

They have the following values:

- $H_{A_1}(f_1)$ - $H(f)$ for path $A_1 \rightarrow RX$ and frequency f_1,
- $H_{A_1}(f_2)$ - $H(f)$ for path $A_1 \rightarrow RX$ and frequency f_2,
- $H_{A_2}(f_1)$ - $H(f)$ for path $A_2 \rightarrow RX$ and frequency f_1,
- $H_{A_2}(f_2)$ - $H(f)$ for path $A_2 \rightarrow RX$ and frequency f_2.

The antenna A_1 transmits the carrier f_1-th in state U, while antenna A_2 in state $-V^*$:

$$\text{Carrier}(f_1): \quad Y(f_1) = H_{A_1}(f_1) \cdot U - H_{A_2}(f_1) \cdot V^*. \tag{24.23}$$

The same time the antenna A_1 transmit the carrier f_2 in state V, while the antenna A_2 in state U^*:

$$\text{Carrier}(f_2): \quad Y(f_2) = H_{A_1}(f_2) \cdot V + H_{A_2}(f_2) \cdot U^*. \tag{24.24}$$

After complex conjugation of the last equation one obtains

$$\text{Carrier}(f_2): \quad Y^*(f_2) = H_{A_1}^*(f_2) \cdot V^* + H_{A_2}^*(f_2) \cdot U. \tag{24.25}$$

Now we combine Eqs. (24.23), (24.25) into a matrix formula:

$$\begin{bmatrix} Y(f_1) \\ Y^*(f_2) \end{bmatrix} = \underbrace{\begin{bmatrix} H_{A_1}(f_1) & -H_{A_2}(f_1) \\ H_{A_2}^*(f_2) & H_{A_1}^*(f_2) \end{bmatrix}}_{\mathbf{H}} \begin{bmatrix} U \\ V^* \end{bmatrix} \tag{24.26}$$

Let multiply the matrix $\mathbf{H}$ from Eq. (24.26) by its conjugated transposition:

$$\mathbf{H}\left(\mathbf{H}^T\right)^* = \begin{bmatrix} H_{A_1}(f_1)H_{A_1}^*(f_1) + H_{A_2}(f_1)H_{A_2}^*(f_1), & H_{A_1}(f_1)H_{A_2}(f_2) - H_{A_1}(f_2)H_{A_2}(f_1) \\ H_{A_1}^*(f_1)H_{A_2}^*(f_2) - H_{A_1}^*(f_2)H_{A_2}^*(f_1), & H_{A_1}(f_2)H_{A_1}^*(f_2) + H_{A_2}(f_2)H_{A_2}^*(f_2) \end{bmatrix} \tag{24.27}$$

If for each antenna, the channel frequency response is similar for the frequencies f_1 and f_2 (it should be because the frequencies are close to each other):

$$H_{A_1}(f_1) \approx H_{A_1}(f_2), \quad H_{A_2}(f_1) \approx H_{A_2}(f_2), \tag{24.28}$$

then the following simplification can be done for Eq. (24.27):

$$\mathbf{H}\left(\mathbf{H}^T\right)^* \approx C \begin{bmatrix} 1 & 0 \\ 0 & 1 \end{bmatrix}, \quad C = H_{A_1}(f_{12})H_{A_1}^*(f_{12}) + H_{A_2}(f_{12})H_{A_2}^*(f_{12}), \tag{24.29}$$

where denotation f_{12} is used only for reminding that the $H(f)$ is assumed to be valid for both frequencies f_1 and f_2. The last result tells us that the matrix $\mathbf{H}$ is almost orthogonal. Therefore its inverse is equal to its appropriately scaled conjugated transposition, i.e. Hermitian transposition $(.)^H$, and the Eq. (24.26) can be solved easily:

$$\begin{bmatrix} U \\ V^* \end{bmatrix} = \frac{1}{\sqrt{C}} \begin{bmatrix} H_{A_1}(f_{12}) & -H_{A_2}(f_{12}) \\ H_{A_2}^*(f_{12}) & H_{A_1}^*(f_{12}) \end{bmatrix}^H \begin{bmatrix} Y(f_1) \\ Y_2^*(f_2) \end{bmatrix} = \frac{1}{\sqrt{C}} \begin{bmatrix} H_{A_1}^*(f_{12}) & H_{A_2}(f_{12}) \\ -H_{A_2}^*(f_{12}) & H_{A_1}(f_{12}) \end{bmatrix} \begin{bmatrix} Y_1 \\ Y_2^* \end{bmatrix} \tag{24.30}$$

In the LTE standard, all carriers used by PBCH in 4 consecutive time slots (see Fig. 24.2), excluding the antenna pilots (CRS signals), are put together into a long vector—one-by-one slot-after-slot. This is shown in Fig. 24.16. In 4 time slots (TS) we have 6 resource blocks (RB) with 12 sub-carriers, i.e. together 288. Excluding $2TS \cdot 6RB \cdot 4 = 48$ pilots, 240 sub-carriers remain: 120 carrier pairs using the above-described frequency/space diversity coding scheme which originates from the Alamouti method.

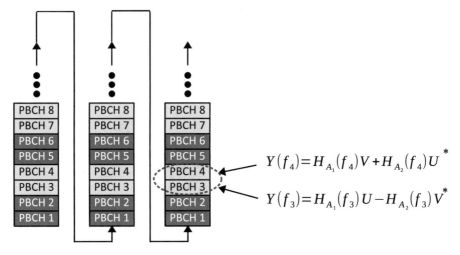

Fig. 24.16: Graphical illustration of exploiting antenna space–frequency diversity during PBCH decoding: PBCH data are transmitted as complex-values U and V by pairs of two neighboring PBCH sub-carriers of PBCH time slots, scanned down-up, and slot-by-slot

To Remember In fact the above-described new signal processing functionality of the LTE digital telephony, the multi-antenna support, is a characteristic mark of the standard and new telecommunication era that is coming. In this direction goes further the 4G LTE successor, the 5G New Radio, allowing better support and usage even up to 64 transmitting and 64 receiving antennas in the most advanced beam-forming option (64 in base-stations, a phone has typically 1–2 antennas). Thanks to it a 5G user will have a specially designed for him/her, optimal, directional radio beam improving the signal reception.

Exercise 24.8 (Testing Alamouti Transmission Scheme). Write program that implements Eqs. (24.23), (24.24), and (24.30). Assume some complex-values $H_{A_k}(f_l), k = 1, 2, l = 1, 2, U$ and V, and calculate $Y(f_1)$ and $Y(f_2)$ from the first two equations. Then recover U and V from the third equation. Try to fulfill conditions (24.28).

Observing PBCH Signal Constellation Points And now, the final view of our risky LTE climbing. Somebody will say: our *Show-Time* or *Happy-Hours*. Obtained PBCH scatter plots are presented in Fig. 24.17. On the left—without fractional CFO estimation and correction, on the right—with it. In the correct right plot we can distinguish well separated four carrier states of the QPSK modulation. We can hope that thanks to the added data redundancy and after Viterbi decoder number of errors will be negligible.

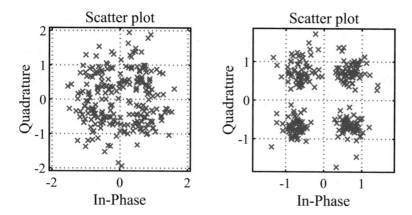

Fig. 24.17: Scatter plots of found PBCH channel QPSK constellation points in the beginning of carrier frequency offset adaptation (left) and after some time (right)

24.4 Experiments Using Synthesized LTE Signals

Till now we were analyzing first algorithmic steps of the LTE decoder, leading to extracting information of the physical broadcast channel (PBCH). This allows a phone to establish a connection with a base-station and be granted system resources, i.e. resource blocks (RBs) of 12 sub-carriers in consecutive 6 or 7 time slots. We were focused on identification of existing problems and their the easiest solutions. At present, having the whole path described and program working, we would like to test robustness of our solution in more sterile conditions and observe PBCH scatter plots for different types of impairments individually. Our intention is comparing quantity of scatter plot degradation for different disturbances and verification of efficiency of algorithmic constellation points repairing. In order to have an unambiguous conclusions, we will use *clear*, synthetic IQ LTE signals which support ADALM-PLUTO SDR modules. The files consist of only one LTE frame which can be transmitted in a loop by the PLUTO hardware in free-access frequency bands. The files include 5, 10, and 10 MHz LTE services (LTE5.mat, LTE10.mat, LTE20.mat) and only QPSK carrier modulation. In order to turn on the signal disturbing, user should set do_disturbing=1 in the LTE program beginning and then choose values of disturbance parameters in the code fragment, presented in Listing 24.11. Scatter plots $Q(k) = f(I(k))$, obtained from LTE transmission simulation for PBCH sub-carriers for different disturbances (impairments), are presented in Fig. 24.18. As we see, fractional carrier frequency offset is the most harmful, however, scatter plot degradation due to multi-path fading channel and incorrect ADC sampling rate is also significant and non-acceptable for 16-QAM and higher-level modulations. The scatter plot for all disturbances occurring the same time is not plotted, since it is very similar to the fractional CFO scatter. Correction algorithms work satisfactory in all cases and sub-carrier states can be recovered even when all disturbances occur the same time (see the plot in left-down corner). Below all plots are shortly commented.

Listing 24.11: Matlab program for PBCH decoding in LTE signal (cont.)—addition of disturbances (distortions)

```
1   % lab24_ex_4g_lte.m - continuation
2
3   if(do_disturbing)
4       % Parameter values
5       cfo=0; npwr=-160; ADCppm=0; G0=1; G1=0; G2=0; D0=0; D1=10; D2=25;
6       npwr=-30;                  % noise power [dB] : -160, ...,-60,-50,-40,-30,-20
7       % cfo=1245;                % carrier frequency offset (CFO) [Hz]
8       % ADCppm=200;              % ADC error [ppm] : from -400 to +400
9       % G0=0.5; G1=0.25; G2=0.1; D0=0; D1=10; D2=25; % Multi-path: gains, delays
10
11      % Disturbing signal
12      NADC = round((1+ADCppm/1000000)*N); step = N/NADC;      % ADC error
13      if( step ~= 1 )
14          disp('ADC ERROR -> generation !');
15          NADC = length(0:step:N-step);
16          y(1,1:NADC)=interp1( [0:N-1], y(1:N), [0:step:N-step], 'spline' );
17          N = NADC; y(NADC+1:end)=[];
18      end
19      y = exp(j*2*pi*rand) * G0*[ y(end-D0+1:end) y(1:end-D0) ] + ... % multi-path
20          exp(j*2*pi*rand) * G1*[ y(end-D1+1:end) y(1:end-D1) ] + ... %
21          exp(j*2*pi*rand) * G2*[ y(end-D2+1:end) y(1:end-D2) ];      %
22      y = y .* exp( j*2*pi*cfo/fs*(0:N-1));                   % CFO
23      y = y + 10^(npwr/20) / sqrt(2) * (randn(1,N)+j*randn(1,N));     % AWGN
24  end
```

Noise Noise is very unwanted. It is not only influencing the FFT output, making the IQ scatter more *cloudy*, but also it makes estimation of the carrier frequency offset and momentum channel frequency response significantly more difficult. In consequence, it causes CFO and channel estimation errors, wrong corrections, and wrong constellation point positions. Amplified thermal noise of a receiver and interference from neighbor physical radio cells are the main noise sources.

Fading Channel The channel attenuates and delays the LTE signal frequency sub-carriers in different manner, according to its frequency response. Original, transmitted amplitudes and phases of sub-carriers are lost. It is absolutely necessary to estimate the channel impulse/frequency response permanently when the channel is dynamically changing. In LTE approximately one per 35 microseconds (due to user mobility). In our experiment the channel has significantly attenuated the sig-

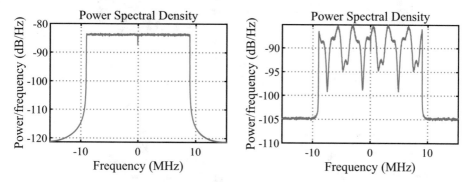

Fig. 24.18: PSD spectrum of the investigated 20 MHz LTE synthetic signal for signal not distorted (left) and distorted by all impairments occurring the same time

nal amplitude. Additionally its multi-path nature causes significant broadening of the scatter plot constellation points. Both effects are very harmful and they cause application of higher order QAM modulations impossible. Therefore, the channel was estimated and compensated (see right plot in the second row in Fig. 24.19): at present the scatter plot is sharper and properly scaled.

Carrier Frequency Offset This is the most dangerous impairment in regard to scatter plot *smearing*. Since all signal components are shifted in frequency, the LTE sub-carriers no longer represent orthogonal DFT harmonics. Therefore, the DFT coefficients capture energy of many LTE sub-carriers and their values and IQ constellation points become *fuzzy*. After precise CFO estimation and correction, this effect can be almost completely canceled.

Wrong ADC Sampling Rate Wrong ADC sampling rate also causes losing sub-carrier orthogonality in DFT, and has to be corrected absolutely. The ADC sampling rate can be estimated by counting number of samples between PSS synchronization signals. If they are too many or too less, the signal can be interpolated/resampled (what is very time-consuming) or sampling frequency has to be properly adjusted in ADC hardware (what is preferred). The same situation was in DAB radio, as you remember.

Exercise 24.9 (Disturbances). In the program `lab24_ex_4g_lte.m` choose an input file with a synthetic IQ LTE signal (`LTE5.mat`, `LTE10.mat` or `LTE20.mat`). Set `do_disturbing=1`. Then, one-by-one, turn ON one disturbance only: noise, fractional carrier offset, channel and ADC ppm error. Run the program, increase the disturbance level, and try to obtain plots shown in Fig. 24.19. Next, check how the situation will change when disturbances exist together which is typical. At the end, choose file `LTE_TM3p1_10MZ_18p22dBFS.txt` with 16-QAM data embedded in weak noise. Observe displayed scatter plots. Start adding disturbances.

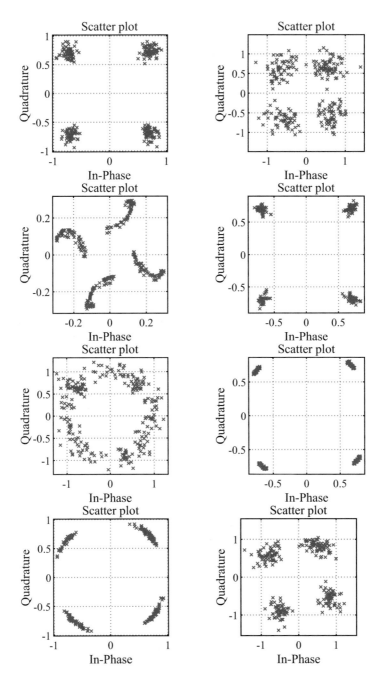

Fig. 24.19: Results from 20 MHz LTE transmission simulation, using synthetic IQ files with only QPSK sub-carrier modulation. PBCH scatter plots for: (first row) noise power -30 dB (left) and -20 dB (right), (second row) fading channel G0=0.5; G1=0.25; G2=0.1; D0=0; D1=10; D2=25 not corrected (left) and corrected (right), (third row) fractional carrier frequency offset 1234 Hz without correction (left) and with correction (right), (fourth row) 400 ppm ADC sampling rate error without correction (left) and all disturbances occurring together (noise -30 dB) with correction (right)

24.5 5G New Radio

The 5G New Radio standard is an attempt of incorporating in one normalization act as many mobile services of the future as possible. We can say that 5G is a common operational shell for different use cases of mobile, wireless data exchange, i.e. it is a supervisor which is managing radio-frequency resources and giving access to them for all users: state, institutional, and public. It is like a computer operation system granting many programs time-multiplexed or parallel access to a processor, memory, and input/output devices.

24.5.1 Introduction to 5G NR

Fulfillment of growing new needs requires new technical solutions. The 5G New Radio standard [10] addresses the following important new use cases: [18, 26]

1. **enhanced Mobile Broadband (eMBB)**—e.g. high-speed Internet access for browsing, high-definition video streaming, and virtual reality—10–20 Gbps peak throughput, 100 Mbps guaranteed, 10000× bigger traffic, 100× network energy saving, support for high user mobility (500 km/h),
2. **ultra Reliable Low Latency Communications (uRLLC)**—e.g. for autonomous vehicles, remote surgery, industry automation—ultra responsive, reliable and available (99.9999%), low latency (below 1 ms in air, 5 ms end-to-end (E2E)), extremely low error rates, low to medium data rates (50 kbps–10 Mbps), high-speed mobility,
3. **massive Machine-Type Communication (mMTC)**—e.g. massive Internet of Things—high density of devices, long range, low data rate (1–100 kbps), machine-to-machine (M2M) ultra low cost, asynchronous access, 10 years battery life.

Apart from them, in between on the cross-roads, smart homes and mission critical applications are located. In general—data exchange everywhere and anytime supervised by one joint wireless telecommunication framework. In Fig. 24.20 a scope of 5G application fields is summarized.

In order to get the requested, challenging, new functionalities, significant new technical solutions had to be applied in the 5G NR. In Table 24.3 the 5G NR is compared with legacy 3G and 4G digital telephony standards. We can briefly conclude that it differs from 4G using:

- new, higher frequency bands in millimeter wave spectrum, up to 37–52 GHz,
- wider channel bandwidths, up to 400 MHz in frequency bands above 6 GHz and up to 100 MHz below 6GHz,
- filtered OFDM [1] generalization of CP-OFDMA and SC-OFDM,
- new channel coding techniques, i.e. LDPC [21] and polar codes [3],

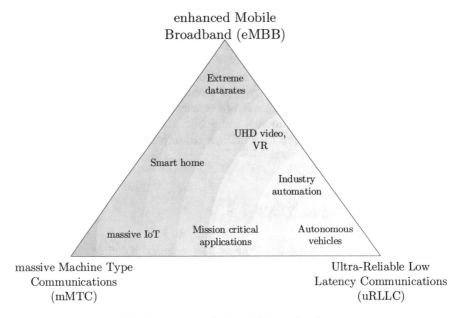

Fig. 24.20: New fields of 5G applications

and offering:

- bigger throughput, up to 20 Gbps in down-link and 10 Gbps in up-link,
- smaller latency down to 1 millisecond.

In Table 24.3 first releases of the 3G (R99) and 4G LTE (R8) standards are included also in order to compare what was offered when both standards appeared. Later 3G offered 2×2 MIMO systems and 4G (LTE) 8×8 and even 32×32 MIMO. In LTE for MIMO bigger than 4×4 user-equipment specific reference signals (UE-RS) are used instead of physical cell-specific ones (CRS), similarly as in 5G.

Since 5G is using filtered OFDM (f-OFDM), in Fig. 24.21 the new modulation method is compared with the classical OFDM used in 4G LTE. Both techniques exploit the cyclic prefix and bandwidth shaping filters. In 4G LTE one IFFT is performed and a wide-band filter is used, while in 5G—multiple IFFTs are calculated which are followed by filters with different non-overlapping pass-bands (arbitrarily wide separating zones are used between them). Thanks to this the harmful influence of out-of-band interference (disturbances) is significantly reduced in 5G. But what is more important: different 5G services can have different sub-carrier spacing and different length of OFDM symbols.

From the signal processing perspective, the 5G is addressing and solving signal processing tasks listed in Fig. 24.22. Below we only comment the most important, new or significantly enhanced ones.

Table 24.3: Comparison of 3G (Release 99), 4G LTE (Release 10), and 5G New Radio (Release 15) digital telephony

	3G WCDMA	3G HSPA+	4G LTE	5G NR
Year of public.	2000–2001	2007	2011	2019
Freq bands (GHz)	0.6–3.5	0.6–3.5	0.6–6	0.6–6, 24–30, 37–40
Channel BW (MHz)	5	5	1.4–20	5–100 (<6 GHz), 50–400 (>6 GHz)
Peak downlink (Mb/s)	0.384	28	900[a]	20,000
Peak uplink (Mb/s)	0.384	11	50	10,000
Latency (ms)	150	50	10	1
Modulation	QPSK	QPSK-64QAM	BPSK-64QAM	BPSK-256QAM
Multiple access	CDMA	CDMA	CP-OFDMA SC-FDMA	CP-OFDMA SC-FDMA,f-OFDMA
Error correction	Turbo, Conv	Turbo, Conv	Turbo, Conv	LDPC, Polar codes
Antenna config.	1 × 1	2 × 2 MIMO	⩽ 8 × 8 MIMO	massive MIMO,ant.⩽ 64 × 64

[a]—valid for aggregation of three 20 MHz services and 64-QAM

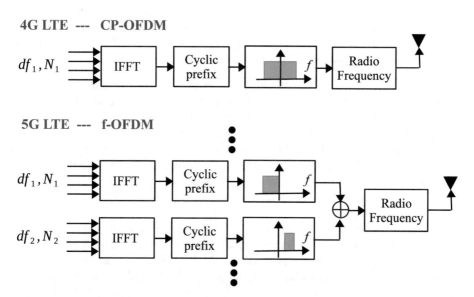

Fig. 24.21: Comparison of 4G CP-OFDM (up) with 5G f-OFDM (down)

- 5G NR physical layer better exploits existing beam-forming possibilities: all synchronization signals are User-Equipment (UE) specific. At present, maximum number of MIMO layers is the same in 4G and 5G. Number of antennas is not restricted by 4G and 5G standards.

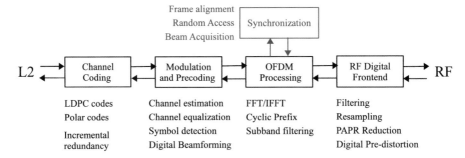

Fig. 24.22: 5G NR signal processing tasks

- Usage of the new f-OFDM modification, in which several IFFT outputs are sub-band filtered and combined with other bandwidth parts (BWP), together with appropriate separating zones, allows integration of different type services inside 5G. This makes the standard more universal/flexible.
- When very high frequencies, millimeter waves, are used, an analog TX hardware is more non-linear and a transmitted signal should be adaptively pre-distorted before the digital-to-analog converter (DAC). This way momentum non-linear characteristic of analog power amplifier is compensated.
- For millimeter waves (> 6 GHz) a phase noise, caused by frequency jitter, is a significant problem which has to be efficiently solved [20].

In 5G many different signals and data transmission channels are used in down-link and up-link. Each of them has precisely defined role in the system. Main 5G signals and data channels are characterized in Table 24.4, which are given: their names, modulation used, coding applied, and purpose of usage. We can observe that: (1) synchronization and network access signals use low-level modulations (BPSK, QPSK), (2) synchronization signals are not coded (there is no sense doing this—we are only interested in analog values of carrier states), (3) control channels are coded with polar codes (in 4G—convolutional codes), while shared channels use LDPC (in 4G—turbo codes).

Finally, in Table 24.5 and in Fig. 24.23, physical 5G down-link and up-link specification is given [15]. We see in the table that the standard provides a support for the 4G set-up: $\Delta f = 15$ kHz, sampling rate 30.72 MHz, time slot duration 1 millisecond and its division into 14 OFDM symbols, OFDM symbol duration 66.67μs and cyclic prefix length equal to 4.69μs. Increase of sub-carrier spacing 2^n times causes the same increase of slots in sub-frame. It is clearly visible in Fig. 24.23 where 5G frames are shown for different Δf values (resulting from sampling frequency used).

24.5.2 Exemplary Data Decoding

The Physical Broadcast Channel (PBCH) data decoding is similar in 4G and 5G. The 5G base-station has to generate the Synchronization Signal Block (SSB) consisting of [15, 17]:

1. the primary PSS synchronization signal, being a modulated maximum length sequence (MLS) having 127 samples (not Zadoff–Chu sequence as in 4G LTE); there are 3 different PSS sequences defined,
2. the secondary SSS synchronization signal, being also a MLS sequence having 127 samples; there are 336 different SSS sequences defined,
3. the BPCH broadcast signal, delivering some necessary 5G NR system information to a user equipment.

Together PSS and SSS defines $1008 = 3 \cdot 336$ different physical radio cell identities. After detection and decoding PSS and SSS, user equipment, a phone, is calculating the physical cell number and using it for generation of specific pilots exploited for PBCH synchronization. In 5G all pilot signals are cell or user specific, similarly like in 3G and 4G. Thanks to them a phone can achieve down-link time and frequency synchronization with OFDM symbols and acquire time instants of the PBCH channel.

The PBCH, similarly to LTE, carries the basic 5G NR system information which is required to connect with the base-station. PBCH has its own pilots. It is send in only one antenna port (no TX diversity is applied, the signal is linearly precoded using beam-forming) and coded by polar codes. It is transmitted many times during radio frame using different beams. As a result, a user phone recognizes the best beam for it during access to the network.

In 5G PBCH makes use of 20 central Physical Resource Blocks (PRBs) of OFDM symbol and $20 \cdot 12 = 240$ sub-carriers. Sequence in the SSB block is as follows: PPS (127 sub-carriers), then one complete PBCH symbol, then SSS in the middle and PBCH around it (4 RBs on both sides), then again one complete PBCH symbol. In separate OFDM time–frequency cells (slots) of PBCH, specific pilot signals are sent by only one antenna port. In general, the 5G PBCH decoding procedure is very similar to this trained by us during analysis of the LTE signal.

In 4G LTE 2×2, 4×4, and 8×8 antenna arrays were used. In 5G choice is significantly bigger. Below 6 GHz maximally 8×8 antenna arrays are allowed, while generally—up to 64×64. It is important to stress that the standards 4G and 5G do not limit number of physical antennas. Some limiting numbers are given only when practical consideration are taken into account. In 64×64 MIMO, the base-station have 64 transmit and receive antennas, while a single user equipment—significantly less, 2 or 4. These very large antenna arrays in the base-station are referred to as massive MIMO. It is considered as one of development [14].

For millimeter, i.e. shorter, waves antenna dimensions are lower and application of bigger antenna arrays is more practical. Gain and phase of each antenna is adjusted, in the most optimistic scenario—even for single user. Thanks to this, RF beams in different directions can be stronger and RF reception power is increased

Table 24.4: Physical channels and signals used in 5G NR

DN/UP	Name	Abbrev.	Modulation	Coding	Application
DOWN	Primary Synchro Signal	PSS	BPSK	–	Initial radio cell access
DOWN	Secondary Synchro Signal	SSS	BPSK	–	Initial radio cell access
DOWN	Physical Broadcast Ch.	PBCH	QPSK	Polar code	System broadcast information
DOWN	Physical Downlink Shared Ch.	PDSCH	QPSK, 16/64/256-QAM	LDPC	User data DOWN
DOWN	Physical Downlink Control Ch.	PDCCH	QPSK	Polar code	Resource scheduling grants, ac-knowledgments
DOWN	CSI Reference Signal	CSI-RS	QPSK	–	Channel measurements, beam acquisition in down-link
UP	Physical Random Access Ch.	PRACH	Zadoff-Chu	–	Initial radio cell access (asyn-chronous) + init. up-link
UP	Physical Uplink Shared Ch.	PUSH	QPSK, 16/64-QAM	LDPC	User data UP
UP	Physical Uplink Control Ch.	PUCCH	QPSK	Polar code	Scheduling requests, ac-knowledgments, measurement reports
UP	Sounding Reference Signal	SRS	QPSK	–	Channel measurements, beam acquisition in up-link

Table 24.5: 5G New Radio physical down-link and up-link specification

Sub-carrier spacing Δf [kHz]	15	30	60	120
OFDM symbol duration [µs]	66.67	33.33	16.67	8.33
Cyclic prefix duration [µs]	4.69	2.34	1.17	0.586
Slot duration (14 symbols) [µs]	1000	500	250	125
Slots in sub-frame	1	2	4	8
Sampling rate	30.72	61.44	122.88	245.76

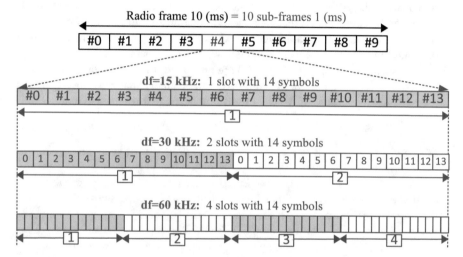

Fig. 24.23: Frame structure used in 5G New Radio

in a particular direction, which allows for a spatial separation of users. In 5G SSB signals of the PBCH are transmitted periodically on multiple beams and the mobile terminal can select the best beam for it during initial physical cell access. Beam-forming is performed in both direction, up-link and down-link, but usually only in the base-station. It consists of antenna signal scaling in amplitude and shifting in phase, therefore it is simple in implementation even for large antenna arrays. In contrary to Alamouti coding, the beam-forming operation is transparent for a phone—it does not have to know that beam-forming is utilized.

It is important to note that MIMO 2×2, 4×4, etc., denote number of parallel data streams on input and output, not the number of antenna used. Both 4G LTE and 5G NR allow usage up to 8 data layers, the bigger amount is not practical. However antenna arrays which are used can have bigger dimensions. It is not restricted by standards, but practical implementation aspects, like resulting antenna array size or hardware design complexity.

Another story is transmission and decoding data in down-link and up-link shared channels. On this web page a Matlab program simulating the physical up-link transmission is given. Its description is presented in this paper [5], available at https://github.com/gc1905/5g-nr-pusch The block diagram, describing all operations performed in the program, is presented in Fig. 24.24. It is only informative and its goal is to give us a general idea what will be going on in our future 5G phones.

PUSCH/PDSCH Decoding The most difficult signal processing part of a 5G phone deals with synchronization, channel estimation and correction as well as efficient error checking and correction issues. In the transmission simulation program used by us, the PUSCH data are recovered using algorithm which block diagram is presented in Fig. 24.25. Result obtained with its use is shown in Fig. 24.26. The following scatter plots of $Q(k) = f(I(k))$ are presented: for input data (left), for uncorrected data (center), and for synchronized and channel equalized data (right). As we see—"nothing is easy." If we want to live on Mars, first we have to reach it.

Exercise 24.10 (5G NR Is Coming). A Matlab program with simulation of up-link shared transmission of user data (physical up-link shared channel (PUSH)) is given at this web page. Two main files have names `run_5gnr_codec.m` (LDPC encoder/decoder without front-end base-band processing layer, modulation mapper and de-mapper) and `run_5gnr_sim_sweep.m` (full simulation). Algorithmic solutions, implemented in the program, were briefly described in this chapter. In wide form, they are presented in the paper [5] available at this ResearchGate page. Analyze code of both programs. Then look inside functions that are called by them, and into sub-functions called by functions... Visit all sub-folders and read names of all subroutines present in them—their names are self-presenting. Then run the programs. Observe figure when they are plotted. Find them in this this paper.

24.6 Summary

Modern digital telephony is a part of complex system of completely connected digital world including: smart home, autonomous vehicles, Internet of Things, intelligent automated industry, sensor networks, mission critical and extremely high data-rate applications. It is a part of human dreams about everything possible "just now and here!" It is a part of yesterday science-fiction becoming reality today. It is realization of our dreams about convenient and safe life. How to conclude all of this? What is the most important?

INPUT BITS   **OUTPUT BITS**

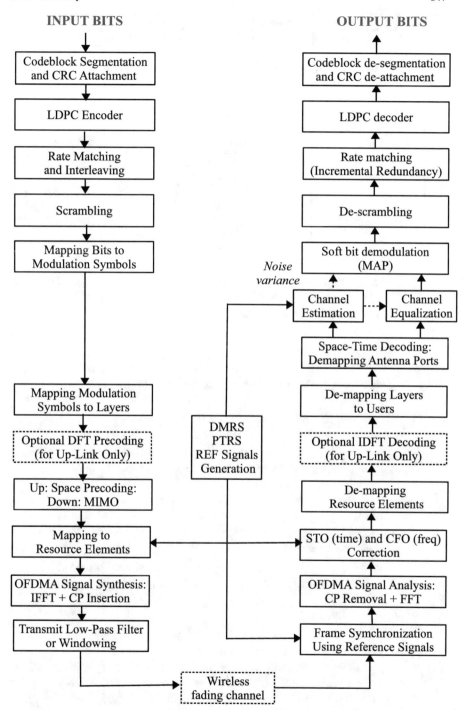

Fig. 24.24: Simplified diagram of physical down-link (PDSCH) and up-link (PUSCH) shared channel data/signal processing in 5G New Radio

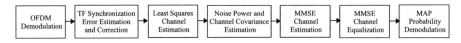

Fig. 24.25: Algorithm of PUSCH channel data recovery implemented in this program

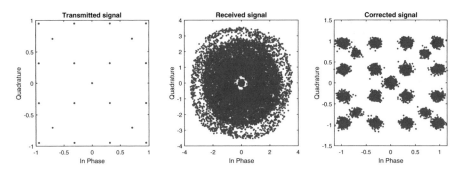

Fig. 24.26: Simulation result of PUSCH data recovery using algorithm from Fig. 24.25 and (left) original scatter, (middle) scatter after LS channel equalization, (right) scatter after MMSE channel scatter equalization

1. Our world is in permanent progress and nowadays solutions will not be used forever. After 5G we will have 6G, 7G, and so on. Therefore it is crucial to have good understanding of existing technical telecommunication problems and wide knowledge about many methodologies of their solving. The methods will change but their philosophy—not necessarily.

2. Advanced contemporary and future telecommunication systems will have many users and layers, options, and different application scenarios. To ensure all their features anytime and anywhere for everybody, front-end signaling will be becoming bigger and bigger. More bricks in the wall will be used. More players will be in the game. We should understand general concepts of all main operations, in order not to be imprisoned by a single equation or a single program line.

3. The 4G LTE and 5G NR technology represent a natural further development of concepts originated from telephone and cable discrete multi-tone modems, having in mind multiple access, user mobility, and channel fast variability.

4. In order to ensure achievement of very ambition objectives, in 4G LTE and 5G NR many control and synchronization signals are transmitted (for example, the physical radio cell-specific reference signals (CRS) in LTE are sent every 35 μs) and channel estimation and equalization is done continuously. There are numerous synchronization signals in 5G and they are different than in DSL and DAB, but general concept of their generation,

detection, and usage remain the same. The same is true for channel estimation: methods are the same, only their realization is different. Channel is estimated in specified points in time–frequency grid, for selected carriers (pilots) and selected time slots, then interpolated and applied to remaining data carriers. Statistical signal processing, with its means and covariances, helps us to cope with many transmission distortions and track the channel changes. Concluding: we have to "un-do" what channel and imperfect hardware "did" with our signals.

5. The MIMO antenna technique, already present in late 3G (HSPA+) and widely applied in 4G LTE telephony, is significantly enhanced in 5G New Radio. Different transmitter antennas sent the same data using different carriers (i.e. the same IQ state is transmitted a few times) and this frequency carrier (space) diversity increases the system noise robustness even when the receiver has one antenna only (MISO—multiple inputs, single output). In case of several RX antennas (MIMO)—the BER is much more reduced. Thanks to advanced MIMO techniques, base-station may offer even a special signal beam-forming for a single user. And this is a completely new DSP story, may be worthy writing another book.

6. The 5G New Radio is coming. But we should not be afraid of. Details are different, but, from signal processing perspective, the techniques applied are similar and should not be difficult for us. In this chapter operations performed by the 5G transmitter and receiver were briefly described and link to a Matlab program [5], simulating the data link was indicated (see https://github.com/gc1905/5g-nr-pusch this web page). We should not worry! We are 5G Ready.

24.7 Private Investigations: Free-Style Bungee Jumps

Exercise 24.11 (*** Going Forward with a Smile). Dear Reader. Everything has an end. Our together DSP journey also. After hundreds of pages, equations, figures, and diagrams. After hours of reading. After writing and running many programs. It is time to go forward alone, to taste sweet fruits of your new knowledge, to formulate your own tasks, to look for treasures hidden in signals around us. It is time for better understanding surrounding world, for being a technical detective and private investigator similar to Sherlock Holmes and Philip Marlowe, for live with passion, for enjoying beauty of work and for searching it.

I liked a lot The Muppet Show News with a charming announcement: "There are no news today." Impossible in our speeding forward world, inventing new and new gadgets. Science and technology rush ahead and our dreams and expectations together with them. This which was impossible yesterday, today is our everyday reality allowing us efficient functioning in cyberspace.

References

1. J. Abdoli, M. Jia, J. Ma, Filtered OFDM: A new waveform for future wireless systems, in *Proc. IEEE 16th Int. Workshop on Signal Processing Advances in Wireless Communications*, Stockholm, July 2015
2. S. Alamouti, A simple transmit diversity technique for wireless communications. IEEE J. Sel. Areas Commun. **16**(8), 1451–1458 (1998)
3. E. Arikan, Channel polarization: A method for constructing capacity-achieving codes for symmetric binary-input memoryless channels. IEEE Trans. Inf. Theory **55**(7), 3051–3073 (2009)
4. D. Chu, Polyphase codes with good periodic correlation properties. IEEE Trans. Inf. Theory 18(4), 531–532 (1972)
5. G. Cisek, T.P. Zieliński, Prototyping software transceiver for the 5G new radio physical uplink shared channel, in *Conference: Signal Processing Symposium*, Cracow, pp. 17–19, Sep. 2019
6. J. Demel, *Empfang von LTE-Signalen in GNU Radio*, M.Sc. Thesis, Karlsruhe Institute of Technology, Communications Engineering Lab, 2012
7. J. Demel, S. Koslowski, F.K. Jondral, A LTE receiver framework using GNU radio. J. Sig. Process. Syst. **78**(3), 313–320 (2015)
8. A. Donarski, T. Lamahewa, J. Sorensen, Downlink LTE synchronization: A software defined radio approach, in *Proc. 8th Int. Conf. on Signal Processing and Communication Systems*, Gold Coast, Dec. 2014
9. ETSI 4G-LTE Standards, Online: https://www.etsi.org/deliver/etsi_ts/136200_136299/
10. ETSI 5G NR Standards, Online: https://www.etsi.org/deliver/etsi_ts/138100_138199/
11. ETSI TS 136 211, LTE: Evolved Universal Terrestrial Radio Access (E-UTRA); Physical channels and modulation, Online: https://www.etsi.org/deliver/etsi_ts/136200_136299/136211/
12. R. Frank, S. Zadoff, R. Heimiller, Phase shift pulse codes with good periodic correlation properties. IRE Trans. Inf. Theory **8**(6), 381–382 (1962)
13. A. Golnari, M. Shabany, A. Nezamalhosseini, G. Gulak, Design and implementation of time and frequency synchronization in LTE. IEEE Trans. Very Large Scale Integr. (VLSI) Syst. **23**(12), 2970–2982 (2015)
14. E.G. Larsson, O. Edfors, F. Tufvesson, T.L. Marzetta, Massive MIMO for next generation wireless systems. IEEE Commun. Mag. **52**(2), 186–195 (2014)

15. S.-Y. Lien, S.-L. Shieh, Y. Huang, B. Su, Y.-L. Hsu, H.-Y. Wei, 5G New Radio: Waveform, frame structure, multiple access, and initial access. IEEE Commun. Mag. **55**(6), 64–71 (2017)

16. H. Myung, J. Lim, D. Goodman, Single carrier FDMA for uplink wireless transmission. IEEE Veh. Technol. Mag. **1**(3), 30–38 (2006)

17. A. Omri, M. Shaqfeh, A. Ali, H. Alnuweri, Synchronization procedure in 5G NR systems. IEEE Access **7**, 41286–41295 (2019)

18. S. Parkvall, E. Dahlman, A. Furuskar, M. Frenne, NR: The new 5G radio access technology. IEEE Commun. Stand. Mag. **1**(4), 24–30 (2017)

19. J.G. Proakis, M. Salehi, G. Bauch, *Modern Communication Systems Using Matlab* (Cengage Learning, Stamford, 2004, 2013)

20. Y. Qi, M. Hunukumbure, H. Nam, H. Yoo, S. Amuru, On the phase tracking reference signal (PT-RS) design for 5G new radio (NR), in *Proc. IEEE 88th Vehicular Technology Conference (VTC-Fall)*, Chicago, Aug. 2018

21. T. Richardson, S. Kudekar, Design of low-density parity check codes for 5G new radio. IEEE Commun. Mag. **56**(3), 28–34 (2018)

22. N. Rupasinghe, Í. Güvenc, Capturing, recording, and analyzing LTE signals using USRPs and Lab VIEW, in *Proc. IEEE SoutheastCon*, Fort Lauderdale, April 2015

23. M.A.N. Shukur, K. Pahwa, H.P. Sinha, Implementing primary synchronization channel in mobile cell selection 4G LTE-A network. Proc. Int. J. Sci. Res. Eng. Technol. (IJSRET) **3**(1), 19–26 (2014)

24. M.R. Sriharsha, D. Sreekanth, K. Kiran, A complete cell search and synchronization in LTE. EURASIP J. Wirel. Commun. Netw. **101**, 1–14 (2017). Online: https://link.springer.com/content/pdf/10.1186%2Fs13638-017-0886-3.pdf

25. Y. Yu, Q. Zhu, A novel time synchronization for 3GPP LTE cell search, in *Proc. 8th Int. Conf. on Communications and Networking in China*, Guilin, Aug. 2013, pp. 328–331

26. A.A. Zaidi, R. Baldemair, V. Moles-Cases, N. He, K. Werner, A. Cedergren, OFDM numerology design for 5G new radio to support IoT, eMBB, and MB-SFN. IEEE Commun. Stand. Mag. **2**(2), 78–83 (2018)

27. H. Zarrinkoub, *Understanding LTE with Matlab. From Mathematical Modeling to Simulation and Prototyping* (Wiley, Chichester, 2014)

Correction to: Starting Digital Signal Processing in Telecommunication Engineering

Tomasz P. Zieliński

Correction to:
T. P. Zieliński, *Starting Digital Signal Processing in Telecommunication Engineering*, Textbooks in Telecommunication Engineering,
https://doi.org/10.1007/978-3-030-49256-4

This book was inadvertently published with the incorrect copyright year. This has now been amended throughout the book and the correct copyright year of the book is 2021.

The updated online version of this book can be found at
https://doi.org/10.1007/978-3-030-49256-4

Index

Printed in the United States
by Baker & Taylor Publisher Services